Metal Nanocluster Chemistry

Metal Nanocluster Chemistry
Ligand-Protected Metal Nanoclusters With Atomic Precision

Manzhou Zhu
Changjiang Chair Professor of Chemistry, Anhui University, Hefei, Anhui, China

Elsevier
Radarweg 29, PO Box 211, 1000 AE Amsterdam, Netherlands
The Boulevard, Langford Lane, Kidlington, Oxford OX5 1GB, United Kingdom
50 Hampshire Street, 5th Floor, Cambridge, MA 02139, United States

Copyright © 2023 Elsevier Inc. All rights reserved.

No part of this publication may be reproduced or transmitted in any form or by any means, electronic or mechanical, including photocopying, recording, or any information storage and retrieval system, without permission in writing from the publisher. Details on how to seek permission, further information about the Publisher's permissions policies and our arrangements with organizations such as the Copyright Clearance Center and the Copyright Licensing Agency, can be found at our website: www.elsevier.com/permissions.

This book and the individual contributions contained in it are protected under copyright by the Publisher (other than as may be noted herein).

Notices
Knowledge and best practice in this field are constantly changing. As new research and experience broaden our understanding, changes in research methods, professional practices, or medical treatment may become necessary.

Practitioners and researchers must always rely on their own experience and knowledge in evaluating and using any information, methods, compounds, or experiments described herein. In using such information or methods they should be mindful of their own safety and the safety of others, including parties for whom they have a professional responsibility.

To the fullest extent of the law, neither the Publisher nor the authors, contributors, or editors, assume any liability for any injury and/or damage to persons or property as a matter of products liability, negligence or otherwise, or from any use or operation of any methods, products, instructions, or ideas contained in the material herein.

ISBN: 978-0-323-90474-2

For information on all Elsevier publications
visit our website at https://www.elsevier.com/books-and-journals

Publisher: Susan Dennis
Acquisitions Editor: Charles Bath
Editorial Project Manager: Andrea R. Dulberger
Production Project Manager: Kumar Anbazhagan
Cover Designer: Christian J. Bilbow

Typeset by STRAIVE, India
Transferred to Digital Printing 2023

Contents

Preface — ix

1. **Introduction to metal nanoclusters—Concepts and prospects**
   Manzhou Zhu and Xi Kang
   1.1 Why to research metal nanoclusters—A historical overview — 1
      1.1.1 From the size point of view — 2
      1.1.2 From the structure point of view — 2
      1.1.3 From the property point of view — 3
   1.2 How to study metal nanoclusters—Where are we? — 4
      1.2.1 The research status of the controllable preparation — 5
      1.2.2 The research status of the structure elucidation — 5
      1.2.3 The research status of the property regulation — 6
      1.2.4 The research status of the potential application — 6
   1.3 What are included in this book — 7
   References — 7

2. **Controllable preparation of metal nanoclusters**
   Manzhou Zhu and Sha Yang
   2.1 Introduction — 11
   2.2 Vacuum synthesis — 11
      2.2.1 Sputtering — 12
      2.2.2 Fast atom bombardment — 13
      2.2.3 Liquid metal ion source — 14
   2.3 Gas phase nanoclusters — 15
      2.3.1 Evaporation and gas condensation — 15
      2.3.2 Laser vaporization — 17
   2.4 Chemical synthesis in solution — 19
      2.4.1 Metal complex reduction method — 19
      2.4.2 Brust synthesis and beyond — 20
      2.4.3 Kinetically controlled synthesis method — 25
      2.4.4 One-pot synthesis — 26
      2.4.5 Controlled conversion method — 30
      2.4.6 Inorganic anions as synthetic templates — 33
   2.5 Synthesis of alloy nanoclusters — 34
      2.5.1 Co-reduction method — 34
      2.5.2 Ligand exchange method — 35
      2.5.3 Method exchange method — 36
      2.5.4 Intercluster reaction method — 37
   2.6 Other methods of synthesis — 38
   References — 39

3. **Characterizations and atomically precise structures of metal nanoclusters**
   Manzhou Zhu and Qinzhen Li
   3.1 Characterization techniques for nanoclusters — 45
      3.1.1 UV-vis absorption spectroscopy — 45
      3.1.2 Photoluminescence spectroscopy — 45
      3.1.3 Fourier transform infrared spectrometer — 47
      3.1.4 Circular dichroic spectroscopy — 48
      3.1.5 Transient absorption spectroscopy — 49
      3.1.6 Mass spectrometry — 50
      3.1.7 Transmission electron microscope — 52
      3.1.8 Dynamic light scattering — 53
      3.1.9 Thermogravimetric analysis — 54
      3.1.10 Nuclear magnetic resonance spectroscopy — 54
      3.1.11 Electron paramagnetic resonance — 55
      3.1.12 X-ray photoelectron spectroscopy — 56

3.1.13 X-ray absorption fine structure ... 57
3.1.14 Single crystal X-ray diffraction ... 58
3.2 **Atomically precise structures** ... 59
　3.2.1 Introduction ... 59
　3.2.2 Ligands and metal-ligand interfaces ... 59
　3.2.3 Metallic kernels in nanoclusters ... 65
References ... 73

4. Mechanism of size conversion and structure evolution of metal nanoclusters

Manzhou Zhu and Haizhu Yu

4.1 **Concept of size-conversion and structural evolution** ... 79
4.2 **Mechanism of size-conversion** ... 79
　4.2.1 General understanding on the size-growth of metal nanoclusters ... 79
　4.2.2 Molecular-level mechanism of size-conversion of metal nanoclusters ... 82
4.3 **Insights into the size-evolution principals** ... 137
　4.3.1 Electronic structure evolution of metal nanoclusters ... 137
　4.3.2 Size and optical absorption property evolution of the icosahedral metal nanoclusters ... 140
　4.3.3 Size and excited-state dynamics evolution of the FCC-packed gold clusters ... 140
　4.3.4 Size-evolution of the FCC-packed silver clusters ... 142
　4.3.5 Size-evolution principals ... 144
4.4 **Conclusion and perspective** ... 146
References ... 146

5. Physical-chemical properties of metal nanoclusters

Manzhou Zhu and Shuang Chen

5.1 **Optical absorption of metal nanoclusters** ... 153
5.2 **Photoluminescence of metal nanoclusters** ... 155
　5.2.1 Inherent geometry and electronic structure ... 156
　5.2.2 Ligand effect on PL of metal nanoclusters ... 158
　5.2.3 Metal doping and alloying ... 162
　5.2.4 Aggregation-induced emission ... 165
　5.2.5 Others ... 168
5.3 **Magnetism of metal nanoclusters** ... 170
　5.3.1 Magnetic properties of Au_{25}-based metal nanoclusters ... 170
　5.3.2 Magnetic properties of other metal nanoclusters ... 172
5.4 **Chirality of metal nanoclusters** ... 175
　5.4.1 Identify the chirality through crystal structure ... 175
　5.4.2 Chiral separation of metal nanoclusters ... 178
　5.4.3 Chiral ligands/compounds induce chirality for metal nanoclusters ... 179
　5.4.4 Combination of chirality and other properties ... 182
5.5 **Electrochemical property of metal nanoclusters** ... 184
　5.5.1 Nanoelectrochemistry ... 184
　5.5.2 Electrochemical analysis for metal nanoclusters ... 185
　5.5.3 Electrochemiluminescence ... 187
　5.5.4 Photoelectric conversion ... 191
References ... 191

6. Theoretical simulations on metal nanocluster systems

Manzhou Zhu and Haizhu Yu

6.1 **Foundation of theoretical simulation methods** ... 201
　6.1.1 Theoretical basis of DFT functionals ... 201
6.2 **Application of density functional theory simulations** ... 205
　6.2.1 Single point energy calculations ... 205
　6.2.2 Geometry optimization ... 206
　6.2.3 Frequency calculation ... 212
　6.2.4 Electronic state elucidation ... 213
　6.2.5 Reaction mechanism analysis ... 216
　6.2.6 Theoretical simulation on optical properties ... 221
　6.2.7 Calculations on NMR spectra ... 223
　6.2.8 Structure-property correlation ... 224
6.3 **Application of quantum mechanics-molecular mechanics calculations** ... 226
6.4 **Conclusion and prospect** ... 228
References ... 228

7. Assembly of metal nanoclusters

Manzhou Zhu and Shan Jin

7.1	Introduction—A brief introduction to the assembly of nanomaterials	233
7.2	Cluster-based host-gust nanosystem for assemble nanomaterial	235
	7.2.1 Supramolecular interactions of NCs as hosts with cyclodextrins (CDs)	235
	7.2.2 Supramolecular interactions of NCs as hosts with fullerene	237
	7.2.3 Supramolecular interactions of NCs as hosts with other molecules	239
	7.2.4 Clusters as guests	240
7.3	Assembly in crystal lattice for assemble nanomaterial via supramolecular interactions	245
7.4	Cluster-based metal-organic framework via linker	263
7.5	Perspective	280
	References	281

8. Practical applications of metal nanoclusters

Manzhou Zhu and Yuanxin Du

8.1	Introduction	289
8.2	Sensors	289
	8.2.1 Chemical sensor	289
	8.2.2 Biological sensor	306
	8.2.3 Other sensor	322
8.3	Biological application	323
	8.3.1 Biolabeling/imaging	323
	8.3.2 Disease diagnostics and therapy	329
	8.3.3 Self-vaccine	333
	8.3.4 Antimicrobial agents	333
8.4	Catalysis	335
	8.4.1 Electrocatalysis	335
	8.4.2 Photocatalysis	340
	8.4.3 Catalytic selective oxidation	342
	8.4.4 Catalytic selective reduction	348
	8.4.5 Catalysis of coupling reactions	352
	8.4.6 Other catalytic reactions	355
8.5	Conclusions and outlooks	357
	References	358

9. Summary and perspectives

Manzhou Zhu and Xi Kang

9.1	Summary of this book	373
9.2	Personal perspectives to the nanocluster science	373
	9.2.1 New synthetic methodologies	373
	9.2.2 Preparing more novel alloy nanoclusters	373
	9.2.3 Ligand effect on nanoclusters	374
	9.2.4 Cocrystallization of metal nanoclusters	374
	9.2.5 Periodicity of nanoclusters	374
	9.2.6 Seeing the transformation of nanoclusters	374
	9.2.7 Fixing the boundary between quantum-sized nanoclusters and metallic-state nanoparticles	375
	9.2.8 Mechanisms of properties of nanoclusters	375
	9.2.9 Future theoretical works	375
	9.2.10 Nanocluster-based assemblies	375
	9.2.11 Nanocluster crystals as materials	376
	9.2.12 Grant applications of cluster-based nanomaterials	376
	References	376

Index 379

Preface

Metal nanocluster chemistry remains a thriving research field. Metal nanoclusters, also known as ultra-small metal nanoparticles, occupy the gap between discrete atoms and plasmonic nanomaterials and are an emerging class of nanomaterials. Owing to their atomically precise structures, prominent quantum size effect, and discrete electronic energy levels, metal nanoclusters have progressed tremendously in the exploration of accurate structures and structure-property correlations at the atomic level. Based on the new knowledge acquired, the researchers can customize the nanomaterials more rationally, and atomically precise cluster-based nanomaterials will find new opportunities in a wide range of applications.

Since Frank Albert Cotton proposed the ground-breaking concept of cluster, the research of metal clusters constantly flutters high at the forefront of nanoscience. However, how to break through the bottleneck of this research field, the efficient and directional preparation of metal nanoclusters, always remains a pressing academic task. Thanks to my master's supervisor Prof. Zheng Cui at Shenyang Pharmaceutical University, my doctoral supervisor Prof. Qingxiang Guo, and my postdoctoral supervisor Prof. Jianguo Hou at the University of Science and Technology of China, I mastered basic skills of natural product separation, organic synthesis, and molecular surface self-assembly, which I believe are the fundamental solutions aiming at the abovementioned bottleneck of metal nanocluster chemistry. Therefrom, during my postdoctoral period under the guidance of Prof. Rongchao Jin at Carnegie Mellon University, I achieved the high-yield synthesis and X-ray crystal structure determination of the $Au_{25}(SR)_{18}$ nanocluster. From here, the preparation of metal nanoclusters stepped from random to precise, and then several downstream explorations, including the determination of cluster structures, the investigation of their structure-property correlations, and the application of cluster-based materials, have gradually become a reality. Thereafter, I witnessed dramatic progress in the metal nanocluster research field. I also realized that there were several unknowns in metal nanocluster chemistry, which greatly inspired me to explore further. In this context, after returning to China to work at Anhui University, I took the initiative to organize the lab of "Centre for Atomic Engineering of Advanced Materials" for continuously studying the metal nanocluster chemistry.

After tremendous research efforts, the metal nanocluster chemistry has been booming in the past two decades. Meanwhile, a comprehensive and timely book that introduces and summarizes the preparations, the characterizations, the structure determinations, the property investigations, and the practical applications of metal nanoclusters is still urgently needed. In this context, my colleagues (all from my lab in Anhui University) and I wrote this book: Chapters 1 and 9 (Dr. Xi Kang); Chapter 2 (Dr. Sha Yang); Chapter 3 (Dr. Qinzhen Li); Chapters 4 and 6 (Dr. Haizhu Yu); Chapter 5 (Dr. Shuang Chen); Chapter 7 (Dr. Shan Jin); and Chapter 8 (Dr. Yuanxin Du). By combining all these contributions, we presented this book titled *Metal Nanocluster Chemistry*, pointing out what is metal nanocluster, why we study nanocluster, and how to study nanocluster.

This book provides audiences with a comprehensive synthetic toolbox and insightful researching fundamentals of metal nanoclusters. I hope that this book attracts the attention of graduate students, young scientists, and experienced researchers across many disciplines in this ever-growing field.

Manzhou Zhu
Anhui University, Hefei, China

Chapter 1

Introduction to metal nanoclusters—Concepts and prospects

Manzhou Zhu and Xi Kang

1.1 Why to research metal nanoclusters—A historical overview

In retrospect to the evolution of human social civilization, the development of materials has been regarded as progressive milestones. Indeed, the ability to understand and use materials determines the form of society and the quality of human life. Among the developments of materials, the properties of metals have fascinated human beings since thousands of years ago, and the employment of metal products has promoted the evolution of human society from the Stone Age to the Bronze Age. Metal is the most important material basis of human civilization, especially for the agricultural civilization and the industrial civilization. Metal has been one of the most important materials for human beings to manufacture tools and weapons since the Bronze Age, and the production and application of metals determine the development of the modern social productivity.

During thousands of years' practice, the science of metal materials has been a significant subject and has become a hot area in research and application. Accompanied by the development of science technology, metal researchers have shifted their focuses from "large (bulk)" to "small (nano)" by exploiting the nanotechnology. It is widely accepted that the chemical-physical performances of metals or their compounds are dependent on their interior structures. The application of nanotechnology contributes to the atomic, molecular, and supramolecular insights into the structure-performance correlations of metal compounds, which is of great critical for grasping the origin of their chemical-physical properties and further guiding the preparation of more metal compounds with customized structures and performances. Among these compounds, the metal-based nanoparticles have received the most impressive researching focus [1–7].

A metal nanoparticle (or colloidal nanoparticle) is usually defined as a particle of metals that is between 1 and 100 nm (distinguished from microparticles with sizes of 1–1000 μm) in diameter [1–7]. In retrospect, scientific research on colloidal nanoparticles has a long history, especially for gold colloids, and one of the first seminal studies of the modern era on metal nanoparticles was carried out by Faraday in 1857 [8]. In this work, he performed systematic studies on metal colloids (including gold, platinum, palladium, and silver) made by a two-phase reaction between metal salts (in water) and phosphorus (in ether), that products was called "potable gold" with vivid colors [8]. After that, systematic studies on inorganic colloids of metals and their inorganic compounds were carried out, initiating a rapid development of metal colloid science [8–12]. Prior to the 1930s that the transmission electron microscopy (TEM) technology was invented, the particle sizes of these colloids were primarily determined by the ultramicroscope approach or the ultracentrifugation approach using Stokes' law of sedimentation [8]. Besides, in 1918, Scherrer reported the estimation of the sizes of gold nanoparticles by X-ray diffraction, which has been widely used till today [8].

However, the atomic-level study of these colloidal nanoparticles remains challenging because of their two major defects: (i) polydispersity and heterogeneity of particle sizes and (ii) uncertainty of surface structures [8]. Nano chemists are often frustrated by the well-known fact that no two colloidal particles are the same, which impedes the deep understanding of their fundamentally chemical-physical performances in which the total structures must be known [8]. For example, the citrate-capped Au nanoparticles are sensitive to salt and will aggregate in the presence of NaCl; by comparison, these Au nanoparticles are extremely stable when stabilized by thiol-terminated DNA [13]. However, without the atomic-level understanding of these nanoparticles, it remains challenging to investigate the root cause of the huge difference between these gold colloids. Besides, although major efforts have been made for the shape-controlled synthesis of nanoparticles, the atomic-level mechanism for the formation of nanoparticles with different facets and how ligands are bonded on these facets are still largely unclear. For understanding these unclear issues at the atomic level, more precise nanomolecular entities are required, as model nanosystems, and precise molecular tools. In this context, the nanocluster science emerges.

Owing to their atomically precise structures (including the core metallic structure, the surface ligand structure, and the bonding between the metal core and the ligands), metal nanoclusters provide an ideal platform for addressing most of these fundamentally important issues remaining in colloidal nanoparticles [8,14–31]. The most reliable approach for the determination of total structures of metal nanoclusters is single-crystal X-ray diffraction. Besides, with these nanocluster structures of different sizes and shapes in hand, it will be possible to study the core-ligand interactions, the mechanism of material properties and their correlations with the nanocluster structure, and the possible pathways of the shape control, size growth, and so on. Then, the differences between metal nanoclusters and colloidal nanoparticles are listed below from three aspects (size, structure, and property).

1.1.1 From the size point of view

Metal nanoclusters bridge between organometallic complexes (with sizes below 1 nm) and plasmonic nanoparticles (with sizes between several to 100 nm). As depicted in Fig. 1.1, for simplicity, the Au-SR (where SR is thiolate) complexes, $Au_{25}(SR)_{18}$ and $Ag_{44}(SR)_{30}$ nanoclusters, and large-sized FCC (face-centered-cubic) nanoparticles are selected as representatives for metal molecules, metal nanoclusters, and metal nanoparticles, respectively. Compared with conventional colloidal nanoparticles, the sizes of metal nanoclusters are smaller (1–3 nm), and thus metal nanoclusters are usually regarded as ultra-small nanoparticles [8]. Probably, the higher crystallinity of metal nanoclusters relative to large-sized colloidal nanoparticles originate from their smaller sizes, and the determination of their atomically precise structures becomes a reality. Besides, compared with Au-SR complexes wherein all metals are in oxidation states (i.e., M^{n+}), metal nanoclusters contain several free electrons, and these free electrons are considered to locate in/on the metallic kernel of metal nanoclusters, rendering these metals with partially zero-valence (i.e., $M^{\delta+}$). Indeed, thiolated Au nanoclusters are always synthesized by reducing Au-SR complexes with reductants, and this operation provides free electrons into the nanocluster framework.

1.1.2 From the structure point of view

One of the most attractive characterizations of metal nanoclusters compared with colloidal nanoparticles is their atomically precise structures. First, the precise structures of metal nanoclusters indeed make them "compounds" because they can be assigned definite formulas (i.e., $Au_n(SR)_m$, where n and m are determined), rather than a mere size range (e.g., 10 ± 1 nm). Generally, the structure of a defined nanocluster contains several parts, including the innermost metallic kernel, the metal-ligand interface, the main chain of ligands, and the outermost terminal of ligands (Fig. 1.2). Benefiting from the precise structures of nanoclusters, the manipulation of each part of a nanocluster can be thoroughly tracked, which is of great significance for analyzing the structure evolutions of nanoclusters. Second, it is inadequate to consider the morphology of metal nanoclusters because of their ultrasmall sizes. In this context, the TEM graphics of nanoclusters show scarcely

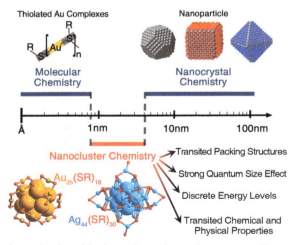

FIG. 1.1 Illustration of the scope of nanocluster chemistry. Metal nanoclusters bridge the organometallic complexes and plasmonic metal nanoparticles. These nanoclusters exhibit several intriguing characterizations, such as transited packing structures, strong quantum size effects, discrete energy levels, and transited chemical-physical properties. *(No permission required.)*

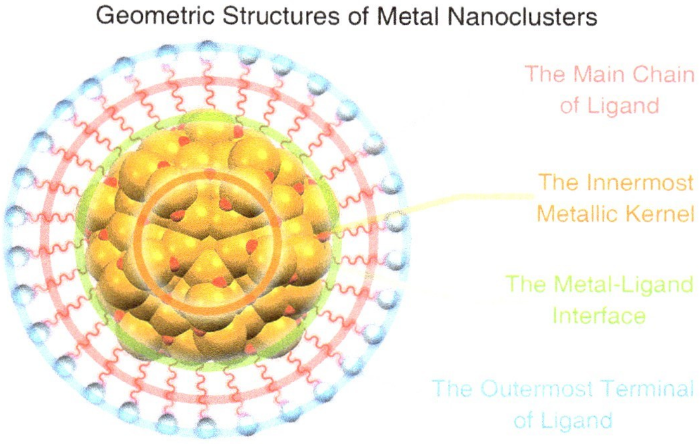

FIG. 1.2 Geometric structures of metal nanoclusters. Geometric structures of metal nanoclusters contain several parts, mainly including the innermost metallic kernel, the metal-ligand interface, the main chain of ligand, and the outermost terminal of ligand. *(No permission required.)*

any sign of crystalline facet and lattice fringe, which, however, are two of the most important parameters to represent the structures of large-sized colloidal nanoparticles [8].

1.1.3 From the property point of view

When metal nanoparticles enter the ultrasmall size regime (i.e., being considered as metal nanoclusters), strong quantum size effects come into play, and these metal nanoclusters indeed become sensitive to their structures and sizes [32,33]. As a result, even a single-atom engineering of a nanocluster framework will significantly influence its electronic structure as well as chemical-physical properties. By comparison, such strong effects are hardly manifest in large-sized colloidal nanoparticles. Besides, the atomically precise structures of metal nanoclusters render them ideal nanomodels for investigating the mechanism of their material properties. For instance, accompanied by the substitution of the innermost Ag atom by an introduced Au, remarkably enhancement on the emission of $Ag_{29}(SSR)_{12}(PPh_3)_4$ has been observed, and such an enhancement is assigned to the relativistic effect of the introduced Au heteroatom [34,35]. Furthermore, the small-sized metal nanoclusters exhibit discrete electronic energy levels, whereas large-sized colloidal nanoparticles are more like bulk metal with continuously electronic bands (Fig. 1.3) [36–39]. Accordingly, metal nanoclusters are believed to display abundant physical-chemical properties, and these properties are easy to dictate by controlling the structures and compositions of nanoclusters. Such a characterization helps to establish a sufficient sample database for the future application of cluster-based nanomaterials.

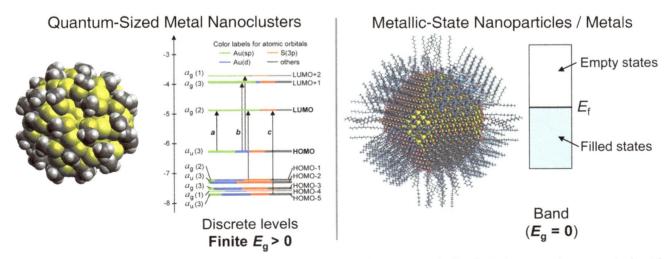

FIG. 1.3 Comparison between nanocluster and large-sized colloidal nanoparticles. (A) Molecule-like electronic structure in quantum-sized metal nanoclusters (where, HOMO=highest occupied molecular orbital, LUMO=lowest unoccupied molecular orbital, E_g=HOMO-LUMO gap). (B) Continuous band electronic structure of metallic-state nanoparticles and bulk metals (where E_f=Fermi level/energy). *(No permission required.)*

Collectively, benefit from their atomically precise structures, easily controllable frameworks/compositions, and abundant physical-chemical properties, metal nanoclusters are attracting increasing attention in both fundamental researches and practical applications. The nanocluster science has been showing its unique charm among all nanomaterials.

1.2 How to study metal nanoclusters—Where are we?

In retrospect, the research of phosphine ligand stabilized metal nanoclusters is much earlier than that of the thiolated nanoclusters. Research on the topic of phosphine ligand-protected nanoclusters could be traced back to the late 1960s when the crystal structure of the Au_{11} nanocluster was reported [40]. By comparison, inspired by works on self-assembled monolayers (SAMs) of thiols on bulk gold surfaces, researchers started to explore thiolated metal nanoclusters in the 1990s [41–43]. After decades of development, the study of metal nanoclusters has become a research hotspot, and several efforts have been made in this researching field, together to contribute to a prosperous discipline—the nanocluster science. In this subsection, we will provide an overview of the research status of metal nanocluster from four aspects: the controllable preparation, the structure elucidation, the property regulation, and the potential application (Fig. 1.4). Besides, the perspectives on existing issues and future efforts will be disclosed as well with these aspects.

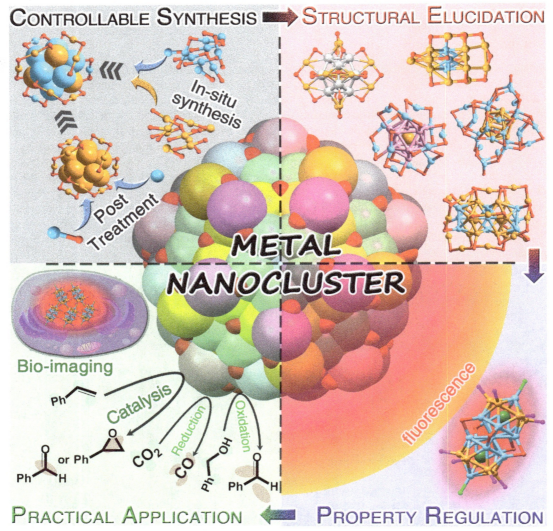

FIG. 1.4 Researching efforts of metal nanoclusters. Researching efforts on (A) the controllable preparation, (B) the structure elucidation, (C) the property regulation, and (D) the potential application of metal nanoclusters. *(No permission required.)*

1.2.1 The research status of the controllable preparation

The preparation of metal nanoclusters with high-yield and -purity is the basis of their fundamental researches and further applications. After decades of development, researchers now can increase the synthetic yield of nanoclusters to satisfy the subsequent investigations on structures and properties. Here, we take one of the most well-studied classical nanocluster systems as an example, namely, the thiolated $Au_{25}(SR)_{18}$ nanocluster. In 1994, Brust and coworkers reported the preparation of gold nanoparticles by exploiting a two-phase system, leading to a new period for synthesizing nanoparticles with homogeneous sizes [44]. The Whetten group reported the isolation and selected properties of a series of glutathione (GSH)-stabilized gold nanoclusters [33]. The Murray group reported the synthesis of $Au_{25}(S-C_2H_4Ph)_{18}$ nanocluster by modifying the "Brust-Schiffrin two-phase method" [45]. By using GSH as the stabilizer, the Tsukuda group reported the synthesis of hydrophilic $Au_n(SG)_m$ nanoclusters, and further extracted $Au_{25}(SG)_{18}$ from the $Au_n(SG)_m$ mixture by polyacrylamide gel electrophoresis [46]. However, the in situ synthesis of $Au_{25}(SR)_{18}$ nanocluster with high yield and purity remains challenging at that time. On this issue, in 2008, Zhu et al. reported the kinetically controlled, thermodynamically selective synthesis of the $Au_{25}(S-C_2H_4Ph)_{18}$ nanocluster with a high yield (cal. 40%) [47]. Indeed, delicately regulating the synthetic route is of great significance for the high-yield and -purity preparation of nanoclusters, which is important for the subsequent study of these nanoclusters.

Up to the present, aside from the in situ synthetic procedures, several posttreatment approaches have been reported to prepare nanoclusters with novel formulas and structures, such as ligand exchange [48], antigalvanic reduction [49], metal exchange [27], heat treatment [50], redox treatment [51], and so on. The presynthesized nanoclusters with a high purity are required for these posttreatment approaches.

However, we should note that the controllable preparation of giant nanoclusters containing more than 150 metal atoms remains challenging at present. Besides, nanocluster researchers have a dream to develop the nanocluster family more regular and comprehensive, as the periodic table of elements; however, there are also several vacancies in the member list of the nanocluster family, which calls for more efforts to discovery and determination.

1.2.2 The research status of the structure elucidation

One of the most attractive characterizations of metal nanoclusters is the atomically precise structure. For the determination of total structures of metal nanoclusters, the most reliable approach is the single-crystal X-ray diffraction (SC-XRD), with which help hundreds of structures of nanoclusters have been reported up to the present. The precise structures of nanoclusters enable the elucidation of many issues that remain mysteries in the nanoparticle research field, such as the size-growth modes, the metal-ligand interactions, the origins of material properties, and so on.

There are some limitations on the structure determination of nanoclusters by SC-XRD. For example, to differentiate the Pt locations from Au nanoclusters, or the Pd locations from Ag nanoclusters, remains challenging depend only on SC-XRD by considering the extremely similar electron clouds between Pt and Au (or Pd and Ag). Besides, several metal-hydride nanoclusters have been reported, but the hydride locations in these nanoclusters are impossible to define by SC-XRD (unless using the neutron diffraction). In such cases, the extended characterization means are needed to ascertain the structures of metal nanoclusters, including mass spectrometry, density functional theory (DFT) calculation, Fourier transform infrared, and so on. For the current researches, a combination of modern analytical techniques is required to characterize the precise structures of nanoclusters at the atomic level.

For those metal nanoclusters whose crystals are hard to culture, such as SG ligand stabilized nanoclusters, the X-ray absorption spectroscopy (XAS) provides an efficient way to enable the atomic site-specific analysis of local structures and electronic characters [52]. For example, the Zhang group reported the structural determination of $Au_{18}(SG)_{14}$ by exploiting the extended X-ray absorption fine structure (EXAFS), which provided an ideal platform together with the previously reported $Au_{18}(SR)_{14}$ to analyze the effect of hydrophobic and hydrophilic ligands on the structure of metal nanoclusters [53].

At present, although hundreds of crystal structures of metal nanoclusters have been obtained, the structure determination of metastable nanoclusters remains challenging, which impedes the in-depth understanding of their structural evolutions. The ligand-exchange or alloying processes of nanoclusters have been thoroughly tracked by the mass spectrometry, which, however, only provides information on the composition variation of nanoclusters but the information on the structure variation lays behind. The future structure determination of metal nanoclusters should make more efforts on these metastable nanoclusters (or reaction intermediates). In situ EXAFS may provide more precise information on this.

1.2.3 The research status of the property regulation

Owing to their strong quantum size effect and discrete electron energy levels, metal nanoclusters display structure-dependent chemical-physical properties, and even a single-atom variation will remarkably influence their properties. In this context, metal nanoclusters are prone to exhibit abundant material properties, such as fluorescence, chirality, magnetism, and electrochemical and electroluminescence properties, and so on.

Several approaches have been proposed to control the properties of metal nanoclusters, including the molecular and the supramolecular approaches (or the intracluster control and the intercluster control). Specifically, the intracluster control of nanoclusters refers to the manipulation over their metal compositions, ligand types, and metal-ligand interactions at the single molecular level (such as heterometal alloying, ligand exchange, molecular charge regulation, etc.), whereas the intercluster control mainly focuses on the manipulation over their assembled patterns among several cluster molecules in amorphous or crystallographic forms (such as the aggregation-induced emission of nanoclusters, the cluster-based metal-organic frameworks, the hierarchical structural complexity of nanoclusters, the intercluster reactions, etc.). All these operations and the corresponding property regulations result from the variations on geometric/electronic structures of metal nanoclusters.

The atomically precise structures of metal nanoclusters render them ideal nanomodels for the investigation of structure-property correlations at the atomic level. Taking photoluminescence as an example, the DFT calculations have demonstrated that the emission of nanoclusters originates from the LMCT (ligand-to-metal charge transfer) process [54]. Aikens and coworkers theoretically investigated the origin of emission of Au_{25} and Au_{13} nanoclusters, and proposed that the photoluminescence originated from the superatomic $P \leftarrow D$ transitions of nanoclusters, which belonged to core-based orbitals [55]. Zhou et al. suggested that the visible and near-infrared emissions of $Au_{25}(SR)_{18}$ nanoclusters have different lifetimes and arose from the core-shell charge transfer state and the Au_{13} core state, respectively [56]. All these explanations are proposed to explicate the origin of nanocluster emission from different perspectives.

The future efforts of metal nanoclusters in terms of their chemical-physical properties should lay on two aspects: (i) to customize the properties more delicately via controlling the structures of nanoclusters and (ii) to explicate the mechanism of properties more precise via analyzing the structure-property correlations of nanoclusters. Of note, both aspects are based entirely on the atomically precise structures of metal nanoclusters.

1.2.4 The research status of the potential application

The abundant chemical-physical properties provide metal nanoclusters and cluster-based nanomaterials with more potential in applications relative to small-sized metal complexes and large-sized colloidal nanoparticles [8].

As for the catalysis aspect, first, metal nanoclusters possess relatively large specific surface areas compared with large-sized colloidal nanoparticles. Second, through controlling their surface structures, metal nanoclusters can exhibit specific surface-active sites. Third, the controllable metallic kernel@surface ligands framework of metal nanoclusters endows fine control over their catalytic properties. In this context, metal nanoclusters have shown their unique charm in catalysis-related applications, such as click reaction, oxidation of benzyl alcohol, oxidation of carbon monoxide, electrocatalytic redox, and so on.

As for the emission-related applications, due to the intriguing absorption and luminescence properties, metal nanoclusters have been widely used in environmental monitoring and chemical sensing. Several hydrophilic noble metal nanoclusters with high emission have been reported, and their specific responses on emission to Hg^{2+}, Ag^+, and Pb^{2+} metal ions make them promising chemical sensors in both environment and biology [57–59]. Besides, together with the strong luminescence, the high biocompatibility, good photostability, low toxicity, and anticancer activity of such nanoclusters make them highly promising in cell labeling, phototherapy, bioimaging, and biotherapy.

As for the porous structure-related applications, the nanocluster-based assemblies with optimized performances and specific pore structures have been selected for gas adsorption-related applications, such as gas storage, gas separation, gas sensing, and so on. On one hand, the controllable nanocluster structures, as well as their assembling modes, point out multilevel approaches to customize the pore sizes and structures, as well as their adsorption capacity to specific gas molecules, of the assembled materials. On the other, the combination of catalysis and gas adsorption paves a promising way to adsorb harmful gases and then convert them into harmless ones (e.g., the absorb and photodegradation of sulfur mustard by Ag_{12} assemblies) [60].

Of note, for each type of nanomaterials, the researching evolution should contain three stages at least: the discovery of materials, the manipulation of materials, and the application of materials. After decades of development, the control over the structures and properties of metal nanoclusters is getting mature, and it is time for these precise nanomaterials to move towards the application stage (or the real commercial stage).

In a word, the era of atomically precise nanocluster research has arrived, and calls for more efforts for leading to a better future for these promising nanomaterials.

1.3 What are included in this book

In this book, we review the past, the present, the future of metal nanoclusters, and introduces the preparations, the characterizations, the structure determinations, the property investigations, and the practical applications of metal nanoclusters, completing the fundamental view and the comprehensive view. Besides, the researching advances in the very recent years, as well as some personal perspectives, are included in this book, completing the insightful view and the timely view of this book. The overview of each chapter are listed below.

In Chapter 2, we will introduce the synthetic methods of metal nanoclusters, including the general chemical synthesis, the microwave-assisted synthesis, the ligand exchange, the metal exchange, the antigalvanic reduction, and so on. The size and structure controls of metal nanoclusters along with the synthesis are disclosed as well.

In Chapter 3, we will introduce the approaches to analyze and determine the structures of nanoclusters, including single-crystal X-ray diffraction (SC-XRD), transmission electron microscopy (TEM), Fourier transform infrared (FTIR), X-ray photoelectron spectroscopy (XPS), mass spectrometry (MS), thermogravimetric analysis (TGA), nuclear magnetic resonance (NMR), and so on. The atomically precise structures of these nanoclusters are to be presented.

In Chapter 4, we will introduce the mechanisms of how specific nanoclusters are generated. Specifically, the previously proposed mechanisms of size growth, structure evolution, and nanocluster conversion are summarized.

In Chapter 5, we will introduce the chemical-physical properties of metal nanoclusters, including their optical absorption, photoluminescence, magnetism, chirality, electrochemical property, and so on. Besides, the approaches to manipulate these properties by controlling the structures of nanoclusters are disclosed.

In Chapter 6, we will introduce the calculation methods in evaluating the structures and properties of nanoclusters. These theoretical results can provide a better understanding of the electronic structures and the structure-property correlations of metal nanoclusters.

In Chapter 7, we will introduce the aggregation/assembly of nanoclusters in solutions, micelles, and crystals, including the hierarchical assembly, the cluster-based metal-organic frameworks, aggregation-induced emission, and so on. Besides, the effects of assembling on nanocluster structures and properties are summarized.

In Chapter 8, we will introduce the applications of metal nanoclusters, including chemical sensing in the environment and biology, bioimaging, therapeutic application, catalysis, and so on. An outlook for the applications of nanoclusters and cluster-based nanomaterials are also presented.

We hope that this book can point out what is the metal nanocluster, why we study the nanocluster, and how to study the nanocluster. And we believe this book can serve as a versatile and powerful tool for the different target audience(s) such as graduate students in physics, chemistry, and materials sciences, the initiate researches, and professional researches.

References

[1] M.R. Langille, J. Zhang, M.L. Personick, S. Li, C.A. Mirkin, Stepwise evolution of spherical seeds into 20-fold twinned icosahedra, Science 337 (6097) (2012) 954–957, https://doi.org/10.1126/science.1225653.

[2] R. Jin, Y. Cao, C.A. Mirkin, K.L. Kelly, G.C. Schatz, J.G. Zheng, Photoinduced conversion of silver nanospheres to nanoprisms, Science 294 (5548) (2001) 1901–1903, https://doi.org/10.1126/science.1066541.

[3] N.L. Rosi, C.A. Mirkin, Nanostructures in biodiagnostics, Chem. Rev. 105 (4) (2005) 1547–1562, https://doi.org/10.1021/cr030067f.

[4] X. Yang, M. Yang, B. Pang, M. Vara, Y. Xia, Gold nanomaterials at work in biomedicine, Chem. Rev. 115 (19) (2015) 10410–10488, https://doi.org/10.1021/acs.chemrev.5b00193.

[5] A. Gole, C.J. Murphy, Seed-mediated synthesis of gold nanorods: role of the size and nature of the seed, Chem. Mater. 16 (19) (2004) 3633–3640, https://doi.org/10.1021/cm0492336.

[6] K.D. Gilroy, A. Ruditskiy, H.-C. Peng, D. Qin, Y. Xia, Bimetallic nanocrystals: syntheses, properties, and applications, Chem. Rev. 116 (18) (2016) 10414–10472, https://doi.org/10.1021/acs.chemrev.6b00211.

[7] M.C. Daniel, D. Astruc, Gold nanoparticles: assembly, supramolecular chemistry, quantum-size-related properties, and applications toward biology, catalysis, and nanotechnology, Chem. Rev. 104 (1) (2004) 293–346, https://doi.org/10.1021/cr030698+.

[8] R. Jin, C. Zeng, M. Zhou, Y. Chen, Atomically precise colloidal metal nanoclusters and nanoparticles: fundamentals and opportunities, Chem. Rev. 116 (18) (2016) 10346–10413, https://doi.org/10.1021/acs.chemrev.5b00703.

[9] G. Frens, Particle size and sol stability in metal colloids, Kolloid-Zeitschrift und Zeitschrift für Polym. 250 (7) (1972) 736–741, https://doi.org/10.1007/BF01498565.

[10] G. Frens, Controlled nucleation for the regulation of the particle size in monodisperse gold suspensions, Nat. Phys. Sci. 241 (105) (1973) 20–22, https://doi.org/10.1038/physci241020a0.

[11] N. Uyeda, M. Nishino, E. Suito, Nucleus interaction and fine structures of colloidal gold particles, J. Colloid Interface Sci. 43 (2) (1973) 264–276, https://doi.org/10.1016/0021-9797(73)90374-3.

[12] M.P.A. Viegers, J.M. Trooster, Mössbauer spectroscopy of small gold particles, Phys. Rev. B 15 (1) (1977) 72–83, https://doi.org/10.1103/PhysRevB.15.72.

[13] R. Jin, G. Wu, Z. Li, C.A. Mirkin, G.C. Schatz, What controls the melting properties of DNA-linked gold nanoparticle assemblies? J. Am. Chem. Soc. 125 (6) (2003) 1643–1654, https://doi.org/10.1021/ja021096v.

[14] B. Bhattarai, Y. Zaker, A. Atnagulov, B. Yoon, U. Landman, T.P. Bigioni, Chemistry and structure of silver molecular nanoparticles, Acc. Chem. Res. 51 (12) (2018) 3104–3113, https://doi.org/10.1021/acs.accounts.8b00445.

[15] I. Chakraborty, T. Pradeep, Atomically precise clusters of noble metals: emerging link between atoms and nanoparticles, Chem. Rev. 117 (12) (2017) 8208–8271, https://doi.org/10.1021/acs.chemrev.6b00769.

[16] A. Fernando, K.L. Dimuthu, M. Weerawardene, N.V. Karimova, C.M. Aikens, Quantum mechanical studies of large metal, metal oxide, and metal chalcogenide nanoparticles and clusters, Chem. Rev. 115 (12) (2015) 6112–6216, https://doi.org/10.1021/cr500506r.

[17] Z. Gan, N. Xia, Z. Woo, Discovery, mechanism, and application of antigalvanic reaction, Acc. Chem. Res. 51 (11) (2018) 2774–2783, https://doi.org/10.1021/acs.accounts.8b00374.

[18] T. Higaki, M. Zhou, K.J. Lambright, K. Kirschbaum, M.Y. Sfeir, R. Jin, Sharp transition from nonmetallic Au-246 to metallic Au-279 with nascent surface plasmon resonance, J. Am. Chem. Soc. 140 (17) (2018) 5691–5695, https://doi.org/10.1021/jacs.8b02487.

[19] X. Kang, M. Zhu, Cocrystallization of atomically precise nanoclusters, ACS Mater. Lett. 2 (10) (2020) 1303–1314, https://doi.org/10.1021/acsmaterialslett.0c00262.

[20] X. Kang, M. Zhu, Intra-cluster growth meets inter-cluster assembly: the molecular and supramolecular chemistry of atomically precise nanoclusters, Coord. Chem. Rev. 394 (2019) 1–38, https://doi.org/10.1016/j.ccr.2019.05.015.

[21] X. Kang, M. Zhu, Metal nanoclusters stabilized by selenol ligands, Small 15 (43) (2019), https://doi.org/10.1002/smll.201902703.

[22] X. Kang, M. Zhu, Structural isomerism in atomically precise nanoclusters, Chem. Mater. 33 (1) (2020) 39–62, https://doi.org/10.1021/acs.chemmater.0c03979.

[23] X. Kang, M. Zhu, Tailoring the photoluminescence of atomically precise nanoclusters, Chem. Soc. Rev. 48 (8) (2019) 2422–2457, https://doi.org/10.1039/c8cs00800k.

[24] X. Kang, M. Zhu, Transformation of atomically precise nanoclusters by ligand-exchange, Chem. Mater. 31 (24) (2019) 9939–9969, https://doi.org/10.1021/acs.chemmater.9b03674.

[25] X. Kang, H. Chong, M. Zhu, Au-25(SR)(18): the captain of the great nanocluster ship, Nanoscale 10 (23) (2018) 10758–10834, https://doi.org/10.1039/c8nr02973c.

[26] X. Kang, Y. Li, M. Zhu, R. Jin, Atomically precise alloy nanoclusters: syntheses, structures, and properties, Chem. Soc. Rev. 49 (17) (2020) 6443–6514, https://doi.org/10.1039/c9cs00633h.

[27] S. Wang, Q. Li, X. Kang, M. Zhu, et al., Customizing the structure, composition, and properties of alloy nanoclusters by metal exchange, Acc. Chem. Res. 51 (11) (2018) 2784–2792, https://doi.org/10.1021/acs.accounts.8b00327.

[28] S. Takano, S. Hasegawa, M. Suyama, T. Tsukuda, Hydride doping of chemically modified gold-based superatoms, Acc. Chem. Res. 51 (12) (2018) 3074–3083, https://doi.org/10.1021/acs.accounts.8b00399.

[29] J. Yan, B.K. Teo, N. Zheng, Surface chemistry of atomically precise coinage-metal nanoclusters: from structural control to surface reactivity and catalysis, Acc. Chem. Res. 51 (12) (2018) 3084–3093, https://doi.org/10.1021/acs.accounts.8b00371.

[30] Q. Yao, X. Yuan, T. Chen, D.T. Leong, J. Xie, Engineering functional metal materials at the atomic level, Adv. Mater. 30 (47) (2018), https://doi.org/10.1002/adma.201802751.

[31] L. Zhang, E. Wang, Metal nanoclusters: new fluorescent probes for sensors and bioimaging, Nano Today 9 (1) (2014) 132–157, https://doi.org/10.1016/j.nantod.2014.02.010.

[32] R. Jin, Quantum sized, thiolate-protected gold nanoclusters, Nanoscale 2 (3) (2010) 343–362, https://doi.org/10.1039/b9nr00160c.

[33] T.G. Schaaff, G. Knight, M.N. Shafigullin, R.F. Borkman, R.L. Whetten, Isolation and selected properties of a 10.4 kDa gold: glutathione cluster compound, J. Phys. Chem. B 102 (52) (1997) 10643–10646, https://doi.org/10.1021/jp9830528.

[34] G. Soldan, M.A. Aljuhani, M.S. Bootharaju, et al., Gold doping of silver nanoclusters: a 26-fold enhancement in the luminescence quantum yield, Angew. Chem.-Int. Ed. 55 (19) (2016) 5749–5753, https://doi.org/10.1002/anie.201600267.

[35] X.-Y. Xie, P. Xiao, X. Cao, W.-H. Fang, G. Cui, M. Dolg, The origin of the photoluminescence enhancement of gold-doped silver nanoclusters: the importance of relativistic effects and heteronuclear gold-silver bonds, Angew. Chem. Int. Ed. 57 (31) (2018) 9965–9969, https://doi.org/10.1002/anie.201803683.

[36] T. Higaki, M. Zhou, K.J. Lambright, K. Kirschbaum, M.Y. Sfeir, R. Jin, Sharp transition from nonmetallic Au246 to metallic Au279 with nascent surface plasmon resonance, J. Am. Chem. Soc. 140 (17) (2018) 5691–5695, https://doi.org/10.1021/jacs.8b02487.

[37] R. Jin, T. Higaki, Open questions on the transition between nanoscale and bulk properties of metals, Commun. Chem. 4 (1) (2021), https://doi.org/10.1038/s42004-021-00466-6.

[38] C. Zeng, Y. Chen, K. Kirschbaum, K.J. Lambright, R. Jin, Emergence of hierarchical structural complexities in nanoparticles and their assembly, Science 354 (6319) (2016) 1580–1584, https://doi.org/10.1126/science.aak9750.

[39] M. Zhou, C. Zeng, Y. Chen, et al., Evolution from the plasmon to exciton state in ligand-protected atomically precise gold nanoparticles, Nat. Commun.s 7 (2016), https://doi.org/10.1038/ncomms13240.

[40] L.H. Pignolet, M.A. Aubart, K.L. Craighead, R.A.T. Gould, D.A. Krogstad, J.S. Wiley, Phosphine-stabilized, platinum-gold and palladium-gold cluster compounds and applications in catalysis, Coord. Chem. Rev. 143 (1995) 219–263, https://doi.org/10.1016/0010-8545(94)07009-9.

[41] C.D. Bain, E.B. Troughton, Y.T. Tao, J. Evall, G.M. Whitesides, R.G. Nuzzo, Formation of monolayer films by the spontaneous assembly of organic thiols from solution onto gold, J. Am. Chem. Soc. 111 (1) (1989) 321–335, https://doi.org/10.1021/ja00183a049.

[42] M.H. Dishner, J.C. Hemminger, F.J. Feher, Scanning tunneling microscopy characterization of organoselenium monolayers on Au(111), Langmuir 13 (18) (1997) 4788–4790, https://doi.org/10.1021/la970397t.

[43] S.W. Han, S.J. Lee, K. Kim, Self-assembled monolayers of aromatic thiol and selenol on silver: comparative study of adsorptivity and stability, Langmuir 17 (22) (2001) 6981–6987, https://doi.org/10.1021/la010464q.

[44] M. Brust, M. Walker, D. Bethell, D.J. Schiffrin, R. Whyman, Synthesis of thiol-derivatised gold nanoparticles in a two-phase Liquid–Liquid system, J. Chem. Soc. Chem. Commun. (7) (1994) 801–802, https://doi.org/10.1039/C39940000801.

[45] R.L. Donkers, D. Lee, R.W. Murray, Synthesis and isolation of the molecule-like cluster Au-38(PhCH2CH2S)(24), Langmuir 20 (5) (2004) 1945–1952, https://doi.org/10.1021/la035706w.

[46] Y. Negishi, K. Nobusada, T. Tsukuda, Glutathione-protected gold clusters revisited: bridging the gap between gold(I)-thiolate complexes and thiolate-protected gold nanocrystals, J. Am. Chem. Soc. 127 (14) (2005) 5261–5270, https://doi.org/10.1021/ja042218h.

[47] M. Zhu, E. Lanni, N. Garg, M.E. Bier, R. Jin, Kinetically controlled, high-yield synthesis of Au-25 clusters, J. Am. Chem. Soc. 130 (4) (2008) 1138, https://doi.org/10.1021/ja0782448.

[48] X. Kang, L. Huang, W. Liu, et al., Reversible nanocluster structure transformation between face-centered cubic and icosahedral isomers, Chem. Sci. 10 (37) (2019) 8685–8693, https://doi.org/10.1039/c9sc02667c.

[49] Z. Wu, Anti-galvanic reduction of thiolate-protected gold and silver nanoparticles, Angew. Chem. Int. Ed. 51 (12) (2012) 2934–2938, https://doi.org/10.1002/anie.201107822.

[50] T. Chen, Q. Yao, X. Yuan, R.R. Nasaruddin, J. Xie, Heating or cooling: temperature effects on the synthesis of atomically precise gold nanoclusters, J. Phys. Chem. C 121 (20) (2016) 10743–10751, https://doi.org/10.1021/acs.jpcc.6b10847.

[51] T. Higaki, C. Liu, Y. Chen, et al., Oxidation-induced transformation of eight-electron gold nanoclusters: [Au23(SR)16]- to [Au28(SR)20]0, J. Phys. Chem. Lett. 8 (4) (2017) 866–870, https://doi.org/10.1021/acs.jpclett.6b03061.

[52] P. Zhang, X-ray spectroscopy of gold-thiolate nanoclusters, J. Phys. Chem. C 118 (44) (2014) 25291–25299, https://doi.org/10.1021/jp507739u.

[53] D.M. Chevrier, L. Raich, C. Rovira, et al., Molecular-scale ligand effects in small gold-thiolate nanoclusters, J. Am. Chem. Soc. 140 (45) (2018) 15430–15436, https://doi.org/10.1021/jacs.8b09440.

[54] Z. Wu, R. Jin, On the ligand's role in the fluorescence of gold nanoclusters, Nano Lett. 10 (7) (2010) 2568–2573, https://doi.org/10.1021/nl101225f.

[55] K.L.D.M. Weerawardene, P. Pandeya, M. Zhou, Y. Chen, R. Jin, C.M. Aikens, Luminescence and electron dynamics in atomically precise nanoclusters with eight superatomic electrons, J. Am. Chem. Soc. 141 (47) (2019) 18715–18726, https://doi.org/10.1021/jacs.9b07626.

[56] M. Zhou, Y. Song, Origins of visible and near-infrared emissions in Au-25(SR)(18) (-) nanoclusters, J. Phys. Chem. Lett. 12 (5) (2021) 1514–1519, https://doi.org/10.1021/acs.jpclett.1c00120.

[57] J. Xie, Y. Zheng, J.Y. Ying, Highly selective and ultrasensitive detection of Hg2+ based on fluorescence quenching of Au nanoclusters by Hg2+-Au+ interactions, Chem. Commun. 46 (6) (2010) 961–963, https://doi.org/10.1039/b920748a.

[58] P. Nath, M. Chatterjee, N. Chanda, Dithiothreitol-facilitated synthesis of bovine serum albumin-gold nanoclusters for Pb(II) Ion detection on paper substrates and in live cells, ACS Appl. Nano Mater. 1 (9) (2018) 5108–5118, https://doi.org/10.1021/acsanm.8b01191.

[59] W. Jiang, B. Rao, Q. Li, et al., Fluorescence signal amplification of gold nanoclusters with silver ions, Anal. Methods 10 (43) (2018), https://doi.org/10.1039/c8ay01955j.

[60] M. Cao, R. Pang, Q.-Y. Wan, et al., Porphyrinic silver cluster assembled material for simultaneous capture and photocatalysis of mustard-gas simulant, J. Am. Chem. Soc. 141 (37) (2019) 14505–14509, https://doi.org/10.1021/jacs.9b05952.

Chapter 2

Controllable preparation of metal nanoclusters

Manzhou Zhu and Sha Yang

2.1 Introduction

"Noble metals" mainly refer to gold, silver, and platinum group metals (ruthenium, rhodium, palladium, osmium, iridium, platinum) and other metals in the VIII main group. Most of these metals have beautiful color and luster, with strong chemical stability, under general conditions is not easy to react with other chemical substances. Because of their poor abundance as well as good for human body and appear to be precious. At present, noble metals are widely used in the fields of aviation, aerospace, national defense industry, automotive industry and electronic information technology, new energy technology and environmental protection technology, and others. Due to the beautiful color is one of their common features, all of them are widely used in the production of jewelry. Other major application areas are electronic industry and medicine, tableware and other utensils followed by their excellent electrical conductivity and bactericidal and pharmacological effects, respectively. For example, gold and silver have strong bactericidal effect. *Escherichia coli* is not easy to survive in the gold and silver vessels, so the food in the gold and silver vessels is not easy to corrupt. In addition, gold can be used in medicine, including the Chinese patent medicine auranofin, which contains gold to treat rheumatic heart disease and the use of gold isotopes to kill cancer cells. Therefore, gold and silver metals have drawn tremendous interest from the scientific community because of their versatile applications. With the gradual weakening of the monetary function of gold, silver, and other precious metals, the number of industrial precious metals has increased sharply. One of the reasons is that the combination of high and new technologies such as nanotechnology with the traditional deep processing technology of precious metals has greatly expanded the scope and number of applications of precious metals in industry.

Among them, metal nanoclusters with the ultrasmall size (≤ 3 nm) have become an important and emerging branch of nanoscience with many unusual properties. Nanoclusters have precise atomic composition and structure, which can be expressed by molecular-like "molecular formulas" such as $[M_n(L)_m]^q$ (m, n, and q are the number of metal atom M, ligand L, and electric charge in the nanocluster) [1]. In addition, due to their ultrasmall scale, metal nanoclusters also have unique molecular-like physicochemical properties, such as discrete electron energy levels, strong luminescence, magnetism, and intrinsic optical rotation [2]. The physical and chemical properties of the molecules show a strong structural correlation: the metal nanoclusters of isomers also show significant differences. This sensitive structural correlation provides an effective means for understanding the structure-activity relationship and further rational design and synthesis of nanoclusters. Therefore, metal nanoclusters can provide an ideal platform for the research of inorganic functional nanomaterials.

To make good use of this kind of material, it is imperative to synthesize monodisperse clusters with varying atomic composition on a large scale. Development of metal clusters synthetic methodology, to synthesize atomic precision metal clusters in high yield. Further reveal a series of metal clusters of crystal structure, and then to explore the design, synthesis, and assembly rules of metal clusters with different properties, are the basis for studying the relationship between the stacking mode and properties in metal nanoclusters. Moreover, it has important scientific significance for the development of synthetic chemistry and structural chemistry. To date, there are three categories of synthetic method for preparing metal nanoclusters: physical methods, chemical methods, as well as comprehensive synthetic method. According to the formation conditions of nanoclusters, these three methods can be further divided into (a) vacuum process, (b) gas phase method, and (c) liquid phase synthesis strategy.

2.2 Vacuum synthesis

Since 1984, Professor Knight, from the University of California, observed the magic number properties of Na_n clusters. Then C_{60} was discovered and a series of methods have been developed to break through a large scale with simple synthesis methods [1]. The studying of nanoclusters has been rapidly developed in worldwide, and the corresponding meetings also have been gradually increased. For example, the international conference on nanotechnology in engineering and the international conference on C_{60} and its derivatives to discuss the electronic, atomic structures, and properties, theoretical research of clusters also including experimental methods and techniques to prepare or synthesize nanoclusters. The main strategy at that time focused on the vacuum synthesis method of nanoclusters.

2.2.1 Sputtering

In the early stages of the cluster, the ion bombardment and sputtering method regarded as the intriguing approach to producing metal clusters [2–4]. The energetic-particle incident on the surface of solid target which are coated with samples will sputter various kinds of secondary particles, such as electrons, ions, atoms, and clusters. Among them, when the energy of particles was increases from a few to dozens and particles change into fast atom, the resulting clusters are those with electric charges and characterized by mass spectrometry. Fig. 2.1 show the schematic device of the synthesizer for the formation of clusters in ultra-vacuum which is made of ion source, sputtering chamber, mass analyzing system, and detector [5]. The inert gas atoms or ions, such as Ar, Xe, and Kr are often used as molecular beam, due to their inertness of the noble gases does not cause any chemical interference to the sample. Clusters with corresponding charges also can be obtained by adjusting the positive or negative voltages of the corresponding poles in the target chamber. In this device, the sputtering chamber is connected to an external gas tank through the trachea to inflate. The resulting secondary ions reach the target chamber B_2 through an electromagnet which has an average radius of 30 cm and an angle of 60 degrees. Among them, the target chamber B_2 can also be filled with a certain amount of gas to make the generated clusters collide and disintegrate. Then, the ion cluster was passed through a 127 degrees electrostatic analyzer with a radius of 120 mm. An electron multiplier is installed behind the electrostatic analyzer to collections. The detection system was mounted on a removable shelf. Most importantly, the whole device is sealed in a cover, keeping the vacuum of 10^{-9} Torr.

Clusters were prepared by this device, and a large proportion of the neutral clusters were generated by sputtering. An atom escapes from the surface of the solid target in a neutral state after sputtering from noble gases bombardment beam and then in an excited state of free ionization, that is M^x. Then, the process of Auger type de-excitation was talked place and in turn form clusters with charge, $M^x \rightarrow M^+ + e^-$. Of note, the number of atoms in the sputtering clusters and their charge states are also related to the mass, energy and other experimental parameters of the incident particles. Taking synthetic Al clusters as an example, a solid aluminum target is bombarded with an Ar^+ bombardment beam to produce ion clusters [6]. When the target temperature is up to 330°C, it is not only observed that the emission intensity of Al_n^+ clusters was decreased with the increase of n but also found other clusters such as Ar_n^+ and $ArAl_n^+$ clusters. This indicates that the preparation of clusters is also affected by the target temperature and lattice orientation. The formation of these clusters of Ar_n^+ and $ArAl_n^+$ is likely due

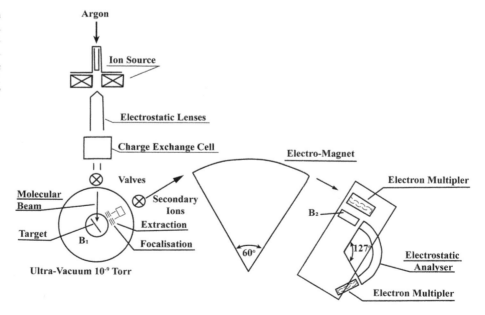

FIG. 2.1 Schema of device in vacuum synthesis. *(From F.M. Devienne, R. Combarieu, M. Teisseire, Action of different gases, specially nitrogen, on the formation of uranium clusters; comparison with niobium and tantalum clusters, Surf. Sci. 106 (1981) 204–221, https://doi.org/10.1016/0039-6028(81)90202-8.)*

to the diffusion of Ar^+ in aluminum which has the high temperature, then they clump together and emitted. As mentioned above, the inert gas argon cannot form clusters in the free state of general equilibrium.

In details, for the synthesis of clusters of type M^q, the surface of solid which is M was bombarded using molecular or ion beam of inert gas with high energy from some keV up to 20 keV. This indicates that the sputtering ion clusters produced by this device are related to the composition and structure of the bombardment target. For example, the ion beam bombarded on the target of copper, aluminum or tantalum, will be generated accordingly Cu_n^+, Al_n^+, and Ta_n^+ (the n is the number of metallic atoms in clusters), respectively. Devienne et al. using this device to obtain the tantalum clusters, while the clusters obtained in ultra-vacuum is rather complex. Moreover, the ratio of clusters is closely related to reaction conditions such as temperature and type of bombardment beam. Compared with temperature, it is found that the bombardment beam has a decisive effect on the proportion of clusters. In addition, the nature of the beam is more important relative to the energy which is attached to the bombardment beam. The oxygen using as bombardment beam and Ta oxides are observed to replace Ta_n^+ clusters. Moreover, with the increase in oxygen bombardment beam pressure, the content of Ta atoms in the Ta cluster are decreases. The Ta cluster contain one or two Ta atoms as main component with the oxygen pressure reaching 7×10^{-7} Torr. While the bombardment beam was change to nitrogen, we can find that the N atoms are always less than two in the Ta cluster and Ta cluster which contain two Ta atoms as main component.

Wang et al. using the neutral argon atoms as bombardment beam to sputter the surface of copper solid target for preparation of Cu clusters [7]. In this work, distinguishing different isotopic clusters from the same isotopic clusters is first proposed and Wang et. al. found the isotopic effect of sputtering ion clusters for the first time in the experiments. In the detailed experiment, the device of VG ZAB-HS was used. The polycrystal copper with the high purity was selected as solid target and the formula of products were determined by mass analyzed ion kinetic energy spectrometry. For example, a series of Cu clusters were found and the number of Cu atoms in the Cu cluster ranges from 1 to 8. In these Cu clusters, two isotopes of ^{63}Cu and ^{65}Cu have been found. The results of mass spectrometry showed that in the Cu-1 clusters, the ratio of ^{63}Cu and ^{65}Cu is close to it in their natural abundances. They also displayed the ratio of ^{63}Cu and ^{65}Cu and kinds of cluster ions which can be divided into two broad categories including the homoisotopic clusters and heteroisotopic clusters. The proportions of heteroisotopic clusters in the Cu clusters which have same Cu atoms also are summarized and we can find that the proportions of heteroisotopic clusters is has a great correlation with the parity of number of Cu atoms. For example, the heteroisotopic clusters had a high proportion in Cu_1, Cu_3, Cu_5, Cu_7, and Cu_9 clusters indicating that the isotopic effect is important in the formation of metal cluster ions by sputtering.

2.2.2 Fast atom bombardment

The fast atom bombardment on the surface of solid target is an important method for the production of metal cluster ions [8]. In the work reported by Sharpe et al., the silver metal and silver salts using as solid target to generate silver cluster ions, that is Ag_x^+, $x=1–5$ [9]. The preparation of the silver cluster is performed on Bruker CMS 47X FT-ICR mass spectrometer used the xenon atoms and ions with 8^{-10} kV beam as bombardment beam. In the initial state of the synthesis process, only clean silver foil was used as solid target, the weak signals of clusters were found after bombardment from ion beam. In this series of silver clusters, the odd atoms of silver than for even with a huge percentage due to the spin-paired electronic structure of odd atoms clusters. The signals of the clusters are increasing with the silver salts (AgBr) used as solid target and the number of silver atoms is range from 1 to 9. Similar to that of silver foil solid target, the most abundant of these ions have $x=1, 3$, and 5. In this case, the silver bromide clusters $Ag_xBr_y^+$ and stoichiometry $[Ag_x(AgBr)_y]^+$ as the products. When the silver salts (AgBr) change into AgO, the product of Ag_x^+ clusters with $x=1–9$ and $Ag_xO_y^+$, as well as Ag_xHO^+ clusters were found.

To prepare alloy clusters, alloy targets can be used as bombardment objects. It is worth noting that the effects of target characteristics on sputtering ion clusters are different. In general, the stoichiometry of the compounds in the target does not favor clusters that produce similar compositions. That is the proportion of compounds in the target is a regulatory factor to prepare different clusters. In addition, the different phases in the metal have a great influence on the formation of clusters. For example, copper and aluminum alloy is a solid solution. After annealing at 330°C, copper is deposited in the small particles with the size of about 1 nm, forming the θ phase with the component is Al_2Cu. In this alloy phase, the emission intensity of the $Al_nCu_2^+$ clusters are much greater than that of in the solid solution phase. While there is little difference between the two the emission intensity of Al_nCu^+ clusters in the alloy phase and the solid solution phase, respectively.

In order to study the characteristics of sputtering cluster, the relationship between the cluster size and their dynamical properties were explored in charged clusters. Wöste et al. designed and built an experimental apparatus for sputtering mass spectrometry and shown in Fig. 2.2 [10]. Based on this device, they achieved the explore on the collision-induced dissociations, chemical reactions and photofragmentation experiments at single cluster level. Similarly, clusters are produced by ion beam bombardment of a solid target. The primary ions are produced in a cold reflective discharge ion source (CORDIS).

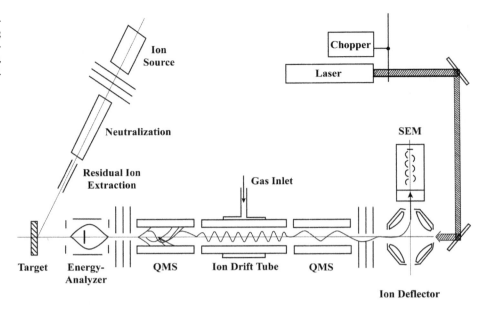

FIG. 2.2 Experimental apparatus for generating nonejection scattering by sputtering cluster ions. *(From P. Fayet, L. Wöste, Production and study of metal cluster ions, Surf. Sci. 156 (1985) 134–139, https://doi.org/10.1016/0039-6028(85)90566-7.)*

The ion source is placed in a gradient vacuum chamber and the distance from the target is 50 cm. The incident angle of the ion source is 50 degrees with the energy of 15 keV. The diameter of beam spot was controlled at 10 mm and the intensity of beam is 3 mA. Neutral beams are employed to prevent charge accumulation in cases where the target is not a conductor. The specific operation is to set up an argon grid of about 10 cm (internal argon pressure should be noted) to obtain neutral particles from the primary ions through the resonance charge exchange. After that the 10% argon ions come out from the ion source turn into neutral ions, and other ions are then ejected from the capacitor. In a vacuum target chamber, two flanges are arranged on both sides facing the target to ionize the sputtering neutral ions and enlarge the cluster beam by photoionization. Cluster ions are generated from sputtering and the ionized neutral particle beam after the selection of energy and mass through the axial energy analyzer and quadrupole lens (QMS). Then the cluster ions come in ion drift tube which has 60 cm long and the cross section of ion trap is similar to that of QMS. By adjusting the voltage of the quadrupole to reduce the kinetic energy of the incoming particles, they can be retained in the ion trap for up to 10 ms. In addition, the vacuum of the ion trap should be as high into 10^{-12} Pa, but it has a gas inlet to allow the collision reaction or chemical reaction between the gas and the cluster ions. The second quadrupole lens in the device is installed at the exit of the ion trap to analyze inelastic scattering events. Finally, the ions clusters are transmitted to the electron multiplier by electrostatic deflection and recorded.

In addition, the device can actually work in a variety of ways. If only the first quadrupole lens is adjusted as the mass spectrum, the mass spectrum of sputtering ion clusters is obtained without the second quadrupole lens moving. If the cluster ion which selected by the first quadrupole lens is received by the ion trap, the collision fragmentation or chemical reaction will occur with the additional of gas atom or molecule in the reaction chamber. In this case, the reaction product is analyzed by the second quadrupole projection which as mass spectrometry. Clusters can also be photo-dissociated by the modulated laser, and the kinetic properties of the clusters can be explored. The photoinduced fragment spectra of Ag_3^+, Ag_5^+, Ag_7^+, and Ag_9^+ clusters as examples. These curves indicate that the signal peaks of the parent cluster (the negative signal) are gradually consumed and the fragmented signal peaks of the products were appeared (the positive signal). The experimental details in this case are that the laser power is 1.6 W with a wavelength of 488 nm, and the kinetic energy of cluster ions keep in 17 keV. The interaction time of them is approximately in 6 ms. It is worth noting that the strong fission reaction of the cluster is induced by a small laser power. This indicated that the cross section of the cluster in the excited state which induced by electron excitation is quite large. This is beneficial to the study of the interaction force between atoms in the cluster. Therefore, the experimental design of three quadrupole lenses not only can be used to study the formation of ions in sputtered clusters but also helpful to study the decay and reaction kinetics of clusters.

2.2.3 Liquid metal ion source

Liquid metal ion source (LMIS) is also a common method of preparing clusters [11]. Ion source is one of the key components in ion beam equipment who is the device to produce ion beam. As one kind of charged material, ion beam shows variety of excellent features. It can be concentrated into thin beams or parallel beams under the action of electric and magnetic fields. And the ion beam can adjust the interaction energy with the target by accelerating or decelerating. Therefore, in the design and manufacture of ion beam equipment, the development of high-performance ion source is very important. In the late of 1970s, the liquid metal ion source was successfully developed, which has small size, high height and stable emission [12]. Besides, using liquid metal ion source make the beam spot of the aggregation ion beam as small as submicron magnitude. The liquid metal ion source is made of a fine tungsten wire with a tip of a few microns in radius at one end, which is inserted into a liquid metal tube. In the manufacture of a liquid metal ion source, the tungsten wire with a diameter of 0.5 mm is first dealt with electrochemical etched to obtain a tungsten needle with a tip of only 5–10 µm. The molten liquid metal is then attached to the tip of a tungsten needle. Under the action of an external reinforcing electric field, the liquid monitor forms a tiny tip, that is Taylor cone. It is worth noting that the electric field intensity at the tip of the liquid metal ion source can be as high as 1010 V/m. So, at such high electric fields, metal ions on the surface of liquid metal will be escape by means of evaporation to create a beam of ions. The variety of particles, mainly monatomic ions, but also including a considerable number of ionized clusters and charged metal droplets, are emitted by intense field evaporation.

The electric field at the emission point of the ion cluster is controlled by the extraction voltage, that is V_{ex}. The ions will be emitted with the extraction voltage increased to a range of 5–7 kV. Using this device, Walle et al. prepared a series of Ge_n^p ($n/p \leq 25$, $1 \leq p \leq 4$) ions clusters and their molecular formula were characterized by mass spectrometry. From the mass spectrum, we can observe that many Ge clusters with the high charge, for example Ge_{25}^{3+}, Ge_{31}^{3+}, Ge_{64}^{3+}, and Ge_{76}^{3+} which have great abundance. It is worth noting that the number of clusters in these clusters presents a scale, mainly with a periodicity of six. For example, in the case of N < 45, each additional 6 atoms result in exactly 1 additional sp^3 closed ring of 8 atoms. For N > 45, the number of atoms added to satisfy the periodicity is $43 + 3 + m6$, where m is an integer ≥ 2. In the structure of 43 atomic clusters, the 6 hanging bonds are in such a position that only 3 atoms are needed to form a new 6-membered ring. Add another 12 atoms to make 58, which is a pretty tight structure. And then follow the scale of 6 until the n up to 88. This periodicity is also observed in Si clusters, suggesting that it is related to cluster geometry and bonding mode. Should it be noted that in order for the liquid metal ion source to operate continuously, the pressure of the metal at the time of melting is required to be sufficient to maintain the extraction voltage. The Liquid metal ion source (LMIS) are mainly applied to metals such as gold, gallium, indium, and tin [13].

2.3 Gas phase nanoclusters

The molecules or clusters produced in the real air phase are ideal isolated model systems. The conditions for the formation of meteorological clusters include (1) the generation of large quantities of gases (single or diatomic molecules); (2) the monomer is cooled by a quenching process (by collision with an inert gas atom or by adiabatic expansion); and (3) monomers gather into clusters or grow by colliding with each other. After summarizing, it can be found that there are two main ways of cluster generation: the one is the monomers gather in an atmosphere of inert atoms and the another is the growth of monomers into clusters by their own cooling.

2.3.1 Evaporation and gas condensation

The basic process of evaporation and gas condensation is that the original material (element or compound) is placed in an evaporating pan which in inert gas chamber with a low pressure (usually <1 atmosphere) and heated at high temperature until gasification. The evaporated atoms or molecules collide with inert atoms or molecules and it cools down rapidly as it loses energy. Subsequently, a supersaturated region is formed near the evaporation source, which leads to nucleation and growth into clusters. After the formation of clusters, they quickly move away from the supersaturated region through convection (because of the temperature difference) to avoid further agglomeration and growth. Therefore, the use of flowing inert gas is beneficial for controlling cluster size and improving cluster generation efficiency. There have three basic factors for the controlled synthesis of clusters using this approach: (1) provides energy for monomer atoms to aggregate to form supersaturated regions; (2) the cooling medium (inert gas) can carry away the energy of the evaporated atoms; and (3) the rate at which clusters are able to depart from the supersaturated region after formation. At this time, the metal atoms close to the evaporation source have a low degree of supersaturation, and the energy removed by the evaporated atoms is slow, but the convection is fast. Therefore, for given source material, low evaporation rate, light inert gas, and low pressure will result in small cluster size. The technique was pioneered by Pfund, Burger, and van Cittert, and by Harris et al. which make ultrafine metal particles by well controlled evaporation from an oven.

The work reported by Granqvist et al. presents this novel and versatile technique for the production of clusters by evaporation from a temperature-regulated oven containing a reduced atmosphere of an inert gas [14]. In this work, the authors gained insights into the synthesis of Al clusters, and then extended to the synthesis and preparation of other metal cluster systems including the Mg, Zn, and Sn also including the particles of Cr, Fe, Co, Ni, Cu, and Ga, as well as larger Al particles. In addition, the author explored the relationship between the size of several metal clusters formed by thermal evaporation and the type of inert gas, the pressure and the evaporation rate during the synthesis process. The process is that, during large amounts of evaporation, the clusters are exposed to the microscope grid, which is usually located in the center of the cooling plate. They also did a series of experiments to verify that the size of the cluster was independent of its location on the cooling plate. However, the size of the clusters increases linearly with increases of pressure of the inert gas. And the inert gases with larger atomic masses helps to produce the larger clusters. The size cluster of is proportional to the logarithm of the vapor pressure of metal atoms under the first order approximation.

Thermal evaporation and gas condensation were first used to produce and study atomic clusters, and played an important role in the study of cluster formation conditions, mechanisms and agglomeration characteristics. Thermal evaporation and gas condensation methods are widely used in the preparation of large size nanoclusters and solid nanomaterials. However, the clusters obtained by direct evaporation under inert gas still have a wide distribution by using these methods even when the experimental conditions are strictly controlled. It is usually a lognormal distribution of volume shown as Eq. (2.1).

$$f_{LN}(x) = \frac{1}{(2\Pi)^{1/2} \ln \sigma} \exp\left(-\frac{(\ln x - \ln \bar{x})^2}{2 \ln^2 \sigma}\right) \tag{2.1}$$

In the Eq. (2.1), the $f_{LN}(x)$ is the normalized normal distribution function and x is the size of the cluster. Among them, x is the statistical median and σ is the geometric standard deviation.

$$\ln \bar{x} = \sum n_i \ln x_i \left(\sum n_i\right)^{-1}; \quad \ln \sigma = \left[\sum n_i (\ln x_i - \ln \bar{x})^2 \left(\sum n_i\right)^{-1}\right]^{1/2}$$

where n is the number of clusters of diameter x. As $\ln \to 0$, Eq. (2.1) approaches the Gaussian. The size distribution of the evaporated aluminum clusters with the pressure of Ar is 3.5 Torr. The average diameter of the clusters is 4.1 nm. However, the size range is from 2 to 10 nm. This wide size distribution indicates the aggregation between clusters is occurred due to the slow migration of clusters from the vicinity of the evaporation source.

In previous work, the formation mechanism including the nucleation and growth of cluster which between atoms and solids have been solved by scientists. As a platform for exploring physical and chemical properties, clusters with a single atom or fixed size are ideal models. However, only the clusters with broad size range have been obtained. Therefore, how to obtain clusters with narrow size range is a major problem faced by researchers. In 1980, Sattler et al. gives the first report which is realized reduce the size of clusters and narrow the size distribution by using forced convection in an inert atmosphere [15]. In this work, the authors have for the first time synthesized full-scale clusters between atoms and small particles and characterized their molecular formulas by mass spectrometry. They found the intensity of clusters beam are surprisingly high in the presence of inert gas condensation. In addition, the intensity of cluster beam will gradually decrease with the increase in the size until they are faint and invisible. The authors used this equipment to prepare antimony clusters using He as an inert gas, and investigated the effects of various factors on cluster synthesis. The detailed process is He gas added into the condensing cell, which contains metallic vapor (Antimony steam) and then the metal clusters ejected through the upper hole of O_1. Next, the clusters pass through the differential vacuum chamber and hole O_2 into the mass spectrometer. The holes of O_1 and O_2 perform the dual functions of collimation and differential vacuum. The inlet of inert gas is G and inert gas is pumped out by a differential pump system.

The intensity of antimony clusters decreased gradually with the increased pressure of He gas until the strength is almost zero. The author thinks that the main reason is that antimony is dispersed in the He gas and then deposited on the wall of the condensing chamber. Then the influential factors of cluster formation were explored and indicated that the type and purity of the inert gas are important (other types or mixed with the small amount other gases in He will lead to the failure of the experiment). Besides, the temperature of the He gap flow is also an important determinant (even a slight change in temperature cannot synthesize antimony clusters). From the Mass spectrogram of antimony clusters, we can find that there have a series of Sb_n clusters, where n ranges from 1 to 100. Start with Sb_{16} cluster, the clusters of Sb_{4n} have higher intensity than others. Although the clusters with $n > 100$ can be detectable, but they are generally of low intensity. To prove the universality of this method, the authors synthesized two other kinds of clusters of Bi and Pb clusters. In the case of Bi_n clusters, the n is concentrated between 1 and 21, and Bi_5, as well as Bi_7 have the larger intensity than other clusters. In Pd_n clusters, the Pb_7 clusters with the highest intensity was found in the series of clusters between Pb_3 and Pb_{13}.

2.3.2 Laser vaporization

Vaporizing metal materials with pulsed laser is an effective mean to obtain the metallic cluster which metal is refractory materials. The basic principle is focusing the laser to a small area by using the optical system, is the joule level of energy acting on the solid target surface and resulting energy applied to the solid target surface is of the joule order. Therefore, the temperature of the fixed area on the surface of the metal target will be extremely high, which is prone to thermionic emission and neutral particle evaporation. Then, ion or neutral particle are cooled by ultrasonic expansion or inert gas collisions to form clusters. So, we can know that the laser evaporation method is actually using pulse laser instead of furnace heating [16–21].

In 1982, Smalley and co-workers report the produce ultracold beams of copper clusters within a pulsed supersonic nozzle has been used [22]. Prior to this work, this group designed laser-vaporization supersonic nozzle source for forming cluster beams of aluminum. However, the above method has two defects poor stability and difficulty in reproducibility. In this work, the author made the pulse-to-pulse reproducible by moving the target to a new position prior to each vaporization laser pulse. The schematic diagram of this device for generating clusters by laser evaporation is shown in the Fig. 2.3. Cluster preparation with this device has the following distinct characteristics: (a) the excited metal takes the shape of a rod and rotates and shifts under the light of a laser to maintain sufficient evaporation; (b) the evaporation of the target is done with a concentrated pulsed laser beam; and (c) metallic clusters were formed in high pressure conduit areas, and the high-pressure chambers are derived from inert gas beam. To prepare copper clusters, a 0.65-cm diameter copper rod was selected and rotated as well as translated by a screw mechanism which operates continuously during the pulsed vaporization. Eventually, they synthesized a series of copper clusters in the range from 1 to 13 copper atoms.

Rohlfing et al. reported the combination of the above laser evaporation with ultrasonic expansion, laser ionization, and time of flight mass spectrometry to design a combined device [23]. Using this device (Fig. 2.4), they explored the relationship between the distribution of Ni clusters with the laser intensity and frequency. Then the relationship between the ionization potential and the size of the clusters is revealed by them. The evaporation of laboratory metal target was performed by the second order resonant output of Q switched Nd:YAG laser. The photoionization process is carried out on two laser systems. Although the variety of gas phase clusters is very rich, the synthesis of these devices has been reported endlessly. We will give more attention to the gas phase synthesis of noble metal nanoclusters and mainly in Au, Ag, and Cu. The earliest studies of gas phase Au and Ag clusters date back to the 1960s [24].

Although sputtering, evaporation, and other methods are used to prepare the gas phase cluster, but it is not easy to control and cannot be separated or stabilized. Coating a layer of organic ligands on the metal surface may be an effective way to solve the above problems. In Faraday's time, colloidal chemistry began to sprout and colloidal particles were synthesized gradually, among which the gold sol gained more attention because of the gorgeous color and fascinating performance. The main reason is that Faraday published a pioneering work in which gold colloids were prepared from gold salt solution and phosphine which dissolved in ether by a two-phase reaction (Eq. 2.2).

$$HAuCl_4 + P \rightarrow Au(sol) \tag{2.2}$$

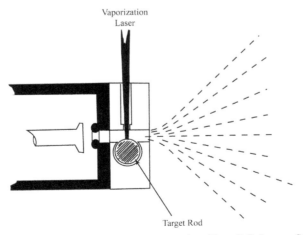

FIG. 2.3 The schematic diagram of device for generating clusters by laser evaporation. *(From D.E. Powers, S.G. Hansen, M.E. Geusic, A.C. Puiu, R.E. Smalley, Supersonic metal cluster beams: laser photoionization studies of copper cluster (Cu_2), J. Phys. Chem. 86(14) (1982) 2556–2560, https://doi.org/10.1021/j100211a002.)*

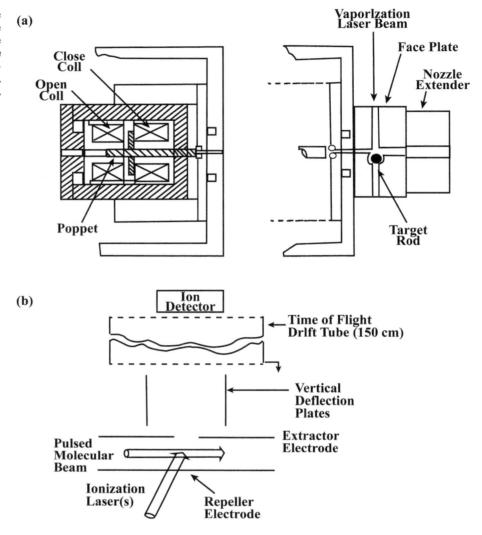

FIG. 2.4 A schematic diagram of (A) the pulsed nozzle and laser vaporization zone and (B) the photoionization region of the time-of-flight mass spectrometer. *(From E.A. Rohlfing, D.M. Cox, A. Kaldor, Photoionization of isolated nickel atom clusters, J. Phys. Chem. 88(20) (1984) 4497–4502, https://doi.org/10.1021/j150664a011.)*

Since Faraday's work laid the foundation for colloidal science, researchers have carried out a series of studies and explorations to prepare colloidal clusters. Colloid clusters can actually be divided into two broad categories, one is semiconductor cluster and the another is solution of clusters. Semiconductor clusters have attracted much attention because of their applications in microelectronics industry. With suitable energy band structure, large specific surface area and unique surface effect, semiconductor nanomaterials can be used to produce new energy (hydrogen energy) and deal with environmental pollution problems. In general, liquid phase deposition reactions for compound semiconductor clusters depend on homogeneous nucleation forming seed nuclei that can be stabilized as colloid. To achieve the target semiconductor cluster with sizes <5 nm, the proper control of the reaction temperature, concentration, and solvent is required. The semiconductor clusters with thus small size will exhibit the electronic states of discrete molecular orbitals rather than continuous band structures. Ionomers are a class of copolymers containing ion edge-chain groups that aggregate into domains similar to negative ion traps formed in colloidal clusters. The metal ions such as Cd^{2+} and Pd^{2+} can easily be exchanged into these ionic domains to synthesize ideal semiconductor clusters.

The solution of clusters are common synthetic clusters, which consist of hydrophilic and hydrophobic clusters. The overall method is relatively simple and that is the metal ions are dissolved in an aqueous solution and a suitable reducing agent is added and then the cluster is formed. It is worth mentioning that the ligands are considered for the stability of isolated clusters. Although the variety of metallic clusters have been reported and their synthesis methods are largely similar. We will give the more attention on the synthesis of noble metal nanoclusters and mainly in Au, Ag, and Cu. Although several efforts were made to synthesize colloids of noble metals, but most of them resulted in nanoparticles with a broad size distribution. Colloidal gold is a typical hydrophobic colloid solution and has attracted much more attention.

However, the size of the gold sol has not been characterized due to the limitation of the electron microscopy at that time. With the development of the electron microscope, a preliminary characterization of the synthesized colloidal gold clusters was carried out by electron microscopy reported by Turkevich et al. in 1952. In this work, the author investigated the nucleation and growth of gold in colloidal solution and revealed the role of reducing agents in the identification of nucleating agents. A series of experimental results proved the effect of the amount of gold, nucleation mechanism, and growth characteristics on the final products. In 1972, Frens et al. inspired by the large *London-van der Waals* forces on the surface of metal to consider the attractive forces between clusters. A new idea about the stability of colloidal clusters is put forward and the conjecture is verified experimentally. The authors show that the size of the clusters plays a key role in the stability of the colloids, and the small size of the clusters contributes to the stability of the colloids. The influence is due to *van der Waals* forces between the small clusters.

Cluster with the monodisperse that is to synthesize truly uniform nanoclusters at the ultimate atomic level, namely, atomically precise nanoclusters is an ideal platform for study of nanoscience. However, creation of highly monodisperse nanoparticles was a great challenge in those early days. In the field of colloidal clusters, several synthetic methods have been developed to prepare atomically accurate nanoclusters. For example, the chemical synthesis, microwave-assisted, sonochemical, hydrothermal, photoreductive, etching-based, kinetically controlled, electrostatically induced reversible phase-transfer method, template controlled, and so on. Among them, chemical synthesis as the mainstream method has gained more attention. In this book, we take gold, silver, and copper as an example to introduce the above several synthetic strategies.

2.4 Chemical synthesis in solution

2.4.1 Metal complex reduction method

In gold nanocluster, the phosphine was first selected to enclose the gold core as ligands to maintain cluster stability. The halogens contained in metal salts are often wrapped on the surface of gold clusters which protected by phosphine ligands as in situ ligands [25]. In order to synthesize these kinds of clusters, metal salts react with the ligand and form M-L complexes in a suitable solvent. Then, the M-L complexes reacted with reducing agent (sodium borohydride, carbon monoxide) giving nanoclusters. Using this method, a series of gold(I) triphenylphosphine derivatives were prepared and reported by Malatesta et al. [26] They also found that the product of AuCl was prepared from the precursors of AuCl salt and the number of ligands on the metal can be controlled by the amount of ligand added.

After a series of reports on gold(0)-gold(I) polynuclear complexes, the first gold nanocluster of $Au_{11}(PPh_3)_7(SCN)_3$ was reported by Mason et al. [27] On the basis of analytical data, the authors describe the geometric structure of the cluster and report the structural information of this cluster. From structural analysis that Au_{11} cluster has an incomplete icosahedral geometric construction with an Au(I) kernel which is surrounded by ten gold atoms and Au_{11} further protected by ligands. Not only that, the authors also predicted the existence of other clusters such as $Au_{10}L_7X_2$, and $Au_{12}L_7X_4$. Since then, a series of metal clusters protected by phosphine ligands or phosphine-halogen mixed ligands have been studied and reported. Of note, the phosphine ligand not limited to monodentate phosphine. Among the reported gold cluster, Au_{13} cluster with completed icosahedral structure has attracted more attention due to 8-electron stable closed electron shell structure. The previous reports of simulations for the bond patterns in clusters showed that the clusters with higher nuclear were present. Therefore, the Au_{13} cluster is considered to be the initial state of a polyhedral cluster due to its regular icosahedral structure. In 1981, Brian et al. confirmed this prediction and giving the report of the synthesis and structure for $[Au_{13}(PMe_2Ph)_{10}Cl_2](PF_6)_3$ cluster [28]. Prior to this report, there have two methods of metal evaporation and reduction of gold complex to prepared gold clusters or gold compounds. In this work, the additives are considered in order to increase yield and synthesize new substances. The first thing examined was the use of $Ti(\eta-C_7H_8)_2$ as an additive and added into the solution which included the precursors of gold complex and found that the yield of the product has been greatly improved and up to 80% base on the gold complex. The author thinks that the main reason is that in toluene solvent, the additive reaction with gold complex and the combination form an insoluble substance of $[Au_3(PR_3)_2TiCl_3]_n$. However, in the solvent of ethanol, the intermediates of $[Au_3(PR_3)_2TiCl_3]_n$ which insoluble in toluene form several gold clusters and trace mononuclear complexes. With the help of excessive counterions, the $[Au_9(PPh_3)_8]Y_3$ and $[Au_{11}(PR_3)_{10}]Y_3$ were gained with the high purity and yield. To prepared $[Au_{13}(PMe_2Ph)_{10}Cl_2](PF_6)_3$ cluster, the $[Au_{11}(PPhMe_2)_{10}]^{3+}$ was chose as precursors and dissolve in ethanol reaction with the NEt_4Cl. After the X-ray crystallographic analysis, the $[Au_{13}(PMe_2Ph)_{10}Cl_2](PF_6)_3$ is determined to have the center of symmetry and represents a marked contrast with the centered cuboctahedral geometry established for Rh_{13}. After that, many groups have studied the cluster of Au_{13} extensively.

Since then, nanoclusters with the high nuclear have gained a lot of attention and They aim to reveal the geometry structure and explore their bonding patterns and crystal stacking model. In the field of gold/silver nanoclusters protected by phosphine ligands, Teo' group has carried out a series of work. Pure gold cluster of $[(Ph_3P)_{14}Au_{39}Cl_6]^{2+}$ coprotected by phosphine and chlorine ligands has been reported by Teo et al. [29] The synthesis of $[(Ph_3P)_{14}Au_{39}Cl_6]^{2+}$ is done in a similar way, with the addition of Ph_3P, $HAuCl_4$, and $NaBH_4$. While Au_{39} has a high nuclear structure with an unprecedented 1:9:9:1:9:9:1 layered hep structure. Another important cluster with high nuclei is $Au_{55}(PPh_3)_{12}Cl_6$ due to it has a high tendency to self-assemble forming one- (1D), two- (2D), and three-dimensional (3D) organized structures. Because of this self-organization, Au_{55} cluster has been used in several applications. In order to synthesize $Au_{55}(PPh_3)_{12}Cl_6$ cluster, Ph_3PAuCl complex mix with B_2H_6 and reduced to get $Au_{55}(PPh_3)_{12}Cl_6$ reported by Schmid et al. [30].

Based on the synergistic interaction between difference metals, the alloy nanoclusters exhibit more excellent physical and chemical properties. In order to synthesize the alloy nanoclusters, two metal salts were selected to complexing with the ligands and then reduced to form the targeted alloy clusters. In 1984, the famous cluster of 25 was synthesized and reported by Teo et al. which is an assembly cluster consisting of two icosahedral 13 by sharing a vertex [31]. Au-Ag alloy nanocluster of $[(Ph_3P)_{12}Au_{13}Ag_{12}Cl_6]^{m+}$ synthesized by means of metal complex reduction method. The specific steps are generally described the gold salt is mixed with silver salt and reaction with the Ph_3P ligand and to form Ph_3PAuCl and $(Ph_3P)_4Ag_4Cl_4$ complex. The Ph_3PAuCl and $(Ph_3P)_4Ag_4Cl_4$ complex were further reduction by $NaBH_4$ in ethanol and gain the novel $[(Ph_3P)_{12}Au_{13}Ag_{12}Cl_6]^{m+}$ nanocluster. Then, single crystals were cultured by liquid-liquid diffusion. Therefore, a combination of Au and Ag might be a fruitful way to produce large metal alloy clusters. The synthesis and reporting of 25-atoms cluster is expected to provide important insights into the synthesis and structure of clusters with high nuclear and also give the benefit to the development of new materials.

In Teo' reports, the vertex-sharing poly-icosahedral clusters not just limited to 25-atoms clusters, the 38-atoms and 37-atoms are also the members of the "cluster of clusters" which is a series of clusters that follows a well-defined growth pathway. In those years, there are two broad categories of highly symmetrical high-nuclearity metal clusters which both can be ascribed to "cluster of clusters" [32]. The one is V_n polyhedral clusters is defined as a cluster with $(n+1)$ atoms on each edge of the polyhedron. Another is the S_n supra-cluster is defined as a cluster of n smaller cluster units fused together via vertex-, edge-, or face-sharing. These clusters are particularly interesting because they have a completely closed shell structure that contributes to stability. Both the 38-atoms and 37-atoms have three icosahedral 13-atoms and sharing three vertices in a cyclic manner with the one or two capping atoms plus. In addition, they have been confirmed to idealized C_3 symmetry. To synthesize these two clusters, the $(p\text{-Tol})_3PAuCl$ mixes with $(p\text{-Tol})_3PAgCl$ complex in ethanol and then the $NaBH_4$ was added. Among them, the neutral cluster of $(p\text{-tolyl}_3P)_{12}Au_{18}Ag_{20}Cl_{14}$ was first to be reported which is believed to be the largest bimetallic cluster of gold and silver at the time [33]. This nanocluster was isolated from the reduction of mixed solution including the $p\text{-tol}_3P$, $HAuCl_4$, and $AgAsF_6$ by $NaBH_4$ in the ethanol. Then the cationic cluster of $[(Ph_3P)_{14}Au_{18}Ag_{20}Cl_{12}]^{2+}$ which is closely related to the neutral ones.

2.4.2 Brust synthesis and beyond

With the development of the field of metallic colloidal solution, various preparation techniques have been established. Particles in colloidal solutions prepared by different methods tend to agglomerate and eventually leads to the loss of dispersity and the formation of large incompatible materials [34,35]. In addition, solvent removal usually results in the complete loss of the ability of the metal particles to reforming the colloid. In an earlier report of the preparation of gold colloidal solution by Faraday, the two-phase system was employed, in which gold salt is dissolved in water and phosphine is dissolved in ether and result is a ruby-colored aqueous solution consisting of dispersed gold particles. In 1994, Brust et al. developed a method to combine the two-phase system with ion extraction technique, using alkane mercaptan as monolayer protective ligand expected to produce a unique metal nanoparticle (Fig. 2.5) [36]. In the first step, metal atoms or ions gather together to form a metal core, which is then complexed with a self-assembled mercaptan monolayer to protect the stability of the metal core. Then the cluster grows and stabilizes with the simultaneous growth of the metal core and the protective ligand. The two-phase system can promote the metal nucleus to react with the ligand surface during the growth process. It is worth noting that in order to promote the REDOX reaction between the two phases, the appropriate oxidation reducing agent should be selected. In this reaction, the tetraoctylammonium bromide chose as phase-transfer catalyst transfers $AuCl_4^-$ from aqueous solution to toluene (as shown in Eq. 2.3). The mercaptan is added to the system and reacts with the metal salt in the same phase (toluene) reduces Au^{3+} to Au^{1+} and to form the Au-S complex (as shown in Eq. 2.4).

$$H^+AuCl_4^-(aq) + N(C_8H_{17})_4^+Br^- \rightarrow N(C_8H_{17})_4^+AuX_4^- + HX \quad (2.3)$$

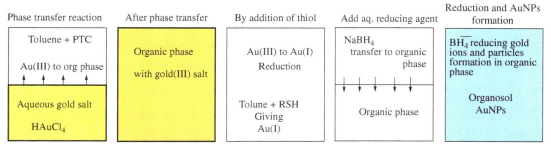

FIG. 2.5 Overview of the synthesis of thiol capped gold nanoparticles by the method of Brust et al. *(From M. Brust, M. Walker, D. Bethell, D.J. Schiffrin, R. Whyman, Synthesis of thiol-derivatised gold nanoparticles in a two-phase liquid–liquid system, J. Chem. Soc. Chem. Commun. 7 (1994) 801–802, https://doi.org/10.1039/c39940000801.)*

$$N(C_8H_{17})_4^+ AuX_4^- + 3RSH \rightarrow {}^-(AuSR)n^- + RSSR + N(C_8H_{17})_4^+ + 4X^- + 3H^+ \tag{2.4}$$

Then, the Au-S complex (-(AuSR)n-) was reduced by sodium borohydride (dissolved in aqueous solution) accompanied by the color of the solution varies from colorless to dark (as shown in Eq. 2.4). The sodium borohydride is the main reducing agent and reduces the part of Au^{1+} to Au^0. These reactions for the biphasic were initially considered to occur at the organic-aqueous interface. The conditions of the reaction should be considered such as the ratio of Au and S, the reaction solvent, as well as reaction temperature. The final product of this reaction was characterized by transmission electron microscope (TEM), infrared spectroscopy (IR), and X-ray photoelectron spectroscopy (XPS). The size of them are determined to be in 1–3 nm range.

$$-(AuSR)_n^- + BH_4^- + RSH + RSSR \rightarrow Au_x(SR)_y$$

$$N(C_8H_{17})_4^+ AuX_4^- + BH_4^- + RSH + RSSR \rightarrow Au_x(SR)_y$$

Since then, this method has attracted extensive attention because it promotes the synthesis of highly stable functionalized nanoclusters which has small size at room temperature. This method give a major breakthrough in the field of the stable nanoclusters. This method was later named as Brust-Schiffrin method (BSM) and has been widely used in the synthesis of multifunctional nanoclusters. The importance of this approach described here is threefold. The first one is that the Brust-Schiffrin method can be regarded as one simple method for preparing surface-functionalized nanoclusters. Using this method, the plentiful stable nanoclusters were synthesized and reported [37]. The stability studies show that the extraordinary stability of these nanoclusters is can be attributed to the strong coordination ability between mercaptan and metal. That is the mercaptan forms a strong and stable protective shell on the surface of the metal core because of the strong interaction of M—S bonds. This passivates the surface of the metal core and prevents it from cracking or further growing. The second point is that the size of clusters can be controlled by reaction conditions due to the kinetics of cluster growth are determined by surface cover. In the previous report of colloidal metallic nanoclusters, the kinetics of reaction is usually controlled by the metal ion reduction. Therefore, the controllable synthesis of clusters can be achieved by adjusting the conditions. A series of reports suggests that the size of nanoclusters can be controllable regulation by the different types or amounts of thiol. According to the reports, the size of nanoclusters can be reduced from 8 to 2 nm by changing the ratio of thiol to gold. And last but not least, the nanoclusters as a new kind of metal nanomaterials show the unique physical and chemical properties due to their quantum confinement effect. Therefore, their intrinsic interests make them have great potential in a wide range of application which attract researchers.

Despite the Brust-Schiffrin method has been widely used and studied, there are still some issues that are not clear. For example, the amount of reducing agent is usually an excellent control condition for controlling cluster size in other synthesis methods. While using the Brust-Schiffrin method, how the thiol ligands regulate the size of nanocluster during the synthesis process is unknown. In addition, the resulting cluster products still show polydispersity even though the thiol ligands and metal surface have strong binding energy. The most important is the lack of understanding of the synthesis process and the exploration of the mechanism of this method. Researchers initially believed that this reaction occurred at the interface between the oil phase and the water phase in this method. However, the results of the investigation show that the factors affecting the response are mainly concentrated in intensity of mixing, vessel size suggesting that the reactions do not occur on the interface. In the traditional approach, the ratio of S with Au is set to be <3 and believed that only a small part of the Au^{3+} is reduced to Au^+. After reduction, metal nanoclusters of a specific size are obtained, which has nothing to do with the type of thiol. However, the reports of larger size nanoclusters using the very bulky thiol as ligands

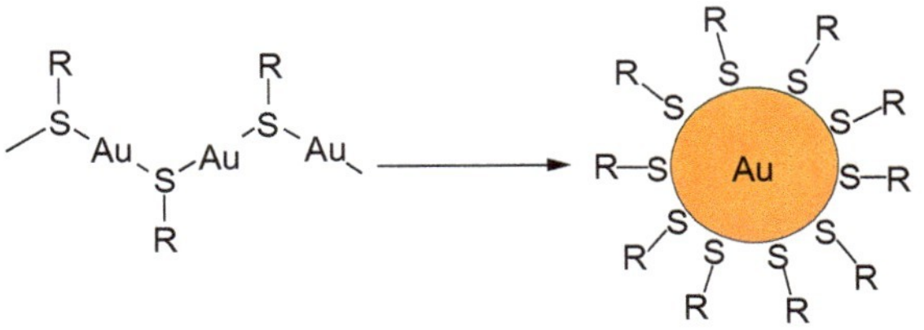

FIG. 2.6 Schematic representation of thiol-capped gold nanoparticles synthesis route of soluble Au(I)-thiolates. *(From M.K. Corbierre, R.B. Lennox, Preparation of Thiol-capped gold nanoparticles by chemical reduction of soluble Au(I)–thiolates, Chem. Mater. 17(23) (2005) 5691–5696. https://doi.org/10.1021/cm051115a.)*

indicate the emergence of special cases. Therefore, the in situ formed Au(I)-thiolates complex is the precursor for the synthesis of nanoclusters have been putted forward. In 2005, the soluble Au-S complexes were prepared and separated by Lennox et al. [38] and then the Au-S complexes were further reduced by superhydride. In this work, they demonstrated that the solubility of the gold(I)-thiolate complex in solvent could be controlled by bulky thiol (Fig. 2.6). Avoiding its insolubility in solvents by changing the interaction between Au⋯Au. This work confirms that the complex is indeed the precursor of the synthetic nanocluster. In addition, nothing that were produced just by the two-phase reactions without additional of phase transfer catalyst. However, with the addition of phase transfer catalyst, nanoparticles are rapidly formed. In other words, the phase transfer catalyst promotes the formation of clusters in the presence of both mercaptan and metal. The above results and discussion indicate that the reaction is carried out in organic phase because of the aqueous solution is remain colorless during the reaction process.

In order to further explore the role of Au(I)-thiolates complex in cluster synthesis, Lee et al. [39] went on to explore the reaction process using the dioctadecyl disulfides to replace the octadecanethiols. In this work, the dioctadecyl disulfides was selected as a ligand or covering to stabilize the nanoclusters. The size of the product nanoclusters was measured to evaluate the influence of the Au(I)-thiolates complex on it. Their TEM images show that the two nanoclusters are roughly spherical and have the comparable size. It is important to note that the ratio of gold to S is the same in above two reactions. This indicated that the constant ratio of sulfur to gold used in the synthesis process (1:3) is a major factor affecting the growth of gold nanoclusters. In 2001, Murray et al. [40] further explored the influence of complexes on the synthesis of nanocluster. Three kinds of thiol or disulfide including the unsymmetrical disulfides and thiol mixtures were considered. In the first part of this work, the authors propose several factors affecting the synthesis process and products. Using the alkanethiol as ligands, the Au(III) which in $Oct_4N^+AuCl_4^-$ complex is reduced to Au^{1+} with a guaranteed ratio of S:Au greater than 3 to form the Au(I)-thiolates complex as well as dialkyl disulfides. In the second step, the Au(I)-thiolates complex was reduced by BH_4^- in the presence of dialkyl disulfides. Therefore, both the Au(I)-thiolates complex and dialkyl disulfides may affect the nucleation and passivation of nanoclusters. Moreover, the authors speculated that the Au (III) could not be reduced by dialkyl disulfides and it reduced to Au^0 by the BH_4^-. However, the experimental results show that the nanoclusters synthesized by using alkanethiol or dialkyl disulfides as ligands have considerable size. This indicated that the Au with 3+ or 1+ may have the similar nucleation and passivation processes/mechanism.

Then, the mixed monolayers on the surface of the nanocluster also have been explored. The main mixed objects in this report are the mixture of different alkanethiol (the ratio is 1:1) and unsymmetrical disulfides. The final product was characterized by NMR, and the results showed that the proportion of the product was not equal to the proportion of the alkanethiol. One of the thiol-protected nanoclusters has an absolute advantage and occupies a large proportion. This result can indicate that nanoclusters reflects a strong preferential binding to certain kind of thiol in the mixed thiol system. The authors propose a variety of possible influencing factors such as thermodynamic effects, the composition of nanoclusters, the relative salvations of thiol, or the functional groups on the surface of the ligand. To explore the major influencing factors and the kinetic reactivity of the Au(I)-thiolate complex, the Au(I)-thiolate complexes were isolated and characterized by NMR. The final result shows that all sulfur-containing substances in the reaction medium act on the passivation product of the final product. That is, the main factor in the composition of the nanoclusters may come from thermodynamics and/or the type of alkanethiol. The number of ligands on the surface of the cluster is varies due to the different binding capacities of different kinds of thiol to the metal. Therefore, the unsymmetrical dialkyl disulfides was taken into consideration due to the equal amounts of −SR and −SR′ can be brings on the surface of the Au. The results of NMR characterization show that the ratio of gold nanoclusters protected by −SR and −SR′ ligand is almost one indicates that this is a kinetically determined product.

The above series of reports indicate that it is an effective means to achieve controlled synthesis of the final products of clusters by changing the factors which effect on the kinetics. In addition, thermodynamics is also an indispensable factor in the process of nanocluster synthesis. Kinetics plays an important role in the matter of nucleation and growth nanocluster, while their thermodynamics is reflected in the stabilization of nanoclusters make them have a particular size or shape. For example, Moiseev et al. prepared and reported a family of Pd_n nanoclusters base on this scheme indicating that it can be used for the preparation of nanoclusters with single size or mono-dispersity [41]. Then, to explore the influence factors of the specific size and shape of nanoclusters in solution, Gelbart and Whetten developed a model to compare the dispersity nanoclusters which stabilized by thermodynamically and surfactant, respectively [42]. The results show that the average size and morphology of clusters are mainly determined by ratio of surfactant to nanoclusters. In their system, they assume that all the clusters are in statistical thermodynamic equilibrium with each other.

In 1999, Gelbart et al. explored the influence of thermodynamics on the size of metal nanoclusters by means of experiments and theoretical combination [43]. In this work, the gold nanoparticles were synthesized firstly based on the Brust-Schiffrin method. In Brust' work, the size of nanoclusters is relatively concentrated and small (about 2 nm). While this work shows that nanoclusters with a wider range of sizes (1.5–20 nm) be synthesized using this method, and the main inducer is the ratio of gold to thiol. In addition, the thermodynamic theory of metal nanoclusters is proposed.

The detailed synthesis steps are consistent with those reported by Brust et al. with alter the molar ratio of gold to sulfur from 3:1 to multiple variables including 1:1, 2:1, 3:1, 3.5:1, 4:1, and 6:1. And they found that the nanoparticles did not exhibit outstanding stability, especially in the atmosphere of air and light. These nanoparticles were characterized by X-ray powder diffraction and the particle size is decrease gradually with the increase of the thiol ratio. The smallest size of nanoparticles in thermodynamic stability is about 1.5 nm and the required ratio of Au with S is 2.5:1. Then the author put forward the thermodynamic stability theory which lay a foundation for the rational design and synthesis of nanoclusters with the given size and morphology.

The model mentioned above not only finds out the basis of particle size variation but also predicts that particle size will increase dramatically to infinity with the mercaptan thiol approaches to zero. In 1995, Schiffrin group prepared gold nanoclusters using Brust-Schiffrin method to substitute thiol with organic disulfide [44]. And the insoluble dithiol cross-linked nanoclusters with about 2 nm were obtain and the size are similar to that of thiol stabilization. Then, the effect of reaction solvent on the size of the nanoclusters was investigated by replacing toluene with diethylether. The authors found that the size of the clusters could increase to 8 nm without the addition of thiol. In addition, the stability nanocluster is also difficult to maintain unless the clusters remain in the original solution without any isolate or purification.

Metal nanoclusters with the desired size or shape show the unique and attractive performance provides the application potential in various fields. Therefore, it is valuable to investigate the relationship between nanocluster size with their additional nanocluster properties. To address the above the issues, the ability to synthesize a large number of clusters in a simple way is necessary. In this regard, it is important to determine the reaction factors that may affect the product in the synthesis reaction. Following the insight into the types of thiol, the ratio of thiol to metal, thermodynamics, as well as kinetics, Murray et al. [45] have investigated the size trends of nanocluster are based on temperature variations in the first step. In this work, the authors present the synthesis and characterization of a series of unfractionated gold nanoclusters which protected by dodecanethiolate. In this work, the Brust-Schiffrin method were still used and the ratio of thiol to chloroauric acid are extend on a much larger scale. Most importantly, the author changed the temperature and rate of addition for reducing agent. And, the size range of the nanoclusters is determined to be from 1.5 to 5.2 nm by combined results of various characterization techniques. More importantly, each kind of nanoclusters has good dispersion and high size concentration. According to the above analysis, the size of the cluster mainly depends on its synthesis path.

Compared with the others method, Brust-Schiffrin method has attracted more attention due to its tremendously influential in the field of metal nanomaterials. Even though many groups have been joined in the research of Brust-Schiffrin method with the various metal nanoclusters have also been synthesized, there remain significant questions regarding their detailed mechanisms [46]. Or more accurately, it is not clear in our current understanding for the precursor species before the addition of reducing agents (i.e., BH_4^-). In previous reports, the M-L complex has been assumed as the precursor of two-phase reactions (aqueous and organic phase), and the influence of various reaction conditions on the size and morphology of the final product has been explored. As a result, determining the composition of precursor is important because it directly affects the size and properties of the nanocluster. Therefore, the Lennox' group address the above issues via the identification and quantification the composition of precursor in Au, Ag, as well as Cu nanoclusters under the Brust-Schiffrin method [47]. In this work, the authors track and monitor the reaction process in typical two phases Brust-Schiffrin method base on the modern analytical techniques (Fig. 2.7).

The transformation of Au^{3+} to Au^{1+} was first determined by the ultraviolet-visible (UV-vis) absorption spectra. Through experiments, the authors determined that the conversion of Au^{3+} to Au^{1+} was followed by the addition of thiol, and the thiol has to be added in 2 equiv of to ensure the complete conversion. The composition of the material in the reaction process was determined by nuclear magnetic resonance spectroscopy (NMR) spectroscopy. The NMR spectroscopy of the organic phase which separated after sufficient stirring were employed. Contrary to previous conjecture, the peak attributed to

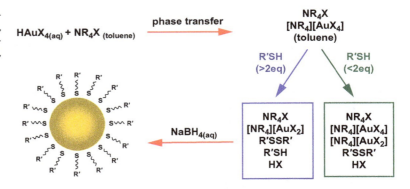

FIG. 2.7 Schematic representation of two-phase Brust-Schiffrin Au nanoparticle synthesis. *(From P.J.G. Goulet, R.B. Lennox, New insights into Brust–Schiffrin metal nanoparticle synthesis, J. Am. Chem. Soc. 132(28) (2010) 9582–9584, https://doi.org/10.1021/ja104011b.)*

Au-tetraalkylammonium complexes were detected instead of Au-S complex. So that means that the intermediate process of the cluster is going to form Au-tetraalkylammonium complex. That is, Au-tetraalkylammonium complexes is the precursor to synthetic nanocluster (reaction 2.5). Then the authors verified the above conclusion by adding excessive thiol reaction with Au-tetraalkylammonium complexes.

$$[NR_4][AuX_4] + 2R'SH \rightarrow [NR_4][AuX_2] + R'SSR' + 2HX \qquad (2.5)$$

However, the authors go on to discuss the composition of the intermediate in the case of the water in the system. The author puts forward the water in the reaction contributes to the formation of Au-thiolate complexes. Therefore, there is a high probability that Au-thiolate complexes is a precursor to the reaction due to the it is difficult to completely remove the water in the two-phase system and the water will also be added with the addition of sodium borohydride. In this case, the synthesis of nanoclusters follows the reaction (2.6).

$$[NR_4][AuX_4] + 3R'SHf\,[AuSR']n + R'SSR' + NR_4X + 3HX \qquad (2.6)$$

Since the water-soluble phase is always colorless, previous work assumed that the reaction would take place in the organic phase. However, Lennox' group showed that the precursor of the reaction was Au-tetraalkylammonium complex before the addition of the reducing agent by NMR characterization technology in the case of without water. On account of this, Tong' group [48] suggested that the reaction takes place in the inverse micelles with water core in the organic solvent. Then, the metal ion inside the inverse micelle is reduced by thiol and then further reduced by reducing agents to form nanoclusters. The final step is the ligands exchange of thiol to replace the ammonium bromide. However, there are several problems with this path that cannot be explained clearly. For example, how the thiol enters into the interior of the micelle to reduce the Au^{3+} to Au^{1+}. Whereas inside the micelle is a water core, most of kinds of thiol are hydrophobic and difficult to get into. Previous reports have also suggested that the thiol dissolved in toluene could not reaction with the gold in the aqueous solution phase in the absence of a phase transfer catalyst. Then a series experiments proved that the process is still the metal ion is brought from the aqueous phase to the organic phase by complexation with phase transfer catalyst.

The aforesaid series of studies shows the nanoclusters are different synthesized by Brust-Schiffrin method compare to the previous method such as gas-phase cluster formation or use the other small organic molecules that are not thiol as ligands. The most significant difference is that the thiolate ligand-protected nanoclusters have the ability to repeatedly isolated from and re-dissolved in common solvents just as organic molecules. Then, Murray et al. summarizes the synthesis, composition, structure, performance, and functional applications of these monolayer-protected nanoclusters [49]. As the extensive exploration and research for monolayer-protected nanoclusters, the Brust-Schiffrin method have been identified to have three important steps including nucleation, growth and passivation. Among them, the passivation rate of the particles is determined by the concentration of thiol. That is, the high thiol concentration makes the faster passivation of the nanocluster and resulting in small size. On the other side, the rapid addition of reducing agents such as sodium borohydride after the nucleation formed by burst clusters is an effective strategy for the synthesis of small size. In fact, the size of clusters and their particle size distribution will be very little affected from the experimental point of view. Although several mechanisms that may affect the size of nanocluster are suggested in this review, they are in fact not clear because the synthesis process is not understanding with comprehensive enough. For example, to rationally design some other thiol ligands, continue to regulate its reaction solvent or some other modification on the reaction conditions.

In 2013, Kumar et al. insight into the formation mechanism of particle in the Brust-Schiffrin method use a detailed population balance model [50]. Their results indicate that the nanoclusters were formed in Brust-Schiffrin method adopts a new synthetic route, including nucleation, generation and covering (Fig. 2.8). Although the synthesis of gold nanoclusters

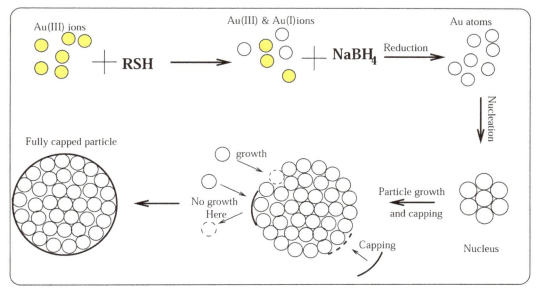

FIG. 2.8 Schematic representation of synthetic route for Brust-Schiffrin method. *(From M. Brust, M. Walker, D. Bethell, D.J. Schiffrin, R. Whyman, Synthesis of thiol-derivatised gold nanoparticles in a two-phase liquid–liquid system, J. Chem. Soc. Chem. Commun. 7 (1994) 801–802, https://doi.org/10.1039/c39940000801.)*

has achieved remarkable success under Brust-Schiffrin method. However, the nanoclusters synthesized by the current method is polydisperse. Most importantly, the yields of nanocluster obtained under the current method are often low. However, the separation and purification of clusters is quite difficult, which hinders its further application in the fields of nanomaterials.

2.4.3 Kinetically controlled synthesis method

Total synthesis is a milestone discovery in human efforts to uncover the mystery of nature. It refers to an ideal synthesis method of generating desired organic and/or biological molecules with atomic precision from simple and readily available precursors. It contributed to the centuries-long prosperity of organic chemistry as not only a science of expression but also an art. This fine line of reaction is also desirable in nanoscience. Recently, atomically precise Au_{25} nanoclusters has been synthesized and reported by Donkers et al. using the Brust-Schiffrin method. However, the reported synthetic yield of Au_{25} was quite low, and requiring multistep solvent extraction to remove other large and multidispersed clusters. In 2007, Zhu et al. established a kinetically controlled method with low temperature aggregation base on the Brust-Schiffrin method (Fig. 2.9) [51]. This synthesis strategy was first applied to synthesize Au_{25} clusters protected by thiol ligands with high yield. In this reaction, the two-phase method is used as the model system, the regulation of reaction kinetics is used to realize the controllable synthesis of nanoclusters with single-size and high yield. This process usually consists of two steps: (i) Au(III) (such as $HAuCl_4$) is reduced to Au(I) by thiol to form intermediate Au(I)-S complex; (ii) further reduction of Au(I) to Au (0) by a strong reducing agent (i.e., $NaBH_4$). An important finding of this work is that the kinetics of formation of Au(I)-S intermediates is crucial for the synthesis of Au_{25} clusters by two-phase method with high yield. The experimental

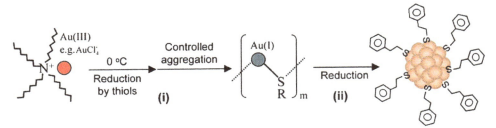

FIG. 2.9 Schematic representation of synthetic route for kinetically controlled method. *(From M. Zhu, E. Lanni, N. Garg, M.E. Bier, R. Jin, Kinetically controlled, high-yield synthesis of Au_{25} clusters, J. Am. Chem. Soc. 130(4) (2008) 1138–1139, https://doi.org/10.1021/ja0744302.)*

results show that controlling the reaction temperature (0°C) and stirring conditions (slow) can produce the aggregation state of specific Au(I)-SR intermediates, which leads to the enrichment and formation of Au_{25}. The low temperature method successfully eliminated the formation of clusters other than Au_{25}.

In order to further understand the dynamic control process, especially the correlation between the aggregation state of Au(I)-S intermediates and the high yield of Au_{25} clusters. Dynamic light scattering (DLS) measurements have been performed on Au(I)-S intermediates. DLS measurements showed that Au(I)-S formation was mainly distributed at 100–400 nm at 0°C and very slow stirring conditions. While at room temperature, Au(I)-SR intermediate formation showed peaks of multiple distributions (including small sizes distributed at 100–400 nm and other macroscopic deposits with larger sizes). These results clearly indicate that these two synthesis strategies result in significant differences in size and internal structure of Au(I)-S polymers. Subsequently, Zhu et al. further investigated the effect of Au(I)-S aggregation time on the final Au_{25} yield. $NaBH_4$ solution was added at different times to reduce Au(I) aggregation into gold clusters. The UV-vis spectrum of the final product showed that the optimal aggregation state of Au(I) was the polymerization duration of Au(I)-S complex of 1 h, corresponding to the highest yield of Au_{25}.

After the high-yield synthesis of $[Au_{25}(SC_2H_4Ph)_{18}]^-TOA^+$ clusters, this synthesis strategy has been applied to the synthesis of other gold clusters, especially novel clusters. In 2009, two novel phenethyl mercaptan-protected gold clusters including the $Au_{20}(SC_2H_4Ph)_{16}$ and $Au_{24}(SC_2H_4Ph)_{20}$ [52] were synthesized with high yield by adjusting the adding rate and method of reducing agent and using the low-temperature aggregation kinetically controlled synthesis method. In order to synthesize $Au_{20}(SC_2H_4Ph)_{16}$ cluster, [Au(I)-S] aggregates were formed in the reaction system and reduced by adding $NaBH_4$ (about one time equivalent of gold) aqueous solution slowly within 30 min. During this period, the reaction mixture was still in the state of continuous slow stirring. The crude products of $Au_{20}(SC_2H_4Ph)_{16}$ clusters were extracted with mixed solvent, and the very pure Au_{20} nanoclusters were obtained. The gold clusters collected were analyzed by particle size exclusion chromatography (SEC) to verify their purity. For the synthesis of $Au_{24}(SC_2H_4Ph)_{20}$, the reduction agent $NaBH_4$ (about one equivalent of gold) was dropped into the reaction system within 15 min to reduce [Au(I)-S] aggregates. UV-vis spectroscopic characterization showed that the crude reaction product before separation showed a distinct absorption band (about 700 nm) and a shoulder peak (about 780 nm). In combination with other detailed analyses, the cluster composition was determined to be $Au_{24}(SC_2H_4Ph)_{20}$ by electrospray ionization mass spectrometry (ESI-MS).

Except for gold nanoclusters, a series of silver and copper nanoclusters were synthesized by kinetically controlled synthesis method. For example, one Cu_{25} (Cu-H) nanoclusters [53] and two Cu(I) clusters (Cu_{11} [54] and Cu_{13} [55]) protected by mixed ligands have been synthesized. Among them, the Cu(I) have been found own the strong fluorescence emission properties. The synthesis strategy can be briefly described as: metal salt reaction with ligands (two kinds of ligands) to forms M-L complex; the M-L complex is transformed into final product reduced by reducing agent. By means of a series of characterization methods such as hydrogen nuclear magnetic resonance spectroscopy, ESI mass spectrometry, thermogravimetric analysis, XPS, as well as other methods to characterize their molecular formula. Moreover, their precise atomic structures were also determined by single crystal X-ray diffraction.

The design and optimization ligands on the surface of metal nanoclusters is an effective mean to adjust their size, structure, and properties at the atomic level. There are three types of thiols that have been reported including the aliphatic, aromatic, and bulk thiol ligands. Based on the kinetic control synthesis method, Zou et.al designed and synthesized a pair of silver nanoclusters by manipulating the type of thiol ligands [56]. The compositions of the two silver nanoclusters were determined by single crystal X-ray diffraction (SC-XRD) and X-ray photoelectron spectroscopy (XPS). Notably, the asymmetric distribution from the three ligands (thiolate, phosphines, and halogens) which on the surface of the clusters results in the chirality of the nanoclusters. The structural analysis of them shows that the two kinds of nanoclusters have certain similarity and uniqueness. The differences between the two nanoclusters indicate that the structure of the silver clusters is very sensitive to the composition of thiolates and halogen ligands. These findings demonstrate the key principle of ligand-shell anchoring tri-ligand protection for silver clusters. This work will provide further insights into the synthesis of chiral metal clusters by tailoring surface ligands.

2.4.4 One-pot synthesis

Because the nanoclusters synthesized by Brust-Schiffrin method show the excellent stability promotes exploration of their performance and application. With the development of nanoclusters, nanoclusters have become a new type of material and have attracted wide attention and have been inspired a number of related approaches. In 1995, Brust et al. developed a related method of Brust-Schiffrin method and different is the reaction is performed in a single phase [57]. To synthesize the nanoclusters with similar stability, the bifunctional stabilizing thiol ligand (i.e., *p*-mercaptophenol) considered. However, the *p*-mercaptophenol is soluble in aqueous solution and cannot complete the reaction with gold salt in the organic phase. The reaction was therefore performed in a single phase with polar methanol as the solvent. In addition, methanol and water are mutually soluble which contribute to the subsequent reaction of water-soluble sodium borohydride with Au-S complex in the system. The following steps are similar to that of Brust-Schiffrin method, and the final products are characterized by transmission electron microscope. The detailed synthesis steps can be described as (a) the $HAuCl_4$ and p-mercaptophenol were co-dissolved in the methanol solvent. To avoid deprotonation of thiol, the acetic acid was added to the above mixture. Subsequently, a freshly prepared aqueous solution of sodium borohydride was added and continue stirring for 30 min; (b) to purify the final product, the solvent purification method is mentioned, in which excess thiol and other by-products are extracted by diethyl ether and removed the water by anhydrous Na_2SO_4. The final product was characterized by TEM and IR which indicated that the range size of particle is from 2.4 to 7.6 nm. In this work, the authors expected that his method will be a new way to synthesize clusters. Over the next few decades with the rapid development in cluster science, single-phase method has indeed become the mainstream strategy in the field of nanocluster for the preparation of monodisperse and atomically precise nanoclusters.

By combining previous reports and the mechanism of nanocluster synthesis, the implementation of single-phase or one pot method is mainly in the choice of polar solvent, such as the one pot method common use the method and tetrahydrofuran as the reaction solvent. Furthermore, different nanoclusters can be synthesized by adjusting the reaction conditions such as temperature, solvent, concentration, reducing agent, and so on.

Except for thiol-ligand protected Au_{25}, individual clusters were difficult to synthesize directly in 2009. Despite the synthesis of nanoclusters having been greatly developed, the final product is usually multiple nanoclusters. Therefore, simple synthesis methods should be developed for precise control over particle size, that is, one pot for one size. The single-phase method is considered to have great potential for development. In 2009, the Jin' group developed a one pot method for the preparation of Au_{25} compounds which protected by functional thiol [58]. Based on the one pot method, the functional thiol reaction directly with metal salts without the phase transfer catalyst. Another reason is that most functional mercaptans have a water-soluble phase that makes it difficult to reach the metal to complexed in the two-phase process. Therefore, the synthesis of functional Au_{25} clusters by one-pot method will further promote the application of the nanoclusters in the field of materials. The general synthesis steps are as follows: $HAuCl_4$ dissolves in tetrahydrofuran solvent and the system ice bath for 30 min. Then, the thiol is added to the system and reacts with gold salt at a particularly low rotational speed with color of the reaction solution changes gradually from yellow to colorless. The sodium borohydride dissolved in ice water is added to the system rapidly and the reaction is going to stir faster. The product was purified by solvent method and the most cases it is done several times with methanol to remove excess thiol. The UV-vis spectrum of the crude product is very similar to that of pure Au_{25}. In the mass spectrum of this product, only a single peak around 7395 Da was found. From the UV-vis spectrum and mass spectrum, we can classify the product as Au_{25}, and it is pure Au_{25}. This indicated that the one pot is a suitable and simple method is used to prepare monodisperse nanoclusters. In addition, in order to determine the universality of this method, several other functional ligands were extended to synthesize a series of Au_{25} nanoclusters including water-soluble phase clusters.

Au_{144} is an important gold nanocluster because it bridges the famous Au_{25} cluster and gold nanoparticles with surface plasmon resonance in the FCC structure. In previous reports, Au_{144} was synthesized by a two-phase method. Purification requires the further size-focused under the high temperature and excess of thiol. Since the one pot method is a simple, convenient, and efficient synthetic methods for the synthesis of the nanocluster with a single size, the synthesis of Au_{144} should be able to be implemented by regulated the reaction conditions. Therefore, after the series of conditions explored, Qian et al. synthesized the Au_{144} [59]. The general synthesis procedure is that the gold salt reacts with the thiol ligand in the presence of tetraoctylammonium bromide to form Au(I)-SR complex. And then the sodium borohydride dissolve in aqueous solution comes in and reduces Au(I)-SR complex to nanoclusters. To synthesize Au_{144} nanoclusters, the resulting products need to be further combined with size focusing method. Of note, the final product consists of two types of nanoclusters the major part is Au_{144} and the minor of Au_{25} nanoclusters. The separation of the two is based on solvent extraction (Fig. 2.10). Most importantly, the author gives insight into the influences of the gold-to-thiol ratio, TOABr, and O_2 on the synthesis of Au_{144} nanoclusters. The results of a series of controlled experiments show that a large number of thiols are critical conditions to facilitate the synthesis of accurate atomic Au_{144} nanoclusters with monodisperse. The main reason is that in the size etching stage excessive thiol promotes the decomposition of other unstable nanoclusters, and it also can improve the stability of the Au_{25} and Au_{144} nanoclusters that have been formed. For the role of TOABr, after the analysis of the results from a series of comparative experiments, the authors suggest that the effect of Br^-/Cl^- in TOABr is minimal, but the targeted clusters cannot be synthesized without TOA^+. The above results indicate that TOABr is essential for the synthesis of Au_{144} nanocluster. The author believes that the reason is that TOABr promotes the formation of Au(I)-SR complex and modifies its structure to make it suitable for the synthesis of Au_{144} nanocluster. Finally, it was found that oxygen is also the main influencing factor of this reaction.

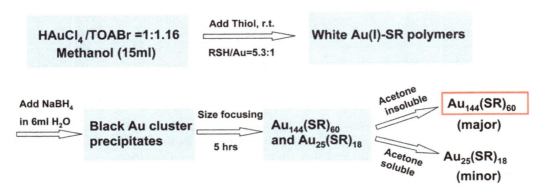

FIG. 2.10 Synthetic procedure for synthetic the $Au_{144}(SR)_{60}$ nanocluster via one-pot method. *(From H. Qian, R. Jin, Ambient synthesis of $Au_{144}(SR)_{60}$ nanoclusters in methanol, Chem. Mater. 23(8) (2011) 2209–2217, https://doi.org/10.1021/cm200143s.)*

Not only the reaction solvent but also the selection of reducing agent in the reaction is very important for the synthesis of nanoclusters of different sizes. The selection of reducing agent mainly depends on its different reducing capability, such as the strong and mild. As previously reported, sodium borohydride is a common and major reductant, and in fact the reductant of the nanocluster is not limited to sodium borohydride. Sodium cyanoborohydride, borane-tert-butylamine complex, triphenylsilane, hydrazine hydrate, as well as CO are the common reducing agent for the synthesis of the nanoclusters. Compared with typical sodium borohydride which is a relatively strong reducing agent, sodium cyanoborohydride with the mild reducing capability in organic chemical reactions and often selected as a selective reducing agent. Because the nucleation and growth rate of nanoclusters directly affect their size, controlling the growth rate is an effective means to synthesize nanoclusters with the controlled size. For example, Pradeep's group chose the sodium cyanoborohydride as the reducing agent in order to obtain smaller clusters by regulating the size of clusters with its mild reducing capability [60]. The smaller nanocluster, that is $Au_{18}SG_{14}$ (SG is the abbreviation of glutathione which in thiolate form), shows a very strong photoluminescence performance (with the quantum yield is 0.053) with a high purity. It emits red light in both aqueous and solid state under UV illumination. In addition, the author also verified the universality of the method and synthesized several nanoclusters with clear composition.

Recently, Xie' group report the detailed synthetic process of Au_{25} nanoclusters under the well-controlled reduction using CO as reducing agent [61]. To address the issues of the mechanism in synthesis process, the UV-vis absorption and ESI-MS characterization techniques were employed and point out the all stable intermediate species in the process from the Au(I)-thiolate complexes to Au_{25} nanoclusters. This new synthetic strategy is based on the one pot method and can be divided into two main steps. As shown in Fig. 2.11A, the first step is the kinetically controlled growth of soluble Au(I)-thiolate complexes which form from the reaction of gold salt to thiol by CO and to form a narrow size distribution of nanoclusters [62]. The second step is the thermodynamically controlled size focusing to obtain the thiolated Au_{25} nanocluster which is carried out under 25°C about 3 days. The final product is characterized by UV-vis absorption and

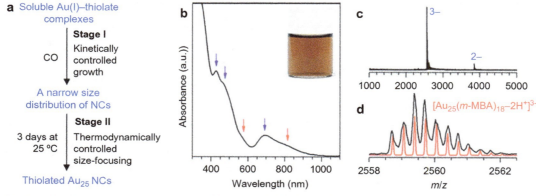

FIG. 2.11 CO-directed synthesis of the atomically precise thiolated Au_{25} NCs. *(From Z. Luo, V. Nachammai, B. Zhang, N. Yan, D.T. Leong, D. Jiang, J. Xie, Toward understanding the growth mechanism: tracing all stable intermediate species from reduction of Au(I)–thiolate complexes to evolution of Au_{25} nanoclusters, J. Am. Chem. Soc. 136(30) (2014) 10577–10580, https://doi.org/10.1021/ja505429f.)*

ESI-MS as Au_{25} nanocluster (Fig. 2.11B–D) [62]. This work provides evidence that gold nanoclusters protected by thiol ligands have well-controlled growth in a unique environment, that is, CO-directed synthesis. The different color changes of the products during the reaction indicate the formation of nanoclusters of different sizes and they are further determined by UV-vis absorption and ESI-MS.

Since it has been reported that kinetics is an effective means of controlling the size of synthetic nanoclusters, it was certainly useful to change the redox potentials of reactants in the reaction. Because the reduction potential of some reactants is sensitive to pH in the reaction environment, pH is often considered as an influencing factor in the synthesis of nanocluster. For example, the reducing capability of CO can be enhanced by increasing the pH of synthesis leading to the increase in the size of nanocluster. Therefore, Xie's group give the report of the systematic study of the pH effects on the formation of thiolated Au NCs in the reaction of CO-reduction [63,64]. As shown in Fig. 2.12, the Au^{3+} reaction with the GSH to form the Au(I)-SR complexes which is the precursor of the nanocluster. Then, the pH adjusting and CO-reduction were performed to adjusting the reaction pH from 7 to 11. The pH in reaction is brought by the NaOH which is often to adjusting the reaction pH in the synthesis of nanoclusters. With the injection of CO, color of the reaction solution is constantly changing, and the color change at different pH is also very different (Fig. 2.12). The crude products synthesized under different pH display different colors, implying that they have different sizes. The resulting nanoclusters were characterized by UV-vis absorption, ESI-MS, as well as 1D and 2D NMR analysis. This work not only presents a simple one-pot method which can produce different sized Au nanoclusters but also shows that pH adjusting is an effective method in tailoring the nanocluster size.

In addition to the solvents, reducing agents and pH, the order in which raw materials are added and the temperature in the synthesis process are also the important factors [65,66]. In 2014, Song et al. reported the synthesis of phenylselenol ligand-protected Au_{24} [67] and Au_{25} [68] nanoclusters using the one-phase method (Fig. 2.13). The two clusters were synthesized by adjusting the order in which the reactants were added. The detailed synthesis steps are (i) the transfer of gold salt ($HAuCl_4 \cdot 3H_2O$) from aqueous phase to toluene organic phase under the action of tetraoctylammonium bromide (TOABr); (ii) cooling Au(III) of the organic phase to 0°C (ice bath); and (iii) synchronous reduction is added on top of the original low temperature aggregation. The changes in step were as follows: PhSeH was dissolved in glacial toluene and $NaBH_4$ was dissolved in 1 mL glacial water. They synchronous added to Au(III) coolant drop by drop. Using this synthesis strategy, the synthesis yield of Au_{25} clusters was increased to 55%, and the reaction time was shorter. Furthermore, the overall atomic structure of $[Au_{25}(SePh)_{18}]^- ToA^+$ nanoclusters was revealed by modern characterization methods, and the results show that the framework structure and atomic stowage pattern of $[Au_{25}(SePh)_{18}]^- ToA^+$ nanoclusters are similar to that of the thiol-ligand protected Au_{25} nanoclusters.

However, the change of ligand has a great influence on the electronic structure and properties of the nanocluster. By controlling the reaction conditions, adjust ligand and reducing agent in the reaction system and gold, the proportion of phenol was prepared with a novel structure of benzene selenium protected Au_{24} nanoclusters. This new nanocluster was determined by MALDI-TOF mass spectra, thermogravimetric analysis (TGA), nuclear magnetic resonance (NMR), element analysis, and X-ray single crystal diffraction. Then according to the concept of different ligand coordination ways, phenylselenol and triphenylphosphine as the common ligand to be considered. After the screening and optimization of reaction conditions, a new nanocluster was synthesized. Through the regulation of ligand and reducing agent, a

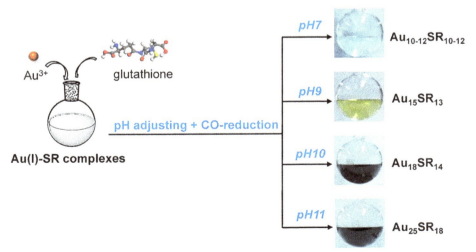

FIG. 2.12 Schematic illustration of a series of gold nanoclusters were synthesized via the CO-reduction method by adjusting PH. *(From Y. Yu, X. Chen, Q. Yao, Y. Yu, N. Yan, J. Xie, Scalable and precise synthesis of thiolated Au_{10-12}, Au_{15}, Au_{18}, and Au_{25} nanoclusters via pH controlled CO reduction, Chem. Mater. 25(6) (2013) 946–952, https://doi.org/10.1021/cm304098x.)*

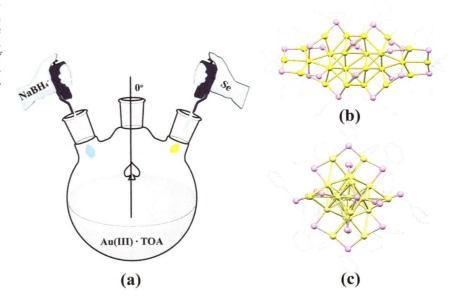

FIG. 2.13 Schematic illustration of phenylselenol ligand-protected Au_{24} and Au_{25} nanoclusters using the one-phase method. *(From Y. Song, S. Wang, J. Zhang, X. Kang, S. Chen, P. Li, H. Sheng, M. Zhu, Crystal structure of selenolate-protected $Au_{24}(SeR)_{20}$ nanocluster, J. Am. Chem. Soc. 136(8) (2014) 2963–2965, https://doi.org/10.1021/ja4131142.)*

cyclic $[Au_{60}Se_2(Ph_3P)_{10}(PhSe)_{15}]^+$ nanocluster was obtained by one pot method [69]. It shows a palm-shaped Se-Au_5 structure that connects five Au_{13} monomers together by means of common vertices, so the cluster is also regarded as a "cluster self-assembly material". In addition, two kinds of nanoclusters with different charges $[Au_{25}(PPh_3)_{10}(SePh)_5Cl_2]^{+2/+1}$ were obtained by solvent screening and temperature control, and their precise structures were characterized by ESI mass spectrometry, thermogravimetric analysis, and X-ray single crystal diffraction [70].

2.4.5 Controlled conversion method

Nanoclusters have been considered to be an ideal model system for studying the structure-activity relationship of nanomaterials at the atom level due to their molecular-like properties and well-defined structure. In recent years, the research on the transformation of nanoclusters has attracted much attention. The postmodification or transformation of the nanoclusters with clear structures can help to create the application-oriented functional materials. Therefore, a series of single dispersed clusters has been synthesized by means of cluster transformation strategy. Among them, organic ligand and reducing agent can regard as the driving force of transformation. In addition, the transformation can be achieved in single phase or two phases, and under the further induction of alkali, counterions, and other inorganic ions can also achieve the cluster transformation under control.

2.4.5.1 Two-phase ligand exchange method

As early as 2005, the smallest $Au_{15}(SG)_{13}$ and $Au_{18}(SG)_{14}$ water-soluble clusters were protected by glutathione (GSH) have been reported by Tsukuda's group [71]. The crystal structure of this cluster is very important for understanding the nucleation mechanism of thiol-ligand-protected gold clusters and the growth mechanism of the nanoclusters. However, due to the flexibility of GS ligands, it is difficult to obtain the crystal structure of these transition-size nanoclusters. Replacing flexible GS ligands with rigid mercaptans ligands is expected to solve the problem of nanocluster crystallization. In 2015, our research group replaced the flexible glutathione ligand of $Au_{18}(SG)_{14}$ with rigid cyclohexanethiol by a controlled two-match exchange method, synchronously realizing the transfer of water-soluble phase clusters to oil-soluble phase (Fig. 2.14) [72]. Then we determined the X-ray structure of $[Au_{18}(C_6H_{11}S)_{14}]$ nanoclusters protected by cyclohexanethiol. In 2018, we used $Au_{15}(SG)_{13}$ as the precursor and employed the controlled two-match exchange method again. By adjusting the types of organic mercaptan ligands, we transformed $Au_{15}(SG)_{13}$ into $Au_{16}(SAdm)_{12}$ clusters which is the smallest gold nanocluster [73]. Furthermore, we realized the transformation of water-soluble phase $Au_{15}(SG)_{12}^-$ to oil-soluble phase cluster $Cd_1Au_{14}(StBu)_{12}$ by adding Cd ions. Combined with the precise atomic structures of the above three transitional size clusters, we determined the formation mode of small-size mercaptan ligand gold clusters as "gold insertion and mercaptan collapse," which provides a reliable scientific reference for the basic research on understanding the structural evolution of gold complex to nanoclusters.

FIG. 2.14 Schematic illustration of Au_{18} nanoclusters using the two-phase ligand exchange method. *(From S. Chen, S. Wang, J. Zhong, Y. Song, J. Zhang, H. Sheng, Y. Pei, M. Zhu, The structure and optical properties of the [$Au_{18}(SR)_{14}$] nanocluster, Angew. Chem. Int. Ed. 54(10) (2015) 3145–3149, https://doi.org/10.1002/anie.201410295.)*

In two-phase ligand exchange method, the addition of inductive additives into the system is an effectively regulate the reaction conditions and realize the controlled synthesis of nanoclusters. For example, the proper amount of base (e.g., NaOH) added into aqueous solution can regulate the types of precursors and synthesize the nanoclusters with different sizes, especially in the synthesis of silver nanoclusters. In addition, the counterion (i.e., $NaBF_6$ and Ph_4PBr) was also an essential inducer in the synthesis. In the synthesis of the silver cluster, Chen et.al investigated the role of $NaBF_6$. In the presence of $NaBF_6$, the color of the solution remained dark brown during the synthesis of water-soluble precursors. However, without the addition of $NaBF_6$, the solution gradually becomes colorless. The results of electrospray ionization mass spectrometry (ESI-MS) showed that, the water-soluble precursors exhibited multiple mass spectrum peaks in the presence of $NaBF_6$. However, the signal displayed by the colorless precursor ESI-MS was in the small size range without the $NaBF_6$. In addition, the products were change with the $NaBF_6$ is replaced by other counterions (i.e., $NaBPh_4$, KPF_6, $NaAlF_6$). These results suggest that $NasBF_6$ is crucial for determining the size distribution of the precursors of Ag nanoclusters and the structure of the final products.

By means of the two-phase ligands exchange method, the water-soluble phase clusters can transfer to the organic phase, and then reveal their precise structure by X-ray diffraction of single crystal. However, the organic ligands and organic solvents are not suitable for their applications in the biological field due to their hydrophobicity. Therefore, it is meaningful to transfer the precise organic phase nanoclusters to the aqueous phase through the ligands exchange of functional ligands under two-phase ligands exchange method. As a first example, Bakr's group achieved the transformation of Ag_{44} nanoclusters from organic phase to aqueous solution based on 4-fluorothiophenol ligands [74]. The authors emphasized the importance of 4-fluorothiophenol ligand which is can be regarded as promising ligand for charge transport applications and helps for the stability of Ag_{44} nanocluster. Moreover, the different of this ligand to the original ligands is also an important consideration. The unique characterization of the final product in the water-soluble phase is determined by UV-vis spectroscopy, which shows nearly the same UV-vis absorption peaks as the original nanocluster in the organic phase. More encouragingly, the authors found that Ag_{44} nanocluster exhibited excellent photoluminescent property in aqueous solution. The authors suggest that the luminescence of this nanoclusters is due to the dipole moment interaction between the clusters which are in the charge-transfer excited state and water.

Based on the simplicity and convenience of one-pot method and the effectiveness of two-phase ligands exchange method, a synthetic strategy of in situ two-phase ligand exchange method was developed (Fig. 2.15). The detail detailed synthesis steps are (i) the metal salt is mixed with glutathione in the aqueous solution and (ii) the reducing agent and an organic mercaptans ligand codissolvent in organic solvent and added into the above mixture. In this in situ system, the ligands exchange and reduction of the Au-S complex are occurred synchronously. Therefore, the reaction time of nanoclusters synthesized by this method is short. The reaction system is more complex, which is conducive to the synthesis of novel clusters that have not been reported [75].

To explore the reaction process, the time dependent UV-vis spectroscopy combined with in situ mass spectrometry were used to trace the targeted reaction process. The results showed that the reaction system of in situ method was more complex and the precursors were more dispersed, which was beneficial to the selective exchange of organic ligands in the next step. These results indicate that in situ method show the different reaction mechanism from traditional synthesis method and is beneficial to synthesize new clusters. The infrared spectra of the precursor were also employed to explore the influence of the coexistence of ligand and reducing agent (in situ) in the reaction. The C—C stretching vibration peak which is belonging to S*t*Bu was found, and is not found in the precursors under traditional methods. These results indicate that the glutathione ligands in the aqueous phase replaced by tert-butyl thiol ligands is synchronous with the reduction of Au(I)-S complex. Therefore, it is concluded that the reaction environment reflected by in situ method is different from that of traditional one-pot method and ligand exchange method due to the coexistence of ligand and reducing agent.

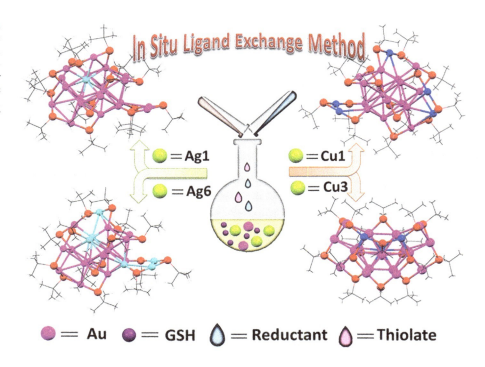

FIG. 2.15 Schematic illustration of in situ two-phase ligand exchange method. *(From S. Yang, J. Chai, Y. Song, J. Fan, T. Chen, S. Wang, H. Yu, X. Li, M. Zhu, In situ two-phase ligand exchange: a new method for the synthesis of alloy nanoclusters with precise atomic structures, J. Am. Chem. Soc. 139(16) (2017) 5668–5671, https://doi.org/10.1021/jacs.7b00668.)*

2.4.5.2 Ligand exchange method

The synthesis of nanocluster induced by ligand is not only applicable to in two phases but also in single phase. Nanoclusters with known atomically precise structures are selected to use as precursors and transformed into other nanoclusters with different sizes/structures by effective ligand regulation. This strategy has been widely used in the synthesis of nanoclusters. For example, the famous nanocluster of $Au_{25}(PET)_{18}$ as a precursor is converted to $Au_{28}(TBBT)_{20}$ after the ligand exchange with excess TBBT thiol [76]. While the $Au_{24}(SCH_2Ph)_{20}$ was synthesized after the $Au_{25}(PET)_{18}$ reaction with excess benzyl mercaptan. In 2013, Jin et al. [77] gives a particularly important report on the transformation from $Au_{38}(PET)_{24}$ to $Au_{36}(TBBT)_{24}$ by ligand induction. Most important of all is that the reaction kinetics are studied to unravel the total transformation process under time depend UV-vis and ESI mass. The results show a clear transformation process, which begins with ligand substitution and then induces structural transformation which provided a guiding significance for the study of the later cluster transformation. In addition, Wu's group found that Au_{38} can be transformed into another novel $Au_{60}S_6(SCH_2Ph)_{36}$ nanocluster with benzyl mercaptan as the incoming ligand [78]. Interestingly, continuous transformation has also been found in the field of nanoclusters.

2.4.5.3 Others

In addition to the transformation from the one to another, the interconversion between two clusters, as well as linkage transformation can also be achieved by ligand induce. In 2017, Dass reported the core size interconversions of $Au_{30}(StBu)_{18}$ and $Au_{36}(SPhX)_{24}$ nanoclusters [79]. With the HSPhX induced, the $Au_{30}(StBu)_{18}$ was transformed into $Au_{36}(SPhX)_{24}$ nanocluster with the help of thermochemical etching. The reverse process, that is transform $Au_{36}(SPhX)_{24}$ to $Au_{30}(StBu)_{18}$ nanocluster, is obtained by the reaction of tert-butyl thiol with $Au_{36}(SPhX)_{24}$ nanocluster. For linkage transformation, the $Au_{99}(SPh)_{42}$ nanocluster was synthesized after $Au_{144}(PET)_{60}$ reaction with benzenethiol which reported by Dass [80]. Then, the $Au_{99}(SPh)_{42}$ nanocluster use as precursors and transform into $Au_{133}(TBBT)_{52}$ after treatment with excess 4-tert-butylbenzenethiol at 80°C for 4 days [81].

Previous reports show that the oxidation-reduction reaction in the fields of nanoclusters is interesting phenomena are worth to explored. For example, the different charge states formed from their oxidation or reduction reaction will affect the geometric and electronic structures. what's more, nanocluster with the different charge states may induce the nanocluster to produce magnetic, photoluminescence and other unique properties, which makes them have a great application prospect in catalysis and other fields. In the recent study, the conversion of $([Ag_{62}S_{13}(StBu)_{32}]^{4+})$ clusters to $([Ag_{62}S_{12}(StBu)_{32}]^{2+})$ has been achieved by electrochemical reduction reported by our group [82]. A key intermediate was obtained at a voltage of -0.6 V, and its atomic structure was determined as $[Ag_{62}S_{13}(StBU)_{32}]^{2+}$ by single crystal X-ray crystallography.

The detailed structural comparison shows that, based on the principle of overenergy advantage, the centroid-S atom in the cubic core of $Ag_{14}S$ can be extruded from the cluster through the cavity during the reduction process, which is also the first reported phenomenon. In addition, we also study the specific conversion process and the accompanying changes in optical properties. Based on the electrochemical reduction strategy, $[Au_{18}Cu_{32}(SPhCl)_{36}]^{2-}$ can be reduced into $[Au_{18}Cu_{32}(SPhCl)_{36}]^{3-}$ clusters, which have the same frame structure and different charge valence states [83]. And this reduction reaction can be achieved not only by electrochemistry.

Light is a kind of green energy, and it is inexhaustible. Recently, Tang et al. provided the report on the insight of the light-induced transformation of gold nanoclusters [84]. In this work, the $[Au_{23}(S\text{-}c\text{-}C_6)_{16}]^-(TOA)^+$ nanocluster was selected as the reactant due to it can excited to $[Au_{23}(S\text{-}c\text{-}C_6)_{16}]^-(TOA)^+$ by photons. When $[Au_{23}(S\text{-}c\text{-}C_6)_{16}]^-(TOA)^+$ with the negatively charged is easily to oxidized to be the neutral one. Because the $[Au_{23}(S\text{-}c\text{-}C_6)_{16}]^0$ with the neutral charge is unstable and this neutral cluster is unstable and decomposed into smaller nanocluster and finally reassembled into the $Au_{28}(S\text{-}c\text{-}C_6)_{20}$ nanocluster. The reaction process is monitored by means of several analysis methods. This work provides new insights in the terms of the transformation of nanoclusters at the atomic level.

Oscillation is an interesting phenomenon in nature. However, no structural oscillation has been found in semiconductor nanoparticles, which is mainly due to the difficulty of structural resolution at the atomic level. The emergence of gold nanoclusters (ultrafine nanoparticles) provides an excellent opportunity to solve some challenging problems in the field of nanoparticles. In 2020, two $Au_{28}(CHT)_{20}$ (CHT: cyclohexyl thiol) structural isomers ($Au_{28}i$ and $Au_{28}ii$) were simultaneously synthesized by using the quasireverse current method [85]. Driven by the process of dissolution and crystallization, at least reversible transformation between them was 10 Cycles. The transition from $Au_{28}ii$ to $Au_{28}i$ is solvent medium constant dependent, and the deuteration effect of dichloromethane is significant. The photoluminescence values of these two isomers are significantly different, which are not only important for structure-property correlation but also have potential applications in transformation and sensing.

Recently, our group found that a metastable nanocluster of $Au_{22}(SAdm)_{16}$ can be spontaneous transformed into $Au_{21}(SAdm)_{15}$ nanocluster [86]. The transformation from $Au_{22}(SAdm)_{16}$ to $Au_{21}(SAdm)_{15}$ is spontaneous in any state including the nanocluster dissolved in a solvent and in a solid state. The transformation process was monitored via time dependent UV-vis and ESI-MS. In addition, the transformation process of them in different solvent was also investigated and make clear that the solvent effect has occurred. Both the unitary solvent system and binary solvent systems are executed to complete the transformation of nanocluster. The results show that the affinity between the solvent molecules and Au atoms may account for the boost in conversion rates.

2.4.6 Inorganic anions as synthetic templates

Inorganic anions as synthetic templates are often used in the synthesis of silver nanoclusters because the silver clusters are not sufficiently stable. The above synthesis strategies can also be used for controllable synthesis of silver nanoclusters based on the optimization of synthesis conditions. As early as 1999, Mak and Gou reported the concept that anions have the highest ligation number. They suggest that the coordination of silver to various anions contributes to increased stability because of the argyrophilic interactions of the anions. In recent years, ligand-protected high-nuclearity silver nanoclusters have attracted extensive attention due to their potential applications in the fields of catalysis, photoelectron and luminescent materials despite it have great challenge in their synthesis. The main challenge is that the formation of high-nuclearity silver nanoclusters requires a very complex process involving multiple components, so it is very difficult to achieve the ordered assembly of atoms in the nanoclusters. Recently, the results of Wang's group and Sun's group show that the inorganic anions as synthetic templates is an effective method for the controllable synthesis of high-nuclearity silver clusters in an ordered way.

In the early stage of inorganic anions as synthetic templates use to synthesize the silver nanocluster are main in Ag(I) complex and the kind of inorganic anions that are used are a relatively simple type. For example, A cage-like silver compound $[Ag_{14}(tBuC\equiv C)_{12}Cl]OH_{14}^-$ nanocluster was synthesized and reported by Mingos et al. in 2009 in which the kernel of the cluster was occupied by a chloride ion [87]. In addition, the importance of anions in the synthesis of caged silver nanoclusters is emphasized. Besides halogen anions, carbonates have been found have an important templating effect on the synthesis of silver clusters. Series of silver nanoclusters include the $[Ag_{17}(tBuC\equiv C)_{14}CO_3]OTf$, $[Ag_{19}(tBuC\equiv C)_{16}CO_3]BF_4$, $[Ag_{22}(tBuC\equiv C)_{18}CrO_4](BF_4)_2$, and $[Ag_{21}(tBuC\equiv C)_{18}SO_4]BF_4$ nanoclusters are prepared by using the carbonates as synthetic templates.

The polyoxometalates (POMs) which has the large oxo anions and they can bind silver ions with high affinity. Therefore, polyoxometalates has recently been developed as an anion to synthesize the silver nanoclusters under inorganic anions as synthetic templates. Moreover, the synthetized silver nanoclusters use the polyoxometalates as inorganic anions are high-nuclearity silver clusters. For example, the giant silver alkynyl cluster, $[Ag_{60}(Mo_6O_{22})_2(tBuC{\equiv}C)_{38}]$ $(CF_3SO_3)_6$ was synthesized with $Mo_6O_{22}^{8-}$ as templates and reported by Wang's group [88]. The complicated silver nanoclusters of $\{(NO_3)_2@Ag_{16}(C{\equiv}CPh)_4[(tBuPO_3)_4V_4O_8]_2(DMF)_6(NO_3)_2\}{\cdot}DMF{\cdot}H_2O$ and $\{[(O_2)V_2O_6]_3@Ag_{43}$ $(C{\equiv}CPh)_{19}[(tBuPO_3)_4V_4O_8]_3(DMF)_6\}{\cdot}5DMF{\cdot}2H_2O$ were gain with the $[(O_2)V_2O_6]^{4-}$ template anions. Among them, the $[(tBuPO_3)_4V_4O_8]^{4-}$ unit are revealed as integral shell to stabilize the nanocluster reported by Mar's group. It is worth mentioning that Sun' group have been performed the comprehensive studies on the silver nanoclusters protected by organic ligands. Most of the high-nuclearity silver nanoclusters were synthesized by use of inorganic template anions. In their work, the chromate and tungstate are used as anion template combines with alkynyl to form silver alkynyl nanoclusters. In thiolated silver nanoclusters, the chromate and polyoxometalates are commonly used as anion templates. Recent studies have shown that anion templates play an important role in controlling the size, shape, stability, and properties of silver nanoclusters.

2.5 Synthesis of alloy nanoclusters

In metallurgy, the properties of metals can be changed by adding other metals and this change usually improves the properties of materials which is more conducive to their practical application. In the field of metal nanoclusters, alloy nanoclusters are also functional materials with great application potential due to their properties different from those of single metal nanoclusters. In order to further investigate the synergistic effect between metals in alloy nanoclusters, large-scale synthesis of alloy nanoclusters is necessary [89–97].

With the development of science and technology, a variety of alloy nanoclusters have been synthesized. Most of these alloy nanoclusters are clustered with gold-based and silver-based alloy nanoclusters. They include Au-Ag, Au-Cu, Au-Pt, Au-Pd, Au-Cd, Ag-Cu, Ag-Pt, and Ag-Pd bimetallic nanoclusters. Recently, the crystal structures of Au-Ag-Cu, Au-Ag-Pt, and other three metal nanoclusters have also been revealed. There are many interesting scientific questions in the field of alloy nanoclusters, including how many heteroatoms can be doped into the parent cluster and how the doping of these heteroatoms is distributed. Two kinds of doping behaviors have been found in the reported alloy nanoclusters: (a) some bimetallic nanoclusters have high precision heteroatom doping numbers and positions, which are usually concentrated in Pd, Pt, and Cd heteroatom doping and (b) the number of heteroatoms in some bimetallic nanoclusters is distributed, but the total number of dopants remains at the parent number within the order.

Au-Ag bimetallic nanoclusters have been widely reported in thiol-protected gold-based alloy nanoclusters. The silver atoms are doped into gold nanoclusters and seem to prefer doped onto the surface of the nanoclusters. For example, the $Ag_xAu_{25-x}(SR)_{18}$ [98,99] and $Ag_xAu_{38-x}(SR)_{24}$ [100,101] nanoclusters reported by Kumara; The Jin's group [102] found that $Au_{23}(SR)_{16}^-$ converted into $Ag_xAu_{25-x}(SR)_{18}$ with the doping number of silver atoms was up to 20 and the silver atoms could occupy not only the surface but also the central position of the clusters. Similar to silver, copper easily occupies the surface of Au-Cu nanoclusters and is mostly distributed doped [103]. The Pd atoms all selectively occupy the center of the cluster and have precise doping number [104,105]. Although the molecular weight of Pt is similar to that of Au, the crystal structure of Au-Ag-Pt three-metal nanoclusters indicates that Pt is easy to occupy the center of nanoclusters [106].

For alloy nanocluster, there have a number of mainstream synthesis methods and some of which are similar to single cluster synthesis methods which will not be highlighted here. The other part is about the interesting synthesis of alloy clusters and this section will focus on it.

2.5.1 Co-reduction method

Among the traditional synthesis methods, one-pot or direct method is the most widely used for the synthesis of nanoclusters, which is also suitable for alloy nanoclusters. In alloy nanoclusters, this method is also known as the co-reduction method [107]. The direct synthesis of thiol-protected alloy nanoclusters can be summarized as three steps: the first step is to mix two or more kinds of metal ions and dissolve them into a suitable solvent; In the second step, excessive thiol or other ligands were added to reduce the metal ions to a lower state. At this time, the metal ions reaction with ligands to form the complexes. Then, the third step is the reduction of the above complexes formed into alloy nanoclusters by adding a reducing agent. Using the above synthesis method, a large number of bimetallic nanoclusters and a few trimetallic nanoclusters have been reported.

It is worth noting that any change in conditions will have a great effect on the type of the final alloy nanoclusters. Therefore, the metal salt type and the type of solvent, including the single solvent and mixed solvent, ligands, reducing agent, proportion of materials and order of addition are very important factors to be considered. In 1984, the first phosphine-protected bi-icosahedral Au and Ag alloy nanocluster was synthesized following co-reduction method by

Teo and Keating [108]. The obtained alloy NC contained 25 metal atoms with the molecular composition of $[(PPh_3)_{12}Au_{13}Ag_{12}Cl_6]^{m+}$. This work was followed by the synthesis of $[(Ph_3P)_{10}Au_{13}Ag_{12}Br_8](PF_6)$. Then, the same group found new tri-icosahedral and tetra-icosahedral alloy NCs, $[(p\text{-}Tol_3P)_{12}Au_{18}Ag_{20}Cl_{14}]$ and $[(Ph_3P)_{12}Au_{22}Ag_{24}Cl_{10}]$, respectively [109]. In 2009, Murray et al. [110] synthesized $PdAu_{24}(PET)_{18}$, which was found to have different electrochemical properties than its monometallic analog. Ag, Cu, and Pt atoms were incorporated into $Au_{25}(PET)_{18}$ to make $Au_{25-x}Ag_x(PET)_{18}$ [111], $Au_{25-x}Cu_x(PET)_{18}$ [112], and $PtAu_{24}(PET)_{18}$ [113] via this method. Ni-doped bi- and trimetallic clusters such as Ag_4Ni_2 [114] and $Ag_{12}Au_{12}Ni$ [115] were achieved by this method, which can exhibit interesting magnetic properties. Various other alloys with interesting properties can be synthesized easily. Such as chiral alloy nanocluster, larger-sized, Ag-rich, and with exceptional stability alloy nanoclusters.

2.5.2 Ligand exchange method

Ligand exchange is an efficient method for the synthesis of atomically precise NCs in which one NC is converted to another, with the same or different nuclearity in the presence of foreign ligands. It is the dream of nano-chemistry researchers to control the size and appearance of nanoparticles and even to achieve the precision control on the structure of nanoclusters. A new strategy for controlling the size and structure of nanoclusters, known as ligand-induced strategy, is described in 2015 and reported by Jin's group [116]. In this work, several groups of nanoclusters have been shown to be transformed by ligands. The author studied the transformation process and suggested that it be the same as the transformation in organic chemistry. The subsequent mechanism exploration indicated the role of ligands in inducing the size and structure of nanoclusters. This work shows that ligands play a major role in the control of nanoclusters with certain size and structure. In other words, ligand induction will be an effective means to expand the type and number of nanoclusters. In alloy nanoclusters, the $Pt_2Ag_{23}(PPh_3)_{10}Cl_7$ was converted to $PtAg_{24}(DMBT)_{18}$ [117] and $PtAg_{28}\text{-}(BDT)_{12}(PPh_3)_4$ [118] with the induction by DMBT and BDT ligands, respectively. $PtAg_{28}(S\text{-}Adm)_{18}(PPh_3)_4$ was synthesized by the conversion of $PtAg_{24}(DMBT)_{18}$ after addition of Adm-SH and PPh_3 [105]. Similarly, $MAg_{28}(BDT)_{12}(PPh_3)_4$ (M=Ni/Pd/Pt) was synthesized via the LEIST method starting from $MAg_{24}(DMBT)_{18}$ (M=Ni/Pd/Pt) (see Fig. 2.16) [119]. Moreover, Zhu et al. showed the synthesis of highly luminescent $Pt_1Ag_{12}(dppm)_5(DMBT)_2$ from feebly luminescent $Pt_2Ag_{23}(PPh_3)_{10}Cl_7$ by introducing dppm along with DMBT [117]. It has been proposed that the drastic change in the nuclearity of the NC is due to the change in the electron-withdrawing and electron-donating effect of ligands, which depends on the position of the functional groups. The noncovalent interactions and steric hindrance are also crucial factors for this transformation. However, the effect of ligand on the nuclearity and structure of the NC is unclear.

Metal exchange method is an effective strategy for the synthesis of alloy nanoclusters. In addition, the alloy nanoclusters were synthesized by metal exchange can maintain the framework unchanged, which is beneficial to the understanding of metal synergies at the atomic level. Therefore, it has received extensive attention as a universal synthesis strategy in the past few years. The method is briefly described as follows: the monometallic nanoclusters are prepared as templates and doped with the second metal to prepared alloy nanoclusters. The metal exchange method is different from the metal replacement reaction in that the exchange between metals is not restricted by the electrode potential. According to the classic galvanic theory, a metal ion of higher reduction potential in solution gets reduced and replaces a metal atom present in a material and the latter subsequently enters the solution after being oxidized. In the field of nanoclusters, the exchange of metals may or

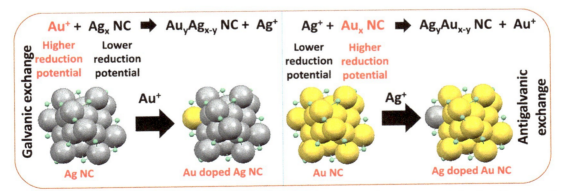

FIG. 2.16 Schematic representation of galvanic and antigalvanic exchange reaction processes. *(From E. Khatun, T. Pradeep, New routes for multicomponent atomically precise metal nanoclusters, ACS Omega 6(1) (2021) 1–16, https://doi.org/10.1021/acsomega.0c04832.)*

2.5.2.1 Galvanic synthesis

In 2015, the galvanic synthesis in the atomically precise alloy nanoclusters was reported by Bakr's group in the [Ag$_{24}$Au(SR)$_{18}$]$^-$ nanocluster [121]. The [Ag$_{25}$(SR)$_{18}$]$^-$ nanocluster was employed as precursor and reaction with AuClPPh$_3$ complex. After that the Au$^+$ replaced the one Ag atom in [Ag$_{25}$(SR)$_{18}$]$^-$ nanocluster and form a [Ag$_{24}$Au(SR)$_{18}$]$^-$ nanocluster and signifies the completion of the exchange of metals. Followed by this report, Zhu's group demonstrates a metal exchange method for obtaining alloy NCs using the Au$_{25}$(SR)$_{18}^-$ NCs as the template [122]. A series of gold-based alloy nanoclusters including the Ag$_x$Au$_{25-x}$, Cu$_x$Au$_{25-x}$, Cd$_1$Au$_{24}$, and Hg$_1$Au$_{24}$, were prepared by reacting with a series of metal mercaptan complexes. In fact, the author not only tried to prepare the abovementioned alloy nanoclusters but also considered the reaction between Au$_{25}$(SR)$_{18}^-$ and Ni-S, Pd-S, and Pt-S complexes, but failed to get Ni$_x$Au$_{25-x}$, Pd$_x$Au$_{25-x}$, and Pt$_x$Au$_{25-x}$ alloy nanoclusters. The above results indicate that in the metal exchange process, the Au atoms in the Au$_{25}$(SR)$_{18}^-$ NC can be exchanged by metals with different activities (e.g., Cu, Ag, Cd, Hg, etc.) to produce corresponding alloy nanoclusters. Therefore, the authors suggest that the metal exchange method for alloy NCs is, to a large extent, associated with electron shell closing and the NC's structural stability, but less on the metal activity. After that Kang et al. reported shape-altered synthesis of alloy NCs using this method [123]. The incorporation of Au atoms in PtAg$_{24}$(DMBT)$_{18}$ resulted in the formation of shape-unaltered-trimetallic PtAu$_x$Ag$_{24-x}$(DMBT)$_{18}$ when Au-DMBT was used as the precursor, while the use of AuBrPPh$_3$ led to the formation of shape-altered trimetallic Pt$_2$Au$_{10}$Ag$_{13}$(PPh$_3$)$_{10}$Br$_7$. In this case, Br$^-$ peeled away the PtAg$_{12}$ core from the staple motifs and transformed it into a bi-icosahedron, which was stabilized by PPh$_3$ ligand instead of DMBT. Trimetallic PtCu$_x$Ag$_{28-x}$(BDT)$_{12}$(PPh$_3$)$_4$ and tetrametallic Pt$_1$Ag$_{12}$Cu$_{12}$Au$_4$(SAdm)$_{18}$(PPh$_3$)$_4$ NCs were synthesized by galvanic replacement procedure [124].

2.5.2.2 Antigalvanic reduction method

The antigalvanic reduction reaction in metal exchange was also found and reported by several groups. In 2010, Murray's group report the reactivity of the gold nanoparticle [TOA$^+$] [Au$_{25}$(SC$_2$Ph)$_{18}$]$^{1-}$ with Ag$^+$, Cu^{2+}, and Pb^{2+} ions [125]. Among them, added the Ag$^+$ into the dichloromethane solution containing the [Au$_{25}$(SC$_2$Ph)$_{18}$]$^{1-}$ nanoclusters would change the forcing of the absorption peaks of the nanoclusters and lead to changes in their absorption peaks. The results of mass spectrometry showed that a series of Ag$_x$Au$_{25-x}$(SR)$_{18}$ nanoclusters were formed. In 2012, the concept of the antigalvanic reduction reaction is mentioned by Wu in reaction of Au$_{25}$(PET)$_{18}$ with Ag ions [126]. The final product is characterized as Ag$_2$Au$_{25}$(SR)$_{18}$ nanocluster by using laser desorption ionization (LDI) MS. To prove that antigalvanic exchange reaction is not a unique property of Au$_{25}$(PET)$_{18}$ NC, approximately 2–3 nm nanoparticles were treated with AgNO$_3$ solution. The incorporated Ag was detected by XPS and, after the reaction, the binding energy of Ag indicated the incorporation of neutral Ag, which confirmed the reduction of Ag ions by Au nanoparticles. A similar experiment was performed on 3 nm sized Ag nanoparticles using Cu salt and incorporation of neutral Cu in Ag nanoparticles was observed using XPS. This proved that Cu ions can be reduced by more noble Ag atoms. Antigalvanic reduction reaction is feasible due to the enhanced reducing ability when the metal is in nanoscale form. One of the important driving factors for Antigalvanic reduction is the protective ligands on the surface of nanoparticles or NCs. It was proposed that the partial negative charge present on the surface ligands plays a crucial role in the reduction of more reactive ions. After that, Wu's group showed the incorporation of bivalent Cd/Hg in Au$_{25}$(SR)$_{18}$ NCs.

2.5.3 Method exchange method

The alloy nanoclusters which synthesized using the metal exchange method can be mainly classified into two categories: (i) the obtained alloy nanoclusters are consistent with the structure framework of the parent with only a partial metal substitution and (ii) the obtained alloy nanoclusters are inconsistent with the parent structure frame, and the replacement of metals induces structural changes. Based on the metal-exchange method, most of the obtained alloy nanoclusters do not change their structure which can be classified as the first type. However, in some NC systems, the replacement with a foreign metal atom can lead to structural transformations. In 2014, Zhu's research group synthesized two "rod-like" [Ag$_x$Au$_{25-x}$(PPh$_3$)$_{10}$(SR)$_5$Cl$_2$]$^{2+}$ clusters using polydisperse gold nanoparticles and monodisperse Au$_{11}$ as precursors base on the metal-exchange method [99]. The two have the same frame structure and similar silver doping sites, but the difference lies in the doping degree of silver. Using polydisperse gold nanoparticles with triphenylphosphine ligands as precursors, two kinds of "assembled" clusters were obtained by the reaction of Cu-S complex with different ligands that is

$Cu_xAu_{25-x}(PPh_3)_{10}(PhC_2H_4S)_5Cl_2^{2+}$ and $Cu_3Au_{34}(PPh_3)_{13}(tBuPhCH_2S)_6S_2^{3+}$, respectively [127]. On the other hand, doping of Ag atoms in $Au_{23}(CHT)_{16}$ NCs first led to the formation of $Au_{23-x}Ag_x(CHT)_{16}$, which then got converted to $Au_{25-x}Ag_x(CHT)_{18}$ [102]. Using $[Pt_1Ag_{24}(SPhMe_2)_{18}]$ as precursor, the "rod-like" $Pt_2Au_{10}Ag_{13}(Pph_3)_{10}Br_7$ cluster was obtained after the reaction of $[Pt_1Ag_{24}(SPhMe_2)_{18}]$ with Au-P complex. The authors hypothesize that the change inducers come from changes in chemical composition and morphological changes in structure [124].

2.5.4 Intercluster reaction method

In 2016, Pradeep's group present the first example of intercluster reactions between atomically precise of $Au_{25}(SR)_{18}$ and $Ag_{44}(SR)_{30}$ nanocluster and suggested that this is a new method for the synthesis of alloy nanoclusters (Fig. 2.17) [128]. In this work, the authors revealed that the reaction is spontaneous and driven by the exchange of metal atoms and ligands between the two nanoclusters. With $Ag_{44}(SR)_{30}$ nanocluster as the silver source, $Au_{25}(SR)_{18}$ is transformed into $Au_{25-x}Ag_x(SR)_{18}$ in which the X is up to 20, which cannot be achieved in co-reduction and other silver source doping. The total number of atoms and the overall charge state remain preserved in the formed alloy NCs during this kind of substitution reaction. It is worth noting that in this work the amount of $Au_{25}(SR)_{18}$ was always greater than the amount of $Ag_{44}(SR)_{30}$ nanocluster which result in the product is gold $Au_{25-x}Ag_x(SR)_{18}$. With the ratio of $Ag_{44}(SR)_{30}$ to $Au_{25-x}Ag_x(SR)_{18}$ is up to 2.8, the main product is $[Au_{12}Ag_{32}(FTP)_{30}]^{4-}$ after the $Ag_{44}(SR)_{30}$ nanocluster reaction with $Au_{25}(SR)_{18}$ nanocluster. In this case, the author observed a systematic trend in the Au atom substitution into $Ag_{44}(SR)_{30}$, reflecting the geometric shell structure of the reactant and product clusters. We showed that the metal atom substitution chemistry of $Ag_{44}(SR)_{30}$ proceeds through an unexpected from superatom to nonsuper-atom transition, resulting from a change in the overall charge state of the $Au_xAg_{44-x}(SR)_{30}$ alloy clusters formed.

With the precursor of $Ag_{44}(SR)_{30}$ is replaced by $Ag_{25}(SR)_{18}$ nanocluster which has the same number of atoms and similar geometry structure to $Au_{25}(SR)_{18}$ nanocluster, the adduct of two nanoclusters were formed. After the reaction about 5 min, the precursor peaks disappear and a new peak can be described as $Au_xAg_y(SR)_{18}$ appears. This adduct formation suggested that the intercluster reaction is bimolecular. The atomic structure of this adduct was predicted by DFT calculations and found to be different from that of the original cluster in bond length and bond angle. Except for the silver nanoclusters, the $Ir_9(PET)_6$ nanocluster was also employed as reactant to react with $Au_{25}(SR)_{18}$ to create new alloy nanocluster [129]. Base on the intercluster reaction between $Au_{25}(PET)_{18}$ and $Ir_9(PET)_6$, the new alloy nanocluster of $Au_{22}Ir_3(PET)_{18}$ was prepared. The formula of $Au_{22}Ir_3(PET)_{18}$ nanocluster was determined by mass spectrometry and other spectroscopic techniques. The DFT study revealed that the Ir atoms occupy the center and surface of the icosahedral core. The intercluster reactions were then extended to silver nanoclusters protected by mixed ligands. The results showed that the metal exchange was slow and ligand exchange could not occur because of the indicated rigidity the surface of nanocluster. Not only the bimetallic alloy nanoclusters but also the intercluster reactions can be applied to the preparation of trimetallic nanoclusters. Recently, the intercluster reactions between the bimetallic $MAg_{28}(BDT)_{12}(PPh_3)_4$ (where M=Ni/Pd/Pt) with monometallic $Au_{25}(PET)_{18}$ was performed [119]. Hence, based on this, a mixture of trimetallic $MAu_xAg_{28-x}(BDT)_{12}(PPh_3)_4$ and bimetallic $Ag_xAu_{25-x}(PET)_{18}$ nanoclusters were obtained. In addition to the interactions between the clusters, they also used $Au_{25}(PET)_{18}$ and $Au_{38}(PET)_{24}$ NCs to react with Ag, Cu, and Cd foils before and after functionalization with thiols. The doping rate was observed to be different for the treated and untreated foils. Treated Ag foils exhibited a faster reaction rate,

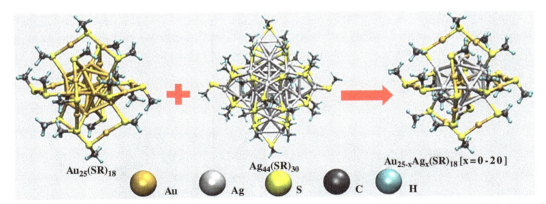

FIG. 2.17 Schematic representation of intercluster reaction method. *(From K.R. Krishnadas, A. Ghosh, A. Baksi, I. Chakraborty, G. Natarajan, T. Pradeep, Intercluster reactions between $Au_{25}(SR)_{18}$ and $Ag_{44}(SR)_{30}$, J. Am. Chem. Soc. 138(1) (2016) 140–148, https://doi.org/10.1021/jacs.5b09401.)*

which decreased with time, while the untreated one showed a lower reaction rate at first and slowly increased after a certain time. This observation indicated that prefunctionalized thiols on the foils play a crucial role in the reaction, which indeed emphasized the importance of metal-ligand interfaces during the reaction between NCs. In the case of Cu and Cd foils, the reaction did not occur with the bare metal foils.

2.6 Other methods of synthesis

Even though silver and gold belong to the same group, they exhibit slightly different physical and chemical properties. For example, silver is much more reactive and easily oxidizable than gold in multiple states, making it difficult to controllable synthetic the stable silver nanoclusters. The synthetic methods which are applicable to both gold and silver have been described in the preceding paragraph. This part, we introduce a novel synthesis strategy which is mainly suitable for the synthesis of silver nanoclusters.

In 2010, Pradeep's group developed a novel synthesis method for preparing silver nanoclusters [130]. The novel of this method is that the operation is performed not in the conventional gas and liquid phases, but in the solid-state route (Fig. 2.18). The implementation of this method requires the use of a grinding bowl as a reaction vessel. The solid silver nitrate is added and ground into fine particles. Then the solid thiol ligands are added into the system and continue to be ground uniformly with fine particles of silver nitrate. The changing trend of solid color in the system can be used to detect the uniformity of grinding. After that, the sodium borohydride in the solid-state was added and the grinding continues until the solid in the system change to darkens. The resulting crude product is washed with ethanol to remove excess thiol or other byproducts. And the final product was extracted with the corresponding solvent to obtain the pure nanoclusters.

Based on this method, Ag_9 nanocluster was first synthesized. The detailed synthesis process is divided into three steps: (1) The $AgNO_3$ is mixture with thiol in a molar ratio of 1:5 and ground in the solid state with the color change from white to orange indicates the completion of the reaction; The possible product of this process is silver thiolate. (2) Then, the sodium borohydride with the ratio is more than 5:1 to $AgNO_3$ was added also in the solid and further ground make it react with silver thiolate. In this process, the silver thiolate was reduced by sodium borohydride and the color changed from orange to brownish black. (3) The final product exhibits strong hydrophilicity and the water was added into the system. The strong effervescence was found indicating the release of hydrogen. It is worth noting that the rate of water addition directly affects the yield of nanoclusters, the slow addition of water leads to the formation of a large number of nanoclusters. The final product is precipitated by the addition of ethanol. The nanocluster was characterized by transmission electron microscopy (TEM) and electrospray ionization mass spectrometry (ESI-MS; Fig. 2.18).

After that, the solid-state method was then used to synthesize $Ag_{44}(SePh)_{30}$ nanocluster which is the famous one in silver nanoclusters [131]. In the previously report, the Ag_{44} nanocluster was protected by thiol ligands and $Ag_{44}(SePh)_{30}$ nanocluster was the first silver nanocluster protected by selenolate ligand. In the synthesis of $Ag_{44}(SePh)_{30}$ nanocluster, the silver nitrate is mixed with phenylselenol first and fully ground until the color of the system turns into yellowish orange. After the silver selenolate is formed, the sodium borohydride added into the system in the solid state and then mixed. The system continues grinding until the color changes from yellowish orange to dark yellow. After the ethanol was added into the grinding bowl and wash away the excess thiol ligands. In addition, the color of system goes from dark yellow to deep brown with the ethanol added into and reserved for a certain amount of time. The product is collected by centrifugation to remove the supernatant to obtain a solid product. The solid product was dissolved in acetone and removed by vacuum rotary evaporator and gain the solid product again. Then the solid product continues to wash away excess thiol or other byproducts through ethanol. Repeat the above steps at least 3 times to ensure the purity of the product. Deep pink colored Ag_{44} cluster was obtained in acetone and it was stored in a refrigerator at ∼4°C. Several clusters such as Ag_{32} and Ag_{152} have been synthesized using this procedure.

FIG. 2.18 Photographs representing the changes during the cluster synthesis in the solid-state route. *(From T.U.B. Rao, B. Nataraju, T. Pradeep, Ag_9 quantum cluster through a solid-state route, J. Am. Chem. Soc. 132(46) (2010) 16304–16307, https://doi.org/10.1021/ja909321d.)*

References

[1] H.W. Kroto, J.R. Heath, S.C. O'Brien, R.F. Curl, R.E. Smalley, C60: buckminsterfullerene, Nature 318 (1985) 162–163, https://doi.org/10.1038/318162a0.

[2] O. Kylián, J. Kratochvíl, J. Hanuš, O. Polonskyi, P. Solař, H. Biederman, Fabrication of Cu nanoclusters and their use for production of Cu/plasma polymer nanocomposite thin films, Thin Solid Films 550 (2014) 46–52, https://doi.org/10.1016/j.tsf.2013.10.029.

[3] V. Pelenovich, X. Zeng, W. Zuo, et al., Enhanced sputtering yield of nanostructured samples under Ar+ cluster bombardment, Vacuum 172 (2020), 109096, https://doi.org/10.1016/j.vacuum.2019.109096.

[4] W.A. de Heer, The physics of simple metal clusters: experimental aspects and simple models, Rev. Mod. Phys. 65 (3) (1993) 611–676, https://doi.org/10.1103/RevModPhys.65.611.

[5] F.M. Devienne, R. Combarieu, M. Teisseire, Action of different gases, specially nitrogen, on the formation of uranium clusters; comparison with niobium and tantalum clusters, Surf. Sci. 106 (1981) 204–221, https://doi.org/10.1016/0039-6028(81)90202-8.

[6] G. Slodzian, Some problems encountered in secondary ion emission applied to elementary analysis, Surf. Sci. 48 (1) (1975) 161–186, https://doi.org/10.1016/0039-6028(75)90315-5.

[7] G. Wang, L. Dou, Z. Liu, Isotopic effect in the formation of copper-ion clusters by neutral-argon-atom bombardment, Phys. Rev. B 37 (1988) 9093, https://doi.org/10.1103/physrevb.37.9093.

[8] K.L. Rinehart, Fast atom bombardment mass spectrometry, Science 218 (4569) (1982) 254–260, https://doi.org/10.1126/science.218.4569.254.

[9] P. Sharpe, C.J. Cassady, Gas-phase reactions of silver cluster ions produced by fast atom bombardment, Chem. Phys. Lett. 191 (1–2) (1992) 111–116, https://doi.org/10.1016/0009-2614(92)85378-n.

[10] P. Fayet, L. Wöste, Production and study of metal cluster ions, Surf. Sci. 156 (1985) 134–139, https://doi.org/10.1016/0039-6028(85)90566-7.

[11] M.L. Mandich, V.E. Bondybey, W.D. Reents Jr., Reactive etching of positive and negative silicon cluster ions by nitrogen dioxide, J. Chem. Phys. 86 (7) (1987) 4245–4257, https://doi.org/10.1063/1.451885.

[12] R.L. Seliger, J.W. Ward, V. Wang, R.L. Kubena, A high-intensity scanning ion probe with submicrometer spot size, Appl. Phys. Lett. 34 (5) (1979) 310–312, https://doi.org/10.1063/1.90786.

[13] J.V. Van de Walle, P. Joyes, Remarkable periodicity of Ge$_n$p ions (n/p $\leq$ 25, 1 $\leq$ p $\leq$ 4) formed by the liquid-metal ions-source technique, Phys. Rev. B 32 (12) (1985) 8381–8383, https://doi.org/10.1103/physrevb.32.8381.

[14] C.G. Granqvist, R.A. Buhrman, Ultrafine metal particles, J. Appl. Phys. 47 (1976) 2200–2219, https://doi.org/10.1063/1.322870.

[15] K. Sattler, J. Mühlbach, E. Recknagel, Generation of metal clusters containing from 2 to 500 atoms, Phys. Rev. Lett. 45 (10) (1980) 821–824, https://doi.org/10.1103/PhysRevLett.45.821.

[16] T.G. Dietz, M.A. Duncan, D.E. Powers, R.E. Smalley, Laser production of supersonic metal cluster beams, J. Chem. Phys. 74 (11) (1981) 6511, https://doi.org/10.1063/1.440991.

[17] A. Hoareau, B. Cabaud, P. Melinon, Time-of-flight mass spectroscopy of supersonic beam of metallic vapours; intensities and appearance potentials of Mx aggregates, Surf. Sci. 106 (1-3) (1981) 195–203, https://doi.org/10.1016/0039-6028(81)90201-6.

[18] J. Li, H. Zhong, M. Xu, T. Li, L. Wang, Boosting the activity of Fe-Nx moieties in Fe-N-C electrocatalysts via phosphorus doping for oxygen reduction reaction, *Sci. China Mater.* 63 (2020) 965–971, https://doi.org/10.1007/s40843-019-1207-y.

[19] J.C. Rienstra-Kiracofe, G.S. Tschumper, H.F. Schaefer, S. Nandi, G.B. Ellison, Atomic and molecular electron affinities: photoelectron experiments and theoretical computations, Chem. Rev. 102 (1) (2002) 231–282, https://doi.org/10.1021/cr990044u.

[20] A. Sanov, R. Mabbs, Photoelectron imaging of negative ions, Int. Rev. Phys. Chem. 27 (1) (2008) 53–85, https://doi.org/10.1080/01442350701786512.

[21] Y. Zhao, J. Cui, M. Wang, D.Y. Valdivielso, A. Fielicke, Dinitrogen fixation and reduction by $Ta_3N_3H_{0,1}^-$ cluster anions at room temperature: hydrogen-assisted enhancement of reactivity, J. Am. Chem. Soc. 141 (32) (2019) 12592–12600, https://doi.org/10.1021/jacs.9b03168.

[22] D.E. Powers, S.G. Hansen, M.E. Geusic, A.C. Puiu, R.E. Smalley, Supersonic metal cluster beams: laser photoionization studies of copper cluster (Cu_2), J. Phys. Chem. 86 (14) (1982) 2556–2560, https://doi.org/10.1021/j100211a002.

[23] E.A. Rohlfing, D.M. Cox, A. Kaldor, Photoionization of isolated nickel atom clusters, J. Phys. Chem. 88 (20) (1984) 4497–4502, https://doi.org/10.1021/j150664a011.

[24] I. Katakuse, T. Ichihara, Y. Fujita, T. Matsuo, T. Sakurai, H. Matsuda, Mass distributions of copper, silver and gold clusters and electronic shell structure, Int. J. Mass Spectrom. Ion Process. 67 (2) (1985) 229–236, https://doi.org/10.1016/0168-1176(85)80021-5.

[25] J. Chen, Q. Zhang, P.G. Williard, L. Wang, Synthesis and structure determination of a new Au_{20} nanocluster protected by tripodal tetraphosphine ligands, Inorg. Chem. 53 (8) (2014) 3932–3934, https://doi.org/10.1021/ic500562r.

[26] L. Malatesta, L. Naldini, G. Simonetta, F. Cariati, Triphenylphosphine-gold(0)/gold(I) compounds, *Coord. Chem. Rev.* 1 (1-2) (1966) 255–262, https://doi.org/10.1016/s0010-8545(00)80179-4.

[27] M. McPartlin, R. Mason, L. Malatesta, Novel cluster complexes of gold(0)–gold(I), J. Chem. Soc. D 7 (1969) 334, https://doi.org/10.1039/c29690000334.

[28] C.E. Briant, B.R.C. Theobald, J.W. White, L.K. Bell, D.M.P. Mingos, Synthesis and X-ray structural characterization of the centred icosahedral gold cluster compound $[Au_{13}(PMe_2Ph)_{10}Cl_2](PF_6)_3$; the realization of a theoretical prediction, J. Chem. Soc. Chem. Commun. 5 (1981) 201–202, https://doi.org/10.1039/c39810000201.

[29] B.K. Teo, X. Shi, H. Zhang, Pure gold cluster of 1:9:9:1:9:9:1 layered structure: a novel 39-metal-atom cluster $[(Ph_3P)_{14}Au_{39}Cl_6]Cl_2$ with an interstitial gold atom in a hexagonal antiprismatic cage, J. Am. Chem. Soc. 114 (7) (1992) 2743–2745, https://doi.org/10.1021/ja00033a073.

[30] G. Schmid, R. Pfeil, R. Boese, F. Bandermann, J.W.A. van der Velden, Au$_{55}$[P(C$_6$H$_5$)$_3$]$_{12}$Cl$_6$—ein goldcluster ungewöhnlicher Größe, *Chem Ber Recl.* 114 (11) (1981) 3634–3642, https://doi.org/10.1002/cber.19811141116.

[31] B.K. Teo, K. Keating, Novel triicosahedral structure of the largest metal alloy cluster: hexachlorododecakis(triphenylphosphine)-gold-silver cluster [(Ph$_3$P)$_{12}$Au$_{13}$Ag$_{12}$Cl$_6$]$_m^+$, J. Am. Chem. Soc. 106 (7) (1984) 2224–2226, https://doi.org/10.1021/ja00319a061.

[32] B.K. Teo, H. Zhang, Clusters of clusters: self-organization and self-similarity in the intermediate stages of cluster growth of Au-Ag supraclusters, Proc. *Natl. Acad. Sci. U. S. A.* 88 (12) (1991) 5067–5071, https://doi.org/10.1073/pnas.88.12.5067.

[33] B.K. Teo, M. Hong, H. Zhang, D. Huang, X. Shi, Cluster of clusters: structure of a novel 38-atom cluster (p-tolyl3P)$_{12}$Au$_{18}$Ag$_{20}$Cl$_{14}$, J. Chem. Soc. Chem. Commun. 2 (1988) 204–206, https://doi.org/10.1039/c39880000204.

[34] G. Frens, *Kolloid-Z. Z. Polym.* 250 (1972) 736–741, https://doi.org/10.1007/BF01498565.

[35] J. Turkevich, P.C. Stevenson, J. Hillier, *Discuss Faraday Soc.* 11 (1951) 55–75, https://doi.org/10.1039/DF9511100055.

[36] M. Brust, M. Walker, D. Bethell, D.J. Schiffrin, R. Whyman, Synthesis of thiol-derivatised gold nanoparticles in a two-phase liquid–liquid system, J. Chem. Soc. Chem. Commun. 7 (1994) 801–802, https://doi.org/10.1039/c39940000801.

[37] H. Qian, M. Zhu, Z. Wu, R. Jin, Quantum sized gold nanoclusters with atomic precision, Acc. Chem. Res. 45 (9) (2012) 1470–1479, https://doi.org/10.1021/ar200331z.

[38] M.K. Corbierre, R.B. Lennox, Preparation of thiol-capped gold nanoparticles by chemical reduction of soluble Au(I)−thiolates, Chem. Mater. 17 (23) (2005) 5691–5696, https://doi.org/10.1021/cm051115a.

[39] L.A. Porter, D. Ji, S.L. Westcott, et al., Gold and silver nanoparticles functionalized by the adsorption of dialkyl disulfides, Langmuir 14 (26) (1998) 7378–7386, https://doi.org/10.1021/la980870i.

[40] Y.S. Shon, C. Mazzitelli, R.W. Murray, Unsymmetrical disulfides and thiol mixtures produce different mixed monolayer-protected gold clusters, Langmuir 17 (25) (2001) 7735–7741, https://doi.org/10.1021/la015546t.

[41] I.P. Stolyarov, Y.V. Gaugash, G.N. Kryukova, D.I. Kochubei, I.I. Moiseev, New palladium nanoclusters. Synthesis, structure, and catalytic properties, Russ. Chem. B 53 (2004) 1194–1199, https://doi.org/10.1023/b:rucb.0000042273.76439.7b.

[42] R.L. Whetten, W.M. Gelbart, Nanocrystal microemulsions: surfactant-stabilized size and shape, J. Phys. Chem. 98 (13) (1994) 3544–3549, https://doi.org/10.1021/j100064a042.

[43] D.V. Leff, P.C. Ohara, J.R. Heath, W.M. Gelbart, Thermodynamic control of gold nanocrystal size: experiment and theory, J. Phys. Chem. 99 (18) (1995) 7036–7041, https://doi.org/10.1021/j100018a041.

[44] M. Brust, D.J. Schiffrin, D. Bethell, C.J. Kiely, Novel gold-dithiol nano-networks with non-metallic electronic properties, Adv. Mater. 7 (9) (1995) 795–797, https://doi.org/10.1002/adma.19950070907.

[45] M.J. Hostetler, J.E. Wingate, C. Zhong, J.E. Harris, R.W. Murray, Alkanethiolate gold cluster molecules with core diameters from 1.5 to 5.2 nm: core and monolayer properties as a function of core size, Langmuir 14 (1) (1998) 17–30, https://doi.org/10.1021/la970888d.

[46] M.K. Corbierre, N.S. Cameron, M. Sutton, S.G.J. Mochrie, R.B. Lennox, Polymer-stabilized gold nanoparticles and their incorporation into polymer matrices, Am. Chem. Soc. 123 (42) (2001) 10411–10412, https://doi.org/10.1021/ja0166287.

[47] P.J.G. Goulet, R.B. Lennox, New Insights into Brust−Schiffrin metal nanoparticle synthesis, J. Am. Chem. Soc. 132 (28) (2010) 9582–9584, https://doi.org/10.1021/ja104011b.

[48] Y. Li, O. Zaluzhna, B. Xu, Y. Gao, J.M. Modest, Y.J. Tong, Mechanistic Insights into the Brust−Schiffrin two-phase synthesis of organo-chalcogenate-protected metal nanoparticles, J. Am. Chem. Soc. 133 (7) (2011) 2092–2095, https://doi.org/10.1021/ja1105078.

[49] A.C. Templeton, W.P. Wuelfing, R.W. Murray, Monolayer-protected cluster molecules, Acc. Chem. Res. 33 (1) (2000) 27–36, https://doi.org/10.1021/ar9602664.

[50] S.R.K. Perala, S. Kumar, On the mechanism of metal nanoparticle synthesis in the Brust−Schiffrin method, Langmuir 29 (31) (2013) 9863–9873, https://doi.org/10.1021/la401604q.

[51] M. Zhu, E. Lanni, N. Garg, M.E. Bier, R. Jin, Kinetically controlled, high-yield synthesis of Au$_{25}$ clusters, J. Am. Chem. Soc. 130 (4) (2008) 1138–1139, https://doi.org/10.1021/ja0744302.

[52] M. Zhu, H. Qian, R. Jin, Thiolate-protected Au$_{24}$(SC$_2$H$_4$Ph)$_{20}$ nanoclusters: superatoms or not? J. Phys. Chem. Lett. 1 (6) (2010) 1003–1007, https://doi.org/10.1021/jz100133n.

[53] T.D. Nguyen, Z.R. Jones, B.R. Goldsmith, et al., A Cu$_{25}$ nanocluster with partial Cu(0) character, J. Am. Chem. Soc. 137 (41) (2015) 13319–13324, https://doi.org/10.1021/jacs.5b07574.

[54] H. Li, H. Zhai, C. Zhou, et al., Atomically precise copper cluster with intensely near-infrared luminescence and its mechanism, J. Phys. Chem. Lett. 11 (12) (2020) 4891–4896, https://doi.org/10.1021/acs.jpclett.0c01358.

[55] J.-H. Liao, S. Kahlal, Y.-C. Liu, et al., [Cu$_{13}${S$_2$CNnBu$_2$}$_6$(acetylide)$_4$]$^+$: a two-electron superatom, Angew. Chem. Int. Ed. 55 (47) (2016) 14704–14708, https://doi.org/10.1002/anie.201608609.

[56] X. Zou, S. Jin, W. Du, et al., Multi-ligand-directed synthesis of chiral silver nanoclusters, Nanoscale 9 (43) (2017) 16800–16805, https://doi.org/10.1039/C7NR06338E.

[57] M. Brust, J. Fink, D. Bethell, D.J. Schiffrin, C. Kiely, Synthesis and reactions of functionalised gold nanoparticles, J. Chem. Soc. Chem. Commun. 16 (1995) 1655–1656, https://doi.org/10.1039/C39950001655.

[58] Z. Wu, J. Suhan, R. Jin, One-pot synthesis of atomically monodisperse, thiol-functionalized Au$_{25}$ nanoclusters, J. Mater. Chem. 19 (5) (2009) 622–626, https://doi.org/10.1039/B815983A.

[59] H. Qian, R. Jin, Ambient synthesis of Au$_{144}$(SR)$_{60}$ nanoclusters in methanol, Chem. Mater. 23 (8) (2011) 2209–2217, https://doi.org/10.1021/cm200143s.

[60] A. Ghosh, T. Udayabhaskararao, T. Pradeep, One-step route to luminescent $Au_{18}SG_{14}$ in the condensed phase and its closed shell molecular ions in the gas phase, J. Phys. Chem. Lett. 3 (15) (2012) 1997–2002, https://doi.org/10.1021/jz3007436.

[61] Y. Yu, Z. Luo, Y. Yu, J.Y. Lee, J. Xie, Observation of cluster size growth in CO-directed synthesis of $Au_{25}(SR)_{18}$ nanoclusters, ACS Nano 6 (9) (2012) 7920–7927, https://doi.org/10.1021/nn3023206.

[62] Z. Luo, V. Nachammai, B. Zhang, et al., Toward understanding the growth mechanism: tracing all stable intermediate species from reduction of Au(I)−thiolate complexes to evolution of Au_{25} nanoclusters, J. Am. Chem. Soc. 136 (30) (2014) 10577–10580, https://doi.org/10.1021/ja505429f.

[63] Q. Yao, Y. Yu, X. Yuan, Y. Yu, J. Xie, J.Y. Lee, Two-phase synthesis of small thiolate-protected Au_{15} and Au_{18} nanoclusters, Small 9 (16) (2013) 2696–2701, https://doi.org/10.1002/smll.201203112.

[64] Y. Yu, X. Chen, Q. Yao, Y. Yu, N. Yan, J. Xie, Scalable and precise synthesis of thiolated Au_{10-12}, Au_{15}, Au_{18}, and Au_{25} nanoclusters via pH controlled CO reduction, Chem. Mater. 25 (6) (2013) 946–952, https://doi.org/10.1021/cm304098x.

[65] J. Yan, H. Su, H. Yang, et al., Asymmetric synthesis of chiral bimetallic $[Ag_{28}Cu_{12}(SR)_{24}]^{4-}$ nanoclusters via ion pairing, J. Am. Chem. Soc. 138 (39) (2016) 12751–12754, https://doi.org/10.1021/jacs.6b08100.

[66] H. Yang, Y. Wang, H. Yang, et al., All-thiol-stabilized Ag_{44} and $Au_{12}Ag_{32}$ nanoparticles with single-crystal structures, J. Am. Chem. Soc. 138 (39) (2016) 12751–12754, https://doi.org/10.1021/jacs.6b08100.

[67] Y. Song, S. Wang, J. Zhang, et al., Crystal structure of selenolate-protected $Au_{24}(SeR)_{20}$ nanocluster, J. Am. Chem. Soc. 136 (8) (2014) 2963–2965, https://doi.org/10.1021/ja4131142.

[68] Y. Song, J. Zhong, S. Yang, et al., Crystal structure of $Au_{25}(SePh)_{18}$ nanoclusters and insights into their electronic, optical and catalytic properties, Nanoscale 6 (22) (2014) 13977–13985, https://doi.org/10.1039/C4NR04631E.

[69] Y. Song, F. Fu, J. Zhang, et al., The magic Au_{60} nanocluster: a new cluster-assembled material with five Au_{13} building blocks, Angew. Chem. Int. Ed. 54 (29) (2015) 8430–8434, https://doi.org/10.1002/anie.201501830.

[70] Y. Song, S. Jin, X. Kang, et al., How a single electron affects the properties of the "non-superatom" Au_{25} nanoclusters, Chem. Mater. 28 (8) (2016) 2609–2617, https://doi.org/10.1021/acs.chemmater.5b04655.

[71] Y. Negishi, K. Nobusada, T. Tsukuda, Glutathione-protected gold clusters revisited: bridging the gap between gold(I)−thiolate complexes and thiolate-protected gold nanocrystals, J. Am. Chem. Soc. 127 (14) (2005) 5261–5270, https://doi.org/10.1021/ja042218h.

[72] S. Chen, S. Wang, J. Zhong, et al., The structure and optical properties of the $[Au_{18}(SR)_{14}]$ nanocluster, Angew. Chem. Int. Ed. 54 (10) (2015) 3145–3149, https://doi.org/10.1002/anie.201410295.

[73] S. Yang, S. Chen, L. Xiong, et al., Total structure determination of $Au_{16}(S-Adm)_{12}$ and $Cd_1Au_{14}(StBu)_{12}$ and implications for the structure of $Au_{15}(SR)_{13}$, J. Am. Chem. Soc. 140 (35) (2018) 10988–10994, https://doi.org/10.1021/jacs.8b04257.

[74] L.G. AbdulHalim, S. Ashraf, K. Katsiev, et al., A scalable synthesis of highly stable and water dispersible $Ag_{44}(SR)_{30}$ nanoclusters, J. Mater. Chem. A 1 (35) (2013) 10148–10154, https://doi.org/10.1039/C3TA11785E.

[75] S. Yang, J. Chai, Y. Song, et al., In situ two-phase ligand exchange: a new method for the synthesis of alloy nanoclusters with precise atomic structures, J. Am. Chem. Soc. 139 (16) (2017) 5668–5671, https://doi.org/10.1021/jacs.7b00668.

[76] R. Jin, Atomically precise metal nanoclusters: stable sizes and optical properties, Nanoscale 7 (5) (2015) 1549–1565, https://doi.org/10.1039/C4NR05794E.

[77] D.M. Chevrier, A. Chatt, P. Zhang, C. Zeng, R. Jin, Unique bonding properties of the $Au_{36}(SR)_{24}$ nanocluster with FCC-like core, J. Phys. Chem. Lett. 4 (19) (2013) 3186–3191, https://doi.org/10.1021/jz401818c.

[78] Z. Gan, J. Chen, J. Wang, et al., The fourth crystallographic closest packing unveiled in the gold nanocluster crystal, Nat. Commun. 8 (2017) 14739, https://doi.org/10.1038/ncomms14739.

[79] M. Rambukwella, A. Dass, Synthesis of $Au_{38}(SCH_2CH_2Ph)_{24}$, $Au_{36}(SPh-tBu)_{24}$, and $Au_{30}(S-tBu)_{18}$ nanomolecules from a common precursor mixture, Langmuir 33 (41) (2017) 10958–10964, https://doi.org/10.1021/acs.langmuir.7b03080.

[80] P.R. Nimmala, A. Dass, $Au_{99}(SPh)_{42}$ nanomolecules: aromatic thiolate ligand induced conversion of $Au_{144}(SCH_2CH_2Ph)_{60}$, J. Am. Chem. Soc. 136 (49) (2014) 17016–17023, https://doi.org/10.1021/ja5103025.

[81] P.R. Nimmala, S. Theivendran, G. Barcaro, et al., Transformation of $Au_{144}(SCH_2CH_2Ph)_{60}$ to $Au_{133}(SPh-tBu)_{52}$ nanomolecules: theoretical and experimental study, J. Phys. Chem. Lett. 6 (11) (2015) 2134–2139, https://doi.org/10.1021/acs.jpclett.5b00780.

[82] S. Jin, S. Wang, L. Xiong, et al., Two electron reduction: from quantum dots to metal nanoclusters, Chem. Mater. 28 (21) (2016) 7905–7911, https://doi.org/10.1021/acs.chemmater.6b03472.

[83] Q. Li, J. Chai, S. Yang, et al., Multiple ways realizing charge-state transform in Au-Cu bimetallic nanoclusters with atomic precision, Small 17 (27) (2021) 1907114, https://doi.org/10.1002/smll.201907114.

[84] L. Tang, X. Kang, S. Wang, M. Zhu, Light-induced size-growth of atomically precise nanoclusters, Langmuir 35 (38) (2019) 12350–12355, https://doi.org/10.1021/acs.langmuir.9b01527.

[85] N. Xia, J. Yuan, L. Liao, et al., Structural oscillation revealed in gold nanoparticles, J. Am. Chem. Soc. 142 (28) (2020) 12140–12145, https://doi.org/10.1021/jacs.0c02117.

[86] Q. Li, S. Yang, T. Chen, et al., Structure determination of a metastable $Au_{22}(SAdm)_{16}$ nanocluster and its spontaneous transformation into $Au_{21}(SAdm)_{15}$, Nanoscale 12 (46) (2020) 23694–23699, https://doi.org/10.1039/d0nr07124b.

[87] F. Gruber, M. Schulz-Dobrick, M. Jansen, Structure-directing forces in intercluster compounds of cationic $[Ag_{14}(C \equiv CtBu)_{12}Cl]^+$ building blocks and polyoxometalates: long-range versus short-range bonding interactions, Chem. Eur. J. 16 (5) (2010) 1464–1469, https://doi.org/10.1002/chem.200902538.

[88] Z. Wang, H.-F. Su, X.-P. Wang, et al., Johnson solids: anion-templated silver thiolate clusters capped by sulfonate, Chem. Eur. J. 24 (7) (2017) 1640–1650, https://doi.org/10.1002/chem.201704298.

[89] H. Akbarzadeh, M. Abbaspour, E. Mehrjouei, Effect of systematic addition of the third component on the melting characteristics and structural evolution of binary alloy nanoclusters, J. Mol. Liq. 249 (2018) 412–419, https://doi.org/10.1016/j.molliq.2017.11.075.

[90] A.I. Ayesh, Size-selected fabrication of alloy nanoclusters by plasma-gas condensation, J. Alloys Compd. 745 (15) (2018) 299–305, https://doi.org/10.1016/j.jallcom.2018.02.219.

[91] X. Kang, Y. Li, M. Zhu, R. Jin, Atomically precise alloy nanoclusters: syntheses, structures, and properties, Chem. Soc. Rev. 49 (17) (2020) 6443–6514, https://doi.org/10.1039/c9cs00633h.

[92] T. Kawawaki, Y. Imai, D. Suzuki, et al., Atomically precise alloy nanoclusters, Chem. Eur. J. 26 (69) (2020) 16150–16193, https://doi.org/10.1002/chem.202001877.

[93] X. Mao, L. Wang, Y. Xu, P. Wang, Y. Li, J. Zhao, Computational high-throughput screening of alloy nanoclusters for electrocatalytic hydrogen evolution, NPJ Comput. Mater. 7 (2021) 46, https://doi.org/10.1038/s41524-021-00514-8.

[94] S. Sharma, K.K. Chakrahari, J.-Y. Saillard, C.W. Liu, Structurally precise dichalcogenolate-protected copper and silver superatomic nanoclusters and their alloys, Acc. Chem. Res. 51 (10) (2018) 2475–2483, https://doi.org/10.1021/acs.accounts.8b00349.

[95] A.G. Walsh, P. Zhang, Thiolate-protected bimetallic nanoclusters: understanding the relationship between electronic and catalytic properties, J. Phys. Chem. Lett. 12 (1) (2021) 257–275, https://doi.org/10.1021/acs.jpclett.0c03252.

[96] S. Wang, Q. Li, X. Kang, M. Zhu, Customizing the structure, composition, and properties of alloy nanoclusters by metal exchange, Acc. Chem. Res. 51 (11) (2018) 2784–2792, https://doi.org/10.1021/acs.accounts.8b00327.

[97] J. Yang, F. Muckel, W. Baek, et al., Chemical synthesis, doping, and transformation of magic-sized semiconductor alloy nanoclusters, J. Am. Chem. Soc. 139 (19) (2017) 6761–6770, https://doi.org/10.1021/jacs.7b02953.

[98] R. Jin, S. Zhao, C. Liu, et al., Controlling Ag-doping in $[Ag_xAu_{25-x}(SC_6H_{11})_{18}]^-$ nanoclusters: cryogenic optical, electronic and electrocatalytic properties, Nanoscale 9 (48) (2017) 19183–19190, https://doi.org/10.1039/c7nr05871c.

[99] S. Wang, X. Meng, A. Das, et al., A 200-fold quantum yield boost in the photoluminescence of silver-doped Ag_xAu_{25-x} nanoclusters: the 13th silver atom matters, Angew. Chem. Int. Ed. 53 (9) (2014) 2376–2380, https://doi.org/10.1002/anie.201307480.

[100] C. Kumara, A. Dass, AuAg alloy nanomolecules with 38 metal atoms, Nanoscale 4 (14) (2012) 4084–4086, https://doi.org/10.1039/c2nr11781a.

[101] C. Kumara, K.J. Gagnon, A. Dass, X-ray crystal structure of $Au_{38-x}Ag_x(SCH_2CH_2Ph)_{24}$ alloy nanomolecules, J. Phys. Chem. Lett. 6 (7) (2015) 1223–1228, https://doi.org/10.1021/acs.jpclett.5b00270.

[102] Q. Li, S. Wang, K. Kirschbaum, K.J. Lambright, A. Das, R. Jin, Heavily doped $Au_{25-x}Ag_x(SC_6H_{11})_{18}^-$ nanoclusters: silver goes from the core to the surface, Chem. Commun. 52 (29) (2016) 5194–5519, https://doi.org/10.1039/c6cc01243d.

[103] J. Zhou, S. Yang, Y. Tan, H.S. Cheng, J. Chai, M. Zhu, Cu doping-induced transformation from $[Ag_{62}S_{12}(SBut)_{32}]_2^+$ to $[Ag_{62-x}Cu_xS_{12}(SBut)_{32}]_4^+$ nanocluster, Chem. Asian J. 16 (19) (2021) 2973–2977, https://doi.org/10.1002/asia.202100739.

[104] M.S. Bootharaju, S.M. Kozlov, Z. Cao, et al., Tailoring the crystal structure of nanoclusters unveiled high photoluminescence via ion pairing, Chem. Mater. 30 (8) (2018) 2719–2725, https://doi.org/10.1021/acs.chemmater.8b00328.

[105] J. Yan, H. Su, H. Yang, et al., Total structure and electronic structure analysis of doped thiolated silver $[MAg_{24}(SR)_{18}]_2^-$ (M = Pd, Pt) clusters, J. Am. Chem. Soc. 137 (37) (2015) 11880–11883, https://doi.org/10.1021/jacs.5b07186.

[106] H. Xu, B. Yan, S. Li, et al., Highly open bowl-like PtAuAg nanocages as robust electrocatalysts towards ethylene glycol oxidation, J. Power Sources 384 (30) (2018) 42–47, https://doi.org/10.1016/j.jpowsour.2018.02.067.

[107] L. Xu, Q. Li, T. Li, J. Chai, S. Yang, M. Zhu, Construction of a new $Au_{27}Cd_1(SAdm)_{14}(DPPF)Cl$ nanocluster by surface engineering and insight into its structure–property correlation, Inorg. Chem. Front. 8 (22) (2021) 4820–4827, https://doi.org/10.1039/d1qi01015h.

[108] G.K. Wertheim, J. Kwo, B.K. Teo, K.A. Keating, XPS study of bonding in ligated Au clusters, Solid State Commun. 55 (4) (1985) 357–361, https://doi.org/10.1016/0038-1098(85)90623-4.

[109] B.K. Teo, Cluster of clusters: a new series of high nuclearity Au-Ag clusters, Polyhedron 7 (22–23) (1988) 2317–2320, https://doi.org/10.1016/s0277-5387(00)86348-2.

[110] J.F. Parker, C.A. Fields-Zinna, R.W. Murray, The story of a monodisperse gold nanoparticle: $Au_{25}L_{18}$, Acc. Chem. Res. 43 (9) (2010) 1289–1296, https://doi.org/10.1021/ar100048c.

[111] D.R. Kauffman, D. Alfonso, C. Matranga, H. Qian, R. Jin, A quantum alloy: the ligand-protected $Au_{25-x}Ag_x(SR)_{18}$ cluster, J. Phys. Chem. C 117 (15) (2013) 7914–7923, https://doi.org/10.1021/jp4013224.

[112] S. Hossain, D. Suzuki, T. Iwasa, et al., Determining and controlling Cu-substitution sites in thiolate-protected gold-based 25-atom alloy nanoclusters, J. Phys. Chem. C 124 (40) (2020) 22304–22313, https://doi.org/10.1021/acs.jpcc.0c06858.

[113] M. Suyama, S. Takano, T. Nakamura, T. Tsukuda, Stoichiometric formation of open-shell $[PtAu_{24}(SC_2H_4Ph)_{18}]^-$ via spontaneous electron proportionation between $[PtAu_{24}(SC_2H_4Ph)_{18}]_2$ and $[PtAu_{24}(SC_2H_4Ph)_{18}]_0$, J. Am. Chem. Soc. 141 (36) (2019) 14048–14051, https://doi.org/10.1021/jacs.9b06254.

[114] S.R. Biltek, S. Mandal, A. Sen, A.C. Reber, F. Pedicini Anthony, S.N. Khanna, Synthesis and structural characterization of an atom-precise bimetallic nanocluster, $Ag_4Ni_2(DMSA)_4$, J. Am. Chem. Soc. 135 (1) (2013) 26–29, https://doi.org/10.1021/ja308884s.

[115] L. Zhan, M. Zhu, L. Liu, J. Wang, C. Xie, J. Zhang, Synthesis of MAuAg (M = Ni, Pd, or Pt) and NiAuCu heterotrimetallic complexes ligated by a tritopic carbanionic N-heterocyclic carbene, Inorg. Chem. 60 (21) (2021) 16035–16041, https://doi.org/10.1021/acs.inorgchem.1c01964.

[116] Y. Chen, C. Liu, Q. Tang, et al., Isomerism in $Au_{28}(SR)_{20}$ nanocluster and stable structures, J. Am. Chem. Soc. 138 (5) (2016) 1482–1485, https://doi.org/10.1021/jacs.5b12094.

[117] X. Kang, L. Xiong, S. Wang, Y. Pei, M. Zhu, De-assembly of assembled Pt_1Ag_{12} units: tailoring the photoluminescence of atomically precise nanoclusters, Chem. Commun. 53 (93) (2017) 12564–12567, https://doi.org/10.1039/C7CC05996E.

[118] M.S. Bootharaju, S.M. Kozlov, Z. Cao, et al., Direct versus ligand-exchange synthesis of $[PtAg_{28}(BDT)_{12}(TPP)_4]^{4-}$ nanoclusters: effect of a single-atom dopant on the optoelectronic and chemical properties, Nanoscale 9 (27) (2017) 9529–9536, https://doi.org/10.1039/C7NR02844J.

[119] E. Khatun, P. Chakraborty, B.R. Jacob, et al., Intercluster reactions resulting in silver-rich trimetallic nanoclusters, Chem. Mater. 32 (1) (2020) 611–619, https://doi.org/10.1021/acs.chemmater.9b04530.

[120] E. Khatun, T. Pradeep, New routes for multicomponent atomically precise metal nanoclusters, ACS Omega 6 (1) (2021) 1–16, https://doi.org/10.1021/acsomega.0c04832.

[121] S. Bootharaju Megalamane, C.P. Joshi, M.R. Parida, O.F. Mohammed, O.M. Bakr, Templated atom-precise galvanic synthesis and structure elucidation of a$[Ag_{24}Au(SR)_{18}]^-$ nanocluster, *Angew. Chem.* 128 (3) (2016) 934–938, https://doi.org/10.1002/ange.201509381.

[122] S. Wang, Y. Song, S. Jin, et al., Metal exchange method using Au_{25} nanoclusters as templates for alloy nanoclusters with atomic precision, J. Am. Chem. Soc. 137 (12) (2015) 4018–4021, https://doi.org/10.1021/ja511635g.

[123] X. Kang, L. Xiong, S. Wang, et al., Shape-controlled synthesis of trimetallic nanoclusters: structure elucidation and properties investigation, *Chem. Eur. J.* 22 (48) (2016) 17145–17150, https://doi.org/10.1002/chem.201603893.

[124] X. Kang, X. Wei, S. Jin, et al., Rational construction of a library of M29 nanoclusters from monometallic to tetrametallic, Proc. Natl. Acad. Sci. U. S. A. 116 (38) (2019) 18834–18840, https://doi.org/10.1073/pnas.1912719116.

[125] J.-P. Choi, C.A. Fields-Zinna, R.L. Stiles, et al., Reactivity of $[Au_{25}(SCH_2CH_2Ph)_{18}]^{1-}$ nanoparticles with metal ions, J. Phys. Chem. C 114 (38) (2010) 15890–15896, https://doi.org/10.1021/jp9101114.

[126] Z. Wu, Anti-galvanic reduction of thiolate-protected gold and silver nanoparticles, Angew. Chem. 124 (12) (2012) 2988–2992, https://doi.org/10.1002/ange.201107822.

[127] S. Yang, J. Chai, T. Chen, et al., Crystal structures of two new gold–copper bimetallic nanoclusters: $Cu_xAu_{25-x}(PPh_3)_{10}(PhC_2H_4S)_5Cl_2^+$ and $Cu_3Au_{34}(PPh_3)_{13}(tBuPhCH_2S)_6S_2^+$, Inorg. Chem. 56 (4) (2017) 1771–1774, https://doi.org/10.1021/acs.inorgchem.6b02016.

[128] K.R. Krishnadas, A. Ghosh, A. Baksi, I. Chakraborty, G. Natarajan, T. Pradeep, Intercluster reactions between $Au_{25}(SR)_{18}$ and $Ag_{44}(SR)_{30}$, J. Am. Chem. Soc. 138 (1) (2016) 140–148, https://doi.org/10.1021/jacs.5b09401.

[129] S. Bhat, A. Baksi, S.K. Mudedla, G. Natarajan, V. Subramanian, T. Pradeep, $Au_{22}Ir_3(PET)_{18}$: an unusual alloy cluster through intercluster reaction, J. Phys. Chem. Lett. 8 (13) (2017) 2787–2793, https://doi.org/10.1021/acs.jpclett.7b01052.

[130] T.U.B. Rao, B. Nataraju, T. Pradeep, Ag_9 quantum cluster through a solid-state route, J. Am. Chem. Soc. 132 (46) (2010) 16304–16307, https://doi.org/10.1021/ja909321d.

[131] I. Chakraborty, W. Kurashige, K. Kanehira, et al., $Ag_{44}(SeR)_{30}$: a hollow cage silver cluster with selenolate protection, J. Phys. Chem. Lett. 4 (19) (2013) 3186–3376, https://doi.org/10.1021/jz401879c.

Chapter 3

Characterizations and atomically precise structures of metal nanoclusters

Manzhou Zhu and Qinzhen Li

3.1 Characterization techniques for nanoclusters

3.1.1 UV-vis absorption spectroscopy

Both ultraviolet and visible absorption spectra belong to molecular spectra, which are generated by the transition of valence electrons. The composition and structure of a substance can be analyzed, measured and inferred by the UV-vis spectrum and absorption intensity generated by the absorption of molecules or ions of substances to ultraviolet and visible light. In organic compound molecules, there are σ electrons forming single bonds, π electrons forming double bonds, and lone pair n electrons forming no bonds. When a molecule absorbs a certain amount of radiant energy, these electrons will jump to higher energy levels. At this time, the orbital occupied by electrons is called antibonding orbital, and this kind of electronic transition is closely related to the internal structure. There are four types of electron transitions in ultraviolet absorption spectrum: $\sigma \rightarrow \sigma^*$, $n \rightarrow \sigma^*$, $\pi \rightarrow \pi^*$, and $n \rightarrow \pi^*$.

The UV-vis spectrum has the following characteristics: (1) the solution under test with the same concentration has different absorbances to different wavelengths of light; (2) for the same solution to be tested, the greater the concentration, the greater the absorbance; and (3) for the same substance, regardless of the concentration, the wavelength corresponding to the maximum absorption peak (λ_{max}) is the same, and the shape of the curve is also exactly the same.

Due to the ultrasmall size of metal nanoclusters (<2.2 nm), quantum confinement becomes an important characteristic of them. The quantum confinement effect endows the nanoclusters with molecule-like properties, so the electron transitions can occur within the cluster molecules to form a distinct molecular spectrum, just like an organic molecule. Therefore, UV-vis spectroscopy is a useful tool to identify nanoclusters. Different with metal nanoparticles which usually show a broad surface plasmon resonance (SPR) absorption peak, the appearance of step-like features with spectral broadening in their absorption profiles due to a molecule-like HOMO-LUMO transition. This kind of conversion from electronic band structure to distinct energy levels leads to well-defined optical bands of nanoclusters. UV-vis spectroscopy is often used as a convenient method of preliminary nanocluster identification. The UV-vis spectra of metal nanoclusters show significant structure-dependent characteristic. For instance, Negishi et al. separated a mix of polydisperse GS-protected Au NCs with PAGE and found that different components showed distinct UV-vis spectra (Fig. 3.1) [1].

This result demonstrates that the UV-vis spectrum of a nanocluster is highly correlative to its molecular structure. Even two nanoclusters are structural isomers, their UV-vis spectrum are totally different. For example, $Au_{38}(SR)_{24}$ has been found to have two structures successively, that is, Au_{38Q} and Au_{38T} [2]. The UV-vis spectra of these two isomers are shown in Fig. 3.2. As we can see, the two nanoclusters show significant differences in their UV-vis absorption, allowing them to be easily distinguished by UV-vis spectrum.

Moreover, UV-vis spectroscopy can also be used to reflect the purity of clusters, monitor the process of nanocluster reaction or transformation, and reflect their stability under different conditions. With the help of DFT calculations, the absorption peaks in the UV-vis spectrum can be finely identified and assigned to different types of orbital transitions. This enables us to further understand the relationship between cluster structure and UV-vis spectrum.

3.1.2 Photoluminescence spectroscopy

Photoluminescence (PL) is a kind of cold luminescence, which refers to the process of photons being reradiated by a substance after absorbing them. According to the theory of quantum mechanics, this process can be described as a process in which a matter absorbs photons and transitions to an excited state with a higher energy level, then returns to a lower energy

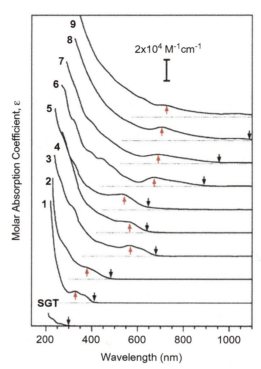

FIG. 3.1 UV-vis spectra of a series of GS-protected Au NCs separated by PAGE. *(From Y. Negishi, K. Nobusada, T. Tsukuda, T. Glutathione-protected gold clusters revisited: bridging the gap between gold(I)–thiolate complexes and thiolate-protected gold nanocrystals, J. Am. Chem. Soc. 127(14) (2005) 5261–5270, https://doi.org/10.1021/ja042218h.)*

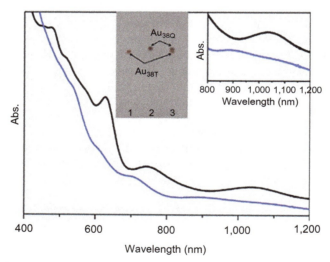

FIG. 3.2 UV-vis spectra of Au_{38Q} and Au_{38T}. Insets are the photo of thin-layer chromatography. *(From S. Tian, Y.-Z. Li, M.-B. Li, J. Yuan, J. Yang, Z. Wu, R. Jin, Structural Isomerism in gold nanoparticles revealed by X-ray crystallography, Nat. Commun. 6 (2015) 8667, https://doi.org/10.1038/ncomms9667.)*

state and emits photons at the same time. Photoluminescence can be divided into fluorescence or phosphorescence according to delay time.

Fluorescence spectrophotometer is the main instrument for characterizing the photoluminescence properties of nanoclusters. According to the different types, it can provide many physical parameters, including excitation spectrum, emission spectrum, fluorescence intensity, quantum yield, fluorescence lifetime, fluorescence polarization and so on.

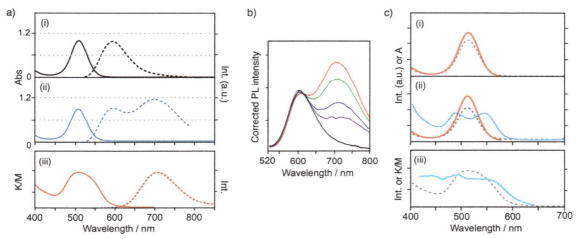

FIG. 3.3 (A) Absorption/diffuse reflectance *(solid lines)* and PL *(dotted lines, $\lambda_{ex}=510\,nm$)* spectra of Au_8 clusters. (B) Concentration-corrected PL spectra of Au_8 clusters. (C) PL excitation spectra of Au_8 clusters in (i) CH_2Cl_2 (12.5 μM), (ii) MeOH (12.5 μM), and (iii) the solid state upon monitoring at 596 nm *(red lines)* and 708 nm *(blue lines)*. *(From M. Sugiuchi, J. Maeba, N. Okubo, M. Iwamura, K. Nozaki, K. Konishi, Aggregation-induced fluorescence-to-phosphorescence switching of molecular gold clusters, J. Am. Chem. Soc. 139(49) (2017) 17731–17734, https://doi.org/10.1021/jacs.7b10201.)*

Fluorescence spectrophotometer used for measuring fluorescence intensity is composed of excitation light source, monochromator, slit, sample chamber, signal detection and amplification system, signal readout and recording system. The excitation source provides the source of incident light used to excite the sample. The monochromator is used to separate the desired monochromatic light. The signal detection and amplification system are used to convert the fluorescent signal into electrical signal, and the fluorescent signal can be displayed or recorded with the readout device on the amplification system.

The other two important parameters of fluorescent nanoclusters, quantum yield and fluorescence lifetime, can be also recorded by fluorescent instrument. The instrument used to measure the fluorescence quantum yield is usually equipped with an integrating sphere. The instrument determines the quantum yield of the sample by calculating the ratio of the number of photons emitted by the sample to the number of photons absorbed by the sample. The obtained quantum yield is called the absolute quantum yield, which does not need the reference material. TCSPC (time-correlated single photon counting) is a commonly used fluorescence lifetime measurement technology. Fluorescence lifetime is usually measured on the order of ps ∼ μs, and TCSPC is the most mature and accurate test method in such a short time. The use of a synchronous signal source driving laser, the output optical pulse irradiation sample pool, taking advantage of the photon detection device (mostly PMT) the fluorescence signal detection, each photon counting signal will fall into a corresponding time window, after a certain time after statistical stacking get fluorescence life curve. After hundreds of thousands of repetitions, different time channels accumulate different numbers of photons. The fluorescence delay curve can be obtained by plotting the number of photons against time.

Fluorescence spectrophotometer is usually used to characterize the fluorescence intensity of nanoclusters and record their excitation spectra and emission spectra. Meanwhile, the fluorescence quantum yield and fluorescence lifetime of clusters are also characterized by corresponding types of fluorescence spectrophotometer. Moreover, it can also be used to compare the fluorescence intensity of different nanoclusters qualitatively or quantitatively, to monitor the response of nanoclusters' fluorescence to external factors (such as temperature, small organic molecules, inorganic ions, solvent polarity, etc.), and even to monitor the generation of fluorescent nanoclusters and optimize their synthesis conditions.

In addition, by careful comparison and analysis, fluorescence spectrum can also be used to predict the emission type and mechanism of nanoclusters. As shown in Fig. 3.3, the emission wavelength of Au_8 cluster was found to be different in solution and solid state, also in different solvent. Together with other evidence, an aggregation-induced fluorescence-to-phosphorescence switching phenomenon of the Au_8 was proposed [3].

3.1.3 Fourier transform infrared spectrometer

Fourier transform infrared (FTIR) spectrometer is mainly composed of infrared light source, beam splitter, interferometer, sample pool, detector, computer data processing system and record system, etc., it is a typical representative of the

48 Metal nanocluster chemistry

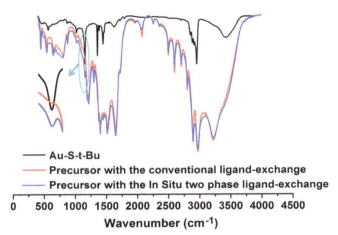

FIG. 3.4 The FTIR spectra of Au(I)-S*t*Bu complex *(red line)*, and the precursors with in situ *(blue line)* and the conventional ligand-exchange *(black line)*. (From S. Yang, J. Chai, Y. Song, J. Fan, T. Chen, S. Wang, H. Yu, X. Li, M. Zhu, In situ two-phase ligand exchange: a new method for the synthesis of alloy nanoclusters with precise atomic structures, J. Am. Chem. Soc. 139(16) (2017) 5668–5671, https://doi.org/10.1021/jacs.7b00668.)

interferometric infrared spectrometer. Different from the working principle of the dispersion-type infrared apparatus, it has no monochromator and slit. Michelson interferometer is used to obtain the interferogram of the incident light and then by Fourier mathematical transformation, the time domain function for frequency domain interference figure transformation function.

The main principle of Michelson interferometer is that the light emitted by the light source is divided into two beams to form a certain optical path difference, and then combined to produce interference. The obtained interferogram function contains all the frequency and intensity information of the light source. The frequency distribution of the intensity of the original light source can be calculated by Fourier transform of the interferogram function with a computer. Compared with traditional spectrometers, FTIR spectrometer has the advantages including fast scanning speed, high resolution, high precision, extreme sensitivity and wide spectral range.

In organic molecules, the atoms that make up the chemical bonds or functional groups are in a state of constant vibration at a frequency comparable to that of infrared light. Therefore, when an organic molecule is irradiated with infrared light, the chemical bonds or functional groups in the molecule can vibrate and absorb. Different chemical bonds or functional groups have different absorption frequencies and will be at different positions in the infrared spectrum, thus obtaining the information of what kind of chemical bonds or functional groups are contained in the molecule.

Since the metal nanoclusters are protected by organic ligands, FTIR can be used to characterize ligands on the surface of nanoclusters. That is, identifying the existence of the certain chemical bonds or functional groups in the nanoclusters, and revealing the change of the certain chemical bonds or functional groups after their combining with metal core in nanoclusters (e.g., the breaking of the S—H bond). Besides, considering the interaction between the nanoclusters and organic molecule, for instance, a catalytic reaction in which nanoclusters serve as the catalysts or the functionalization of nanoclusters with functional groups or molecules, FTIR spectrometer will be the useful tool for identifying or monitoring the reaction process. In a recent work, Yang et al. developed an in situ two phase ligand-exchange method, in which GSH was used as ligand in the synthesis of water-soluble precursors for further ligand exchange with HS*t*Bu. To reveal the difference in precursors prepared by conventional ligand-exchange and in situ two phase ligand-exchange method, the FTIR spectra of Au-SR, precursor prepared with conventional or in situ method were recorded for comparison. As shown in Fig. 3.4, The C-H stretching vibration (2956 cm^{-1}) of methyl for S*t*Bu was found in both Au-S*t*Bu complex and the in situ precursors [4]. These results indicate that some GS-ligands have been replaced by the S*t*Bu ligands in the water phase, when the Au(I)SR complex was converted into the precursors under the reductive environment.

3.1.4 Circular dichroic spectroscopy

The circular dichromatic (CD) spectrometer consists of a light source, a monochromator, a polarizing mirror, a photoelectric modulator and a detector. The light from the light source changes into monochromatic light after passing through the monochromator, and then becomes linearly polarized light after passing through the polarizer. The linearly polarized light is divided into left and right circularly polarized light through the photoelectric modulator, and then passes through the sample with optical activity. The sample has different absorption of left and right circularly polarized light, so as to synthesize elliptical polarized light, showing circular dichroism, and showing circular dichroism on the detector.

Plane polarized light through a medium with optical activity, due to the medium of the same kind of chiral optical activity of molecules exist different two configurations, their plane polarized light down into the right-hand and left-hand circularly polarized light absorption is different, the amplitude of the output when the electric field vector, synthetic again not circularly polarized light, polarized light but elliptically polarized light, resulting in a circular dichroism. It can be divided into electron circular dichroism (ECD) and vibration circular dichroism (VCD) according to the wavelength range of measurement.

It can be known from the above that the measurement of circular dichroism depends on the optical absorption of the sample itself. Since nanoclusters have the molecule-like UV-vis absorption, nanoclusters with optical activity can also show CD signals. CD spectroscopy is one of the main methods to characterize chiral nanoclusters. Similar to the UV-vis spectrum, the CD spectrum of nanoclusters also show high degree of structural correlation, so CD spectrum can be used for the identification of chiral nanoclusters. Besides, a pair of enantiomers of a nanocluster have a symmetric CD spectrum with respect to the x-axis. Therefore, two different chiral configurations of a nanocluster can be distinguished with CD spectrum, which is not possible with UV-vis spectrum. For the same reason, CD spectrum can be used to monitor the chiral transition process of nanoclusters and reflect the change of optical purity.

3.1.5 Transient absorption spectroscopy

Transient absorption (TA) spectroscopy is a common ultrafast laser pump-probe technique, which is a powerful tool to study the relaxation process of excited states in luminescent or nonradiative recombination processes. The so-called pump-probe technique refers to the technique of using optical pump pulse to excite the sample to the excited state, and then using probe pulse to monitor the relaxation process back to the ground state. In this technology, two femtosecond pulses with time delay need to be used. The one with higher energy and earlier time is used as pump light, and the one with lower energy and later time is used as probe light to excite and detect the sample respectively.

Its basic process is, first, the pump pulse and probe pulse are generated by the laser device. Then, the pump pulse passes through a certain volume of sample, resonates with the electron transition in the sample, and induces a certain number of molecules to transition to the excited state. After a certain relaxation time, the lower energy probe pulse passes through the same sample. The probe pulse is also generated by the laser, which can be separated from the strong pump pulse by the beam splitter. In each time relaxation period, the probe pulse intensity with or without pump pulse is recorded. In this process (differential absorption spectroscopy) absorbance changes are recorded to assess signals associated with excited states and photoinduced speciation. With the continuous change of the time delay between the pump light and the probe light, the received signal intensity of the probe light through the sample will change, which indicates that the number of particles in the excited state has changed. In this way, we can get the dynamic information of the excited state decay of the sample.

There are three main types of TA spectral signals: ground state bleaching (GSB), excited state absorption (ESA) and stimulated emission (SE). After irradiating by the pump pulse, only a small part of the molecules can be excited to the excited state, while the rest of the ground state molecules will absorb part of the probe pulse, showing a negative signal in the TA spectrum, which is called ground state bleaching. On the other hand, molecules excited by pump pulse can absorb probe light again and be excited again to a higher energy level, which is also shown as a negative signal in TA spectrum, called excited state absorption. Besides, after irradiating by the pump pulse and being excited to the excited state, the excited molecules can return to a lower energy level and generate emission, which is shown as a positive signal in TA spectrum, called stimulated emission.

The molecular or metallic state of a nanocluster can be well defined by the TA spectrum based on its laser power-dependent behavior. As shown in Fig. 3.5, a laser power-independent dynamic was found in $Au_{246}(SR)_{80}$. On the contrary, all the relaxation dynamics in $Au_{279}(SR)_{84}$ are dependent on the laser fluence. That is, a laser power-independent dynamic indicates that the nanocluster is in the nonmetallic state and the nanoclusters in metallic state will show laser power-dependent dynamics. The transition from the excitonic state in $Au_{246}(SR)_{80}$ to the nascent plasmonic state in $Au_{279}(SR)_{84}$ can be further demonstrated by comparing their TA spectra. The TA spectrum of $Au_{279}(SR)_{84}$ probed at 0.3 ps shows a single ground state bleaching at ~530 nm and excited state absorptions on both wings. Such features of $Au_{279}(SR)_{84}$ are similar to those of plasmonic gold nanoparticles, indicating the metallic/plasmonic state of the $Au_{279}(SR)_{84}$ nanocluster. In contrast, the excitonic $Au_{246}(SR)_{80}$ shows multiple ground state bleaching bands overlapped with the broad excited state absorptions, similar to smaller sized gold nanoclusters in excitonic state [5].

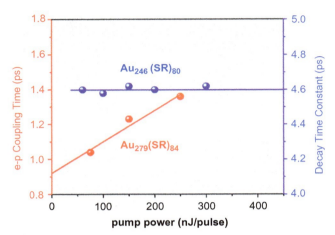

FIG. 3.5 Extracted τe-ph of $Au_{246}(SR)_{80}$ (in *blue*, 470 nm pump) and $Au_{279}(SR)_{84}$ (in *red*, 360 nm pump) as a function of pump fluence. *(From T. Higaki, M. Zhou, K.J. Lambright, K. Kirschbaum, M.Y. Sfeir, R. Jin, Sharp transition from nonmetallic Au_{246} to metallic Au_{279} with nascent surface plasmon resonance, J. Am. Chem. Soc. 140(17) (2018) 5691–5695, https://doi.org/10.1021/jacs.8b02487.)*

3.1.6 Mass spectrometry

Mass spectrometry (MS) is an analytical method for measuring mass-to-charge ratio (m/z) of ions. Its basic principle is to ionize each component in the sample in the ion source, generate charged ions with different mass-to-charge ratio, and form ion beam by accelerating electric field, and then enter the mass analyzer. In the mass analyzer, the electric field and magnetic field are used to cause the opposite velocity dispersion, and the mass spectrum is obtained by focusing them respectively. Mass spectrometer generally consists of sample introduction system, ion source, mass analyzer, detector and data processing system. Common mass spectrometry analyzers include fan-shaped magnetic fields, time-of-flight, quadrupole, quadrupole ion traps and ion cyclotron resonance mass analyzers.

Compared with other mass analyzers, the time-of-flight mass analyzer (TOF) has the advantages of simple construction, high sensitivity and wide mass range (it is easier to measure macromolecular ions), so it is suitable for the detection of nanoclusters since they molecular weights are usually in the thousands to tens of thousands Da. The principle of TOF is that the ions accelerate to fly through the flight pipeline under the action of electric field and are detected according to the different flight time of arriving at the detector, that is, the m/z of the measured ions is proportional to the flight time of the ions, and the ions are detected. TOF-MS commonly used for detecting nanoclusters can be divided into matrix assisted laser desorption ionization (MALDI) TOF-MS and electrospray ionization (ESI) according to different ion sources.

3.1.6.1 MALDI-TOF-MS

MALDI-TOF-MS is a new type of soft ionization mass spectrometry developed in recent years. It is very simple and efficient both in theory and design. The principle of MALDI is to irradiate the co-crystalline film formed by the sample and the matrix with laser. The matrix absorbs energy from the laser and transfers it to the sample. In the process of ionization, protons are transferred to the sample or get protons from the sample and ionize the sample. Therefore, it is a soft ionization technique, which is suitable for the determination of mixtures and macromolecules. In nanocluster studies, commonly used matrixes include trans-2-[3-(4-tert-butylphenyl)-2-methyl-2-propenylidene]malononitrile (DCTB) for organically soluble nanoclusters, α-cyano-4-hydroxycinnamic acid (α-CHCA) and 2,5-dihydroxybenzoic acid (DHB) for water soluble nanoclusters and sinapinic acid (SA) for protein-coated nanoclusters.

In the synthesis of nanoclusters, the mixture with multiple components is often obtained. If someone wants to analyze the composition of nanoclusters, MALDI-TOF-MS is a very convenient method. It does not need ones to separate different cluster components, just need to wash away the high boiling point substances, then it can directly load the mass spectrum, and get the mass distribution of the mixture, which is very useful in the exploration of nanocluster synthesis methods and the synchronous monitoring of some important reaction process.

Earlier in 2009, Dass et al. demonstrated the capabilities of MALDI TOF-MS in studying size evolution in the synthesis of $Au_{25}(SCH_2CH_2Ph)_{18}$ at room temperature. As shown in Fig. 3.6, a mixture of Au_{25}, Au_{38}, $Au_{\sim44}$, Au_{68}, and Au_{102} is

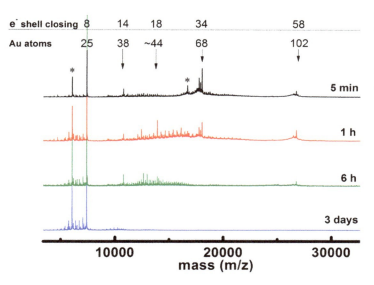

FIG. 3.6 MALDI-TOF mass spectra of the as-prepared nanoclusters using DCTB matrix and operating at threshold laser fluence. Nanocluster synthesis conditions: gold/phenylethanethiol mol ratio 1:6; room temperature; fast stirring (~500 rpm); reaction solvent, THF. *(From A.C. Dharmaratne, T. Krick, A. Dass, Nanocluster size evolution studied by mass spectrometry in room temperature $Au_{25}(SR)_{18}$ synthesis, J. Am. Chem. Soc. 131(38) (2009) 13604–13605, https://doi.org/10.1021/ja906087a.)*

formed at the first 5 min in the synthesis of Au_{25} and the size of the nanoclusters gradually focus into monodisperse Au_{25} after 3 days. This study emphasizes the importance of MALDI-TOF-MS as a tool to monitor the synthesis progress of nanocluster [6].

3.1.6.2 ESI-TOF-MS

ESI is a kind of ionization technology, which transforms ions in solution into gas phase ions for mass spectrometry analysis. The process of electrospray can be simply described as: the sample solution is sprayed into a mist like charged droplet under the action of electric field and auxiliary air stream, and the volatile solution gradually evaporates at high temperature. The density of the charge body on the droplet surface increases with the decrease of radius, and the Coulomb explosion of the final droplet produces a smaller charged droplet. The above process is repeated, and finally the sample is ionized. Because there is no direct external energy acting on the molecule in this process, the molecular structure is less damaged, which is a typical "soft ionization" mode. Compared with MALDI using laser as energy source, ESI has better retention of molecular integrity, which is very important for nanoclusters' characterization. In most cases, people only care about the complete molecular weight of the nanocluster, comparing whether its mass corresponds to the molecular formula obtained from the single crystal structure, or inferring its molecular formula from its molecular weight (mainly in the characterization of water-soluble nanoclusters). On the other hand, based on the high-resolution detector, the isotope peaks of the signal peaks of nanoclusters can be distinguished, which makes it possible to further improve the reliability of the molecular formula according to the degree of consistency between the experimental and calculated isotopic patterns.

ESI-TOF-MS is usually used to determine the molecular formula. In addition, because it uses solution injection, it can reflect some reaction processes in the solution more truly, so it is often used to track the formation of nanoclusters, the transformation and reaction between nanoclusters. After the structural determination by single crystal X-ray diffraction, the molecular composition of a nanocluster is need to be further confirm by ESI-TOF-MS since it can usually provide the molecular ion peak and isotopic pattern owing to its "soft" ionization method and high resolution.

For example, Fig. 3.7A shows the ESI-TOF-MS of $Cd_1Au_{24}(SR)_{18}$ nanoclusters. Two major peaks were found in the range of 3000–10,000 Da which were subsequently assigned into $[Cd_1Au_{24}(SR)_{18}Cs_2]^{2+}$ and $[Cd_1Au_{24}(SR)_{18}Cs]^{+}$. The isotopic patterns of these two peaks are consistent with the calculated patterns which further support analysis results of these two peaks. Of note, the isotopic pattern of a MS peak has two pieces of important information. The first one is the spacing between the two adjacent isotopic peaks. This spacing is caused by the isotope element in the sample and the two adjacent isotopic peaks are differed with a neutron (ca. 1 Da), so the value of the spacing can be calculated to be $1/z$ since the x-axis of MS is m/z. As a result, one can learn about the charge state of a certain peak according to the isotope pattern spacing. For example, a set of isotopic peaks exhibiting a spacing of ~0.5 Da indicates that this species bearing two charges. The second one is the relative intensity of the isotopic peaks. The theoretical relative intensity of the isotopic peaks can be calculated with the isotopic natural abundance of different element, it means that a definite molecular formula corresponds to a definite isotope pattern. It is useful in distinguishing two species (except for isomers) that display the same molecular weight [7].

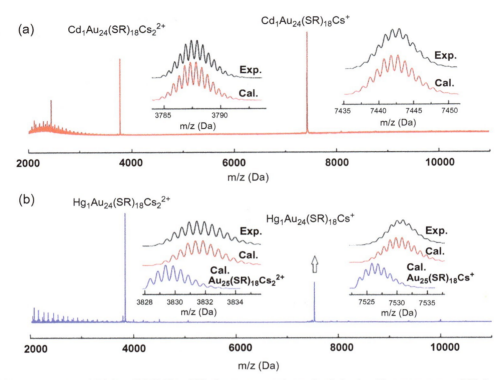

FIG. 3.7 (A) ESI mass spectrum of $Cd_1Au_{24}(SC_2H_4Ph)_{18}$ NCs. Insets are experimental and simulated isotope patterns of $Cd_1Au_{24}(SC_2H_4Ph)_{18}Cs_2^{2+}$ and $Cd_1Au_{24}(SC_2H_4Ph)_{18}Cs^+$, respectively. (B) ESI-MS spectrum of $Hg_1Au_{24}(PhC_2H_4S)_{18}$ NCs. Insets are experimental and simulated isotope patterns of $Hg_1Au_{24}(SC_2H_4Ph)_{18}Cs_2^{2+}$ (together with simulated $Au_{25}(SC_2H_4Ph)_{18}Cs_2^{2+}$) and $Hg_1Au_{24}(SC_2H_4Ph)_{18}Cs^+$ (together with simulated $Au_{25}(SC_2H_4Ph)_{18}Cs^+$), respectively. *(From S. Wang, Y. Song, S. Jin, X. Liu, J. Zhang, Y. Pei, X. Meng, M. Chen, P. Li, M. Zhu, Metal exchange method using Au_{25} nanoclusters as templates for alloy nanoclusters with atomic precision, J. Am. Chem. Soc. 137(12) (2015) 4018–4021, https://doi.org/10.1021/ja511635g.)*

As shown in Fig. 3.7B, the m/z value and the isotopic pattern of the two signal peaks correspond well with the formula $[Hg_1Au_{24}(SR)_{18}Cs_2]^{2+}$ and $[Hg_1Au_{24}(SR)_{18}Cs]^+$ and differ with the both the calculated molecular weight and isotopic pattern with $Au_{25}(SR)_{18}$.

ESI-MS can also be used for understanding the reaction mechanisms. For example, Xie et al. employed ESI-MS to capture the intermediates during the synthesis of $Au_{25}(SR)_{18}$ in the aqueous solution. The growth of $[Au_{25}(SR)_{18}]^-$ from Au(I)-thiolate complex precursors was monitored at an interval of 2 min and a $2e^-$-reduction pathway was proposed for the growth of the final product $[Au_{25}(SR)_{18}]^-$. Different from the MALDI-TOF-MS, a solution of the sample can be used directly for ESI-TOF-MS test; it allows one to monitor the reaction process at real time [8].

3.1.7 Transmission electron microscope

In the transmission electron microscope (TEM), electron beams are accelerated and concentrated onto a very thin sample. The electrons collide with the atoms in the sample and change direction, resulting in solid angle scattering. The size of the scattering angle is related to the density and thickness of the sample, so different light and dark images can be formed. The image will be displayed on the imaging device (such as fluorescent screen, film, and photosensitive coupling assembly) after amplification and focusing.

Because the de Broglie wavelength of the electron is very short, the resolution of the transmission electron microscope is much higher than that of the optical microscope, which can reach 0.1~0.2 nm, and the magnification is tens of thousands~millions of times. Thus, the use of a transmission electron microscope can be used to observe the fine structure of a sample, even a single list of atoms, tens of thousands of times smaller than the smallest structure that can be seen with a light microscope. TEM is an important analytical method in many scientific fields related to physics and biology, such as cancer research, virology, materials science, as well as nanotechnology, semiconductor research, and so on.

The general working principle of TEM is the electron beam emitted by the electron gun passes through the condenser along the optical axis of the lens body in the vacuum channel, and then converges into a sharp, bright and uniform light spot through the condenser, illuminating the sample in the sample chamber; the electron beam passing through the sample carries the internal structure information of the sample. The electron beam passing through the sample is less in the dense part and more in the sparse part. After the converging focusing and primary magnification of the objective lens, the electron beam enters the intermediate lens and the first and second projection lenses at the lower level for comprehensive magnification and imaging, and finally the amplified electronic image is projected on the fluorescent screen in the observation room. Fluorescent screens convert electronic images into visible light for the user to view.

In TEM test, the contrast of metal core of nanoclusters is much greater than that of their peripheral ligands, so in general, only the metal core of nanoclusters can be observed by TEM. The size of the metal cores in the nanoclusters is generally about 0.8–2 nm. In cluster studies, TEM is used to characterize the size monodispersity of nanoclusters, especially for water-soluble nanoclusters. Besides, TEM is a useful tool for investigating the distribution of the nanoclusters loaded on the supports when constructing a heterogeneous catalyst. In a study reported by Tsukuda et al., porous carbon-supported $Au_{25}(SC_{12})_{18}$ were used as the catalyst for the aerobic alcohol oxidation [9]. TEM images showed that Au_{25} were highly dispersed on the porous carbon supports with an average size of 1.2 nm. In addition, TEM can also be used to reveal the interaction or assemble form between nanoclusters. For example, a cationic polymer poly(allylamine hydrochloride) (PAH) was used to coat the GSH-protected Au NCs, forming the Au-GSH-PAH assembly. Revealed from the TEM images, Au NCs were evenly dispersed in the PAH nanogel, no obvious intercluster aggregation was found in the assembly [10]. In another work reported by Yao et al., amphiphilic Au NCs were construct via patching hydrophilic NCs with hydrophobic cetyltrimethylammonium ion to about half of a monolayer coverage. The TEM was used to investigate the assembling form of this amphiphilic Au NCs at the two-phase interface, showing that a single-NC-thick sheet which was folded in several places. The high-resolution TEM image confirms the sheet was composed of small NCs with the size below 2 nm [11].

3.1.8 Dynamic light scattering

Dynamic light scattering (DLS), also known as photon correlation spectroscopy (PCS), quasielastic scattering, is used for measuring the fluctuation of light intensity with time and finally obtaining the hydrodynamic diameter of the nanoparticles in solvent.

The specific principle is: when the light passes through the liquid containing nanoparticles, the particles will scatter the light, and the light signal can be detected at a certain angle. The detected signal is the result of the superposition of multiple scattered photons, which has statistical significance. The instantaneous light intensity is not a fixed value and fluctuates under a certain average value, but the fluctuation amplitude is related to the particle size. The light intensity at one time is the same as that at another time in a very short time. We can think that the correlation is 1. After a long time, the similarity of light intensity decreases. When the time is infinite, the light intensity is completely different from that before, and the correlation is 0. According to the optical theory, the light intensity correlation equation can be obtained. The velocity of a particle in Brownian motion is related to the particle size (Stokes Einstein equation). Large particles move slowly, small particles move fast. If large particles are measured, the intensity of scattering spot will fluctuate slowly because of their slow motion. Similarly, if small particles are measured, the density of the scattered spot will fluctuate rapidly due to their fast motion. Finally, the particle size and its distribution were calculated by light intensity fluctuation and light intensity correlation function.

Different from the diameter of metal core measured by TEM, the hydrodynamic diameter obtained by DLS is the diameter of the whole nanoparticles containing ligand layer and solvent layer. DLS technology has the advantages of accuracy, rapidity and repeatability, and has become a more conventional characterization method in nanotechnology. With the update of the instrument and the development of data processing technology, the current dynamic light scattering instrument not only has the function of measuring particle size, but also has the ability of measuring zeta potential and molecular weight of macromolecules.

The sample should be well dispersed in the liquid medium. Under ideal conditions, the dispersant should have the following conditions: transparent; the dispersant and solute particles have different refractive index; it should match the solute particles (i.e., it will not cause swelling, resolution or association); master the accurate refractive index and viscosity, the error is less than 0.5%; it is clean and can be filtered.

DLS technology is useful in analyzing the aggregation state of nanoclusters in the solution since the measurement was carried out using the solution directly, which makes it possible for probing the "nondestructive" state of the nanoclusters in their solution. The most common application scenario of DLS about nanoclusters is in the size measurement of AIE-type nanoclusters. For instance, reported by Yahia-Ammar et al., fluorescent GSH-protected Au NCs were embedded into the PAH polymer, which made its size in solution changed from ~5 nm (individual Au NC) to 120 nm (AIE-type Au NC). Further, by regulating the pH value of the solution, the size of the assembly would change accordingly due to the weakening or strengthening of the electrostatic interaction between GSH and PAH [10]. In another work reported by Xie et al., GSH-Au NCs was inserted into the chitosan nanogel. Similarly, the size of this composite nanoparticle was measured by DLS in water, showing an approximate size of 202 nm, while the size of the particles revealed by the TEM is smaller than that measured by DLS, because the sample used for TEM analysis should be desolvated, leading to the shrinkage of the nanoparticles [12].

3.1.9 Thermogravimetric analysis

Thermogravimetric analysis (TGA) is a method to measure the relationship between the mass of a substance and temperature or time under programmed temperature control. By analyzing the thermogravimetric curve, we can know the composition, thermal stability, thermal decomposition and products of the sample and its possible intermediate products.

Derivative thermogravimetric analysis (DTG), also known as derivative thermogravimetric analysis, can be derived from thermogravimetric analysis. The experimental results are derivative thermogravimetry curve, which takes mass change rate as ordinate and represents decrease from top to bottom; abscissa is temperature or time and represents increase from left to right.

The main characteristic of thermogravimetric analysis is that it is quantitative and can accurately measure the mass change and the rate of change. According to this characteristic, it can be said that as long as the mass of a substance changes when heated, it can be studied by thermogravimetric analysis. Thermogravimetric analysis can be used to detect the physical and chemical changes in the process. We can see that these physical and chemical changes are mass changes, such as sublimation, vaporization, adsorption, desorption, absorption and gas-solid reaction.

The instrument for thermogravimetric analysis, called thermogravimeter, is mainly composed of three parts: temperature control system, detection system and recording system.

TGA can be used as an additional method to verify the molecular composition of the nanoclusters. Metal nanoclusters are composed of a metal core and an organic ligand layer, in the measure of TGA, the ligand layer with lower dissociation temperature will escape and lead to the mass loss of the sample and remains the metal element with high boiling temperature (in some cases, part of the sulfur may be left in form of metal sulfide). Generally, a nanocluster is first structural determined by X-ray crystallography, and with the crystal structure, the metal proportion can be calculated. Further, the TGA result was collected and checked its match degree of the calculated data. TGA can also used for comparison in nanoclusters with similar molecular composition. As reported, $Au_{25}(PET)_{18}$ and $Au_{24}Hg_1(PET)_{18}$ showed a slight difference in the weight loss. The percentage of mass loss of $Au_{24}Hg_1(PET)_{18}$ is $\sim 3\%$ higher than $Au_{25}(PET)_{18}$, which can be attributed to the further loss of Hg in the temperature range from 220°C to 510°C [13].

3.1.10 Nuclear magnetic resonance spectroscopy

Nuclear magnetic resonance (NMR) spectrometer, refers to the absorption of radiofrequency radiation by atomic nuclei, is one of the most powerful tools for qualitative analysis of the composition and structure of various organic and inorganic substances, and sometimes can be used for quantitative analysis. Its working principle is in the strong magnetic field, the nucleus energy level split, when the absorption of foreign electromagnetic radiation, will occur nuclear energy level transition, that is, the so-called NMR phenomenon. When the frequency of the applied RF field is the same as the frequency of the spin precession of the nucleus, the energy of the RF field can be effectively absorbed by the nucleus to provide assistance for the energy level transition. So a particular nucleus, in a given external magnetic field, absorbs only the energy from a particular radiofrequency field at a particular frequency, thus forming an NMR signal. NMR studies the absorption of radiofrequency radiation by atomic nuclei in a strong magnetic field. There are two types of NMR spectrometer: high resolution NMR spectrometer and wide line NMR spectrometer. The former can only measure liquid samples and is mainly used for organic analysis. The latter can directly measure the solid sample and is often used in the field of physics. According to the working mode of the spectrometer, it can be divided into continuous wave NMR spectrometer (ordinary spectrometer) and Fourier transform NMR spectrometer.

NMR is a powerful tool to reveal the chemical environment of the ligands protecting the nanoclusters, which can be used for structural determination. Considering the ligand type used for protecting nanoclusters, ^1H, ^{13}C, and ^{31}P NMR are the most commonly used.

For instance, Liu and co-workers measured the temperature-dependent ^{31}P NMR spectra of $Pt_2Ag_{33}(dtp)_{17}$ (dtp: dipropyl dithiophosphate). This clusters showed a sharp peak at 101.1 ppm and 2 broad bands centered at 95.7 and 104.9 ppm at 20°C, which can be subsequently resolved into 8 peaks at −60°C. Authors assigned the sharp peak at 101.1 ppm to the 2 dtp ligands located at both ends of this cluster. The 5 intense peaks with similar integral areas in range of 110–104 ppm are corresponding to the 10 dtp ligands, 5 on each side. Further, the 3 peaks at ~ 95 ppm with integration ratios 1:2:2 can be assigned to the 5 dtp ligands located at the waist of this rod-like nanocluster [14].

Besides, NMR can be used to monitor the change of size/structure or even electronic configuration of nanoclusters since these changes will lead to the rearrangement or distortion of the outer organic ligand layers of nanoclusters. $Au_{25}(PET)_{18}$

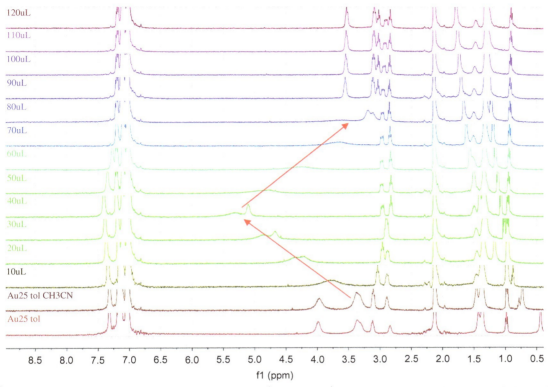

FIG. 3.8 ^{1}H NMR monitoring of the reaction of $Au_{25}(SC_2H_4Ph)_{18}^-TOA^+$ and $TEMPO^+BF_4^-$. *(From Z. Liu, M. Zhu, X. Meng, G. Xu, R. Jin, Electron transfer between [Au$_{25}$(SC$_2$H$_4$Ph)$_{18}$]$^-$TOA$^+$ and oxoammonium cations, J. Phys. Chem. Lett. 2(17) (2011) 2104–2109, https://doi.org/10.1021/jz200925h.)*

have found to exist three kind of charge state, that is, −1, 0, and +1. Fig. 3.8 shows the ^{1}H NMR spectra of [Au$_{25}$(PET)$_{18}$]$^-$ added with different amount of TEMPO$^+$BF$_4^-$ oxidant. With the with addition of 1 eq. TEMPO$^+$, the CH$_2$ signal shifts downfield from 3.79 to 5.1 ppm, corresponding to change of charge state of Au$_{25}$ from −1 to 0. Further, addition of 1 more equivalent of TEMPO$^+$ makes the peak at 5.1 ppm upshifted back to 3.5 ppm, indicating the charge state have been oxidized to +1 [15].

NMR has been used for determining the presence of a certain element in nanoclusters. For example, in the characterization of hydride-contained nanoclusters, ^{2}H NMR is usually used to confirm the existence of hydride ligands and investigate the how many chemical environments there are. For example, [Cu$_{25}$H$_{22}$(PPh$_3$)$_{12}$]Cl contains 22 H$^-$ ligand, by the replacement of NaBH$_4$ with NaBD$_4$, [Cu$_{25}$D$_{22}$(PPh$_3$)$_{12}$]Cl was obtained. In this way, the hydride ligands in clusters can be detected with ^{2}H NMR without interference of the H atom on PPh$_3$ ligands. Its ^{2}H NMR showed 3 D-signals in the range of −1.5 to 2.5 ppm with integration ratios about 3:6:2, indicating there are 3 kind of hydride ligand in Cu$_{25}$ with 6, 12, and 4 H$^-$ for each group respectively [16].

3.1.11 Electron paramagnetic resonance

Electron paramagnetic resonance (EPR) is a magnetic resonance phenomenon caused by electron spin, so it is also called electron spin resonance (ESR). The research objects of the EPR analysis are the materials with unpaired electrons, such as atoms and molecules with odd numbers of electrons (or unpaired electrons), ions with unfilled inner electron shells, free radicals generated by radiation, semiconductors, metals, etc. The principle is that by placing a paramagnetic crystal in a constant and strong magnetic field, the element magnetic moments of the paramagnetic body precession around the constant magnetic field. At this point, an alternating magnetic field (microwave band) is added in the direction perpendicular to the constant magnetic field. When the frequency of the alternating magnetic field is equal to the precession frequency of the element magnetic moment, paramagnetic resonance occurs, that is, paramagnetic body absorbs energy from the alternating field. The function between microwave absorption and changing parameters can be measured by changing the frequency of microwave or changing the field strength of a constant magnetic field while other parameters remain constant. This

56 Metal nanocluster chemistry

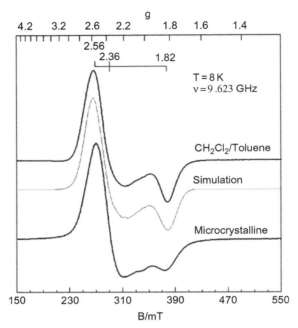

FIG. 3.9 EPR spectra of $[Au_{25}(SR)_{18}]^0$ for the conditions listed. Simulation parameters: **g** =(2.556, 2.364, 1.821), g strain (σ_g)=0.03, 13 equivalent I=3/2 nuclei with **A** =(50, 71, 142) MHz. *(From M. Zhu, C.M. Aikens, M.P. Hendrich, R. Gupta, H. Qian, G.C. Schatz, R. Jin, Reversible switching of magnetism in thiolate-protected Au_{25} superatoms, J. Am. Chem. Soc. 131(7) (2009) 2490–2492, https://doi.org/10.1021/ja809157f.)*

functional relationship is called electron paramagnetic resonance spectroscopy. This map can be used to identify paramagnetic substances and to infer the properties of intramolecular chemical bonds from the interaction of the magnetic moments of the elements with their external environment.

Free electrons have been found to exist in the nanoclusters, so when the number of electrons is odd, the EPR signal can also appear in the nanoclusters. So far, no nanocluster with an even number of free electrons has been found to exhibit an EPR signal (similar to the two unpaired electrons in O_2 molecule). EPR measurements of nanoclusters are usually performed at ultra-low temperatures (<10K) to enhance their resonance signals. In addition, the strength of the signal is proportional to the amount of sample, so it is usually necessary to prepare a larger amount of nanocluster samples (usually in solid state) to obtain a better EPR profile.

The first found nanoclusters with EPR signal is the $Au_{25}(SR)_{18}^0$, obtained by the oxidization of $Au_{25}(SR)_{18}^-$. The $Au_{25}(SR)_{18}^0$ was calculated to possess 7 free electrons in which an unpaired electron is expected. This hypothesis is subsequently verified by the EPR measurement. The sample of $Au_{25}(SR)_{18}^0$ show an S=1/2 signal with g=(2.56, 2.36, 1.82) (Fig. 3.9). EPR quantification indicates that $Au_{25}(SR)_{18}^0$ has one unpaired spin per particle [17].

3.1.12 X-ray photoelectron spectroscopy

X-ray photoelectron spectroscopy (XPS) is an advanced analytical technique in the microanalysis of electronic materials and components. Because it can measure the inner electron binding energy and chemical shift more accurately than Auger electron spectroscopy, it can not only provide the information of molecular structure and atomic valence state for chemical research, but also provide the information of element composition and content, chemical state, molecular structure and chemical bond for electronic material research. In the analysis of electronic materials, it can provide not only the overall chemical information, but also the surface, micro area and depth distribution information.

The energy of X-ray photons is between 1000 and 1500eV, which can not only ionize the valence electrons but also excite the inner electrons. The energy level of the inner electrons is little affected by the molecular environment. The binding energies of the inner electrons of the same atom in different molecules vary very little, so it is characteristic. In addition to the different values of the binding energy of the same inner shell electron of different elements, the binding energy of a given inner shell electron of a given atom is also related to the chemical binding state of the atom and its chemical environment. With the different molecules of the atom, the photoelectron summit of the given inner shell electron has a shift, which is called chemical shift. This is because the binding energy of the inner shell electrons is not only determined by the nuclear charge, but also affected by the surrounding valence electrons. The atom with higher electronegativity tends to pull the valence electron of the atom to the side, which makes the nucleus combine with its 1s electron firmly, thus increasing the binding energy.

As a modern analytical method, XPS has the following characteristics:

(1) All elements except H and He can be analyzed, and the sensitivity to all elements has the same order of magnitude.
(2) The spectral lines of the same energy level of adjacent elements are far away from each other with less interference, and the qualitative identification of elements is strong.
(3) It can observe chemical shift. The chemical shift is related to the atomic oxidation state, atomic charge and functional group. Chemical shift information is the basis of XPS for structural analysis and chemical bond research.
(4) It can be used for quantitative analysis. It can measure the relative concentration of elements and the relative concentration of different oxidation states of the same element.
(5) It is a highly sensitive and ultra-micro surface analysis technique. The depth of sample analysis is about 2 nm, the signal comes from several atomic layers on the surface, the sample size can be as small as 10^{-8} g, and the absolute sensitivity can reach 10^{-18} g.

In the cluster-related domain, XPS generally has three kind of application. The first one is to determine the type of elements in the samples. For example, the counterions of the nanoclusters are sometime hard to be find in their crystal structure due to the disorder. At that time, XPS can act as a tool to confirm the existence of the certain type of counterion (e.g., Cl^-, P in the PPh_4^+, Sb and F in the SbF_6^-).

The second one is to reveal the valence of the certain element, in most cases, the valence of the metal element in the nanoclusters. The valence of the metal element in nanoclusters usually lies between M(0) and M(I) (M=Au, Ag). Negishi et al. reported the PAGE separation of the GSH-Au NCs synthesized in which nine separated bands were found [1]. The XPS results of these bands along with the Au(0) film and Au(I)-SG complex. As a result, Au(0) film shows an Au $4f_{7/2}$ binding energy located at 84.0 eV and the binding energy of Au $4f_{7/2}$ in Au(I)-SG complex lies at ~86 eV. Of note, different from the Au(0), the binding energy of Au(I) species is variable, depending on the specific form of the Au(I) species. As we can see, the binding energies of all the nine binds are located between the Au(0) and Au(I) species, demonstrating that the average valances of all these GSH-Au NCs lie between 0 and +1. In addition, the binding energies of different bind showed slight shifts, which are caused by the different Au(0)/Au(I) ratios in these nanoclusters.

The third application of XPS is used to investigate the change of valence of a certain element in a nanocluster after some treatments (e.g., after loaded on a support). In this way, it is useful to reveal the electron transfer direction. For example, $Au_{25}(L-Cys)_{18}$ shows the Au $4f_{7/2}$ located at 84.6 eV, after loaded on the ZIF-8 and further coated with ZIF-67, a sandwich cluster composite nanocatalysts was form. In this state, the binding energy of Au $4f_{7/2}$ shifted to 84.2 eV, which is caused by the electron transfer between $Au_{25}(L-Cys)_{18}$ and ZIF-8 or ZIF-67 [18].

3.1.13 X-ray absorption fine structure

Depending on the range of energy, X-ray absorption fine structure (XAFS) can be divided into two parts: X-ray absorption near edge structure (XANES) and extended X-ray absorption fine structure.

X-ray excited photoelectrons can be scattered by the surrounding coordination atoms, which causes the X-ray absorption intensity to oscillate with the energy. The electron and geometric local structure of the studied system can be obtained by studying these oscillating signals. XANES is formed by low energy photoelectrons scattering from coordination atoms and then returning to absorption atoms to interfere with the emitted wave, which is characterized by strong oscillation. EXAFS are formed by the single scattering of the coordination atoms around the absorbed atoms back to the absorption atoms and the interference of the emitted wave, which is characterized by small amplitude, like sinusoidal wave. The appearance of synchrotron radiation source greatly shortens the measurement time of XAFS and improves the signal-to-noise ratio of XAFS spectrum, which lays a foundation for the application of XAFS. XAFS has the following characteristics: atomic selectivity; It can provide the information of the local structure around the absorption atom with subatomic resolution; All atoms respond to XAFS; There is no special requirement for the state of the sample, it can be solid and solution, or gas, crystal or noncrystal. XANES can determine valence states, characterize d-band characteristics, measure coordination charges, and provide structural information including orbital hybridization, coordination number, and symmetry. The generation of EXAFS is related to the scattering of absorbing atoms and other atoms around them, that is, both are related to the structure. Therefore, the nearest structure around the absorption atom can be studied by measuring EXAFS, and parameters such as atomic spacing, coordination number and mean azimuth shift of atoms can be obtained. The main feature of EXAFS method is that different kinds of atoms can be measured separately, the nearest neighbor

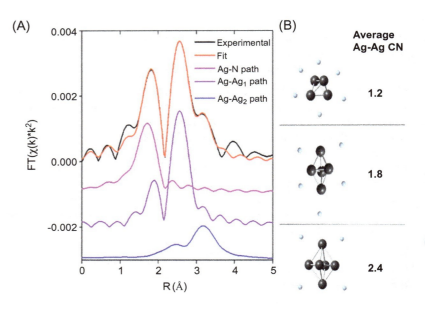

FIG. 3.10 (A) Ag K-edge EXAFS of Ag-DNA conjugates fitted with three individual scattering paths shown separately and (B) proposed Ag core structures and corresponding theoretical CN values (*gray* atoms and *faded blue* atoms represent Ag^0 and Ag^+, respectively). *(From J.T. Petty, O.O. Sergev, M. Ganguly, I.J., Rankine, D.M. Chevrier, P. Zhang, A segregated, partially oxidized, and compact Ag_{10} cluster within an encapsulating DNA host, J. Am. Chem. Soc. 138(10) (2016) 3469–3477, https://doi.org/10.1021/jacs.5b13124.)*

structure of the specified element atom can be given, and the types of the nearest neighbor atoms can be distinguished. The use of strong X-ray sources can also be used to study the structure of the nearest neighbors of atoms in small quantities and can be used to study both ordered and disordered matter. Thus, EXAFS can be used to solve some structural problems of substances that are difficult or impossible to solve by other methods.

XANES can be used for determining the average oxidation state of the sample of a certain element by analyzing the white-line integration area. For example, Petty et al. measured the oxidation state of a Ag_{10}-DNA cluster using Ag L_3-edge XANES spectra [19]. Using Ag(0) foil and silver acetate (Ag^+) as references, linearly interpolating the integrated area from the spectrum of the cluster gave ~35% Ag(0) and ~65% Ag(I). Thus, the oxidation state of Ag in Ag_{10}-DNA was determined to be 0.65. Further, EXAFS spectra were employed to reveal the inner structure of the Ag_{10} core. (Fig. 3.10) These spectra identified three single scattering paths in a multishell fitting analysis. The associated bond distances, coordination numbers, and Debye-Waller factors for each scattering path were determined by the theoretical modeling. The first peak exhibiting a bond length of 2.24 Å is ascribed to a silver-DNA interaction, and this bond length is corresponded with the Ag—N bond in Ag^+-nucleobase, Ag^+-amine, and Ag-pyridine complexes. The second peak with a 2.743 Å bond length is assigned to the Ag—Ag bond in the kernel of the cluster considering its comparable bond length with the cubic closest packed crystals in bulk Ag. The third peak is ascribed to long-range Ag-Ag interactions since the average bond distance of 3.37 Å is comparable to the combined van der Waals radii of 3.44 Å for 2 silver atoms. These two types of Ag—Ag bond length indicated that the Ag_{10} clusters contained a compact inner core and a loose core-shell interaction. The EXAFS spectra can also reveal the average coordination number (CN) of the Ag atoms in the clusters. The CN of Ag-nucleobase is 1.02, indicating that one Ag was combined with one nucleobase on the DNA. For the Ag_{10} metal core, a 2.2 CN was revealed by the EXAFS spectrum for the second peak. Authors established three packing mode of the inner core of Ag_{10} as shown in Fig. 3.10B. Among them, an octahedral configuration Ag_6 core exhibiting a 2.4 CN is most consisted with the EXAFS result.

3.1.14 Single crystal X-ray diffraction

In 1912, Laue et al. confirmed the periodic arrangement of atoms within tens to hundreds of pm apart in crystalline materials according to theoretical predictions. The periodic arrangement of the atomic structure can become the "diffraction grating" of X-ray diffraction; X-ray has the characteristic of wave, the wavelength is tens to hundreds of pm, and has the ability of diffraction. The experiment marked the first milestone in X-ray diffraction. When a beam of monochromatic X ray incident to crystal, because the crystal is made up of atomic rules are arranged into a cell, the rules of the distance between the atoms and the incident X-ray wavelengths in the same order of magnitude with single crystal X-ray diffraction (SC-XRD) analysis, so by the different atomic scattering X-ray mutual interference, stronger effects in some special directions X-ray diffraction, the diffraction lines in the spatial distribution of location and intensity, is closely related to the crystal structure, each crystal diffraction patterns produced by reflect the internal atomic distribution regularity of the crystal. This is the basic principle of X-ray diffraction.

A common instrument used to measure the intensity of single crystal X-ray diffraction is a four-circle diffractometer. The four-circle diffractometer uses a counter to measure the intensity of the diffracted beam. A single crystal can have hundreds to tens of thousands of diffraction points. The counter is always kept on the horizontal plane, but each diffraction point to be measured can fall on the horizontal plane by rotating the orientation of the crystal on the goniometer head. The azimuth of goniometer and counter is determined by four Euler angles ($\chi, \omega, 2\theta, \varphi$), so it is called four circle diffractometers. χ angle can adjust the direction of the rotation axis of the goniometer in the range of 0–360 degrees, while the φ angle is the rotation angle on the goniometer. It can further adjust the orientation of the crystal and its diffraction point in the range of 0–360 degrees. The angle between the diffracted beam measured by the counter and the incident X-ray can be adjusted by 2θ angle, i.e., Bragg angle. The angle ω can adjust the rotation of the angle measuring head around the axis of rotation perpendicular to the horizontal plane. At the same time, when the diffraction intensity is measured, the scanning range of the angle is generally very small, about 0.5–5 degrees. The four-circle diffractometer has a fixed optical center, and the four rotation axes of Euler angle intersect at the optical center. Therefore, when the crystal is precisely adjusted in the optical center, any Euler angle rotation in the experiment of measuring diffraction intensity will not move the space position of the crystal. This ensures that the incident X-ray always passes through the crystal.

The overall working process is as follows: the characteristic X-ray acts on the single crystal mounted on the goniometer to generate the diffraction point, and the intensity data of the diffraction point is recorded through the detector. Through the computer control to complete the automatic peak diffraction, determination of cell parameters, collecting diffraction intensity data, statistics system extinction law, determine the space group and then calculate the crystal structure.

SC-XRD can get all the three-dimensional structure information, including the bond length and bond angle between atoms, the packing mode in the crystal, the molecular interaction in the crystal, the hydrogen bond and the π-π interaction and other useful information. SC-XRD is the most powerful tool for the structure determination of nanoclusters. In addition, intermolecular interaction between clusters and packing mode of nanoclusters can also been revealed by SC-XRD. With the atomically precise structures of nanoclusters, the structure-property relationship can be well established and many important processes can be well understood. Noble metal nanoclusters with crystal structures revealed by SC-XRD will discuss in the next section.

3.2 Atomically precise structures

3.2.1 Introduction

X-ray structures have been seen as the "holy grail" of nanocluster research. Revealing the overall structure of the nanoclusters is crucial for understanding their stability, binding mode of metal-ligand interface and physicochemical properties [20,21]. Many methods introduced in the last chapter have been developed to paves the way for the rational synthesis of noble metal nanoclusters with molecular purity, which makes it possible to determine the crystallization of nanoclusters and their total structures. Since the first structural determination of the gold nanoclusters $Au_{102}(SR)_{44}$ in 2007 [22], and then $Au_{25}(SR)_{18}^-$ in 2008 [23,24], and the first silver nanocluster $Ag_{44}(SR)_{30}^{4-}$ in 2013 [25], more and more nanoclusters have been structurally solved. As the increase of the crystal structures of nanoclusters, their structural features can be now well-described. That is, the core-shell structures of metal nanoclusters. Metal nanoclusters are most composed of a core-shell structure, in which the inner core shows a more impact metal-metal interaction, forming some characteristic kernel such as icosahedral M_{13}, face-centered cubic M_{13}, octahedral M_6, decahedral M_7 and so on. The inner cores of nanoclusters are further protected by a series of metal-ligand polymer, also called as motifs, through covalent bond and metal-metal interaction. Likewise, these protecting motifs have shown their structural features based on the type of the metal and ligand. For example, staple-like $Au_n(SR)_{n+1}$ motifs are most usually found in the thiolate-protected Au NCs, and the M-PR_3 interaction is usually formed as mono-coordination metal-ligand bond. In this section, we will discuss the ligand types used for protecting nanoclusters, metal-ligand interfaces and the inner cores of the nanoclusters, including gold, silver, copper and the alloying nanoclusters.

3.2.2 Ligands and metal-ligand interfaces

Most of the metal nanoclusters are prepared in the liquid phase, different with the clusters generated in the gas phase; a naked metal core is hard to survive in the liquid phase due to its high surface free energy. To solve this problem, ligands were introduced to protect the metal core, thus forming the metal-ligand interfaces. This is an effective way to reduce the surface free energy which greatly improves the stability of the nanoclusters in solution, and makes it more convenient for ones to study their structures, physicochemical properties, applications, etc. Selection of suitable ligands is a key step in the synthesis of nanoclusters, which requires that there must be functional groups that can form strong interactions (generally covalent or coordination bonds) with metals in the ligands, the ligands themselves must be stable, and that the ligands must have good solubility in solvents. As the outermost layer of the nanoclusters, the ligands will directly interact with the external environment (e.g., solvents, molecules and cells) influencing nanoclusters in various applications; and the interfacial chemistry between ligands and metal atoms can determine the structures, as well as the physicochemical properties of nanoclusters. When ligands participate in the process of forming metal nanoclusters, ligands first form metal-ligand complexes with metal ions. With the addition of reducing agents, metals begin to nucleate. At this time, metal-ligand complexes will wrap around the exterior of metal cores and form the so-called core-shell structure. The metal-ligand structure involved in the formation of the ligand varies with the ligand, and this structure is also called "motif" in the nanocluster study. Moreover, the same ligand combined with different metal elements can produce different types of surface motifs, and even the motif formed by the same ligand and the same metal also has multiple forms. As an important component of metal nanoclusters, the surface motif of nanocluster not only plays a role in stabilizing the inner core of nanocluster, but also profoundly influences the overall structure of nanocluster, the way of interaction between molecules, basic physical properties (such as solubility and polarity), physicochemical properties (such as optical properties, chirality, catalytic activity), etc. In addition, different functional groups on the ligands can also give the clusters different properties and surface modifiability. So far, many types of ligands have been used in the synthesis of clusters, such as thiols [22–24,26], phosphines [27–29], alkynes [30–33], selenols [34–37], N-heterocyclic carbenes (NHCs) [38–41], etc. The next section will discuss the bonding types, bonding methods and bonding characteristics of different types of ligands. The characteristics of the metal-ligand structural units formed with metal and the binding mode of the structural units with metal core will be also introduced.

3.2.2.1 Thiolate ligands

Due to the strong metallophilic properties of sulfhydryl groups (—SH) in the molecules, thiols can combine with metal atoms in nanoclusters to form M-S covalent bonds. Thiol is one of the most commonly used ligands in the synthesis of nanoclusters because of the strong covalent interaction between metal and sulfur so that the nanoclusters can be well stabilized. Based on their solubility in solvent, thiolate ligands can be divided into hydrophilic and hydrophobic types. Hydrophilic ligands usually contain the hydrophilic groups in their molecular structures, such as carboxyl (—COOH), sulfonate (—SO_3H) and amino (—NH_2), which make the as-protected nanoclusters water soluble. Unfortunately, most of the nanoclusters protected by hydrophilic ligands have not been structural solved due to the very high difficulty in crystallization. This may cause by the high flexibility of the carbon skeleton in the ligand and difficulty in purification of water-soluble nanoclusters. So, more effects are needed in the development of crystallization as to reveal the atomically precise structures of the water-soluble nanoclusters. But interestingly, the first reported thiolate-protected Au NC and Ag NC are both protected by the hydrophilic ligand p-mercaptobenzoic acid (p-MBA), that is, the $Au_{102}(p\text{-MBA})_{44}$ [22] and $Ag_{44}(p\text{-MBA})_{30}$ [25]. Compared with other common hydrophilic like GSH, p-MBA ligands show a more rigid molecular structure. Besides, these two nanoclusters can be finely purified; these two reasons make the as-coated nanoclusters easier for crystallizing.

Even so, subsequent nanoclusters are mostly prepared using rigid hydrophobic ligands that are relatively easy to be crystallized, such as adamantanethiol (HSAdm) and 4-(tert-butyl)benzenethiol (TBBT). Moreover, after years of development, a set of relatively mature methods have been developed for the synthesis, purification and crystallization of nanoclusters protected by hydrophobic ligands, and more and more crystal structures have been obtained, which provides opportunities to further study the structure-property relationship.

Though differing in solubility, hydrophilic and hydrophobic thiols have the same coordination mode with metals essentially. As mentioned earlier, thiolates can bind to the metal in the nanocluster by covalent bonds, but the way it binds varies according to the type of metal. Most of the thiolate ligands adopt the μ_2-coordination mode to the gold atoms, forming the staple-like $Au_n(SR)_{n+1}$ motifs, bridging thiolates anchoring on the core, and ring $Au_n(SR)_n$ motifs (Fig. 3.11). For the staple-like $Au_n(SR)_{n+1}$ motifs, they will cap on the kernel of the nanoclusters by two terminal SRs, and the adjacent SRs will link with each other via an Au(I) atom in a linear fashion. This type of motifs is the most common found in the Au NCs. For example, six $Au_2(SR)_3$ staple motifs were found in the crystal structure of $Au_{25}(SR)_{18}$ [23,24]. For the bridging thiolates, they are preferred to cap in the <110> direction on the Au{100} surface of the FCC core in Au NCs, such as $Au_{92}SR_{44}$ [42] and $Au_{279}SR_{84}$ [5,43]. The ring-like $Au_n(SR)_n$ motifs are rarely found in Au NCs, and can only found in the $Au_{20}(TBBT)_{16}$ [44] and $Au_{13}(SAdm)_8(DPPB)_2^+$ [45] yet, which both exhibit a $Au_8(SR)_8$ motif. In some cases, thiolate ligand can also adopt μ_3-coordination mode in the Au NCs. For example, in the $Au_{21}(SAdm)_{15}$ or

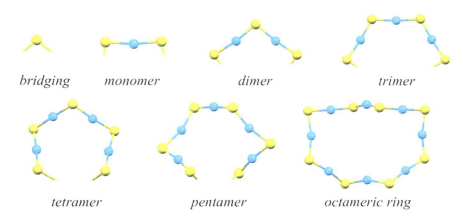

FIG. 3.11 Different surface-protecting motifs in $Au_n(SR)_m$ nanoclusters, including the bridge thiolate -SR, staple-like $Au_x(SR)_{x+1}$ motifs, and ring motifs such as $Au_8(SR)_8$. *(From R. Jin, C. Zeng, M. Zhou, Y. Chen, Atomically precise colloidal metal nanoclusters and nanoparticles: fundamentals and opportunities, Chem. Rev. 116(18) (2016) 10346–10413, https://doi.org/10.1021/acs.chemrev.5b00703.)*

bridging   *monomer*   *dimer*   *trimer*

tetramer   *pentamer*   *octameric ring*

$Au_{22}(SAdm)_{16}$ nanoclusters [46–48], a μ_3-thiolate ligand was observed in middle of the long $Au_8(SR)_9$ motif which surrounded the Au_{10} core of the Au_{21}/Au_{22} nanoclusters.

The bind modes between silver and thiolate ligand are more diverse than that in Au-SR, the coordination number of the thiolate ligand to the silver can be μ_2, μ_3 and μ_4 in the Ag NCs, which makes the structures of Ag NCs a great difference with Au NCs. At present, only one case of Ag NC with the same structural framework as Au NC has been found, that is, the $Ag_{25}(SR)_{18}^-$ [49] nanoclusters. In Ag NCs, a "Y" type $Ag(SR)_3$ construction unit is most commonly found. Moreover, two or more $Ag(SR)_3$ motif can further assemble into more complex structures via sharing the SR ligand. For example, six V-shaped $Ag_2(SR)_5$ motifs formed by two $Ag(SR)_3$ units via SR-sharing can be found in the $Ag_{44}(SR)_{30}^{4-}$ nanoclusters [25]. In addition, a μ_3 and μ_4-thiolate ligand anchoring on the kernel or linking up the structure units are also very common modes to find in a Ag NC. The high coordination number between thiolate ligands and silver atoms makes the combination of structural units very diverse, which endows Ag NCs with various structural characteristics and is different from Au NCs. The Ag-S shells derived from these combinations are basically different in each Ag NCs. Multiple structural units are often connected with each other to form a network or cage-like shell wrapped with the metal core. Although there are some researches on the structural evolution of Ag NCs, the research on the binding mode and law of the external ligand metal interface is still very limited. To solve this problem, more crystal structure of Ag NCs and theoretical calculation of Ag NCs are needed.

The coordination modes between thiolate ligands and Cu atoms are similar to that between Ag and SR. μ_2, μ_3, and μ_4-SR have also been found in the Cu NCs or M-Cu alloyed NCs. For example, the $[Cu_{32}(PET)_{24}H_8Cl_2](PPh_4)_2$ nanocluster exhibits all the three types of SR (μ_2, μ_3, and μ_4) in its outer shell [50]. The rod-shaped $Cu_{14}H_8$ core is coated by two complex $Cu_7(PET)_{11}Cl$ and two Cu_2PET metal ligand frameworks, constructing the total structure of Cu_{32} nanoclusters.

3.2.2.2 Phosphines and NHCs

Phosphine ligands can be divided into monodentate, bidentate, and polydentate phosphine ligands. Unlike thiolate ligands, phosphine ligands usually bind to metals in a monocoordination mode. Regardless of the type of the phosphines or the targeting metal, they are capping on the kernel surface of nanoclusters by M—P bonds between each terminal phosphorus atom and one metal atom. Of note, though all the phosphines adopt the mono-coordination mode with metal, there are some differences between monodentate and bidentate or polydentate phosphines. That is, bidentate or polydentate phosphines can bind with more than one metal atom in the nanocluster, so a staple-like mode can be formed on the surface of the metal kernel, similar to that of $Au_n(SR)_{n+1}$ staple motif in Au NCs. In fact, the $Au_1(SR)_2$ motifs on the $Au_{23}(SR)_{16}^-$ have been proved that they can be replace by the 1,2-bis(diphenylphosphino) methane (DPPM) ligands [51]. For bidentate ligands, the distance of the metal atoms bond to the bisphosphine ligand in the nanocluster is increased by the influence of the length of the carbon chain between the two phosphines in the ligand. For example, the distance between two metals attached to a DPPM is usually less than 3.2 Å, while in the Au_{22} cluster protected by the 1,8-bis(diphenylphosphino) octane (DPPO), the distance between the gold atoms attached to the same phosphine ligand can reach as much as 8 Å [52]. Interestingly, in the recently reported $M@Ag_{21}(DPPF)_3(SAdm)_{12}(BPh_4)_2$ (M=Au/Ag; DPPF=1,1'-bis(diphenylphosphino)ferrocene), the two P atoms in the DPPF were found to bind with the same Ag atom, which may cause by the special electronic structure of DPPF [53].

Because phosphine ligands form coordination bonds with metals, phosphine-gold bonds exhibit weaker binding ability than metal-sulfur bonds. This phenomenon results in the partial detachment of phosphine ligands from the surface of the clusters, which can be observed by mass spectrometry. It was observed in the ESI-MS of $Ag_{29}(BDT)_{12}(Ph_3P)_4$ that the four Ph_3P can dissociate in the solution [54]. Further, Kang et al. found that by adding excess Ph_3P into the solution of Ag_{29}, the PL intensity of the nanoclusters could be significantly enhanced due to the restricted Ph_3P dissociation-aggregation process [55].

In recent years, NHCs have been developed as a new kind of protecting ligands to construct the metal nanoclusters. The coordination mode of NHCs is similar as the phosphines, only a single M—C bond is formed between the metal atoms and NHC ligand. Due to the stronger M—C bond than M—P bond, compared with nanoclusters protected by phosphines, NHCs-protected nanoclusters display better thermal stability. For example, it was reported that a PPh_3 ligand on the Au_{11} nanocluster can be exchanged by an NHC molecule due to its stronger affinity, producing the first NHC-protected Au NCs [39]. The as-anchored NHC on the Au_{11} nanoclusters adopt the μ_1-Au-C coordination bond same as the Au—P bond and it was revealed that the stability of this Au_{11} clusters was enhanced significantly. Further, $[Au_{13}(NHC)_9Cl_3]^{2+}$ and rod-like $[Au_{25}(NHC)_{10}Br_7]^{2+}$ co-protected by NHCs and halogen were successfully synthesized [40,41]. The crystal structures of these two nanoclusters were revealed to be similar to their analogues protecting by phosphine ligands. Interestingly, in addition to the excellent thermal stability, NHC-Au_{13} showed a significant red emission compared with that of PPh_3-Au_{13}, and the NHC-Au_{25} are displayed high catalytic activity in the cycloisomerization of alkynyl amines to indoles. An NHC-protected Au NC with new structural framework, $Au_{44}(NHC)_9(C\equiv CPh)_6Br_8$, was also reported later, showing high thermostability and can be used as heterogeneous catalyst for phenylacetylene hydration [56]. In addition to gold nanoclusters, NHC-protected silver nanoclusters $[Ag_{12}(PSiMe_3)_6(NHC)_6]$ and $[Ag_{26}P_2(PSiMe_3)_{10}(NHC)_8]$ have also been reported in recent years [57]. More recently, an NHC-contained Cu NC $Cu_{31}(RS)_{25}(NHC)_3H_6$ was obtained. The NHCs used in this work is a bidentate ligand, as revealed by the crystal structure, the bidentate NHCs bind with two Cu atoms via two Cu—C bonds [58]. The as-prepared Cu_{31} exhibits excellent stability due to the protection of NHC, this provides a new idea for improving the stability of Cu NCs with poor stability.

3.2.2.3 Alkynyl ligands

Similar as thiolate ligand, α-alkynyl used for protecting nanoclusters contains a terminal H atom that can be removed while bond to the metal atom ($RC\equiv C^-$ vs RS^-). So, to some extent, alkynyl ligands and thiolate ligands have similar coordination patterns with metals. Alkynyl can bind to the metal atom via μ_1, μ_2 and μ_3 bind modes. For instance, alkynyl ligand can form staple-like $Au_n(C\equiv CR)_{n+1}$ motifs and ring-like $Au_n(C\equiv CR)_n$ motifs like the surface motifs formed between thiolate ligands and Au atoms. But differently, alkynyl ligands can adopt either σ- or π-coordination modes, while the thiolate ligands only adopt σ-coordination, and this difference can dramatically affect the structure. In addition, the binding atoms in alkynyl ligands are two carbon atoms, while the binding atoms in thiolate ligands have only one sulfur atom, which makes the motif types involving alkynyl ligands more diverse. For example, in the Au_2L_3 motifs (L=SR or $C\equiv CR$), thiolate ligands have only one form, while alkynyl ligands can produce several different $Au_2(C\equiv CR)_3$ forms by changing the coordination modes between the two carbon atoms and gold. In details, one alkynyl ligand in the $Au_2(C\equiv CR)_3$ binds to two Au atoms, that is, the μ_2-form. Alkynyl ligand can bind to both of these two Au atoms via the same terminal carbon, forming two σ-Au bonds (μ_2-η_1, η_1). In another way, the alkynyl ligand can bind to these two gold atoms via a σ bonds and a π bond (μ_2-η_1, η_2). The staple motifs do not simply resemble the surface structures of thiolate-protected nanoclusters, because the incorporation of alkynyl ligands may significantly alter properties of nanoclusters. Compared with thiolate-protected gold nanoclusters, alkynyl-protected ones with identical metal cores exhibit distinctly different absorption profiles and show much improved catalytic activities for semihydrogenation of alkynes. In addition, the participation of alkynyl ligands could profoundly affect the luminescent properties of nanoclusters. These "ligand effects" are mainly attributed to the different nature of alkynyl ligands, as electronic perturbation through π-conjugated units may largely modulate the electronic structure of the whole nanocluster. For example, due to the similarity in binding mode, several alkynyl-protected Au nanoclusters with a similar structural framework to thiolate-protected Au NCs were obtained. That is, the structural determination of $Au_{36}(C\equiv CPh)_{24}$, $Au_{44}(C\equiv CPh)_{28}$ and $Au_{144}(C\equiv CC_6H_4$-2-F$)_{60}$ revealed that they can be viewed as the counterparts of $Au_{36}(SR)_{24}$, $Au_{44}(SR)_{28}$, and $Au_{144}(SR)_{60}$ [30,33]. Though similar in the structure, the UV-vis spectra between these analogues in pair are distinct.

3.2.2.4 Halogens and oxygen group elements as ligands

Halogen ions and sulfur ions usually appear in nanoclusters as auxiliary ligands. Since they have only a single atom, they can bind not only on the surface of the cluster, but also into the voids of the metal core to form multiligand bonds (usually more than four bonds). When the halogen ion binds to the surface of the cluster, it may bind to the metal in the form of single coordination, double coordination and multiple coordination (less common in gold cluster, more common in silver cluster).

When bound in a mono-coordination manner, it acts similar to the mono-dentate phosphine ligand, usually in metals that protrude from the cluster, such as the chlorine atoms at the both ends of rod-like $[Au_{25}(PPh_3)_{10}(SR)_5Cl_2]^{2+}$ [59]. When conjugated in a two-coordination manner, it acts like the thiolate ligands with μ_2-coordination mode and is used to bridge and stabilize two relatively close metal atoms. In contrast to halogen ions, sulfur ions as ligands binding on the surface of nanoclusters often appear as μ_3 bonds, such as the sulfur ions in $Au_{30}S(SR)_{18}$ [60,61] and $Au_{38}S_2(SR)_{20}$ [62]. In addition, this kind of μ_3-S can further form a Au_4S_4 ring on the surface of the metal core, which was found in the $Au_{108}S_{24}(PPh_3)_{16}$ and $Au_{70}S_{20}(PPh_3)_{12}$ [63,64]. Furthermore, the μ_4-S and the μ_5-Se have also been observed on the surface of the nanoclusters, that is, in the $Au_{60}S_6(SR)_{36}$ and $[Au_{60}Se_2(PPh_3)_{10}(SeR)_{15}]^+$ respectively [36,65]. When halogens or sulfur ions are used to treat the internal voids of nanoclusters, the coordination number is usually greater than three. They are connected with metal atoms around them by covalent bonds and have three forms. One is to enter into the innermost part of the nanocluster and form the core together with metal atoms. The other is to participate in the formation of network or cage-like shell structure at the periphery of the nanocluster. The last one is to bridge the core and shell of the nanocluster as the intermediate layer between the core and shell. Most of these kinds of halogen or oxygen group element ligand are found in silver clusters. For example, a single S^{2-} was found in the center of the FCC Ag_{14} kernel of $Ag_{62}S_{13}(SBu^t)_{32}$ clusters, forming the $Ag_{14}S_1$ kernel of this cluster. It is also in $Ag_{62}S_{13}(SBu^t)_{32}$ that 12 S^{2-} are combined with the $Ag_{14}S_1$ kernel and further connected the $Ag_{48}(SR)_{32}$ outer shell of which is composed of Ag atoms and thiolate ligands, forming the hierarchical structure of $S@Ag_{14}@S_{12}@Ag_{48}(SR)_{32}$ [66]. More recently, Wei et al. report two Cl-rich Ag NCs $Ag_{48}Cl_{14}(S-Adm)_{30}$ and $Ag_{50}Cl_{16}(S-Adm)_{28}(DPPP)_2$ [67]. These two Ag NCs show strong structural dependence that they both exhibit the Ag_8 core which is coated with 14 Cl^-, the as-formed Ag_8Cl_{14} core is further cap by the outer shell. Of note, in the shell of the Ag_{50}, two Cl^- was observed to participate the formation of $Ag_{42}(SR)_{28}(DPPP)_2Cl_2$ with μ_2-coordination mode.

3.2.2.5 Hydride ions

Hydride ions can be used as auxiliary ligands in metal clusters, which are usually derived from the addition of reducing agents in the nanocluster synthesis process, such as $NaBH_4$. Of note, the position of the hydride ions in nanoclusters cannot be determined by the SC-XRD. Instead, neutron diffraction, MS, ^2H NMR and DFT calculations can be combined to reveal its position in the clusters. The coordination mode of hydrides is quite diverse. A hydride ion can connect with a metal atom to form a single M—H bond, or bridging two metal atoms by μ_2 type, and it can also cap on the triangle M_3 or quadrilateral M_4 through higher μ_3 and μ_4 coordination modes. Moreover, because of their small size, hydride ion can not only bind to the outside of the nanocluster, but also enter the spaces between metal atoms. In this case, the coordination number of the hydrides depends on the type of void it is in, for instance, a μ_4-H in tetrahedral M_4 or a μ_6-H in octahedral M_6. Hydrides have been found in Au NCs, Ag NCs and Cu NCs. For Au NCs, the hydrides were first found in the $[Au_9H(PPh_3)_8]^{2+}$ clusters reported by Tsukuda's group [68]. In this case, the existence of the single hydride ion was confirmed by ESI-MS and NMR, and the position of it was determined by DFT calculation that this H atom bind with the center Au atom in Au_9 with μ_1 Au—H bond. Subsequently, they obtained the hydride-doped bimetallic nanoclusters $[HPdAu_{10}(PPh_3)_7Cl_2]^{3+}$ which contains a μ_2-H binding with one Au atom and the central Pd atom [69]. Crystal structural analysis and DFT calculations on $(HPdAu_{10})^{3+}$ showed that the interstitially doped H atom induced a notable deformation of the core. Recently, Yuan et al. reported two ultrastable hydride-contained Au_{20} NCs, which are composed of a Au_{11} and Au_9 units [70]. These two units connect with each other via three Au—Au bonds. Considering that there are three hydrides in the Au_{20} revealed by ESI-MS, authors proposed that each Au—Au bond between these two structural units was anchored by a hydride with μ_2 coordination. Similarly, four μ_2-H atoms were found bridging the four pair of Au-Au atom between the two Au_{11} units in the $[Au_{22}H_4(DPPO)_6]^{2+}$ nanoclusters [71].

The first structurally characterized homometallic polyhydrido silver cluster is the $[Ag_6H_4(DPPM)_4(OAc)_2]$, in which the four hydrides were anchored on four of the Ag_3 triangle of the octahedral Ag_6 core via μ_3 coordination mode [72]. As mentioned earlier, the hydrides in the clusters can also appear in the interstitial space between the metal atoms. For example, Liu et al. reported a hydride-centered heptanuclear silver clusters, $[Ag_7(H)\{E_2P(OR)_2\}_6]$, in which the hydride ion was imbedded into the center of the tetrahedral Ag_4 unit [73]. Recently, Yuan et al. also reported a $[Ag_{40}(DMBT)_{24}(PPh_3)_8H_{12}]^{2+}$ in which twelve hydrides have the same chemical environment revealed by the ^2H NMR [74]. By combining the crystal structure and DFT calculations, these twelve hydrides are proposed to anchor on the Ag_8 inner core and linking up the core with the $Ag_{32}(SR)_{24}(PPh_3)_8$ shell. As a result, these twelve hydride ions can be regarded as the μ_4-H in twisted Ag_4 tetrahedrons.

In 2015, a Cu_{25} nanocluster with partial Cu(0) was structurally determined, formulated as $[Cu_{25}H_{22}(PPh_3)_{12}]Cl$ [16]. The 22 hydrides can be divided into three types in the ratios of 12:6:4. The first and third types of hydrides are both in the

μ_3 coordination mode but in different chemical environments, and the 6 second type of hydrides are in the μ_4 coordination mode in the Cu_4 unit. In fact, the position of hydrides in the nanoclusters can be experimentally confirmed via the single-crystal neutron diffraction. Using this method, Liu et al. determined the exact location of the hydrogen anion in the Cu_{11} cluster, and the results showed that the two hydrogen atoms in the cluster molecules were inside two Cu_4 tetrahedrons, represented as two μ_4 hydrogens. Subsequently, they reported another Cu NC with the molecular formula $[Cu_{30}H_{18}\{E_2P(OR)_2\}_{12}]$ (Cu_{30} for short) [75]. The structure of Cu_{30} except for hydrides was first determined by SC-XRD, and then single-crystal neutron diffraction was used to further determine the position of hydride ions in the nanocluster. The results show that there are three kinds of hydrogen anions with different coordination types in Cu_{30} molecule, namely μ_3-H, interstitial μ_4-H and a square pyramidal type μ_5-H. In addition, a μ_6-H has been found in a trimethyltriazacyclohexane-protected octahedral Cu_6 [76].

3.2.2.6 Other ligands

In addition to the common ligands mentioned above, there are also a few nanoclusters protected by other ligands. Some of them are either isolated cases or their use is emerging and not yet systematic, which will be described in this section. Wang's group used amine ligands to protected the nanoclusters and have successfully solved the structure of several nanoclusters. The first case using the amine ligands in nanoclusters is the $[Au_{32}(Ph_3P)_8(dpa)_6][SbF_6]_2$ (Hdpa=2,2′-dipyridylamine) [77], however, in this case, the dpa ligands anchor on the metal core only through the two pyridine-N binding with two Au atom, the amine group have not form the covalent bond with Au. After this pioneering work, two Ag NCs protected by dpa were successfully prepared, that is, $[Ag_{21}(dpa)_{12}]SbF_6$ and $[Ag_{22}(dpa)_{12}](SbF_6)_2$ [78]. The coordination modes of dpa in this work showed some different to that in Au_{32}. A dpa molecule has three binding sites to Ag atom, the two pyridine-N can bind to the Ag via coordinate bond, and the amine-N in the middle can form the σ bonds with one or two Ag atoms via μ_1 or μ_2 configuration. More recently, $[Au_{23}(Ph_3P)_{10}(dpa)_2Cl](SO_3CF_3)_2$ have been reported which bearing two dpa ligand in the nanocluster [79]. These two dpa bind with two Au atoms via one pyridine-N and one amine-N leaving the other pyridine-N unbonded.

Recently, an all-carboxylate-protected Ag_8 cluster was synthesized using perfluoroglutarate ligands [80]. This ligand contains two carboxyl groups in the molecular structure, and each of the carboxyl group bind with two Ag atoms through the Ag—O bonds. So the carboxylic group can be thought of as a bidentate ligand, that is, two oxygen atoms on the carboxylic group are connected to two metals, but it is worth noting that in the anion of the carboxylic acid, due to electron delocalization, the two oxygens within the molecule are actually completely equivalent. In fact, carboxylic acids have previously been reported to protect clusters as auxiliary ligands. For example, an Ag_{12} clusters co-protected by CF_3COO^- and thiolate ligands was reported [81] and $^nPrCOO^-$ and $PhCOO^-$ were used as auxiliary ligands for constructing the Ag_{84} nanoclusters [82]. In these cases, the carboxylic group all bond to two metal atoms via μ_1 coordination mode. Specially, amino acids can also act as the protecting ligand for nanoclusters. In the $Ag_{47}(_L/_D\text{-valine})_{12}(C{\equiv}C^tBu)_{16}]^+$ nanoclusters reported by Liu et al. [83], the thiol-free amino acid valine showed its ability to stabilize the Ag NCs. In essence, the amino acid also binds to the nanocluster as a carboxylic acid ligand, but in contrast, the amino acid molecule also contains amino groups that can form coordination bonds with Ag atoms, giving the amino acid three active sites. This difference profoundly affects the binding mode of amino acids to nanoclusters and thus resulting in different arrangement of surface metal-ligand structures. In addition, using the amino acid as the chiral template, the chiral arrangement of the Ag NCs can be finely controlled.

In 2016, Guan et al. used calixarene as protecting ligand to synthesize the Ag_{35} $(H_2L)_2(L)(C{\equiv}CBu^t)_{16}]$ $(SbF_6)_3$ (L=p-tert-butylthiacalix[4]-arene) nanoclusters [84]. Calixarenes are good candidates for binding metal ions because of their preorganized multidentate coordination sites. The calixarenes used in this study contain four phenolic hydroxyl group and four thioether group, which are available for anchoring the metal atoms. There are two types of motifs between calixarenes and Ag atoms found in this Ag_{35} nanocluster. The first one is that the four O atoms bind with five Ag atoms as μ_3-coordination mode, while the four S atoms bind to four of these Ag atoms. The second type of motif revealed by the crystal structure is that one calixarene molecule bind with four Ag atoms, two of the Ag atoms are connected with two adjacent O atoms and one S atom and the other two Ag atoms bind to the left two S and O atoms, forming two O-Ag-S structures. Further, two new nanoclusters with calixarene ligands, $[Ag_{34}(BTCA)_3(C{\equiv}CBu^t)_9(tfa)_4(CH_3OH)_3]SbF_6$ and $[AuAg_{33}(BTCA)_3(C{\equiv}CBu^t)_9(tfa)_4(CH_3OH)_3]SbF_6$ were reported [85]. In these two nanoclusters, the calixarene only have the first one type of the bind mode to Ag atoms, which was found that more stable than the Ag_{35} nanoclusters bearing the second motif type.

Recently, Zhang et al. have induced the inorganic ions as capping ligands to protect the Ag NCs. They reported the successful synthesis of $Ag_{48}(C{\equiv}CBu^t)_{20}(CrO_4)_7$ [86]. Different with the previously reported inorganic oxo anion templated Ag(I) clusters, the CrO_4^{2-} ions was anchored on the surface of the nanoclusters acts as protecting ligands. Revealed by the crystal structure, there are two binding modes of CrO_4^{2-} in this Ag_{48} nanoclusters. For the first type, three O atoms in the CrO_4^{2-} bind with the pentagonal Ag_5 through two μ_2-O and one μ_1-O atoms leaving the fourth O atom unbonded. In the second mode, two of the O atoms in CrO_4^{2-} bind with both the Ag atoms in the kernel and outer shell via μ_3-coordination mode, similar to the coordination mode of Cl^- or S^{2-} bridging the core and the shell. The other two O atoms bind to two Ag atoms on the shell via μ_1-coordination mode.

In a more recent work, a series of Ag clusters encapsulated in the Ti-based organic cages were prepared. Taking one of them for instance, a Ag_6 cluster was encapsulated in the $Ti_6(Sal)_6$ ($SalH_2$ = salicylic acid) metal-organic cage [87]. As revealed by X-ray crystallography, the cage binds with the six Ag atom via the twelve Ag—O bonds. This work developed a new form of protection for Ag NCs and provides a feasible method to construct stable Ag NCs.

3.2.3 Metallic kernels in nanoclusters

The kernels of the nanoclusters refer to the metal cores formed by the tightly packed metal atoms inside the nanoclusters. Due to their tight packing pattern, the average metal-to-metal bond length in the nanoclusters' core is usually comparable to, or even smaller than, the bond length in the bulk metal (e.g., the average Au—Au bond length is about 2.88 Å in the bulk gold, and the minimum bond length in the kernel of Au NCs can be <2.7Å). On the other hand, in the nanoclusters containing free electrons, it is generally believed that most of the free electrons in the ground state are delocalized in the kernel. Therefore, compared with the metal atoms in the surface motifs, the valence states of the metal atoms in the kernel of the nanoclusters are more inclined to the M(0) valence state. Similar to bulk metals, the kernels of nanoclusters also have a variety of stacking modes, such as face-centered cubic (FCC), hexagonal close-packed (HCP) and body-centered cubic (BCC) modes. In addition, cluster cores also have nucleation modes not found in bulk metals, such as icosahedral M_{13}, M_{12}, and decahedral M_7, as well as a variety of irregular cores, and nucleation assembly between different types of core units. Besides, in the larger nanoclusters, the core often shows a multilayer structure, that is, starting from the most central core, the metal atoms further pile up and wrap around it, forming a kernel similar to "Russian dolls." It has been proved that the inner core of the nanocluster has a major influence on the intrinsic properties of the nanoclusters, and even a small change in the core may lead to a significant change in the physicochemical properties of the overall nanocluster. For example, the single crystal analysis results of a pair of reported Ag_{40} and Ag_{46} clusters showed that the two nanoclusters had the same shell but different kernels (simple cubic Ag_8 in Ag_{40} while FCC-Ag_{14} core in Ag_{46}), which led to significant differences in UV-vis absorption and oxygen reduction reaction (ORR) electrocatalytic reactions between the two clusters [88]. In this section, we will summarize the different types of metal kernels in the nanoclusters. The packing modes of metal atoms in different kind of basic kernels, the shell-by-shell kernels and the core-shell interfaces will be discussed.

3.2.3.1 FCC kernels in nanoclusters

In the bulk gold, silver and copper, the metal atoms are deposited in the form of FCC. When the size is reduced to several hundred atoms, this kind of FCC stacking mode is still retained, such as the FCC stacking in gold nanoparticles. The problem is whether the FCC structure can exist when the size of the metallic condensed state is further reduced to nanoclusters, and the number of metal atoms in the core of the nanoclusters is sharply reduced. If so, how is it presented? We know that the construction of a complete FCC framework needs 14 metal atoms, eight atoms at the vertex positions and six atoms at the face center positions. At present, M_{14} with this structure has been found in nanoclusters (see below). But in addition, the interesting thing is that there is not only one type of FCC kernel in nanoclusters but many other ways to present FCC in nanoclusters, which may be the incomplete FCC framework or the transformed structure of FCC. They can be regarded as a part of the FCC lattice, and the kernel with transformed structure of FCC latter also retains the symmetrical features of FCC. As shown in Fig. 3.12, after extending the FCC cell by a half-cell and drop the left half of the initial cell, an M_{13} structure containing a central metal atom can be constructed, which is also called cuboctahedron containing eight triangular facets and six square facets. It retains all the symmetric elements of the FCC cell, such as the C_2, C_3 and C_4 symmetry, and the 12-coordinate number of the central metal. When observed along the C_3 axis, this M_{13} structure can be divided into three layers with 3:7:3 atom, which do not overlap each other, corresponding to the three packing layers of A/B/C in the FCC structure.

The first Au NCs exhibiting an FCC kernel is $Au_{36}(SR)_{24}$ [89]. It has FCC Au_{28} kernel, which can be regarded as the fusion of four cuboctahedra M_{13}, exhibiting a truncated tetrahedral shape. Along the C_3 axis, this Au_{28} kernel showed a four-layer 3:7:12:6 structures, consisted with the A/B/C/A stacking sequence of FCC structures. This Au_{28} kernel exposes four {111} facets and six {100} facets, and each {111} facet is protected by an $Au_2(SR)_3$ motif, while the {100} facets are capped by 12 bridging thiolate ligands.

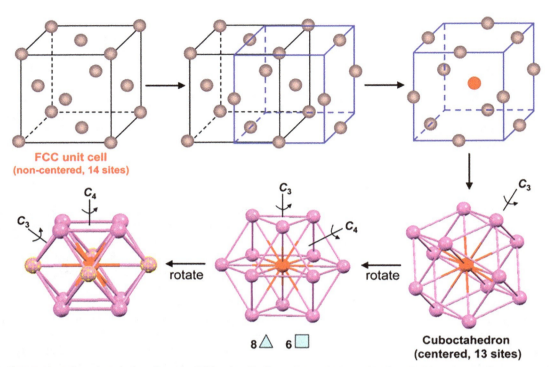

FIG. 3.12 Construction of a cuboctahedron from the FCC unit cell. *(From R. Jin, C. Zeng, M. Zhou, Y. Chen, Atomically precise colloidal metal nanoclusters and nanoparticles: fundamentals and opportunities, Chem. Rev. 116(18) (2016) 10346–10413, https://doi.org/10.1021/acs.chemrev.5b00703.)*

Interestingly, with the expansion of the crystal structure library, more and more nanoclusters with FCC kernels have been found, which also gives ones the opportunity to study the structural evolution relationship between FCC cores. That is, a novel evolution pattern of FCC kernel was found in a magic series of Au NCs, including $Au_{28}(SR)_{20}$, $Au_{36}(SR)_{24}$, $Au_{44}(SR)_{28}$ and $Au_{52}(SR)_{32}$ [89–92]. This series of Au NCs show a uniform progression of $Au_8(SR)_4$ units. When focusing on the relationship between their core structures, we can see that their kernels grow in groups of eight gold atoms, i.e., Au_{20}, Au_{28}, Au_{36}, Au_{44}, respectively. As mentioned above, the Au_{28} kernel in $Au_{36}(SR)_{24}$ can be regarded as four fused Au_{13} units. By analyzing the cores of the other three nanoclusters in this way, it can be found that the Au_{20} kernel in $Au_{28}(SR)_{20}$ is composed of two cuboctahedra Au_{13} units, the Au_{36} kernel in $Au_{44}(SR)_{28}$ is composed of six Au_{13} units, and eight Au_{13} units in Au_{44} kernel of $Au_{52}(SR)_{32}$. In details, if the two Au_{13} units added each time are regarded as a rod-shaped Au_{20} unit, then the newly added Au_{20} units are added to the existing kernel in the direction of crossing each other. In these series of Au NCs, the surface motifs combine with the kernel in the same manner of the as-mentioned in Au_{36}, that is, $Au_2(SR)_3$ capping the exposed {111} facets and bridging thiolate ligands anchoring on the {100} facets. So, these series of FCC Au NCs can be formulated as $Au_{8n+4}(Au_2SR_3)_4(SR)_{4n}$ (n is the number of Au_{13} units). Unexpectedly, Liao et al. reported an $Au_{56}(SR)_{34}$ nanocluster with highly structural correlation with the as-mentioned FCC type Au NCs [93]. When the series of clusters are observed along the {100} plane, it is found that the nanoclusters mentioned earlier grow in a pattern of eight gold atoms per layer. But in $Au_{56}(SR)_{34}$, it adds only four gold atoms in one layer compared with $Au_{52}(SR)_{32}$, which is equivalent to half a layer. This finding provides a new perspective for understanding the core growth pattern of FCC nanoclusters.

In addition to this series of nanoclusters, the core of the FCC M_{13} is also found in $Au_{23}(SR)_{16}$ [94]. The core of $Au_{23}(SR)_{16}$ is a cuboctahedron of M_{13} plus two additional capping atoms along the C_4 axis, forming a Au_{15} inner core. This Au_{15} core was protected by two $Au_3(SR)_4$, two $Au_1(SR)_2$ and four simple μ_2-SR.

As mentioned above, gold nanoparticles retain the FCC packing pattern in bulk gold. With the structural determination of the plasmonic $Au_{279}(SR)_{84}$, the specific packing pattern and the binding pattern of FCC core and its interfaces between kernel and external thiolate ligands and motifs are revealed [43]. Specifically, the crystal structure shows that the $Au_{279}(SR)_{84}$ exhibit a shell-by-shell Au_{249} kernel ($Au@Au_{12}@Au_{42}@Au_{92}@Au_{54}@Au_{48}$). In details, the central gold atom is coated by 12 gold atoms to form the cuboctahedron Au_{13}, and then the gold atoms on the outside of it are packed in the same way to form the larger cuboctahedron Au_{55} and then the Au_{147}. Six {100} facets of Au_{147} are covered by 9×6 gold atoms, forming a truncated octahedron Au_{201}. Finally, the 8 {111} facets of Au_{201} are covered by 6×8 gold atoms to form the final Au_{249} kernel. Subsequently, Sakthivel et al. obtained the structure of a $Au_{191}(SR)_{66}$ nanoclusters which showed a monotwinned/stacking faulted FCC Au_{89} inner core. The stacking sequence of the Au_{89} can be represented as ABCBA with atom ratios as 12:19:27:19:12.

The FCC patterns in Ag NCs are quite different from that of Au NCs. The most representative ones are a series of FCC Ag NCs protected by thiolate and phosphine ligands. They are Ag_{14}, Ag_{23}, Ag_{38} and Ag_{63} respectively [95–97]. They show obvious structural correlation, and FCC Ag_{14} of can be assembled into other FCC Ag NCs with FCC Ag_{14} as the smallest component unit. Different from the gold nanoclusters, the FCC structure of this series of clusters contains the silver atoms in the core and the outer shell, that is to say, the silver atoms in the whole cluster contribute to the complete FCC framework. Taking the smallest Ag_{14} as an example, its complete molecular formula is $Ag_{14}(SR)_{12}(PR_3)_8$. This cluster has an octahedral Ag_6 kernel, surrounded by cubes composed of eight Ag atoms, which combine to form a complete structural framework of Ag_{14} FCC, that is, six silver atoms in the kernel are at the face centers, while eight silver atoms in the outer are at the vertices. The twelve thiolate ligands in Ag_{14} bridge the two silver atoms at the vertices and one silver atom with μ_3-coordination mode, and the eight Ag atoms at the vertices were further capped by eight PR_3 ligands. Taking the FCC type of Ag_{14} as the basic unit, and when two Ag_{14} units are combined in the form of face sharing, the structure of $Ag_{23}(SR)_{18}(PR_3)_8$ is formed. Because of the face sharing, the two Ag_{14} units share five silver atoms, so the number of silver atoms increases by only nine. Observing the ligand coordination pattern on its surface, there are some differences from Ag_{14}. The 18 thiolate ligands in Ag_{23} can be divided into two kinds. The first kind is the same as that in Ag_{14}, which connect two silver atoms on the edge and a face-center silver atom of FCC framework by the μ_3 coordination mode. The other two thiolate ligands are located in the middle of the shared edges of the two Ag_{14} faces, connecting the silver atoms of the two silver atoms on the edge and the two face-center silver atoms by the coordination mode of μ_4. Similarly, the eight Ag atoms at the vertices are capped by eight phosphine ligands. Based on this one-dimensional growth rule, the next FCC Ag NCs should be the $Ag_{32}(SR)_{24}(PR_3)_8$ with three Ag_{14} units. Obviously, the assembly of Ag_{14} is not limited to one-dimensional growth. When it grows in a two-dimensional or three-dimensional way, a series of clusters with FCC structure can be derived. For instance, the reported structures of Ag_{38} and Ag_{63} can be viewed as the combination of four Ag_{14} in $2 \times 2 \times 1$ mode and eight Ag_{14} in $2 \times 2 \times 2$ mode respectively. In these cases, the thiolate ligands can be also divided into the edge-type and face-type which binding in μ_3 and μ_4 coordination modes, respectively. With this packing mode, the structures of Ag_{53} ($2 \times 3 \times 1$), Ag_{74} ($3 \times 3 \times 1$), Ag_{88} ($3 \times 3 \times 2$) and Ag_{174} ($3 \times 3 \times 3$) can be predicted (Fig. 3.13) [98].

This type of FCC structure has not been found in Au NCs, which may be due to the difference in the way the gold and silver atoms coordinating with the ligands. It has been mentioned before that the thiolate ligands in Au NCs are generally μ_2-SR, while the thiolate ligands in such Ag NCs are μ_3-SR (edge) and μ_4-SR (facet). As proof, the coordination mode of copper is similar to that of silver, and FCC Cu_{23} with similar structure to Ag_{23} has been reported to be successfully synthesized [99].

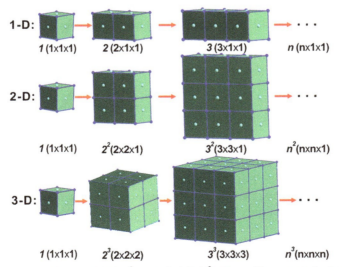

FIG. 3.13 Schematic representations of (A) 1-D (**n**), (B) 2-D (**n²**), and (C) 3-D (**n³**) arrays of face-centered cubes. *(From B.K. Teo, H. Yang, J. Yan, N. Zheng, Supercubes, supersquares, and superrods of face-centered cubes (FCC): atomic and electronic requirements of $[M_m(SR)_l(PR'_3)_8]^q$ nanoclusters (M = coinage metals) and their implications with respect to nucleation and growth of FCC metals, Inorg. Chem. 56(19) (2017) 11470–11479, https://doi.org/10.1021/acs.inorgchem.7b00427.)*

There is another type of evolution pattern of FCC Ag NCs which can be summarized from the crystal structures of Ag_{46} and Ag_{67} clusters [88,100,101]. If the as-mentioned growth mode above is called the complete cubic evolution model, this new kind of evolution model is the cubic evolution model ended with half-cubic blocks. In details, the Ag_{46} cluster exhibits a standard FCC-Ag_{14} kernel, and this FCC kernel is allowed to grow one more single layer of Ag atom, that is, four Ag atoms at each facet of this Ag_{14}, forming the FCC Ag_{38} framework. In the case of Ag_{67}, the kernel of the cluster is the FCC Ag_{23}, each facet of this kernel grows one more layer of Ag atoms, that is, four Ag atoms on the square facet and seven Ag atoms on the rectangular facets, forming the FCC Ag_{59} framework. The eight corners of the Ag_{38} or Ag_{59} are capped by eight $Ag_1S_3P_1$ tetrahedral motifs. In the Ag_{46}, all the 24 thiolate ligands in the nanoclusters belong to the 8 $Ag_1S_3P_1$ motifs, but in the Ag_{67}, except for the 24 thiolate ligands in the corners, 2 addition μ_3-thiolate ligands are found on each rectangular facet. According to this growth rule, the structure of next Ag NCs is the Ag_{88} nanoclusters with an FCC Ag_{32} kernel [102].

More recently, an Ag_{100} nanocluster with a new FCC packing mode differed with the above two packing mode was reported [103]. It exhibits a rhombicuboctahedron structure. There are 92 sliver atoms within this rhombicuboctahedron. First, an FCC Ag_{14} is found in the center of this Ag NC. Then each facet of this grows with 2 FCC layers of Ag atoms, that is, 9 silver atoms for each facet, forming the Ag_{68} framework which can be viewed as 7 FCC Ag_{14} combining via facet-sharing. Every two Ag_{14} units outside are connected by two silver atoms. Thus, 6 outer Ag_{14} units are connected by 24 Ag atoms, forming the FCC Ag_{92} pattern. This Ag_{92} is capped by 8 $Ag_1S_3P_1$ motifs at the corners and the left thiolate ligands can be divided into 2 types. One is the μ_3-SR connecting two Ag atoms at the edges and one Ag atom at the face center of the outer Ag_{14} units, there are two thiolate ligands of this type for each outer Ag_{14} unit. The other type of thiolate ligands connecting two Ag atoms at the edges of outer Ag_{14} unit and two bridging Ag atoms with μ_4-coordination mode.

3.2.3.2 HCP kernels in nanoclusters

Compared with FCC, HCP, another close packing form, have also been observed in nanoclusters. In bulk metals, HCP stacking behaves in the ABAB stacking mode, in which the atomic layers cycle in two layers. Thus, the construction of the simplest HCP model requires at least three layers, to be stacked in an ABA-like manner.

Earlier in 1992, Teo et al. reported a phosphine-ligated $Au_{39}(PPh_3)_{14}Cl_6$ [29]. The 39 Au atoms were found stacked in 7 layers as 1:9:9:1:9:9:1. In one-half of the molecule, the layered structure 1:9:9:1 can be observed as nearly HCP stacking, except that the center atom in the third layer is displaced into forth layer. The left half 1:9:9:1 adopts the same arrangement as HCP stacking. These two halves of HCP layers combined after twisted by 30 degrees, forming the final HCP/HCP' structure of Au_{39} nanocluster.

In thiolate-protected Au NCs, the smallest Au NCs containing an HCP kernel was reported to be the $Au_{18}(SR)_{14}$ nanoclusters [104,105]. It contains an HCP Au_9 kernel with three Au_3 layer. The Au_9 kernel is protected by one $Au_4(SR)_5$ motif at the bottom, one $Au_2(SR)_3$ motif at the top and three $Au_1(SR)_2$ at the waist.

Subsequently, a larger $Au_{30}(SR)_{18}$ nanocluster was discovered exhibiting an HCP Au_{18} kernel [106]. This Au_{18} kernel is composed of four layers containing 3, 6, 6, and 3 Au atoms respectively in an ABAB manner. The layer 1 and 3 are further capped and bridge by three $Au_2(SR)_3$ motifs in a C_3 symmetry and layer 2 and 4 are also protected by three $Au_2(SR)_3$ in the same manner.

Different with Au NCs, simple HCP kernels have not been found in silver and copper nanoclusters, they were only found in some mix-packed kernels in these nanoclusters. Diecke et al. reported a highly unstable $Ag_{64}(P^nBu_3)_{16}Cl_6$ nanocluster which exhibit a close-packed metal core [107]. In this Ag_{64} metal core, both FCC and HCP packing mode of Ag atoms can be found, thus, a mix-packed kernel in this Ag NC. The Ag_{64} metal core is capped by 16 phosphine ligands with μ_1 coordinate bond and 6 Cl^- with μ_2-coordination mode, leaving lots of uncoordinated silver atoms on the surface, which accounts for its poor stability.

In the Cu NCs family, a $Cu_{53}(C\equiv CPhPh)_9(dppp)_6Cl_3(NO_3)_9$ NCs containing part of HCP kernel was recently reported [108]. In details, the Cu_{53} cluster contains six-layer Cu_{41} kernel with atom ratios of 6:7:12:10:6. The first four layers stack by the ABAB manner, that is, the HCP mode. But the fifth layer shows a different mode with the first four layers, along with the third and fourth layers, an ABC packing mode with FCC feature is formed. Thus, the total kernel shows an ABABC packing mode which can be regarded as a mixed FCC and HCP structure.

3.2.3.3 BCC kernels in nanoclusters

As the third type of packed mode, BCC stacking has also been found in the nanoclusters. Structurally, the BCC stacking can be derived from the FCC stacking by the shrinking of metal-metal distance along the C_4 axis in FCC framework. The squeezing force possibly comes from the ligands. The coordination number of metal atoms in BCC lattice is 8, which is smaller than that in FCC or HCP lattice. Also, the space utilization of BCC is ~68%, lower than that in FCC or

HCP ($\sim$74%). BCC kernels are rarely found in nanoclusters, and the BCC stacked kernels have only been found in $Au_{38}S_2(SR)_{20}$ nanoclusters until now [62]. Its kernel consists of 30 gold atoms, which can be contained in a $3 \times 3 \times 2$ multicell BCC lattice. There are two intact BCC units in this Au_{30} kernel, forming an Au_{14} via facet-sharing, and this Au_{14} is extended by the addition 16 Au atoms following the BCC manner, forming the final BCC Au_{30} kernel. This Au_{30} kernel is further protected by six $Au_2(SR)_3$ motifs and eight simple bridging thiolate ligands. In addition, two more μ_3-sulfido atoms were anchored on the Au_{30} kernel. Of note, the structure of the Au_{30} kernel deviates slightly from that of the ideal BCC framework, due to the interaction between the outer $Au_2(SR)_3$ motifs and the kernel.

3.2.3.4 Icosahedron kernels in nanoclusters

The icosahedron consists of twelve vertices and twenty triangular facets. According to the presence of central atoms, the icosahedron can be divided into M_{13} and hollow M_{12} structures. Along the C_5 axis of M_{13}, it can be divided into five layers with atom ratios 1:5:1:5:1 (four layers 1:5:5:1 for M_{12}). The M_{13} icosahedron can be obtained by twisting peripheral 6 metal atoms in the middle layer of the FCC M_{13} unit. The reported conversion from $Au_{23}(SR)_{16}^-$ to $Au_{25-x}Ag_x(SR)_{18}$ proved that this transformation process can be experimentally realized [109].

The icosahedron is perhaps the most widely observed structure in nanoclusters. The existence of icosahedron in nanoclusters can be varied. In addition to simple individual icosahedral kernel, icosahedron can also assemble into larger kernels via vertex, face sharing or interpenetration. Further, in larger nanoclusters, the kernels will show as the shell-by-shell structure in which the icosahedron can serve as the center of these multilayer kernels. Icosahedron can be also found as part of some complicated kernels. Next, we will introduce and discuss each of these cases.

Isolated icosahedral kernels

Icosahedral M_{13} can appear as a core alone, and even individual M_{13} nanoclusters can be obtained by appropriate ligand and synthesis methods. However, individual hollow M_{12} icosahedral nanoclusters have not been reported, indicating that the central metal atom is essential for stabilizing the structure of the icosahedron. Earlier in 1981, Briant et al. had successfully prepared and solved the structure of $[Au_{13}(PMePh)_{10}Cl_2](PF_6)_3$ [110]. The 13 Au atoms in this cluster form an intact icosahedron which is further protected by phosphines and Cl^-. One icosahedron has 12 vertices outside for the anchoring of ligands, which can be regarded as 4 layers in atoms ratios 1:5:5:1 along the C_5 axis. In this case, the top and bottom Au atoms were capped by 2 Cl^- and the left 10 Au atoms at the waist were protected by 10 phosphine ligands. The Au_{13} kernel was calculated to bear 8 free electrons, which can be interpreted in terms of the superatom complex model (SACM) [111]. Benefit from its 8-electron configuration, icosahedral Au_{13} shows excellent stability so that the nanoclusters containing icosahedral kernels will be more stable, this may explain why there are so many nanoclusters with icosahedral kernel.

In 2008, the structure of $Au_{25}(SR)_{18}^-$, which has received the most research and attention, is reported [23,24]. An icosahedral Au_{13} kernel was found in the molecular structure of $Au_{25}(SR)_{18}^-$. The Au_{13} kernel is capped by three pairs of $Au_2(SR)_3$ motifs, each pair of $Au_2(SR)_3$ binds on the Au_{13} in a near-plane manner and the Au atoms in the $Au_2(SR)_3$ motifs occupy at the Au_3 facets of the icosahedron, leaving 8 Au_3 facets uncovered. Such kind of uncovered Au_3 facets is potentially interesting in catalysis.

As a kind of common kernel, icosahedral kernels have also found in Ag NCs and Cu NCs and alloyed NCs. $Ag_{25}(SR)_{18}^-$, which exhibits similar structural framework of $Au_{25}(SR)_{18}^-$, have been reported by Joshi et al. [49] That is, $Ag_{25}(SR)_{18}^-$ contains an icosahedral Ag_{13} kernel and six $Ag_2(SR)_3$ motifs. In another work, $Ag_{29}(SSR)_{12}(PPh_3)_4$ with an icosahedral Ag_{13} was reported [54]. In this case, the Ag_{13} kernel is caged by an outer shell consisting of four Ag_3S_3 rings and four $Ag_1S_3P_1$ motifs. These eight construction units cap on eight Ag_3 facets of the icosahedron, and the left 12 facets are covered by the 12 phenyl groups in the 12 thiolate ligands. Besides, the $[Ag_{21}\{S_2P(O^iPr)_2\}_{12}]^+$ clusters reported by Dhayal et al. also contained an icosahedron Ag_{13} kernel [112].

For Cu NCs, the first reported Cu(0) containing Cu_{25} NCs was observed to possess an icosahedral Cu_{13} kernel [16]. The Cu_{13} core is protected by four triangular $Cu(PPh_3)_3$ motifs, which cap the icosahedron in a tetrahedral arrangement. After a ligand-exchange process, the Cu_{25} can transform into Cu_{29} NCs without changing the icosahedral Cu_{13} kernel [113].

Icosahedron-based assembly kernels

As early as the 1980s, Teo et al. found that icosahedral M_{13} could be used as a building block to construct the larger clusters, which were called as "clusters of clusters." In their reported $Au_{13}Ag_{12}(PPh_3)_{12}Cl_6$ nanocluster, the $Au_{13}Ag_{12}$ metal core was found to be composed of two icosahedral M_{13} combined via vertex sharing [114]. Furthermore, the structures of $Au_{18}Ag_{20}(PPh_3)_{12}Cl_{14}$ and $Au_{18}Ag_{19}(PPh_3)_{12}Br_{11}$ both contain the M_{36} kernels formed by three icosahedral M_{13} via vertex sharing [115,116]. Similarly, the subsequently reported nanoclusters such as $[Au_{25}(PPh_3)_{10}(SR)_5Cl_2]^{2+}$ [59],

$[Au_xAg_{25-x}(PPh_3)_{10}(SR)_5Cl_2]^{2+}$ [117], $[Cu_3Au_{34}(PPh_3)_{13}(SR)_6S_2]^{3+}$ [118], $[Pt_2Ag_{23}(PPh_3)_{10}(SR)_5Cl_2]^{2+}$ [119], $Pd_2Au_{23}(PPh_3)_{10}Br_7$ [120] and $[Pt_3Ag_{33}(PPh_3)_{12}Cl_8]^+$ [121] also displayed the similar assembly mode between icosahedral M_{13} units. In addition, Li et al. recently showed a $Cd_2Au_{29}(SR)_{17}(dppf)_2$ nanocluster which also exhibits a rod-like Au_{25} kernel composed of two icosahedral Au_{13} units [122]. Interestingly, Jin et al. found that three icosahedrons can assemble linearly via sharing two vertices, forming the rod-like $[Au_{37}(PPh_3)_{10}(SC_2H_4Ph)_{10}X_2]^+$ clusters [123]. The waist of these clusters is bridged by 10 μ_2-SR, and both ends of this cluster are capped by 2 Cl^-. The left 10 Au atoms are bonded by 10 PPh_3. This type of linear assembly was also recently found in a series of Pt-Ag alloyed nanoclusters reported by Chiu et al. That is, $Pt_2Ag_{33}(dtp)_{17}$ and $Pt_3Ag_{44}(dtp)_{22}$ with two and three icosahedral Pt_1Ag_{12} units respectively [14]. More icosahedrons assembling in a nanocluster via vertex sharing is found in the $[Au_{60}Se_2(PPh_3)_{10}(SeR)_{15}]^+$ [36], in which five icosahedral Au_{13} assemble end to end by sharing five vertex Au atoms, forming the Au_{13}-based ring-type nanocluster. More recently, Yuan et al. reported the novel $Ag_{61}(dpa)_{27}(SbF_6)_4$ (Hdpa = dipyridylamine) nanoclustesr which exhibits a unique Ag_{49} kernel composing by four icosahedral Ag_{13} unit via vertex-sharing. This cluster showed strong electron coupling between the icosahedral units [124].

All the above are the M_{13} cores assembled by vertex sharing. In addition, M_{13} can also be assembled by other ways, including face sharing and interpenetration. The face sharing mode was first observed in the $Au_{38}(SR)_{24}$ reported by Qian et al. [26]. The Au_{23} kernel in this nanocluster can be regarded as two icosahedral Au_{13} combined via sharing a Au_3 facet. Subsequently, this type of M_{23} kernel was also found in Ag NCs. In the $[Ag_{45}(dppm)_4(SR)_{16}Br_{12}]^{3+}$ nanoclusters, a Ag_{23} kernel composed of two icosahedral Ag_{13} units is found to be caged by an $Ag_{22}S_{16}P_8Br_{12}$ shell [125]. Similarly, in the alloyed nanocluster $[Pt_2Ag_{51}(SR)_{28}(PPh_3)_2Cl_7]^{2+}$, two icosahedral Pt_1Ag_{12} units face fuse to form the Pt_2Ag_{21} kernel, and further protected by a cage-like $Ag_{30}(SR)_{28}(PPh_3)_2$ motif [126]. Interestingly, Li et al. reported a $Au_{36}Ag_2(SAdm)_{18}$ alloy nanoclusters recently, which was found to exhibit a $Au_{30}Ag_2$ kernel composing by three icosahedral M_{13} units via facet-sharing between each of two, and the two Ag atoms were found to dope into the center of this kernel [127].

When two or more icosahedrons are assembled and share more metal atoms (shared atoms >3), an interpenetration will occur. This situation is more common in Au-Ag alloy nanoclusters. In the first time, Jin et al. found that the Au_8Ag_{41} kernel in the $[Au_8Ag_{57}(dppp)_4(C_6H_{11}S)_{32}Cl_2]Cl$ and $[Au_8Ag_{55}(dppp)_4(C_6H_{11}S)_{34}][BPh_4]_2$ clusters was formed by eight icosahedral M_{13} through 3D assembly [128]. In this case, two icosahedral M_{13} are first combined to form a M_{19} unit by sharing seven metal atoms, and then the M_{19} units assemble through further interpenetration, forming the final Au_8Ag_{41} kernel. In another case, $[Au_3Ag_{48}(SR)_{28}Cl_7]^{2+}$ exhibits a Au_3Ag_{22} which is composed by three Au_1Ag_{12} icosahedrons via interpenetration (sharing 7 metal atoms).

Icosahedron-centered shell-by-shell kernels

The icosahedrons and their assemblies can be found in the center of the multishell kernel. At this time, the kernel of the nanocluster will grow outwards to form a shell-by-shell structure. There are three main forms of existence, the first is the icosahedral M_{13} as the center, the second is the hollow M_{12} as the center, and the third is the M_{19} assembled by two M_{13} units.

The construction of a smallest icosahedron needs 12 metal atoms, in this case, each facet of the icosahedron has three metal atoms. When this icosahedron is caged by a larger icosahedron, the construction of this second icosahedral would need 42 metal atoms and each facet will contain 6 atoms, and so on a third shell with 92 atoms (10 atoms for each facet). That is, the total metal atoms of a complete icosahedron will occur as 13, 55 and 147 with one, two or three layers respectively (or 12, 54 and 146 for a hollow kernel).

For example, in the reported $Au_{133}(SR)_{52}$, a central Au_{13} icosahedron was first coated by a larger Au_{42} icosahedral shell, forming the ideal Au_{55} Mackay icosahedron [129,130]. Then 16 Au_3 triangles was found to cap on the facets of this icosahedron in an ABCB manner, other four single Au atoms capped on the other four facets in a ABCA manner, forming the final Au_{107} ($Au_{13}@Au_{42}@Au_{52}$) kernel. For the Ag NCs, Hu et al. reported a $[Ag_{112}Cl_6(C\equiv CAr)_{51}]^{3-}$ exhibiting an icosahedron-containing shell-by-shell Ag_{103} kernel [131]. This kernel can be divided into three shells, that is, an Ag_{13} coated by a larger icosahedral Ag_{42} shell, forming an ideal Ag_{55} icosahedron. The third shell Ag_{48} does not have a regular shape, but can be viewed as a barrel formed by cutting off two Ag_6 fragments from the anti-Mackay icosahedron (or rhombicosidodecahedron) Ag_{60}. Excitingly, the complete three-shell icosahedral M_{147} ($M_{13}@M_{42}@M_{92}$) was recently found in the Au-Ag alloyed nanocluster $Au_{267-x}Ag_x(SR)_{80}$ [132]. The 120 Ag atoms in the outermost $Ag_{120}(SR)_{80}$ shell form an anti-Mackay shell coating this M_{147} icosahedron.

Differently, when the M_{13} (M = Au/Ag) is coated by a Cu-shell, due to the smaller atomic radius of Cu atoms, the Cu cage out of the M_{13} did not show an icosahedral framework, but usually as an icosidodecahedral Cu_{30} cage. That is, 30 Cu atoms of the Cu_{30} cage capped on the 30 edges of the M_{13} icosahedron, it can be derived from the icosahedral M_{42} cage by removing the 12 vertex atoms. For instance, in the $[Ag_{61}Cu_{30}(SR)_{38}S_3]^+$ nanoclusters, it has an icosahedral Ag_{13} inner most kernel and a second Cu_{30} shell, forming the $Ag_{13}Cu_{30}$ kernel which was further capped by a $Ag_{48}(SR)_{38}S_3$ outer shell [133]. In another work, the $[Au_{19}Cu_{30}(C{\equiv}CR)_{22}(Ph_3P)_6Cl_2]^{3+}$ cluster also exhibits a Au_{13} inner core but caged by a Au_6Cu_{30} shell. In this case, the icosahedral M_{42} cage has been only removed 6 vertical atoms, and the left 6 Au atoms on the Au_6Cu_{30} shell adopt a chair conformation.

Multilayer kernels with a hollow icosahedron as the core have also been reported. They share some similarities with the M_{13}-base multilayer kernels but also have their own unique features. It was first found in the $Ag_{44}(SR)_{30}^{4-}$ nanocluster, which exhibits a Ag_{32} kernel composted of a hollow icosahedral Ag_{12} inner core and a dodecahedral Ag_{20} outer shell [25]. Each Ag atom of the Ag_{20} cap on the Ag_3 triangles of the Ag_{12} icosahedron. In addition, the Ag_{12} inner core can be changed into Au_{12} by the addition of Au precursor during the synthesis. Further, via a seed-growth method, the $Ag_{44}(SR)_{30}$ can be transformed into the $Ag_{50}(SR)_{30}(dppp)_6$ nanoclusters which still exhibits the $Ag_{12}@Ag_{20}$ kernel [134]. More recently, Yuan et al. reported the successful synthesis of the "golden fullerene" Au_{32} nanoclusters, of which the same Au_{32} metal core was also reported by Kenzler et al. [77,135] Similarly, this Au_{32} is composed of a hollow icosahedral Au_{12} inner core and a dodecahedral Au_{20} outer shell. Recently, the long-pursued structure of $Au_{144}(SR)_{60}$ and its alkynyl-protected analogue $Au_{144}(C{\equiv}CR)_{60}$ have been revealed [30,136]. They show similar structures that are composed of a three-shell Au_{114} ($Au_{12}@Au_{42}@Au_{60}$) kernel. Similar as that in Au_{13}-centered kernel, the icosahedral Au_{12} is caged by a larger icosahedral Au_{42} shell. This icosahedral Au_{54} is then further coated by an outer rhombicosidodecahedron Au_{60} shell, forming the final shell-by-shell kernel with totally 114 Au atoms. A rhombicosidodecahedron has 20 triangular, 30 quadrate and 12 pentagonal facets on its surface. The left $Au_{30}(SR/C{\equiv}CR)_{60}$ forms 30 $Au_1(SR/C{\equiv}CR)_2$ staple motifs capping on the 30 quadrate facets on the Au_{60} shell to form the entire $Au_{144}(SR/C{\equiv}CR)_{60}$ framework. This kind of $M_{12}@M_{42}@M_{60}$ kernels was also found recently in two Au-Ag alloyed nanoclusters $[Au_{78}Ag_{66}(C{\equiv}CR)_{48}Cl_8]^{q-}$ and $[Au_{74}Ag_{60}(C{\equiv}CR)_{40}Br_{12}]^{2-}$ [137]. They all exhibit the $Au_{12}@Au_{42}@Ag_{60}$ kernels.

The third kind of icosahedron-base inner cores in the multilayer kernels of nanoclusters displays as an M_{19} framework that can be viewed as the interpenetration of two icosahedral M_{13} (by sharing a Au_7 decahedron) [138]. This kind of M_{19} inner core was found in the $[Ag_{141}(SR)_{40}Br_{12}]^{3+}$ nanoclusters. In this cluster, the Ag_{19} was further caged by a second outer Ag_{52} shell and a $Ag_{70}(SR)_{40}Br_{12}$ outermost layer.

Other icosahedron-containing kernels

In some cases, the entire kernel of the nanoclusters may not exhibit a regular framework, or in other work, it's not made up entirely of icosahedron. The icosahedral M_{13} or its assemblies make up only part of the kernel of the clusters. For example, $Au_{44}(SR)_{26}$ reported by Liao et al. consists of a Au_{29} kernel capped by an exterior shell including two μ_2 thiolate ligands, three $Au_1(SR)_2$, and six $Au_2(SR)_3$ staple motifs [139]. The Au_{29} kernel can be divided into an Au_{23} formed by two icosahedral Au_{13} via facet sharing (like the kernel in Au_{38Q}) [26] and a special Au_6 cap. In another work reported by Zhao et al, a $Au_{40}(SR)_{22}$ cluster was found to consist of an irregular Au_{29} kernel [140]. This Au_{29} kernel contains an icosahedral Au_{13}, and this Au_{13} further combine with an irregular Au_{16} framework to form the final Au_{29} kernel. Due to the influence of the addition Au_{16} fragment, the Au_{13} unit in the kernel is slightly disordered and deviates from the ideal icosahedral shape.

3.2.3.5 Decahedral kernels in nanoclusters

The decahedron in the kernels of the nanoclusters generally refers to the pentagonal bipyramid frameworks with C_5 axis of symmetry. Another characteristic of this decahedron is that it has a σ_h symmetry plane, that is, a symmetry plane perpendicular to the axis of C_5 symmetry, which gives it a structure of upper and lower symmetry. The simplest decahedron consists of seven metal atoms, the pentagonal bipyramid shape of M_7. It has 15 edges and 10 triangular facets. In addition, there are two kind of variants of the decahedron in the kernels of the clusters, namely the Ino decahedron and Marks decahedron. The simplest Ino-decahedron has a total of 12 metal atoms (considering the central atom, there are 13 atoms), which can be obtained by rotating the 5 metal atoms in the waist of the icosahedron M_{13} by 36 degrees. At this point the middle of M_{13} will change from 10 triangular facets to 5 rectangular facets. An Ino decahedron contains 25 edges, 10 triangular faces and 5 rectangular faces. Thus, the Ino decahedron can be thought of as a pentagonal bipyramid with an elongated waist. Of note, the icosahedral and Ino decahedral M_{13} can be regarded as two decahedral M_7 sharing a vertex, while the two Au_5 pentagons are in the staggered and eclipsed conformation respectively along the C_5 axis. Since the icosahedral M_{13} kernels have been introduced in the previous section, the following discussion will not involve them. The Marks decahedron can be derived from the regular decahedron with the 5 vertical atoms at the waist removed. So the construction of a Marks decahedron requires at least 49 atoms which is derived from the two-shell $Au_7@A_{47}$ regular decahedron.

Decahedral M_7-centered shell-by-shell kernels

The final shape of the decahedral M_7-centered shell-by-shell kernels depends on the different packing mode of the shell outside the decahedral M_7. Earlier in 2007, when the thiol monolayer-protected Au NCs was structurally solved for the first time, a decahedral Au_7 was found as the innermost core of this nanocluster [22]. The Au_7 is further caged by an Au_{42} outer shell. This Au_{42} shell can be view as a complete Au_{47} decahedron with five vertices at the waist removed, that is, a Marks decahedron. Further, each of the {111} facets of this Marks decahedral $Au_7@Au_{42}$ were capped by an Au_3 triangle, forming the entire Au_{79} kernel. This kernel can be also viewed as the multiply twinned ABCB stacking from the center to exterior shell with atom ratios 1:3:6:3. The subsequently reported $Au_{103}S_2(SR)_{41}$ was also found to exhibit a similar Au_{49} kernel inside [141]. Furthermore, the Marks decahedral M_{49} kernel was also recently found in the $Au_{130-x}Ag_x(SR)_{55}$ alloyed nanoclusters [142]. At the innermost locations, the Ag-rich sites show a decahedral M_7 with the addition of two more metal atoms for each vertex forming an M_{17} pentagram. It was further covered by an Au-rich M_{32} cage, thus forming the M_{49} Marks decahedron.

In addition to the Marks decahedron, an M_7 core can be also covered by an Ino decahedron shell. As reported by Zeng et al., the $Au_{246}(SR)_{80}$ consist of a Au_{116} Ino decahedron in the center of the kernel. In details, the Au_{116} can be divided into three shells [143]. The innermost part is an Au_7 decahedron. Then the second shell displays as an Au_{32} Ino decahedron, it can be also derived from a complete Au_{47} decahedron with a circle of Au atoms in the waist removed. A third shell of a large Ino decahedron with 77 Au atoms further covers the inner core, forming the three-shell Au_{116} Ino decahedron. Each {111} facet (totally 10 facets) of this Au_{116} Ino decahedron is capped by a Au_6 triangle and each {100} facet (totally 5 facets) of it is capped by a quasirectangular (3 × 2) Au_6, that is, a 90-gold-atom forth shell. Thus, the entire kernel of $Au_{246}(SR)_{80}$ has 206 Au atoms that can be represented as $Au_7@Au_{32}@Au_{77}@Au_{90}$. This kind of M_7-centered Ino decahedral kernels have also been observed in Ag NCs. Yan et al. reported the structural determination of $[Ag_{206}(SR)_{68}F_2Cl_2]^q$ which exhibit a $Ag_7@Ag_{32}@Ag_{77}@Ag_{90}$ shell-by-shell framework [144]. This metallic framework is quite similar to that of the Au_{206} kernel in $Au_{246}(SR)_{80}$ with only some slight twist on the outer shell. Interestingly, these two cases unambiguously reveal the different binding modes of metal-ligand interfaces in Au and Ag NCs. That is, the protection of the Au_{206} kernel require the participation of staple-like $Au_n(SR)_{n+1}$ motif, and the bridging thiolates on the surface show the μ_2 coordination mode, while the protection of Ag_{206} can be done just by the multiply coordinate ligands. Song et al. also found a two-shell Ino decahedral $Ag_7@Ag_{32}$ inner core in the $Ag_{146}(SR)_{80}Br_2$ cluster, which was surrounded by 12 addition Ag atoms at the waist, forming the Ag_{51} kernel [145]. This kernel was further caged by an $Ag_{95}S_{80}Br_2$ shell.

The truncated decahedral cores (Marks ore Ino decahedron) are more likely to form in Au NCs due to the higher surface energy of {111} facets in Au NCs and the truncation can reduce the surface energy. As a result, the shell-by-shell kernel with a regular decahedral shape is rarely found in the Au NCs. For the Ag NCs, complete decahedral kernels have been observed in the $[Ag_{136}(SR)_{64}Cl_3Ag_{0.45}]^-$ [146]. This nanocluster exhibits a two-shell $Ag_7@Ag_{47}$ with the regular decahedral framework. This inner core is further cap by 10 Ag_6 triangle on the {111} facets, capping with 30 thiolate ligands and 2 Cl^-. An addition $Ag_{22}SR_{34}Cl_1$ framework further surrounds on the equator of this kernel to form the entire nanocluster. Interestingly, in a recently reported Au-Cu alloyed nanocluster $Au_{52}Cu_{72}(SR)_{55}$ [147], a complete decahedral Au_{47} shell was found to cage the decahedral Au_5Cu_2 innermost core. Through structural analysis, the complete Au_{47} decahedron can be formed in this Au-Cu cluster is due to the shrinking of the distance between the copper atoms at the two poles in the Au_7Cu_2 innermost core, thus reducing the internal stress of the whole structure, and finally making the complete decahedron can be formed. Finally, this $Au_7Cu_2@Au_{47}$ decahedral kernel is further cover by a $Cu_{70}(SR)_{55}$ cage.

Ino decahedral M_{13}-centered shell-by-shell kernels

The configuration of the smallest M_{13} Ino decahedron has been mentioned above. When it serves as the innermost core of the nanocluster, the metal atoms will further pile up on its surface to form a multilayer Ino decahedron structure. Since the Ino decahedron can be derived from the icosahedron, as a result, the shell-by-shell Ino decahedrons will have the same atomic number with the icosahedrons with the same number of shells. Thus, a two-layer Ino decahedron has a total of 55 atoms ($M_{13}@M_{42}$) and a three-layer one will have 147 atoms ($M_{13}@M_{42}@M_{92}$). In the Au NCs, $Au_{130}(SR)_{50}$ has been found to possess the Ino decahedral Au_{13} innermost core [148]. Outside the M_{13}, a second Ino decahedral Au_{42} shell was found. Further, each of the ten {111} facets of the Au_{55} decahedron is capped with a triangular Au_3 in an HCP manner, and each of the five {100} facets is capped with an Au_4 square. Thus, the third shell has totally 50 Au atoms. 25 $Au_1(SR)_2$ staple motifs further protect this Au_{105} kernel to form the final structural framework. In a large $Ag_{374}(SR)_{113}Br_2Cl_2$ nanocluster, the inner Ino decahedral kernel in it exhibits three shells, that is, $Ag_{13}@Ag_{42}@Ag_{92}$ [146]. The core is further surrounded by a pentagonal-cylinder layer of 60 Ag atoms at the waist, forming the Ag_{207} ($Ag_{13}@Ag_{42}@Ag_{92}@Ag_{60}$) kernel in this cluster. Differently, Zhang et al. reported a $[Ag_{78}(SR)_{30}(dppp)_{10}Cl_{10}]^{4+}$ nanocluster which shows a core-shell structure comprised of a Ag_{53} kernel surrounded by an Ag_{25} discontinuous metal-organic shell [149]. The Ag_{53} kernel is composed by an Ag_{13} Ino decahedron and an Ag_{40} drum-like outer shell. Two $Ag_{10}S_{10}P_{10}Cl_5$ ring-like motifs further cap on the two poles of this kernel and five Ag_1S_2 staples were found to cover on its waist.

Other decahedron-containing kernels

In addition to the nanocluster with decahedral-M_7/M_{13}-centered kernel as the core, there are some other types of decahedron-contained nanoclusters, which have been found as individual cases. Here we will briefly introduce them.

The reported $Ag_{210}(SR)_{71}(PPh_3)_5Cl$ and $Ag_{211}(SR)_{71}(PPh_3)_6Cl$ were found to exhibit the Ino decahedral Ag_{19} innermost kernel constructing by two interspersed Ino decahedral Ag_{13} (sharing a Ag_7 decahedron) [150]. The second shell for covering the Ag_{19} is an Ino decahedral Ag_{52} and then the third shell as an Ag_{45} pentagonal cylinder surrounding the $Ag_{19}@Ag_{52}$ kernel at the waist. An outermost $Ag_{89}(SR)_{71}(PPh_3)_5Cl$ finally cage the $Ag_{19}@Ag_{52}@Ag_{45}$ kernel to form the total $Ag_{210}(SR)_{71}(PPh_3)_5Cl$ cluster (or an $Ag_{90}(SR)_{71}(PPh_3)_6Cl$ shell for $Ag_{211}(SR)_{71}(PPh_3)_6Cl$). Zhang et al. synthesized an $Ag_{48}(C{\equiv}CR)_{20}(CrO_4)_7$ nanocluster with an Ag_{23} kernel and a drum-like Ag_{25} shell [86]. The Ag_{23} kernel is composed of an Ag_{13} Ino decahedron in the middle and two Ag_5 pentagons at the two poles. More recently, Wang et al. first observed the bi-decahedral Ag_{11} kernel in the Ag_{54} and Ag_{55} nanoclusters [151]. This novel Ag_{11} kernel is formed by two Ag_7 decahedron via sharing an Ag_3 triangle.

References

[1] Y. Negishi, K. Nobusada, T. Tsukuda, Glutathione-protected gold clusters revisited: bridging the gap between gold(I)−thiolate complexes and thiolate-protected gold nanocrystals, J. Am. Chem. Soc. 127 (14) (2005) 5261–5270, https://doi.org/10.1021/ja042218h.

[2] S. Tian, Y.-Z. Li, M.-B. Li, et al., Structural isomerism in gold nanoparticles revealed by X-ray crystallography, Nat. Commun. 6 (2015) 8667, https://doi.org/10.1038/ncomms9667.

[3] M. Sugiuchi, J. Maeba, N. Okubo, M. Iwamura, K. Nozaki, K. Konishi, Aggregation-induced fluorescence-to-phosphorescence switching of molecular gold clusters, J. Am. Chem. Soc. 139 (49) (2017) 17731–17734, https://doi.org/10.1021/jacs.7b10201.

[4] S. Yang, J. Chai, Y. Song, et al., In situ two-phase ligand exchange: a new method for the synthesis of alloy nanoclusters with precise atomic structures, J. Am. Chem. Soc. 139 (16) (2017) 5668–5671, https://doi.org/10.1021/jacs.7b00668.

[5] T. Higaki, M. Zhou, K.J. Lambright, K. Kirschbaum, M.Y. Sfeir, R. Jin, Sharp transition from nonmetallic Au_{246} to metallic Au_{279} with nascent surface plasmon resonance, J. Am. Chem. Soc. 140 (17) (2018) 5691–5695, https://doi.org/10.1021/jacs.8b02487.

[6] A.C. Dharmaratne, T. Krick, A. Dass, Nanocluster size evolution studied by mass spectrometry in room temperature $Au_{25}(SR)_{18}$ synthesis, J. Am. Chem. Soc. 131 (38) (2009) 13604–13605, https://doi.org/10.1021/ja906087a.

[7] S. Wang, Y. Song, S. Jin, et al., Metal exchange method using Au25 nanoclusters as templates for alloy nanoclusters with atomic precision, J. Am. Chem. Soc. 137 (12) (2015) 4018–4021, https://doi.org/10.1021/ja511635g.

[8] T. Chen, V. Fung, Q. Yao, Z. Luo, D. Jiang, J. Xie, Synthesis of water-soluble $[Au_{25}(SR)_{18}]^-$ using a stoichiometric amount of $NaBH_4$, J. Am. Chem. Soc. 140 (36) (2018) 11370–11377, https://doi.org/10.1021/jacs.8b05689.

[9] T. Yoskamtorn, S. Yamazoe, R. Takahata, et al., Thiolate-mediated selectivity control in aerobic alcohol oxidation by porous carbon-supported Au_{25} clusters, ACS Catal. 4 (10) (2014) 3696–3700, https://doi.org/10.1021/cs501010x.

[10] A. Yahia-Ammar, D. Sierra, F. Mérola, N. Hildebrandt, X. Le Guével, Self-assembled gold nanoclusters for bright fluorescence imaging and enhanced drug delivery, ACS Nano 10 (2) (2016) 2591–2599, https://doi.org/10.1021/acsnano.5b07596.

[11] Q. Yao, X. Yuan, Y. Yu, Y. Yu, J. Xie, J.Y. Lee, Introducing amphiphilicity to noble metal nanoclusters via phase-transfer driven ion-pairing reaction, J. Am. Chem. Soc. 137 (5) (2015) 2128–2136, https://doi.org/10.1021/jacs.5b00090.

[12] N. Goswami, F. Lin, Y. Liu, D.T. Leong, J. Xie, Highly luminescent thiolated gold nanoclusters impregnated in nanogel, Chem. Mater. 28 (11) (2016) 4009–4016, https://doi.org/10.1021/acs.chemmater.6b01431.

[13] L. Liao, S. Zhou, Y. Dai, et al., Mono-mercury doping of Au_{25} and the HOMO/LUMO energies evaluation employing differential pulse voltammetry, J. Am. Chem. Soc. 137 (30) (2015) 9511–9514, https://doi.org/10.1021/jacs.5b03483.

[14] T.-H. Chiu, J.-H. Liao, F. Gam, et al., Homoleptic platinum/silver superatoms protected by dithiolates: linear assemblies of two and three centered icosahedra isolobal to Ne_2 and I_3, J. Am. Chem. Soc. 141 (33) (2019) 12957–12961, https://doi.org/10.1021/jacs.9b05000.

[15] Z. Liu, M. Zhu, X. Meng, G. Xu, R. Jin, Electron transfer between $[Au_{25}(SC_2H_4Ph)_{18}]^-$ TOA+ and oxoammonium cations, J. Phys. Chem. Lett. 2 (17) (2011) 2104–2109, https://doi.org/10.1021/jz200925h.

[16] T.-A.D. Nguyen, Z.R. Jones, B.R. Goldsmith, et al., A Cu_{25} nanocluster with partial Cu(0) character, J. Am. Chem. Soc. 137 (41) (2015) 13319–13324, https://doi.org/10.1021/jacs.5b07574.

[17] M. Zhu, C.M. Aikens, M.P. Hendrich, et al., Reversible switching of magnetism in thiolate-protected Au_{25} superatoms, J. Am. Chem. Soc. 131 (7) (2009) 2490–2492, https://doi.org/10.1021/ja809157f.

[18] Y. Yun, H. Sheng, K. Bao, et al., Design and remarkable efficiency of the robust sandwich cluster composite nanocatalysts ZIF-8@Au_{25}@ZIF-67, J. Am. Chem. Soc. 142 (9) (2020) 4126–4130, https://doi.org/10.1021/jacs.0c00378.

[19] J.T. Petty, O.O. Sergev, M. Ganguly, I.J. Rankine, D.M. Chevrier, P. Zhang, A segregated, partially oxidized, and compact Ag_{10} cluster within an encapsulating DNA host, J. Am. Chem. Soc. 138 (10) (2016) 3469–3477, https://doi.org/10.1021/jacs.5b13124.

[20] I. Chakraborty, T. Pradeep, Atomically precise clusters of noble metals: emerging link between atoms and nanoparticles, Chem. Rev. 117 (12) (2017) 8208–8271, https://doi.org/10.1021/acs.chemrev.6b00769.

[21] R. Jin, C. Zeng, M. Zhou, Y. Chen, Atomically precise colloidal metal nanoclusters and nanoparticles: fundamentals and opportunities, Chem. Rev. 116 (18) (2016) 10346–10413, https://doi.org/10.1021/acs.chemrev.5b00703.

[22] P.D. Jadzinsky, G. Calero, C.J. Ackerson, D.A. Bushnell, R.D. Kornberg, Structure of a thiol monolayer-protected gold nanoparticle at 1.1 Å resolution, Science 318 (5849) (2007) 430–433, https://doi.org/10.1126/science.1148624.

[23] M.W. Heaven, A. Dass, P.S. White, K.M. Holt, R.W. Murray, Crystal structure of the gold nanoparticle $[N(C_8H_{17})_4][Au_{25}(SCH_2CH_2Ph)_{18}]$, J. Am. Chem. Soc. 130 (12) (2008) 3754–3755, https://doi.org/10.1021/ja800561b.

[24] M. Zhu, C.M. Aikens, F.J. Hollander, G.C. Schatz, R. Jin, Correlating the crystal structure of a thiol-protected Au_{25} cluster and optical properties, J. Am. Chem. Soc. 130 (18) (2008) 5883–5885, https://doi.org/10.1021/ja801173r.

[25] A. Desireddy, B.E. Conn, J. Guo, et al., Ultrastable silver nanoparticles, Nature 501 (2013) 399–402, https://doi.org/10.1038/nature12523.

[26] H. Qian, W.T. Eckenhoff, Y. Zhu, T. Pintauer, R. Jin, Total structure determination of thiolate-protected Au_{38} nanoparticles, J. Am. Chem. Soc. 132 (24) (2010) 8280–8281, https://doi.org/10.1021/ja103592z.

[27] Y. Kamei, Y. Shichibu, K. Konishi, Generation of small gold clusters with unique geometries through cluster-to-cluster transformations: octanuclear clusters with edge-sharing gold tetrahedron motifs, Angew. Chem. Int. Ed. 50 (32) (2011) 7442–7445, https://doi.org/10.1002/anie.201102901.

[28] Y. Shichibu, M. Zhang, Y. Kamei, K. Konishi, $[Au_7]3+$: a missing link in the four-electron gold cluster family, J. Am. Chem. Soc. 136 (37) (2014) 12892–12895, https://doi.org/10.1021/ja508005x.

[29] B.K. Teo, X. Shi, H. Zhang, Pure gold cluster of 1:9:9:1:9:9:1 layered structure: a novel 39-metal-atom cluster $[(Ph_3P)_{14}Au_{39}Cl_6]Cl_2$ with an interstitial gold atom in a hexagonal antiprismatic cage, J. Am. Chem. Soc. 114 (7) (1992) 2743–2745, https://doi.org/10.1021/ja00033a073.

[30] Z. Lei, J.-J. Li, X.-K. Wan, W.-H. Zhang, Q.-M. Wang, Isolation and total structure determination of an all-alkynyl-protected gold nanocluster Au_{144}, Angew. Chem. Int. Ed. 57 (28) (2018) 8639–8643, https://doi.org/10.1002/anie.201804481.

[31] P. Maity, H. Tsunoyama, M. Yamauchi, S. Xie, T. Tsukuda, Organogold clusters protected by phenylacetylene, J. Am. Chem. Soc. 133 (50) (2011) 20123–20125.

[32] X.-K. Wan, X.-L. Cheng, Q. Tang, et al., Atomically precise bimetallic $Au_{19}Cu_{30}$ nanocluster with an icosidodecahedral Cu_{30} shell and an alkynyl–Cu interface, J. Am. Chem. Soc. 139 (28) (2017) 9451–9454, https://doi.org/10.1021/jacs.7b04622.

[33] X.-K. Wan, Z.-J. Guan, Q.-M. Wang, Homoleptic alkynyl-protected gold nanoclusters: $Au_{44}(PhC{\equiv}C)_{28}$ and $Au_{36}(PhC{\equiv}C)_{24}$, Angew. Chem. Int. Ed. 56 (38) (2017) 11494–11497, https://doi.org/10.1002/anie.201706021.

[34] W. Kurashige, S. Yamazoe, K. Kanehira, T. Tsukuda, Y. Negishi, Selenolate-protected Au_{38} nanoclusters: isolation and structural characterization, J. Phys. Chem. Lett. 4 (18) (2013) 3181–3185, https://doi.org/10.1021/jz401770y.

[35] Y. Negishi, W. Kurashige, U. Kamimura, Isolation and structural characterization of an octaneselenolate-protected Au_{25} cluster, Langmuir 27 (20) (2011) 12289–12292, https://doi.org/10.1021/la203301p.

[36] Y. Song, F. Fu, J. Zhang, et al., The magic Au_{60} nanocluster: a new cluster-assembled material with five Au_{13} building blocks, Angew. Chem, Int. Ed. 54 (29) (2015) 8430–8434, https://doi.org/10.1002/anie.201501830.

[37] Y. Song, S. Wang, J. Zhang, et al., Crystal structure of selenolate-protected $Au_{24}(SeR)_{20}$ Nanocluster, J. Am. Chem. Soc. 136 (8) (2014) 2963–2965, https://doi.org/10.1021/ja4131142.

[38] Y. Jiang, Y. Yu, X. Zhang, et al., N-heterocyclic carbene-stabilized ultrasmall gold nanoclusters in a metal-organic framework for photocatalytic CO_2 reduction, Angew. Chem. Int. Ed. 60 (32) (2021) 17388–17393, https://doi.org/10.1002/anie.202105420.

[39] M.R. Narouz, K.M. Osten, P.J. Unsworth, et al., N-heterocyclic carbene-functionalized magic-number gold nanoclusters, Nat. Chem. 11 (5) (2019) 419–425, https://doi.org/10.1038/s41557-019-0246-5.

[40] M.R. Narouz, S. Takano, P.A. Lummis, et al., Robust, highly luminescent Au_{13} superatoms protected by N-heterocyclic carbenes, J. Am. Chem. Soc. 141 (38) (2019) 14997–15002, https://doi.org/10.1021/jacs.9b07854.

[41] H. Shen, G. Deng, S. Kaappa, et al., Highly robust but surface-active: an N-heterocyclic carbene-stabilized Au_{25} nanocluster, Angew. Chem. Int. Ed. 58 (49) (2019) 17731–17735, https://doi.org/10.1002/anie.201908983.

[42] C. Zeng, C. Liu, Y. Chen, N.L. Rosi, R. Jin, Atomic structure of self-assembled monolayer of thiolates on a tetragonal Au_{92} nanocrystal, J. Am. Chem. Soc. 138 (28) (2016) 8710–8713, https://doi.org/10.1021/jacs.6b04835.

[43] N.A. Sakthivel, S. Theivendran, V. Ganeshraj, A.G. Oliver, A. Dass, Crystal structure of faradaurate-279: $Au_{279}(SPh-tBu)_{84}$ plasmonic nanocrystal molecules, J. Am. Chem. Soc. 139 (43) (2017) 15450–15459, https://doi.org/10.1021/jacs.7b08651.

[44] C. Zeng, C. Liu, Y. Chen, N.L. Rosi, R. Jin, Gold–thiolate ring as a protecting motif in the $Au_{20}(SR)_{16}$ nanocluster and implications, J. Am. Chem. Soc. 136 (34) (2014) 11922–11925, https://doi.org/10.1021/ja506802n.

[45] Q. Li, B. Huang, S. Yang, et al., Unraveling the nucleation process from a Au(I)-SR complex to transition-size nanoclusters, J. Am. Chem. Soc. 143 (37) (2021) 15224–15232, https://doi.org/10.1021/jacs.1c06354.

[46] S. Chen, L. Xiong, S. Wang, et al., Total structure determination of $Au_{21}(S-Adm)_{15}$ and geometrical/electronic structure evolution of thiolated gold nanoclusters, J. Am. Chem. Soc. 138 (34) (2016) 10754–10757, https://doi.org/10.1021/jacs.6b06004.

[47] Q. Li, S. Yang, T. Chen, et al., Structure determination of a metastable Au$_{22}$(SAdm)$_{16}$ nanocluster and its spontaneous transformation into Au$_{21}$(SAdm)$_{15}$, Nanoscale 12 (46) (2020) 23694–23699, https://doi.org/10.1039/d0nr07124b.

[48] Y. Li, M.J. Cowan, M. Zhou, et al., Atom-by-atom evolution of the same ligand-protected Au$_{21}$, Au$_{22}$, Au$_{22}$Cd$_1$, and Au$_{24}$ nanocluster series, J. Am. Chem. Soc. 142 (48) (2020) 20426–20433, https://doi.org/10.1021/jacs.0c09110.

[49] C.P. Joshi, M.S. Bootharaju, M.J. Alhilaly, O.M. Bakr, [Ag$_{25}$(SR)$_{18}$]$^-$: the "golden" silver nanoparticle, J. Am. Chem. Soc. 137 (36) (2015) 11578–11581, https://doi.org/10.1021/jacs.5b07088.

[50] S. Lee, M.S. Bootharaju, G. Deng, et al., [Cu$_{32}$(PET)$_{24}$H$_8$Cl$_2$](PPh$_4$)$_2$: a copper hydride nanocluster with a bisquare antiprismatic core, J. Am. Chem. Soc. 142 (32) (2020) 13974–13981, https://doi.org/10.1021/jacs.0c06577.

[51] Q. Li, T.-Y. Luo, M.G. Taylor, et al., Molecular "surgery" on a 23-gold-atom nanoparticle, Sci. Adv. 3 (5) (2017), e1603193, https://doi.org/10.1126/sciadv.1603193.

[52] J. Chen, Q.-F. Zhang, T.A. Bonaccorso, P.G. Williard, L.-S. Wang, Controlling gold nanoclusters by diphospine ligands, J. Am. Chem. Soc. 136 (1) (2014) 92–95, https://doi.org/10.1021/ja411061e.

[53] X. Zuo, S. He, X. Kang, et al., New atomically precise M$_1$Ag$_{21}$ (M = Au/Ag) nanoclusters as excellent oxygen reduction reaction catalysts, Chem. Sci. 12 (10) (2021) 3660–3667, https://doi.org/10.1039/d0sc05923d.

[54] L.G. AbdulHalim, M.S. Bootharaju, Q. Tang, et al., Ag$_{29}$(BDT)$_{12}$(TPP)$_4$: a tetravalent nanocluster, J. Am. Chem. Soc. 137 (37) (2015) 11970–11975, https://doi.org/10.1021/jacs.5b04547.

[55] X. Kang, S. Wang, M. Zhu, Observation of a new type of aggregation-induced emission in nanoclusters, Chem. Sci. 9 (11) (2018) 3062–3068, https://doi.org/10.1039/c7sc05317g.

[56] H. Shen, Z. Xu, M.S.A. Hazer, et al., Surface coordination of multiple ligands endows N-heterocyclic carbene-stabilized gold nanoclusters with high robustness and surface reactivity, Angew. Chem. Int. Ed. 60 (7) (2021) 3752–3758, https://doi.org/10.1002/anie.202013718.

[57] B.K. Najafabadi, J.F. Corrigan, N-heterocyclic carbene stabilized Ag–P nanoclusters, Chem. Commun. 51 (4) (2015) 665–667, https://doi.org/10.1039/c4cc06560c.

[58] H. Shen, L. Wang, O. López-Estrada, et al., Copper-hydride nanoclusters with enhanced stability by N-heterocyclic carbenes, *Nano Res.* 14 (9) (2021) 3303–3308, https://doi.org/10.1007/s12274-021-3389-9.

[59] Y. Shichibu, Y. Negishi, T. Watanabe, N.K. Chaki, H. Kawaguchi, T. Tsukuda, Biicosahedral gold clusters [Au$_{25}$(PPh$_3$)$_{10}$(SC$_n$H$_{2n+1}$)$_5$Cl$_2$]$^{2+}$ (n = 2 − 18): a stepping stone to cluster-assembled materials, J. Phys. Chem. C 111 (22) (2007) 7845–7847, https://doi.org/10.1021/jp073101t.

[60] D. Crasto, S. Malola, G. Brosofsky, A. Dass, H. Häkkinen, Single crystal XRD structure and theoretical analysis of the chiral Au$_{30}$S(S-t-Bu)$_{18}$ Cluster, J. Am. Chem. Soc. 136 (13) (2014) 5000–5005, https://doi.org/10.1021/ja412141j.

[61] H. Yang, Y. Wang, A.J. Edwards, J. Yan, N. Zheng, High-yield synthesis and crystal structure of a green Au$_{30}$ cluster co-capped by thiolate and sulfide, Chem. Commun. 50 (92) (2014) 14325–14327, https://doi.org/10.1039/c4cc01773k.

[62] C. Liu, T. Li, G. Li, et al., Observation of body-centered cubic gold nanocluster, Angew. Chem. Int. Ed. 54 (34) (2015) 9826–9829, https://doi.org/10.1002/anie.201502607.

[63] S. Kenzler, C. Schrenk, A. Schnepf, Au$_{108}$S$_{24}$(PPh$_3$)$_{16}$: a highly symmetric nanoscale gold cluster confirms the general concept of metalloid clusters, Angew. Chem. Int. Ed. 56 (1) (2017) 393–396, https://doi.org/10.1002/anie.201609000.

[64] S. Kenzler, C. Schrenk, A.R. Frojd, H. Häkkinen, A.Z. Clayborne, A. Schnepf, Au$_{70}$S$_{20}$(PPh$_3$)$_{12}$: an intermediate sized metalloid gold cluster stabilized by the Au$_4$S$_4$ ring motif and Au-PPh$_3$ groups, Chem. Commun. 54 (3) (2018) 248–251, https://doi.org/10.1039/c7cc08014j.

[65] Z. Gan, J. Chen, J. Wang, et al., The fourth crystallographic closest packing unveiled in the gold nanocluster crystal, Nat. Commun. 8 (2017) 14739, https://doi.org/10.1038/ncomms14739.

[66] G. Li, Z. Lei, Q.-M. Wang, Luminescent molecular Ag−S nanocluster [Ag$_{62}$S$_{13}$(SBut)$_{32}$](BF$_4$)$_4$, J. Am. Chem. Soc. 132 (50) (2010) 17678–17679, https://doi.org/10.1021/ja108684m.

[67] X. Wei, H. Shen, C. Xu, et al., Ag$_{48}$ and Ag$_{50}$ nanoclusters: toward active-site tailoring of nanocluster surface structures, Inorg. Chem. 60 (8) (2021) 5931–5936, https://doi.org/10.1021/acs.inorgchem.1c00355.

[68] S. Takano, H. Hirai, S. Muramatsu, T. Tsukuda, Hydride-doped gold superatom (Au$_9$H)$^{2+}$: synthesis, structure, and transformation, J. Am. Chem. Soc. 140 (27) (2018) 8380–8383, https://doi.org/10.1021/jacs.8b03880.

[69] S. Takano, H. Hirai, S. Muramatsu, T. Tsukuda, Hydride-mediated controlled growth of a bimetallic (Pd@Au$_8$)$^{2+}$ superatom to a hydride-doped (HPd@Au$_{10}$)$^{3+}$ superatom, J. Am. Chem. Soc. 140 (39) (2018) 12314–12317, https://doi.org/10.1021/jacs.8b06783.

[70] S.-F. Yuan, J.-J. Li, Z.-J. Guan, Z. Lei, Q.-M. Wang, Ultrastable hydrido gold nanoclusters with the protection of phosphines, Chem. Commun. 56 (51) (2020) 7037–7040, https://doi.org/10.1039/d0cc02339f.

[71] J. Dong, Z.-H. Gao, Q.-F. Zhang, L.-S. Wang, The synthesis, bonding, and transformation of a ligand-protected gold nanohydride cluster, Angew. Chem. Int. Ed. 60 (5) (2021) 2424–2430, https://doi.org/10.1002/anie.202011748.

[72] A.W. Cook, T.-A.D. Nguyen, W.R. Buratto, G. Wu, T.W. Hayton, Synthesis, characterization, and reactivity of the group 11 hydrido clusters [Ag$_6$H$_4$(dppm)$_4$(OAc)$_2$] and [Cu$_3$H(dppm)$_3$(OAc)$_2$], Inorg. Chem. 55 (23) (2016) 12435–12440, https://doi.org/10.1021/acs.inorgchem.6b02385.

[73] C.W. Liu, Y.-R. Lin, C.-S. Fang, C. Latouche, S. Kahlal, J.-Y. Saillard, [Ag$_7$(H){E$_2$P(OR)$_2$}$_6$] (E = Se, S): precursors for the fabrication of silver nanoparticles, Inorg. Chem. 52 (4) (2013) 2070–2077, https://doi.org/10.1021/ic302482p.

[74] X. Yuan, C. Sun, X. Li, et al., Combinatorial identification of hydrides in a ligated Ag$_{40}$ nanocluster with noncompact metal core, J. Am. Chem. Soc. 141 (30) (2019) 11905–11911, https://doi.org/10.1021/jacs.9b03009.

[75] S.K. Barik, S.-C. Huo, C.-Y. Wu, et al., Polyhydrido copper nanoclusters with a hollow icosahedral core: [Cu$_{30}$H$_{18}${E$_2$P(OR)$_2$}$_{12}$] (E=S or Se; R=nPr, iPr or iBu), Chem. Eur. J. 26 (46) (2020) 10471–10479, https://doi.org/10.1002/chem.202001449.

[76] R.D. Köhn, Z. Pan, M.F. Mahon, G. Kociok-Köhn, Trimethyltriazacyclohexane as bridging ligand for triangular Cu_3 units and C–H hydride abstraction into a Cu_6 cluster, Chem. Commun. 11 (2003) 1272–1273, https://doi.org/10.1039/b302670a.

[77] S.-F. Yuan, C.-Q. Xu, J. Li, Q.-M. Wang, A ligand-protected golden fullerene: the dipyridylamido Au_{328}^+ nanocluster, Angew. Chem. Int. Ed. 58 (18) (2019) 5906–5909, https://doi.org/10.1002/anie.201901478.

[78] S.-F. Yuan, Z.-J. Guan, W.-D. Liu, Q.-M. Wang, Solvent-triggered reversible interconversion of all-nitrogen-donor-protected silver nanoclusters and their responsive optical properties, Nat. Commun. 10 (2019) 4032, https://doi.org/10.1038/s41467-019-11988-y.

[79] S.-F. Yuan, Z. Lei, Z.-J. Guan, Q.-M. Wang, Atomically precise preorganization of open metal sites on gold nanoclusters with high catalytic performance, Angew. Chem. Int. Ed. 60 (10) (2021) 5225–5229, https://doi.org/10.1002/anie.202012499.

[80] K.-G. Liu, X.-M. Gao, T. Liu, M.-L. Hu, D. Jiang, All-carboxylate-protected superatomic silver nanocluster with an unprecedented rhombohedral Ag_8 core, J. Am. Chem. Soc. 142 (40) (2020) 16905–16909, https://doi.org/10.1021/jacs.0c06682.

[81] R.-W. Huang, Y.-S. Wei, X.-Y. Dong, et al., Hypersensitive dual-function luminescence switching of a silver-chalcogenolate cluster-based metal–organic framework, Nat. Chem. 9 (7) (2017) 689–697, https://doi.org/10.1038/nchem.2718.

[82] Z. Wang, H.-T. Sun, M. Kurmoo, et al., Carboxylic acid stimulated silver shell isomerism in a triple core–shell Ag_{84} nanocluster, Chem. Sci. 10 (18) (2019) 4862–4867, https://doi.org/10.1039/c8sc05666h.

[83] W.-D. Liu, J.-Q. Wang, S.-F. Yuan, X. Chen, Q.-M. Wang, Chiral superatomic nanoclusters Ag_{47} induced by the ligation of amino acids, Angew. Chem. Int. Ed. 60 (20) (2021) 11430–11435, https://doi.org/10.1002/anie.202100972.

[84] Z.-J. Guan, J.-L. Zeng, Z.-A. Nan, X.-K. Wan, Y.-M. Lin, Q.-M. Wang, Thiacalix[4]arene: new protection for metal nanoclusters, Sci. Adv. 2 (8) (2016), e1600323, https://doi.org/10.1126/sciadv.1600323.

[85] Z.-J. Guan, F. Hu, S.-F. Yuan, Z.-A. Nan, Y.-M. Lin, Q.-M. Wang, The stability enhancement factor beyond eight-electron shell closure in thiacalix[4]arene-protected silver clusters, Chem. Sci. 10 (11) (2019) 3360–3365, https://doi.org/10.1039/c8sc03756f.

[86] S.-S. Zhang, F. Alkan, H.-F. Su, C.M. Aikens, C.-H. Tung, D. Sun, $[Ag_{48}(C{\equiv}CtBu)_{20}(CrO_4)_7]$: an atomically precise silver nanocluster co-protected by inorganic and organic ligands, J. Am. Chem. Soc. 141 (10) (2019) 4460–4467, https://doi.org/10.1021/jacs.9b00703.

[87] X.-M. Luo, C.-H. Gong, X.-Y. Dong, L. Zhang, S.-Q. Zang, Evolution of all-carboxylate-protected superatomic Ag clusters confined in Ti-organic cages, *Nano Res.* 14 (7) (2020) 2309–2313, https://doi.org/10.1007/s12274-020-3227-5.

[88] J. Chai, S. Yang, Y. Lv, et al., A unique pair: Ag_{40} and Ag_{46} nanoclusters with the same surface but different cores for structure–property correlation, J. Am. Chem. Soc. 140 (46) (2018) 15582–15585, https://doi.org/10.1021/jacs.8b09162.

[89] C. Zeng, H. Qian, T. Li, et al., Total structure and electronic properties of the gold nanocrystal $Au_{36}(SR)_{24}$, Angew. Chem. Int. Ed. 51 (52) (2012) 13114–13118, https://doi.org/10.1002/anie.201207098.

[90] C. Zeng, Y. Chen, K. Iida, et al., Gold quantum boxes: on the periodicities and the quantum confinement in the Au_{28}, Au_{36}, Au_{44}, and Au_{52} magic series, J. Am. Chem. Soc. 138 (12) (2016) 3950–3953, https://doi.org/10.1021/jacs.5b12747.

[91] C. Zeng, Y. Chen, C. Liu, K. Nobusada, N.L. Rosi, R. Jin, Gold tetrahedra coil up: kekulé-like and double helical superstructures, Sci. Adv. 1 (9) (2015), e1500425, https://doi.org/10.1126/sciadv.1500425.

[92] C. Zeng, T. Li, A. Das, N.L. Rosi, R. Jin, Chiral structure of thiolate-protected 28-gold-atom nanocluster determined by X-ray crystallography, J. Am. Chem. Soc. 135 (27) (2013) 10011–10013, https://doi.org/10.1021/ja404058q.

[93] L. Liao, C. Wang, S. Zhuang, et al., An unprecedented kernel growth mode and layer-number-odevity-dependent properties in gold nanoclusters, Angew. Chem. Int. Ed. 59 (2) (2020) 731–734, https://doi.org/10.1002/anie.201912090.

[94] A. Das, T. Li, K. Nobusada, C. Zeng, N.L. Rosi, R. Jin, Nonsuperatomic $[Au_{23}(SC_6H_{11})_{16}]^-$ nanocluster featuring bipyramidal Au_{15} kernel and trimeric $Au_3(SR)_4$ motif, J. Am. Chem. Soc. 135 (49) (2013) 18264–18267, https://doi.org/10.1021/ja409177s.

[95] C. Liu, T. Li, H. Abroshan, et al., Chiral Ag_{23} nanocluster with open shell electronic structure and helical face-centered cubic framework, Nat. Commun. 9 (2018) 744, https://doi.org/10.1038/s41467-018-03136-9.

[96] H. Yang, J. Lei, B. Wu, et al., Crystal structure of a luminescent thiolated Ag nanocluster with an octahedral Ag_6^{4+} core, Chem. Commun. 49 (3) (2013) 300–302, https://doi.org/10.1039/c2cc37347e.

[97] H. Yang, J. Yan, Y. Wang, et al., Embryonic growth of face-center-cubic silver nanoclusters shaped in nearly perfect half-cubes and cubes, J. Am. Chem. Soc. 139 (1) (2017) 31–34, https://doi.org/10.1021/jacs.6b10053.

[98] B.K. Teo, H. Yang, J. Yan, N. Zheng, Supercubes, supersquares, and superrods of face-centered cubes (FCC): atomic and electronic requirements of $[M_m(SR)l(PR'_3)_8]^q$ nanoclusters (M = coinage metals) and their implications with respect to nucleation and growth of FCC metals, Inorg. Chem. 56 (19) (2017) 11470–11479, https://doi.org/10.1021/acs.inorgchem.7b00427.

[99] R.-W. Huang, J. Yin, C. Dong, et al., $[Cu_{23}(PhSe)_{16}(Ph_3P)_8(H)_6] \cdot BF_4$: atomic-level insights into cuboidal polyhydrido copper nanoclusters and their quasi-simple cubic self-assembly, ACS Mater. Lett. 3 (1) (2021) 90–99, https://doi.org/10.1021/acsmaterialslett.0c00513.

[100] M.J. Alhilaly, M.S. Bootharaju, C.P. Joshi, et al., $[Ag_{67}(SPhMe_2)_{32}(PPh_3)_8]^{3+}$: synthesis, total structure, and optical properties of a large box-shaped silver nanocluster, J. Am. Chem. Soc. 138 (44) (2016) 14727–14732, https://doi.org/10.1021/jacs.6b09007.

[101] M. Bodiuzzaman, A. Ghosh, K.S. Sugi, et al., Camouflaging structural diversity: co-crystallization of two different nanoparticles having different cores but the same shell, Angew. Chem. Int. Ed. 58 (1) (2019) 189–194, https://doi.org/10.1002/anie.201809469.

[102] Y. Lv, X. Ma, J. Chai, H. Yu, M. Zhu, Face-centered-cubic Ag nanoclusters: origins and consequences of the high structural regularity elucidated by density functional theory calculations, Chem. Eur. J. 25 (61) (2019) 13977–13986, https://doi.org/10.1002/chem.201903183.

[103] X. Ma, Y. Bai, Y. Song, et al., Rhombicuboctahedral Ag_{100}: four-layered octahedral silver nanocluster adopting the Russian nesting doll model, Angew. Chem. Int. Ed. 59 (39) (2020) 17234–17238, https://doi.org/10.1002/anie.202006447.

[104] S. Chen, S. Wang, J. Zhong, et al., The structure and optical properties of the [Au$_{18}$(SR)$_{14}$] nanocluster, Angew. Chem. Int. Ed. 54 (10) (2015) 3145–3149, https://doi.org/10.1002/anie.201410295.

[105] A. Das, C. Liu, H.Y. Byun, et al., Structure determination of [Au$_{18}$(SR)$_{14}$], Angew. Chem. Int. Ed. 54 (10) (2015) 3140–3144, https://doi.org/10.1002/anie.201410161.

[106] T. Higaki, C. Liu, C. Zeng, et al., Controlling the atomic structure of Au$_{30}$ nanoclusters by a ligand-based strategy, Angew. Chem. Int. Ed. 55 (23) (2016) 6694–6697, https://doi.org/10.1002/anie.201601947.

[107] M. Diecke, C. Schrenk, A. Schnepf, Synthesis and characterization of the highly unstable metalloid cluster Ag$_{64}$(PnBu$_3$)$_{16}$C$_{16}$, Angew. Chem. Int. Ed. 59 (34) (2020) 14418–14422, https://doi.org/10.1002/anie.202006454.

[108] M. Qu, F.-Q. Zhang, D.-H. Wang, H. Li, J.-J. Hou, X.-M. Zhang, Observation of non-FCC copper in alkynyl-protected Cu$_{53}$ nanoclusters, Angew. Chem. Int. Ed. 59 (16) (2020) 6507–6512, https://doi.org/10.1002/anie.202001185.

[109] Q. Li, S. Wang, K. Kirschbaum, K.J. Lambright, A. Das, R. Jin, Heavily doped Au$_{25-x}$Ag$_x$(SC$_6$H$_{11}$)$_{18}^-$ nanoclusters: silver goes from the core to the surface, Chem. Commun. 52 (29) (2016) 5194–5197, https://doi.org/10.1039/c6cc01243d.

[110] C.E. Briant, B.R.C. Theobald, J.W. White, L.K. Bell, D.M.P. Mingos, A.J. Welch, Synthesis and X-ray structural characterization of the centred icosahedral gold cluster compound [Au$_{13}$(PMe$_2$Ph)$_{10}$Cl$_2$](PF$_6$)$_3$; the realization of a theoretical prediction, J. Chem. Soc. Chem. Commun. 5 (1981) 201–202, https://doi.org/10.1039/c39810000201.

[111] M. Walter, J. Akola, O. Lopez-Acevedo, et al., A unified view of ligand-protected gold clusters as superatom complexes, Proc. Natl. Acad. Sci. U. S. A. 105 (27) (2008) 9157–9162, https://doi.org/10.1073/pnas.0801001105.

[112] R.S. Dhayal, J.-H. Liao, Y.-C. Liu, et al., [Ag$_{21}${S$_2$P(OiPr)$_2$}$_{12}$]$^+$: an eight-electron superatom, Angew. Chem. Int. Ed. 54 (12) (2015) 3702–3706, https://doi.org/10.1002/anie.201410332.

[113] T.-A.D. Nguyen, Z.R. Jones, D.F. Leto, G. Wu, S.L. Scott, T.W. Hayton, Ligand-exchange-induced growth of an atomically precise Cu$_{29}$ nanocluster from a smaller cluster, Chem. Mater. 28 (22) (2016) 8385–8390, https://doi.org/10.1021/acs.chemmater.6b03879.

[114] B.K. Teo, K. Keating, Novel triicosahedral structure of the largest metal alloy cluster: hexachlorododecakis(triphenylphosphine)-gold-silver cluster [(Ph$_3$P)$_{12}$Au$_{13}$Ag$_{12}$Cl$_6$]$^{m+}$, J. Am. Chem. Soc. 106 (7) (1984) 2224–2226, https://doi.org/10.1021/ja00319a061.

[115] B.K. Teo, M.C. Hong, H. Zhang, D.B. Huang, Cluster of clusters: structure of the 37-atom cluster [(p-Tol3P)$_{12}$Au$_{18}$Ag$_{19}$Br$_{11}$]$_2$⊕ and a novel series of supraclusters based on vertex-sharing icosahedra, Angew. Chem. Int. Ed. 26 (9) (1987) 897–900, https://doi.org/10.1002/anie.198708971.

[116] B.K. Teo, H. Zhang, X. Shi, Cluster of clusters: a modular approach to large metal clusters. structural characterization of a 38-atom cluster [(p-Tol3P)$_{12}$Au$_{18}$Ag$_{20}$Cl$_{14}$] based on vertex-sharing triicosahedra, J. Am. Chem. Soc. 112 (23) (1990) 8552–8562, https://doi.org/10.1021/ja00179a046.

[117] S. Wang, X. Meng, A. Das, et al., A 200-fold quantum yield boost in the photoluminescence of silver-doped Ag$_x$Au$_{25-x}$ nanoclusters: the 13th silver atom matters, Angew. Chem. Int. Ed. 53 (9) (2014) 2376–2380, https://doi.org/10.1002/anie.201307480.

[118] S. Yang, J. Chai, T. Chen, et al., Crystal structures of two new gold–copper bimetallic nanoclusters: Cu$_x$Au$_{25-x}$(PPh$_3$)$_{10}$(PhC$_2$H$_4$S)$_5$Cl$_{22}^+$ and Cu$_3$Au$_{34}$(PPh$_3$)$_{13}$(tBuPhCH$_2$S)$_6$S$_{23}^+$, Inorg. Chem. 56 (4) (2017) 1771–1774, https://doi.org/10.1021/acs.inorgchem.6b02016.

[119] X. Kang, L. Xiong, S. Wang, Y. Pei, M. Zhu, De-assembly of assembled Pt$_1$Ag$_{12}$ units: tailoring the photoluminescence of atomically precise nanoclusters, Chem. Commun. 53 (93) (2017) 12564–12567, https://doi.org/10.1039/c7cc05996e.

[120] X. Kang, J. Xiang, Y. Lv, et al., Synthesis and structure of self-assembled Pd$_2$Au$_{23}$(PPh$_3$)$_{10}$Br$_7$ nanocluster: exploiting factors that promote assembly of icosahedral nano-building-blocks, Chem. Mater. 29 (16) (2017) 6856–6862, https://doi.org/10.1021/acs.chemmater.7b02015.

[121] S. Yang, J. Chai, Y. Lv, et al., Cyclic Pt$_3$Ag$_{33}$ and Pt$_3$Au$_{12}$Ag$_{21}$ nanoclusters with M$_{13}$ icosahedra as building-blocks, Chem. Commun. 54 (85) (2018) 12077–12080, https://doi.org/10.1039/c8cc06900j.

[122] T. Li, Q. Li, S. Yang, et al., Surface engineering of linearly fused Au$_{13}$ units using diphosphine and Cd doping, Chem. Commun. 57 (38) (2021) 4682–4685, https://doi.org/10.1039/d1cc00577d.

[123] R. Jin, C. Liu, S. Zhao, et al., Tri-icosahedral gold nanocluster [Au$_{37}$(PPh$_3$)$_{10}$(SC$_2$H$_4$Ph)$_{10}$X$_2$]$^+$: linear assembly of icosahedral building blocks, ACS Nano 9 (8) (2015) 8530–8536, https://doi.org/10.1021/acsnano.5b03524.

[124] S.-F. Yuan, C.-Q. Xu, W.-D. Liu, J.-X. Zhang, J. Li, Q.-M. Wang, Rod-shaped silver supercluster unveiling strong electron coupling between substituent icosahedral units, J. Am. Chem. Soc. 143 (31) (2021) 12261–12267, https://doi.org/10.1021/jacs.1c05283.

[125] X. Zou, S. Jin, W. Du, et al., Multi-ligand-directed synthesis of chiral silver nanoclusters, Nanoscale 9 (43) (2017) 16800–16805, https://doi.org/10.1039/c7nr06338e.

[126] Y. Liu, S. Wang, X. Kang, et al., Heterogeneous metal alloy engineering: embryonic growth of M$_{13}$ icosahedra in Ag-based alloy superatomic nanoclusters, Chem. Commun. 56 (91) (2020) 14203–14206, https://doi.org/10.1039/d0cc05575a.

[127] Y. Li, S. Li, A.V. Nagarajan, et al., Hydrogen evolution electrocatalyst design: turning inert gold into active catalyst by atomically precise nanochemistry, J. Am. Chem. Soc. 143 (29) (2021) 11102–11108, https://doi.org/10.1021/jacs.1c04606.

[128] S. Jin, M. Zhou, X. Kang, et al., Three-dimensional octameric assembly of icosahedral M$_{13}$ Units in [Au$_8$Ag$_{57}$(Dppp)$_4$(C$_6$H$_{11}$S)$_{32}$Cl$_2$]Cl and its [Au$_8$Ag$_{55}$(Dppp)$_4$(C$_6$H$_{11}$S)$_{34}$][BPh$_4$]$_2$ derivativethree-dimensional octameric assembly of icosahedral M$_{13}$ units in [Au$_8$Ag$_{57}$(Dppp)$_4$(C$_6$H$_{11}$S)$_{32}$Cl$_2$]Cl and its [Au$_8$Ag$_{55}$(Dppp)$_4$(C$_6$H$_{11}$S)$_{34}$][BPh$_4$]$_2$ derivative, Angew. Chem. Int. Ed. 59 (10) (2020) 3891–3895, https://doi.org/10.1002/anie.201914350.

[129] A. Dass, S. Theivendran, P.R. Nimmala, et al., Au$_{133}$(SPh-tBu)$_{52}$ Nanomolecules: X-ray crystallography, optical, electrochemical, and theoretical analysis, J. Am. Chem. Soc. 137 (14) (2015) 4610–4613, https://doi.org/10.1021/ja513152h.

[130] C. Zeng, Y. Chen, K. Kirschbaum, K. Appavoo, M.Y. Sfeir, R. Jin, Structural patterns at all scales in a nonmetallic chiral Au$_{133}$(SR)$_{52}$ nanoparticle, Sci. Adv. 1 (2) (2015), e1500045, https://doi.org/10.1126/sciadv.1500045.

[131] F. Hu, J.-J. Li, Z.-J. Guan, S.-F. Yuan, Q.-M. Wang, Formation of an alkynyl-protected Ag_{112} silver nanocluster as promoted by chloride released in situ from CH_2Cl_2, Angew. Chem. Int. Ed. 59 (13) (2020) 5312–5315, https://doi.org/10.1002/anie.201915168.

[132] J. Yan, S. Malola, C. Hu, et al., Co-crystallization of atomically precise metal nanoparticles driven by magic atomic and electronic shells, Nat. Commun. 9 (2018) 3357, https://doi.org/10.1038/s41467-018-05584-9.

[133] X. Zou, Y. Li, S. Jin, et al., Doping copper atoms into the nanocluster kernel: total structure determination of $[Cu_{30}Ag_{61}(SAdm)_{38}S_3](BPh_4)$, J. Phys. Chem. Lett. 11 (6) (2020) 2272–2276, https://doi.org/10.1021/acs.jpclett.0c00271.

[134] W. Du, S. Jin, L. Xiong, et al., $Ag_{50}(Dppm)_6(SR)_{30}$ and its homologue $Au_xAg_{50-x}(Dppm)_6(SR)_{30}$ alloy nanocluster: seeded growth, structure determination, and differences in properties, J. Am. Chem. Soc. 139 (4) (2017) 1618–1624, https://doi.org/10.1021/jacs.6b11681.

[135] S. Kenzler, F. Fetzer, C. Schrenk, et al., Synthesis and characterization of three multi-shell metalloid gold clusters $Au_{32}(R_3P)_{12}Cl_8$, Angew. Chem. Int. Ed. 58 (18) (2019) 5902–5905, https://doi.org/10.1002/anie.201900644.

[136] N. Yan, N. Xia, L. Liao, et al., Unraveling the long-pursued Au_{144} structure by X-ray crystallography, Sci. Adv. 4 (10) (2018), https://doi.org/10.1126/sciadv.aat7259. eaat7259.

[137] X. Yuan, S. Malola, G. Deng, et al., Atomically precise alkynyl- and halide-protected AuAg nanoclusters $Au_{78}Ag_{66}(C\equiv CPh)_{48}Cl_8$ and $Au_{74}Ag_{60}(C\equiv CPh)_{40}Br_{12}$: the ligation effects of halides, Inorg. Chem. 60 (6) (2021) 3529–3533, https://doi.org/10.1021/acs.inorgchem.0c03462.

[138] L. Ren, P. Yuan, H. Su, et al., Bulky surface ligands promote surface reactivities of $[Ag_{141}X_{12}(S\text{-}Adm)_{40}]_3^+$ (X = Cl, Br, I) nanoclusters: models for multiple-twinned nanoparticles, J. Am. Chem. Soc. 139 (38) (2017) 13288–13291, https://doi.org/10.1021/jacs.7b07926.

[139] L. Liao, S. Zhuang, C. Yao, et al., Structure of chiral $Au_{44}(2,4\text{-}DMBT)_{26}$ nanocluster with an 18-electron shell closure, J. Am. Chem. Soc. 138 (33) (2016) 10425–10428, https://doi.org/10.1021/jacs.6b07178.

[140] Y. Zhao, S. Zhuang, L. Liao, et al., A dual purpose strategy to endow gold nanoclusters with both catalysis activity and water solubility, J. Am. Chem. Soc. 142 (2) (2020) 973–977, https://doi.org/10.1021/jacs.9b11017.

[141] T. Higaki, C. Liu, M. Zhou, T.-Y. Luo, N.L. Rosi, R. Jin, Tailoring the structure of 58-electron gold nanoclusters: $Au_{103}S_2(S\text{-}Nap)_{41}$ and its implications, J. Am. Chem. Soc. 139 (29) (2017) 9994–10001, https://doi.org/10.1021/jacs.7b04678.

[142] T. Higaki, C. Liu, D.J. Morris, et al., $Au_{130-x}Ag_x$ nanoclusters with non-metallicity: a drum of silver-rich sites enclosed in a marks-decahedral cage of gold-rich sites, Angew. Chem. Int. Ed. 58 (52) (2019) 18798–18802, https://doi.org/10.1002/anie.201908694.

[143] C. Zeng, Y. Chen, K. Kirschbaum, K.J. Lambright, R. Jin, Emergence of hierarchical structural complexities in nanoparticles and their assembly, Science 354 (6319) (2016) 1580–1584, https://doi.org/10.1126/science.aak9750.

[144] J. Yan, J. Zhang, X. Chen, et al., Thiol-stabilized atomically precise, superatomic silver nanoparticles for catalysing cycloisomerization of alkynyl amines, Natl. Sci. Rev. 5 (5) (2018) 694–702, https://doi.org/10.1093/nsr/nwy034.

[145] Y. Song, K. Lambright, M. Zhou, et al., Large-scale synthesis, crystal structure, and optical properties of the $Ag_{146}Br_2(SR)_{80}$ nanocluster, ACS Nano 12 (9) (2018) 9318–9325, https://doi.org/10.1021/acsnano.8b04233.

[146] H. Yang, Y. Wang, X. Chen, et al., Plasmonic twinned silver nanoparticles with molecular precision, Nat. Commun. 7 (2016) 12809, https://doi.org/10.1038/ncomms12809.

[147] Y. Song, Y. Li, H. Li, et al., Atomically resolved $Au_{52}Cu_{72}(SR)_{55}$ nanoalloy reveals marks decahedron truncation and penrose tiling surface, Nat. Commun. 11 (1) (2020) 478, https://doi.org/10.1038/s41467-020-14400-2.

[148] Y. Chen, C. Zeng, C. Liu, et al., Crystal structure of barrel-shaped chiral $Au_{130}(p\text{-}MBT)_{50}$ nanocluster, J. Am. Chem. Soc. 137 (32) (2015) 10076–10079, https://doi.org/10.1021/jacs.5b05378.

[149] W.-J. Zhang, Z. Liu, K.-P. Song, et al., A 34-electron superatom Ag_{78} cluster with regioselective ternary ligands shells and its 2D rhombic superlattice assembly, Angew. Chem. Int. Ed. 60 (8) (2021) 4231–4237, https://doi.org/10.1002/anie.202013681.

[150] J.-Y. Liu, F. Alkan, Z. Wang, et al., Different silver nanoparticles in one crystal: $Ag_{210}(iPrPhS)_{71}(Ph_3P)_5Cl$ and $Ag_{211}(iPrPhS)_{71}(Ph_3P)_6Cl$, Angew. Chem. Int. Ed. 58 (1) (2019) 195–199, https://doi.org/10.1002/anie.201810772.

[151] Z. Wang, H.-F. Su, G.-L. Zhuang, et al., Carbonate–water supramolecule trapped in silver nanoclusters encapsulating unprecedented Ag_{11} Kernel, CCS Chem. 2 (1) (2020) 663–672, https://doi.org/10.31635/ccschem.019.201900058.

Chapter 4

Mechanism of size conversion and structure evolution of metal nanoclusters

Manzhou Zhu and Haizhu Yu

4.1 Concept of size-conversion and structural evolution

The ultrasmall metal nanoclusters (<2nm) comprising several to hundreds of metal atoms are in medium size between the metal complexes and nanoparticles. Their precise atomic structures (elucidated by single crystal X-ray diffraction, mass spectra, density functional theory calculations, etc.) have open opportunities to unravel the inherent correlations of different sized nanoclusters in the quantum size regime. Specifically, the recent size-conversion (including size-growth, size-maintained conversion, and size-reduction) of the atomically precise metal nanoclusters has contributed to the preliminary understanding on dynamics of metal clusters under certain circumstances (such as changes in solvent, pH, temperature, etc.), providing unprecedent understanding on the structure-size-evolution/conversion activity correlations at molecular level. Herein, it is noteworthy that the concepts of size-evolution and size-conversion are not strictly defined in the reported literatures. But for clarity reasons, in this section, the size-conversion is specifically used to depict the chemical reactions in which one cluster changes to another one, under the specific conditions (such as heating or changes in solvents). By contrast, size-evolution refers to the structural and property correlations between a series of structurally related clusters. In other words, the size-evolution and conversion could be understood from the kinetic and static correlations between different clusters, respectively.

4.2 Mechanism of size-conversion

The preliminary mechanistic understanding on how the monomers or oligomers nucleate into nanoclusters/particles could date back to 1950s, when Lamer and co-workers first reported the growth mechanism of colloidal particles. On the basis of the pioneering understanding on the formation mechanism of metal nanoclusters (and particles), varies strategies, such as the ligand engineering, temperature control, etc., have recently developed to improve the synthetic methods, with a primary goal to control the size-distribution. With the continuous progress in synthetic methods (such as the modified Brust synthesis, ligand-exchange, metal exchange method, etc.), a variety of metal nanoclusters with controlled size and morphology has been prepared with high purity and mono-dispersity. Specifically, the atomic precision of these ultrasmall metal nanoclusters (size <2nm, with several to a few hundreds of metal atoms) could be precisely determined by multiple strategies, which is fundamentally important to elucidate the mechanism of the size-conversion processes at molecular level.

Due to the limited space, the following sections mainly focus on the recent achievements in mechanistic studies relating to the group IB metal nanoclusters protected by thiolates or mixed ligands of thiolates and phosphine, chlorides, etc. For clarity, the general and molecular-level mechanistic investigations, depending on whether the formed clusters are atomically precise or not, were given separately. Meanwhile, this chapter mainly refers the molecular-level mechanism, and thus only selected models for the general mechanism of size-growth are include in the following Section 4.2.1.

4.2.1 General understanding on the size-growth of metal nanoclusters

4.2.1.1 Lamer growth

Lamer model (also known as instantaneous/burst nucleation, and "diffusion-controlled" growth) represents a typical, and the most cited qualitative theory to explain how the monodispersed colloidal particles and nanoparticles were formed by a

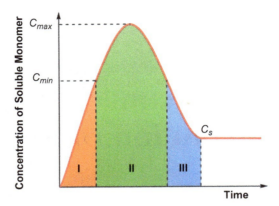

FIG. 4.1 Illustrative diagram for the classic Lamer model.

dilution method. This model was first reported by Lamer and co-workers in 1950 [1] focusing on the mechanism for formation of the mono-dispersed sulfur hydrosols. But in recent decades, this model has been extensively applied to deduce the formation of different types of nanomaterials, such as quantum dots, semiconduction quantum wells, and metal nanoparticles.

In a typical Lamer model, the size growth from the soluble monomers is divided into three stages, and Fig. 4.1 shows qualitatively the variation of monomer concentration versus time. During stage I, the concentration of the soluble monomer builds up until reaches saturation (at the concentration of C_s), and then to the supersaturated state at C_{min} (a critical concentration when nucleation starts to occur). In stage II, an instantaneous nucleation (i.e., formation of nuclei from the monomers, and C_{max} is the hypothetical upper limit of the concentration of monomers) partially relieves the supersaturation. In stage III, the diffusion induced size-growth of particles proceeds when the monomers are remains supersaturated (with concentration in the between of C_s and C_{min}). In this context, the concentration of the monomers represents a balance between the rate of size-growth via monomer addition (i.e., consumption of the excessive monomers) and the rate of sinks/precipitation. Herein, the rapid self-nucleation in stage II corresponds to an instantaneous homogeneous, "burst" nucleation, and this is also the reason why this model was referred to as the instantaneous/burst nucleation in some other cases. Meanwhile, the entire Lamer growth are driven by the diffusion state of the monomers, and therefore "diffusion-controlled" growth was also used in some literatures. According to Lamer theory, the burst nucleation in stage II occurs in a narrow time interval, the growth rates of early nuclei are similar, and therefore the particles are produced with narrow size distribution.

Despite the high academic value of Lamer model in explaining the size-growth of colloidal nanoparticles, the potential interference of dust, the possibility of heterogenous nucleation (size-growth of nanoparticles by interacting with different agents), and agglomeration were omitted in the classic theory. Meanwhile, there is still a lack of proof experimental evidences for the mechanism of size-growth from the monomer to the particles to date. Hopefully, the time-dependent spectral (with UV-vis, MS, NMR, etc.) monitoring technique and skills could be helpful to solve these problems. The interested authors are suggested to refer to works by Richard G. Finke et al. [2] for more detailed comments on the origins, assumptions, and recent progress of the Lamer model.

4.2.1.2 Aggregative growth

Distinct from the nucleation-growth (with monomers) type of Lamer growth, the most critical species in aggregative growth are the smaller primary nanocrystal formed in the early stage of the synthesis. In 2010, Buhro and co-workers tracked the changes in nanocrystal size distributions (with TEM tests) during the thermal coarsening of the decanethiolate-capped Au nanocrystals in presence of various concentration of tetraoctylammonium bromide (TBABr) at 180°C [3]. Similar to the Lamer growth theory, the aggregative growth also follows a typical nucleation-growth process, and the overall size-growth was divided into three stages. Nevertheless, in aggregative growth mechanism, smaller, or primary nanocrystals was suggested to be formed in the first stage (Fig. 4.2). This process corresponds to the primary nucleation process, and was proposed to be rapid. The smaller-sized nanocrystals were also known as the critical-sized aggregates, which play a crucial role in the overall size-growth. In stage II, the thermodynamically unstable, smaller-sized nanocrystals aggregate to form relatively larger crystals. This step is also known as the second nucleation process. Due to the slow rate, the formed larger-sized crystals may possibly undergo the reversible reaction (i.e., decomposition) to produce the monomer or smaller sized nanocrystals. Finally, size-growth occurs on the remaining larger-sized

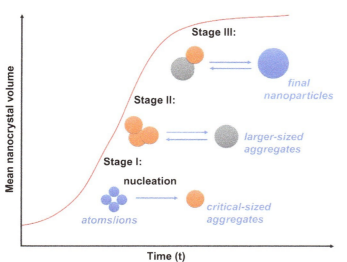

FIG. 4.2 Illustrative diagram for the aggregative growth model.

nanoparticles and the primary nanocrystals to produce the supercritical nanoparticles, until the consumption of all the primary nanocrystals. In this context, the overall size-growth is controlled by the aggregative-nucleation process. Quantitative description of aggregative growth was deduced, via fitting the kinetic data with the Gaussian profile. The height of the profile is the maximum nucleation rate, while the 2σ width is the time window for nucleation.

According to the aforementioned descriptions, the main difference of the Lamer growth and the aggregative growth model lies in the dominating factor: the concentration of the soluble **monomers** is pivotal in Lamer model, while the **smaller-sized aggregates** dominate the aggregative growth.

4.2.1.3 Mechanism of the Brust synthesis

Since Brust and co-workers reported the controllable synthesis of 1–3 nm thiolated gold particles via the two-phase (water-to-toluene) reduction of $AuCl_4^-$ by $NaBH_4$ and in the presence of alkanethiol in 1994 [4], Brust synthesis (also known as the Brust-Schiffrin two-phase synthesis) has been extensively explored (and not limited to the biphasic synthesis) to prepare chalcogen protected group IB metal nanoclusters with size-control. Albeit the promising synthetic advantages, the mechanism of Brust synthesis has been under debate for long time.

One typical mechanism for Brust synthesis includes three stages [5,6]: phase-transfer of a hydrophilic gold salt (commonly $HAuCl_4$) with the aid of the phase-transfer agents (such as TOAB: tetrabutylammonium bromide, Eq. 4.1); reduction of Au(III) to Au(I) with thiolate (Eq. 4.2); and the reduction of Au(I) to Au(0) with extra-reductant (e.g., $NaBH_4$, Eq. 4.3)

$$HAuCl_4 \text{ (aq)} + TOAB \text{ (os)} \rightarrow [AuX_4][TOA] \text{ (os)} + HX \text{ (aq)} \tag{4.1}$$

$$[AuX_4][TOA] \text{ (os)} + HSR \rightarrow [Au(I)SR]_n \text{ complexes (os)} \tag{4.2}$$

$$[Au(I)SR]_n \text{ (os)} + NaBH_4 \text{ (aq)} \rightarrow Au_n(SR)_m \tag{4.3}$$

where, aq and os denote aqueous and organic solvents, and X = Br/Cl.

In contrast to the formation of $[Au(I)SR]_n$ complexes, another typical mechanism proposed for Brust synthesis involves a reduction induced metal nucleation (formation of metal particles) prior to the formation of metal-chalcogen bonds. In the pioneering study on the two-phase (e.g., water: toluene) reaction system of TOAB, $HAuCl_4$ with thiols (dodecanethiol/phenylethanethiol) reported by Lennox and co-workers [7], $[TOA]^+[M(I)X_2]^-$ (M = Au, Ag, Cu; X = Br, Cl) were found the key metal precursors according to ^1H NMR monitoring analysis. Interestingly, thiolates are necessary for the formation of tetraalkylammonium metal complexes in gold and copper systems (Eqs. 4.4, 4.5), while $[TOA][AgBr_2]$ is formed directly via the co-dissolving TOAB and $AgNO_3$. Compared with the intermediacy of $[M\text{-}SR]_n$ polymers, $[TOA][MX_2]$ are proposed the precursors for reaction systems in absence of polar solvent (such as water, methanol, ethanol, etc.). In other words, $[TOA][MX_2]$ are dominant precursors in the one-phase reaction systems in nonpolar solvents, while $[M\text{-}SR]_n$ are dominating metal precursors for the two-phase Brust synthesis when larger quantities of polar solvents (or $NaBH_4$) are present, or one-phase synthesis in polar solvent.

$$[TOA]^+[AuX_4]^- + 2\,RSH \rightarrow [TOA]^+[AuX_2]^- + RS-SR + 2\,HX \tag{4.4}$$

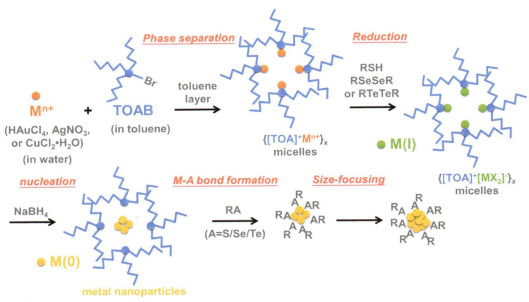

FIG. 4.3 Illustrative diagram for the mechanism of the Brust synthesis via a prior nucleation step to the M—A bond formation steps [8].

$$[TOA]^+[CuX_4]^-[TOAB] + RSH \rightarrow [TOA]^+[CuX_2]^- + 1/2\ RS-SR + TOAB + HX \quad (4.5)$$

Carrying out a similar $NaBH_4$-absent reaction by mixing $HAuCl_4$, TOAB, and dodecanethiol, Lennox and co-workers [8] further verify the formation of $[TOA]^-[M(I)X_2]^+$ (M=Au, Ag, Cu) precursors by the Raman spectral characterizations. In addition, to figure out the details for the formation of metal-ligand bond, a reversed synthetic route by addition of $NaBH_4$ earlier than thiols was devised, and explored with the detailed Raman spectra, ^1H NMR, and surface plasmon resonance (SPR) spectroscopic characterizations. Due to the consistency between the reverse protocol and the typical Brust synthesis, the two systems were suggested to occur via the same mechanism. Briefly, the overall size-growth occurs via the following steps (see illustrative diagram in Fig. 4.3): formation of the tetraalkylammonium metal (M^{n+}, oxidation state) complexes in organic phase, the reduction of M^{n+} to M(I) to form the key intermediate of $[TOA]^-[M(I)X_2]^+$ (by chalcogen ligand in most cases), formation of metal nucleation centers (metal nanoparticles) via $NaBH_4$ reduction, M-A (A=S/Se/Te) bond formation via the ligand etching processes, and the final size-focusing process to generate the mono-dispersed metal nanoparticles with size-control. According to this mechanism, regulating the stirring time before the addition of thiol was predicted an efficient strategy to control the size of the target metal particles (longer stirring step in inverse protocols induces larger-sized particles).

4.2.2 Molecular-level mechanism of size-conversion of metal nanoclusters

With the great progress in the spectral monitoring skills of metal nanoclusters in recent years, the mechanism, and especially the dynamics of conversion of group IB metal nanoclusters have been widely explored. Herein, it is noteworthy that various size-conversions, including the size-growth, size-maintained transformation (such as the generalized isomerization with the unchanged number and type of metal atoms and ligands, and alloying processes with maintained number of metal atoms), and size-reduction reactions have all been reported lately. The preliminary mechanistic investigation on these systems provides important insights into the structure-activity correlations of the cluster systems, and especially the structural response of the target clusters to certain environmental changes. Therefore, the following text refers to the mechanistic study of all these types of size-conversions.

4.2.2.1 From polydisperse nanoparticles to monodisperse gold nanoclusters

The conversion of polydisperse Au nanoparticles (1–3.5 nm, prepared by the two-phase reduction of $HAuCl_4$ with $NaBH_4$ in the presence of PPh_3) to the monodisperse, rod-like, bi-icosahedral Au_{25} clusters via the one-phase thiol etching has recently been reported by Jin and co-workers [9]. Associated with the black crystals of $[Au_{25}(PPh_3)_{10}(PET)_5Cl_2]^{2+}$ (verified by SCXRD and MALDI-MS, and MALDI-MS is short for matrix-assisted laser desorption ionization mass-spectrometry), some orange crystals have been separated and determined to be $[(PPh_3)_2Au_2(PET)]^+$ by SCXRD and

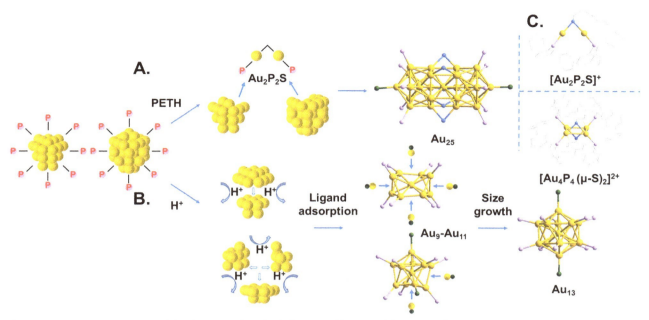

FIG. 4.4 The proposed mechanism of the size-conversion of polydisperse gold nanoparticles to monodisperse Au_{25} (A, via thiolate etching method) or Au_{13} clusters (B, via HCl etching method), and the single crystal structure of $[(PPh_3)_2Au_2(PET)]^+$ (**$[Au_2P_2S]^+$**) and $[(PPh_3)_4Au_4(\mu\text{-}SC_6H_4CH_3)_2]^{2+}$ (**$[Au_4P_4(\mu\text{-}S)_2]^+$**) (C).

ESI-MS characterization [10]. In view of the rapid formation of the Au_2 complexes (dominant signal of Au_2 on ESI-MS after reacting for 1 h), the size-conversion was suggested to be driven by the strong Au—S bonding affinity. An illustrative diagram for the mechanism of size-conversion is shown in Fig. 4.4A: the thiol etching of surface atoms on nanoparticles occurs first, and then the by-product $[Au_2(PPh_3)_2(SC_2H_4Ph)]^+$ is released and a thermodynamically controlled size-focusing proceeds to produce the rod-like Au_{25} (i.e., thiol etching-size-focusing mechanism). Herein, it is noteworthy that the structure of $[Au_2(PPh_3)_2(SC_2H_4Ph)]^+$ looks like half that of the tetranuclear structure of $[(PPh_3)_4Au_4(\mu\text{-}SC_6H_4CH_3)_2]^{2+}$ [11] (prepared by the mild oxidation of $Ph_3PAu(SC_6H_4CH_3)$ with $(Cp_2Fe)PF_6$), but each of them is thermodynamically stable under solution and sold states (unlikely to convert to each other spontaneously). The results indicate the dominating effect of the type of thiolate ligands (aliphatic or aromatic thiol) on the nucleation of the gold clusters, and more efforts are clearly necessary to elucidate the origin of such ligand effect.

Distinct from the aforementioned thiol etching initiated size-conversion, the HCl-etching induced size-conversion of polydisperse Au_n clusters (n ranges from 15 to 65, prepared via reducing $Au_2(dppp)_2Cl_2$ with $NaBH_4$ [12]) to a monodisperse Au_{13} cluster (i.e., $Au_{13}(dppp)_4Cl_4$, dppp is short for 1,3-bis(diphenylphosphino)propane) was suggested to occur via a fragmentation-first mechanism (in accordance with the top-down synthesis) [13]. According to the time-dependent UV-vis, XANES (X-ray absorption near-edge structure), and MALDI-MS tracking on the reaction system, the smaller-sized Au_8-Au_{13} clusters (associated with the formation of Au_{30}-Au_{40} mixtures) appears after about 1 h. The abundance of Au_{13} then increases with the consumption of Au_{30}-Au_{40} and Au_{11}-Au_{12} clusters successively. After about 10 h, Au_{13} was left as the sole product. In this context, the overall size-conversion were deduced to occur via a two-stage mechanism (Fig. 4.4B). In the first stage (first 2 h), decomposition of the polydisperse clusters occurs to generate the intermediate clusters of Au_8-Au_{13}, presumably initiated by the interaction of H^+ with the surface Au atoms that are not fully protected by ligands. In the second stage (2–10 h), the $Au^I Cl$ species (possibly formed in the first stage) induces the size growth of Au_8-Au_{12} intermediates into the thermodynamically more stable Au_{13} clusters (which is suggested to originate from the geometric facility of the closed, icosahedral structure in Au_{13}).

4.2.2.2 Reduction/oxidation induced size-conversion of atomically precise metal nanoclusters

As mentioned in Section 4.2.1, reduction of the oxidation state of metal salts is pivotal for size-control in the typical synthesis of metal nanoclusters. With the continuous progress on atomic precision of metal nanoclusters, pioneering progress on the mechanism of reduction induced conversion from one atomically precise metal nanocluster to another one has been reported in recent years. Interestingly, some of these reactions are reversible, and oxidation with O_2, H_2O_2 could possibly

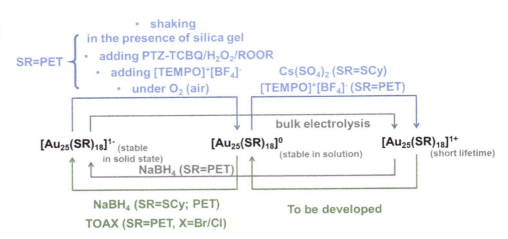

FIG. 4.5 The SET conversions between the series of $[Au_{25}(SR)_{18}]^n$ ($n = -1, 0, +1$) clusters under different redox conditions.

induce the reverse size-conversion of the reduction induced one. In addition, the redox species (reductant or oxidant) could either function as a simple electron-delivery species to induce the single-electron oxidation/reduction processes, or play a multiple role to induce a comprehensive structural change of the precursor clusters. In the following, we will briefly introduce the recent progress on both categories, and start with the redox-induced electron-transfer transformations.

Electron-transfer transformations

Among the various single electron-transfer (SET) reactions of the atomically precise metal nanoclusters, the SETs induced conversion among the series of $[Au_{25}(SR)_{18}]^n$ ($n = -1, 0, +1$) represents one of the most extensively studied topic. Refer Fig. 4.5 for a brief summary on the reported SET reactions between the series of Au_{25} clusters.

In 2007, Tsukuda and co-workers found that the $[Au_{25}(SCy)_{18}]^x$ is stable in different charge state (with $x = 1-, 0, 1+$, and HSCy denotes cyclohexanethiol) [14]. The oxidation of the neutral $[Au_{25}(SCy)_{18}]^0$ with strong oxidant of $Ce(SO_4)_2$ generates the cationic $[Au_{25}(SCy)_{18}]^{1+}$, while the reduction of the neutral state with $NaBH_4$ generates the anionic $[Au_{25}(SCy)_{18}]^{1-}$ (the formula of all these Au_{25} clusters were determined by ESI-MS characterization). In view of the same atomic composition but distinct number of free electrons (6, 7 and 8 for the cationic, neutral and anionic Au_{25} clusters), Tsukuda and co-workers suggested that the geometric stability (instead of the electronic stability such as the magic number principals [15]) is responsible for the stability of the series of Au_{25} clusters. This proposal was quickly verified by the atomic precision of $[Au_{25}(PET)_{18}]^{1-}$ [15] and the spectral characterizations on its air oxidized system in 2008 (HPET denotes 2-phenylethanethiol) [16]. According to Jin's study, the UV-vis absorption spectra of the neutral state is different from that of the anionic one mainly in the peak intensity at 400, 450, 630, and 800 nm, despite they share the same framework and atomic composition (the structural distortion mainly lies in the staples). The control experiments with bubbling N_2 into the solvent (no reaction) or cover the solution surface (significantly slower reaction rate) indicate the positive role of the solvated O_2 during oxidation. Meanwhile, the solid state of the anionic Au_{25} cluster is fairly stable (no changes occur within 2 months), but could be transforms to the neutral state when extra water was added. All these results demonstrate that the tiny amount of the solvated O_2 could stimulate the one electron oxidation of the anionic Au_{25} cluster, and the easiness of the conversion from $[Au_{25}(PET)_{18}]^{1-}$ to $[Au_{25}(PET)_{18}]^0$ under oxidation conditions. For this reason, some other oxidation conditions, including the addition of H_2O_2 [17], the donor-acceptor complex PTZ-TCBQ (phenothiazine: tetrachloro-p-benzoquinone = 1:1) [18], or by simply shaking in the presence of silica gel [19], have been developed to regulate the single electron oxidation of $[Au_{25}(PET)_{18}]^{1-}$ to $[Au_{25}(PET)_{18}]^0$ (Fig. 4.5). Meanwhile, the reverse reaction of $[Au_{25}(PET)_{18}]^0$ to $[Au_{25}(PET)_{18}]^{1-}$ could be easily regulated by the addition of reductant (e.g., $NaBH_4$), NaBr (the in situ formed $[Au_{25}(PET)_{18}]^-[Na]^+$ is instable in presence of excessive Na^+) [20], TOAX (X = Br, Cl). Specifically, on the basis of a group of control experiments and the MALDI-MS characterization on the intermediate state [20] of the reaction system of $[Au_{25}(PET)_{18}]^0$ with NaBr demonstrates the ligand exchange of thiolate(s) with bromide(s) in the early state of the reaction. In view of the one electron reduction from $[Au_{25}(PET)_{18}]^0$ to $[Au_{25}(PET)_{18}]^{1-}$ and the presence of 7 free electrons in the neutral precursor, the stoichiometry of $[Au_{25}(PET)_{18}]^0$ with the salt was proposed to be 8:7 (the conversion could be described as Eq. 4.6).

$$8\left[Au_{25}(PET)_{18}\right]^0 + 7NaBr \rightarrow 7\left[Au_{25}(PET)_{18}\right]^{1-} + 7AuBr + 18AuSR \qquad (4.6)$$

Meanwhile, with the aid of bulk electrolysis, high quality of the cationic $[Au_{25}(PET)_{18}]^{1+}$ could be formed directly from the crystallized $[Au_{25}(PET)_{18}]^{1-}$. Nevertheless, an immediate crystallization treatment is necessary to settle the instability issues of $[Au_{25}(PET)_{18}]^{+}$ [19].

In addition, a cascade one electron-one electron oxidation of $[Au_{25}(PET)_{18}]^{1-}$ to $[Au_{25}(PET)_{18}]^{0}$ and then to $[Au_{25}(PET)_{18}]^{1+}$ has been recently by Jin and co-workers, using the as-prepared $[TEMPO]^{+}[BF_4]^{-}$ (abbreviated for 2,2,6,6-tetramethylpiperidin-1-oxoammonium tetrafluoroborate) as the oxidant [21]. According to the UV-vis and EPR (electron paramagnetic resonance) monitoring, after adding an equivalent amount of $[TEMPO]^{+}$, the $[Au_{25}]^{1-}$ precursor is oxidized to its neutral form quantitatively, and $[Au_{25}]^{1+}$ becomes the final product when two or more equivalent $[TEMPO]^{+}$ is added. The other oxidized product, such as $[Au_{25}]^{2+}$, was not observed, even in presence of largely excessive $[TEMPO]^{+}$. The UV-vis spectral monitoring demonstrates a kinetic plateau after the formation of the cationic $[Au_{25}]^{+}$. This conclusion correlates with the short lifetime of the $[Au_{25}(PET)_{18}]^{2+}$ and $[Au_{25}(PET)_{18}]^{3+}$ determined by Maran and Gascón et al. [22] In their study, the conversion of the neutral $[Au_{25}(PET)_{18}]^{0}$ to its analogues with a relatively larger charge state distribution (from −2 to +3) has been accomplished under the well-designed redox conditions. Specifically, the cationic $[Au_{25}(PET)_{18}]^{1+}$ was found to be a singlet-state species, according to the combination of continuous-wave electron paramagnetic resonance (cw-EPR), NMR analysis and DFT simulation, while the species with −2, +2 and +3 charge states are fairly instable with very short lifetimes.

In another study of Maran and co-workers [23], the UV-vis absorption spectroscopy was used to monitor the oxidation of the neutral $[Au_{25}(PET)_{18}]^{0}$ or anionic $[Au_{25}(PET)_{18}]^{1-}$ upon addition of different amount of peroxide (bis(pentafluorobenzoyl)peroxide, abbreviated as $(RCOO)_2$). Compared with the monotonically varied absorption intensity of the characteristic peaks along with the addition of different amount of peroxide in the Au_{25}^{0} system (Fig. 4.6A), the gradual addition of peroxide results in a much more comprehensive changes (Fig. 4.6B) and an enhancing-weakening process of the 390 nm peak (Fig. 4.6B inset). The observation correlates with a stepwise 1e-1e oxidation of Au_{25}^{1-} (instead of a simultaneous 2e oxidation). Meanwhile, the oxidation of Au_{25}^{0} terminates after adding 1.3 equiv. peroxide, demonstrating that the Au_{25}^{1+} (formed via one electron oxidation of Au_{25}^{0}) is unlikely to undergo a further oxidation to produce Au_{25}^{2+}.

In addition to the extensively reported single electron oxidation/reduction of the popular Au_{25} clusters, the SETs of other reaction systems have also been reported. For example, the redox induced conversion between the neutral $[Au_{133}(TBBT)_{52}]^{0}$ (with one unpaired electron, the paramagnetic character is evidenced by EPR. TBBT is short for 4-tert-butylbenzenethiolate) and the cationic $[Au_{133}(TBBT)_{52}]^{+}$ (diamagnetic) has been recently reported by Jin and co-workers [24]. Addition of H_2O_2 to the CH_2Cl_2 solution of $[Au_{133}(TBBT)_{52}]^{0}$ leads to the generation of $[Au_{133}(TBBT)_{52}]^{+}$, while the reverse reaction could be regulated by adding $NaBH_4$ into the CH_2Cl_2 solution of $[Au_{133}(TBBT)_{52}]^{+}$ (Fig. 4.7A). The SCXRD analysis shows that $[Au_{133}(TBBT)_{52}]^{0}$ contains the Mackay icosahedral $Au@Au_{12}@Au_{42}$ core (Fig. 4.7B). The 16 facets of this Au_{55} core surface are each capped by three gold atoms, while the remaining four facets is each capped by only one gold atom. This four-layered $Au@Au_{12}@Au_{42}@Au_{52}$ structure is further protected by 26 AuS_2 staples. As the UV-vis absorption spectra of $[Au_{133}(TBBT)_{52}]^{+}$ is exactly the same as that of $[Au_{133}(TBBT)_{52}]^{0}$, the cationic Au_{133} cluster was suggested to adopt the same framework with that of the neutral one.

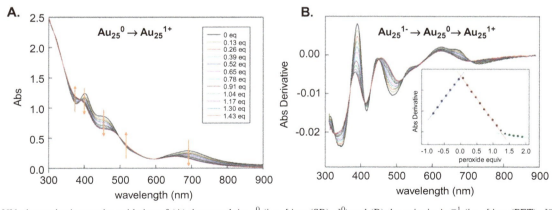

FIG. 4.6 UV-vis monitoring on the oxidation of (A) the neural Au_{25}^{0} (i.e., $[Au_{25}(SR)_{18}]^{0}$) and (B) the anionic Au_{25}^{-1} (i.e., $[Au_{25}(PET)_{18}]^{-1}$) upon the stepwise addition of peroxide, inset: the changes in the absorption intensity of the 390 nm peak along with the addition of different amount of peroxide. *(Reproduced from A. Venzo, S. Antonello, J.A. Gascón, I. Guryanov, R.D. Leapman, N.V. Perera, A. Sousa, M. Zamuner, A. Zanella, F. Maran, Effect of the charge state (z = −1, 0, +1) on the nuclear magnetic resonance of monodisperse $Au_{25}[S(CH_2)_2Ph]18^z$ clusters, Anal. Chem. 83 (2011) 6355–6362, with the permission from American Chemical Society (Copyright 2018).)*

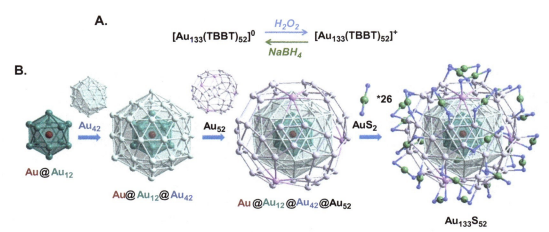

FIG. 4.7 The redox induced single electron oxidation/reduction of the pair of $[Au_{133}(TBBT)_{52}]^{0/+1}$ clusters (A), and the structural anatomy of the framework of the $Au_{133}(TBBT)_{52}$ cluster (B).

Similar to the aforementioned SET of the thiolate protected spherical Au_{25} nanoclusters, the SET oxidation of the selephenol, phosphine, and chloride co-protected, rod-like Au_{25} clusters, i.e., $[Au_{25}(SePh)_5(PPh_3)_{10}Cl_2]^{1+}$ with H_2O_2 has also been reported lately (Eq. 4.7) [25]. By contrast, the reverse reaction, i.e., the reduction of $[Au_{25}(SePh)_5(PPh_3)_{10}Cl_2]^{2+}$ to $[Au_{25}(SePh)_5(PPh_3)_{10}Cl_2]^{1+}$ is unlikely to be regulated by the addition of $NaBH_4$. Herein, the distinct optical absorption at 870 nm of the precursor and product clusters was used to determine whether the SET conversion occurs or not.

$$[Au_{25}(SePh)_5(PPh_3)_{10}Cl_2]^{1+} + H_2O_2 \rightarrow [Au_{25}(SePh)_5(PPh_3)_{10}Cl_2]^{2+} \quad (4.7)$$

In addition to the aforementioned SET among the monometallic systems, the SET conversions have also been observed in alloy metal cluster systems. In the recent study of Zhu and co-workers, the reversible SET conversion of $[Au_{18}Cu_{32}(SPh^{4-}Cl)_{36}]^{2-}$ with $[Au_{18}Cu_{32}(SPh^{4-}Cl)_{36}]^{3-}$ has been regulated by either an electrochemical or a chemo-redox strategy ($HSPh^{4-}Cl$ = 4-chlorobenzenethiol, Fig. 4.8A) [26]. The electrochemical reduction is based on the DPV analysis of $[Au_{18}Cu_{32}(SPh^{4-}Cl)_{36}]^{2-}$, using a voltage of -0.23 V (vs Ag/AgCl, correspond to the first reduction peak of the $[Au_{18}Cu_{32}(SPh^{4-}Cl)_{36}]^{2-}$ precursor). Meanwhile, the electrochemical oxidation of $[Au_{18}Cu_{32}(SPh^{4-}Cl)_{36}]^{3-}$ is accomplished using a voltage of $+0.3$ V (vs Ag/AgCl, correspond to the first oxidation peak of the $[Au_{18}Cu_{32}(SPh^{4-}Cl)_{36}]^{3-}$ precursor). The distinct absorption intensity of these two clusters lies the foundation for a UV-vis spectral monitoring, while an isosbestic point (at 585 nm, see Column I for the tips on the UV-vis spectroscopic analysis) has been observed in both the oxidation and reduction systems. Specifically, the atomic structure of both clusters has been elucidated by SCXRD analysis, which indicate the same framework but the slightly different configuration of the orientation of two groups of thiolate ligands (Fig. 4.8B). Of note, the mutual conversion between $[Au_{18}Cu_{32}(SPh^{4-}Cl)_{36}]^{2-}$ and $[Au_{18}Cu_{32}(SPh^{4-}Cl)_{36}]^{3-}$ could also be mediated by adding $NaBH_4$ to $[Au_{18}Cu_{32}(SPh^{4-}Cl)_{36}]^{2-}$ solution or adding H_2O_2 to $[Au_{18}Cu_{32}(SPh^{4-}Cl)_{36}]^{3-}$ solution.

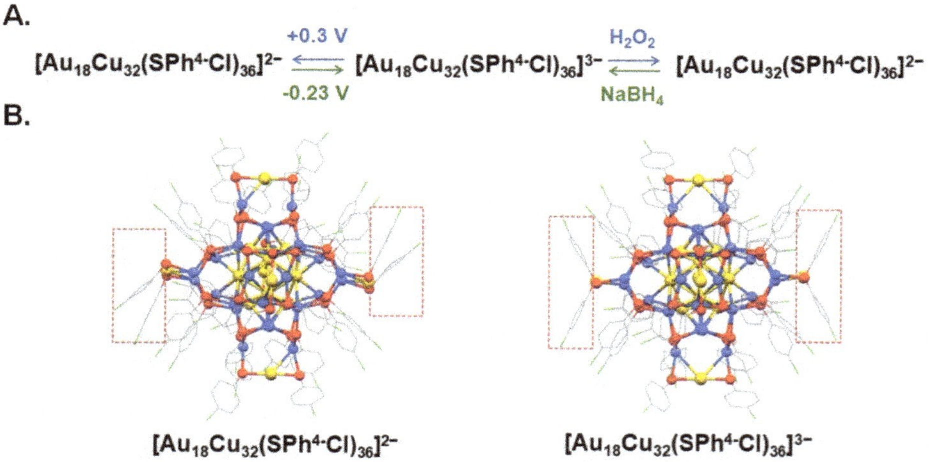

FIG. 4.8 The mutual conversion between $[Au_{18}Cu_{32}(SPh^{4-}Cl)_{36}]^{2-}$ and $[Au_{18}Cu_{32}(SPh^{4-}Cl)_{36}]^{3-}$ under electro- and chemo-redox conditions (A), and the single crystal structure of the two clusters, wherein the main structural differences are labeled with the dashed square (B). *((B) Reproduced from Q. Li, J. Chai, S. Yang, Y. Song, T. Chen, C. Chen, H. Zhang, H. Yu, M. Zhu, Multiple ways realizing charge-state transform in Au-Cu bimetallic nanoclusters with atomic precision, Small, 17 (2021) 1907114, with the permission from WILEY-VCH (Copyright 2020).)*

COLUMN I Tips on UV-vis spectroscopic analysis

Thanks to the molecular-like, discrete energy levels, the atomically precise metal nanoclusters commonly exhibit a series of distinctive absorptive peaks on the UV-vis spectra. The reported studies have frequently indicated that the optical absorption could be largely influenced by the number/type of metal atoms, charge state, and the morphology of the cluster structures. In addition, UV-vis spectrometer is inexpensive (compared with other equipment such as mass spectrometer) and easy to operate. Therefore, time-dependent UV-vis spectra have been extensively used as a powerful strategy to track the size changes during the cluster formation/conversion processes.

The optical absorbance of a certain species could be determined by Lambert-Beer Law (also called Beer's Law) shown in Eq. (4.8), wherein A, I^0, I, ε, c, and l denote the absorbance, incident intensity, transmitted intensity, molar absorption coefficient (in $M^{-1} cm^{-1}$), molar concentration (in M) and optical path length (in cm). In this context, the linear correlation of absorbance and the molar concentration (in a certain equipment) lay the foundation to determine the changes in the concentration of a concerned cluster via monitoring the changes of its characteristic absorbance in UV-vis spectra.

$$A = \lg(I^0/I) = \varepsilon c l \tag{4.8}$$

In the premise of the well-characterized UV-vis of the cluster precursor and the products, the UV-vis spectra monitoring could be used to illustrate qualitatively the consumption of one atomically precise cluster, and the formation of another one along different reaction time (Fig. 4.9A [27]) or after the addition of certain amount of additives (Fig. 4.9B [28]). Specifically, the absorption derivative could also be used, and it is especially convenient to figure out the change of shoulder peaks that is difficult to identify visually [29].

In addition to qualitative description, the UV-vis absorbance at a characteristic wavelength of the precursor could be possibly used to calculate the conversion rate constant (k), in the premise of pseudo zero/first/second order reactions (Fig. 4.10). For example, size-conversion from $Au_{23}(p\text{-}MBA)_{16}^-$ to $Au_{25}(p\text{-}MBA)_{18}^-$ was assumed a pseudo first-order reaction (p-MBA = para-mercaptobenzoic acid, Fig. 4.9A). In this context, the reaction rate constant could be deduced via fitting the time-dependent intensity of the characteristic peak of $Au_{23}(p\text{-}MBA)_{16}^-$ at 589 nm with the pseudo first-order decay profile of $OD = a \cdot e^{-kt} + b$ (OD: normalized optical density setting the initial absorbance as the reference, and b is the final absorbance at the end state of reaction) [27].

In some size-conversion reactions, one or more isosbestic point(s) could be figured out (see Fig. 4.9B as an example), wherein the optical absorbance at one/several given wavelength (or frequency/wavenumber, etc.) maintained throughout the monitoring time. According to the IUPAC gold book [30], an isosbestic point does not necessarily mean a quantitative reaction from one species to another one or an equilibrium between only two species. Instead, the observation of isosbestic point(s) implies the maintained stoichiometry of the target reaction and the absence of secondary reaction throughout the testing time. According to Lambert-Beer Law in Eq. (4.8), the optical absorbance at wavelength of λ follows the equation of $A_\lambda = \sum_{i=1}^{n} \varepsilon_{i,\lambda}[c_i]l$, in which the $\varepsilon_{i,\lambda}$ and $[c_i]$ denote the molar absorption coefficient at λ position and molar concentration of the ith species in the target reaction system. For a reaction of a A + b B → c C + d D + e E, an isosbestic point appears at every wavelength where the equation of $\varepsilon_{A,\lambda}[c_A] + \varepsilon_{B,\lambda}[c_B] = \varepsilon_{C,\lambda}[c_C] + \varepsilon_{D,\lambda}[c_D] + \varepsilon_{E,\lambda}[c_E]$ holds. In this context, the observation of the isosbestic point in Fig. 4.9A and B illustrate a maintained stoichiometry during size-conversion of target clusters, for example $Au_{23}(p\text{-}MBA)_{16}^- + 2 [Au p\text{-}MBA] \rightarrow Au_{25}(p\text{-}MBA)_{18}^-$ in Fig. 4.9A.

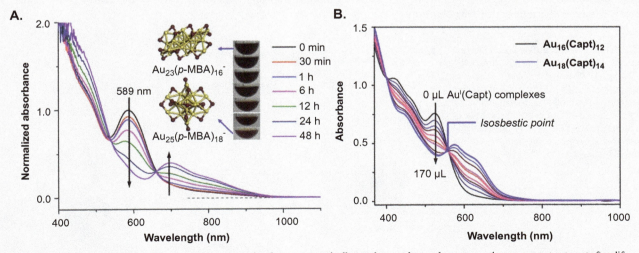

FIG. 4.9 UV-vis spectroscopic tailoring on size-conversion from one atomically precise metal nanocluster to another one upon treatment after different time (A) or addition of different amount of captopril ligand (B). *(Reproduced from Q. Yao, V. Fung, C. Sun, S. Huang, T. Chen, D.-E. Jiang, J.Y. Lee, J. Xie, Revealing isoelectronic size conversion dynamics of metal nanoclusters by a noncrystallization approach, Nat. Commun. 9 (2018) 1979, with permission that is licensed under the Creative Commons Attribution 4.0 International License, copyright the authors 2018, and W. Jiang, Y. Bai, Q. Li, X. Yao, H. Zhang, Y. Song, X. Meng, H. Yu, M. Zhu, Steric and electrostatic control of the pH-regulated interconversion of $Au_{16}(SR)_{12}$ and $Au_{18}(SR)_{14}$ (SR: deprotonated captopril), Inorg. Chem. 59 (2020) 5394–5404, with the permission from American Chemical Society (Copyright 2020).)*

Continued

COLUMN I Tips on UV-vis spectroscopic analysis—cont'd

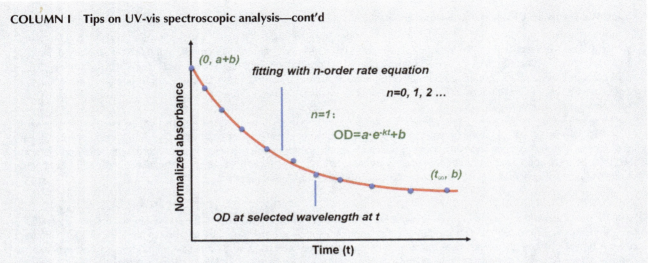

FIG. 4.10 Deduction of the rate constant k via the kinetic analysis on the size conversion from one atomically precise cluster to another one by fitting the decay profile at selected wavelength with the rate equations.

Redox induced ligand addition/removal reactions on metal clusters

In strong contrast to the aforementioned single electron oxidation/reduction conversions (with maintained framework), an intriguing redox induced single ligand addition/elimination has been recently reported in several cases. For example, the exclusion of the innermost S atom from the quantum dots of $[Ag_{62}S_{13}(StBu)_{32}]^{4+}$ (classified into quantum dots due to the absence of free electrons) to from the 4e cluster of $[Ag_{62}S_{12}(StBu)_{32}]^{2+}$ has been recently mediated by the electrochemical reduction method [31] Starting from $[Ag_{62}S_{13}(StBu)_{32}]^{4+}$ and setting the voltage in the amperometric i–t curve parameters at -0.6 V for 180 min, a reduced state, i.e., the 2e cluster of $[Ag_{62}S_{13}(StBu)_{32}]^{2+}$, is formed and separated as the intermediate species (Fig. 4.11A). After that, keeping the solution of $[Ag_{62}S_{13}(StBu)_{32}]^{2+}$ at a voltage of -1.2 V for another 180 min, the 4e cluster, i.e., $[Ag_{62}S_{12}(StBu)_{32}]^{2+}$ is formed as the final product. According to SCXRD characterization [31–33], all these three Ag_{62} species adopt the multilayer structure of $Ag_{14}@Ag_{12}S_{12}@Ag_{36}StBu_{32}$ (Fig. 4.11B), while the main difference lies in the presence (in $[Ag_{62}S_{13}(StBu)_{32}]^{4+}$ and $[Ag_{62}S_{13}(StBu)_{32}]^{2+}$) or absence of the central S atom (in $[Ag_{62}S_{12}(StBu)_{32}]^{2+}$). Meanwhile, the Ag_{14} shell is significantly shrunk with the removal of the central S atom and the increasement in the number of free electrons.

FIG. 4.11 The electrochemical reduction induced conversion of $[Ag_{62}S_{13}(StBu)_{32}]^{4+}$ to $[Ag_{62}S_{13}(StBu)_{32}]^{2+}$ and then to $[Ag_{62}S_{12}(StBu)_{32}]^{2+}$ (A) and an illustrative diagram for the framework of $[Ag_{62}S_{13}(StBu)_{32}]^{2+}$ (note: the framework of $[Ag_{62}S_{12}(StBu)_{32}]^{2+}$ is mainly differentiated from this structure in the removal of the central S atom).

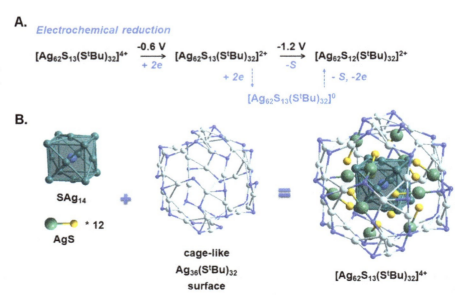

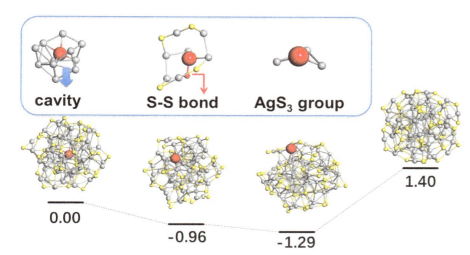

FIG. 4.12 The proposed mechanism for extruding the central S atom from the theoretically predicted intermediate of $[Ag_{62}S_{13}(StBu)_{32}]^0$, and the inset: zoom-in diagram for the selected region (reaction center) of the related intermediates. All relative energies (data below the bars) are given in eV. (Reproduced from S. Jin, S. Wang, Y. Song, M. Zhou, J. Zhong, J. Zhang, A. Xia, Y. Pei, M. Chen, P. Li, M. Zhu, Crystal structure and optical properties of the $[Ag_{62}S_{12}(SBut)_{32}]^{2+}$ nanocluster with a complete face-centered cubic kernel, J. Am. Chem. Soc. 136 (2014) 15559–15565, with the permission from American Chemical Society (Copyright 2014).)

In view of the correlations of these three Ag_{62} species in the atomic structure and the electronic state, a neutral charged $[Ag_{62}S_{13}(StBu)_{32}]^0$ with four free electrons is proposed as a plausible intermediate (formed after the 2e reduction of the $[Ag_{62}S_{13}(StBu)_{32}]^{2+}$, Fig. 4.11A). Indeed, in the predicted structure of $[Ag_{62}S_{13}(StBu)_{32}]^0$ by DFT calculations (Fig. 4.12), strong structural distortion was observed in the S@Ag_{14} kernel, resulting in a structural cave to facilitate the exclusion of the central S atom. After that, the movement of the central S atom to the $Ag_{12}S_{12}$ layer occurs to form a di-sulfur bond. The subsequent exclusion of S to the surface forms a SAg_3 group with the Ag atoms on the outmost $Ag_{36}StBu_{32}$ layer, following with the final removal of S atom. The aforementioned S migration steps from the kernel position to the outmost layer has been predicted to be thermodynamically highly feasible by DFT calculations.

Another study for the redox-induced ligand removal/addition has been recently reported by Xie and co-workers [34]. Under air oxidation and in the presence/absence of excessive MHA (i.e., 6-mercaptohexanoic acid) ligand, the pH = 9 aqueous solution of $[Au_{25}(MHA)_{18}]^-$ cluster is converted to the $[Au_{25}(MHA)_{19}]^0$ (evidenced by ESI-MS characterization) after 24 h (Fig. 4.13A). Meanwhile, the reverse conversion could be regulated by bubbling CO into the pH = 12 aqueous solution of the neutral $[Au_{25}(MHA)_{19}]^0$ for 2 min, and then proceed the reaction for 12 h airtight. In the oxidation induced ligand addition system ($[Au_{25}(MHA)_{18}]^- \rightarrow [Au_{25}(MHA)_{19}]^0$), the intermediacy of the neutral $[Au_{25}(MHA)_{18}]^0$ and the cationic $[Au_{25}(MHA)_{18}]^+$ (formed via single electron oxidation of the anionic precursors) was elucidated by the ESI-MS characterization, and the control experiments with CO/H_2O_2. Meanwhile, the two isosbestic points in the UV-vis absorption spectra at 536 and 627 nm (Fig. 4.13B) demonstrates a maintained stoichiometry throughout the transformation. In addition, the tandem mass spectrometry demonstrates the similar fragmentation pattern of the two Au_{25} clusters,

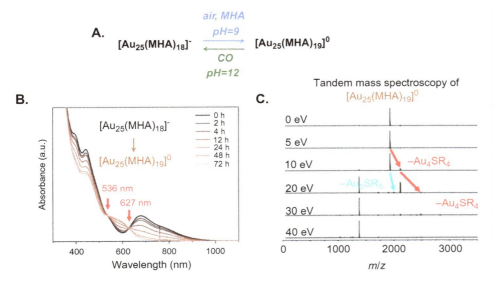

FIG. 4.13 The redox induced interconversion of $[Au_{25}(MHA)_{18}]^-$ and $[Au_{25}(MHA)_{19}]^0$ under oxidation (single ligand addition) and reduction (elimination) conditions (A), the UV-vis monitoring on the oxidation induced single-ligand addition process (B), and the tandem mass spectroscopy of $[Au_{25}(MHA)_{19}]^0$ (C). ((B and C) Reproduced from Y. Cao, V. Fung, Q. Yao, T. Chen, S. Zang, D.-E. Jiang, J. Xie, Control of single-ligand chemistry on thiolated Au_{25} nanoclusters, Nat. Commun. 11 (2020) 5498, with permission that is licensed under the Creative Commons Attribution 4.0 International License, copyright the author(s) 2020.)

as each of them dissociative two Au_4S_4 blocks successively (the tandem ESI-MS of $[Au_{25}(MHA)_{19}]^0$ was provided in Fig. 4.13C). The fragmentation peak corresponding to the dissociation of one Au_5S_5 structure has also been observed in the tandem MS of $[Au_{25}(MHA)_{19}]^0$, demonstrating the lengthened motif after the ligand exchange. This proposal is further supported by the DFT calculations on the possible structures of the $[Au_{25}(MHA)_{19}]^0$, because the isomeric structure with one Au_3S_4 motif (extruding one Au atom on the icosahedral surface) is remarkably more stable than all other examined ones.

Reduction induced 2e⁻ hopping mechanism

Due to the extra-stability, fantastic physicochemical properties (such as discrete energy levels, fluorescence, etc.), and the tolerance of the M-S framework to various types of thiolate ligands and metal components, the spherical $[M_{25}(SR)_{18}]^-$ (M = Au/Ag/Cu) clusters have attracted extensive research interest [35]. In this context, breakthrough mechanistic understanding on both the formation of $[Au_{25}(SR)_{18}]^-$ (from $Au^I(SR)$ salts) and its size-growth to other larger-sized clusters (such as $[Au_{44}(SR)_{26}]^{2-}$) under reduction circumstance has been reported. Both systems were proposed to occur via the 2 e⁻ (i.e., two-electrons) hopping mechanism, wherein the 2e reduction dominates the overall size-growth. The following are some experimental details.

Based on the general understanding on a prior formation of Au^I complexes with a subsequent reduction of Au^I to Au^0 in Brust synthesis (Eqs. 4.1–4.3), Xie and co-workers recently proposed the stoichiometry for formation of $[Au_{25}(p\text{-MBA})_{18}]^-$ from $Au^I(p\text{-MBA})$ complexes with the aid of the electrospray ionization mass spectrometry (abbreviated as ESI-MS) monitoring (see Column II for the tips on ESI-MS analysis) [37]. As shown in Eq. (4.9), $[Au^I(p\text{-MBA})]_x$ was used to depict the oligomeric Au^I reactant, which could be prepared by the reaction of p-MBA and $HAuCl_4$ (with a subsequent adjustment of pH to 11.5). As one $NaBH_4$ could provide eight electrons, the molar ratio of $NaBH_4$ to Au atom (i.e., $NaBH_4:HAuCl_4$ in synthesis) is determined to be 1:32. Specifically, a series of intermediates have been figured out by a real-time ESI-MS characterization (Fig. 4.14). All the identified intermediates carry an even number of free electrons ($N^e = 2, 4, 6, 8, 10$, see Chapter 6 for more detailed calculation of N^e), suggesting a 2 e⁻ hopping mechanism in a bottom-up size-growth pattern. The rapid appearance of the cluster peaks at about 2 min and the slow consumption processes suggested the overall size-growth to occur via a two-stage mechanism: a fast reduction-growth stage with a subsequent, slow size-focusing stage.

$$32/x \left[Au^I(p-MBA)\right]_x + 8\, e^- \rightarrow \left[Au_{25}(p-MBA)_{18}\right]^- + 7 \left[Au(p-MBA)_2\right]^- \quad (4.9)$$

Upon treating the $[Au_{25}(p\text{-MBA})_{18}]^-$ with a mild reductant (CO, via bubbling) and $Au^I(p\text{-MBA})$ complexes, the size-growth could further occur to generate a larger cluster, i.e., $[Au_{44}(p\text{-MBA})_{26}]^{2-}$. Several intermediate species has been identified via the time-dependent UV-vis spectral and ESI-MS monitoring [38]. The overall size-growth starts with the accumulation of $[Au_{25}(p\text{-MBA})_{18}]^-$ by CO reduction of $Au^I(p\text{-MBA})$ complexes (evidenced by the enhanced intensity of characteristic peaks

---

COLUMN II Tips on ESI-MS characterization

In recent years, ESI-MS has become an appealing strategy to determine the formula of the target metal clusters (commonly with molecular weight within 100,000) and to track the changes along a size-conversion process. Compared with the "hard" desorption/ionization source of laser in MALDI-MS, the electrospray is a remarkably softer ionization source. For some labile clusters (such as the metal nanoclusters bearing phosphine ligands), it might be difficult to identify the intact cluster peak in MALDI-MS, while the fragmentation peaks caused by losing one or more ligands/groups have been frequently figured out. In some cases, the intact cluster peak (without losing any ligands/groups) could be identified with the aid of ESI-MS. For example, in the recent study by Jin and co-workers [36], the intact cluster peak of $Au_{36}(TBBT)_{24}$ has been observed in ESI-MS characterization, while only the fragmentation peaks corresponding to the losing of one thiolate ligand from $Au_{36}(TBBT)_{24}$ has been identified in MALDI-MS characterization.

Due to the high possibility of ESI-MS in identifying the cluster peaks, the lifetime of a certain intermediate could be determined by ESI-MS to gain a more detailed mechanistic understanding. This analysis relies on a time-dependent ESI-MS characterization of the target reaction system, and could be achieved via either a real-time or an offline tracking mode (depending on whether the reaction could be conducted directly in the ESI-MS spectrometer or not). In a typical lifetime analysis, the peak intensity along the reaction time was first recorded, and is then normalized by setting the strongest peak intensity as the reference. Of note, for a complex system with multiple testing samples (Fig. 4.14), the peak intensity of each sample should be normalized independently. To this end, the formation and consumption of each sample along the reaction time could be tracked precisely, and the detailed transformation could be deduced, accordingly.

COLUMN II   Tips on ESI-MS characterization—cont'd

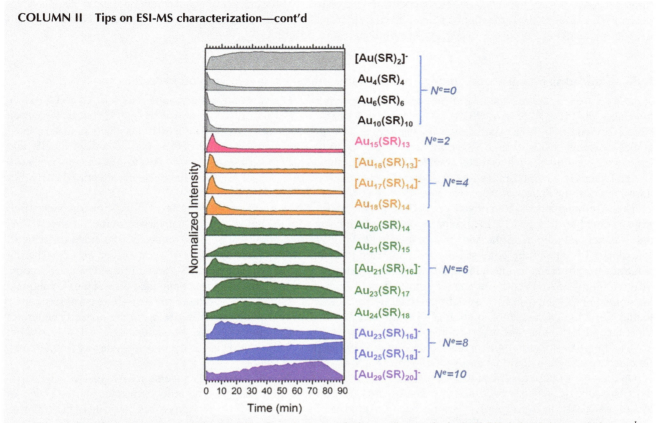

FIG. 4.14 Lifetime analysis (by time-dependent ESI-MS monitoring) on the intermediate species in the NaBH$_4$ induced size growth from Au^I (p-MBA) complexes to [Au$_{25}$(p-MBA)$_{18}$]$^-$ (SR = p-MBA, NaBH$_4$:HAuCl$_4$ = 1:32). *(Reproduced from T. Chen, V. Fung, Q. Yao, Z. Luo, D.-E. Jiang, J. Xie, Synthesis of water-soluble [Au$_{25}$(SR)$_{18}$]$^-$ using a stoichiometric amount of NaBH$_4$, J. Am. Chem. Soc. 140 (2018) 11370–11377, with permission from the ACS Publications (Copyright 2018).)*

of [Au$_{25}$(p-MBA)$_{18}$]$^-$ on UV-vis and ESI-MS spectra), and this process is denoted as the "pregrowth stage". This stage occurs immediately after bubbling CO into the reaction mixture, and lasts for about 1 h. Thereafter, stepwise size-growth occurs, and ESI-MS indicate the co-existence of a monotonic size-growth (Fig. 4.15A) and a volcano-shaped growth (Fig. 4.15B) pattern. As the intermediate clusters feature the increasement of even number of free electrons during the size-growth, both pathways correlate with the 2 e$^-$ hopping mechanism. At the end of the size-growth (~48 h), [Au$_{44}$(p-MBA)$_{26}$]$^{2-}$ has been formed as the predominant species. Finally, a slow size-focusing (need about 4 days) from the other sized clusters such as [Au$_{40}$(p-MBA)$_{25}$]$^+$ and [Au$_{46}$(p-MBA)$_{27}$]$^-$ to the thermodynamically stable product of Au$_{44}$ occurs.

Of note, the homogeneous and heterogeneous nucleation-growth mechanisms have also been used to explain the size-growth of metal nanoclusters/particles. These two patterns refer to the size growth from Au^I precursors to stable clusters

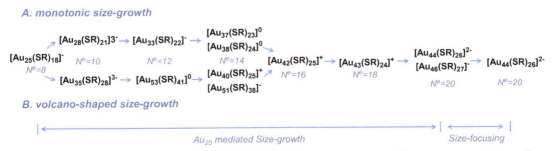

FIG. 4.15 The main size-growth and size-focusing processes during the size-growth from [Au$_{25}$(p-MBA)$_{18}$]$^-$ to [Au$_{44}$(p-MBA)$_{26}$]$^{2-}$ via monotonic size-growth (A) or volcano-shape size-growth pattern (B) [38].

(homogeneous) or from seed cluster (heterogenous) to larger-sized clusters, respectively. According to the molecular-level mechanistic explorations as mentioned above, it can be seen that the homogeneous and heterogeneous nucleation-growth mechanisms could occur competitively in one size-conversion system.

Redox-induced size-conversion among a series of phosphine protected gold nanoclusters

In the past decade, the mechanism of the redox regulated comprehensive conversions among Au_6 (i.e., $[Au_6(PR_3)_8]^{2+}$), Au_8 (i.e., $[Au_8(PR_3)_8]^{2+}$ or $[Au_8(PR_3)_8Cl_2]^{2+}$), Au_9 (i.e., $[Au_9(PPh_3)_8]^{3+}$) and Au_{11} (i.e., $[Au_{11}(PR_3)_{10}]^{3+}$) clusters have been extensively studied by both experimental and theoretical strategies. For clarity, we provide a brief introduction on the related studies separately in this section. Nevertheless, it is noteworthy that the interconversions of these clusters are quite facile, and some other conditions, such as metal complex (e.g., Au^I complex induced size-growth from Au_6 to Au_8 clusters [39]) and foreign ligands (e.g., diphosphine ligands induced size-conversion from Au_9 to Au_8 [39]) have also been reported lately. For clarity reasons, the details of those reactions are discussed more fully later in this chapter.

The size-maintained interconversion between $[Au_8(dppp)_4](NO_3)_2$ (dppp is short for 1,3-bis(diphenylphosphino) propane) and $[Au_8(dppp)_4Cl_2](PF_6)_2$ represents a pioneering study for the redox-induced conversion of metal nanoclusters, and was reported by Konishi and co-workers in 2011 [39]. According to SCXRD analysis, the metallic core of $[Au_8(dppp)_4]^{2+}$ adopts the edge-sharing tri-tetrahedral structure, while $[Au_8(dppp)_4Cl_2]^{2+}$ consists of an edge-sharing bi-tetrahedral structure, in which the opposite side of the shared edge in each tetrahedron is further bridged by an additional Au atom (Fig. 4.16A). In both Au_8 clusters, the eight gold atoms were interlocked by four dppp ligands, and the two outmost Au atoms in $[Au_8(dppp)_4Cl_2]^{2+}$ is each protected by an additional chloride. According to the experimental observations, adding NEt_4Cl to the methanol solution of $[Au_8(dppp)_4]^{2+}$ under aerobic condition at room temperature instantly generates $[Au_8(dppp)_4Cl_2]^{2+}$, while the reverse conversion occurs when ethanol solution of $NaBH_4$ was added to the CH_2Cl_2 solution of $[Au_8(dppp)_4Cl_2]^{2+}$ at room temperature. Due to the distinct optical characteristics of the two Au_8 clusters (Fig. 4.16B), the cluster conversion could be easily verified from UV-vis characterization.

On the basis of the experimental observations, the mechanism of the redox-induced conversion between $[Au_8(dppp)_4]^{2+}$ and $[Au_8(dppp)_4Cl_2]^{2+}$ has been recently investigated by DFT calculations [40]. The reduction induced conversion of $[Au_8(dppp)_4Cl_2]^{2+}$ to $[Au_8(dppp)_4]^{2+}$ mainly occurs via the elementary steps of two SET processes, removal of one chloride on one corner Au atom, Au-Au dissociation of the other corner Au atom, Au—Au bond formation of the two corner Au atoms to reconstruct the metallic core, and the final dissociation of the bridging chloride (Fig. 4.17). Herein, it is noteworthy that the addition of $NaBH_4$ results in a spontaneous ligand exchange to form the intermediate of $[Au_8(dppp)_4(BF_4)_2]^{2+}$, which is a resting state of the overall transformation (Fig. 4.17A). Compared with this mechanism, the other pathways are mainly precluded by the difficulty in Au-Cl dissociation steps, which is sensitive to the coordination environment of the target Au atoms. That is, the Au-Cl dissociation is relatively easier for Au atom(s) with higher electron density (e.g., reduction state of the cluster precursor) or with higher coordination number.

The oxidation induced conversion of $[Au_8(dppp)_4]^{2+}$ to $[Au_8(dppp)_4Cl_2]^{2+}$ starts with the 2e oxidation, and the outermost Au—Au bonds are significantly lengthened in the formed $[Au_8(dppp)_4]^{4+}$ (compared with the related ones in

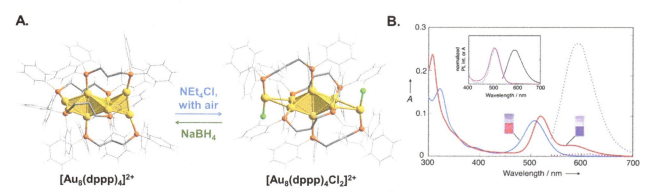

FIG. 4.16 Redox induced interconversion between $[Au_8(dppp)_4]^{2+}$ and $[Au_8(dppp)_4Cl_2]^{2+}$ (A), and the distinct optical property of $[Au_8(dppp)_4Cl_2]^{2+}$ (*blue solid line*: UV-vis absorption spectra; *blue dotted line*: optical emission) and $[Au_8(dppp)_4]^{2+}$ (*red solid line*: UV-vis absorption spectra; *red dotted line*: optical emission) (B). Inset: Normalized optical spectra of $[Au_8(dppp)_4Cl_2]^{2+}$ at 509 nm (*red*: optical absorption curve; *dashed black*: excitation curve) and 600 nm (emission curve). *(Reproduced from Y. Kamei, Y. Shichibu, K. Konishi, Generation of small gold clusters with unique geometries through cluster-to-cluster transformations: octanuclear clusters with edge-sharing gold tetrahedron motifs, Angew. Chem. Int. Ed. 50 (2011) 7442–7445, with permission from the Wiley-VCH (Copyright 2011).)*

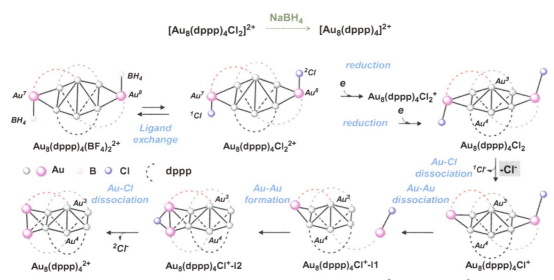

FIG. 4.17 The most plausible mechanism for the reduction induced conversion of $[Au_8(dppp)_4Cl_2]^{2+}$ to $[Au_8(dppp)_4]^{2+}$ by DFT calculations. *(Reproduced from Y. Bai, S. He, Y. Lv, M. Zhu, H. Yu, Redox-induced interconversion of two Au$_8$ nanoclusters: the mechanism and the structure–bond dissociation activity correlations, Inorg. Chem. 60 (2021) 5724–5733, with permission from the ACS Publications (Copyright 2021).)*

$[Au_8(dppp)_4]^{2+/3+}$ precursors). To this end, the coordination of chloride on the outermost Au—Au bond is potentially facilitated by both the lessened steric hindrance and the improved electrostatic attraction. After that, the bridging chloride further weakens the adjacent Au—Au bond (i.e., Au—Au bond was further activated), and thus the incorporation of the second chloride results in a spontaneous Au-Au dissociation. After a kinetically facile isomerization on the monocoordinated biphosphine ligand and Au—Au bond formation, the core-reconstruction is accomplished, and the $[Au_8(dppp)_4Cl_2]^{2+}$ was formed as the product (Fig. 4.18).

Distinct from the aforementioned interconversion between the two Au$_8$ clusters, treating $[Au_8(dppp)_4Cl_2]^{2+}$ with excessive NaBH$_4$ in Huang's recent study results in an immediate size-reduction, generating $[Au_6(dppp)_4]^{2+}$ as the sole product (Fig. 4.19A, the formation of Au$_6$ cluster was first evidenced by UV-vis spectra) [41]. UV-vis spectroscopic

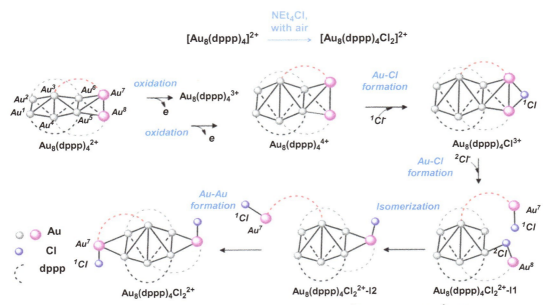

FIG. 4.18 The proposed mechanism for the oxidation induced conversion from $[Au_8(dppp)_4]^{2+}$ to $[Au_8(dppp)_4Cl_2]^{2+}$. *(Partially reproduced from Y. Bai, S. He, Y. Lv, M. Zhu, H. Yu, Redox-induced interconversion of two Au$_8$ nanoclusters: the mechanism and the structure–bond dissociation activity correlations, Inorg. Chem. 60 (2021) 5724–5733, with permission from the ACS Publications (Copyright 2021).)*

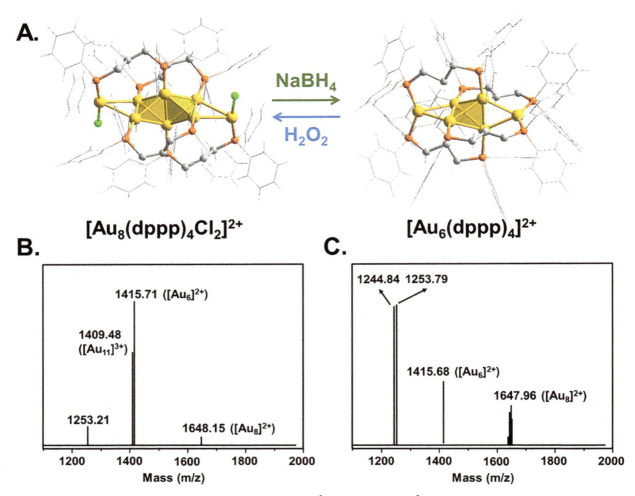

FIG. 4.19 Redox induced reversible size-conversion between [Au$_6$(dppp)$_4$]$^{2+}$ and [Au$_8$(dppp)$_4$Cl$_2$]$^{2+}$ (A), the ESI-MS characterization on the intermediate state during the size-conversion from the reaction of [Au$_8$(dppp)$_4$Cl$_2$]$^{2+}$ with NaBH$_4$ (B) and the reaction of [Au$_6$(dppp)$_4$]$^{2+}$ with H$_2$O$_2$ (C). *Parts (B) and (C) reproduced from X. Yang, X. Lin, C. Liu, R. Wu, J. Yan, J. Huang, Reversible conversion between phosphine protected Au6 and Au8 nanoclusters under oxidative/reductive conditions, Nanoscale, 9 (2017) 2424–2427, with permission from the Royal Society of Chemistry (Copyright 2017).*

tracking on the size-conversion via a drop-by-drop addition of NaBH$_4$ indicates that the size-conversion from [Au$_8$(dppp)$_4$Cl$_2$]$^{2+}$ to [Au$_6$(dppp)$_4$]$^{2+}$ is dependent on the amount of NaBH$_4$. Meanwhile, the ESI-MS characterization on the intermediate state implies multiple cluster peaks corresponding to [Au$_8$(dppp)$_4$Cl$_2$]$^{2+}$, [Au$_6$(dppp)$_4$]$^{2+}$, [Au$_{11}$(dppp)$_5$]$^{3+}$ and [Au$_2$(dppp)$_2$Cl]$^+$ (Fig. 4.19B). Herein, the distinct reduction induced size-conversion of the [Au$_8$(dppp)$_4$Cl$_2$]$^{2+}$ precursors (to [Au$_8$(dppp)$_4$]$^{2+}$ in Konishi's study and [Au$_6$(dppp)$_4$]$^{2+}$ in Huang's study) could be possibly caused by the different counter-anion (PF$_6^-$ vs Cl$^-$) or different types of solvents (CH$_2$Cl$_2$ vs EtOH) of the Au$_8$ cluster precursors in the two reaction systems. More efforts are necessary to elucidate the details on the reason for the distinct chemo-selectivity.

In addition to the reduction induced size-conversion from [Au$_8$(dppp)$_4$Cl$_2$]$^{2+}$ to [Au$_6$(dppp)$_4$]$^{2+}$, Huang and co-workers also reported the reverse conversion, which occurs via adding H$_2$O$_2$ to the ethanol solution of [Au$_6$(dppp)$_4$]$^{2+}$ (Fig. 4.19A). The size-conversion occurs quickly within 2 min, and is completely finished to generate [Au$_8$(dppp)$_4$Cl$_2$]$^{2+}$ after 120 min. The ESI-MS characterization on the intermediate state (after reacting for 50 min) shows the cluster peaks of [Au$_6$(dppp)$_4$]$^{2+}$, [Au$_8$(dppp)$_4$Cl$_2$]$^{2+}$ and [Au$_2$(dppp)$_2$Cl]$^+$. Referring to the single crystal structure of [Au$_6$(dppp)$_4$]$^{2+}$ and [Au$_8$(dppp)$_4$Cl$_2$]$^{2+}$ [39], Huang et al. suggested the involvement of an octahedral-like Au$_6$ intermediate in the two-stage mechanism for both the size-conversion of Au$_6$ to Au$_8$ and its reverse reaction.

Similar to the size-reduction from Au$_8$ to Au$_6$, the size-reduction from a dppp protected Au$_{11}$ (i.e., [Au$_{11}$(dppp)$_5$]$^{3+}$) to Au$_6$ cluster ([Au$_6$(dppp)$_4$]$^{2+}$) has also been regulated by the reduction environments [42]. According to Huang, Liu and

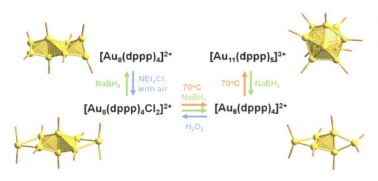

FIG. 4.20 The comprehensive interconversion between the series of Au_6, Au_8, and Au_{11} clusters. Only one C atom on the aromatic ligands of PPh_3/dppp were left, and all H atoms were omitted for a better view.

Yan's experimental observation, adding an alcoholic solution of $NaBH_4$ to $[Au_{11}(dppp)_5]^{3+}$ the results in the formation of $[Au_6(dppp)_4]^{2+}$. Time-dependent UV-vis and ESI-MS tracking indicates the gradual formation of Au_6 clusters during the step by step addition of $NaBH_4$, and Au_6 becomes the sole cluster product when excessive $NaBH_4$ was added.

For clarity, the inter-conversion network among Au_6, Au_8, and Au_{11} clusters is given separately in Fig. 4.20 (both redox and heating induced conversions are given, while the details on heating induced conversion is provided in Section 4.2.2.5).

Reduction induced size-conversion between Au_{25} and Au_{23} clusters

In addition to the reduction induced size-growth of $[Au^I(p\text{-MBA})]_x \rightarrow [Au_{25}(p\text{-MBA})_{18}]^-$ and $[Au_{25}(p\text{-MBA})_{18}]^- \rightarrow [Au_{44}(p\text{-MBA})_{26}]^{2-}$ mentioned above, the size-reduction of $[Au_{25}(SePh)_{18}]^- \rightarrow [Au_{23}(SePh)_{16}]^-$ upon adding aqueous solution of $NaBH_4$ has also been reported recently (Fig. 4.21) [43] The composition and structure of the formed $[Au_{23}(SePh)_{16}]^-$ was verified by a combination of UV-vis spectra, MALDI-MS, TD-DFT calculations, and the control experiments (see Section 6.2.2.1 for the details). According to the time-dependent UV-vis absorption spectra and MALDI-MS characterizations, the characteristic peaks of the Au_{25} precursor are attenuated, and the characteristic peaks of the Au_{23} product is enhanced after reacting for 2 h. After 10 h, Au_{23} becomes the sole product in the reaction system. Mechanistic exploration with DFT calculations on the modeling $[Au_{25}(SeH)_{10}]^n$ ($n=-1, -2$ and -3) indicates that the electronic structure of $[Au_{25}(SeH)_{10}]^{-3}$ is relatively unstable, because both the HOMO (highest occupied molecular orbital) and LUMO (lowest unoccupied molecular orbital) energy levels are positive (note: the HOMO and LUMO energies of $[Au_{25}(SeH)_{10}]^{1-}$ and its 1e-oxidized $[Au_{25}(SeH)_{10}]^{2-}$ are all negative). Meanwhile, Ab initio molecular dynamics (AIMD) simulation on the two-hydride bearing structure of $[Au_{25}(SeH)_{10}]^{1-}$ (i.e., $[Au_{25}(SeH)_{10}H_2]^{3-}$) results in a remarkable structural distortion, with the removal of two Au-H moiety and the formation of a $[Au_{23}(SeH)_{18}]^{3-}$. After that, dissociation of two anionic selenate and structural tautomerization occur to finally generate the $[Au_{23}(SeH)_{16}]^{1-}$ clusters.

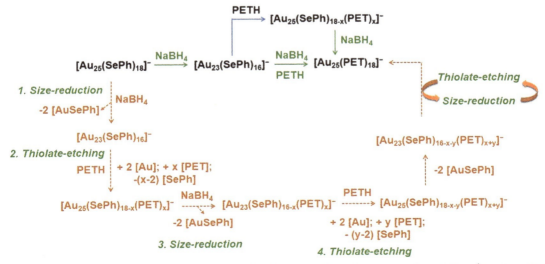

FIG. 4.21 Illustrative diagram for the thiolate (PETH) and reductant ($NaBH_4$) regulated ligand exchange of $[Au_{25}(SePh)_{18}]^{1-}$ to $[Au_{25}(PET)_{18}]^{1-}$. Note: the solid line connects the species evidenced by experimental and DFT strategies, while the dashed line links the deduced intermediates in the overall size-conversion.

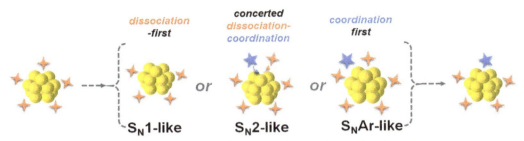

FIG. 4.22 Illustrative diagram for the three typical mechanisms (S_N1-like, S_N2-like and S_NAr-like) of the size-maintained ligand exchange.

Interestingly, adding PETH into the solution of $[Au_{23}(SePh)_{16}]^{1-}$ results in distinct products [43], depending on whether $NaBH_4$ is added or not (Fig. 4.21). In the presence of $NaBH_4$, $[Au_{25}(PET)_{18}]^{-1}$ is obtained (evidenced by UV-vis spectra and MALDI-MS analysis), while $[Au_{25}(SePh)_x(PET)_{18-x}]^{-1}$ with variable ligand components is obtained in the absence of $NaBH_4$. In other words, Au_{25} is obtained in both systems, but the presence of $NaBH_4$ results in a complete removal of the -SePh ligand, while some -SePh ligands are preserved in the absence of $NaBH_4$. Of note, a post addition of $NaBH_4$ in the second reaction system could finally generate $[Au_{25}(PET)_{18}]^{-1}$ as the product. In this context, the overall transformation mechanism for the thiolate and reductant induced ligand exchange of $[Au_{25}(SePh)_{18}]^{-1}$ to $[Au_{25}(PET)_{18}]^{-1}$ was suggested to occur via four stages (see Fig. 4.21 for an illustrative diagram): (1) Size-reduction of $[Au_{25}(SePh)_{18}]^{1-}$ to $[Au_{23}(SePh)_{16}]^{1-}$ by $NaBH_4$; (2) Thiolate etching of $[Au_{23}(SePh)_{16}]^{1-}$ into $[Au_{25}(SePh)_{18-x}(PET)_x]^{1-}$ by the added PETH; (3) Size-reduction of $[Au_{25}(SePh)_{18-x}(PET)_x]^{1-}$ into $[Au_{23}(SePh)_{16-x}(PET)_x]^{1-}$ upon reduction (i.e., formally removal of two -AuSePh groups); and (4) Ligand-etching of $[Au_{23}(SePh)_{18-x-y}(PET)_{x+y}]^{1-}$ into $[Au_{25}(SePh)_{16-x-y}(PET)_{x+y}]^{1-}$. In other words, the $NaBH_4$ induced $Au_{25} \rightarrow Au_{23}$ size-reduction (stage 3) occurs via removing two Au(SePh) groups, while the ligand exchange induces $Au_{23} \rightarrow Au_{25}$ conversion (stage 4) occurs via incorporating two Au(PET) groups each time. The 2*Au(SePh) removal and 2*Au(PET) addition occur repeatedly occurs until all -SePh groups were finally removed, and PET become the only type of ligand on the Au_{25} cluster product.

The mechanism in Fig. 4.21 demonstrates that even for the size-maintained ligand exchange, size-conversion might be involved to tentatively adjust the core-shell interference (and thus the spacing for the ligand shell), and to accommodate versatile arrangement of ligands. Such mechanistic possibility could be easily neglected in these days, as the dissociative mechanism (dissociation of one ligand occurs prior to the addition of a foreign one, analogous to the typical S_N1 reaction); associate pathway (dissociation of the original ligand occurs simultaneously with the coordination of the foreign ligand, analogous to the typical S_N2 reaction); or addition-first mechanism (coordination of the foreign ligand occurs prior to the dissociation of the original one, analogous to the typical S_NAr reaction) are more-commonly proposed. For clarity reasons, an illustrative diagram for these three types of mechanism is provided in Fig. 4.22. These proposals commonly fit the simplicity principal (Nature always acts by the shortest course-Pierre de Fermat), and might be responsible for most reaction systems. But due to the "softness" of the core-shell structure of metal nanoclusters, the structural tautomerization between different sized clusters, might also be responsible for certain reaction systems.

Oxidation induced size-growth from $[Au_{23}(SR)_{16}]^-$ clusters

Compared with the extensively reported reduction-induced size-conversion of atomically precise metal nanoclusters, the oxidation induced conversion is relatively rare. In addition to the aforementioned $Au_6 \rightarrow Au_8$ and $Au_8 \rightarrow Au_8$ transformations, the size-growth of a relatively larger Au cluster, i.e., $[Au_{23}(SCy)_{16}]^-$ (counteraction: TOA^+) to $[Au_{28}(SCy)_{20}]^0$ has been reported lately (Fig. 4.23A) [44]. The reaction has been conducted in a biphasic system by mixing the CH_2Cl_2 solvent of $[Au_{23}(SCy)_{16}]^-$ with an aqueous solution of H_2O_2. Stirring the solution for 2h generates the target $[Au_{28}(SCy)_{20}]^0$, the structure of which was determined by SCXRD, ESI-MS, etc. As shown in Fig. 4.23B, the Au_{23} precursor and the Au_{28} product shows great structural similarity. Au_{23} and Au_{28} comprise a cuboctahedral Au_{15} and a bicuboctahedral Au_{20} (via sharing six Au atoms) core structure, respectively. Outside the metallic core structure, four and eight bridging thiolates are present on Au_{23} and Au_{28}. Aside from that, each core structure is further protected by two AuS_2 and two Au_3S_4 motifs.

According to the UV-vis monitoring (Fig. 4.23C), the characteristic absorption peak at 570nm of the Au_{23} precursor clusters attenuates along with the reaction processes, and four isosbestic points at 355, 468, 526, and 641 nm were observed after 1h (when the spectral change was almost completed). The results indicate a consistent mechanism throughout the size-conversion, without the formation of a metastable intermediate such as a charge-neutral Au_{23} cluster (note: metastable intermediate commonly refers to the intermediate species that can be possibly figured out via spectroscopic analysis or

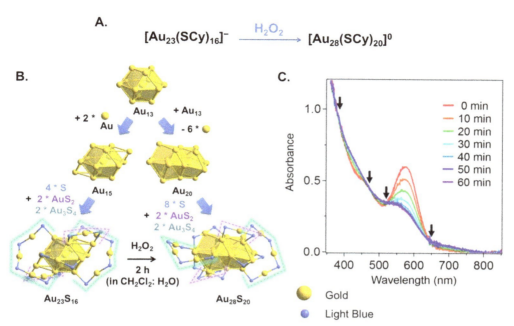

FIG. 4.23 The oxidation induced size-growth of $[Au_{23}(SCy)_{16}]^-$ (A), structural anatomy for the framework of $[Au_{23}(SCy)_{16}]^-$ and $[Au_{28}(SCy)_{20}]$ (B); and time-dependent tracking on the reaction solutions with UV-vis spectra (C). *((C) Reproduced from T. Higaki, C. Liu, Y. Chen, S. Zhao, C. Zeng, R. Jin, S. Wang, N.L. Rosi, R. Jin, Oxidation-induced transformation of eight-electron gold nanoclusters: $[Au_{23}(SR)_{16}]^-$ to $[Au_{28}(SR)_{20}]^0$, J. Phys. Chem. Lett. 8 (2017) 866–870, with permission from the ACS Publications (Copyright 2017).)*

separated from the reaction system). This proposal is further supported by the combination of DPV (differential pulse voltammetry), bulk electrolysis, and control experiments (by replacing the oxidant of H_2O_2 with O_2, $Ce(SO_4)_2$).

Interestingly, the size-growth of the same Au_{23} precursor (i.e., $[Au_{23}(SCy)_{16}]^-[TOA]^+$) to Au_{28} clusters could also be regulated by photo oxidation (Fig. 4.24A) [45]. The control experiments indicate that both light excitation and O_2 (in air) are requisite. In addition, the formation of the anionic dioxygen ($O_2-\bullet$) is verified by ESR analysis, demonstrating the electron transfer from the Au_{23} precursor to dioxygen. Besides, the reaction proceeds smoothly upon excitation with UV-vis light at 427, 620, and 730 nm, but no conversion occurred when 940 nm infrared light is used. To this end, Zhu and co-workers suggested that the photoexcitation energy should exceed the HOMO-LUMO gap of the Au_{23} precursor (1.55 eV, determined by UV-vis spectrum). Meanwhile, the fragmentation peak corresponding to $Au_{18}(SCy)_{14}$ is identified in the time-dependent MALDI-TOF-MS throughout the solid-state reaction (note: the solution phase reaction is too rapid to track with MS analysis), while the characteristic peak corresponding to $[Au_{18}(SCy)_{14}+Au_{23}(SCy)_{16}]/2$ was observed in the latter stage. In accordance with these results, the overall size-conversion was suggested to occur via a mechanism (Fig. 4.24B) consisting of photo excitation of the Au_{23} precursor, electron transfer from the excited Au_{23} to dioxygen (with the possible formation of neutral form of Au_{23}), formation of Au_{18} and Au_nS_{n+1} ($n=2, 3, 4$) via fragmentation of Au_{23}, and the re-assembly of the fragments with the remaining Au_{23} precursor to form the final Au_{28} products.

The aforementioned studies draw out a general mechanism on the oxidation induced size-conversion from Au_{23} to Au_{28} clusters. On this basis, the mechanistic issues mainly lie in the fragmentation pattern of the Au_{23} precursor, the origin of the site-preference and the inherent structure-activity (for bond dissociation) correlations, and the reason for the thermodynamic stability of the Au_{28} product (over the other re-assembly possibilities).

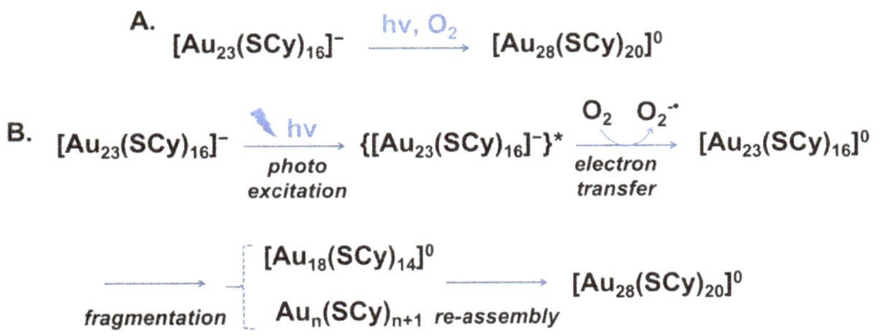

FIG. 4.24 The photo-oxidation induced size-growth of $[Au_{23}(SCy)_{16}]^-$ (A), and the proposed mechanism for the size-conversion (B).

4.2.2.3 Mechanistic insights into the ligand exchange

As a powerful and practical synthetic method, ligand exchange (and the two-phase ligand exchange in some cases) has been extensively explored to prepare metal nanoclusters with atomic precision. Specifically, important mechanistic insights have been gained in the ligand exchange of one atomically precise metal cluster with foreign ligands. According to the reported literatures, the ligand-exchange of metal nanoclusters could occur via different modes, i.e., size-growth, size-maintain (with retained framework or not), or size-reduction. Accordingly, we mainly categorize the ligand etching of the atomically precise metal nanoclusters into three groups: (i) size growth of one atomically precise metal nanoclusters to a relatively larger sized cluster (such as the seed growth), (ii) size-maintained ligand exchange of metal nanoclusters (with another same-sized metal nanocluster, or organic ligand that is different from the preexistent ones), and (iii) size-reduction of a relatively larger-sized cluster into a smaller one (associated with heating conditions in some cases).

Size-growth induced by ligand etching

As mentioned in Section 4.2.1.1, the small nanoparticles play a crucial role in the typical Lamer growth, and these species were also called "seed NPs" in some early studies. Thanks to the atomic precision of the numerical metal nanoclusters, the size-growth from one atomically precise cluster to another one has been widely reported in recent decades. In some of the size-conversion reactions, the reactant and product clusters share great structural similarity, and the formation of the larger-sized cluster product could be viewed as the addition of certain metal-ligand motifs on the precursor clusters. The observation correlates with the traditional concept of "seed growth," albeit different reaction conditions might be involved in the two-stage reaction system (i.e., formation of the cluster precursors, and the size-growth from the precursor to the product). In view of the largely maintained framework (determined by SCXRD), these reactions are also designated as the seed-growth reactions. The main mechanistic progress, such as the insights into the kinetic correlation of the target clusters, and the inherent driving force for size-growth, have been given below.

The size growth from $Au_{18}(SCy)_{14}$ to a relatively larger sized $Au_{21}(SAdm)_{15}$ has recently been accomplished by Zhu and co-workers via mixing the Au_{18} clusters with the foreign ligand of HSAdm (short for adamantane mercaptan) in CH_2Cl_2 solvent (Fig. 4.25A) [46]. According to SCXRD analysis, both clusters comprise a $Au_6@Au_4S_5$ structure (Fig. 4.25B). On this basis, the framework of $Au_{18}(SCy)_{14}$ could be viewed as the coating of the face-fusing bi-octahedral Au_9 core with one Au_4S_5, three AuS_2 and one Au_2S_3 staple motifs. Compared with the Au_{18} cluster precursor, the single crystal structure of $Au_{21}(SAdm)_{15}$ comprises a similar bi-octahedral core, but in an edge-sharing pattern. Therefore, there are ten Au atoms in the core structure of Au_{21} cluster, which is protected by one Au_2S_3, one AuS_2, and one Au_8S_9 (i.e., the connection of the Au_4S_5 block with the Au_4S_4 block, Fig. 4.25B), and a bridging thiolate ligand.

Density functional theory calculations were conducted to shed light on the kinetics of the size-growth, and the proposed mechanism is shown in Fig. 4.26. First, an exterior Au atom is inserted into the Au_2S_3 staple motif of the modeling $Au_{18}(SR)_{14}$ cluster (**C1 → C2** in Fig. 4.26). After that, a structure tautomerization occurs, which could be viewed as the insertion of the newcome Au atom into the Au—S bond of one of the AuS_2 motifs (**C2 → C3**). In this context, a new

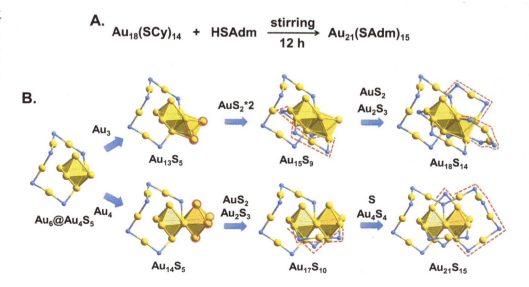

FIG. 4.25 The H-SAdm etching induced size-growth of $Au_{18}(SCy)_{14}$ to $Au_{21}(SAdm)_{15}$ (A), and the structural analysis on the Au-S framework of the precursor and product clusters (B).

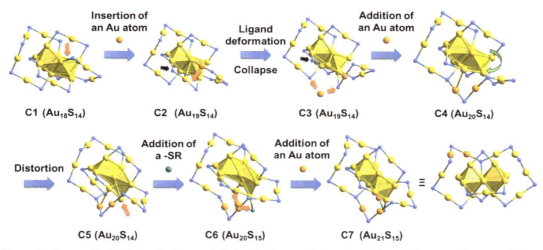

FIG. 4.26 Illustrative diagram for the changes in framework during the size-growth from $Au_{18}(SCy)_{14}$ (**C1**) to $Au_{21}(SAdm)_{15}$ (**C7**).

trimetric motif is formed, and the remaining thiolate ligand on the AuS_2 motif becomes a bridging thiolate (labeled with black arrow). Thereafter, incorporation of the second Au atom occurs to generate the intermediate **C4**, in which a relatively long staple motif of Au_8S_9 is formed via the ligation of the Au_4S_5 motif and the newly formed Au_3S_4 motif with the newcome Au atom. From **C4**, isomerization occurs, and both octahedral blocks are significantly distorted in the formed intermediate **C5**. Finally, the successive addition of a thiolate and the third Au atoms occurs to generate the thermodynamically stable product of **C7** (theoretical modeling structure of $Au_{21}(SR)_{15}$). According to this mechanism, the overall size-growth from Au_{18} to Au_{21} clusters was concluded to undergo a three-stage transformation: (i) core-growth via the insertion of single Au atom into the surface motif; (ii) Collapse of surface motif and growth of the core structure; and (iii) growth of new ligand motifs on activate sites of the cluster intermediates (the exposed thiolate ligand and/or metal core atoms).

In the aforementioned example, a thiolate-to-thiolate exchanges occurs, associated with the structural tautomerization and a size-growth of the precursor cluster. Herein, it is noteworthy that the incorporation of different types of ligands could also be achieved via the ligand exchange reactions. For example, in the recent study by Zhu and co-workers, the thiolate protected Ag_{44} cluster (i.e., $Ag_{44}(p\text{-MBA})_{30}$) [47] has been converted to a thiolate and phosphine co-protected Ag_{50} cluster (i.e., $Ag_{50}(dppm)_6(TBBT)_{30}$) via the addition of $AgNO_3$, dppm, TBBM, and $NaBH_3CN$ (Fig. 4.27A, and p-MBAH, dppm,

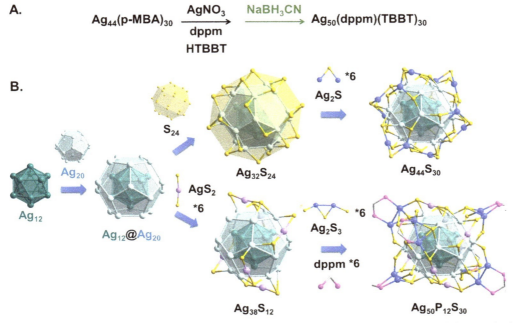

FIG. 4.27 The seed growth from $Ag_{44}(p\text{-MBA})_{30}$ to $Ag_{50}(dppm)_6(TBBT)_{30}$ (A), and the structural analysis of the Ag-S framework of $Ag_{44}(p\text{-MBA})_{30}$ and $Ag_{50}(dppm)(TBBM)_{30}$ (B). Note that the $Ag_{12}@Ag_{20}$ (i.e., Ag_{32}) structure are not the same (but highly similar) in Ag_{44} and Ag_{50} clusters, but for clarity only one group of Ag_{32} is shown.

and HTBBT are short for p-mercaptobenzoic acid, bis(diphenylphosphino)methane and [4-(tert-butyl)phenyl]methanethiol). Albeit the utilization of multiple conditions (e.g., foreign ligand and reduction), the most prominent structural change lies in the size-growth associated with the replacement of the original p-MBA ligands with the mixture ligands of dppm and TBBT. Therefore, the comprehensive reaction system of the $Ag_{44} \rightarrow Ag_{50}$ conversion is categorized into the type of ligand etching induced size-growth. As shown in Fig. 4.27B, the Ag-S/P framework of the formed Ag_{50} cluster shows high structural similarity with that of Ag_{44}, as them both comprise a double-layered $Ag_{12}@Ag_{20}$ core (the inner icosahedral Ag_{12} kernel is protected by an exterior pentagonal dodecahedron Ag_{20}). The $Ag_{12}@Ag_{20}$ kernel is further protected by 24 bridging-thiolate ligands (with an orientation of a slightly distorted rhombicuboctahedron) and six Ag_2S blocks (in a tautomerized octahedral orientation) in Ag_{44} (Fig. 4.25A); while is protected by six AgS_2 and six Ag_2S_3 blocks (each group adopts a hexagonal prism-like orientation) in Ag_{50} cluster. In addition, each of the six bidentate dppm ligands in Ag_{50} functions as a bridging ligand to connect two adjacent Ag atoms in the Ag_2S_3 motifs. Albeit the great structural correlation, a detailed mechanistic understanding on the size-growth, such as the kinetic profiles, and the possible intermediates, remain to be elucidated.

Compared with the size-conversion from one atomically precise cluster to another one, the ligand exchange induces a cascade size-growth from $[Au_{23}(SCy)_{16}]^-$ to $Au_{28}(TBBT)_{20}$ and then to $Au_{36}(TBBT)_{24}$ in the recent study by Mandal and co-workers (Fig. 4.28A) [48]. Of note, the Au_{28} cluster could also be viewed as a metastable intermediate (i.e., stable for only a certain period of time) during the size-conversion from Au_{23} to Au_{36} clusters, and the similar transformation of $[Au_{23}(SCy)_{16}]^- \rightarrow Au_{28}(SCy)_{20}$ has also been accomplished under oxidation condition (see details in Section 4.2.2.2). According to the time-dependent MALDI-MS characterization, the entire ligand-etching induced conversion is initiated by the size-maintained ligand change ($[Au_{23}(SCy)_{16}]^- \rightarrow [Au_{23}(SCy)_x(TBBT)_{16-x}]^-$, $x=0-14$), which occurs rapidly within 1 min (Fig. 4.28A). After that, the components of $[Au_{23}(SCy)_x(TBBT)_{16-x}]^-$ ($x=0, 1$) gradually increased within the following 4 min, associated with the gradual formation of $Au_{28}(SCy)_x(TBBT)_{19-x}$ ($x=0-2$). The results are suggested to be caused by the tolerance of the primary Au_{23} framework to maximal 14 TBBT ligands, while the incorporation of more TBBT significantly perturbs the precursor structure to form a relatively larger Au_{28} clusters. After that, the Au_{23} clusters were continuously consumed, with the incremental formation of $Au_{28}(TBBT)_{19}$. Herein, $Au_{28}(TBBT)_{19}$ was suggested a fragment of $Au_{28}(TBBT)_{20}$ due to the loss of one thiolate ligand (upon laser irradiation), and some other fragments corresponding to losing of [Au(SR)] components were also observed on MALDI-MS. In this context, a two-stage mechanism was proposed to account for the ligand exchange induced size-conversion from Au_{23} to Au_{28} clusters. The first stage is ligand exchange and structural distortion, consisting of both size-maintained ligand exchange and structural transformation (via either size-conversion-ligand exchange or the ligand exchange-size-conversion pattern, Fig. 4.28B). The second stage

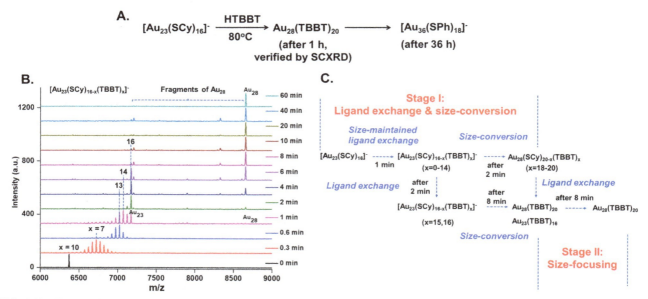

FIG. 4.28 The ligand exchange induced cascade size-growth from $[Au_{23}(SCy)_{16}]^-$ to $Au_{28}(TBBT)_{20}$ and then to $Au_{36}(TBBT)_{24}$ (A), time-dependent MALDI-MS monitoring on the size-conversion from $[Au_{23}(SCy)_{16}]^-$ to $Au_{28}(TBBT)_{20}$ clusters during 1 h (B); and the proposed mechanism (C). ((B) *Reproduced from M.P. Maman, A.S. Nair, H. Cheraparambil, B. Pathak, S. Mandal, Size evolution dynamics of gold nanoclusters at an atom-precision level: ligand exchange, growth mechanism, electrochemical, and photophysical properties, J. Phys. Chem. Lett. 11 (2020) 1781–1788, with permission from the ACS Publications (Copyright 2020).)*

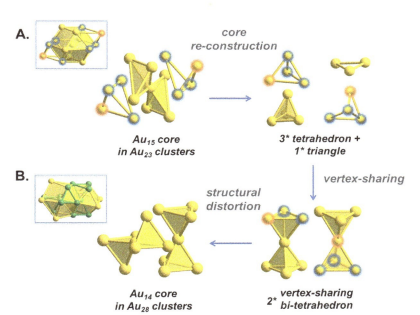

FIG. 4.29 Illustrative diagram for the core-evolution mechanism from the Au_{15} core of $[Au_{23}(SCy)_{16}]^-$ to the Au_{14} core of $Au_{28}(TBBT)_{20}$. Insets: An alternative structural anatomy of the bi-capped cuboctahedron Au_{15} (upper) and bi-cuboctahedron Au_{20} (lower) core structure of Au_{23} and Au_{28} clusters used in Fig. 4.23.

corresponds to the size-focusing process, wherein the Au_{28} clusters were accumulated with prolonged time, and the conversion finished after about 1 h.

With the combination of super atom network model (SANM-Section 6.2.4), the grand unified model (GUM-Section 4.3.1) and preliminary DFT calculations, a plausible core-evolution mechanism for the size-conversion from Au_{23} to Au_{28} was proposed (Fig. 4.29). The core-evolution mainly occurs via three steps: (i) fragmentation of the bicapped cuboctahedral Au_{15} core (or vertex-sharing bi-tetrahedron $Au_7+2*Au_3+2Au^{hub}$) into three tetrahedra and one triangle (core re-construction stage), (ii) re-assembly of the elementary blocks into two bi-tetrahedron Au_7 (vertex-sharing stage), and (iii) the final structural tautomerization (structural distortion stage).

According to the time-dependent MALDI-MS monitoring, $Au_{28}(TBBT)_{20}$ is relatively stable for 6 h (no spectral changes occur). After reacting for 12 h, both the cluster peaks of Au_{28} and the fragmentation peaks of Au_{36} clusters were observed, while the characteristic peak of Au_{28} disappeared after 24 h. In combination of these observation and an earlier proposal on the size-evolution via increasing the tetrahedral Au_4 blocks [36] (see Section 4.3.2 for the details), Mandal et al. suggested the Au_{20} kernel of Au_{36} clusters to be formed by incorporating 6 additional Au atoms (i.e., two triangles) into the Au_{14} core of the Au_{28} precursor.

Interestingly, using the same Au_{23} precursor ($[Au_{23}(SCy)_{16}]^-$, evidenced by MALDI-MS and UV-vis spectra), but replacing the HTBBT ligand with the HSPh, and heating the reaction system to 40°C for 1 h, another cluster product, i.e., $[Au_{25}(SCy)_{18}]^-$ was gained (Fig. 4.30A) [49]. According to the time-dependent UV-vis monitoring, the characteristic peaks of the Au_{23} precursors (at 460, 570 nm) are attenuated within 0.25 min, and completely disappeared after 10 min. The characteristic peaks of the Au_{25} clusters at 425, 460, and 698 nm appeared at 0.5 min and gradually enhanced during the time interval of 0.5–60 min. Due to the fast reaction kinetics and the close mass of the incoming HSPh ligand with the preexisted HSCy, it is difficult to track the detailed size-conversion with MALDI-MS. To reduce reaction rate and gain deep insights into the reaction kinetics, another aromatic thiolate with lower aromaticity (2-naphthalenethiol, abbreviated as 2-NPT), and lower temperature when compared with the aforementioned system (HSPh, 25°C) were used. The time-dependent MALDI-MS and UV-vis spectra characterization demonstrates a three-stage mechanism. In the first stage (0–3 min), the size-maintained ligand exchange occurs, forming $Au_{23}(2\text{-}NPT)_{11}(SCy)_5$ after 3 min. After that, the replacement of one more SCy with 2-NPT (forming $Au_{23}(2\text{-}NPT)_{12}(SCy)_4$) occurs with a simultaneous formation of $Au_{25}(2\text{-}NPT)_{15}(SCy)_3$ after 5 min, demonstrating the core-conversion, and the possible intermediacy of $Au_{23}(2\text{-}NPT)_{13}(SCy)_3$ (correlating with $Au_{23}(2\text{-}NPT)_{12}(SCy)_4$ via ligand exchange, and with $Au_{25}(2\text{-}NPT)_{15}(SCy)_3$ via adding two [Au(2-NPT)]). After 10 min, the formation of another intermediate $Au_{23}(2\text{-}NPT)_{14}(SCy)_2$ was observed, and the cluster peaks of $Au_{25}(2\text{-}NPT)_{16}(SCy)_2$ becomes prominent. The results indicate that $Au_{23}(2\text{-}NPT)_{14}(SCy)_2$ is a plausible intermediate during the size-conversion. According to these results and discussions, the time interval of 3–10 min was denoted as stage II, and its main characteristic lies in the heavily incorporated (>12) 2-NPT ligands induced core-conversion of $Au_{23} \rightarrow Au_{25}$. Stage III covers the time interval of 30–60 min, when the complete conversion of Au_{23} to Au_{25} clusters occur. The thermodynamic stability theory recently proposed by Mpourmpakis and co-workers (Section 6.2.5.1 for details) [50]

FIG. 4.30 The ligand induced size-conversion of [Au$_{23}$(SCy)$_{16}$]$^-$ to [Au$_{25}$(SAr)$_{18}$]$^-$ (SAr = SPh or 2-NPT) (A) and the proposed atomic level mechanism for the size-conversion (B). ((B) Partially reproduced from M.P. Maman, A.S. Nair, A.M. Abdul Hakkim Nazeeja, B. Pathak, S. Mandal, Synergistic effect of bridging thiolate and hub atoms for the aromaticity driven symmetry breaking in atomically precise gold nanocluster, J. Phys. Chem. Lett. 11 (2020) 10052–10059, with permission from the ACS Publications (Copyright 2020).)

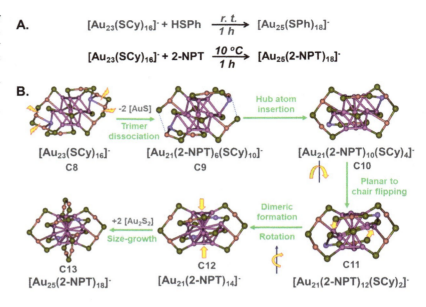

was used to comprehend the relatively inferior stability of the Au$_{23}$ precursor compared with that of the Au$_{25}$ product. The DFT calculation results indicate that the preliminary -SCy ligand(s) on the AuS$_2$ and Au$_3$S$_4$ motifs are more labile to undergo dissociation than those on the Au$_2$S$_3$ motifs. Meanwhile, the optimized geometry of the [Au$_{23}$(SCy)$_{16}$]$^-$ precursor is different from that of the completely exchanged analog of [Au$_{25}$(2-NPT)$_{18}$]$^-$: the central planar Au$_7$ structure distorted into a chair-like one after the complete ligand exchange, correlating with the structural distortion of a cuboctahedron to an icosahedron (Fig. 4.30B). On the basis of the aforementioned results and discussions, the overall size-conversion occurs via the successive [AuS] dissociation on the trimetallic Au$_3$S$_4$ motifs (**C8→C9**), hub atom insertion (**C9→C10**, Fig. 4.30B), planar to chair tautomerization of the central Au$_7$ structure (**C10→C11**), dimeric formation (**C11→C12**), and the final size-growth via adding two Au$_2$S$_2$ structures (**C12→C13**). Referring to the aforementioned three-stage mechanism, the steps prior to the formation of **C10** could be ascribed to stage I, the steps covering **C10-C12** correspond to the core-conversion in stage II, while the transformation from **C12** to **C13** correspond to the stage III.

Interestingly, using a similar HTBBT and heating reaction (to 80°C) system, the ligand exchange of the icosahedral [Au$_{25}$(PET)$_{18}$]$^-$ cluster with HTBBT by Jin and co-workers also generates Au$_{28}$(TBBT)$_{20}$ after 2h (Fig. 4.31) [51]. This reaction is distinct to the size-maintained ligand exchange of [Au$_{25}$(SR)$_{18}$]$^-$ (SR = PET, SBut) systems (vide supra, Fig. 4.34). The yield of Au$_{28}$(TBBT)$_{20}$ is higher than 90% (Au atom basis), without the observation of other byproducts. The structure of the formed Au$_{28}$(TBBT)$_{20}$ is evidenced by SCXRD, and was found to comprise a Au$_{20}$ kernel adopting the face-centered-cubic (FCC) structure. The structure of this Au$_{28}$ shows strong correlation with other thiolate protected Au clusters, such as Au$_{36}$(TBBT)$_{24}$, Au$_{44}$(TBBT)$_{28}$, and Au$_{52}$(TBBT)$_{32}$. In a later study by Wu and co-workers, a similar size-conversion of Au$_{25}$ to Au$_{28}$ cluster was reported [52]. But distinct to the aforementioned reaction, the neutral [Au$_{25}$(PET)$_{18}$]^0 was used as the precursor in Wu's study (Fig. 4.31), and the reaction was conducted by adding both foreign ligand (4-tertbutylphenylmethanethiol, abbreviated as HTBPM) and NaBH$_4$. After reacting at 80°C for 3 h, Au$_{28}$(TBPM)$_{22}$ (with two more thiolate ligands than that of Au$_{28}$(TBBT)$_{20}$ mentioned above) was isolated as the main product, and its atomic structure was evidenced by SCXRD and ESI-MS tests. The control experiments by replacing NaBH$_4$ with ascorbic acid [53] generates the same Au$_{28}$ product after similar reaction conditions, while no Au$_{28}$ was formed when NaBH$_4$ is

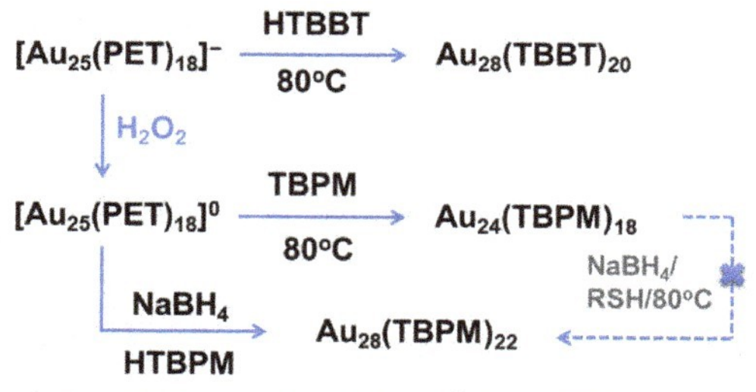

FIG. 4.31 The detailed transformations among [Au$_{25}$(PET)$_{18}$]$^-$, [Au$_{25}$(PET)$_{18}$]^0, Au$_{28}$(TBBT)$_{20}$, Au$_{24}$(TBPM)$_{28}$, and Au$_{28}$(TBBT)$_{22}$.

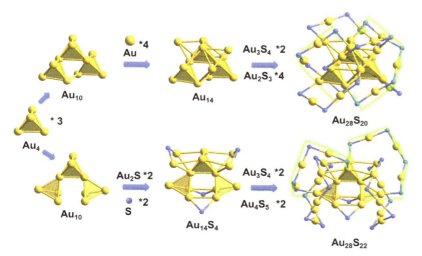

FIG. 4.32 Illustrative diagram for the framework of $Au_{28}(TBPM)_{20}$ and $Au_{28}(TBPM)_{22}$.

changed to NaOH/HOAc. The results indicates that $NaBH_4$ mainly acts as a reductant, instead of a strong base, during the size-conversion. Interesting, without the addition of $NaBH_4$, only $Au_{24}(TBPM)_{20}$ was generated, while the selectivity of Au_{24}: Au_{28} could be well regulated by controlling the amount of $NaBH_4$. In addition, the $Au_{24}(TBPM)_{20}$ was not able to produce target Au_{28} cluster under various examined conditions, such as the addition of HTBPM, $NaBH_4$, or mixture of HTBPM and $NaBH_4$ under 80°C. The results exclude the intermediacy of Au_{24} during the size-conversion from $[Au_{25}(PET)_{18}]^0$ to $Au_{28}(TBPM)_{22}$. Instead, the formation of Au_{24} is competitive with that of Au_{28} under the experimental conditions, and reduction is a driving force for the latter size-conversion. The detailed transformations among $[Au_{25}(PET)_{18}]^-$, $[Au_{25}(PET)_{18}]^0$, $Au_{28}(TBBT)_{20}$, $Au_{24}(TBPM)_{28}$, and $Au_{28}(TBBT)_{22}$ is given in Fig. 4.31 for clarity.

For comparison, the structure anatomy of the frameworks of $Au_{28}(TBPM)_{20}$ and $Au_{28}(TBPM)_{22}$ is given in Fig. 4.32. Each of them consists of a vertex-sharing tri-tetrahedra Au_{10} kernel structure, while the arrangement of the three blocks is slightly different. In $Au_{28}(TBPM)_{20}$, the Au_{10} core is further capped by four Au atoms, constituting the Au_{14} structure with five tetrahedral blocks (assembled via sharing the vertex or edge). The Au_{14} structure is further protected by two Au_3S_4 and four Au_2S_3 staple motifs (note: the structure of $Au_{28}(TBPM)_{20}$ could also be comprehend from the general structural volution principal of the FCC-packed gold clusters, Fig. 4.82). By contrast, the Au_{10} core of $Au_{28}(TBPM)_{22}$ is ligated by two Au_2S blocks and two bridging S atoms (of TBPM ligand) to form the $Au_{14}S_4$ structure, which is further protected by two Au_3S_4 and two Au_4S_5 staple motifs.

In addition to the detailed mechanistic study on the ligand induced size-growth of metal nanoclusters mentioned above, there are also some interesting mechanistic insights on the size-growth of phosphine and halide co-protected Au_{11} clusters. It takes over 10 years to elucidate some of the mechanistic details on the size-conversion from Au_{11} to Au_{25} clusters.

In 2005, the size-growth from Au_{11} to Au_{25} has been accomplished via the ligand exchange of the undecagold Au_{11} with GSH [54]. In view of the cluster peaks of $[Au_{11}(PPh_3)_8Cl_2]^+$ and $[Au_{11}(PPh_3)_8Cl]^{2+}$ on ESI-MS, and the same experimental outcome of the as-prepared Au_{11} clusters with the commercial available $Au_{11}(PPh_3)_8Cl_3$, Tsukuda and co-workers suggested the main components of the Au_{11} cluster precursor to be $Au_{11}(PPh_3)_8Cl_3$ (Fig. 4.33A. Herein, it is noteworthy that the undecagold Au_{11} used here is determined to be a mixture of $Au_{11}(PPh_3)_7Cl_3$ and $[Au_{11}(PPh_3)_8Cl_2]Cl$, and predominantly $Au_{11}(PPh_3)_7Cl_3$ in a latter study [55]). Upon mixing the $CHCl_3$ solution of Au_{11} precursors with the aqueous solution of excessive GSH, the ligand exchange occurs smoothly, and could be easily identified from the changed color of the upper aqueous phase (from colorless to red, Fig. 4.33B). The UV-vis adsorption spectra, ESI-MS and PAGE (polyacrylamide gel electrophoresis) analysis indicates the predominant formation of $Au_{25}(SG)_{18}$ clusters (associated with some smaller sized clusters). Preliminary mechanistic explorations were conducted. The TEM and diffuse reflectance spectrum analysis on the size distribution of the dark-brown floccules at the medium reaction stage (2.5 h, see Fig. 4.33B for an example) indicates the average core-size of 1.5 nm (Fig. 4.33C), demonstrating the ligand-exchange induced aggregation of the Au_{11} precursor clusters. Meanwhile, the composition of the aqueous phase is relatively stable under N_2 atmosphere, while the smaller sized $Au_n(SG)_m$ clusters (with characteristic peaks <600nm on UV-vis spectroscopy) gradually changes to $Au_{25}(SG)_{18}$ under aerobic conditions (Fig. 4.33B inset vs Fig. 4.33D inset). In view of the remarkably changed core-size, the overall size-growth from Au_{11} was suggested to occur via the core-aggregation (possibly start with the replacement of $(PPh_3)AuCl$ moieties by thiolate [56]), and size-focusing processes to generate the kinetically and/or thermodynamically stable Au_{25} products.

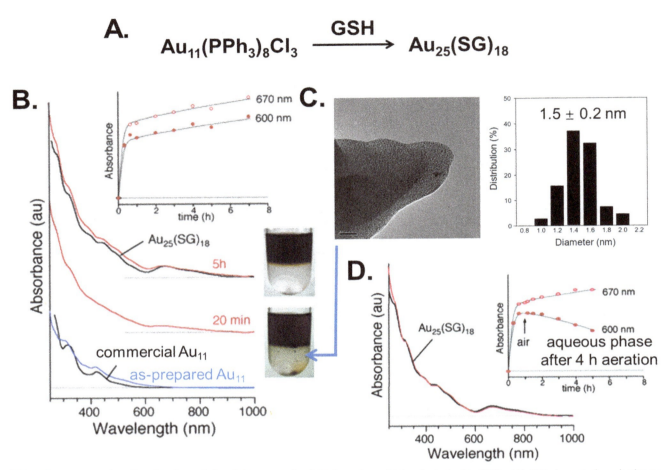

FIG. 4.33 The two-phase ligand exchange induced size-conversion from the undecagold Au_{11} cluster to $Au_{25}(SG)_{18}$ (A), UV-vis spectral monitoring on the ligand exchange of the as-prepared Au_{11} provided (with the UV-vis spectrum of the commercial Au_{11} for comparison) after 20 min and 5 h (B), the digital photo demonstrating the phase-transferred clusters, and the time-dependent changes in absorption intensity at 600 and 670 nm of the aqueous phase (inset); TEM analysis on the dark-brown floccules collected from the biphasic interface after reaction for 2.5 h (C), and the comparison of the UV-vis spectra of the aqueous phase after aeration (D) for 4 h (red line), the $Au_{25}(SG)_{18}$ samples (black line), and time-dependent changes in absorption intensity at 600 and 670 nm of the aqueous phase upon treating with aerobic conditions (inset). ((B—D) *Reproduced from Y. Shichibu, Y. Negishi, T. Tsukuda, T. Teranishi, Large-scale synthesis of thiolated Au_{25} clusters via ligand exchange reactions of phosphine-stabilized Au_{11} clusters, J. Am. Chem. Soc. 127 (2005) 13464–13465, with permission from the ACS Publications (Copyright 2005).)*

In addition to the biphasic ligand exchange induced size-growth of $Au_{11} \rightarrow Au_{25}$ mentioned above, the monophasic ligand exchange of Au_{11} clusters with excessive $C_nH_{2n+1}SH$ ($n=2, 8, 10, 12, 14, 16,$ and 18) in CH_2Cl_2 also produces Au_{25} clusters, but with the co-protection of phosphine, chloride and thiolate ligands [57]. The atomic structure of the produced Au_{25} clusters has been determined to be rod-like, bi-icosahedral $[Au_{25}(PPh_3)_{10}(SC_2H_5)_5Cl_2](SbF_6)_2$ by SCXRD analysis (Fig. 4.34).

Aside from the aforementioned size-conversion of the Au_{11} clusters to the icosahedral/rod-like Au_{25} clusters (with different ligand shell), the ligand exchange induced size-maintained conversion (i.e., $Au_{11} \rightarrow Au_{11}$) has also been reported recently. In an earlier study of Hutchison and co-workers, preliminary experimental outcome on the ligand-exchange of a phosphine protected Au_{11} cluster, $Au_{11}(PPh_3)_8Cl_3$ (as elucidated in a subsequent study, this cluster should be written

FIG. 4.34 Size-maintained ligand exchange of $Au_{11}(PPh_3)_8Cl_3$ with hexanethiol and the related mechanistic proposals.

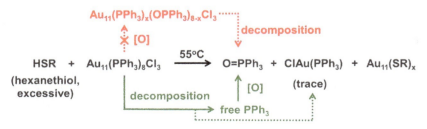

as $[Au_{11}(PPh_3)_8Cl_2]Cl$ [55]), with a series of ω-substituted alkanethiols has been reported [58]. The ligand exchange occurs smoothly via stirring the mixture of the $Au_{11}(PPh_3)_8Cl_3$ precursor with excess thiol at 55°C. The removal of the surface phosphine ligands has been evidenced by XPS and NMR tests. Meanwhile, based on the great similarity in the UV-vis spectra of the ligand-exchange precursor and the product clusters, and the preserved metal core sizes determined by TEM characterizations (transmission electron microscopy, ~0.8 nm), Hutchison et al. proposed a maintained Au_{11} core structure after the ligand exchange.

In a following study of Hutchison's group, the maintained core-size in the ligand exchange reaction of the Au_{11} clusters and thiol ligands were verified by mechanistic studies and trapping experiments [59] Compared with the preliminary study, different types of thiol ligands (including neutral/anionic, organic/water-soluble, alkyl/aromatic thiols) could be used as the exchanging reagents, and all of them are able to undergo complete ligand exchange with the Au_{11} cluster precursors (with slightly modified reaction conditions in some cases). The reaction of $Au_{11}(PPh_3)_8Cl_3$ with hexanethiol was used as the modeling reaction, and the ^{31}P NMR test indicates the predominance of triphenylphosphine oxide (O=PPh_3), trace amount of $ClAu(PPh_3)$, and absence of phosphine ligands bound on the cluster at the end of the reaction. The PPh_3-trapping experiments with 1-azido-2,4-dinitrobenzene [60] indicates the in situ generated free PPh_3 during the ligand exchange (caused by the decomposition of the Au_{11} precursor), precluding the mechanistic possibility for the successive oxidation-dissociation steps (Fig. 4.34). Meanwhile, the removal of free PPh_3 is pivotal to the overall ligand exchange, because the ligand exchange was remarkably inhibited when excessive PPh_3 are present. To this end, the trace $ClAu(PPh_3)$ was suggested to be formed as a by-product during the partial decomposition of the Au_{11} precursor. This mechanism is distinct to the thiol-to-$ClAu(PPh_3)$ exchange mechanism proposed earlier in ligand exchange of $Au_{11}(PPh_3)_{21}Cl_5$ [61].

To this end, distinct experimental outcomes, as whether the atomic size is preserved ($Au_{11} \rightarrow Au_{11}$) or become larger ($Au_{11} \rightarrow Au_{25}$) in the ligand exchange of Au_{11} clusters with thiolate ligands, have been reported. To precisely elucidate the origin for the discrepancy on the ligand exchange of Au_{11} clusters, Hutchison and co-workers reported a comprehensive study in 2014 [55]. First, two types of crystals were identified, and separated mechanistically. SCXRD indicates the red plates and orange needles to be $[Au_{11}(PPh_3)_8Cl_2]Cl$ (note: the Cl inside and outside the bracket of the formular denotes the bound chloride ligand on the metallic core and the chloride counter-anion) and $Au_{11}(PPh_3)_7Cl_3$, respectively. Specifically, each of the two clusters could be readily prepared with high purity via the developed synthetic methods. ^1H and ^{31}P NMR spectra show clear but distinct groups of chemical shifts, verifying the different composition of the two clusters. Meanwhile, the UV-vis absorption spectra of these two clusters show quite similar characteristic peaks, while the main difference lies in the tiny shift of the characteristic peaks and the appearance of smaller peaks on UV-vis spectrum of $Au_{11}(PPh_3)_7Cl_3$ (Fig. 4.35A). The biphasic reaction of the $CHCl_3$ solution of $Au_{11}(PPh_3)_7Cl_3$ and $[Au_{11}(PPh_3)_8Cl_2]Cl$ with the aqueous solution of GSH in biphasic system result in distinct color of the upper aqueous phase (Fig. 4.35B). Meanwhile, the UV-vis absorption spectra are also significantly differentiated by the presence (in the reaction system of $Au_{11}(PPh_3)_7Cl_3$ or the mixture of the two cluster reagents) or absence (reaction of $[Au_{11}(PPh_3)_8Cl_2]Cl$) of the peak at ~670 nm, demonstrating the formation of the spherical $Au_{25}(SG)_{18}$ clusters or not. The distinct experimental outcome was further evidenced by PAGE and ESI-MS characterization. Interestingly, reaction of either $[Au_{11}(PPh_3)_8Cl_2]Cl$ or $Au_{11}(PPh_3)_7Cl_3$ with largely excessive GSH (e.g., ~430 equiv) at 35°C produces $Au_{25}(SG)_{18}$. The experimental outcome of the different systems are given in Fig. 4.35C for clarity reasons.

In view of the same metal core but variation in the ligand shell composition of the two types of Au_{11} clusters, Hutchison and co-workers suggest that the higher reactivity of $Au_{11}(PPh_3)_7Cl_3$ is caused by the lower steric hindrance and the reduced electron density of the metallic core (due to the replacement of one PPh_3 with one chloride). This proposal correlates with the higher stability of $[Au_{11}(PPh_3)_8Cl_2]Cl$ than that of $Au_{11}(PPh_3)_7Cl_3$ in CH_2Cl_2 solvent. To this end, the origin for the discrepancy on the ligand exchange of the undecagold Au_{11} cluster to Au_{25} or smaller sized clusters were finally elucidated.

In addition to the aforementioned systems, there are also some other ligand exchange reactions starting from phosphine and halide co-protected Au_{11} clusters. For example, using $[Au_{11}(PPh_3)_8Cl_2]^+$ as the cluster precursor, the monometallic $[Au_{25}(PPh_3)_{10}(SC_nH_{2n+1})_5Cl_2]^{2+}$ [57] or the bi-metallic alloy $[Ag_xAu_{25-x}(PPh_3)_{10}(PET)_5Cl_2]^{2+}$ ($x=0$–13) [62] has been prepared (Fig. 4.36) via the addition of n-alkanethiol or AgPET. So far, a detailed mechanistic study, especially an atomic level of mechanistic understanding, is still highly challenging.

Size-maintained ligand exchange of atomically precise metal clusters

In the above sections, it can be clearly seen that the combination of spectral (such as UV-vis and MALDI-MS/ESI-MS) monitoring and DFT calculations has become a powerful strategy for mechanistic study in cluster conversion systems. The molecular-like optical absorption of metal clusters lay the foundation to deduce the composition of the intermediate and product clusters, especially when the obtained spectrum is coincident with the reported one. Meanwhile, the changes in

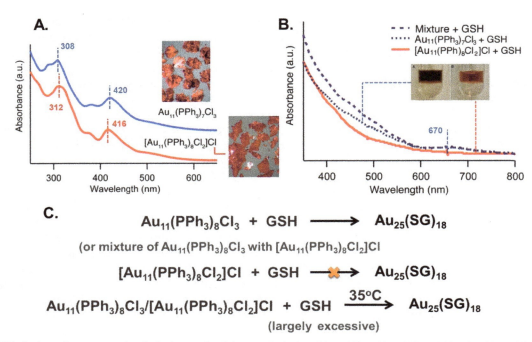

FIG. 4.35 UV-vis absorption spectra and optical micrographs of the crystal solution of $Au_{11}(PPh_3)_7Cl_3$ and $[Au_{11}(PPh_3)_8Cl_2]Cl$ (A), UV-vis absorption spectra of the aqueous phase in two-phase ligand exchange between $Au_{11}(PPh_3)_7Cl_3$, $[Au_{11}(PPh_3)_8Cl_2]Cl$ or the mixture with GSH, associating with the digital photo of the biphasic system after ligand-exchange (B), and the reaction outcome for the biphasic ligand exchange of Au_{11} clusters with GSH under different conditions. *(Reproduced from L.C. McKenzie, T.O. Zaikova, J.E. Hutchison, Structurally similar triphenylphosphine-stabilized undecagolds, $Au_{11}(PPh_3)_7Cl_3$ and $[Au_{11}(PPh_3)_8Cl_2]Cl$, exhibit distinct ligand exchange pathways with glutathione, J. Am. Chem. Soc. 136 (2014) 13426–13435, with permission from the ACS Publications (Copyright 2014).)*

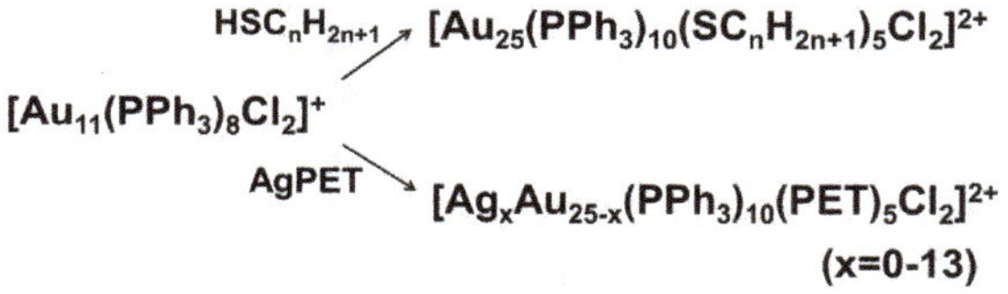

FIG. 4.36 The ligand exchange induced size-growth of the phosphine protected Au_{11} clusters to the rod-like M_{25} (M=Au/Ag) clusters.

molecular formula (e.g., consumption of the precursor clusters, formation and transformation of the intermediate species) along the reaction time could be easily figured out from mass spectral monitoring. Nevertheless, the changes in the atomic packing modes could be hardly deduced from optical tests or MS for reaction systems wherein the spectral characteristics (the group of characteristic peaks) have been unprecedented. To this end, NMR spectra could be used to deduce the changes in the chemical environment of the surface ligands, and especially for the structurally rigid and symmetric cluster systems. For this reason, tailoring the changes in characteristic peaks on NMR could provide important insights into the kinetics of the ligand-exchange reactions. For example, in the recent study by Bürgi and co-workers [63] the time course NMR and MS characterizations have been used to shed light on the mechanism of the ligand exchange of the structurally highly symmetric, icosahedral $[Au_{25}(SR)_{18}]^{-1}$ clusters. Two groups of ligand exchange reactions were conducted, i.e., cluster-ligand reactions wherein the Au_{25} clusters, $[Au_{25}(PET)_{18}]^{-1}$ or $[Au_{25}(SBut)_{18}]^{-1}$ (HPET and HSBut denote 2-phenylethanethiol or butanethiol) react with certain amount of foreign thiolate ligands (HSBut and HPET accordingly, Path I and II in Fig. 4.37); and cluster-cluster reactions wherein the two Au_{25} clusters react with each other (Path III in Fig. 4.37).

In both cluster-ligand reaction systems ($[Au_{25}(PET)_{18}]^-$+HSBut; and $[Au_{25}(SBut)_{18}]^-$+HPET), 18 equiv. of the foreign ligands were added to the Au_{25} cluster solvent to ensure the equivalent amount of two types of ligands. $[Au_{25}(PET)_{12}(SBut)_6]^-$ is the dominate product in both reaction systems, demonstrating the higher affinity of the PET ligands to the Au_{25} framework than that of the SBut ligands. Specifically, conducting the ligand exchange with largely excessed foreign ligand in the two-stage procedure (via a partial etching with a subsequent complete etching stages) [64] under room temperature (to avoid the high temperature induced core-size transformation) could generate the Au_{25} clusters with completely exchanged ligands.

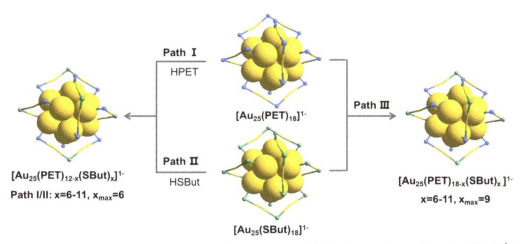

FIG. 4.37 Illustrative diagram for the cluster-ligand (Path I, II) and cluster-cluster (Path III) ligand exchange of the $[Au_{25}(SR)_{18}]^{-1}$ clusters. The x_{max} denotes the number of x in the main product, and only the M-S framework are shown for clarity reasons.

In addition to the cluster-ligand exchange, the intercluster ligand exchange (between $[Au_{25}(SBut)_{18}]^-$ and $[Au_{25}(PET)_{18}]^-$) were conducted. The time-dependent MALDI-MS clearly shows maintained atomic size but varied ligand contents. The molecular ratio of $[Au_{25}(PET)_{18}]^-$ vs $[Au_{25}(SBut)_{18}]^-$ dominates the ratio of PET and SBut ligands in the formed $Au_{25}(PET)_{18-x}(SBut)_x$, while x ranges from 6 to 11 for a 1:1 reaction, and 2–13 for a 1:0.8 reaction after reacting for 48 h. The statistical distribution of ligands (fitting binomial distribution) in the cluster products demonstrates that there is neither preferred ligand composition nor neighboring preference of the two types of ligands in the intercluster reactions. Meanwhile, the NMR tracking on the intercluster ligand exchange system did not show signals of free ligands, and therefore the reaction was suggested to occur via a collision driven pathway, which is distinct from the decomposition-driven pathway in the cluster-ligand exchange reactions.

In addition to the ligand exchange of the homometallic Au_{25} clusters, the doping effect of the central Pd atom in the alloy Pd_1Au_{24} clusters in both cluster-ligand and intercluster ligand exchange reactions has been recently studied by Negishi and co-workers [65]. According to the time-dependent MALDI-MS characterization, the number of the incoming HSR ligand (largely excessive in the reaction system) in the formed Au_{25} cluster increases with the prolonged reaction time. Meanwhile, the average number of the exchanged ligands of the $[Pd_1Au_{24}(SC_{12}H_{25})_{18}]^0$ system is ~1.5–4.2 times higher than that of $[Au_{25}(SC_{12}H_{25})_{18}]^{1-}$ system (Fig. 4.38A vs B), demonstrating the accelerated ligand exchange after the Pd-doping. The reason is mainly attributed to the decreased number of free electrons in the $[Pd_1Au_{24}(SC_{12}H_{25})_{18}]^0$ precursor compared with that of the $[Au_{25}(SC_{12}H_{25})_{18}]^{1-}$ (6 vs 8), and this proposal is further supported by the relatively faster ligand exchange of the neutral $[Au_{25}(SC_{12}H_{25})_{18}]^0$ than that of the anionic one. Meanwhile, the intercluster ligand exchange between the two Au_{25} clusters occurs faster than that between the two Pd_1Au_{24} clusters (Fig. 4.38C vs D), indicating that the doping of the central Pd suppresses the unimolecular dissociation of the ligand/metal-ligand moiety from the ligand shell.

In addition to the framework-maintained ligand exchange of the aforementioned $M_{25} \to M_{25}$ systems, the size-maintained interconversion of $Au_{28}(TBBT)_{20}$ with $Au_{28}(SCy)_{20}$ represents a rare example of ligand-exchange induced

A. $[Au_{25}(SC_{12}H_{25})_{18}]^{1-/0} + HSR \longrightarrow [Au_{25}(SC_{12}H_{25})_{18-x}(SR)_{18-x}]^0$

B. $[PdAu_{24}(SC_{12}H_{25})_{18}]^0 + HSR \longrightarrow [PdAu_{24}(SC_{12}H_{25})_{18-x}(SR)_{18-x}]^{1-}$

(HSR=HSC_6H_{13}, HSC_8H_{17}, $HSC_{10}H_{21}$, HPET, $HSC_{16}H_{33}$)

C. $[Au_{25}(SC_{12}H_{25})_{18}]^{1-} + [Au_{25}(SC_{10}H_{21})_{18}]^{1-}$
$\longrightarrow$ $[Au_{25}(SC_{12}H_{25})_{18-x}(SC_{10}H_{21})_x]^{1-}$
$[Au_{25}(SC_{10}H_{21})_{18-x}(SC_{12}H_{25})_x]^{1-}$ (x=0-4)

D. $[Pd_1Au_{24}(SC_{12}H_{25})_{18}]^0 + [Pd_1Au_{24}(SC_{10}H_{21})_{18}]^0$
$\longrightarrow$ $[Pd_1Au_{24}(SC_{12}H_{25})_{18-x}(SC_{10}H_{21})_x]^0$
$[Au_{25}(SC_{10}H_{21})_{18-x}(SC_{12}H_{25})_x]^0$ (x=0-3)

FIG. 4.38 The ligand exchange of the icosahedral M_{25} clusters with the foreign thiol ligands (A and B) the intercluster ligand-exchange reaction between monometallic Au_{25} (C) and the alloy Pd_1Au_{24} clusters (D).

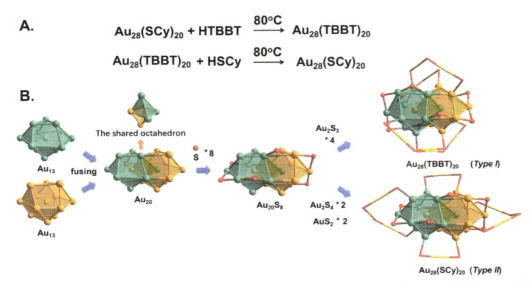

FIG. 4.39 Ligand exchange induced interconversion of the two types of Au_{28} clusters (A), and the structural anatomy of the framework of $Au_{28}(TBBT)_{20}$ and $Au_{28}(SCy)_{20}$ (B).

"isomerization" (generalized isomerization-the number of metal atoms and the type of ligand was preserved) [66]. According to Jin's experiments, adding excessive HSCy/HTBBT to the solution of $Au_{28}(TBBT)_{20}/Au_{28}(SCy)_{20}$ at 80°C quickly yields the isomeric cluster with completely exchanged ligands (Fig. 4.39A). As shown in Fig. 4.39B, the single structure of $Au_{28}(TBBT)_{20}$ (denoted as Type I) and $Au_{28}(SCy)_{20}$ (denoted as Type II) each comprises an octahedron-sharing bi-cubic octahedral FCC-Au_{20} core structure, with eight μ_2-SR protecting ten Au atoms of the Au_{20} core surface. In $Au_{28}(TBBT)_{20}$, the Au_{20} core is further protected by four $Au_2(TBBT)_3$ staple motifs, each of which acts as an interblock ligand, connecting two Au atoms in the two different cubic octahedra. By contrast, in $Au_{28}(SCy)_{20}$, the two $Au_3(SCy)_4$ staple motifs are inner-block ligands (i.e., each protects a pairs of Au atoms in same cubic octahedron), and only the two $Au(SCy)_2$ function as interblock ligands.

Due to the distinct UV-vis spectra of the two isomeric Au_{28} clusters (Fig. 4.40A), the changes in UV-vis spectra could be used conveniently to track the transformation of one Au_{28} to another (Fig. 4.40B and C). The relative stability of Type I structure for $Au_{28}(TBBT)_{20}$ and Type II structure for $Au_{28}(PET)_{20}$ has been well reproduced by DFT calculations, and was

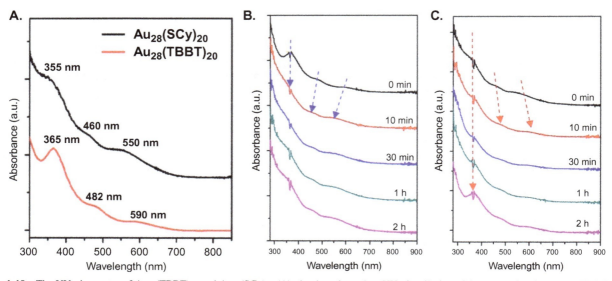

FIG. 4.40 The UV-vis spectra of $Au_{28}(TBBT)_{20}$ and $Au_{28}(SCy)_{20}$ (A), the time-dependent UV-vis tailoring of the conversion from $Au_{28}(TBBT)_{20}$ to $Au_{28}(SCy)_{20}$ (B) and its reverse reaction (C). *(Reproduced from Y. Chen, C. Liu, Q. Tang, C. Zeng, T. Higaki, A. Das, D.-E. Jiang, N.L. Rosi, R. Jin, Isomerism in $Au_{28}(SR)_{20}$ nanocluster and stable structures, J. Am. Chem. Soc. 138 (2016) 1482–1485, with permission from the ACS Publications (Copyright 2016).)*

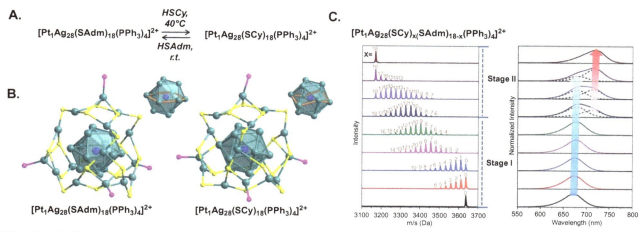

FIG. 4.41 The ligand exchange induced interconversion between [Pt$_1$Ag$_{28}$(SAdm)$_{18}$(PPh$_3$)$_4$]$^{2+}$ and [Pt$_1$Ag$_{28}$(SCy)$_{18}$(PPh$_3$)$_4$]$^{2+}$ (A), the illustrative diagram for the metal framework and the kernel structure in the two clusters (B), and the time-dependent ESI-MS and fluorescence monitoring of the size-conversion from [Pt$_1$Ag$_{28}$(SAdm)$_{18}$(PPh$_3$)$_4$]$^{2+}$ to [Pt$_1$Ag$_{28}$(SCy)$_{18}$(PPh$_3$)$_4$]$^{2+}$ (C). ((C) Reproduced from X. Kang, L. Huang, W. Liu, L. Xiong, Y. Pei, Z. Sun, S. Wang, S. Wei, M. Zhu, Reversible nanocluster structure transformation between face-centered cubic and icosahedral isomers, Chem. Sci. 10 (2019) 8685–8693, with permission that is licensed under the Creative Commons Attribution 3.0 International License, copyright the authors 2019.)

found to originate from the dominance of KS-DFT energy (with both DFT exchange energy and Kohn-sham kinetic energy, favors the Type I-framework) or Van der Walls interactions (favors the Type II-framework).

Compared with the dominating ligand shell rearrangement in the aforementioned interconversion between Au$_{28}$(TBBT)$_{20}$ and Au$_{28}$(SCy)$_{20}$, remarkable change occurs in both the motifs and the core structure in the ligand etching induced isomerization of [Pt$_1$Ag$_{28}$(SAdm)$_{18}$(PPh$_3$)$_4$]$^{2+}$ with [Pt$_1$Ag$_{28}$(SCy)$_{18}$(PPh$_3$)$_4$]$^{2+}$ clusters (Fig. 4.41A) [67]. In Zhu's study, the reaction of HSCy with [Pt$_1$Ag$_{28}$(SAdm)$_{18}$(PPh$_3$)$_4$]$^{2+}$ at 40°C generates the [Pt$_1$Ag$_{28}$(SCy)$_{18}$(PPh$_3$)$_4$]$^{2+}$ in >80% yield after 2h. The reverse reaction was achieved by reacting HSAdm with [Pt$_1$Ag$_{28}$(SCy)$_{18}$(PPh$_3$)$_4$]$^{2+}$ at room temperature for 3min. The atomic precision of [Pt$_1$Ag$_{28}$(SAdm)$_{18}$(PPh$_3$)$_4$]$^{2+}$ by SCXRD analysis [68] indicates an FCC-packed Pt@Ag$_{12}$ structure, which is further protected by a cage-like Ag$_{12}$S$_{18}$ motif, and four AgP groups (Fig. 4.41B). Similarly, the framework of [Pt$_1$Ag$_{28}$(SCy)$_{18}$(PPh$_3$)$_4$]$^{2+}$ could be viewed as a Pt@Ag$_{12}$ core protected by a cage-like Ag$_{12}$S$_{18}$ motif and four AgP groups. However, the Pt@Ag$_{12}$ core in [Pt$_1$Ag$_{28}$(SCy)$_{18}$(PPh$_3$)$_4$]$^{2+}$ adopts an icosahedral structure. The most prominent difference in the FCC- and icosahedral core structures lies in the planar Pt$_1$Ag$_6$ or chair-like Pt$_1$Ag$_6$ in the three-layered Ag$_3$-Pt$_1$Ag$_6$-Ag$_3$ structure (Fig. 4.41B inset). The distinct optical absorption characteristic between this pair of Pt$_1$Ag$_{28}$ clusters lies the foundation for a UV-vis monitoring on the conversion, while the identification of the four isosbestic points at 307, 338, 427, and 454nm indicates an unchanged rate-determining step throughout the isomerization. In addition, the combination of the time-dependent ESI-MS and fluorescence spectral monitoring demonstrates a two-stage mechanism for the conversion from [Pt$_1$Ag$_{28}$(SAdm)$_{18}$(PPh$_3$)$_4$]$^{2+}$ to [Pt$_1$Ag$_{28}$(SCy)$_{18}$(PPh$_3$)$_4$]$^{2+}$. In the first stage (Fig. 4.41C), the fluorescence spectra of the [Pt$_1$Ag$_{28}$(SAdm)$_{18}$(PPh$_3$)$_4$]$^{2+}$ precursors maintain, while the gradual replacement of -SAdm with -SCy results in only slightly redshift in the emission maximum. In the second stage, the fluorescence peak is remarkably broadened, caused by the overlap of the characteristic emission of the precursor with the product clusters. Only the ~720nm emission is preserved after the ligand etching, demonstrating the exclusive formation of [Pt$_1$Ag$_{28}$(SCy)$_{18}$(PPh$_3$)$_4$]$^{2+}$. According to these results, the two stages mentioned above are denoted as the motif-transformation (FCC structure remains) and kernel transformation processes (from FCC to icosahedron, further evidenced by EXAFS analysis), respectively.

With the combination of experimental and theoretical strategies, the mechanism of the ligand exchange of a fairly large gold nanocluster, i.e., Au$_{102}$(p-MBA)$_{44}$ with p-BBA (Fig. 4.42A, and p-BBAH denotes para-bromobenzene thiol), has been recently explored by Ackerson and co-workers [69]. According to the SCXRD analysis on the exchanged product of Au$_{102}$(p-MBA)$_{40}$(p-BBT)$_4$, the incoming p-BBA ligands are favorably incorporated into the Au sites with relatively lower steric hindrance (elucidated by structural analysis and DFT calculations on solvent accessibility surface, see Section 6.2.2.1 for the details). Meanwhile, the theoretical calculations indicate that the direct dissociation of the surface Au—S bond on the Au$_{102}$ cluster precursor is less feasible than the associative pathway. In the associative pathway, the p-BBAH first approaches to the target Au atom on cluster surface in a slightly exothermic pattern. The subsequent H-transfer and removal of one of the original thiolates binding on the attacked Au atom then occur with a remarkably higher energy demand

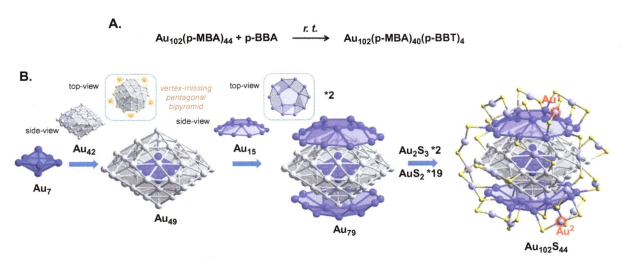

FIG. 4.42 The ligand exchange induced conversion of $Au_{102}(p\text{-MBA})_{44}$ (A) and the structure anatomy of the framework of $Au_{102}(p\text{-MBA})_{44}$ (the C and H atoms are omitted for clarity) and an illustration on the sterically less hindered Au^1/Au^2 sites (undergoing ligand exchange).

(compared with the first binding step). Interestingly, the p-BBAH approaching preference correlates with the steric effect, but is not completely determined by the steric effect. As shown in Fig. 4.42B, ligand exchange occurs favorably on the sterically less hindered Au^1/Au^2 sites (compared with all other surface Au atoms), while the exchange ratio of the Au^2 site is slightly higher than that of the sterically less hindered Au^1 site. The slight deviation of the exchange ratio with the steric effect has been suggested to originate from the different kinetics. This study provides pioneering mechanistic insights into the dynamics of ligand exchange, especially on such large systems (structurally comprehensive and computational costy).

Aside from the aforementioned examples, there are a variety of precedents on the two-phase ligand exchange induced size-maintained conversion (such as the reaction of the water-soluble $[Au_{18}(SG)_{14}]$ with the organic thiolate of HSCy to form the oil-soluble $[Au_{18}(SCy)_{14}]$ [70]). Due to the difficulty in figuring out the detailed interactions on the biphasic water-oil interference, the mechanistic understanding on these reaction systems is still quite limited. The development/utilization of novel analytical strategies (tolerated to the biphasic solution systems) and the more powerful theoretical simulation packages might be helpful in settling these issues.

Ligand etching induced size-reduction of atomically precise metal clusters

In addition to the size-growth and size-maintained conversion from one atomically precise metal nanocluster to another one, the ligand exchange induced size-reduction has also been reported lately [71]. In a recent study of Mandal and coworkers, adding HTBBT to the $Au_{25}(PET)_{18}$ solution under 80°C generates the mixed ligands protected $Au_{22}(SR)_{16}$ (SR = PET or TBBT) as the product (Fig. 4.43A) [72]. As the optical absorption characteristics of the Au_{25} precursor is distinct from that of the Au_{22} cluster products, the size-conversion was first evidenced by UV-vis analysis. In this context, the time-dependent MALDI-MS analysis was then used to determine the changes in chemical compositions along with the reaction procedures. As shown in Fig. 4.43B, the size-maintained ligand exchange ($Au_{25}(PET)_{18} \rightarrow Au_{25}(PET)_{18-x}(TBBT)_x$) occurs in the initial stage of the reaction (0–5 min, Stage I). After that (5 min–7 h, Stage II), a formal "disproportionation" reaction (due to its similarity with the typical disproportionation reactions in traditional chemistry) occurs, with the vanishing of the characteristic peaks of $Au_{25}(PET)_{18-x}(TBBT)_x$, and the appearance of cluster peaks corresponding to a relatively smaller-sized $Au_{22}(PET)_{18-x}(TBBT)_x$, and a larger sized cluster $Au_{28}(PET)_{18-x}(TBBT)_{x+2}$. After 17 h, the cluster peaks of $Au_{28}(PET)_{18-x}(TBBT)_{x+2}$ disappeared completely, while the peaks corresponding to Au_{22} cluster was retained (Stage III). According to the reaction yield analysis, the intermediate Au_{28} clusters convert to Au_{22} within the time interval of 7–17 h (i.e., Stage III), and $Au_{22}(PET)_4(TBBT)_{14}$ becomes the main product after 17 h (no further reaction occurs with prolonged heating time). The observations indicate the relatively higher thermodynamic stability of the Au_{22} cluster compared with that of the Au_{28} clusters.

DFT calculations conducted with Vienna ab initio simulation package (VASP) evidenced the thermodynamic ease of the ligand exchange of the Au_{25} precursor clusters, until 12 TBBT was incorporated. Further ligand exchange is slightly disfavored due to the endothermicity. The observation correlates with the primary size-maintained ligand exchange in MALDI-MS tracking. The latter stage of ligand exchange is mainly precluded by the high steric hindrance among the bulky

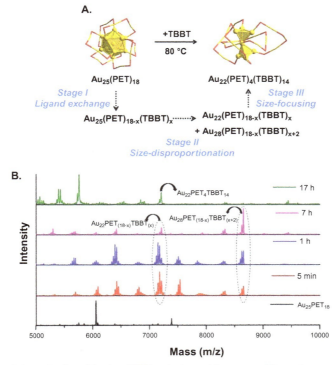

FIG. 4.43 The ligand etching induced size-reduction of the $Au_{25}(PET)_{18}$ cluster, and the proposed three-stage mechanism (linked with the dashed lines) (A), and the time-dependent MALDI-MS of the reaction system (B). *(B) Reproduced from A. George, A. Sundar, A.S. Nair, M.P. Maman, B. Pathak, N. Ramanan, S. Mandal, Identification of intermediate $Au_{22}(SR)_4(SR')_{14}$ cluster on ligand-induced transformation of $Au_{25}(SR)_{18}$ nanocluster, J. Phys. Chem. Lett. 10 (2019) 4571–4576, with permission from the ACS Publications (Copyright 2019).)*

TBBT ligands. Meanwhile, according to DFT calculations, the $Au_{22}(PET)_4(TBBT)_{14}$ product comprises a vertex-sharing bi-tetrahedral Au_7 core, which is protected by one ring-like $Au_6(PET)_4(TBBT)_2$ and three $Au_3(TBBT)_4$ staple motifs (Fig. 4.43A).

Compared with the extensively reported size-conversion between two similar-sized clusters, the size-conversion between clusters with remarkably different atomic sizes is relatively rare. In the recent study of Bakr and co-workers, both the two-phase or the single-phase ligand-exchange has been developed to regulate the size-reduction from $[Ag_{44}(SR)_{30}]^{4-}$ to $[Ag_{25}(SR)_{18}]^-$ (Fig. 4.44A) [73]. In the biphasic reaction system, the aqueous solution of $[Ag_{44}(p\text{-MBA})_{30}]^{4-}$ quickly reacts with the CH_2Cl_2 solution of 2,4-dimethylbenzenethio ($HSPh^{2,4-}Me_2$) in the presence of methanolic solution of PPh_4Br (Fig. 4.44B inset) to generate $[Ag_{25}(SPh^{2,4-}Me_2)_{18}]^-$ within 2 min (evidenced by UV-vis spectra and ESI-MS). In the monophasic ligand exchange, the CH_2Cl_2-soluble $[Ag_{44}(SPh^{4-}F)_{30}]^{4-}$ ($HSPh^{4-}F$ denotes 4-fluorobenzenethiol) has been used as the cluster precursor, and a precisely controlled concentration of reactants could slow down the reaction rates for a better monitoring on the reaction dynamics (the 2-min reaction is too fast to track). According to the time-dependent ESI-MS characterization (Fig. 4.44B), the size-maintained ligand exchange on $[Ag_{44}(SPh^{4-}F)_{30}]^{4-}$ occurs in the first 30 min to generate $[Ag_{44}(SPh^{4-}F)_{12}(SPh^{2,4-}Me_2)_{18}]^{4-}$ as the main product, associating with the formation of $[Ag_{44}(SPh^{4-}F)_1(SPh^{2,4-}Me_2)_{29}]^{4-}$ with almost completely ligand-exchange. In view of the atomic structure of $[Ag_{44}(SR)_{30}]^{4-}$ (protection of the two layered $Ag_{12}@Ag_{20}$ metallic core with six $Ag_2(SR)_5$ motifs) [47,74], the 18 new-coming $-SPh^{2,4-}Me_2$ ligands in the precursor were suggested to distribute far apart to reduce steric hindrance and structural perturbation to the metal framework. In the following 30 min, the relative intensity of the cluster peaks of $[Ag_{44}(SPh^{4-}F)_1(SPh^{2,4-}Me_2)_{29}]^{4-}$ becomes comparable to that of $[Ag_{44}(SPh^{4-}F)_{12}(SPh^{2,4-}Me_2)_{18}]^{4-}$, and some other cluster peaks corresponding to the decomposition species of Ag_{44} clusters, such as $Ag_{43}(SPh^{4-}F)_{10}(SPh^{2,4-}Me_2)_{18}$, and $Ag_{42}(SPh^{4-}F)_{10}(SPh^{2,4-}Me_2)_{17}$, were also observed. After 2 h, the intensity of the cluster peaks of $[Ag_{44}(SPh^{4-}F)_1(SPh^{2,4-}Me_2)_{29}]^{4-}$ overwhelms that of $[Ag_{44}(SPh^{4-}F)_{12}(SPh^{2,4-}Me_2)_{18}]^{4-}$, and some relatively larger-sized clusters, such as Ag_{46}-Ag_{50} were also identified on ESI-MS. After this stage, most $SPh^{4-}F$ on the Ag_{44} precursors were replaced by $SPh^{2,4-}Me_2$, and the cluster framework is destabilized by both the steric hindrance and the interligand π-π interactions. For this reason, the cluster decomposition occurs in the subsequent 1 h, resulting in the significantly strengthened intensity of the cluster peak of $[Ag_{25}(SPh^{4-}F)_1(SPh^{2,4-}Me_2)_{17}]^-$ and also some fragmentation peaks. After 3 h, the characteristic peaks

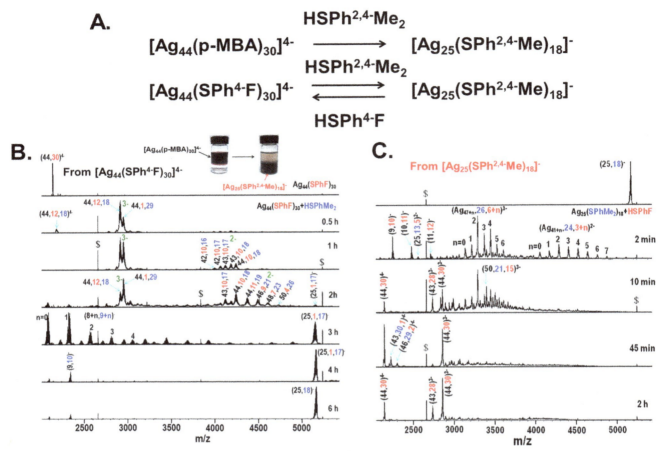

FIG. 4.44 The ligand-exchange induced interconversion of $[Ag_{44}(SR)_{30}]^{4-}$ and $[Ag_{25}(SR)_{18}]^-$ (A), time-dependent ESI-MS tracking on the single-phase size-conversion from $[Ag_{44}(SPh^{4-}F)_{30}]^{4-}$ to $[Ag_{25}(SPh^{2,4-}Me_2)_{18}]^-$ (B, inset: digit photo for the two-phase ligand exchange from aqueous $[Ag_{44}(p-MBA)_{30}]^{4-}$ to the hydrophobic $[Ag_{25}(SPh^{2,4-}Me_2)_{18}]^-$), and the time-dependent ESI-MS tracking on size-conversion from $[Ag_{25}(SPh^{2,4-}Me_2)_{18}]^-$ to $[Ag_{44}(SPh^{4-}F)_{30}]^{4-}$ (C). *(Reproduced from M.S. Bootharaju, C.P. Joshi, M.J. Alhilaly, O.M. Bakr, Switching a nanocluster core from hollow to nonhollow, Chem. Mater. 28 (2016) 3292–3297, with permission from ACS Publications (Copyright 2016).)*

ascribing to the intermediate clusters and the fragments are also vanished, and the intensity of $[Ag_{25}(SPh^{4-}F)_1(SPh^{2,4-}Me_2)_{17}]^-$ is significantly increased, demonstrating the size-focusing process in the latter stage of the ligand exchange. Finally, the excessive $HSPh^{2,4-}Me_2$ in the reaction system results in the formation of the $[Ag_{25}(SPh^{2,4-}Me_2)_{18}]^-$ with completely ligand exchange. The changes in the cluster charge (from −4 in Ag_{44} precursor to −1 in Ag_{25} product) and the formation of the close-shell cluster with 8 free electrons are suggested the main driving force for the size-reduction.

Distinct to the slow size-reduction from $[Ag_{44}(SPh^{4-}F)_{30}]^{4-}$ to $[Ag_{25}(SPh^{2,4-}Me_2)_{18}]^-$ clusters mentioned above, the ligand exchange of $Ag_{25}(SPh^{2,4-}Me_2)_{18}$ with excessive $HSPh^{4-}F$ is finished within 60 min. The time-dependent ESI-MS (Fig. 4.44C) shows comprehensive cluster peaks after only 2 min. The cluster peaks corresponding to $[Ag_{25}(SPh^{2,4-}Me_2)_{13}(SPh^{4-}F)_5]^{2-}$ demonstrates the size-maintained ligand exchange (the first stage: ligand exchange). After that, the Au$_{25}$ clusters undergo rapid dimerization (the second stage) to give the larger-sized $[Ag_{50}(SPh^{2,4-}Me_2)_{26}(SPh^{4-}F)_{10}]^{4-}$ cluster. The removal of one -$SPh^{4-}F$ then occurs to generate a more stable Ag_{50} cluster, i.e., $[Ag_{50}(SPh^{2,4-}Me_2)_{26}(SPh^{4-}F)_9]^{3-}$ cluster (electronically stable, close-shell electron structure with 18 free electrons). From $[Ag_{50}(SPh^{2,4-}Me_2)_{26}(SPh^{4-}F)_9]^{3-}$, a formal size disproportionation (the third stage) could occur to form predominantly two series of clusters, i.e., $[Ag_{47+n}(SPh^{2,4-}Me_2)_{26}(SPh^{4-}F)_{6+n}]^{3-}$ ($n=0–6$), and $[Ag_{41+n}(SPh^{2,4-}Me_2)_{24}(SPh^{4-}F)_{3+n}]^{2-}$ ($n=0–7$), associated with the increased concentration of $[Ag_{44}(SPh^{2,4-}Me_2)_{24}(SPh^{4-}F)_6]^{2-}$. All these three stages occur within 10 min (due to the complexity of the cluster peaks within 2 and 10 min, the details for the size-conversion, such as the sequence of ligand exchange and dimerization are still to be clarified). With prolonged reaction time, the intermediate clusters were all consumed, and the remaining $SPh^{2,4-}Me_2$ on the mixed ligand protected Ag_{44} cluster are completely replaced by $SPh^{4-}F$ ligand, and $[Ag_{44}(SPh^{4-}F)_{30}]^{4-}$ is finally formed as the main product.

According to the aforementioned mechanistic studies on the ligand etching induced size-reduction ($Au_{25} \to Au_{22}$, $Ag_{44} \to Ag_{25}$, $Au_{25} \to Au_{50} \to Au_{44}$), a general mechanism could be deduced. These reactions all start with the rapid size-maintained ligand-exchange (commonly occurs within several minutes). After incorporating certain number of foreign ligands, the original cluster framework becomes fragile and easily decomposes to generate different intermediate species (such as cluster species and metal complexes). Finally, a slow size-focusing with the gradual consumption of the intermediate species occurs to accumulate the thermodynamically most stable cluster products. This proposal corroborates the preliminary understandings on the ligand-exchange and nucleation mechanism of the metal nanoclusters, and more importantly, elucidates the kinetics and inherent structural correlations of different sized nanoclusters at molecular level.

Aside from the aforementioned size-reductions, the thiolate ligand (using HTBBT or HCPT, and HCPT is short for cyclopentanethiol) etching induced size-reduction of $Au_{38}(PET)_{24}$ to $Au_{36}(TBBT)_{24}$ [75]/$Au_{36}(CPT)_{24}$ [76] has also been reported, and both reactions were conducted at heating conditions (80°C, Fig. 4.45A). Time-dependent ESI-MS and UV-vis absorption spectra were used to shed light on the mechanism of HTBBT etching induced size-conversion of $Au_{38}(PET)_{24}$ to $Au_{36}(TBBT)_{24}$.[36]. First, similar to the aforementioned ligand exchange reactions of the Au_{25} systems, ESI-MS tracking on the system of HTBBT with $Au_{38}(PET)_{24}$ indicates an initial size-maintained ligand exchange (Stage I, Fig. 4.45B). Herein, 2–12 PET ligands in the $Au_{38}(PET)_{24}$ precursor are replaced with TBBT within 5 min, and the maintained UV-vis spectra demonstrate the preserved framework. With prolonged time (10–15 min), more PET (6–22) ligands are replaced, associating with the appearance of a new peak on the UV-vis absorption spectrum. The results demonstrate the perturbed framework and the electronic structure of the Au_{38} precursor with the incorporation of more, bulkier TBBT ligands (compared with the PET ligands). Similar to the general mechanism mentioned above, the structural tautomerization caused by increased number of foreign ligands is denoted as Stage II. In Stage III, heavily exchanged Au_{38} clusters with 20–24 TBBT ligands were formed, associating with its disproportionation and the appearance of the cluster peaks corresponding to $Au_{36}(PET)_x(TBBT)_{24-x}$ ($x=22–24$) and $Au_{40}(PET)_x(TBBT)_{26-x}$ ($x=23–26$). The presence of similar number of PET ligands (0–6) in all these three groups of clusters (i.e., Au_{38}, Au_{36}, and Au_{40} clusters) indicate that the transformation occurs preferentially via an internal reconstruction (simply releasing [Au] atoms and its capture by another Au_{38} cluster), instead of a complete decomposition-recombination mechanism. In addition, the identification of the tiny amount of Au_{39} and Au_{41} demonstrate that the size-reduction/growth occurs in a stepwise $[Au_1]$-$[Au_1]$ pattern, rather than a synchronous $[Au_2]$-rearrangement pathway. After 120 min (Stage IV), all Au_{38} clusters are consumed, and a slow size-focusing

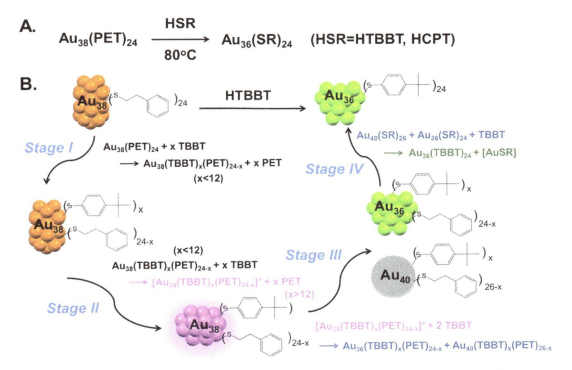

FIG. 4.45 The ligand exchange induced size-reduction of $Au_{38}(PET)_{24}$ (A), and an illustrative diagram for the four-stage size-conversion mechanism (B). *(Part of the diagram was reproduced from C. Zeng, C. Liu, Y. Pei, R. Jin, Thiol ligand-induced transformation of $Au_{38}(SC_2H_4Ph)_{24}$ to $Au_{36}(SPh-t-Bu)_{24}$, ACS Nano, 7 (2013) 6138–6145, with permission from the ACS Publications (Copyright 2013).)*

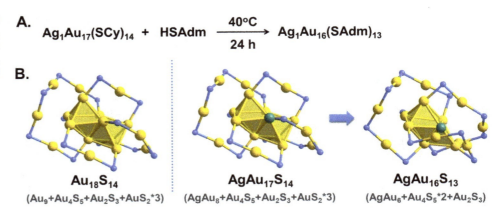

FIG. 4.46 Ligand etching induced size-reduction of $Ag_1Au_{17}(SCy)_{14}$ (A), and comparison of the framework of $Au_{18}(SCy)_{14}$, $Ag_1Au_{17}(SCy)_{14}$ and $AgAu_{16}(SAdm)_{13}$.

(120–300 min) then occurs with the gradual conversion of the metastable Au_{40} clusters into the thermodynamically more stable Au_{36} cluster. This proposal is also verified by the high yield of Au_{36} clusters of ~90% after the ligand exchange. According to the aforementioned analysis, the overall size-conversion comprises four stages, i.e., ligand exchange, structural distortion, disproportionation and size-focusing (Fig. 4.45B). In addition, using the optical absorbance at 550 nm (corresponding to characteristic peak of the structurally distorted Au_{38} clusters in Stage II) as a reference and a first-order kinetic model, the activation energy of Stage II and III is determined 107 and 152 kJ/mol, respectively.

Of note, in Jin's study [36], a control experiment by replacing HTBBT with HSCy was conducted to elucidate the potential ligand effect on the size-conversion. The results indicate a structural perturbation of the $Au_{38}(PET)_{24}$ by HSCy ligand (evidenced by UV-vis spectral analysis), verifying the positive role of the steric bulkiness on the size-conversion. Nevertheless, a mixture of different sized clusters was formed in the HSCy etching system. In this context, the electronic effect (i.e., conjugation of the aromatic group of TBBT ligand) is suggested an important factor for the exclusive formation of Au_{36} cluster products.

Compared with the aforementioned monometallic cluster systems, some interesting alloying effect in the ligand exchange has been reported recently. For example, in the recent study of Zhu and co-workers [77], the reaction of HSAdm with the monometallic $Au_{18}(SCy)_{14}$ was found to produce a relatively larger-sized $Au_{21}(SAdm)_{15}$ (Fig. 4.25) [46]. By contrast, the reaction of HSAdm with $Ag_1Au_{17}(SCy)_{14}$ (alloy analog of $Au_{18}(SCy)_{14}$) gives a smaller $Ag_1Au_{16}(SAdm)_{13}$ as the final product (Fig. 4.46A). According to SCXRD analysis (Fig. 4.46B), the structure of the alloy $Ag_1Au_{17}(SCy)_{14}$ is highly similar to its mono-metallic analog of $Au_{18}(SCy)_{14}$. Each cluster comprises a face-sharing bi-octahedral core structure (Au_9 in $Au_{18}(SCy)_{14}$ and $AgAu_8$ in $AgAu_{17}(SCy)_{14}$), one Au_4S_5, one Au_2S_3 and three AuS_2 motifs. The octahedral $AgAu_8$ block and the Au_4S_5 motif in $AgAu_{17}(SCy)_{14}$ are preserved after size-reduction, while the re-arrangement of the ligand shell and the removal of one [AuS] occurs to generate the framework of $Ag_1Au_{16}(SAdm)_{13}$. The structural distortion in the overall framework of $AgAu_{16}(SAdm)_{13}$ (compared with the structure of $AgAu_{17}(SCy)_{14}$ with higher symmetry) is suggested to originate from the difference in steric bulkiness of the SCy and SAdm ligands.

Another example on the alloying effect in ligand exchange reactions has recently been reported by Wu and co-workers [78]. In their studies, the three cluster analogs, $[Au_{25}(PET)_{18}]^-$, $[Cd_1Au_{24}(PET)_{18}]^0$, and $[Hg_1Au_{24}(PET)_{18}]^0$ bearing the same atomic packing modes but distinct metal composition, was exposed to the addition of extra ligands of $HSCH_2PhtBu$ (4-tert-butylbenzylmercaptan) at 80°C (Fig. 4.47). After ~2 h, the $[Au_{24}(PET)_{20}]^-$ was formed (evidenced by UV-vis and ESI-MS characterizations) in the system of $[Au_{25}(PET)_{18}]^-$ precursor, demonstrating a size-reduction therein. Meanwhile, the reaction of $[Cd_1Au_{24}(PET)_{18}]^0$ with $HSCH_2PhtBu$ generates the completely exchanged $[Cd_1Au_{24}(SCH_2PhtBu)_{18}]^0$ as the main product, and a minor amount of largely exchanged $[Cd_1Au_{24}(SCH_2PhtBu)_{17}(PET)_1]^0$ (evidenced by PTLC and ESI-MS analysis). The preserved characteristic peaks on UV-vis spectra and SCXRD demonstrate the maintained

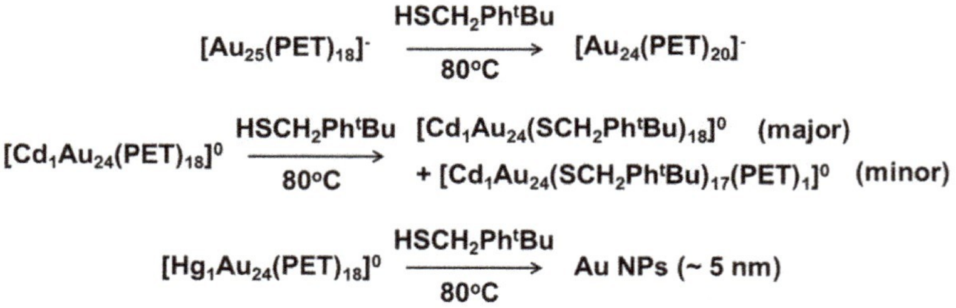

FIG. 4.47 The ligand etching induced conversion of the homometallic $[Au_{25}(PET)_{18}]^-$, and its alloy analogs $[Cd_1Au_{24}(PET)_{18}]^0$, and $[Hg_1Au_{24}(PET)_{18}]^0$.

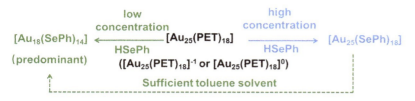

FIG. 4.48 The concentration-dependent conversion of $[Au_{25}(PET)_{18}]^n$ under selenophenol ligand etching.

framework of the icosahedral Cd_1Au_{24} precursor (albeit the structural distortion on the staples) after the ligand exchange. ^{1}H NMR analysis indicates that the remaining PET ligands randomly distribute on the M_2S_3 (M=Au/Cd) staples, and this conclusion is distinct from the preferential ligand exchange of the thiolates nearby the metallic core [79,80]. In contrast to the cluster-to-cluster conversions in the monometallic Au_{25} cluster and the bimetallic Cd_1Au_{24} systems, the reaction of $[Hg_1Au_{24}(PET)_{18}]^0$ with $HSCH_2PhtBu$ results in the formation of $\sim$5 nm sized plasmonic particles (evidenced by TEM analysis) after 2 h. According to the time-dependent MALDI-MS analysis, the Hg_1Au_{24} cluster precursors collapse and aggregate into larger sized particles in the presence of excessive foreign ligands. The reason for the distinct conversion behavior of the three clusters has been ascribed to the differences in the inherent stability (via time-dependent UV-vis-NIR tracking analysis), and could be easily comprehend from the average M—S bond distance in the core-shell interference (the shorter distance induces stronger core-shell interaction, and thus the higher cluster stability). As the preservation and the inversion of the ligand configuration were both observed after the ligand exchange of $[Cd_1Au_{24}(PET)_{18}]^0$, both S_N1 and S_N2-like mechanism was proposed to be plausible (Fig. 4.22).

In addition to the thiolated cluster systems, the selenophenol ligand induced size-reduction from $[Au_{25}(PET)_{18}]^n$ ($n=-1/0$) to $Au_{18}(SePh)_{14}$ (evidenced by ESI-MS and NMR tests, associated with the side product of $Au_{15}(SePh)_{13}$, $Au_{19}(SePh)_{15}$ and $Au_{20}(SePh)_{16}$) has been recently reported by Zhu and co-workers [81]. Of note, the reaction outcome depends on the concentration of the $[Au_{25}(PET)_{18}]^n$ precursors in toluene solvent, as the reaction of selenophenol with $[Au_{25}(PET)_{18}]^n$ gives the size-maintained $Au_{25}(SePh)_{18}$ at high Au_{25}-precursor concentration [82], but gives $Au_{18}(PET)_{14}$ at low precursor concentration (Fig. 4.48). With the aid of the control experiments via diluting certain amount of Au_{25} precursors in different volume of toluene solvents (and tracked the reaction with UV-vis spectra), Zhu and co-workers found that the poor solubility of the selephenol protected Au_{25} cluster is pivotal to the concentration-dependent conversion. In the presence of adequate amount of solvent, the solvated $Au_{25}(SePh)_{18}$ could be easily attacked by the remaining selenophenol ligands to form the $Au_{18}(SePh)_{14}$ as the ligand etching product. By contrast, in the presence of insufficient solvent, the initially formed $Au_{25}(SePh)_{18}$ easily precipitate out from the solution to preclude the future ligand-etching.

Interconversion induced by extra ligands

In addition to the aforementioned size-conversion from one atomically precise cluster to another one, the mutual conversion between two or more clusters could possibly provide more detailed information on the structural and dynamic correlations among different sized clusters. In the recent study of Dass and co-workers [83], the mutual conversion of $Au_{36}(SPhX)_{24}$ (X=H, para-tBu) with $Au_{30}(StBu)_{18}$ has been accomplished by ligand-exchange under heating conditions (Fig. 4.49A).

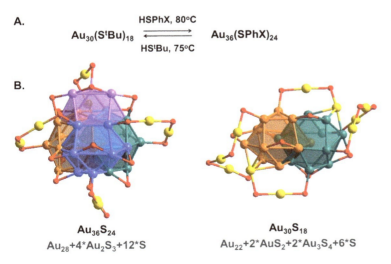

FIG. 4.49 The interconversion of $Au_{36}(SPhX)_{24}$ with $Au_{30}(StBu)_{18}$ (A), and the framework of the precursor and the product clusters (B).

The reaction of $Au_{30}(StBu)_{18}$ with excess HSPhX at 80°C produces the related $Au_{36}(SPhX)_{24}$ after 16 h, mixing $Au_{36}(SPhX)_{24}$ with HStBu at 75°C produces $Au_{30}(StBu)_{18}$ after 30–40 min.

According to the single crystal structure analysis, the Au_{28} core structure of $Au_{36}(SPhX)_{24}$ [84] is assembled by four fused cuboctahedra, and the ligand shell is constituted by four Au_2S_3 staples and twelve bridging thiolate ligands (Fig. 4.49B). Meanwhile, the structure of $Au_{30}(StBu)_{18}$ [85] elucidated by SCXRD demonstrates a bi-capped bi-cuboctahedral Au_{22} core, with the protection of two AuS_2, two Au_3S_4 staples and six bridging thiolate ligands. The time-dependent ESI-MS and MALDI-MS on the intermediate states indicates the formation of the largely exchanged cluster product in both the $Au_{30} \rightarrow Au_{36}$ conversion (such as $Au_{36}(SPhtBu)_{23}(StBu)_1$) and the reverse transformation ($Au_{30}(SPhtBu)_x(StBu)_{18-x}$, $x = 1–4$). In addition, the DFT calculations with the energy fragment decomposition analysis (Section 6.2.5.2) indicate that the different thermodynamic stability of the two clusters mainly originates from the distinct fragmentation energy (the conjugation of the aromatic groups in $Au_{36}(SPhH)_{24}$ weakens the core-shell interactions therein).

Interestingly, in another study by Jin and co-workers, using the similar ligand etching conditions (with HSPhtBu, and heating at 80°C), but starting from a relatively larger-sized $Au_{38}(PET)_{24}$ cluster (compared with the aforementioned Au_{30} precursor), $Au_{36}(SPhtBu)_{24}$ was also obtained as the main product after ~12 h (Fig. 4.50A).[75]. In addition, both the Au_{36} and Au_{38} clusters protected by thiolate ligands (i.e., 2-naphthalenethiol, abbreviated as SNap) could be prepared via the ligand-etching of $Au_{144}(SC_6H_{13})_{60}$ [86] under thermal conditions (80°C, Fig. 4.50B) [87]. According to Li's study, adding excess HSNap to Au_{144} solution at 80°C lead to the formation of $Au_{36}(SNap)_{24}$ (evidenced by UV-vis and MALDI-MS analysis) after 6 h. By contrast, adding pyridine prior to the addition of HSNap in the aforementioned reaction system results in the formation of a distinct cluster product, i.e., $Au_{38}(SNap)_{24}$ (evidenced by UV-vis, MALDI-MS and ESI-MS characterizations). Of note, replacing the organic base of pyridine with the inorganic base, such as K_2CO_3 results in the formation of $Au_{36}(SNap)_{24}$, and the reason was attributed to the low solubility of the inorganic base in the reaction system. In this context, the formation of the Au_{38} clusters was suggested to be promoted by the N(pyridine)···HSNap interactions. Control experiments imply that the $Au_{36}(SNap)_{24}$ and $Au_{38}(SNap)_{24}$ are relatively stable under both heating conditions (80°C, under N_2 atmosphere) or treatment with of extra pyridine additives. The results exclude the intermediacy of either Au_{38} or Au_{36} clusters during the conversion of the Au_{144} precursors (Au_{36} and Au_{38} is formed via different mechanistic pathways). Nevertheless, adding two equivalents of HSNap into the etching system of $Au_{38}(SNap)_{24}$ at 80°C gradually generates $Au_{36}(SNap)_{24}$ as the product (evidenced by UV-vis and MALDI-MS analysis), correlating with the ligand etching induced size-conversion from Au_{38} to Au_{36} clusters mentioned above (Fig. 4.50A). By contrast, the addition of both HSNap and pyridine into the $Au_{38}(SNap)_{24}$ solution results in the decomposition of the Au_{38} precursor.

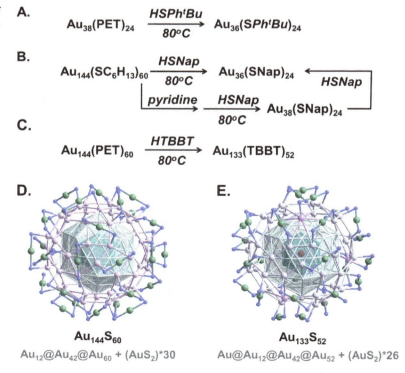

FIG. 4.50 The size-conversion between the series of Au_{144}, Au_{38}, Au_{36} and Au_{133} clusters (A–C), and the framework of Au_{144} (D) and Au_{133} (E) clusters.

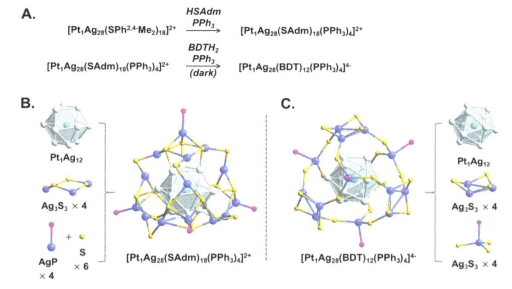

FIG. 4.51 The PPh$_3$ and HSAdm etching induced size conversion of [Pt$_1$Ag$_{24}$(SPh$^{2,4-}$Me$_2$)$_{18}$]$^{2+}$ to [Pt$_1$Ag$_{28}$(SAdm)$_{18}$(PPh$_3$)$_4$]$^{2+}$ (A), PPh$_3$ and BDTH$_2$ induced conversion of [Pt$_1$Ag$_{28}$(SAdm)$_{18}$(PPh$_3$)$_4$]$^{2+}$ to [PtAg$_{28}$(BDT)$_{12}$(PPh$_3$)$_4$]$^{4-}$ (B), and comparison on the framework of [Pt$_1$Ag$_{28}$(SAdm)$_{18}$(PPh$_3$)$_4$]$^{2+}$ and [PtAg$_{28}$(BDT)$_{12}$(PPh$_3$)$_4$]$^{4-}$.

Compared with the aforementioned size-conversion from Au$_{144}$(SC$_6$H$_{13}$)$_{60}$ to Au$_{36}$(SNap)$_{24}$ and Au$_{38}$(SNap)$_{24}$ under HSNap etching, the reaction of Au$_{144}$(PET)$_{60}$ with HTBBT at 80°C generates a remarkably larger-sized cluster product, i.e., Au$_{133}$(TBBT)$_{52}$ (Fig. 4.50C) [88]. The structure of the Au$_{144}$ precursor is deduced from the single crystal structure of its analog of Au$_{144}$(SCH$_2$Ph)$_{60}$, in view of the same ratio of Au:SR and the similar UV-vis characteristic peaks of these two Au$_{144}$ clusters. As shown in Fig. 4.50D, the Au$_{144}$ precursor structure comprise a Au$_{12}$@Au$_{42}$ kernel, which is capped by a C$_{60}$-like icosahedral Au$_{60}$ structure, and 30 AuS$_2$ staples. For comparison, the structure of Au$_{133}$(TBBT)$_{52}$, which correspond to a Au$_{12}$@Au$_{42}$@Au$_{52}$@Au$_{26}$S$_{52}$ structure is given in Fig. 4.50E (see Fig. 4.7 for the detailed structure anatomy).

Aside from the aforementioned examples, it is noteworthy that the size-conversion induced by mixed types of ligands (such as the size-growth from [Pt$_1$Ag$_{24}$(SPh$^{2,4-}$Me$_2$)$_{18}$]$^{2+}$ to [Pt$_1$Ag$_{28}$(SAdm)$_{18}$(PPh$_3$)$_4$]$^{2+}$ in the presence of both HSAdm and PPh$_3$ ligands (Fig. 4.51A) [68]) have also been reported. The single crystal structure of the Pt$_1$Ag$_{24}$ precursor resembles that of the icosahedral Au$_{25}$ cluster, with an icosahedral Pt@Ag$_{12}$ structure protected by six Ag$_2$S$_3$ staples. In a following study by Bakr, Cavallo, Basset, etc., this [Pt$_1$Ag$_{28}$(SAdm)$_{18}$(PPh$_3$)$_4$]$^{2+}$[PPh$_4^-$]$_2$ was used as the precursor to undergo ligand exchange in presence of both PPh$_3$ (sacrificial) and 1,3-benzenedithiol (abbreviated as BDTH$_2$) in a single-phase method. The same-sized, anionic cluster [PtAg$_{28}$(BDT)$_{12}$(PPh$_3$)$_4$]$^{4-}$ was obtained with ~76% yield [89]. Interestingly, the SCXRD analysis indicates the newly formed [PtAg$_{28}$(BDT)$_{12}$(PPh$_3$)$_4$]$^{4-}$ also adopts an Pt@Ag$_{12}$ kernel structure, but in an icosahedral packing mode (distinct to the FCC packed Pt@Ag$_{12}$ kernel in the Pt$_1$Ag$_{28}$ precursor). The icosahedral Pt@Ag$_{12}$ kernel is further protected by four trigonal prisms of Ag$_3$S$_3$ blocks and four tetrahedral AgS$_3$P blocks, both groups adopting orientations of tetrahedral symmetry (Fig. 4.51B). The structural changes result in a remarkable difference in optical emission intensity, while the mechanistic details remain to be elucidated in detail.

Sacrificial ligand induced size-conversion

In aforementioned reactions of this Section, the ligand exchange between an atomically precise metal nanocluster and organic ligands results in structural changes and incorporation of the foreign ligand(s). Compared with these reactions, an uncommon type of cluster-ligand reaction, wherein the ligand is requisite to the size-conversion, but were not incorporated into the cluster products, have been reported recently.

For example, exposing the diphosphine, thiolate and chloride ligands co-protected [Au$_8$Ag$_{57}$(dppp)$_4$(SC$_6$H$_{11}$)$_{32}$Cl$_2$]Cl to phenylacetylene at room temperature produces [Au$_8$Ag$_{55}$(dppp)$_4$(SC$_6$H$_{11}$)$_{34}$](BPh$_4$)$_2$ with ~90% yield after 12h (Fig. 4.52A) [90]. Instead of incorporating the phenylacetylene into the cluster structure, size-reduction occurs exclusively by removing two AgCl moieties, and each vacated position is occupied by an additional thiolate ligand (associated with distortion of the core structure from complete octameric fusion of M$_{13}$ blocks to an incomplete octameric fusion, Fig. 4.52B).

4.2.2.4 Mechanistic insights into the metal doping

In addition to the ligand exchange, the metal doping also represents an efficient strategy to prepare novel metal nanoclusters (and especially alloy clusters), and to improve the physicochemical properties (such as luminescence, catalysis, chirality,

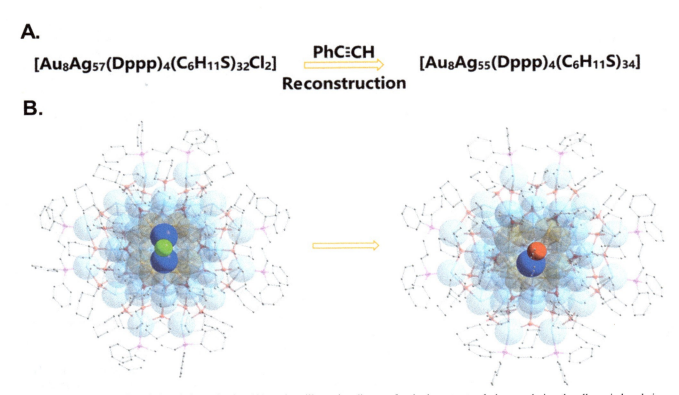

FIG. 4.52 Phenylacetylene induced size-reduction (A), and an illustrative diagram for the key structural changes during the alkyne induced size-conversion from $[Au_8Ag_{57}(dppp)_4(SC_6H_{11})_{32}Cl_2]^+$ (left) to $[Au_8Ag_{55}(dppp)_4(SC_6H_{11})_{34}]^{2+}$ (right) (B). *(Reproduced from S. Jin, M. Zhou, X. Kang, X. Li, W. Du, X. Wei, S. Chen, S. Wang, M. Zhu, Three-dimensional octameric assembly of icosahedral M13 units in $[Au_8Ag_{57}(Dppp)_4(C_6H_{11}S)_{32}Cl_2]$ Cl and its $[Au_8Ag_{55}(Dppp)_4(C_6H_{11}S)_{34}][BPh_4]_2$ derivative, Angew. Chem. Int. Ed. 59 (2020) 3891–3895, with permission from Wiley-VCH, copyright 2020.)*

etc.) of metal clusters via the synergistic effect. Many of the previously reported metal doping reactions result in the formation of a size- and framework-maintained alloy clusters, and such reactions were denoted as the metal-exchange in this section (note: the reported isotope exchange of ^{107}Ag and ^{109}Ag on $[Ag_{25}(SR)_{18}]^-$ is also classified into this category). The other metal doping induced size-conversions could involve the size-growth without a significant structural perturbation on the precursor clusters, or formation of a new cluster with distinct framework. Distinct to these alloying processes, in some of the metal salts regulated size-conversion reactions, such as Ag^+ induced size-growth from Au_6 to Au_7 clusters [91] and the Cu^{2+} induced the size-growth of Au_{25} clusters ($Au_{25}(PET)_{18} \rightarrow Au_{44}(PET)_{18}$) [92], no foreign metal atom is incorporated into the cluster product. Therefore, the mechanism is anticipated to remarkably different from those of the alloying processes. For this reason, the metal salt induced size-conversion reactions (without incorporating the foreign metal atoms) are categorized separately, and the mechanistic details are given in Section 4.2.2.5.

In organic reactions, isotope labeling has become a powerful strategy to elucidate the mechanistic details, such as the kinetics (to deduce the rate-determining step via primary/secondary isotope effect), the possible influence of solvents/other reagents, or the preferential reaction sites. In recent years, isotopic labeling experiments (in combination with mass spectroscopic analysis) have also been used in the metal nanocluster systems and is especially powerful in elucidating the intermediate states and determining their lifetime. For example, with two carefully prepared, isotopically different Ag_{25} clusters, $[^{107}Ag_{25}(SPhMe_2)_{18}]^-$ and $[^{109}Ag_{25}(SPhMe_2)_{18}]^-$ ($HSPhMe_2$ denotes 2,4-dimethyl benzenethiol), the intercluster metal scrambling have been clearly identified and kinetically tracked with ESI-MS by Pradeep and co-workers (Fig. 4.53A) [93]. As shown in Fig. 4.53B, the isotopic pattern of $[^{107}Ag_{25}(SPhMe_2)_{18}]^-$ is slightly different from that of $[^{109}Ag_{25}(SPhMe_2)_{18}]^-$, while the maximal cluster peaks are differentiated by m/z 50 (corresponding to (109–107)*25, m/z denotes the mass-to-charge ratio). Using the mixture of $[^{107}Ag_{25}(SPh^{2,4-}Me_2)_{18}]^-$ and $[^{109}Ag_{25}(SPh^{2,4-}Me_2)_{18}]^-$ (with 1:1 M ratio) as the testing sample, an instantaneous change in the mass spectra occurs. The disappearance of the two series of cluster peaks of the isotopically pure cluster precursors, and the appearance of a new cluster peak with wider distribution of isotopic peaks, demonstrate the metal exchange between the two cluster precursors. In addition, the most intense peak locate at the positions wherein the molar ratio of ^{107}Ag versus ^{109}Ag is close to 1:1. Meanwhile, mixing the

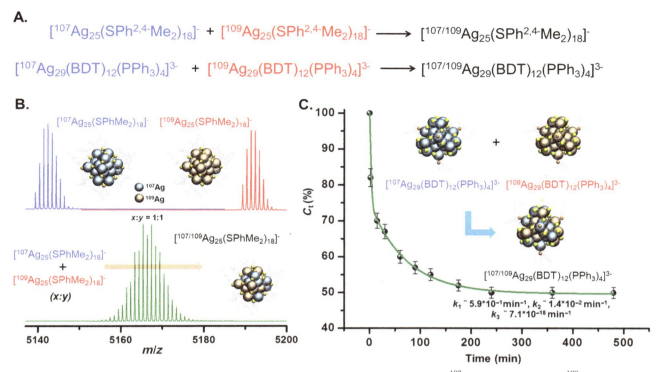

FIG. 4.53 The metal scrambling of the Ag$_{25}$ and Ag$_{29}$ systems (A), ESI-MS of the isotopically pure [^{107}Ag$_{25}$(SPhMe$_2$)$_{18}$]$^-$ and [^{109}Ag$_{25}$(SPhMe$_2$)$_{18}$]$^-$ clusters and the cluster product after mixing equimolar of them at room temperature (B), and kinetic plot of C^t versus time (t) in the reaction system between [^{107}Ag$_{29}$(BDT)$_{12}$(PPh$_3$)$_4$]$^{3-}$ and [^{109}Ag$_{29}$(BDT)$_{12}$(PPh$_3$)$_4$]$^{3-}$ at room temperature. *((B and C) Reproduced from P. Chakraborty, A. Nag, G. Natarajan, N. Bandyopadhyay, G. Paramasivam, K. Panwar Manoj, J. Chakrabarti, T. Pradeep, Rapid isotopic exchange in nanoparticles, Sci. Adv. 5 (2019) eaau7555, Copyright 2019 AAAS.)*

cluster precursor in varied molar ratios results in the rapid metal exchange, and the dominant product fitting binomial mass spectral distribution. These statistical distributions demonstrate the equivalent exchanging possibility of all the 25 metal sites. Of note, lowing the temperature of the testing samples (1:1 mixture of ^{107}Ag$_{25}$ and ^{109}Ag$_{25}$ clusters) and the desolvation temperature (of the ESI-MS equipment) make it possible to track the metal exchange process, and a fast, but stepwise exchange was found to occur within 30s.

Compare to the instantaneous change between the two Ag$_{25}$ clusters, the metal exchange between the two isotopically pure [Ag$_{29}$(BDT)$_{12}$(PPh$_3$)$_4$]$^{3-}$ (with exclusive ^{107}Ag or ^{109}Ag, and BDT is short for benzene dithiol) is much slower (Fig. 4.53C). Depending on the concentration of the precursors and reaction temperature, the exchange may need several hours to several days to finally reach a dynamic equilibrium (higher concentration of the cluster precursors, or elevated temperature could accelerate the reaction rates). For example, the room temperature metal exchange of 1:1 mixture of the isotopically pure [^{107}Ag$_{29}$(BDT)$_{12}$(PPh$_3$)$_4$]$^{3-}$ and [^{109}Ag$_{29}$(BDT)$_{12}$(PPh$_3$)$_4$]$^{3-}$ at a concentration of 1.5 μM reaches a dynamic equilibrium after 3h. The moderate reaction rate enables a convenient spectral tracking on the metal exchange processes (poor data reproducibility might occur for fast reactions finishing within several minutes, while the tracking on long-time reactions is challenging due to the possible perturbation of solution or environmental changes). Specifically, the time-course of the percentage of one isotopically pure Ag$_{29}$ cluster precursor (designate as C^t, the changes of the two precursors are identical in an equimolar reaction, and either one of them could be used) were used to plot a kinetic profile. As shown in Fig. 4.53C, the kinetic data follows a triexponential equation (C^t = $k_1$exp^{-at} + $k_2$exp^{-bt} + $k_3$exp^{-ct}), indicating a three-stage mechanism. In the combination of theoretical simulations (with molecular docking studies and free-energy calculations), the metal exchange was suggested to be initiated by a cluster collision induced fast surface metal exchange (Stage I), following with a relatively slower diffusion of the exchanged metal atoms into the core (Stage II), and the final equilibration in the whole cluster (Stage III). The difference in the first stage has been proposed to account to the relative reaction rates of the two metal-scramming systems. The two isotropic Ag$_{25}$ clusters could quickly bind with each other via the intercluster interaction between the interlocked rings or chain staple motifs on the ligand shell. By contrast, such interactions are unlikely between the Ag$_{29}$ clusters (due to the relatively higher structural rigidity and ligand bulkiness), and Ag$_{29}$ mainly interact with the via the relatively weaker, noncovalent interactions (such as C-H⋯π interactions). Therefore, the reaction rates of the Ag$_{29}$ system is relatively slower than that of the Ag$_{25}$ systems.

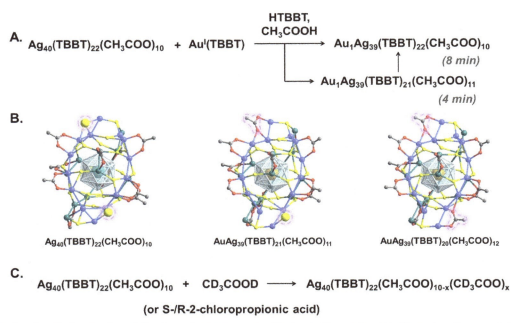

FIG. 4.54 The metal exchange reaction of the Ag_{40} cluster precursor with the Au(I) complex (A), illustrative diagram of the framework of $Ag_{40}(TBBM)_{22}(CH_3COO)_{10}$, $Ag_{40}(TBBM)_{21}(CH_3COO)_{11}$, and $AuAg_{39}(TBBM)_{20}(CH_3COO)_{12}$ (B), and the ligand exchange of the Ag_{40} cluster with different types of acids (C).

In addition to the Ag-Ag scrambling, the Au-Ag exchange has also been reported. For example, the reaction of the atomically precise $Ag_{40}(TBBT)_{22}(CH_3COO)_{10}$ cluster with excessive $Au^I(TBBT)$ complexes (in the presence of HTBBT and CH_3COOH) recently reported by Zhu and co-workers generates the alloy cluster product of $AuAg_{39}(TBBM)_{20}(CH_3COO)_{12}$ with ~95% yield after 8 min (Fig. 4.54A) [94]. Interestingly, with the same reaction procedures (i.e., same amount of reactants, and sequence for adding different reagents), but stopping the reaction in the middle stage (i.e., 4 min), the single crystal of another alloy nanocluster has been separated and identified to be $AuAg_{39}(TBBM)_{21}(CH_3COO)_{11}$. According to the SCXRD characterization on the single crystals, these three clusters share the similar kernel structure, corresponding to an $M@Ag_{12}$ icosahedron, with a central atom to be Ag and Au, respectively (Fig. 4.54B). The changed ratio of TBBM and CH_3COO (22:10 in Ag_{40} and 21:11/20:12 in $AuAg_{39}$) makes little perturbation on the orientation of metal atoms, but changes the coordination environment of two chain-like Ag_7 structure. The intermediate structure of $Ag_{40}(TBBM)_{21}(CH_3COO)_{11}$ bridges those of $Ag_{40}(TBBM)_{22}(CH_3COO)_{10}$ and $AuAg_{39}(TBBM)_{20}(CH_3COO)_{12}$, because the foreign Au atom has been incorporated into the icosahedral core, and a thiolate ligand on one group of chain-like $Ag_7(TBBM)(CH_3COO)_2$ (in Ag_{40}) is replaced by a CH_3COO ligand. Of note, the ligand exchange could also occur in a cluster-ligand reaction pattern via reaction the Ag_{40} cluster with chiral acids (e.g., the deuterium CD_3COOD, or S-/R-2-chloropropionic acid), and the maintained UV-vis spectrum demonstrates that the maintained cluster framework.

Similarly, the single doping by incorporating only one foreign metal atom has been accomplished to produce an alloy cluster of $Au_{17}Ag(SCy)_{14}$ via the reaction of $Au_{18}(SCy)_{14}$ with AgSCy in ice-bath (Fig. 4.55) [77]. The SCXRD analysis indicates that the framework of the Au_{18} precursor and the $Au_{17}Ag$ product adopts the same framework, comprising a face-sharing bi-octahedral FCC-M_9 core (Au_9 in Au_{18} and Au_8Ag in $Au_{17}Ag$, see Fig. 4.46 for the detailed atomic structure), one Au_4S_5, one Au_2S_3 and three AuS_2 staple motifs. Of note, the precise elucidation of the atomic structure of the $AgAu_{17}$ structure verifies a previous theoretical prediction on the doping site [95].

In addition to the aforementioned Au/Ag complex induced size-conversion of Au/Ag clusters, the alloying process induced by other metal complexes has also been reported. In the recent study of Zhu and co-workers [96], the anionic $[Au_{25}(PET)_{18}]^{1-}$ was used as a cluster precursor, and its reaction with the PET coordinated Ag(I), Cu(II), Cd(II) and Hg(II) complexes produce the bimetallic clusters of $Ag_xAu_{25-x}(PET)_{18}$, $Cu_xAu_{25-x}(PET)_{18}$, $Cd_1Au_{24}(PET)_{18}$, and

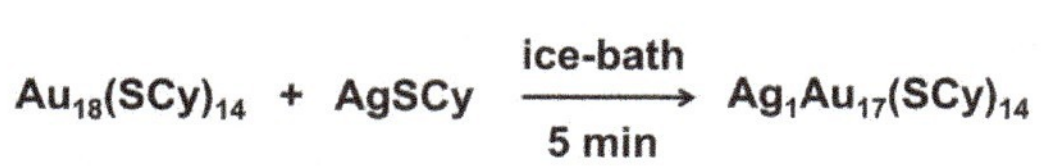

$$Au_{18}(SCy)_{14} + AgSCy \xrightarrow[5\ min]{ice\text{-}bath} Ag_1Au_{17}(SCy)_{14}$$

FIG. 4.55 The single Ag doping reaction of $Au_{18}(SCy)_{14}$ with AgSCy.

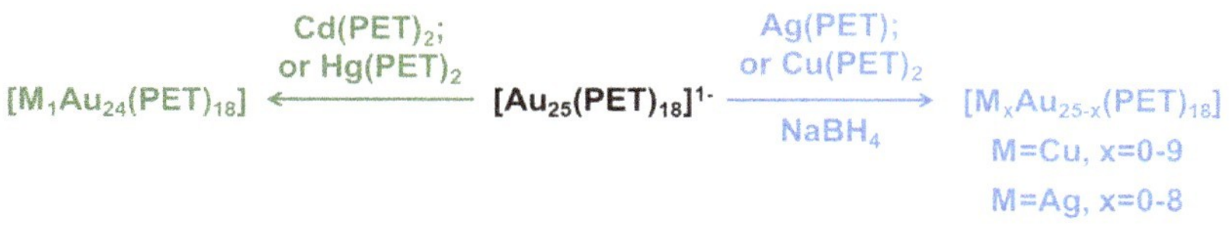

$[M_1Au_{24}(PET)_{18}]$ ⟵[Cd(PET)$_2$; or Hg(PET)$_2$]⟵ $[Au_{25}(PET)_{18}]^{1-}$ ⟶[Ag(PET); or Cu(PET)$_2$ / NaBH$_4$]⟶ $[M_xAu_{25-x}(PET)_{18}]$
M=Cu, x=0-9
M=Ag, x=0-8

FIG. 4.56 Metal exchange via the reaction of $[Au_{25}(PET)_{18}]^{-1}$ with different metal complexes.

Hg$_1$Au$_{24}$(PET)$_{18}$ (evidenced by MALDI-MS ESI-MS, etc.). Interestingly, the reductant of NaBH$_4$ is requisite in the Au-Ag and Au-Cu alloying systems (Fig. 4.56). Compared with the one-pot synthesis, alloy clusters with higher Cu/Ag ratio were obtained with prolonged reaction time. By contrast, the reaction of $[Au_{25}(PET)_{18}]^{1-}$ with the similar Ni(II), Pd(II) and Pt(II) complexes does not produce the alloy cluster product. As the size-maintained alloying occurs only on the metal complexes with d^{10} electronic states, the d^{10}-d^{10} interaction between the foreign metal atom and the Au atom is suggested the driving force of the metal exchange.

4.2.2.5 Mechanism of metal complex induced size-growth

Compared with the ligand induced cluster conversions and the metal complex induced alloying processes, the size-maintained ligand exchange and the size-growth with comprehensive structure tautomerization might also occur in the reaction of the metal clusters with foreign metal complexes. Herein, we mainly focus on these size-conversion reactions.

Ligand exchange/addition via the cluster-complex reaction

Compared with the widely reported ligand exchange of an atomically precise metal nanocluster with foreign ligands (with distinct structure from the original cluster ligands), the size-maintained ligand exchange via the reaction of the atomically precise clusters with a metal complex has been rarely reported.

Interestingly, according to the recent study by Ackerson and co-workers, the ligand exchange between Au$_{25}$(PET)$_{18}$ with phenylacetylene is unlikely under various attempts (such as increasing the amount of the ligands, or elevating the temperature to 60°C). By contrast, the ligand exchange occurs smoothly when using gold(I) or lithium phenylacetylide (i.e., AuPA or LiPA, and PA is short for phenylacetylide) as the alkyne ligand source (Fig. 4.57) [97]. The MALDI-MS characterization on the products indicates that the extent of exchange is directly dependent on the AuPA: Au$_{25}$ ratio: adding 1 or 10 equivalents of AuPA forms the Au$_{25}$(PA)$_x$(PET)$_{18-x}$ components fitting a Gaussian-like distribution; increasing the amount of AuPA into 100 equivalents greatly facilitates the formation of heavily exchanged products, and the cluster peak of Au$_{25}$(PA)$_{18}$ with complete ligand exchange was also figured out in this case. In view of the same experimental outcome of the LiPA and the AuPA systems, Ackerson and co-workers suggest that the gold in the AuPA complex is chemically inert to the ligand exchange, and only functions as a counter cation. Of note, the reverse reaction (i.e., reaction of independently prepared Au$_{25}$(PA)$_{18}$ with free thiol) occurs easily after a simple mixture, and forming the mixed ligands protected Au$_{25}$ clusters (with 1–3 PET ligands) in short time. In addition, the intercluster ligand exchange by reacting Au$_{25}$(PET)$_{18}$ with Au$_{44}$(PA)$_{28}$ occurs within 5 min, indicating the facility of the thiolate-to-acetylide (and the reversed) exchange. DFT calculations on the thermodynamics of the modeling ligand exchange reactions demonstrate that the enthalpic facility is an important driving force for the ligand exchange.

Size-growth via incorporating the extra-metal component(s)

Starting from an atomically precise Au-Cd alloy cluster (i.e., Cd$_1$Au$_{14}$(StBu)$_{12}$, prepared via the reaction of Au$_{15}$(SG)$_{13}$ with CdCl$_2$ and HStBu), the de-alloying by removing the Cd-component has be regulated by the addition of Au(I) complexes (i.e., AuSAdm) [98]. As shown in Fig. 4.58A, adding AuSAdm complex into the Cd$_1$Au$_{14}$(StBu)$_{12}$ cluster solution produces Au$_{16}$(SAdm)$_{12}$ after 20 h (note that Au$_{16}$(SAdm)$_{12}$ could also be independently prepared with the two-phase ligand exchange of Au$_{15}$(SG)$_{13}$ with HSAdm). The single crystal structure of the Cd$_1$Au$_{14}$ precursor cluster (Fig. 4.58B) adopts centrosymmetric-like configuration. The edge-sharing bi-tetrahedral Cd$_1$Au$_5$ core (Fig. 4.58B, left

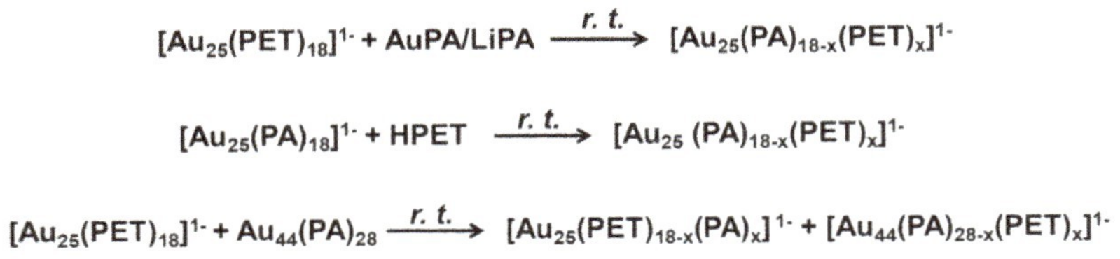

$[Au_{25}(PET)_{18}]^{1-}$ + AuPA/LiPA ⟶[r. t.]⟶ $[Au_{25}(PA)_{18-x}(PET)_x]^{1-}$

$[Au_{25}(PA)_{18}]^{1-}$ + HPET ⟶[r. t.]⟶ $[Au_{25}(PA)_{18-x}(PET)_x]^{1-}$

$[Au_{25}(PET)_{18}]^{1-}$ + Au$_{44}$(PA)$_{28}$ ⟶[r. t.]⟶ $[Au_{25}(PET)_{18-x}(PA)_x]^{1-}$ + $[Au_{44}(PA)_{28-x}(PET)_x]^{1-}$

FIG. 4.57 The ligand exchange by reacting Au$_{25}$ clusters with phenylacetylide salts/thiol ligands or Au$_{44}$ clusters.

FIG. 4.58 The AuSAdm induced size-conversion from $Cd_1Au_{14}(StBu)_{12}$ to $Au_{16}(SAdm)_{12}$, the formation of the precursor and product clusters via the two-phase ligand exchange method (in *gray* color), and the proposed mechanism for the conversion of $Cd_1Au_{14}(StBu)_{12}$ to $Au_{16}(SAdm)_{12}$ by DFT calculations (A), structural anatomy of the framework of $Cd_1Au_{14}(StBu)_{12}$ (B) and the structural anatomy of the framework of $Au_{16}(SAdm)_{12}$ (C).

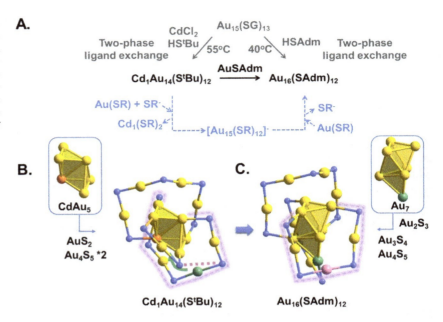

inset) is protected by one AuS_2 and two Au_4S_5 staple motifs. Meanwhile, the atomic structure of the produced $Au_{16}(SAdm)_{12}$ (determined by SCXRD) comprises a mono-capped octahedral Au_7 kernel (Fig. 4.58B, right inset), which is further protected by one Au_2S_3, one Au_3S_4, and one Au_4S_5 motifs (Fig. 4.58C). The overall conversion of the $Cd_1Au_{14}(StBu)_{12}$ to $Au_{16}(SAdm)_{12}$ is suggested to occur via two stages. The first stage includes the conversion of $Cd_1Au_{14}(StBu)_{12}$ to $[Au_{15}(SR)_{12}]^-$ (SR = StBu or SAdm) via replacing the Cd atom with an Au atom, and the second stage corresponds to the size-growth of the $[Au_{15}(SR)_{12}]^-$ to $Au_{16}(SR)_{12}$. The intermediacy of $[Au_{15}(SR)_{12}]^-$ is mainly deduced by the thermodynamic facility, as the DFT calculations indicate the reaction energy for the size-maintained de-alloying step to be -2.83 eV. Meanwhile, the theoretically predicted structure of $[Au_{15}(SR)_{12}]^-$ resembles that of $Cd_1Au_{14}(StBu)_{12}$, with the Au_6 kernel protected by one AuS_2 and two Au_4S_5 staples. From $[Au_{15}(SR)_{12}]^-$, the single "Au atom" size-growth was suggested to occur via the "gold-atom insertion, thiolate-group elimination" (i.e., removal of one anionic SR after the addition of the neutral Au(I)SR complex, see Section 6.2.5.2 for more details).

In Section 4.2.2.2, the interconversion of $[Au_6(dppp)_4]^{2+}$ with $[Au_8(dppp)_4Cl_2]^{2+}$ has successfully regulated by oxidation or reduction (for reverse reaction, Fig. 4.19) conditions. Actually, the size-conversion between this group of Au_6 and Au_8 clusters could date back to 2011, when Konishi and co-workers reported the gold(I) complex (i.e., $Au(PPh_3)Cl$) regulated size-growth from $[Au_6(dppp)_4](NO_3)_2$ to $[Au_8(dppp)_4Cl_2]^{2+}$ [39]. Experimentally, mixing $[Au_6(dppp)_4](NO_3)_2$ with $Au(PPh_3)Cl$ under room temperature gradually generate $[Au_8(dppp)_4Cl_2]^{2+}$ cluster (evidenced by the ESI-MS characterization and SCXRD analysis of $[Au_8(dppp)_4Cl_2](PF_6)_2$). The time-dependent UV-vis absorption spectroscopic monitoring on the diluted reaction system shows three isosbestic points at 297, 453 and 538 nm (Fig. 4.59). The yield of the Au_8 cluster after 5 h was determined to be ~99% according to the molar extinction coefficient. Interestingly, mixing the Au_6 precursor with 2 M equiv. of $Au(PPh_3)Cl$ significantly slows down the reaction rate, and the yield of Au_8 is only ~50% after about 153 h. Meanwhile, replacing $Au(PPh_3)Cl$ with $Au(PPh_3)(NO_3)$ in the reaction system does not generate the Au_8 cluster products (determined from the ESI-MS and UV-vis absorption tests). The results indicate that the self-decomposition during the size-conversion, which correlates with the heating induced size-conversion from Au_6 to Au_{11} clusters (see Section 6.2.5 for the details).

The mechanism of the aforementioned Au(I) complex induced size-growth from $[Au_6(dppp)_4]^{2+}$ to $[Au_8(dppp)_4Cl_2]^{2+}$ has been recently studied by DFT calculations [99]. The most plausible mechanism (Fig. 4.60) was suggested to start with an Au—P(dppp) dissociation on the terminal, bridging Au atom. After that, one Au atom is incorporated via the cascade addition of $Au(PPh_3)Cl$ to the dangling phosphine group of the dppp ligand, dissociation of chloride, and dissociation of PPh_3 (associated with the Au—Au bond formation) steps. Thereafter, the Au—P(dppp) bond dissociates to accommodate an extra- AuCl molecule, via the cascade reaction of adding $Au(PPh_3)Cl$ and dissociating PPh_3. After that, the free chloride in the reaction system easily binds with the outermost Au—Au bonds, and undergoes a quick isomerization via the rapid migration (almost barrierless) to generate the thermodynamic product of $[Au_8(dppp)_4Cl_2]^{2+}$. According to this study, the ease of the Au—P bond dissociation is sensitive to the coordination environment of the Au atom in the precursor. The

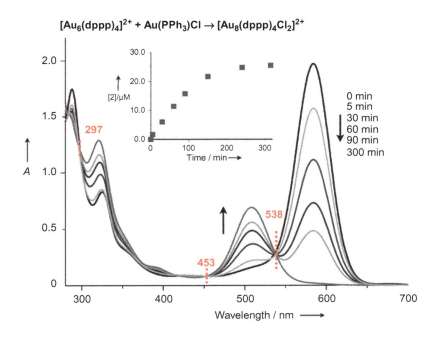

FIG. 4.59 The time-dependent size-growth from $[Au_6(dppp)_4]^{2+}$ to $[Au_8(dppp)_4Cl_2]^{2+}$ and the isosbestic points. Inset: formation of the $[Au_8(dppp)_4Cl_2]^{2+}$ determined by the molar extinction coefficient of the original sample. *(Reproduced from Y. Kamei, Y. Shichibu, K. Konishi, Generation of small gold clusters with unique geometries through cluster-to-cluster transformations: octanuclear clusters with edge-sharing gold tetrahedron motifs, Angew. Chem. Int. Ed. 50 (2011) 7442–7445, from Wiley-VCH, copyright 2011.)*

Au—P bond dissociation is relatively easier on the electron-rich Au atoms, or Au atoms with higher coordination number, and this proposal was latter calibrated by the mechanistic study on the redox induced size-conversion from $[Au_8(dppp)_4]^{2+}$ to $[Au_8(dppp)_4Cl_2]^{2+}$ (Figs. 4.17 and 4.18).

Similar to the aforementioned Au(I) complex induced size-growth of Au_6 with two formal [AuCl] structures (associate with adequate isomerization via changing the coordination of the biphosphine ligands), the recent study by Wang and coworkers reports an Au-alkyne complex induced size-growth from Au_9 to Au_{10} clusters (4–61 A) [100]. As shown in Fig. 4.61B, the structure of $[Au_9(BINAP)_4]^{3+}$ (BINAP = 2,2′-bis-(diphenylphosphino)-1,1′-binaphthyl) elucidated by SCXRD consists of a twisted, face-centered hexagonal Au_7 structure, and half of the Au_7 core is capped by two additional Au atoms to form three face-sharing tetrahedra. Except for the central, bare Au atom, each Au atom is protected by one phosphine group of the BINAP ligand. Mixing the $[Au_9(BINAP)_4](CF_3COO)_3$ solution with the gold-alkyne complex ($C_6H_{11}C\equiv CAu$), $[Au_{10}(BINAP)_4(C_6H_{11}C\equiv C)](CF_3COO)_3$ was quickly formed within 15 min (evidence by UV-vis absorption monitoring). The structure of the Au_{10} cluster (by SCXRD) could be viewed as inserting one Au-alkyne group into an Au—P(BINAP) bond (4–61 B). Of note, the reverse conversion, i.e., from $[Au_{10}(BINAP)_4(C\equiv CR)]^{3+}$ to

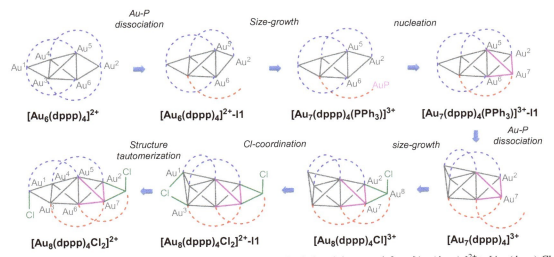

FIG. 4.60 Illustrative diagram for the main transformations in the Au(I) complex induced size-growth from $[Au_6(dppp)_4]^{2+}$ to $[Au_8(dppp)_4Cl_2]^{2+}$ clusters (I denotes isomer). *(Reproduced from Y. Lv, R. Zhao, S. Weng, H. Yu, Core charge density dominated size-conversion from Au_6P_8 to $Au_8P_8Cl_2$, Chem. A Eur. J. 26 (2020) 12382–12387, with the permission from Wiley-VCH, copyright 2020.)*

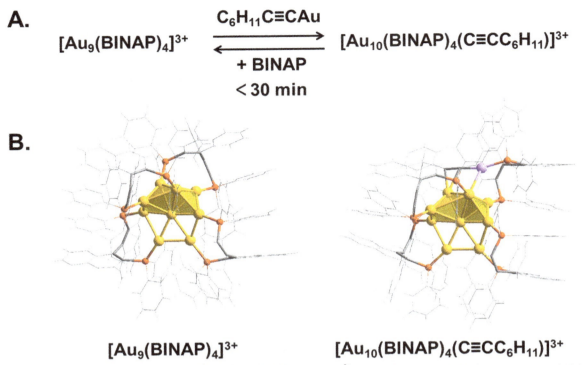

FIG. 4.61 size-conversion between $[Au_9(BINAP)_4]^{3+}$ and $[Au_{10}(BINAP)_4(C\equiv CR)]^{3+}$ clusters induced by Au(I)-alkyne complex and diphosphine ligand (A), and the single crystal structure of the precursor and the product clusters (B).

$[Au_9(BINAP)_4]^{3+}$ cluster, could be regulated by the addition of an equivalent amount of BINAP into the CH_2Cl_2 solution of the aforementioned Au_{10} cluster. Due to the great structural similarity of the two clusters, the interconversion formally corresponds to the *addition* or *elimination* of the Au-alkyne moieties induced by the Au(I) complex and the diphosphine ligands, respectively.

Distinct from the aforementioned size-conversion via incorporating of the monometallic Au(I) complex, the recent mechanistic study on the solvent (e.g., from water/ethanol to pure water) regulated size-growth from $[Au_{23}(p\text{-MBA})_{16}]^-$ to $[Au_{25}(p\text{-MBA})_{18}]^-$ demonstrates an internal driving force for the size-conversion from Au_{23} to Au_{25} (Fig. 4.62A) [27]. The relatively longer Au_3S_4 motifs on $[Au_{23}(p\text{-MBA})_{16}]^-$ were suggested to becomes less stable in strong polar solvent (due to solvent affinity), and thus the $[Au_{25}(p\text{-MBA})_{18}]^-$ with only short $Au_2(SR)_3$ was formed as the thermodynamic product. On the basis of the one-to-one size-conversion (elucidated by UV-vis and ESI-MS monitoring) and the identification of the Au(I) species in low m/z region of the mass spectra, Xie and co-workers conducted the ligand-exchange reaction via reacting $[Au_{23}(p\text{-MBA})_{16}]^-$ with Au(p-NTP) complexes (p-NTP = para-nitrothiophenol). Albeit the consistent formation of Au_{25} clusters, the composition of the formed $[Au_{25}(p\text{-MBA})_x(p\text{-NTP})_{18-x}]^-$ (i.e., number of incorporated foreign ligands) is dependent on the amount of the added Au(p-NTP) complexes. In combination of the spectral analysis and the single crystal structure of the precursor and product clusters, the overall size-conversion was suggested to start with the binding of two $[Au_2S_3]^-$ on the cuboctahedral core of the Au_{23} cluster ($\mathbf{Au_{23} \to Au_{23}\text{-I1}}$, Fig. 4.62B). Thereafter, the surface $[AuS_2]^-$ on the long motif (such as $[Au_3S_4]^-$) are removed to reduce the steric hindrance and electrostatic repulsion in the ligand shell ($\mathbf{Au_{23}\text{-I1} \to Au_{23}\text{-I2}}$). After that, core isomerization ($\mathbf{Au_{23}\text{-I2} \to Au_{23}\text{-I3}}$), re-construction of the core-shell interaction ($\mathbf{Au_{23}\text{-I3} \to Au_{23}\text{-I4}}$), and the final structure relaxation ($\mathbf{Au_{23}\text{-I4} \to Au_{25}}$) occur successively to generate the product Au_{25} clusters. Such surface-motif-exchange (SME) mechanism is also supported by the tentative mass spectral analysis, which indicates the preferential departure of $[AuS_2]^-$ and $[Au_2S]^-$ (compared with other $[Au_nS_{n+1}]^-$ fragments) from the precursor Au_{23} and product Au_{25} clusters, respectively.

In addition to the aforementioned Au(I) complex induced size-growth of $Au_{23} \to Au_{25}$ clusters, silver complex (i.e., Ag(SCy)) has also been successfully utilized to regulate the size-conversion of $[Au_{23}(SCy)_{16}]^-$, forming the alloy clusters of $[Au_xAg_{25-x}(SCy)_{18}]^-$ as a result (Fig. 4.62A) [101]. The framework and the metal composition of the cluster product were elucidated to be $[Au_{5.6}Ag_{19.4}(SCy)_{18}]^-$ with a combination of SCXRD and energy dispersive X-ray spectroscopy (EDX) analysis, and the fractional occupation originate from the compositional disorder of Au/Ag atom. A detailed structural analysis indicate that the doped Ag atoms spread on both the core surface of the icosahedral M_{13} structure, and the

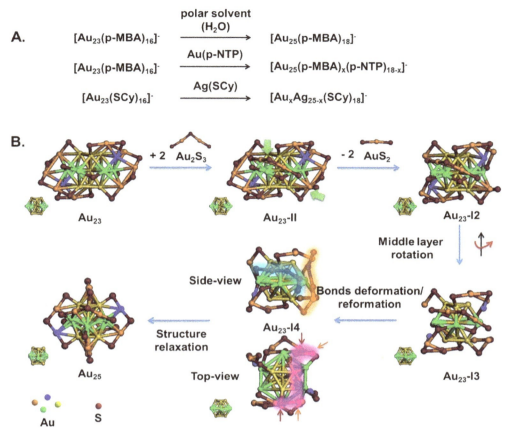

FIG. 4.62 The conversion of the Au$_{23}$ and M$_{25}$ (M=Au/Ag) clusters under different conditions (A), and an illustrative diagram for the conversion mechanism from Au$_{23}$ to Au$_{25}$ clusters (B). ((B) Reproduced from Q. Yao, V. Fung, C. Sun, S. Huang, T. Chen, D.-E. Jiang, J.Y. Lee, J. Xie, Revealing isoelectronic size conversion dynamics of metal nanoclusters by a noncrystallization approach, Nat. Commun. 9 (2018) 1979, with permission that is licensed under the Creative Commons Attribution 4.0 International License, copyright the author(s) 2018.)

M$_2$S$_3$ staples of the alloy M$_{25}$ product. Of note, Ag has been rarely incorporated into the staples/motifs in earlier studies of the Ag-doped Au$_{25}$ clusters. To this end, the Ag(I) complex induced synthetic strategy represents a highly appealing strategy to prepare heavily doped alloy [Au$_x$Ag$_{25-x}$(SCy)$_{18}$]$^-$ clusters, and a detailed mechanistic investigation is also of interest to elucidate the dynamics of the target size-conversion.

Compared with the uncertainty in forming alloy nanoclusters with variable metal compositions, doping specific number of foreign metal atoms (and especially the same group of metal atoms as those of the cluster-precursor) represents a more challenging issue. In this context, the controllable formation of [MAu$_{24}$(PPh$_3$)$_{10}$(PET)$_5$Cl$_2$]$^{2+}$ (M=Ag/Cu) via the reaction of a centrally hollow, rod-like [Au$_{24}$(PPh$_3$)$_{10}$(PET)$_5$Cl$_2$]$^+$ with AgCl/CuCl salts were recently accomplished by Jin and co-workers (Fig. 4.63A) [102]. The atomic structure of both AgAu and CuAu clusters were determined by SCXRD and ESI-MS characterizations. Interestingly, the single Ag atom was found to occupy exclusively the vertex position, while the Cu atom was found possibly occupy both the vertex (30%) and the waiste (70%) positions. The atomic precision excludes the possibility for a direct shuttling of the foreign atom into the central vacancy. In addition, DFT calculations indicates that the locating preference of the foreign metal atom is not determined by the thermodynamic profiles. In view of the stronger M$^+$-ClAu (M=Ag, Cu) interactions compared with that of the M$^+$-SAu, and the relatively larger atomic radii of Ag$^+$ than that of the Cu$^+$, it was suggested that the Ag$^+$-ClAu is predominate in the AuAg alloying system, while both Cu$^+$-ClAu and Cu$^+$-SAu are plausible in the AuCu alloying system. On the basis of this proposal and the additional calculations on the relative energy demands of two possible pathways (differentiated by the mobility of the surface/core metal atoms), the proposed mechanism of Ag$^+$ alloying mainly consists of several steps (Fig. 4.63B): interaction of Ag$^+$ with the vertex AuCl bond (**Au$_{24}$ → AgAu$_{24}$-I1**); insertion of the Ag$^+$ into one of the icosahedral Au$_{12}$ core (**Au$_{24}$-I1 → AgAu$_{24}$-I2**), and the mobility of the surface Au atoms (**AgAu$_{24}$-I2 → AgAu$_{24}$-I3**). Meanwhile, the doping mechanism of the Cu$^+$ in the waist positions of the centrally hollow Au$_{24}$ cluster (Fig. 4.63C) could possibly occur via a first Cu$^+$-SAu interaction step (**Au$_{24}$ → CuAu$_{24}$-I1**), which was then followed with the Cu-Au^{waist} bonding (**CuAu$_{24}$-I1 → CuAu$_{24}$-I2**), and the final

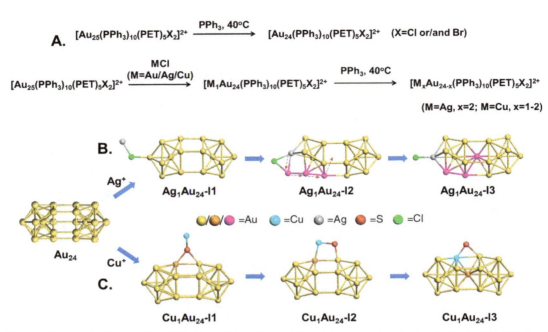

FIG. 4.63 The mutual conversion between the rod-like M$_{25}$ and Au$_{24}$ clusters (A), and the proposed mechanism for the AgCl/CuCl induced size-growth of the hollow [Au$_{24}$(PR$_3$)$_{10}$(SR)$_5$Cl$_2$]$^+$ into [MAu$_{24}$(PR$_3$)$_{10}$(SR)$_5$Cl$_2$]$^{2+}$ (A: M=Ag; B: M=Cu). *(Partially reproduced from S. Wang, H. Abroshan, C. Liu, T.-Y. Luo, M. Zhu, H.J. Kim, N.L. Rosi, R. Jin, Shuttling single metal atom into and out of a metal nanoparticle, Nat. Commun. 8 (2017) 848, with permission that is licensed under the Creative Commons Attribution 4.0 International License, copyright the author(s) 2017.)*

Au^{waist}-shuttling step (**CuAu$_{24}$-I2 → CuAu$_{24}$-I3**). Specifically, etching MAu$_{24}$ (M=Ag, Cu) cluster with PPh$_3$ could further produce the hollow M$_x$Au$_{24-x}$ clusters (evidenced by UV-vis spectra), while the addition of extra CuCl/AgCl results in the formation of the rod-like M$_x$Au$_{25-x}$ clusters with higher doping number ($x=2$ for Ag and $x=1$–2 for Cu, elucidated by ESI-MS). Of note, the PPh$_3$ etching induced size-reduction of the homometallic system, i.e., from the rod-like bi-icosahedral [Au$_{25}$(PPh$_3$)$_{10}$(PET)$_5$X$_2$] (X=Cl/Br) to the centrally hollow, rod-like [Au$_{24}$(PPh$_3$)$_{10}$(PET)$_5$Cl$_2$]$^+$ (evidenced by the UV-vis absorption spectral analysis), has also been reported earlier (Fig. 4.63A) [71]. This is also the synthetic strategy of the Au$_{24}$ cluster in Jin's study.

Similar to the Ag$^+$/Cu$^+$ addition on the rod-like Au$_{24}$ cluster, adding two Ag atoms onto the spherical Au$_{25}$ clusters ([Au$_{25}$(PET)$_{18}$]$^-$) has also been successfully accomplished in Wu's recent study (Fig. 4.64A) [103]. The reaction was carried out by adding AgNO$_3$ solution into the Au$_{25}$ solution drop by drop. The reaction is very rapid, and the cluster product was formed within 2h. Herein, the reduction of the chemically inert Au composition in the cluster precursors with the relatively more activate Ag$^+$ corresponds to the characteristic of the typical AGR reduction. Due to the lack of single crystal structure, and the possibility for both metal exchange and size-growth reactions, Wu and co-workers conducted careful experimental characterization and theoretical simulation to deduce the composition of the product. UV-vis absorption spectrum (similar characteristic peaks of the precursor and product clusters demonstrate the similarity in atomic structure), MALDI-MS (the cluster peaks corresponding to Au$_{25}$Ag$_2$(PET)$_{18}$), TGA (consistency between the experimental weight loss with the theoretical one of Au$_{25}$Ag$_2$(PET)$_{18}$), XPS (absence of counter ion), ICP-AES (undetected Au in the supernatant after reaction, excluding the metal-exchange mechanism), XRD (similar diffraction pattern of the precursor and the product demonstrates the largely maintained framework), and DFT calculations (consistency between the theoretical UV-vis absorption spectra of the modeling Au$_{25}$Ag$_2$(SH)$_{18}$ with the experimental one of Au$_{25}$Ag$_2$(PET)$_{18}$) collaboratively verify the formular of Au$_{25}$Ag$_2$(PET)$_{18}$ of the product. Meanwhile, the possibility for a size-focusing induced mono-dispersity of the alloy product was excluded in view of the short reaction time (size-focusing is commonly time-consuming). But due to the short reaction time, it remains highly challenging to propose an accurate transformation mechanism for this unusual Ag-addition reaction and to elucidate the inherent reason for the preference of "addition" over that of "exchange."

Compared with the aforementioned Ag-addition reaction in Fig. 4.64A, the Ag complex (Ag$_2$(dppm)Cl$_2$, dppm denotes bis(diphenylphosphino)methane) induced size-conversion from [Pt$_1$Ag$_{28}$(SAdm)$_{18}$(PPh$_3$)$_4$]Cl$_2$ to [Pt$_1$Ag$_{31}$(SAdm)$_{16}$(dppm)$_3$Cl$_3$]Cl$_4$ involves both ligand exchange (from monodentate PPh$_3$ to bidentate dppm, and from thiolate to chloride) and size-growth (from Pt$_1$Ag$_{28}$ to Pt$_1$Ag$_{31}$, Fig. 4.64B) [104]. As shown in Fig. 4.64C, the framework of

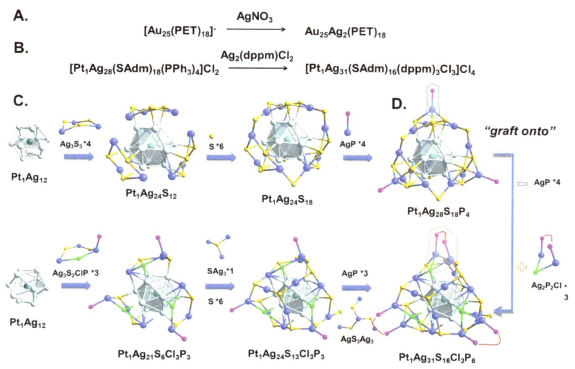

FIG. 4.64 The AgNO$_3$ induced size-growth of Au$_{25}$ cluster (A) and the Ag$_2$ complex induced size-conversion of the alloy Pt$_1$Ag$_{28}$ cluster (B), structural anatomy of the framework of [Pt$_1$Ag$_{28}$(SAdm)$_{18}$(PPh$_3$)$_4$]Cl$_2$ and [Pt$_1$Ag$_{31}$(SAdm)$_{16}$(DPPM)$_3$Cl$_3$]Cl$_4$ clusters from the multilayer (C) or graft-onto (D) point of view.

the Pt$_1$Ag$_{28}$ cluster is similar to that of Pt$_1$Ag$_{31}$. Each of them consists of a Pt@Ag$_{12}$ kernel. Interestingly, the M$_{13}$ adopts an FCC structure in Pt$_1$Ag$_{28}$, but adopts an icosahedral structure in Pt$_1$Ag$_{31}$ cluster. The FCC-packed Pt$_1$Ag$_{12}$ kernel is further protected by four, half-chair-like Ag$_3$S$_3$ motifs, each of which protects one triangular face of the FCC-core surface. The four Ag$_3$S$_3$ blocks adopt a tetrahedral orientation, and the Ag atoms are further cross-linked by six bridging S atoms (on -SAdm ligand). Each of the four Ag$_3$S$_3$ blocks is further capped by an additional AgP group. To this end, the total structure of Pt$_1$Ag$_{28}$ cluster could be depicted as Pt@Ag$_{12}$@(Ag$_{12}$S$_{12+6}$)@(Ag$_4$P$_4$), and the framework exhibits a tetrahedral symmetry. Compared with the FCC-packed Pt@Ag$_{12}$ kernel in Pt$_1$Ag$_{28}$ cluster, the icosahedral Pt@Ag$_{12}$ kernel in Pt$_1$Ag$_{31}$ cluster could be viewed as twisting the medium, planar M$_7$ layer in the tri-layered M$_3$-M$_7$-M$_3$ structure into a chair-like one (Fig. 4.63C). Meanwhile, the monodentate PPh$_3$ ligands were replaced by the bidentate dppm ligands, the dangling P atom of which then binds with an Ag atom on the Ag$_3$S$_3$ block in the third layer. In this context, the thiolate ligands preligated with this Ag atoms was replaced by a chloride. After the replacement of the three thiolate ligands on Ag$_3$S$_3$ with three chlorides, a SAg$_3$ block is incorporated to fill the cavity, resulting the coordination modes of the three, μ$_2$-bridging S atoms (of thiolate) changes to a three-coordinated μ$_3$-S. Compared with the substitution of the PPh$_3$ by dppm ligands in the three Ag$_3$S$_3$@AgP moieties, the structural changes in the fourth Ag$_3$S$_3$@AgP structure is remarkably different. The PPh$_3$ is removed, and the cyclic Ag$_3$S$_3$@Ag block collapses to a trilateral AgS$_3$Ag$_3$ structure. Alternatively, the structural change from Pt$_1$Ag$_{28}$ to Pt$_1$Ag$_{31}$ cluster could also be viewed as replacing three of the four outermost AgP groups with three Ag$_2$(dppm)Cl structures, associated with some other structural distortions, removal of the remaining PPh$_3$ and two thiolate ligands. In view of the direct incorporation of the experimentally prepared Ag complex Ag$_2$(dppm)Cl into the structure of the Pt$_1$Ag$_{28}$ cluster precursor (Fig. 4.63C and D), the size-growth from Pt$_1$Ag$_{28}$ to Pt$_1$Ag$_{31}$ clusters were denoted as a "*graft-onto*" mechanism.

Herein, for each atomically precise metal nanocluster, there are multiple ways to interpret the atomic packing mode. Most of the structural analysis was proposed to fit a certain group of structural evolution principal or experimental procedures. But the readers are suggested to keep in mind that most of the proposed structure anatomy do not correlate directly with the electronic structures or the size-conversion mechanism. Due to the complexity of the size-conversion reaction systems, the atomic level understanding on most of the size-conversion is quite limited to date. Clearly, much more efforts are necessary to develop more powerful theoretical and experimental methods to elucidate the mechanistic details.

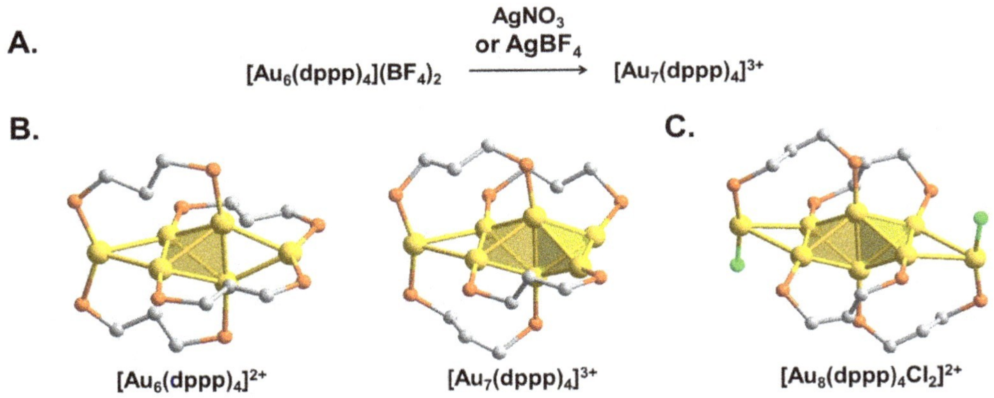

FIG. 4.65 The Ag^+ induced size-growth of Au_6 to Au_7 clusters (A), the framework of the single crystal structure of $[Au_6(dppp)_4]^{2+}$, $[Au_7(dppp)_4]^{3+}$ (B), and $[Au_8(dppp)_4Cl_2]^{2+}$ clusters (C).

Of note, the tautomerization of the FCC-packed $Pt@Ag_{12}$ kernel in the Pt_1Ag_{28} precursor into an icosahedral one has also been reported in several cases, including the PPh_3 and $BDTH_2$ induced conversion of $[Pt_1Ag_{28}(SAdm)_{18}(PPh_3)_4]^{2+}$ to $[PtAg_{28}(BDT)_{12}(PPh_3)_4]^{4-}$ (Fig. 4.51); HSCy induced conversion of $[Pt_1Ag_{28}(SAdm)_{18}(PPh_3)_4]^{2+}$ to $[Pt_1Ag_{28}(S-Cy)_{18}(PPh_3)_4]^{2+}$ (Fig. 4.41).

Metal complex catalyzed size-growth of metal nanoclusters

Compared with the aforementioned size-growth or alloying processes, the metal complex catalyzed size-growth of metal nanocluster has been rarely reported. In this context, the Ag^+ induced size-growth from Au_6 to Au_7 cluster represents an interesting example (Fig. 4.65A) [39]. According to Konishi and co-workers' experimental observation, adding excessive $AgBF_4$ or $AgNO_3$ to $[Au_6(dppp)_4](BF_4)_2$ solution at room temperature results in an immediate formation of $[Au_7(dppp)_4]^{3+}$ cluster (evidenced by the changed solution color from blue to reddish-violate, and ESI-MS characterization). The metallic core structure of the Au_7 cluster could be viewed as bridging one outermost Au—Au bond of the edge-sharing bi-tetrahedral Au_6 structure with an additional Au atom. Except for the bridging Au atom which accommodates two phosphine groups from two dppp ligands, the other Au atoms are each protected by one phosphine hand of the biphosphine ligand. The core structure of Au_7 is distinct from that of $[Au_6(dppp)_4]^{2+}$, but is more similar to the structure of $[Au_8(dppp)_4Cl_2]^{2+}$ (Fig. 4.65B). In a recent DFT study [99] on the $Au(PPh_3)Cl$ complex induced size-conversion from $[Au_6(dppp)_4]^{2+}$ to $[Au_8(dppp)_4Cl_2]^{2+}$, the $[Au_7(dppp)_4]^{3+}$ cluster was found to be a metastable intermediate in the most feasible pathway (Fig. 4.60 for the details). So far, the mechanism for the Ag^+ induced size-growth is remains ambiguous (predominantly due to the rapid reaction rates and difficulty in spectral monitoring), and the ESI-MS tests on the reaction mixture did not show effective signals corresponding to the Ag-containing species.

4.2.2.6 Mechanism of the environmental condition (heating, solvent, etc.) triggered size-conversion

Solvent-triggered isomerization of metal clusters

A pair of isomeric $Au_{23}(C≡CtBu)_{15}$ clusters were recently prepared by Wang and co-workers via the typical one-pot synthesis (i.e., reduction of Me_2SAuCl and $HC≡CtBu/NEt_3$ with $NaBH_4$) [105]. Depending on whether the tetraphenylphosphonium chloride (or tetrabutylam-monium chloride) was added or not, the crystal of **$Au_{23}X_{15}$-1** or **$Au_{23}X_{15}$-2** could be yielded. Each of these two clusters consist of a tri-octahedral Au_{11} structure (Fig. 4.66A and the inset), which is further capped by four Au atoms to constitute the Au_{15} kernel. Each Au_{15} kernel is further protected by three Au_2X_3, two AuX_2, and 2 bridging X ligands (X=—C≡CtBu), while the different orientation of the four capping Au atoms of Au_{15} kernel mainly results in the distinct coordination fashion of the two AuX_2 motifs (Fig. 4.66A). According to the time-dependent UV-vis monitoring on the size-conversion, **$Au_{23}X_{15}$-1** is stable in CH_2Cl_2 for 2 h, and then the characteristic absorption at 540 and 381 nm gradually attenuated, associating with the gradual enhancement of the absorption ranging in 400–500 nm. After 7 h, the characteristic peak of **$Au_{23}X_{15}$-2** (at 412, 598 nm) could be clearly identified, and the size-conversion could also be reflected from the changed solution color (Fig. 4.66B inset). Meanwhile, ^1H NMR tests implies significant changes in the range of 1.0–1.5 ppm (Fig. 4.66C), corresponding to the signals of the tBu groups, and the NMR spectra of the conversion product after 8 h is very similar to that of the target **$Au_{23}X_{15}$-2**. Besides, MALDI-MS characterization on the cluster products shows the cluster peaks corresponding to $Au_{23}X_{15}$, $Au_{23}X_{14}$, $Au_{22}X_{15}$, $Au_{24}X_{15}$, and $Au_{25}X_{16}$. In contrast to the slow conversion of **$Au_{23}X_{15}$-1** to **$Au_{23}X_{15}$-2** at ambient conditions, **$Au_{23}X_{15}$-1** is relatively stable under low temperature

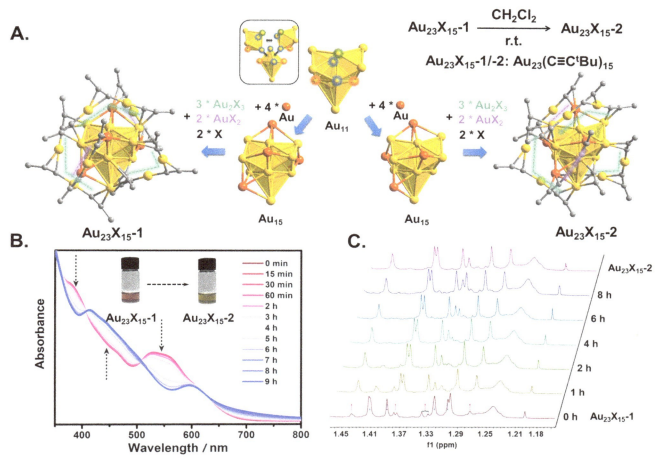

FIG. 4.66 Structural anatomy of the single crystal structure of the two isomeric $Au_{23}X_{15}$ clusters (X = —C≡CtBu) (A) and the inset: illustrative diagram for the face-sharing pattern of the three octahedral blocks in formation of the Au_{11} structure; UV-vis monitoring on the size-conversion from **$Au_{23}X_{15}$-1** to **$Au_{23}X_{15}$-2** (B), inset: the cluster solvent (in CH_2Cl_2) before and after size-conversion; and ^{1}H NMR spectral monitoring on the size-conversion in CD_2Cl_2 (C). *(Reproduced from Z.-J. Guan, F. Hu, J.-J. Li, Z.-R. Wen, Y.-M. Lin, Q.-M. Wang, Isomerization in alkynyl-protected gold nanoclusters, J. Am. Chem. Soc. 142 (2020) 2995–3001, with permission from the ACS Publications (Copyright 2020).)*

(e.g., 4°C) or in the presence of PPh$_4$Cl (at ambient conditions). With the aid of DFT calculations, Wang and co-workers proposed that the conversion is mainly driven by the higher stability of **$Au_{23}X_{15}$-2** than that of **$Au_{23}X_{15}$-1**.

In the recent study of Zang and co-workers [106], ^{1}H NMR was used to evaluate the relative stability of the two Ag_{12} clusters, whose interconversion could be well regulated by changing the solvent conditions (Fig. 4.67A). First, the reaction of Ag(TFA) (TFA = trifluoroacetate) with HS-POSS (i.e., mercaptopropylisobutyl functionalized polyhedral oligomeric silsesquioxane) in a mixed solvent of THF-MeCN (THF is short for tetrahedrofuran) produces the colorless parallelogram crystal, whose structure were determined to be Ag_{12}(S-POSS)$_6$(TFA)$_6$(THF)$_6$ (**Ag_{12}-1**) by SCXRD and Maldi-MS characterization. Either changing the reaction solvent into a mixed CH_2Cl_2-acetone solvent or recrystallizing **Ag_{12}-1** in the mixed CH_2Cl_2-acetone solution could generate another colorless hexagon crystal of Ag_{12} cluster, with the formula of Ag_{12}(S-POSS)$_6$(TFA)$_6$(acetone)$_2$ (**Ag_{12}-2**) as elucidated by SCXRD and Maldi-MS. **Ag_{12}-1** and **Ag_{12}-2** each consists of a three-layered Ag_3-Ag_6-Ag_3 core, and six thiolate ligands (S-POSS) with μ^4-S coordination (Fig. 4.67B). In addition, the six peripheral Ag atoms in Ag_6 layer of **Ag_{12}-1** is each coordinated by two additional oxygen atoms, one from THF solvent, and the other one from the TFA group. In this context, both S-POSS and TFA act as the interlayer ligand, and thus each Ag atom in the upper and lower Ag_3 layer is coordinated by one S atom (of S-POSS) and one O atoms (of TFA). The overall binding modes of the six S-POSS and six TFA to Ag atoms in **Ag_{12}-2** is close to that of **Ag_{12}-1**, while the two acetones coordinate to the two opposite Ag atoms on the Ag_6 layer. Albeit the similar coordination modes of the Ag atoms, thiolate and TFA ligands, the presence of different number of solvent ligands results in the significantly varied structure in both the metallic core and the orientation of the surface POSS groups (Fig. 4.67B). According to the ^{1}H NMR tracking tests, the chemical shift of THF in **Ag_{12}-1** decreased after air dried for 6h, while chemical shift ascribing to acetone in **Ag_{12}-2** is

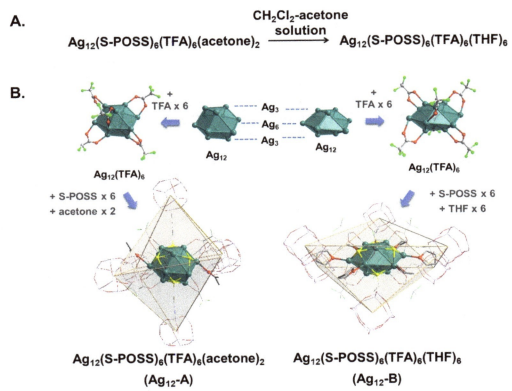

FIG. 4.67 The solvent induced isomerization from one Ag_{12} cluster to another one (A), and an illustrative diagram for the structural analysis of the framework of the two Ag_{12} isomers (B). Note: the Ag-Ag distances in the Ag_6 layer of **Ag_{12}-A** are all over 3.40 Å, which is almost two times the Bondi radii of Ag atoms. Albeit the weak bonding interactions therein, the Ag-Ag atoms are connected for better comparison of the two atomic structures.

largely maintained after almost 2 weeks. The observation indicates the relatively stronger ability of the acetone solvent than that of THF.

Similar to the aforementioned study, Yu, Lang and co-workers recently reported the solvents induced interconversion between a series of Ag_{12} clusters, i.e., $Ag_{12}(SAB)_6(TFA)_{12}(H_2O)_2 \cdot 4CH_3CN$ (**Ag_{12}-3**), $Ag_{12}(SAB)_6(TFA)_{12} \cdot 2DMF$ (**Ag_{12}-4**) and $Ag_{12}(SAB)_6(TFA)_{10}(MeCN)_2 \cdot 2(TFA)$ (**Ag_{12}-5**), with the similar main structure but distinct symmetry elements (HSAB, and DMF denote 4-(trimethylammonio)benzenethiolate and N,N-dimethylformamide, Fig. 4.68A) [107]. The reaction of Ag(I) salt with HSAB in a tertiary solvent system of DMF, CH_3CN and H_2O generates the structurally regulated cluster of **Ag_{12}-3** (evidenced by SCXRD), in which the hollow octahdecahedral Ag_{12} kernel is protected by six SAB ligands via μ_4-S coordination, 12 trifluoroacetate ligands adopting μ_2-/μ_3-O binding modes, and two water molecules with both μ_2—O—Ag bonds and hydrogen bonding with TFA ligands (Fig. 4.68B). Changing the solvent into a binary system of CH_3OH and DMF results in the formation of **Ag_{12}-4**. SCXRD analysis indicates a slightly twisted framework of **Ag_{12}-4** (compared with that of **Ag_{12}-3**) in the absence of two water molecules, and the coordination of the 12 TFA change to four μ_1- and eight μ_2-O modes. Alternatively, conducting the synthesis in MeCN solvent results in the formation of **Ag_{12}-5**, with completely different structural symmetry with that of **Ag_{12}-3** and **Ag_{12}-4**. The 12 Ag atoms adopt a hollow distorted cuboctahedron structure, and is protected by six μ_4-coordinated SAB, ten TFA (with two μ_1- and eight μ_2-coordination modes), and two coordinated MeCN via the Ag—N bonds. Of note, the interconversion of **Ag_{12}-3**, **Ag_{12}-4** and **Ag_{12}-5** could also be regulated by dissolving them in different solvents, associated with re-crystalizing procedures.

Herein, it is noteworthy that the solvent effect could not only induce the isomerization between the same-sized metal nanoclusters bearing different number of coordinated solvents, but is also able to regulate the size-conversion of metal nanoclusters. An intriguing concentration/dilution induced reversible size-conversion between the hexanuclear Au_6 sulfido cluster (i.e., $[Au_6S_2(Ph_2PN(Me)PPh_2)_3]Cl_2$) and a dodecanuclear Au_{12} cluster ($[Au_{12}S_4(Ph_2PN(Me)PPh_2)_6]Cl_4$) was reported recently by Yam and co-workers (Fig. 4.69) [108]. Of note, these two clusters were also categorized into polynuclear Au(I) complexes due to the absence of free electrons. According to SCXRD analysis, these two clusters adopt the cubic-like, and metallamacrobicyclic structures, respectively (Fig. 4.69). From ^1H NMR and $^{31}P\{^1H\}$ NMR spectral characterization (see Column III for details), dissolution of the CD_3OD solution of the Au_{12} cluster to 4×10^{-5} M (or lower) results in the formation of Au_6 as the sole product, while the reverse size-conversion occurs by concentrating the solution

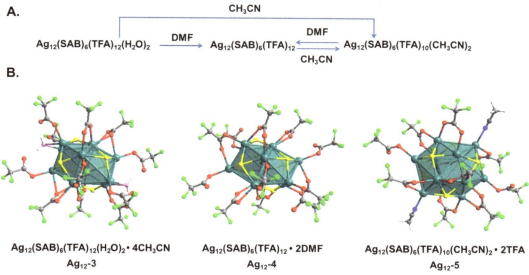

FIG. 4.68 The solvent controlled conversion among the three Ag$_{12}$ clusters (A), and the framework of **Ag$_{12}$-3**, **Ag$_{12}$-4** and **Ag$_{12}$-5** (B). For clarity, the C and H atoms on the thiolate ligands of SAB, and the solvent molecules outside the cluster structure (without metal-ligand coordination) are omitted for clarity.

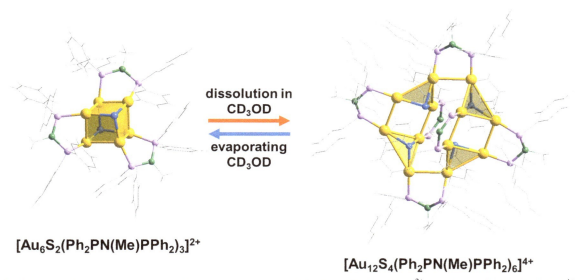

FIG. 4.69 The dissolution/concentration induced size conversion between [Au$_6$S$_2$(Ph$_2$PN(Me)PPh$_2$)$_3$]$^{2+}$ and [Au$_{12}$S$_4$(Ph$_2$PN(Me)PPh$_2$)$_6$]$^{4+}$ clusters. Color legends: *light blue* for S; *lavender* for P; and *sea green* for N.

of Au$_6$ to 8×10^{-4} M via evaporating the CD$_3$OD solvent under reduced pressure (and without heating). In addition to the NMR tests, the interconversion was also verified by UV-vis and high-resolution ESI-MS tests. Similar to the aforementioned reaction system, the concentration regulated size-conversion between the Au$_{12}$ and Au$_6$ clusters bearing PF$_6$ counter-anions ([Au$_6$S$_2$(Ph$_2$PN(Me)PPh$_2$)$_3$](PF$_6$)$_2$ vs [Au$_{12}$S$_4$(Ph$_2$PN(Me)PPh$_2$)$_6$](PF$_6$)$_4$) was achievable in CD$_3$CN solvent. Meanwhile, the concentration of cluster solvents could also be regulated by reducing the solvation via adding bad solvents. In addition, using a control experiment with two different Au$_{12}$ cluster as the reactants (bearing the diphosphine ligand of Ph$_2$PN(Me)PPh$_2$ or Ph$_2$PN(H)PPh$_2$), ligand-exchange between the two Au$_{12}$ cluster precursors occurs, associated with the size-conversion. By contrast, no ligand exchange occurs when [Au$_{12}$S$_4$(Ph$_2$PN(Me)PPh$_2$)$_6$]Cl$_4$ was mixed with the free ligand of Ph$_2$PN(H)PPh$_2$. In this context, Yam and co-workers suggested that the concentration induced size-conversion occurs via a mechanism involving a prior dissociation to the subsequent rearrangement and reorganization steps. In addition, the observation of the characteristic peaks corresponding to [Au$_3$S(Ph$_2$PN(R)PPh$_2$)$_3$]$^+$ in the cryospray-ionization-mass spectroscopy (CSI-MS) of the CH$_3$OH solution of [Au$_{12}$S$_4$(Ph$_2$PN(Me)PPh$_2$)$_6$]Cl$_4$ demonstrates the formation of an intermediate Au$_3$ fragments, which was suggested to be too short-lived to be detected in the NMR spectral tests.

COLUMN III Tips on NMR analysis in mechanistic study

Thanks to the thermodynamic stability and structural rigidity of some atomically precise metal nanoclusters, NMR characterization could be possibly used to elucidate the reaction mechanism by monitoring the consumption/production of certain species, and especially, to determine quantitatively the conversion. For example, in the mechanistic investigation on the concentration regulated size-conversion between the polynuclear Au_6 and Au_{12} clusters, both the Au_{12} cluster precursor and the Au_6 products show clear characteristic peaks on $^{31}P\{^1H\}$ NMR spectroscopy, and it is easy to distinguish the signals corresponding to each cluster species due to the distinct chemical shift (δ, Fig. 4.70). In this context, the increased molar ratio of Au_6 compared with that of the Au_{12} precursors due to the dissolution could be easily quantified by NMR spectroscopy.

Compared with the clear NMR spectra in Fig. 4.70, the NMR spectra in some reaction systems are relatively comprehensive. In this case, comparing NMR spectra of free state of certain reagents with that of the total reaction system could benefit the identification of the characteristic peaks. For example, in the ligand exchange reaction of $[Au_{11}(PPh_3)_8Cl_2]Cl$ and di-isopropyl benzimidazolium hydrogen carbonate (**NHC1**·H_2CO_3, and NHC is short for N-heterocyclic carbene), the ^1H NMR spectra shows the presence of a variety of species in the crude products (Fig. 4.71) [109]. In this context, the ^1H NMR of $OPPh_3$ (oxidative state of the PPh_3 ligands on cluster surface), [Au(**NHC1**)$_2$]Cl helped to rationalize the characteristic peaks in the region of $\delta = 7.4$–7.9 and ~5.4 ppm (Fig. 4.71a, f vs c). Meanwhile, the absence of the effective signals in the region of ~5.5 ppm is potentially helpful to elucidate the presence of [Au(**NHC1**)]Cl in the crude product (Fig. 4.71g vs c). In addition, due to the largely maintained cluster framework and the ligand structure (only one of the eight PPh_3 was replaced by the **NHC1** ligand, and the single crystal X-ray crystallography), the overall ^1H NMR spectra of the $[Au_{11}(PPh_3)_8Cl_2]Cl$ and [Au(**NHC1**)$_2$]Cl greatly overlap in the region of 6.6–7.1 ppm. Nevertheless, the low intensity of the peak at 6.65 ppm (labeled with violet star *) could be used to identify the tiny amount of the remaining cluster precursor. Interestingly, the effect of purification could also be verified with the aid of the comparison between different ^1H NMR spectra. After testing the ^1H NMR spectra of samples before and after purification, $OPPh_3$ was found to be efficiently removed via washing with diethyl ether (Fig. 4.71c vs d), while the subsequent treatment with column chromatography further removes of the [Au(**NHC1**)$_2$]Cl complex (Fig. 4.71c vs d).

In addition to the aforementioned elucidation of the possible components in the crude products, ^{13}C NMR and ^{31}P NMR spectra were also used to characterize the NHC-incorporated cluster products. First, ^{13}C-enriched NHC ligands were used to settle the problem that the natural abundance of ^{13}C is too low to be identified by ^{13}C NMR. Interestingly, ^{31}P NMR spectra of both the cluster precursor $[Au_{11}(PPh_3)_8Cl_2]Cl$ and the product (predominantly $[Au_{11}(PPh_3)_7(\mathbf{NHC1})Cl_2]Cl$) showed a unique characteristic peak at room temperature. However, ^{31}P NMR spectra of $[Au_{11}(PPh_3)_8Cl_2]Cl$ was broadened under lowered temperature ($-70°C$), while ^{31}P NMR spectra of the NHC incorporated cluster products showed peak splitting upon cooling to $-70°C$. To this end, both the characteristic peaks of Au—C bond and the peak splitting corresponding to the distinct chemical environments of the surface phosphine ligands (due to the broken symmetry of the introduced NHC ligand) verifies the ligand exchange.

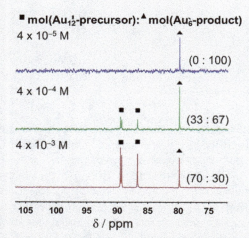

FIG. 4.70 The changes in molar ratios of the Au_{12} precursor and the Au_6 products with the concentration of Au_{12} (in CD_3OD) varied from 4×10^{-3} to 4×10^{-5} M determined by $^{31}P\{^1H\}$ NMR spectroscopy. *(Reproduced from L.-L. Yan, L.-Y. Yao, V.W.-W. Yam, Concentration- and solvation-induced reversible structural transformation and assembly of polynuclear gold(I) sulfido complexes, J. Am. Chem. Soc. 142 (2020) 11560–11568, with permission from the ACS Publications (Copyright 2020).)*

COLUMN III Tips on NMR analysis in mechanistic study—cont'd

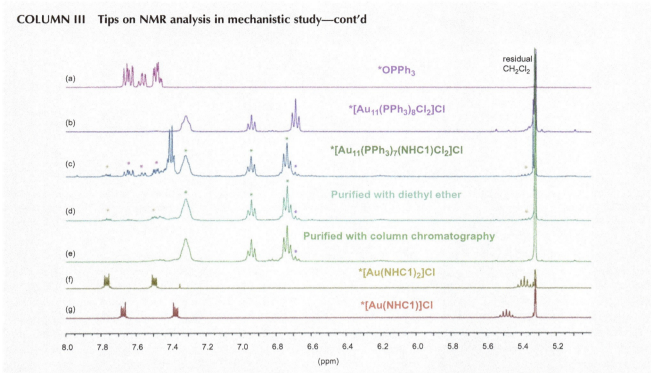

FIG. 4.71 ^{1}H NMR (400 MHz, CD$_2$Cl$_2$, RT) spectra of OPPh$_3$ (a), [Au$_{11}$(PPh$_3$)$_8$Cl$_2$]Cl cluster precursor (b), cluster mixture (crude product) of the reaction system of [Au$_{11}$(PPh$_3$)$_8$Cl$_2$]Cl with **NHC1**·H$_2$CO$_3$ (c), purified crude product after treating with diethyl ether (d), purified crude product by column chromatography (e), [Au(**NHC1**)$_2$]Cl (f), and [Au(**NHC1**)]Cl (g). (*Reproduced from M.R. Narouz, K.M. Osten, P.J. Unsworth, R.W.Y. Man, K. Salorinne, S. Takano, R. Tomihara, S. Kaappa, S. Malola, C.-T. Dinh, J.D. Padmos, K. Ayoo, P.J. Garrett, M. Nambo, J.H. Horton, E.H. Sargent, H. Häkkinen, T. Tsukuda, C.M. Crudden, N-heterocyclic carbene-functionalized magic-number gold nanoclusters, Nat. Chem. 11 (2019) 419–425. Copyright 2019, The Author(s), under exclusive license to Springer Nature Limited.*)

Size-conversion induced by heating

During the synthesis of metal nanoclusters, heating (in the presence or absence of reducing reagents) has been frequently used as an efficient strategy to control the nucleation of metal complexes, and thus the atomic size of the cluster products. With the well elucidation of the atomic structure and physicochemical properties of the large amount of ultrasmall sized metal nanoclusters, the posttreatment with heating of one atomically precise metal nanocluster has been found a plausible strategy to regulate the size-conversion.

In the recent study of Yan, Liu and co-workers, heating the alcoholic solution of the bidentate phosphine protected Au$_6$ cluster, i.e., [Au$_6$(dppp)$_4$]$^{2+}$ at 70°C results in a rapid cluster decomposition, which is evidenced by the expense of the characteristic peak at 585 nm on UV-vis spectra (Fig. 4.20) [42]. Meanwhile, the gradual formation of [Au$_{11}$(dppp)$_5$]$^{3+}$ was verified by the appearance and strengthening of its characteristic peak at 410 nm along reaction time. The time-dependent ESI-MS tracking demonstrates the co-existence of comprehensive species along the reaction times, including Au$_6$ precursors, Au$_{11}$ products, and also some other intermediate species (whose formular are difficult to be identified, and was suggested to be Au(I) complexes). After 460 min, all Au$_6$ and the intermediate species were consumed and the [Au$_{11}$(dppp)$_5$]$^{3+}$ cluster become the only product.

In addition to the heating induced size-conversion from Au$_6$ to Au$_{11}$, heating the alcoholic solution of another cluster precursor, i.e., [Au$_8$(dppp)Cl$_2$]$^{2+}$ also generates Au$_{11}$ as the final product (after several days), while [Au$_6$(dppp)$_4$]$^{2+}$ was formed as an intermediate after about 2 h (evidenced by UV-vis and ESI-MS monitoring on the reaction system). With the combination of the reduction/oxidation induced interconversion of [Au$_6$(dppp)$_4$]$^{2+}$ and [Au$_8$(dppp)Cl$_2$]$^{2+}$, and reduction induced size-reduction from [Au$_{11}$(dppp)$_5$]$^{3+}$ to [Au$_6$(dppp)$_4$]$^{2+}$ (the details of these reactions are omitted because the conditions are beyond the discussion in this section), the inherent size-conversion network of the Au$_6$, Au$_8$, and Au$_{11}$ clusters has been suggested (Fig. 4.20).

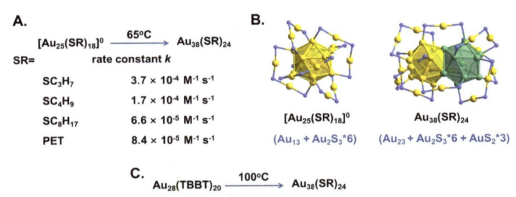

FIG. 4.72 Heating induced size-growth of a series of thiolate ligand protected neutral $[Au_{25}(SC_4H_9)_{18}]^0$ clusters (A), and an illustrative diagram of the framework of the Au_{25} and Au_{38} clusters comprising one icosahedral Au_{13} and face-sharing bi-icosahedral Au_{23} core structure (B), and heating induced size-growth of the $Au_{28}(SC_4H_9)_{20}$ clusters (C) [110].

The heating-controlled fusion of smaller-sized clusters into a relatively larger one has also been reported recently. Using the fusion of the neutral $Au_{25}(SC_4H_9)_{18}$ as a modeling cluster precursor, the size-conversion was conducted at 65°C and tracked with UV-vis absorption spectra. The reaction takes 10–14 days, and the product was determined $Au_{38}(SC_4H_9)_{24}$ by ESI-MS characterization (Fig. 4.72A). On this basis, the ^1H and ^{13}C NMR spectra of the cluster product demonstrate that the Au_{38} adopts a similar framework with that of $[Au_{38}(PET)_{24}]$ [111]. Since the Au_{25} precursor and the Au_{38} product comprise the icosahedral Au_{13} and the face-sharing bi-icosahedral Au_{23} structure (Fig. 4.72B), this size-conversion is also specified as the "fusion" reaction. Meanwhile, the time-dependent UV-vis absorption and DPV (differential pulse voltammetry) demonstrate a second order reaction, with a rate constant k of cal. 1.5×10^{-4} M^{-1} s^{-1} (referring to the concentration of the Au_{25} precursor). Similar size-conversion of $Au_{25} \rightarrow Au_{28}$ was also accomplished in presence of other thiolate ligands (SC_3H_8, PET, etc.), albeit with different reaction rates. In strong contrast to these neutral, paramagnetic $Au_{25}(SC_4H_9)_{18}$ precursors, both the anionic $Au_{25}(SC_4H_9)_{18}$, or the charge neutral $Au_{24}Hg(SC_3H_7)_{18}$ shows no reactivity to the target fusion reaction. In other words, both the charge state and the unpaired electron were suggested pivotal to the size-growth. With the combination of these experimental observation, and the density functional theory calculations, Maran and co-workers proposed that the size-conversion is initiated by a dimerization of two neutral Au_{25} precursors by intercluster van der Walls interaction in the ligand shell and intercluster Au—Au bonding. After that, a slow bond re-organization occurs to finally generate the Au_{38} cluster product. The dimerization could also be viewed as the coupling of two open-shell superatoms to form a supermolecule [110].

Interestingly, with a slightly modified reaction condition (heating at 100°C), the size-growth of the FCC-packed gold cluster of $Au_{28}(TBBT)_{20}$ was achieved, generating the relatively larger $Au_{36}(TBBT)_{24}$ cluster as the product (associating with some insoluble polymeric byproducts and dimer of TBBT ligands) after 36h (Fig. 4.72C) [112]. But compared with the aforementioned core-fusion in the conversion of $Au_{25} \rightarrow Au_{38}$, the size-growth of $Au_{28} \rightarrow Au_{36}$ corresponds to a "layer-by-layer" size growth (Section 4.3.1.1). Meanwhile, the kinetic analysis on the fusion process with time-dependent UV-vis absorption spectral monitoring demonstrate a first-order kinetics (instead of the second order kinetics of the fusion of the neutral Au_{25}). In this context, the rate-determining step is suggested the decomposition of the Au_{28} precursor, in strong contrast to the dimerization of the neutral Au_{25} in the aforementioned system. On the basis of the experimental characterization and the size-evolution principal of the series of FCC-packed Au clusters, Mandal and co-workers suggested that the size-growth involves the rearrangement and decomposition of the core structure, and the removal of the surface thiolates, and then the addition of $Au_8(TBBT)_4$ unit then occurs to generate the target Au_{36} cluster. A balanced equation (2 $Au_{28}(TBBT)_{20} \rightarrow Au_{36}(TBBT)_{24} + 20$ [Au] + 16[TBBT]) was suggested, and the species in square brackets are comprehensive by-products with variable structures.

In addition to the heating induced size-growth mentioned above, a heating regulated isomerization of one $Au_{38}(PET)_{24}$ has also been reported lately (Fig. 4.73) [113]. First, one $Au_{38}(PET)_{24}$ (denoted as $Au_{38}(PET)_{24}$-**1**) was synthesized by the one-pot method, and SCXRD analysis indicates its structure to comprise a Au_{23} core (an icosahedral Au_{13} capped by an Au_{10} structure, Fig. 4.73 left), two Au_3S_4, three Au_2S_3, three AuS_2 staples and one PET ligand. This $Au_{38}(PET)_{24}$-**1** is stable under $-10°C$ for 1 month. But heating its toluene solution to 50°C results in an isomerization after 36h, forming another $Au_{38}(PET)_{24}$ cluster ($Au_{38}(PET)_{24}$-**2**) comprising a bi-icosahedral Au_{23} core, three AuS_2 and six Au_2S_3 staples (Fig. 4.73 right). The UV-vis spectra of these two isomers are slightly different, and two isosbestic points at 360 and 700 nm were observed in the UV-vis monitoring. Meanwhile, the two isomers could be distinguished by PTLC (preparative thin-layer

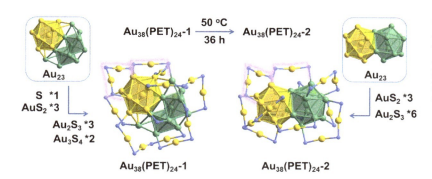

FIG. 4.73 Heating induced isomerization of the two $Au_{38}(PET)_{24}$ clusters, and the illustrative diagram for the framework of the single crystal structure of $Au_{38}(PET)_{24}^{-I1}$ and $Au_{38}(PET)_{24}^{-I2}$ (insets: the different kernel structure in the two isomers). The pink dashed line labels the two blocks that adopting the similar orientation in the two isomers [113].

chromatography) analysis. On the basis of the easy formation and stability of $Au_{38}(PET)_{24}$-**1** under low temperature and its conversion to $Au_{38}(PET)_{24}$-**2** under heating conditions, Jin and co-workers suggest these two clusters to be kinetic and thermodynamic feasible products, respectively.

In addition to the isomerization of the two $Au_{38}(PET)_{24}$ clusters, the isomerization of the two rod-like $[Au_{13}Ag_{12}(PPh_3)_{10}Cl_8]^+$ has also been successfully regulated via alternating the temperature ($-10°C$ or $25°C$, Fig. 4.74A) [114]. According to Li's study, the two $Au_{13}Ag_{12}$ isomers are formed concomitantly during synthesis (reduction of Ph_3PAuCl and $AgSbF_6$ with $NaBH_4$ in an ice bath), and has been separated via the thin-layer chromatography (TLC) treatment. Specifically, crystallization of the product mixture (with both isomers) under different temperatures generates the pure isomer of $\{[Au_{13}Ag_{12}(PPh_3)_{10}Cl_8]^+\}^{-I1}$ (at $25°C$, after 4 weeks) and $\{[Au_{13}Ag_{12}(PPh_3)_{10}Cl_8]^+\}^{-I2}$ (at $-10°C$, after 6 weeks). Meanwhile, conversion of the separated $\{[Au_{13}Ag_{12}(PPh_3)_{10}Cl_8]^+\}$-**1** under $-10°C$, and the reverse conversion of $\{[Au_{13}Ag_{12}(PPh_3)_{10}Cl_8]^+\}$-**2** under $25°C$ have also been accomplished under independent conditions (Fig. 4.74A). SCXRD analysis indicates that both clusters adopt the typical rod-like, vertex-sharing bi-icosahedral structures, with the $Au_{13}Ag_{12}$ metal framework being protected by two chlorides (on outermost Au atoms), ten PPh_3 ligands (on the two shoulder layers), and six bridging chlorides (between the two lumbar layers). Nevertheless, these two isomers are mainly differentiated by the arrangement of the two icosahedral blocks (mirror-like assembly in $\{[Au_{13}Ag_{12}(PPh_3)_{10}Cl_8]^+\}$-**1** and tandem-assembly in $\{[Au_{13}Ag_{12}(PPh_3)_{10}Cl_8]^+\}$-**2**, Fig. 4.74B), and the binding modes of the six bridging chlorides (one μ_4 and five μ_2 chlorides in $\{[Au_{13}Ag_{12}(PPh_3)_{10}Cl_8]^+\}$-**1** and two μ_3 and four μ_2 chlorides in $\{[Au_{13}Ag_{12}(PPh_3)_{10}Cl_8]^+\}$-**2**). Due to the distinct optical absorption characteristics of the two clusters, the isomerization could be easily tracked with UV-vis absorption spectra. The lack of by-products (evidenced by ESI-MS characterization on the independent isomerization systems) demonstrates the thermodynamic stability of both clusters, and thus the isomerization was concluded to reminiscent of the phase transitions. Based on the differential scanning calorimeter (DSC) and TGA analysis, Gibbs free energy was suggested as a driving force for the isomeric conversion.

Size-conversion induced by pH-control

In the synthesis of water-soluble metal nanoclusters (metal clusters protected by water-soluble ligands), pH control has been widely applied to regulate/promote the final size-focusing process. In recent years, pH regulation was also found

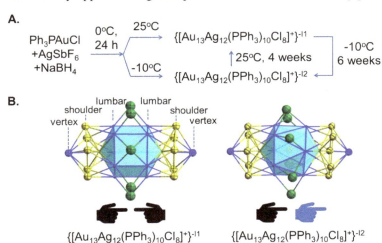

FIG. 4.74 Heating induced interconversion of the two $[Au_{13}Ag_{12}(PPh_3)_{10}Cl_8]^+$ isomers (A), and the structural anatomy of the two isomers (B). The different assembly modes of the two icosahedral cores are illustrated by the approaching of the two hands, and the different color demonstrate the different orientation of gold/silver atoms in the two blocks. *(Part of (B) is reproduced from Z. Qin, J. Zhang, C. Wan, S. Liu, H. Abroshan, R. Jin, G. Li, Atomically precise nanoclusters with reversible isomeric transformation for rotary nanomotors, Nat. Commun., 11 (2020) 6019, with permission that is licensed under the Creative Commons Attribution 4.0 International License, copyright the author(s) 2020.)*

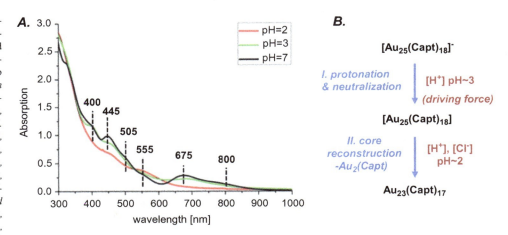

FIG. 4.75 UV-vis spectral monitoring on the pH-regulated size-conversion (A) and the proposed mechanism (B) for the size-conversion from $[Au_{25}(Capt)_{18}]^-$ to $Au_{23}(Capt)_{17}$. ((A) Reproduced from M. Waszkielewicz, J. Olesiak-Banska, C. Comby-Zerbino, F. Bertorelle, X. Dagany, A.K. Bansal, M.T. Sajjad, I.D.W. Samuel, Z. Sanader, M. Rozycka, M. Wojtas, K. Matczyszyn, V. Bonacic-Koutecky, R. Antoine, A. Ozyhar, M. Samoc, pH-induced transformation of pH-induced transformation of ligated Au_{25} to brighter Au_{23} nanoclusters, Nanoscale, 10 (2018) 11335–11341, with permission from the RSC Publications (Copyright 2018).)

a plausible strategy to initiate the postconversion of an as-prepared atomically precise metal nanocluster. For example, Olesiak-Banska and co-workers reported the size-conversion of the captopril (abbreviated as Capt) protected Au_{25} cluster (i.e., $Au_{25}(Capt)_{18}$) into $Au_{23}(Capt)_{17}$ under acidic conditions (pH=2), associating with the remarkably enhanced luminescence [115]. As shown in Fig. 4.75A, the UV-vis absorption spectra of aqueous solution of $[Au_{25}(Capt)_{18}]^-$ (pH=7) features several characteristic peaks at 400, 445, 505, 550, 675, and 800 nm, corresponding to the typical absorption of the anionic, spherical Au_{25} clusters. Most of these bands survived for several hours in pH > 3 buffer solutions, while the tiny changes correlate with the formation of neutral Au_{25} cluster species. The peaks at 800, 670, and 450 nm is significantly attenuated (associated with the enhanced intensity of the 550 nm should peak) in pH=2 solutions after several hours. The product was determined to be $Au_{23}(Capt)_{17}$ by UV-vis absorption spectra and ESI-MS characterization. Interestingly, dissolving the concentrated Au_{25} clusters into HCl or HNO_3 solutions result in distinct experimental outcome: the UV-vis absorption spectra of the HCl-system correlate with that of the buffer systems, while the characteristic peaks of Au_{25} clusters were largely maintained in presence of HNO_3. To this end, chloride was suggested to play a positive role in the size-conversion. In combination of these results and a more comprehensive analysis on the UV-vis spectra, Olesiak-Banska et al. suggested the conversion to occur via a two-stage mechanism: protonation of the anionic $[Au_{25}(Capt)_{18}]^-$ into the neutral $Au_{25}(Capt)_{18}$ under pH > 3 conditions (stage I, Fig. 4.75B), and structural re-construction at pH=2 via losing $Au_2(Capt)$ or $Au_3(Capt)$ moieties (stage II). According to DFT and TDDFT calculations, the theoretical UV-vis absorption profile of $Au_{23}(Capt)_{17}$ deduced by adding a Capt ligand to structure of $Au_{23}(Capt)_{16}$ correlates with the experimental one. By contrast, the theoretical curve significantly deviates the experimental one using the structure of an isomeric $Au_{23}(Capt)_{16}$ deduced by removing two gold atoms and one Capt ligand from $Au_{25}(Capt)_{18}$. The results demonstrate the core-reconstruction associated with the size-conversion from $Au_{25}(Capt)_{18}$ to $Au_{23}(Capt)_{16}$.

In addition to the pH-induced size-conversion from $[Au_{25}(Capt)_{18}]^-$ to $Au_{23}(Capt)_{17}$, the mutual size-conversion between $Au_{16}(Capt)_{12}$ and $Au_{18}(Capt)_{14}$ (Fig. 4.76A) [28]. has also been regulated by changing pH circumstance of the cluster solution. According to UV-vis absorption spectra and ESI-MS characterizations, $Au_{16}(Capt)_{12}$ clusters are relatively stable in alkaline, neutral and weakly acidic solutions (pH ≥ 5), but slowly convert to $Au_{18}(Capt)_{14}$ under more acidic conditions (the size-conversion takes 6h under pH=2.5 circumstance). According to the structural analysis on the precursor and the product clusters, the balance between steric hindrance in the ligand shell and the electrostatic interactions modulate the main composition of the reaction solution. In pH ≥ 5 solutions, the Au_{16} clusters and the Au(I) species (possibly formed during the synthesis procedures of Au_{16} or by decomposition of Au_{16} clusters) are negatively charged due to the acidic dissociation of the surface captopril ligands (Fig. 4.76B). Due to the strong electrostatic repulsion between the anionic ligands in Au_{16} clusters, and that between clusters and Au(I) complexes, Au_{16} cluster with a structurally loose framework is predominant. With enhanced acidity of the cluster solution (especially under pH ≤ 3 conditions (note: pKa of captopril is 3.7 [116]), most surface ligands are protonated, and thus the electrostatic repulsion in the ligand shell of Au_{16} clusters, and between Au_{16} clusters and Au(I) complexes, has been remarkably reduced. In this context, the $Au_{18}(Capt)_{14}$ clusters with relatively denser framework structure and higher stability (induced by the stronger Au—Au and Au—S bonding interactions) are favorably formed. This proposal is further supported by the lowered conversion rate of $Au_{16}(Capt)_{12} \rightarrow Au_{18}(Capt)_{14}$ under lower temperatures; accelerated conversion induced by addition of extra-captopril

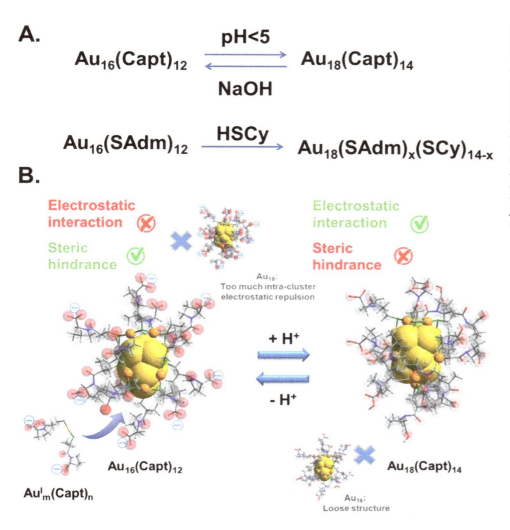

FIG. 4.76 The pH and ligand-exchange induced size-conversion between $Au_{16}(SR)_{12}$ and $Au_{18}(SR)_{14}$ clusters (A), and the proposed mechanism relating to the steric and electrostatic analysis (B). ((B) Reproduced from W. Jiang, Y. Bai, Q. Li, X. Yao, H. Zhang, Y. Song, X. Meng, H. Yu, M. Zhu, Steric and electrostatic control of the pH-regulated interconversion of $Au_{16}(SR)_{12}$ and $Au_{18}(SR)_{14}$ (SR: deprotonated captopril), Inorg. Chem. 59 (2020) 5394–5404, with permission from the ACS Publications (Copyright 2020).)

ligands; alkaline condition induced reverse conversion (i.e., $Au_{18}(Capt)_{12} \rightarrow Au_{16}(Capt)_{14}$), and ligand-exchange induced conversion of $Au_{16}(SAdm)_{12} \rightarrow Au_{18}(SAdm)_x(SCy)_{14-x}$ (replacing the bulky -SAdm ligand with the smaller -SCy ligand promotes the formation of structurally denser Au_{18} cluster).

4.3 Insights into the size-evolution principals

4.3.1 Electronic structure evolution of metal nanoclusters

Earlier before the burst of atomically precise metal nanoclusters (especially of those characterized by SCXRD) reported in these two decades, the electronic structure of gas-phase metal nanoclusters, or colloidal clusters have been extensively studied. Some breakthrough theories, such as the Jellium model, Colloidal model, Superatom complex model, have been successively reported. These studies shed light on the inherent electronic interactions within the metal nanoclusters, and greatly deepen the chemists' understanding on the correlation of the cluster structure and the bonding facility/thermodynamic stability/optical properties, etc. These progresses further enlightened the development of more synthetic strategies to prepare versatile clusters bearing different metal components, ligands, and distinct physicochemical properties. In the following, only selected theories for electronic state analysis of metal nanoclusters were covered, and we suggest the authors to refer to recent reviews for the details of the electronic theories.

4.3.1.1 Jellium model and its updated version

In these decades, the Jellium model and the renewed/extended version of the Jellium model have been frequently used to analyze the electronic structure of metal nanoclusters, and the thermodynamic stabilities. In a typical Jellium model, the

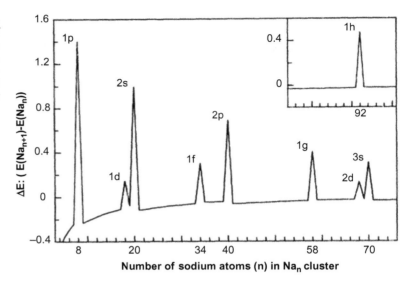

FIG. 4.77 The changes of the energy gap ΔE (i.e., $E(Na_{n+1})$-$E(Na_n)$) along with the increased number of sodium atoms in Na_n cluster. *(Reproduced from ref. W.D. Knight, K. Clemenger, W.A. de Heer, W.A. Saunders, M.Y. Chou, M.L. Cohen, Electronic shell structure and abundances of sodium clusters, Phys. Rev. Lett. 52 (1984) 2141–2143, with permission from the APS (Copyright 1984).)*

positive charges are assumed to be uniformly distributed in the solid space (resemble the atomic nuclei), while concerns are mainly put in the electronic properties of the target sample. In 1980's, the Jellium model has been extensively explored to understand the electronic structure of metal (Na, Al) nanoclusters. For example, with the aid of Local-density approximation (LDA) method [117], spherical jellium model was used by Martins and co-workers to treat sodium clusters, and the orbital degeneracy generated from the spherical effective potential was found to be similar to that of atomic orbitals (e.g., 1 for s, 3 for p, etc.). Latter in 1984, Knight and co-workers probed the mass spectra of sodium clusters (Na_n, $n=2$–100) [118], and found the extra abundance of the clusters with 2, 8, 20, 40, 58, and 92 atoms (compared with the adjacent clusters with less or more atoms). Specifically, solving the electronic structures of each Na_n cluster numerically with Schrödinger equation, the orbital degeneracy of the discrete energy levels correlates with the angular momentum quantum number L in a fashion of $2L+1$ (Fig. 4.77). With the increased number of sodium atoms (n) in the cluster, the energy levels are lowered down regularly. The correlation of the number of n with the energy gap of Na_{n+1} and Na_n (i.e., $\Delta E = E(Na_{n+1})-E(Na_n)$) shows several sharp peaks at $n=8, 18, 20, 34, 68, 70$, etc., indicating that incorporating one additional Na atom significantly perturbs the stability of these clusters. The extra-stability of the metal nanoclusters with certain number of electrons (2, 8, 18 …) were mainly ascribed to the closed-shell electronic structure, and these clusters were also referred to as the magic sized metal nanoclusters in the latter studies.

Due to the similarity of the discrete energy levels (and the degeneracy) of the metal nanoclusters with that of atomic orbitals, and the same valence electron of group IB metal (Au/Ag/Cu) with that of Na, the colloidal metal nanoclusters were also referred as "superatom" in the following studies.

4.3.1.2 A grand unified model (GUM) for gold nanoclusters protected by thiolates, phosphines and halides

To determine a general correlation between the extensively reported metal nanoclusters (with atomic precision), a grand unified model has recently been proposed by Gao and co-workers to comprehend the stability of 71 liganded gold nanoclusters with atomic precision (54 experimentally crystallized structure and 17 theoretically predicted ones) [119]. The metal atoms in each cluster are classified into two groups: the shell gold atoms in $Au_n(SR)_{n+1}$ staples and the core gold atoms. Depending on the coordination environments, the core gold atoms are further categorized into bottom (the gold atom bound with the electron-donating phosphines), middle (the gold atom bound with the one S atom in $Au_n(SR)_{n+1}$ staple motifs or thiolates, or are shared by two different elementary blocks) and top Au (the gold atoms bonded with one nonbridging thiolate or halide ligand) flavors. The number of effective valance electrons for bottom, middle and top Au atoms in each core structure are 1 e, 0.5 e and 0 e (Fig. 4.78A). In this context, the core structure of all the target clusters could be viewed as the combination of certain number of elementary blocks, namely the triangular $Au_3(2e)$ and the tetrahedral $Au_4(2e)$ blocks. Nevertheless, the valence state of the Au atoms and the charge state of each block might be varied, and the symbol of $\triangle_1$–$\triangle_{10}$, and T_1–T_{15} were used to designate the different types of $Au_3(2e)$ or $Au_4(2e)$ blocks (Fig. 4.78A and B), respectively. In view of the duet rule (i.e., both the triangular Au_3 and tetrahedral Au_4 show high

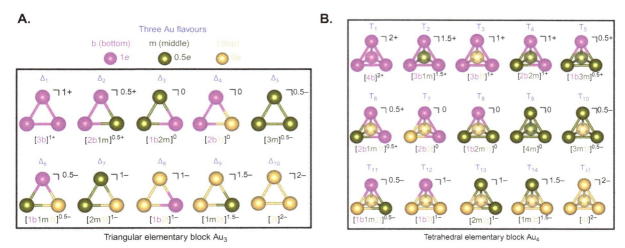

FIG. 4.78 Variants of the triangular $Au_3(2e, A)$ and the tetrahedral $Au_4(2e, B)$ building blocks. *(Reproduced from W.W. Xu, B. Zhu, X.C. Zeng, Y. Gao, A grand unified model for liganded gold clusters, Nat. Commun. 7 (2016) 13574, with permission that is licensed under the Creative Commons Attribution 4.0 International License, copyright the author(s) 2016.)*

tendency to having two valance electrons), the number of elementary blocks for a comprehensive cluster could be first determined by half the number of free electrons in the total cluster.

Herein, the core structure of $[Au_8(PR_3)_6]^{2+}$ (R denotes mesityl) [120] was used as an example to illustrate the application of the aforementioned definitions (Fig. 4.79). Six of the eight Au atoms in $[Au_8(PR_3)_6]^{2+}$ are bonded with the phosphine atoms, and thus each of them is a bottom Au with 1 e. The reminding two Au atoms are each shared by one triangular Au_3 elementary block and a tetrahedral Au_4 block, and is thus a medium Au with 0.5 e. Meanwhile, the six phosphine ligands are charge-neutral ligands, and thus the two positive charges on the cluster are spread exclusively on the metal core. In this context, the Au_8 core could be divided into two Au_3 and a central Au_4 elementary blocks. Each Au_3 block bears two bottoms Au, one medium Au flavors and half positive charge ($\triangle_2$ in Fig. 4.78A). Meanwhile, the Au_4 block comprises two bottom Au, one medium Au flavors and one positive charge (T_4 in Fig. 4.78B). To this end, the 2e counting rule of all these elementary blocks are satisfied.

In addition to a general applicable structural characterization of a large number of reported metal nanoclusters, the GUM also provides a deep insight into the size-evolution of gold nanoclusters via increasing the number of $Au_3(2e)$ and $Au_4(2e)$ elementary blocks in the core structure (while adjusting the charge state of the total cluster to meet the duet rule). Herein, it is noteworthy that the stability of a great number of metal nanoclusters could be well depicted by GUM, but it does not necessarily mean that only the clusters fitting GUM are stable. Meanwhile, GUM provides a convenient way to predict the relative stability of a target structure via simply counting the triangular $Au_3(2e)$ and tetrahedral $Au_4(2e)$ elementary blocks,

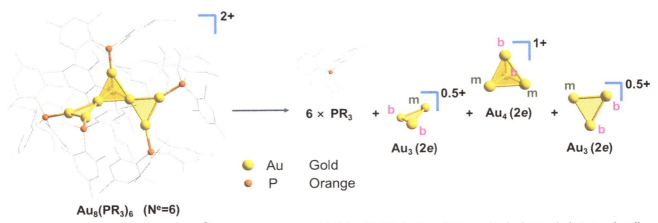

FIG. 4.79 Structural analysis on $[Au_8(PR_3)_6]^{2+}$ via the grand unified model. "b" and "m" in the three elementary blocks denotes the bottom and medium Au flavors, respectively. Color legends: *lavender*: P; *gold*: Au; C: *gray*. All hydrogen atoms are omitted for clarity.

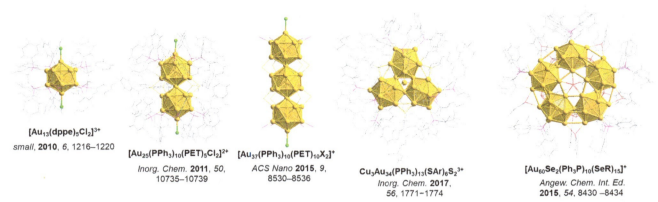

FIG. 4.80 Selected examples for the single crystal structure of icosahedral and assembled icosahedral metal nanoclusters reported in these decades. The *yellow* balls denote Au atoms, except for the Cu_3Au_{34} wherein both Cu and Au are given in *yellow* color. The C-H atoms in the rod-like Au_{37} was not identified in Jin's study, and therefore only the framework was shown (X=Cl/Br).

but such treatment is totally different from the electronic state analysis. To gain deep insights into the electronic structures of target clusters, the readers are suggested to refer some other typical theories, such as the Jellium model, and Mingo's united atom model.

4.3.2 Size and optical absorption property evolution of the icosahedral metal nanoclusters

Icosahedron represents one of the most common building blocks in the organic ligand protected group IB metal nanoclusters. According to the atomic precision elucidated by SCXRD, the icosahedral structure has been identified in hundreds of metal nanoclusters, with different types of ligands (thiolates, phosphines, alkynes, mixed-ligands, etc.) and metal compositions (homometallic or alloy clusters). The extensive studies on structure and property characterizations of these icosahedron-bearing clusters enable the deep understanding on the correlation of different size-clusters, many interesting size-evolution principals, and the associated structure-property correlations.

As early as 1987, Teo and co-workers proposed the common growth mode of "cluster of clusters" to account for the structural correlation of a series of clusters [121], which could be viewed as the assembly of icosahedral M_{13} structures via a vertex sharing pattern (Fig. 4.80). According to this theory, the formula of icosahedra assembled structures (also called supraclusters in their study) is $13n-x$, where n is the number of icosahedral blocks, and x is the number of shared vertexes. Interestingly, this model was first proposed to deduce the molecular formula of relatively large supraclusters with icosahedral blocks arranged in a cyclic fashion [122], but was then quickly generalized to deduce the composition of other icosahedra-assembled clusters.

On the basis of the atomic precision and the structural correlations of the series of icosahedral structures, interesting property evolution was reported in recent years. For example, compared with UV-vis spectra of the icosahedral Au_{13} cluster (i.e., $[Au_{13}(dppe)_5Cl_2]^{3+}$, and dppe is short for 1,2-bis(diphenylphosphino) ethane) [123], the bi-icosahedral, rod-like Au_{25} (i.e., $[Au_{25}(PPh_3)_{10}(PET)_5Cl_2]^{2+}$) exhibits an addition characteristic peak at ~670 nm, while all other optical peaks in these two clusters are similar (Fig. 4.81A). Introducing another icosahedral Au_{13} block results in an extra optical absorption peak at 1230 nm of Au_{37} (i.e., $[Au_{37}(PPh_3)_{10}(PET)_{10}X_2]^{2+}$, X=Cl/Br in Fig. 4.81B). To this end, the HOMO-LUMO transition of the three clusters are determined to be cal. 1.9, 1.7 and 1.2 eV, respectively (insets of Fig. 4.81A and B). The similar red-shift of the HOMO-LUMO transition was also reported in the series of icosahedral Pt_1Ag_{12}, Pt_2Ag_{23} and Pt_3Ag_{33} clusters [124].

4.3.3 Size and excited-state dynamics evolution of the FCC-packed gold clusters

With the extensive characterization of FCC (face-centered-cubic) packed structure in bulk gold and the continuous exploration of icosahedral blocks in gold nanoclusters, the FCC packed gold nanoclusters (in subnanometer scale) have been once predicted to be unstable. But in 2012, Jin and co-workers prepared the first FCC packed $Au_{36}(TBBT)_{24}$ via the ligand etching of $Au_{38}(PET)_{24}$ with HTBBT under 80°C (Fig. 4.50A) [75]. In the following 4 years, the cluster of $Au_{28}(TBBT)_{20}$ [51], $Au_{44}(TBBT)_{28}$ [125], and $Au_{52}(TBBT)_{32}$ [126] clusters are prepared successfully. According to the SCXRD characterization, these four clusters exhibit strong structural correlations (Fig. 4.82), which could be viewed as the addition of the

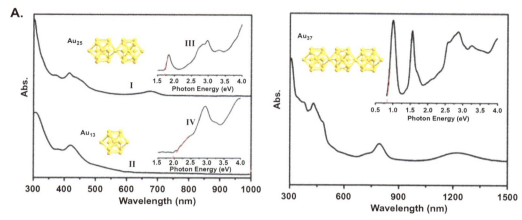

FIG. 4.81 Comparison of the UV-vis-NIR spectra of the icosahedral Au_{13} and the rod-like bi-icosahedral Au_{25} (A) and the rod-like tri-icosahedral Au_{37} clusters (B). *(Reproduced from R. Jin, C. Liu, S. Zhao, A. Das, H. Xing, C. Gayathri, Y. Xing, N.L. Rosi, R.R. Gil, R. Jin, Tri-icosahedral gold nanocluster $[Au_{37}(PPh_3)_{10}(SC_2H_4Ph)_{10}X_2]^+$: linear assembly of icosahedral building blocks, ACS Nano, 9 (2015) 8530–8536, with permission from the ACS Publications (Copyright 2015).)*

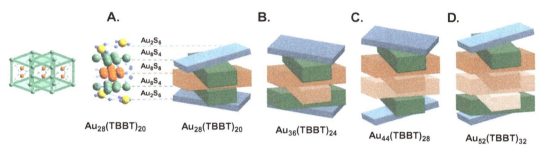

FIG. 4.82 Illustrative diagram for the atomic packing mode of $Au_{28}(TBBT)_{20}$ (A), $Au_{36}(TBBT)_{24}$ (B), $Au_{44}(TBBT)_{28}$ (C), and $Au_{52}(TBBT)_{32}$ (D). The Au_2S_6, Au_8S_4, Au_8S_8 layer are depicted with *blue*, *green*, and *yellow* colors. Each thiolate ligand in the medium layers is shared by the upper and lower two layers, and thus the effect number of thiolate ligands should be halved. For clarity, the two interpenetrating FCC blocks constituted by the 22 Au atoms in $Au_{28}(TBBT)_{20}$ was shown as an example.

hierarchical structure of $Au_8(TBBT)_4$ along the z-direction. The orientation of the metal atoms in the single crystal structure (with slight structural distortion) of $Au_{28}(TBBT)_{20}$ could be viewed as the orthogonal assembly of three layers of Au_8 structures (Fig. 4.82A). Eight bridge thiolates are present to connect the medium layer with the upper and lower layers. The upper and lower layer of the Au_{24} block is each protected by a rectangular Au_2S_6 structure. Compared with the three layered Au_{24} structure in $Au_{28}(TBBT)_{20}$, the single crystal structure of $Au_{36}(TBBT)_{24}$ could be viewed as the insertion of one Au_8 layer within the three-layered Au_{24} structure (of $Au_{28}(TBBT)_{20}$) along the z-direction (Fig. 4.82B). The eight thiolates in this layer are each shared by the Au atoms on the neighboring two layers, and therefore, the efficient number of introduced thiolates is $8*0.5=4$. Similarly, the structure of $Au_{44}(TBBT)_{28}$ and $Au_{52}(TBBT)_{32}$ could be viewed as the incorporation of two and three more $Au_8(TBBT)_4$ layers on $Au_{28}(TBBT)_{20}$ (Fig. 4.82C and D). To this end, the gold clusters shown in Fig. 4.82 could be described with a general formula of $Au_{8n+4}(TBBT)_{4n+8}$ ($n=3$–6). Due to the extensive presence of the FCC-blocks (see Fig. 4.82A as an example), all these clusters were categorized into the FCC-packed structures.

Similar to the observation in the icosahedral packed structures, the series of FCC-packed gold clusters (i.e., Au_{28}, Au_{36}, Au_{44}, and Au_{52}) show systematically red-shift in absorption peaks of UV-vis spectra, and especially in the long-wavelength region [125]. But distinct from the core-shell transitions in the series of icosahedral assembled structures, the photophysics of the four FCC-packed clusters mainly originate from the charge transfer in the metal core [127]. Meanwhile, the real-time and real-space density functional theory (DFT) calculations indicate a denser orbital distribution (near the HOMO and LUMO) of the larger sized clusters than that of the smaller ones, in accordance with the relatively larger freedom of the free electrons (8, 12, 16, and 20 e in Au_{28}, Au_{36}, Au_{44} and Au_{52}) and thus the higher degeneracy of frontier orbitals. In addition, the excited state dynamics on these FCC-packed clusters indicates a slowed relaxation of hot carriers and band-edge carrier recombination with the increased cluster size [127].

4.3.4 Size-evolution of the FCC-packed silver clusters

Similar to the aforementioned gold nanoclusters, a series of FCC-packed silver nanoclusters (co-protected by thiolates and phosphine ligands) have been reported recently, and interesting structure correlations have been elucidated. These studies start with the structural characterization of the simplest FCC-packed $Ag_{14}(SPhF_2)_{12}(PPh_3)_8$, which was reported by Zheng and co-workers in 2013 ($HSPhF_2$ = 3,4-difluoro-benzenethiol) [128]. As shown in Fig. 4.83A, the 14 Ag atoms occupy on the eight vertex and six face-center positions, and the vertex Ag atoms are further protected by a phosphine group. Meanwhile, each of the eight S atoms bisects one edge of six facets. Alternatively, the core structure of Ag_{14} cluster could be viewed as an octahedral Ag^6 block (constituted by the six face-center Ag atoms), each trigonometric facet of which is further capped by a AgS_3P block.

Four years later, Zheng, Häkkinen, and Teo et al. further prepared the single crystal of $Ag_{38}(SPhF_2)_{26}(PR_3)_8$, and $[Ag_{63}(SPhF_2)_{36}(PR_3)_8]^+$ ($HSPhF_2$ = 3,4-difluoro-benzenethiol, and R = Ph/^{n}Bu) [129], and elucidate their atomic structures with SCXRD. Compared with the single FCC structure of Ag_{14} cluster, the metal framework of Ag_{38} and Ag_{63} cluster could be viewed as the assembly of $2 \times 2 = 4$, and $2 \times 2 \times 2 = 8$ FCC building blocks (Fig. 4.83B and C). Herein, it is interesting to note that Ag_{14} and Ag_{38} clusters are prepared with the similar synthetic method, wherein the mixed solvent (CH_2Cl_2:MeOH) of silver tetrafluoroborate ($AgBF_4$) is reduced by $NaBH_4$ in the presence of thiol and phosphine ligand at 0°C. By contrast, during the synthesis of the Ag_{63} cluster, tetrabutylammonium tetraphenylborate (^{n}Bu$_4$N·BPh$_4$) was added as counterion to stabilize the relatively larger-sized charged cluster. In addition to the structural correlation in the metal core, the coordination modes of the thiolate and phosphine ligands in these three clusters are consistent [130]. The thiolate ligands adopt the μ_3- and μ_4-binding modes with Ag atoms on the exterior edges and (100) facets of the cluster surface, and the corner Ag atoms is each protected by an additional phosphine ligand. To this end, the corner Ag atoms all adopt the tautomerized tetrahedral AgS_3P coordination mode, the Ag atoms on the edges adopts the T-shape coordination with the thiolate ligands, while all other Ag atoms on (100) facets show crosslinked μ_4-coordination with four thiolate ligands.

In view of the strong structural correlation of Ag_{14}, Ag_{38}, and Ag_{63} clusters, Zheng and Teo et al. performed systematic investigation on the nucleation and size-evolution principals of the FCC-packed coinage metal [131]. In this context, the 1, 2, and 3-dimensional assembly of the n, n^2, and n^3 FCC- building blocks was proposed to form superrods, supersquares, and supercubes, respectively. The atomic number of metal atoms (coinage metal of Ag, Au, or Cu), thiolate ligands (or other group XI ligands), and phosphine ligands were deduced based on the common structural characteristics of the aforementioned FCC-Ag clusters (i.e., face-sharing FCC-packed Ag atoms, μ_3-S/Ag on edges and μ_4-S/Ag on surface facets). Meanwhile, the number of free electrons were deduced in accordance with the traditional Jellium model (see Section 4.3.1 for more details) [118,132–134], and the recently developed Jelliumatic shell model (by grouping the atomic shells into one or more neighboring electron shells, and the electron count and shape of each shell follows the conventional Jellium model) [135]. The electron counting principal is critical in determining the charge state of the target cluster. To this end, the formular of a series of FCC-packed coinage nanomaterials with the general formula of $[M_m(SR)_l(PR'_3)_8]^q$ (m ranges from 14 to 365) were predicted. Of note, the predicted 1D-assembly of two FCC-blocks with 23 Ag atoms, were quickly verified by the experimental synthesis and characterization of the FCC-Ag clusters, albeit with slight deviation on the number of thiolate ligands (20 vs 18).

In 2018, Jin and co-workers reported the single crystal structure of $Ag_{23}(PET)_{18}(PPh_3)_8$ [136], the metal framework of which correspond to the assembly of two FCC blocks (Fig. 4.83D). Similar to the structural features mentioned above, the thiolate ligands in Ag_{23} also show two types of coordination modes: μ_3-coordination on the exterior edges and μ_4-coordination on the (100) facets. Meanwhile, the corner and Ag atoms adopt the tetrahedral-like AgS_3P, and T-shape AgS_3 coordination modes, similar to the coordination modes of the Ag atoms of all aforementioned FCC-Ag clusters.

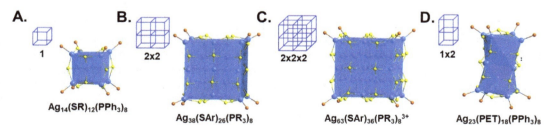

FIG. 4.83 Structure anatomy of Ag_{14} (A), Ag_{38} (B), Ag_{63} (C) and Ag_{23} (D) clusters comprising 1, 4 and 8 FCC building blocks. Color legends: *yellow* for S; *orange* for P, and *blue* for Ag. (*Reproduced from Y. Lv, R. Zhao, S. Weng, H. Yu, Core charge density dominated size-conversion from Au6P8 to Au8P8Cl2, Chem. A Eur. J. 26 (2020) 12382–12387, with the permission from Wiley-VCH, copyright 2019.*)

Nevertheless, distinct from the fairly regular FCC structure of Ag_{14}, Ag_{38}, and Ag_{63} clusters, the metal core of Ag_{23} cluster significantly deviates from FCC structure and results in the core-distortion induced chirality. According to a subsequent study from Yu and co-workers, the chirality of the Ag_{23} cluster in Fig. 4.83D mainly originate from the usage of the alkyl thiolate ligand of PET [137]. By contrast, introduction of the aromatic thiolate ligands (i.e., 4-methoxythiophenol) during synthesis results in the formation of a structurally regulated, double FCC-packed Ag_{23} cluster (evidenced by SCXRD characterization). Meanwhile, the presence of the single unpaired electron in $Ag_{23}(PET)_{18}(PPh_3)_8$ was verified by the electron paramagnetic resonance (EPR) measurement, and the vibrational circular dichroism measurements mainly show characteristic peaks at 1200–1000 cm^{-1}, corresponding to the Ag-P-C and Ag-S-C stretching modes.

In the aforementioned Ag_{14}, Ag_{23}, Ag_{38}, and Ag_{63} clusters, all Ag atoms are incorporated in the complete FCC building blocks (albeit with slight structure distortion). Interestingly, in the recently reported $[Ag_{67}(SPhMe_2)_{32}(PPh_3)_8]^{3+}$ ($HSPhMe_2$ denotes 2,4-dimethylbenzenethiol) by Bakr and Häkkinen et al. [138] the complete structure (determined by SCXRD) could be viewed as the protection of a Ag_{23} metal core with a larger cubic $Ag_{44}SR_{18}P_8$ structure (Fig. 4.84A), while the Ag_{23} core structure is similar to the metal framework of the Ag_{23} cluster shown in Fig. 4.83D (but with a much higher structure regularity). After removing the eight corner AgS_3P motifs, the structure of Ag_{67} could also be viewed as the assembly of two complete FCC blocks with ten half-cubic blocks. In this context, coordination modes of Ag, S, and P atoms are in perfect agreement with the structural features of the aforementioned FCC-Ag clusters, i.e., the μ^3-S/Ag on outermost edges and μ^4-S/Ag on (100) facets and eight AgS_3P structures on the eight corner positions.

Similar to the structural anatomy of the Ag_{67} clusters, two isomeric Ag_{46} clusters, i.e., $[Ag_{46}(SPh^{2,4-}Me_2)_{24}(PPh_3)_8]^{2+}$ [139], and $[Ag_{46}(SPh^{2,5-}Me_2)_{24}(PPh_3)_8]^{2+}$ [140] ($HSPh^{2,4-}Me_2$ and $HSPh^{2,5-}Me_2$ are 2,4- and 2,5-dimethylbenzenethiol) were recently reported by Pradeep and Zhu et al., at almost the same time. According to the SCXRD analysis, the atomic structures of these two clusters are highly alike (except for the difference in the structure of the exterior thiolate ligands). As shown in Fig. 4.84B, the structure of Ag_{46} clusters could be viewed as protecting the six facets of the FCC-Ag_{14} kernel with six half-cubic. Meanwhile, the coordination modes of Ag, S, and P atoms of are Ag_{46} clusters follow the same binding characteristic with all aforementioned FCC-Ag clusters.

On the basis of the newly developed FCC-Ag clusters and the general structural characteristics, an alternative version of the size-evolution principal was recently proposed to comprehend the FCC-Ag clusters, ending with either the complete or the half FCC-elementary blocks [130]. The common structural characteristics are deduced from the reported Ag_{14}, Ag_{23}, Ag_{38}, Ag_{63}, and Ag_{67} clusters, and (111) facets were suggested criminal to size-growth. The linear correlation between the number of building blocks (block A: four edge-sharing octahedra; and block B: five edge-sharing octahedra, Fig. 4.85) and the total binding energy (obtained by density functional theory calculations) evidences the homogeneity in Ag—Ag and Ag—S bonding interactions of the series of 1D FCC-packed Ag clusters, which constitute the basis for the subsequent structural deduction of relatively larger sized Ag clusters. Although the general formula of both types of size-growth were proposed, this theory is limited by the difficulty in figuring out the number of A and B blocks in target clusters, and also the omitted issues on predicting charge states.

Aside from the aforementioned cubic and half-cubic FCC-packed Ag nanoclusters, the synthesis and structural characterization of a large-sized FCC-Ag cluster, i.e., $[Ag_{100}(SPhF_2)(PPh_3)_8]^-$ ($HSPhF_2$ = 3,4-difluorobenzenethiol), adopts a four layered $Ag_6@Ag_{38}@Ag_{48}S_{24}@Ag_8S_{24}P_8$ octahedral structure with consistent symmetry elements (also known as the Russian Nesting Doll Model, Fig. 4.86) [141]. As the synthetic procedures of Ag_{100} is similar to the aforementioned FCC-Ag clusters, the formation of the large Ag_{44} ($Ag_6@Ag_{38}$) core was suggested to originate from the weakened bonding

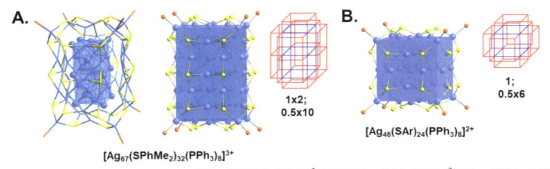

FIG. 4.84 Illustrative diagram for the atomic structure of $[Ag_{67}(SPhMe_2)_{32}(PPh_3)_8]^{3+}$ (A) and $[Ag_{46}(SAr)_{24}(PPh_3)_8]^{2+}$ (SAr = SPhMe_2 or SPhMe_2-2, B). (Reproduced from Y. Lv, X. Ma, J. Chai, H. Yu, M. Zhu, Face-centered-cubic ag nanoclusters: origins and consequences of the high structural regularity elucidated by density functional theory calculations, Chem. A Eur. J. 25 (2019) 13977–13986, with the permission from Wiley-VCH, copyright 2019.)

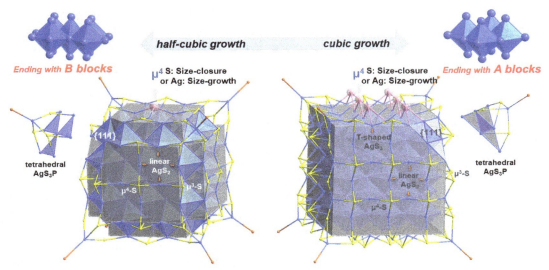

FIG. 4.85 The general structure of FCC-Ag clusters co-protected by thiolate and phosphine ligands. *(Reproduced from Y. Lv, X. Ma, J. Chai, H. Yu, M. Zhu, Face-centered-cubic ag nanoclusters: origins and consequences of the high structural regularity elucidated by density functional theory calculations, Chem. A Eur. J. 25 (2019) 13977–13986, with the permission from Wiley-VCH, copyright 2019.)*

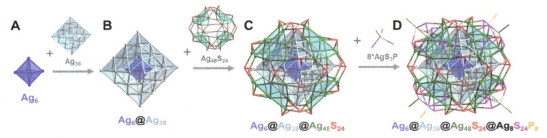

FIG. 4.86 Illustrative diagram for the structure of the four-layered $[Ag_{100}(SPhF_2)(PPh_3)_8]^-$ cluster. *(Reproduced from J. Chai, S. Yang, Y. Lv, T. Chen, S. Wang, H. Yu, M. Zhu, A unique pair: Ag_{40} and Ag_{46} nanoclusters with the same surface but different cores for structure–property correlation, J. Am. Chem. Soc., 140 (2018) 15582–15585, with the permission from Wiley-VCH, copyright 2020.)*

interaction of the Ag atoms with the electron-withdrawing $SPhF_2$ ligands. The pivotal ligand effect in the size-control of FCC-Ag clusters has also been proposed in earlier studies [129].

According to the aforementioned structure characterization, it can be seen that the thiolate and phosphine co-protected FCC-Ag clusters could adopt different nucleation and surface morphology. It has been well documented that choice of reaction conditions (temperature, solvents, etc.), ligands or counterions could play important role in the charge state and atomic structure. So far, the size-evolution principals of the FCC-Ag clusters have been elucidated by both experimental and theoretical methods, which provide important insights into the potential strategies to control the cluster size and charge states. Nevertheless, the size-growth of one FCC-metal cluster to another one, which is anticipated to provide pivotal evidence on the formation mechanism of the relatively large-sized metal nanoparticles, are remain unprecedented. On the other hand, the reactions of one atomically precise metal nanocluster with foreign ligands, metal complexes or reduction reagents have been successfully utilized to generate relatively larger sized metal clusters (such as $Au_{18} \rightarrow Au_{21}$, $Au_{23} \rightarrow Au_{25}$, $Au_{25} \rightarrow Au_{38} \rightarrow Au_{44}$). To this end, the size-conversion between the FCC-Ag clusters (but not limited to the reported ones) might be achievable with the well-designed reaction conditions and reagents.

4.3.5 Size-evolution principals

The size-evolution principals of a series of thiolate protected, small sized gold nanoclusters, $Au_m(SR)_n$ (m, n range from 5 to 12) and their correlation with relatively larger sized clusters were recently deduced by Pei and Ma et al. (Fig. 4.87) [142]. This work starts with the structural analysis on the homoleptic clusters adopting the formula of $Au_m(SR)_m$ ($m=6–12$), in which the number of gold atoms is equal to that of the thiolate ligands. Each of the homoleptic cluster could be viewed as protecting the Au_2 core with two staple motifs, with the formular of $Au_2[Au_{(m-2)/2}(SR)_{m/2}]_2$ when m is an even number, and

COLUMN IV Tips on calculating the connectivity and average binding energy

The connectivity is defined as follows. For Au—Au bond, the cutoff of 3.1 Å was set as the lower limit of a full Au—Au bonding (with connectivity degree of 1), while increasing Au—Au bond distances by <0.1 Å results in a decrease in connectivity by 0.25. In other words, the connectivity of Au—Au bond is set as 1, 0.75, 0.5, 0.25, and 0 when its distance ranges from (≤3.1 Å), (3.1–3.2 Å], (3.2–3.3 Å], (3.3–3.4 Å] and over 3.4 Å. Meanwhile, 2.4 Å was set as the lower limit of a full Au—S bonding, while increasing the bond distance by 0.05 Å results in the decrease in the connectivity by 0.25. To this end, the connectivity of a target Au atom is the sum of the connectivity degree of all Au—Au and Au—S bonds. Since the connectivity degree of the kernel Au atom is remarkably higher than the Au atoms in the core surface and in staple motifs, the higher connectivity correlates with the higher proportion of core Au atoms, and thus the higher nucleation.

Meanwhile, the average binding energy (ABE) of an $Au_m(SR)_n$ cluster is calculated according to the equation of $ABE = \frac{E_b}{m+n}$, wherein E_b is the total binding energy of the target cluster which could be gained directly from the DFT calculations.

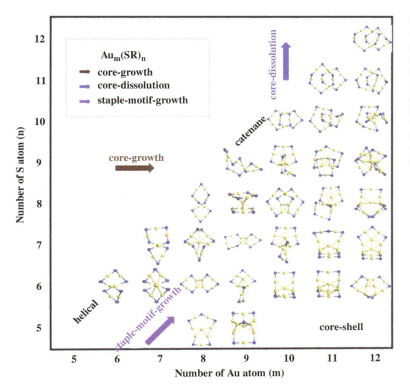

FIG. 4.87 Illustrative diagram for the size evolution of the $Au_m(SR)_n$ (m, n range from 5 to 12) via the core-growth (*brownish red arrow*), core-dissolution (*blue arrow*) and staple-motif-growth (*purple arrow*) pathway. The given structures are generated from the theoretical calculations with a prior basin-hopping algorithm and a subsequent geometry optimization with density functional theory calculation (Dmol³ package: PBE/DND4.4). (*Reproduced from C. Liu, Y. Pei, H. Sun, J. Ma, The nucleation and growth mechanism of thiolate-protected Au nanoclusters, J. Am. Chem. Soc. 137 (2015) 15809–15816, with permission from the ACS Publications (Copyright 2015).*)

$Au_2[Au_{(m-3)/2}(SR)_{(m-1)/2}][Au_{(m-1)/2}(SR)_{(m+1)/2}]$ when m is an odd number. With the addition of gold atoms and thiolate ligands, the size evolution from $Au_m(SR)_m$ to $Au_m(SR)_n$ ($m \neq n$) could possibly occur via three patterns, i.e., core growth, core dissolution, and staple-motif growth pathways. The core-growth (size-evolution along the equatorial direction as shown in Fig. 4.87: brownish red arrow) generates the relatively larger core, including the triangular/linear Au_3, tetrahedral Au_4, trigonal bipyramidal Au_5, octahedral Au_6, and vertex-sharing bi-tetrahedral Au_7. The higher nucleation of the larger sized gold clusters is evidenced by the increased average connectivity (Column IV: Tips on calculation of connectivity). By contrast, the introduction of additional thiolate ligands on homoleptic clusters tends to connect two adjacent Au atoms, resulting in the weakening of the core Au—Au bonds, and increase of the exterior staple motifs. Such process is called the core-dissolution route (size-evolution along the axial direction in Fig. 4.87: blue arrow), similar to the dissolution of the core metal atoms (*solute*) with thiolates (*solvent*). Finally, the addition of equal number of Au atoms and thiolate ligands from a certain $Au_m(SR)_n$ cluster commonly results in the maintained inner core, but lengthened staple motifs. Such size-evolution was denoted as the "staple-motif-growth" pathway (purple arrow in Fig. 4.87). Specifically, the core structure in the series of staple-motif-growth pathway shows great dependency on the difference between m and n. For example, the tetrahedral Au_4, and the bi-tetrahedral Au_7 core could be easily identified in the series of $Au_{n+2}(SR)_n$ and $Au_{n+4}(SR)_n$ clusters. The few examples that do not obey the aforementioned three pathways (such as the structural

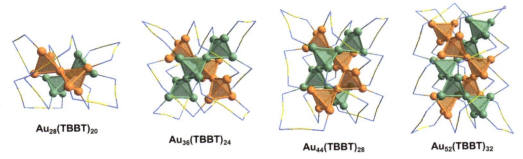

FIG. 4.88 The size-evolution of Au_{28}, Au_{36}, Au_{44}, and Au_{52} clusters from the viewpoint of tetrahedral Au_4-growth pattern.

difference between $Au_8(SR)_8$ to $Au_9(SR)_8$ were suggested to occur via pathways involving the metastable isomers. The relative energy of the isomers might be slightly higher than the lowest-lying minimum (as given in Fig. 4.87), while the structure correlates with the two target clusters before and after the size-evolution.

Interestingly, the average binding energy (ABE, see Column IV for the details) of clusters with 2 or 4 free electrons are relatively larger than that of the other systems. To this end, the formation mechanism of the certain sized nanoclusters from the Au-SR complexes were suggested to occur via a first dissociation of thiolates (upon reduction) and formation of the metal core structure, from which the fusion of the core structures in two clusters then occurs to generate the thermodynamically stable cluster products. This mechanism correlates with the 2e reduction-growth mechanism mentioned in Section 4.2.2.1. In this context, the tetrahedral Au_4 was suggested as one type of elementary building blocks due to both geometric and electronic advantages, and may serve as a superatom unit in the size-evolution of relatively larger-sized gold clusters. This proposal agrees with the structural analysis of the series of FCC-packed Au nanoclusters, including $Au_{28}(TBBT)_{20}$, $Au_{36}(TBBT)_{24}$, $Au_{44}(TBBT)_{28}$, and $Au_{52}(TBBT)_{32}$ (Fig. 4.88, a different view of growth pattern alternative to Fig. 4.82) [125].

4.4 Conclusion and perspective

According to the aforementioned section, it can be seen that the size-conversion and the size-evolution principals between different sized metal nanoclusters have been extensively reported. With the aid of atomic precision, intermediate species characterization/trapping, spectroscopic monitoring, kinetic analysis, and control experiments, a series of interesting structural and property correlations have been elucidated. A variety of mechanistic insights, such as the seed growth, core-dissolution route, thermodynamic/kinetic control, etc., has been proposed, and novel synthetic strategies, and size-conversion reaction have been continuously developed to verify/exclude the previously proposed insights. Nevertheless, it is also noteworthy that the current mechanistic understandings on the cluster systems are still quite limited. There are a lot of challenging issues to settle. For example, in the size-conversion occurring within several minutes (or even several seconds), the determination of the intermediate species will be extremely difficult. Meanwhile, the time-dependent mass spectroscopic analysis has been extensively utilized in mechanistic studies. But the MALDI-MS with strong ionization source always suffer from the fragmentation issue (the observed cluster peaks might be a fake signal caused by the fragmentation of the precursor or even intermediate clusters). Meanwhile, the electronic ionization efficiency of different species in ESI-MS (and also the fragmentation issues) will significantly affect the determination on the relative probability of different intermediates. In addition, in these days, most spectroscopic tests are capable of qualitative description, but not a quantitative determination on the amount of the intermediate species, while the plausibility of DFT calculations is largely affected by the preliminary mechanistic proposal (of the researchers who conducted the mechanistic study). Clearly, much more efforts are necessary to explore more experimental strategies for in situ analysis, as well as the systematic understanding on the structure-activity correlations for reasonable deduction of preliminary mechanisms.

References

[1] V.K. LaMer, R.H. Dinegar, Theory, production and mechanism of formation of monodispersed hydrosols, J. Am. Chem. Soc. 72 (1950) 4847–4854.

[2] C.B. Whitehead, S. Özkar, R.G. Finke, LaMer's 1950 model for particle formation of instantaneous nucleation and diffusion-controlled growth: a historical look at the model's origins, assumptions, equations, and underlying sulfur sol formation kinetics data, Chem. Mater. 31 (2019) 7116–7132.

[3] S.P. Shields, V.N. Richards, W.E. Buhro, Nucleation control of size and dispersity in aggregative nanoparticle growth. A study of the coarsening kinetics of thiolate-capped gold nanocrystals, Chem. Mater. 22 (2010) 3212–3225.
[4] M. Brust, M. Walker, D. Bethell, D.J. Schiffrin, R. Whyman, Synthesis of thiol-derivatised gold nanoparticles in a two-phase liquid–liquid system, J. Chem. Soc. Chem. Commun. (1994) 801–802.
[5] R. Jin, Quantum sized, thiolate-protected gold nanoclusters, Nanoscale 2 (2010) 343–362.
[6] R. Sardar, A.M. Funston, P. Mulvaney, R.W. Murray, Gold nanoparticles: past, present, and future, Langmuir 25 (2009) 13840–13851.
[7] P.J.G. Goulet, R.B. Lennox, New insights into Brust−Schiffrin metal nanoparticle synthesis, J. Am. Chem. Soc. 132 (2010) 9582–9584.
[8] Y. Li, O. Zaluzhna, B. Xu, Y. Gao, J.M. Modest, Y.J. Tong, Mechanistic insights into the Brust−Schiffrin two-phase synthesis of organo-chalcogenate-protected metal nanoparticles, J. Am. Chem. Soc. 133 (2011) 2092–2095.
[9] H. Qian, M. Zhu, E. Lanni, Y. Zhu, M.E. Bier, R. Jin, Conversion of polydisperse Au nanoparticles into monodisperse Au_{25} nanorods and nanospheres, J. Phys. Chem. C 113 (2009) 17599–17603.
[10] H. Qian, W.T. Eckenhoff, M.E. Bier, T. Pintauer, R. Jin, Crystal structures of Au_2 complex and Au_{25} nanocluster and mechanistic insight into the conversion of polydisperse nanoparticles into monodisperse Au_{25} nanoclusters, Inorg. Chem. 50 (2011) 10735–10739.
[11] J. Chen, T. Jiang, G. Wei, A.A. Mohamed, C. Homrighausen, J.A. Krause Bauer, A.E. Bruce, M.R.M. Bruce, Electrochemical and chemical oxidation of gold(I) thiolate phosphine complexes: formation of gold clusters and disulfide, J. Am. Chem. Soc. 121 (1999) 9225–9226.
[12] Y. Shichibu, K. Suzuki, K. Konishi, Facile synthesis and optical properties of magic-number Au13 clusters, Nanoscale 4 (2012) 4125–4129.
[13] L. Yang, H. Cheng, Y. Jiang, T. Huang, J. Bao, Z. Sun, Z. Jiang, J. Ma, F. Sun, Q. Liu, T. Yao, H. Deng, S. Wang, M. Zhu, S. Wei, In situ studies on controlling an atomically-accurate formation process of gold nanoclusters, Nanoscale 7 (2015) 14452–14459.
[14] Y. Negishi, N.K. Chaki, Y. Shichibu, R.L. Whetten, T. Tsukuda, Origin of magic stability of thiolated gold clusters: a case study on $Au_{25}(SC_6H_{13})_{18}$, J. Am. Chem. Soc. 129 (2007) 11322–11323.
[15] M. Walter, J. Akola, O. Lopez-Acevedo, D. Jadzinsky Pablo, G. Calero, J. Ackerson Christopher, L. Whetten Robert, H. Grönbeck, H. Häkkinen, A unified view of ligand-protected gold clusters as superatom complexes, PNAS 105 (2008) 9157–9162.
[16] M.W. Heaven, A. Dass, P.S. White, K.M. Holt, R.W. Murray, Crystal structure of the gold nanoparticle $[N(C_8H_{17})_4][Au_{25}(SCH_2CH_2Ph)_{18}]$, J. Am. Chem. Soc. 130 (2008) 3754–3755.
[17] M. Zhu, W.T. Eckenhoff, T. Pintauer, R. Jin, Conversion of anionic $[Au_{25}(SCH_2CH_2Ph)_{18}]^-$ cluster to charge neutral cluster via air oxidation, J. Phys. Chem. C 112 (2008) 14221–14224.
[18] Z. Liu, Q. Xu, S. Jin, S. Wang, G. Xu, M. Zhu, Electron transfer reaction between Au_{25} nanocluster and phenothiazine-tetrachloro-p-benzoquinone complex, Int. J. Hydrogen Energy 38 (2013) 16722–16726.
[19] M.A. Tofanelli, K. Salorinne, T.W. Ni, S. Malola, B. Newell, B. Phillips, H. Häkkinen, C.J. Ackerson, Jahn–Teller effects in $Au_{25}(SR)_{18}$, Chem. Sci. 7 (2016) 1882–1890.
[20] M. Zhu, G. Chan, H. Qian, R. Jin, Unexpected reactivity of $Au_{25}(SCH_2CH_2Ph)_{18}$ nanoclusters with salts, Nanoscale 3 (2011) 1703–1707.
[21] Z. Liu, M. Zhu, X. Meng, G. Xu, R. Jin, Electron transfer between $[Au_{25}(SC_2H_4Ph)_{18}]^-TOA^+$ and oxoammonium cations, J. Phys. Chem. Lett. 2 (2011) 2104–2109.
[22] S. Antonello, N.V. Perera, M. Ruzzi, J.A. Gascón, F. Maran, Interplay of charge state, lability, and magnetism in the molecule-like $Au_{25}(SR)_{18}$ cluster, J. Am. Chem. Soc. 135 (2013) 15585–15594.
[23] A. Venzo, S. Antonello, J.A. Gascón, I. Guryanov, R.D. Leapman, N.V. Perera, A. Sousa, M. Zamuner, A. Zanella, F. Maran, Effect of the charge state (z = −1, 0, +1) on the nuclear magnetic resonance of monodisperse $Au_{25}[S(CH_2)_2Ph]_{18}^z$ clusters, Anal. Chem. 83 (2011) 6355–6362.
[24] C. Zeng, A. Weitz, G. Withers, T. Higaki, S. Zhao, Y. Chen, R.R. Gil, M. Hendrich, R. Jin, Controlling magnetism of $Au_{133}(TBBT)_{52}$ nanoclusters at single electron level and implication for nonmetal to metal transition, Chem. Sci. 10 (2019) 9684–9691.
[25] Y. Song, S. Jin, X. Kang, J. Xiang, H. Deng, H. Yu, M. Zhu, How a single electron affects the properties of the "non-superatom" Au_{25} nanoclusters, Chem. Mater. 28 (2016) 2609–2617.
[26] Q. Li, J. Chai, S. Yang, Y. Song, T. Chen, C. Chen, H. Zhang, H. Yu, M. Zhu, Multiple ways realizing charge-state transform in Au-Cu bimetallic nanoclusters with atomic precision, Small 17 (2021) 1907114.
[27] Q. Yao, V. Fung, C. Sun, S. Huang, T. Chen, D.-E. Jiang, J.Y. Lee, J. Xie, Revealing isoelectronic size conversion dynamics of metal nanoclusters by a noncrystallization approach, Nat. Commun. 9 (2018) 1979.
[28] W. Jiang, Y. Bai, Q. Li, X. Yao, H. Zhang, Y. Song, X. Meng, H. Yu, M. Zhu, Steric and electrostatic control of the pH-regulated interconversion of $Au_{16}(SR)_{12}$ and $Au_{18}(SR)_{14}$ (SR: deprotonated captopril), Inorg. Chem. 59 (2020) 5394–5404.
[29] T. Dainese, S. Antonello, S. Bogialli, W. Fei, A. Venzo, F. Maran, Gold fusion: from $Au_{25}(SR)_{18}$ to $Au_{38}(SR)_{24}$, the most unexpected transformation of a very stable nanocluster, ACS Nano 12 (2018) 7057–7066.
[30] IUPAC, Compendium of Chemical Terminology, second ed. (the "Gold Book"), Compiled by A. D. McNaught and A. Wilkinson, Blackwell Scientific Publications, Oxford, 1997, ISBN: 0-9678550-9-8, https://doi.org/10.1351/goldbook. Online version (2019-) created by S. J. Chalk.
[31] S. Jin, S. Wang, L. Xiong, M. Zhou, S. Chen, W. Du, A. Xia, Y. Pei, M. Zhu, Two electron reduction: from quantum dots to metal nanoclusters, Chem. Mater. 28 (2016) 7905–7911.
[32] G. Li, Z. Lei, Q.-M. Wang, Luminescent molecular Ag−S nanocluster $[Ag_{62}S_{13}(SBu^t)_{32}](BF_4)_4$, J. Am. Chem. Soc. 132 (2010) 17678–17679.
[33] S. Jin, S. Wang, Y. Song, M. Zhou, J. Zhong, J. Zhang, A. Xia, Y. Pei, M. Chen, P. Li, M. Zhu, Crystal structure and optical properties of the $[Ag_{62}S_{12}(SBu^t)_{32}]^{2+}$ nanocluster with a complete face-centered cubic kernel, J. Am. Chem. Soc. 136 (2014) 15559–15565.
[34] Y. Cao, V. Fung, Q. Yao, T. Chen, S. Zang, D.-E. Jiang, J. Xie, Control of single-ligand chemistry on thiolated Au_{25} nanoclusters, Nat. Commun. 11 (2020) 5498.
[35] X. Kang, H. Chong, M. Zhu, $Au_{25}(SR)_{18}$: the captain of the great nanocluster ship, Nanoscale 10 (2018) 10758–10834.
[36] C. Zeng, C. Liu, Y. Pei, R. Jin, Thiol ligand-induced transformation of $Au_{38}(SC_2H_4Ph)_{24}$ to $Au_{36}(SPh-t-Bu)_{24}$, ACS Nano 7 (2013) 6138–6145.

[37] T. Chen, V. Fung, Q. Yao, Z. Luo, D.-E. Jiang, J. Xie, Synthesis of water-soluble [Au$_{25}$(SR)$_{18}$]$^-$ using a stoichiometric amount of NaBH4, J. Am. Chem. Soc. 140 (2018) 11370–11377.

[38] Q. Yao, X. Yuan, V. Fung, Y. Yu, D.T. Leong, D.-E. Jiang, J. Xie, Understanding seed-mediated growth of gold nanoclusters at molecular level, Nat. Commun. 8 (2017) 927.

[39] Y. Kamei, Y. Shichibu, K. Konishi, Generation of small gold clusters with unique geometries through cluster-to-cluster transformations: octanuclear clusters with edge-sharing gold tetrahedron motifs, Angew. Chem. Int. Ed. 50 (2011) 7442–7445.

[40] Y. Bai, S. He, Y. Lv, M. Zhu, H. Yu, Redox-induced interconversion of two Au$_8$ nanoclusters: the mechanism and the structure–bond dissociation activity correlations, Inorg. Chem. 60 (2021) 5724–5733.

[41] X. Yang, X. Lin, C. Liu, R. Wu, J. Yan, J. Huang, Reversible conversion between phosphine protected Au$_6$ and Au$_8$ nanoclusters under oxidative/reductive conditions, Nanoscale 9 (2017) 2424–2427.

[42] X. Ren, J. Fu, X. Lin, X. Fu, J. Yan, R. Wu, C. Liu, J. Huang, Cluster-to-cluster transformation among Au$_6$, Au$_8$ and Au$_{11}$ nanoclusters, Dalton Trans. 47 (2018) 7487–7491.

[43] Y. Song, H. Abroshan, J. Chai, X. Kang, H.J. Kim, M. Zhu, R. Jin, Molecular-like transformation from PhSe-protected Au$_{25}$ to Au$_{23}$ nanocluster and its application, Chem. Mater. 29 (2017) 3055–3061.

[44] T. Higaki, C. Liu, Y. Chen, S. Zhao, C. Zeng, R. Jin, S. Wang, N.L. Rosi, R. Jin, Oxidation-induced transformation of eight-electron gold nanoclusters: [Au$_{23}$(SR)$_{16}$]$^-$ to [Au$_{28}$(SR)$_{20}$]^0, J. Phys. Chem. Lett. 8 (2017) 866–870.

[45] L. Tang, X. Kang, S. Wang, M. Zhu, Light-induced size-growth of atomically precise nanoclusters, Langmuir 35 (2019) 12350–12355.

[46] S. Chen, L. Xiong, S. Wang, Z. Ma, S. Jin, H. Sheng, Y. Pei, M. Zhu, Total structure determination of Au$_{21}$(S-Adm)$_{15}$ and geometrical/electronic structure evolution of thiolated gold nanoclusters, J. Am. Chem. Soc. 138 (2016) 10754–10757.

[47] A. Desireddy, B.E. Conn, J. Guo, B. Yoon, R.N. Barnett, B.M. Monahan, K. Kirschbaum, W.P. Griffith, R.L. Whetten, U. Landman, T.P. Bigioni, Ultrastable silver nanoparticles, Nature 501 (2013) 399–402.

[48] M.P. Maman, A.S. Nair, H. Cheraparambil, B. Pathak, S. Mandal, Size evolution dynamics of gold nanoclusters at an atom-precision level: ligand exchange, growth mechanism, electrochemical, and photophysical properties, J. Phys. Chem. Lett. 11 (2020) 1781–1788.

[49] M.P. Maman, A.S. Nair, A.M.A.H. Nazeeja, B. Pathak, S. Mandal, Synergistic effect of bridging thiolate and hub atoms for the aromaticity driven symmetry breaking in atomically precise gold nanocluster, J. Phys. Chem. Lett. 11 (2020) 10052–10059.

[50] A. Bhardwaj, J. Kaur, M. Wuest, F. Wuest, In situ click chemistry generation of cyclooxygenase-2 inhibitors, Nat. Commun. 8 (2017) 1.

[51] C. Zeng, T. Li, A. Das, N.L. Rosi, R. Jin, Chiral structure of thiolate-protected 28-gold-atom nanocluster determined by X-ray crystallography, J. Am. Chem. Soc. 135 (2013) 10011–10013.

[52] J. Dong, Z. Gan, W. Gu, Q. You, Y. Zhao, J. Zha, J. Li, H. Deng, N. Yan, Z. Wu, Synthesizing photoluminescent Au$_{28}$(SCH$_2$Ph-tBu)$_{22}$ nanoclusters with structural features by using a combined method, Angew. Chem. Int. Ed. 60 (2021) 17932–17936.

[53] H. Borsook, G. Keighley, Oxidation-reduction potential of ascorbic acid (vitamin C), PNAS 19 (1933) 875–878.

[54] Y. Shichibu, Y. Negishi, T. Tsukuda, T. Teranishi, Large-scale synthesis of thiolated Au$_{25}$ clusters via ligand exchange reactions of phosphine-stabilized Au$_{11}$ clusters, J. Am. Chem. Soc. 127 (2005) 13464–13465.

[55] L.C. McKenzie, T.O. Zaikova, J.E. Hutchison, Structurally similar triphenylphosphine-stabilized undecagolds, Au$_{11}$(PPh$_3$)$_7$Cl$_3$ and [Au$_{11}$(PPh$_3$)$_8$Cl$_2$]Cl, exhibit distinct ligand exchange pathways with glutathione, J. Am. Chem. Soc. 136 (2014) 13426–13435.

[56] R. Balasubramanian, R. Guo, A.J. Mills, R.W. Murray, Reaction of Au$_{55}$(PPh$_3$)$_{12}$Cl$_6$ with thiols yields thiolate monolayer protected Au$_{75}$ clusters, J. Am. Chem. Soc. 127 (2005) 8126–8132.

[57] Y. Shichibu, Y. Negishi, T. Watanabe, N.K. Chaki, H. Kawaguchi, T. Tsukuda, Biicosahedral gold clusters [Au$_{25}$(PPh$_3$)$_{10}$(SC$_n$H$_{2n+1}$)$_5$Cl$_2$]$_{2+}$ (n = 2 − 18): a stepping stone to cluster-assembled materials, J. Phys. Chem. C 111 (2007) 7845–7847.

[58] G.H. Woehrle, M.G. Warner, J.E. Hutchison, Ligand exchange reactions yield subnanometer, thiol-stabilized gold particles with defined optical transitions, J. Phys. Chem. B 106 (2002) 9979–9981.

[59] G.H. Woehrle, J.E. Hutchison, Thiol-functionalized undecagold clusters by ligand exchange: synthesis, mechanism, and properties, Inorg. Chem. 44 (2005) 6149–6158.

[60] H. Staudinger, J. Meyer, Über neue organische phosphorverbindungen III. Phosphinmethylenderivate und phosphinimine, Helv. Chim. Acta 2 (1919) 635–646.

[61] G.H. Woehrle, L.O. Brown, J.E. Hutchison, Thiol-functionalized, 1.5-nm gold nanoparticles through ligand exchange reactions: scope and mechanism of ligand exchange, J. Am. Chem. Soc. 127 (2005) 2172–2183.

[62] S. Wang, X. Meng, A. Das, T. Li, Y. Song, T. Cao, X. Zhu, M. Zhu, R. Jin, A 200-fold quantum yield boost in the photoluminescence of silver-doped Ag$_x$Au$_{25-x}$ nanoclusters: the 13th silver atom matters, Angew. Chem. Int. Ed. 53 (2014) 2376–2380.

[63] G. Salassa, A. Sels, F. Mancin, T. Bürgi, Dynamic nature of thiolate monolayer in Au$_{25}$(SR)$_{18}$ nanoclusters, ACS Nano 11 (2017) 12609–12614.

[64] M. Rambukwella, S. Burrage, M. Neubrander, O. Baseggio, E. Aprà, M. Stener, A. Fortunelli, A. Dass, Au$_{38}$(SPh)$_{24}$: Au$_{38}$ protected with aromatic thiolate ligands, J. Phys. Chem. Lett. 8 (2017) 1530–1537.

[65] Y. Niihori, W. Kurashige, M. Matsuzaki, Y. Negishi, Remarkable enhancement in ligand-exchange reactivity of thiolate-protected Au$_{25}$ nanoclusters by single Pd atom doping, Nanoscale 5 (2013) 508–512.

[66] Y. Chen, C. Liu, Q. Tang, C. Zeng, T. Higaki, A. Das, D.-E. Jiang, N.L. Rosi, R. Jin, Isomerism in Au$_{28}$(SR)$_{20}$ nanocluster and stable structures, J. Am. Chem. Soc. 138 (2016) 1482–1485.

[67] X. Kang, L. Huang, W. Liu, L. Xiong, Y. Pei, Z. Sun, S. Wang, S. Wei, M. Zhu, Reversible nanocluster structure transformation between face-centered cubic and icosahedral isomers, Chem. Sci. 10 (2019) 8685–8693.

[68] X. Kang, M. Zhou, S. Wang, S. Jin, G. Sun, M. Zhu, R. Jin, The tetrahedral structure and luminescence properties of bi-metallic $Pt_1Ag_{28}(SR)_{18}(PPh_3)_4$ nanocluster, Chem. Sci. 8 (2017) 2581–2587.

[69] C.L. Heinecke, T.W. Ni, S. Malola, V. Mäkinen, O.A. Wong, H. Häkkinen, C.J. Ackerson, Structural and theoretical basis for ligand exchange on thiolate monolayer protected gold nanoclusters, J. Am. Chem. Soc. 134 (2012) 13316–13322.

[70] S. Chen, S. Wang, J. Zhong, Y. Song, J. Zhang, H. Sheng, Y. Pei, M. Zhu, The structure and optical properties of the $[Au_{18}(SR)_{14}]$ nanocluster, Angew. Chem. Int. Ed. 54 (2015) 3145–3149.

[71] A. Das, T. Li, K. Nobusada, Q. Zeng, N.L. Rosi, R. Jin, Total structure and optical properties of a phosphine/thiolate-protected Au_{24} nanocluster, J. Am. Chem. Soc. 134 (2012) 20286–20289.

[72] A. George, A. Sundar, A.S. Nair, M.P. Maman, B. Pathak, N. Ramanan, S. Mandal, Identification of intermediate $Au_{22}(SR)_4(SR')_{14}$ cluster on ligand-induced transformation of $Au_{25}(SR)_{18}$ nanocluster, J. Phys. Chem. Lett. 10 (2019) 4571–4576.

[73] M.S. Bootharaju, C.P. Joshi, M.J. Alhilaly, O.M. Bakr, Switching a nanocluster core from hollow to nonhollow, Chem. Mater. 28 (2016) 3292–3297.

[74] H. Yang, Y. Wang, H. Huang, L. Gell, L. Lehtovaara, S. Malola, H. Häkkinen, N. Zheng, All-thiol-stabilized Ag_{44} and $Au_{12}Ag_{32}$ nanoparticles with single-crystal structures, Nat. Commun. 4 (2013) 2422.

[75] C. Zeng, H. Qian, T. Li, G. Li, N.L. Rosi, B. Yoon, R.N. Barnett, R.L. Whetten, U. Landman, R. Jin, Total structure and electronic properties of the gold nanocrystal $Au_{36}(SR)_{24}$, Angew. Chem. Int. Ed. 51 (2012) 13114–13118.

[76] A. Das, C. Liu, C. Zeng, G. Li, T. Li, N.L. Rosi, R. Jin, Cyclopentanethiolato-protected $Au_{36}(SC_5H_9)_{24}$ nanocluster: crystal structure and implications for the steric and electronic effects of ligand, J. Phys. Chem. A 118 (2014) 8264–8269.

[77] X. Kang, L. Xiong, S. Wang, Y. Pei, M. Zhu, Combining the single-atom engineering and ligand-exchange strategies: obtaining the single-heteroatom-doped $Au_{16}Ag_1(S\text{-}Adm)_{13}$ nanocluster with atomically precise structure, Inorg. Chem. 57 (2018) 335–342.

[78] N. Yan, N. Xia, Z. Wu, Metal nanoparticles confronted with foreign ligands: mere ligand exchange or further structural transformation? Small 17 (2021) 2000609.

[79] T.W. Ni, M.A. Tofanelli, B.D. Phillips, C.J. Ackerson, Structural basis for ligand exchange on $Au_{25}(SR)_{18}$, Inorg. Chem. 53 (2014) 6500–6502.

[80] Y. Niihori, Y. Kikuchi, A. Kato, M. Matsuzaki, Y. Negishi, Understanding ligand-exchange reactions on thiolate-protected gold clusters by probing isomer distributions using reversed-phase high-performance liquid chromatography, ACS Nano 9 (2015) 9347–9356.

[81] Q. Xu, S. Wang, Z. Liu, G. Xu, X. Meng, M. Zhu, Synthesis of selenolate-protected $Au_{18}(SeC_6H_5)_{14}$ nanoclusters, Nanoscale 5 (2013) 1176–1182.

[82] X. Meng, Q. Xu, S. Wang, M. Zhu, Ligand-exchange synthesis of selenophenolate-capped Au_{25} nanoclusters, Nanoscale 4 (2012) 4161–4165.

[83] A. Dass, T.C. Jones, S. Theivendran, L. Sementa, A. Fortunelli, Core size interconversions of $Au_{30}(S\text{-}tBu)_{18}$ and $Au_{36}(SPhX)_{24}$, J. Phys. Chem. C 121 (2017) 14914–14919.

[84] P.R. Nimmala, S. Knoppe, V.R. Jupally, J.H. Delcamp, C.M. Aikens, A. Dass, $Au_{36}(SPh)_{24}$ nanomolecules: X-ray crystal structure, optical spectroscopy, electrochemistry, and theoretical analysis, J. Phys. Chem. B 118 (2014) 14157–14167.

[85] A. Dass, T. Jones, M. Rambukwella, D. Crasto, K.J. Gagnon, L. Sementa, M. De Vetta, O. Baseggio, E. Aprà, M. Stener, A. Fortunelli, Crystal structure and theoretical analysis of green gold $Au_{30}(S\text{-}tBu)_{18}$ nanomolecules and their relation to $Au_{30}S(S\text{-}tBu)_{18}$, J. Phys. Chem. C 120 (2016) 6256–6261.

[86] N. Yan, N. Xia, L. Liao, M. Zhu, F. Jin, R. Jin, Z. Wu, Unraveling the long-pursued Au_{144} structure by X-ray crystallography, Sci. Adv. 4 (2018), eaat7259.

[87] Q. Shi, Z. Qin, C. Yu, S. Liu, H. Xu, G. Li, Pyridine as a trigger in transformation chemistry from $Au_{144}(SR)_{60}$ to aromatic thiolate-ligated gold clusters, Nanoscale 12 (2020) 4982–4987.

[88] C. Zeng, Y. Chen, K. Kirschbaum, K. Appavoo, Y. Sfeir Matthew, R. Jin, Structural patterns at all scales in a nonmetallic chiral $Au_{133}(SR)_{52}$ nanoparticle, Sci. Adv. 1 (2015), e1500045.

[89] M.S. Bootharaju, S.M. Kozlov, Z. Cao, A. Shkurenko, A.M. El-Zohry, O.F. Mohammed, M. Eddaoudi, O.M. Bakr, L. Cavallo, J.-M. Basset, Tailoring the crystal structure of nanoclusters unveiled high photoluminescence via ion pairing, Chem. Mater. 30 (2018) 2719–2725.

[90] S. Jin, M. Zhou, X. Kang, X. Li, W. Du, X. Wei, S. Chen, S. Wang, M. Zhu, Three-dimensional octameric assembly of icosahedral M_{13} units in $[Au_8Ag_{57}(Dppp)_4(C_6H_{11}S)_{32}Cl_2]Cl$ and its $[Au_8Ag_{55}(Dppp)_4(C_6H_{11}S)_{34}][BPh_4]_2$ derivative, Angew. Chem. Int. Ed. 59 (2020) 3891–3895.

[91] Y. Shichibu, M. Zhang, Y. Kamei, K. Konishi, $[Au_7]^{3+}$: a missing link in the four-electron gold cluster family, J. Am. Chem. Soc. 136 (2014) 12892–12895.

[92] M.-B. Li, S.-K. Tian, Z. Wu, R. Jin, Cu^{2+} induced formation of $Au_{44}(SC_2H_4Ph)_{32}$ and its high catalytic activity for the reduction of 4-nitrophenol at low temperature, Chem. Commun. 51 (2015) 4433–4436.

[93] P. Chakraborty, A. Nag, G. Natarajan, N. Bandyopadhyay, G. Paramasivam, K. Panwar Manoj, J. Chakrabarti, T. Pradeep, Rapid isotopic exchange in nanoparticles, Sci. Adv. 5 (2019), eaau7555.

[94] W. Du, X. Kang, S. Jin, D. Liu, S. Wang, M. Zhu, Different types of ligand exchange induced by Au substitution in a maintained nanocluster template, Inorg. Chem. 59 (2020) 1675–1681.

[95] B. Molina, A. Tlahuice-Flores, Thiolated Au_{18} cluster: preferred Ag sites for doping, structures, and optical and chiroptical properties, Phys. Chem. Chem. Phys. 18 (2016) 1397–1403.

[96] S. Wang, Y. Song, S. Jin, X. Liu, J. Zhang, Y. Pei, X. Meng, M. Chen, P. Li, M. Zhu, Metal exchange method using Au_{25} nanoclusters as templates for alloy nanoclusters with atomic precision, J. Am. Chem. Soc. 137 (2015) 4018–4021.

[97] C.A. Hosier, I.D. Anderson, C.J. Ackerson, Acetylide-for-thiolate and thiolate-for-acetylide exchange on gold nanoclusters, Nanoscale 12 (2020) 6239–6242.

[98] S. Yang, S. Chen, L. Xiong, C. Liu, H. Yu, S. Wang, N.L. Rosi, Y. Pei, M. Zhu, Total structure determination of $Au_{16}(S-Adm)_{12}$ and $Cd_1Au_{14}(StBu)_{12}$ and implications for the structure of $Au_{15}(SR)_{13}$, J. Am. Chem. Soc. 140 (2018) 10988–10994.

[99] Y. Lv, R. Zhao, S. Weng, H. Yu, Core charge density dominated size-conversion from Au_6P_8 to $Au_8P_8Cl_2$, Chem. A Eur. J. 26 (2020) 12382–12387.

[100] J.-Q. Wang, Z.-J. Guan, W.-D. Liu, Y. Yang, Q.-M. Wang, Chiroptical activity enhancement via structural control: the chiral synthesis and reversible interconversion of two intrinsically chiral gold nanoclusters, J. Am. Chem. Soc. 141 (2019) 2384–2390.

[101] Q. Li, S. Wang, K. Kirschbaum, K.J. Lambright, A. Das, R. Jin, Heavily doped $Au_{25-x}Ag_x(SC_6H_{11})_{18}^-$ nanoclusters: silver goes from the core to the surface, Chem. Commun. 52 (2016) 5194–5197.

[102] S. Wang, H. Abroshan, C. Liu, T.-Y. Luo, M. Zhu, H.J. Kim, N.L. Rosi, R. Jin, Shuttling single metal atom into and out of a metal nanoparticle, Nat. Commun. 8 (2017) 848.

[103] C. Yao, J. Chen, M.-B. Li, L. Liu, J. Yang, Z. Wu, Adding two active silver atoms on Au_{25} nanoparticle, Nano Lett. 15 (2015) 1281–1287.

[104] X. Kang, S. Jin, L. Xiong, S. Wei, M. Zhou, C. Qin, Y. Pei, S. Wang, M. Zhu, Nanocluster growth via "graft-onto": effects on geometric structures and optical properties, Chem. Sci. 11 (2020) 1691–1697.

[105] Z.-J. Guan, F. Hu, J.-J. Li, Z.-R. Wen, Y.-M. Lin, Q.-M. Wang, Isomerization in alkynyl-protected gold nanoclusters, J. Am. Chem. Soc. 142 (2020) 2995–3001.

[106] S. Li, Z.-Y. Wang, G.-G. Gao, B. Li, P. Luo, Y.-J. Kong, H. Liu, S.-Q. Zang, Smart transformation of a polyhedral oligomeric silsesquioxane shell controlled by thiolate silver(I) nanocluster core in cluster@clusters dendrimers, Angew. Chem. Int. Ed. 57 (2018) 12775–12779.

[107] X.-R. Chen, L. Yang, Y.-L. Tan, H. Yu, C.-Y. Ni, Z. Niu, J.-P. Lang, The solvent-induced isomerization of silver thiolate clusters with symmetry transformation, Chem. Commun. 56 (2020) 3649–3652.

[108] L.-L. Yan, L.-Y. Yao, V.W.-W. Yam, Concentration- and solvation-induced reversible structural transformation and assembly of polynuclear gold(I) sulfido complexes, J. Am. Chem. Soc. 142 (2020) 11560–11568.

[109] M.R. Narouz, K.M. Osten, P.J. Unsworth, R.W.Y. Man, K. Salorinne, S. Takano, R. Tomihara, S. Kaappa, S. Malola, C.-T. Dinh, J.D. Padmos, K. Ayoo, P.J. Garrett, M. Nambo, J.H. Horton, E.H. Sargent, H. Häkkinen, T. Tsukuda, C.M. Crudden, N-heterocyclic carbene-functionalized magic-number gold nanoclusters, Nat. Chem. 11 (2019) 419–425.

[110] L. Cheng, C. Ren, X. Zhang, J. Yang, New insight into the electronic shell of $Au_{38}(SR)_{24}$: a superatomic molecule, Nanoscale 5 (2013) 1475–1478.

[111] J.J. Pelayo, R.L. Whetten, I.L. Garzón, Geometric quantification of chirality in ligand-protected metal clusters, J. Phys. Chem. C 119 (2015) 28666–28678.

[112] S. Mukherjee, D. Jayakumar, S. Mandal, Insight into the size evolution transformation process of the fcc-based $Au_{28}(SR)_{20}$ nanocluster, J. Phys. Chem. C 125 (2021) 12149–12154.

[113] S. Tian, Y.-Z. Li, M.-B. Li, J. Yuan, J. Yang, Z. Wu, R. Jin, Structural isomerism in gold nanoparticles revealed by X-ray crystallography, Nat. Commun. 6 (2015) 8667.

[114] Z. Qin, J. Zhang, C. Wan, S. Liu, H. Abroshan, R. Jin, G. Li, Atomically precise nanoclusters with reversible isomeric transformation for rotary nanomotors, Nat. Commun. 11 (2020) 6019.

[115] M. Waszkielewicz, J. Olesiak-Banska, C. Comby-Zerbino, F. Bertorelle, X. Dagany, A.K. Bansal, M.T. Sajjad, I.D.W. Samuel, Z. Sanader, M. Rozycka, M. Wojtas, K. Matczyszyn, V. Bonacic-Koutecky, R. Antoine, A. Ozyhar, M. Samoc, pH-induced transformation of ligated Au_{25} to brighter Au_{23} nanoclusters, Nanoscale 10 (2018) 11335–11341.

[116] A.V. Pereira, A.A. Garabeli, G.D. Schunemann, P.C. Borck, Determination of dissociation constant (Ka) of captopril and nimesulide: analytical chemistry experiments for undergraduate pharmacy, Quim. Nova 34 (2011) 1656–1660.

[117] J.L. Martins, R. Car, J. Buttet, Variational spherical model of small metallic particles, Surf. Sci. 106 (1981) 265–271.

[118] W.D. Knight, K. Clemenger, W.A. de Heer, W.A. Saunders, M.Y. Chou, M.L. Cohen, Electronic shell structure and abundances of sodium clusters, Phys. Rev. Lett. 52 (1984) 2141–2143.

[119] W.W. Xu, B. Zhu, X.C. Zeng, Y. Gao, A grand unified model for liganded gold clusters, Nat. Commun. 7 (2016) 13574.

[120] Y. Yang, P.R. Sharp, New gold clusters $[Au_8L_6](BF_4)_2$ and $[(AuL)_4](BF_4)_2$ (L = P(mesityl)$_3$), J. Am. Chem. Soc. 116 (1994) 6983–6984.

[121] B.K. Teo, M.C. Hong, H. Zhang, D.B. Huang, Cluster of clusters: structure of the 37-atom cluster $[(p-Tol3P)_{12}Au_{18}Ag_{19}Br_{11}]_2^+$ and a novel series of supraclusters based on vertex-sharing icosahedra, Angew. Chem. Int. Ed. 26 (1987) 897–900.

[122] R. Jin, C. Liu, S. Zhao, A. Das, H. Xing, C. Gayathri, Y. Xing, N.L. Rosi, R.R. Gil, R. Jin, Tri-icosahedral gold nanocluster $[Au_{37}(PPh_3)_{10}(SC_2H_4Ph)_{10}X_2]^+$: linear assembly of icosahedral building blocks, ACS Nano 9 (2015) 8530–8536.

[123] Y. Shichibu, K. Konishi, HCl-induced nuclearity convergence in diphosphine-protected ultrasmall gold clusters: a novel synthetic route to "magic-number" Au_{13} clusters, Small 6 (2010) 1216–1220.

[124] S. Yang, J. Chai, Y. Lv, T. Chen, S. Wang, H. Yu, M. Zhu, Cyclic Pt_3Ag_{33} and $Pt_3Au_{12}Ag_{21}$ nanoclusters with M13 icosahedra as building-blocks, Chem. Commun. 54 (2018) 12077–12080.

[125] C. Zeng, Y. Chen, K. Iida, K. Nobusada, K. Kirschbaum, K.J. Lambright, R. Jin, Gold quantum boxes: on the periodicities and the quantum confinement in the Au_{28}, Au_{36}, Au_{44}, and Au_{52} magic series, J. Am. Chem. Soc. 138 (2016) 3950–3953.

[126] C. Zeng, Y. Chen, C. Liu, K. Nobusada, L. Rosi Nathaniel, R. Jin, Gold tetrahedra coil up: Kekulé-like and double helical superstructures, Sci. Adv. 1 (2015) e1500425.

[127] M. Zhou, C. Zeng, M.Y. Sfeir, M. Cotlet, K. Iida, K. Nobusada, R. Jin, Evolution of excited-state dynamics in periodic Au_{28}, Au_{36}, Au_{44}, and Au_{52} nanoclusters, J. Phys. Chem. Lett. 8 (2017) 4023–4030.

[128] H. Yang, J. Lei, B. Wu, Y. Wang, M. Zhou, A. Xia, L. Zheng, N. Zheng, Crystal structure of a luminescent thiolated Ag nanocluster with an octahedral Ag_6^{4+} core, Chem. Commun. 49 (2013) 300–302.

[129] H. Yang, J. Yan, Y. Wang, H. Su, L. Gell, X. Zhao, C. Xu, B.K. Teo, H. Häkkinen, N. Zheng, Embryonic growth of face-center-cubic silver nanoclusters shaped in nearly perfect half-cubes and cubes, J. Am. Chem. Soc. 139 (2017) 31–34.

[130] Y. Lv, X. Ma, J. Chai, H. Yu, M. Zhu, Face-centered-cubic ag nanoclusters: origins and consequences of the high structural regularity elucidated by density functional theory calculations, Chem. A Eur. J. 25 (2019) 13977–13986.

[131] B.K. Teo, H. Yang, J. Yan, N. Zheng, Supercubes, supersquares, and superrods of face-centered cubes (FCC): atomic and electronic requirements of [mm(SR)l(PR′3)8]q nanoclusters (M = coinage metals) and their implications with respect to nucleation and growth of FCC metals, Inorg. Chem. 56 (2017) 11470–11479.

[132] W.A. de Heer, The physics of simple metal clusters: experimental aspects and simple models, Rev. Mod. Phys. 65 (1993) 611–676.

[133] K. Clemenger, Ellipsoidal shell structure in free-electron metal clusters, Phys. Rev. B 32 (1985) 1359–1362.

[134] M. Brack, The physics of simple metal clusters: self-consistent jellium model and semiclassical approaches, Rev. Mod. Phys. 65 (1993) 677–732.

[135] B.K. Teo, S.-Y. Yang, Jelliumatic shell model, J. Clust. Sci. 26 (2015) 1923–1941.

[136] C. Liu, T. Li, H. Abroshan, Z. Li, C. Zhang, H.J. Kim, G. Li, R. Jin, Chiral Ag_{23} nanocluster with open shell electronic structure and helical face-centered cubic framework, Nat. Commun. 9 (2018) 744.

[137] X. Ma, Y. Lv, H. Li, T. Chen, J. Zeng, M. Zhu, H. Yu, Ligand effect on geometry and electronic structures of face-centered cubic Ag_{14} and Ag_{23} nanoclusters, J. Phys. Chem. C 124 (2020) 13421–13426.

[138] M.J. Alhilaly, M.S. Bootharaju, C.P. Joshi, T.M. Besong, A.-H. Emwas, R. Juarez-Mosqueda, S. Kaappa, S. Malola, K. Adil, A. Shkurenko, H. Häkkinen, M. Eddaoudi, O.M. Bakr, $[Ag_{67}(SPhMe_2)_{32}(PPh_3)_8]^+$: synthesis, total structure, and optical properties of a large box-shaped silver nanocluster, J. Am. Chem. Soc. 138 (2016) 14727–14732.

[139] M. Bodiuzzaman, A. Ghosh, K.S. Sugi, A. Nag, E. Khatun, B. Varghese, G. Paramasivam, S. Antharjanam, G. Natarajan, T. Pradeep, Camouflaging structural diversity: co-crystallization of two different nanoparticles having different cores but the same shell, Angew. Chem. Int. Ed. 58 (2019) 189–194.

[140] J. Chai, S. Yang, Y. Lv, T. Chen, S. Wang, H. Yu, M. Zhu, A unique pair: Ag_{40} and Ag_{46} nanoclusters with the same surface but different cores for structure–property correlation, J. Am. Chem. Soc. 140 (2018) 15582–15585.

[141] X. Ma, Y. Bai, Y. Song, Q. Li, Y. Lv, H. Zhang, H. Yu, M. Zhu, Rhombicuboctahedral Ag_{100}: four-layered octahedral silver nanocluster adopting the Russian nesting doll model, Angew. Chem. Int. Ed. 59 (2020) 17234–17238.

[142] C. Liu, Y. Pei, H. Sun, J. Ma, The nucleation and growth mechanism of thiolate-protected Au nanoclusters, J. Am. Chem. Soc. 137 (2015) 15809–15816.

Chapter 5

Physical-chemical properties of metal nanoclusters

Manzhou Zhu and Shuang Chen

5.1 Optical absorption of metal nanoclusters

Metal nanoparticles have continuous band in the metallic state and surface plasmon resonance, and their optical properties can be controlled by adjusting their size and morphology. Differently, ultrasmall metal nanoparticles (diameters <2.2 nm, also called metal nanoclusters) exhibit molecular properties due to the quantum confinement effect, with discrete energy levels and multiple absorption peaks. With the development of synthesis and crystal structure in recent years, significant progress has been made in the study of the optical properties of clusters. Due to the atomic structures, the electronic transitions of metal nanoclusters can be obtained by DFT analysis. In this section, we will discuss the association of experimental optical spectra and DFT-simulated spectra, assign the electronic transitions, and demonstrate the structural effect on optical spectra for metal nanoclusters.

In 2008, Jin et al. determined the crystal structure of the $Au_{25}(SR)_{18}$ cluster and collected its absorption spectrum. Due to strong quantum size effects, the $Au_{25}(SR)_{18}$ cluster displayed multiple molecular-like transitions at 1.8, 2.75, and 3.1 eV. TD-DFT calculations were performed to correlate the electronic structure and optical absorption spectrum of $Au_{25}(SR)_{18}$ cluster [1]. The 6sp atomic orbitals of Au contributed to the HOMO and the lowest three LUMOs, and the 5d10 atomic orbitals of Au contributed to the molecular orbitals from HOMO-1 to HOMO-5. The LUMO ← HOMO transition followed an sp ← sp intraband transition, corresponding to the first excited state at 1.52 eV. The peak at 2.63 eV arises from mixed sp ← sp intraband and sp ← d interband transitions. The peak at 2.91 eV arises principally from an sp ← d interband transition (Fig. 5.1A).

The optical spectra and electronic structures of the cluster are the inherent characterization and vary with the size and geometric structure of metal nanoclusters. In 2015, Chen et al. and Das et al. independently reported the structure of $Au_{18}(SC_6H_{11})_{14}$. Experimental and theoretical results showed that the optical spectrum of $Au_{18}(SC_6H_{11})_{14}$ match well [2,3]. The electronic transitions associated with all absorption peaks were assigned to the d → sp interband transitions. After that, Chen et al. determined the structures of $Au_{21}(S-Adm)_{15}$ and $Au_{16}(S-Adm)_{12}$. DFT simulation showed that the HOMO-LUMO transition of $Au_{21}(S-Adm)_{15}$ is assigned to the Au sp ← sp excitation [4]. For the Au_{16}, the experimental absorption spectrum showed a weak shoulder peak at the 605 nm and theoretical simulation resulted in shoulder peak at and 630 nm, which is resulted from the HOMO → LUMO transitions, but the LUMO is largely contributed by the bare gold atom [5].

Metal doping affects the optical absorption properties of clusters. Aikens studied the effect of Ag doping of $Au_{25}(SH)_{18}^-$. The doping sites and the associated optical spectra were studied. It is found that when $n=1$ in $Au_{25-n}Ag_n(SH)_{18}^-$ ($n=1, 2, 4, 6, 8, 10, 12$), the preferable doping site is on the icosahedral shell of the metal core. When $n \geq 2$, the HOMO-LUMO transition peak exhibited a blue shift, and the peak at 2.5 eV increased as the number of doping Ag atoms increases [6]. Recently, Li et al. reported atom-by-atom manipulation on metal nanoclusters based on the Au_{21}, Au_{22}, $Au_{22}Cd_1$ and Au_{24} nanoclusters. The optical spectra changed with the sizes and structures for these nanoclusters. $Au_{22}Cd_1(SAdm)_{16}$ and $Au_{23}(SR)_{16}^-$ possess the same number of metal atoms [7,8]. $Au_{22}Cd_1(SAdm)_{16}$ showed a red shift in absorption spectrum compared with that of $Au_{23}(SR)_{16}^-$. Similarly, doping heterogeneous metal into Ag nanoclusters can affect the absorption spectra. Zheng et al. prepared $[PdAg_{24}(SR)_{18}]^{2-}$ and $[PtAg_{24}(SR)_{18}]^{2-}$ nanoclusters that possessed the similar structure to that of $Ag_{25}(SR)_{18}^-$. $[PdAg_{24}(SR)_{18}]^{2-}$ displayed two absorption peaks centered at 483 and 628 nm. The optical spectra of $[PtAg_{24}(SR)_{18}]^{2-}$ showed similar profile, and the peaks were blue-shifted to 453 and 564 nm [9].

The optical spectra of metal nanoclusters are related to their structures. $[Au_{24}(PPh_3)_{10}(SC_2H_4Ph)_5X_2]^+$ and $[Au_{25}(PPh_3)_{10}(SC_2H_4Ph)_5X_2]^{2+}$ possessed the similar frame structure. The main difference is a single Au atom in the center

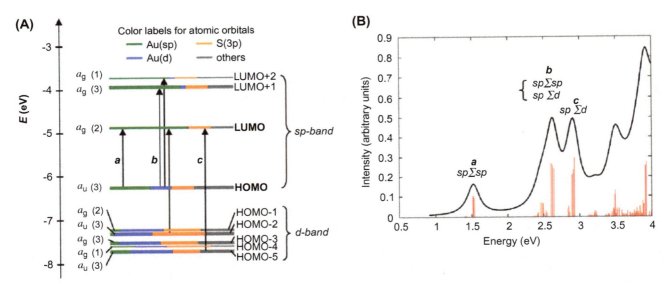

FIG. 5.1 The absorption of $Au_{25}(SH)_{18}$. (A) Kohn-Sham orbital energy level diagram for a model compound $Au_{25}(SH)_{18}^-$. The energies are in units of eV. Each KS orbital is drawn to indicate the relative contributions (line length with color labels) of the atomic orbitals of Au (6sp) in *green*, Au (5d) in *blue*, S (3p) in *yellow*, and others in *gray* (those unspecified atomic orbitals, each with a < 1% contribution). The left column of the KS orbitals shows the orbital symmetry (*g, u*) and degeneracy (in parenthesis); the right column shows the HOMO and LUMO sets. (B) The theoretical absorption spectrum of $Au_{25}(SH)_{18}^-$. Peak assignments: peak *a* corresponds to 1.8 eV observed, peak *b* corresponds to 2.75 eV (observed), and peak *c* corresponds to 3.1 eV (observed). *(From M. Zhu, C.M. Aikens, F.J. Hollander, G.C. Schatz, R. Jin, Correlating the crystal structure of a thiol-protected Au_{25} cluster and optical properties, J. Am. Chem. Soc. 130(18) (2008) 5883–5885, https://doi.org/10.1021/ja801173r.)*

of rod core. The absorption spectrum of rod-like Au_{24} showed prominent bands at 383 and 560 nm, whereas rod-like Au_{25} displayed absorption peaks centered at 415 and 670 nm [10]. Besides, Wang et al. achieved shuttling a single Ag or Cu atom into centrally hollow, rod-shaped Au_{24} nanoclusters, forming Ag_1Au_{24} and Cu_1Au_{24} nanoclusters. It is found that the absorption peak of Cu_1Au_{24} in the high-energy region has a slight blue shift, and the absorption peak of Ag_1Au_{24} in the low-energy region has a slight red shift compared with that of Au_{25} [11].

The number of building unit has an effect on the spectra of metal nanoclusters. The rod-like Au_{25} and Au_{37} are composed of two and three Au_{13} building unit, respectively. The optical energy gap of rod-like Au_{37} is 0.83 eV that is smaller than 1.73 eV of rod-like Au_{25} and 1.96 eV of Au_{13} [12]. On the other hand, the arrangement of the building blocks will affect the electronic transitions of the metal nanoclusters. The $[Au_{60}Se_2(Ph_3P)_{10}(SePh)_{15}]^{2+}$ composed of five Au_{13} units showed a shoulder peak at 835 nm, which is red-shifted compared with that of rod-like Au_{25} [13]. The family of $Au_{28}(TBBT)_{20}$, $Au_{36}(TBBT)_{24}$, $Au_{44}(TBBT)_{28}$ and $Au_{52}(TBBT)_{32}$ nanoclusters further demonstrated the optical change with the number of building units. These nanoclusters is identified with a formula of $Au_{8n+4}(TBBT)_{4n+8}$ ($n = 3$–6), the absorption peaks red shift with the increase in n and also the increase in building units (Fig. 5.2) [14].

Long wavelength absorption is potentially useful for solar thermal conversion and biomedicine [15]. Metal nanoclusters with distinct energy level showed distinct peaks at UV-vis region. Recently, a few metal nanoclusters showed intense near-infrared absorption. Tsukuda et al. employed a slow reduction to prepare $Au_{76}(4\text{-MEBA})_{44}$ clusters that displayed a ($3 \times 10^5 M^{-1} cm^{-1}$) near-infrared absorption band at 1340 nm [16]. Zeng et al. employed a size focusing method to prepared $Au_{64}(\text{S-c-}C_6H_{11})_{32}$ that exhibited an absorption peak centered at 968 nm [17]. In terms of silver nanoclusters with near-infrared absorption, Song et al. reported the gram-scale synthesis of $Ag_{146}Br_2(SR)_{80}$ nanoclusters that displayed a weak absorption band at ∼1200 nm in the NIR region [18].

Femtosecond time-resolved transient absorption spectroscopy has been employed to investigate the electronic properties of metal nanoclusters. Yi et al. studied the electronic energy relaxation of $Au_{144}(SR)_{60}$ nanoclusters and demonstrated the Au_{144} display electronic energy relaxation characteristic of metallic nanosturctures and a quantifiable electron-phonon coupling constant [19]. Zhou et al. investigated the transition of Au_{25}, Au_{38}, Au_{144}, Au_{333}, $Au_{\sim520}$ and $Au_{\sim940}$ nanoparticles by ultrafast spectroscopy and found that the transition from metallic to molecular state is between Au_{333} and Au_{144} (Fig. 5.3) [20]. Further, Higaki et al. collected the transient absorption spectra of Au_{246} and Au_{279}, and revealed that the Au_{279} nanocluster shows a laser power dependence in its excited state lifetime. It is concluded that Au_{279} is metallic but the Au_{246} is nonmetallic [21].

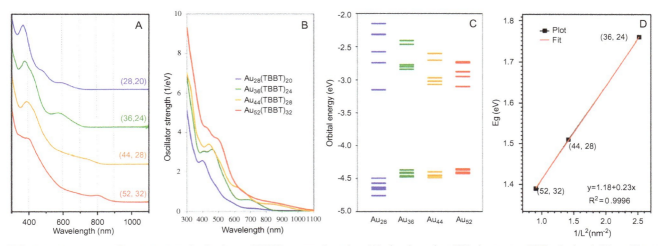

FIG. 5.2 Quantum confinement nature in the $Au_{8n+4}(TBBT)_{4n+8}$ magic series. (A) size-dependent UV-vis spectra; DFT-calculated (B) oscillator strengths and (C) energy levels (HOMO−4 to LUMO+4); and (D) scaling law of energy gap with size. *(From Z. Chenjie, C. Yuxiang, K. Kristin, K. J. Lambright, J. Rongchao, Emergence of hierarchical structural complexities in nanoparticles and their assembly, Science 354(6319) (2016) 1580–1584, https://doi.org/10.1126/science.aak9750.)*

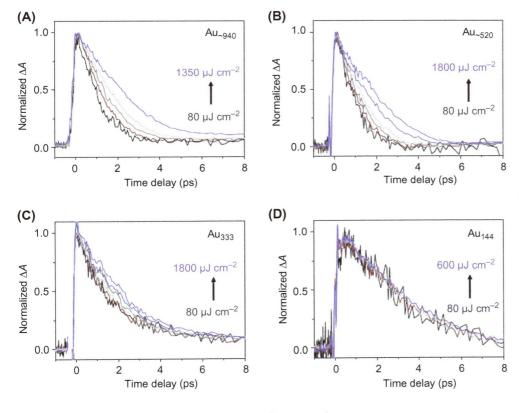

FIG. 5.3 Ultrafast absorption spectra. Pump power-dependent electron dynamics of gold nanoparticles. Normalized decay kinetics as a function of pump fluence for (A) $Au_{\sim 940}$, (B) $Au_{\sim 520}$, (C) Au_{333} and (D) Au_{144}; the kinetics at the maximum of SPR peak of each gold nanoparticle was monitored; for Au_{144}, GSB ~460nm was monitored. *(From M. Zhou, C. Zeng, Y. Chen, S. Zhao, M.Y. Sfeir, M. Zhu, R. Jin, Evolution from the plasmon to exciton state in ligand-protected atomically precise gold nanoparticles, Nat. Commun. 7(1) (2016) 13240, https://doi.org/10.1038/ncomms13240.)*

5.2 Photoluminescence of metal nanoclusters

Photoluminescence (PL) is the phenomenon that objects rely on external light sources to illuminate, thereby obtaining energy, and causing excitation to cause luminescence. It roughly passes through three main stages of absorption, energy transfer and light emission. Light absorption and emission occur between energy levels. All of the transitions go through excited states. The energy transfer is due to the motion of excited states. Ultraviolet radiation, visible light and infrared radiation can all cause photoluminescence. Photoluminescence can be divided into fluorescence (Fluorescence) and phosphorescence (Phosphorescence) according to the energy transition form.

The PL of atomically precise metal nanoclusters have attracted widespread attention due to their excellent photostability, biocompatibility and low toxicity, which have been employed to bioimaging, chemical and biological sensing, and photothermal therapy [22]. The precise composition and structure of the metal nanoclusters provide important support for the study of the PL properties of the nanoclusters [23,24]. Recently, great efforts of metal nanoclusters with PL property have been witnessed, and advanced progresses in terms of the interpretation of PL mechanism and regulation of PL intensity have been made [25–27]. However, the PL quantum yield (QY) of metal nanoclusters is generally still low compared with the widely studied PL materials [24]. It is quite necessary to explore new PL emission mechanisms and develop new methods to increase the PL intensity of nanoclusters.

There are many factors that affect the PL of metal nanoclusters. For example, the core size of the cluster, the rigidity of the structure, and the influence of peripheral ligands. In general, alloying and doping with heterogeneous metal can significantly improve the PL property of nanoclusters, which can lead to tens of times or even hundreds of times of enhancement in PL. In addition, the surrounding environment of the nanocluster also has a strong effect on the luminous intensity of metal nanoclusters, such as temperature, solvent and pH, etc. Generally, the low temperature state is beneficial to the PL intensity enhancement. Under normal circumstances, aggregation induction has a certain effect on the luminescence intensity of metal nanoclusters. By exploring these factors that affect the luminescence of metal nanoclusters and adjusting them appropriately, the luminescence intensity of the metal nanoclusters can be effectively improved, and practical applications related to this can be explored.

In this chapter, we focus on the PL mechanism and regulation (intensity, QY, wavelength) of metal nanoclusters. The following mechanisms and factors of PL for metal nanoclusters, inherent geometry and electronic structure, ligand effect (ligand to metal charge transfer, LMCT), metal doping and alloying, aggregation-induced emission (AIE)/aggregation-induced emission enhancement (AIEE), will be discussed in the following content.

5.2.1 Inherent geometry and electronic structure

Although many metal nanoclusters with fluorescent properties have been reported, it is very difficult to controllably synthesize and prepare metal nanoclusters with inherent fluorescent properties. More often, researchers investigated the PL property of the nanoclusters that have been prepared. Some of these nanoclusters have inherent and intense fluorescence emission due to their unique geometry and electronic structure. In this section, we discuss the nanoclusters with inherent PL, such as Ag_{29}, Ag_{62}, Au_{24}, and Ag_{14} nanoclusters.

In 2002, Whetten et al. reported the PL property of $Au_{28}(SG)_{16}$ where SG is glutathione. This nanoclusters show a broad luminescence in the visible to the infrared with two band of 1.5 and 1.15 eV [28]. The QY of $Au_{28}(SG)_{16}$ is calculated about 3.5×10^{-3} at room temperature. It is found that the visible PL after 500 nm irradiation is assigned to the interband transition from d to sp. band and radiative recombination. The radioactive recombination in the sp. conduction band, across the HOMO-LUMO gap of 1.3 eV, is responsible for the long wavelength emission. On the other hand, this low energy radiation could also be attributed to a triplet state with LMCT process. The whole electronic structure is responsible for the two band emission of $Au_{28}(SG)_{16}$ nanoclusters.

The atomically precise gold nanoclusters usually exhibit relatively weaker fluorescence intensity and lower QY than silver nanoclusters, especially oil-soluble Au nanoclusters. In particular, $Au_{24}(SR)_{20}$ exhibit higher QY. In 2014, Jin group reported the structure and electronic properties of $Au_{24}(SCH_2Ph\text{-}tBu)_{20}$ nanocluster [29]. It is demonstrated that this nanoclusters contain four tetrameric staple motifs and an Au_8 core. Hereafter, Yao et al. identified the $Au_{24}(SC_2H_4Ph)_{20}$ nanoclusters by mass spectrometry, and further studied its fluorescence property. Compared with nanoclusters of $Au_{25}(SC_2H_4Ph)_{18}$, $Au_{38}(SC_2H_4Ph)_{24}$ and $Au_{144}(SC_2H_4Ph)_{60}$ with the same ligand, the QY of $Au_{24}(SC_2H_4Ph)_{20}$ is ~40 times higher than that of $Au_{25}(SC_2H_4Ph)_{18}$ [30]. In their later work, the PL behavior of $Au_{24}(SR)_{20}$ (R=C_2H_4Ph, CH_2Ph, or $CH_2C_6H_4$ tBu) protected by different ligands has been explored. These nanoclusters displayed various emission intensity with the range of $Au_{24}(SePh)_{20} < Au_{24}(SC_2H_4Ph)_{20} < Au_{24}(SCH_2Ph)_{20} < Au_{24}(SCH_2C_6H_4\ tBu)_{20}$ (Fig. 5.4) [31]. It is concluded that the inherently interlocked $Au_4(SR)_5$ staples and the interaction between the core and thiolates are responsible for the extensive red emission. In addition, Au_{24} protected by alkynyl displayed near-infrared (NIR) PL with a maximum emission of 925 nm. DFT calculation demonstrated that this NIR emission can be attributed to the HOMO-LUMO transition.

Silver nanoclusters have received more and more attention due to their rich PL properties. With the advancement of synthesis and separation technology, a series of silver nanoclusters have been synthesized and characterized, and their properties have been explored. The geometry and electronic structure of the peripheral ligands have been shown to play an important role in the adjustment of the PL property of the nanoclusters. $Ag_{29}(BDT)_{12}(TPP)_4$ was reported by Bakr et al. in 2015. This nanocluster possesses four tetrahedrally symmetrical binding site, which is banded to TPP ligands

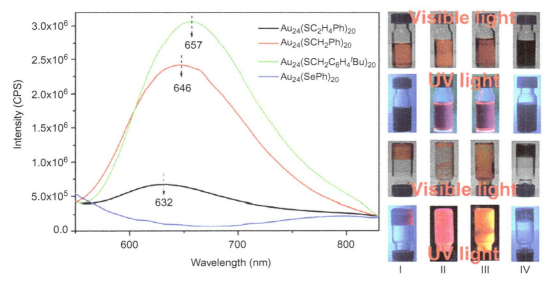

FIG. 5.4 Fluorescence spectra of Au_{24} nanoclusters. Fluorescence spectra of $Au_{24}(SC_2H_4Ph)_{20}$, $Au_{24}(SCH_2Ph)_{20}$, $Au_{24}(SCH_2C_6H_4 \ tBu)_{20}$, and $Au_{24}(SePh)_{20}$ nanoclusters dissolved in DCM (left) and digital photographs (I–IV) under visible and 365nm UV light irradiation (right). *(From Z. Gan, Y. Lin, L. Luo, G. Han, W. Liu, Z. Liu, C. Yao, L. Weng, L. Liao, J. Chen, X. Liu, Y. Luo, C. Wang, S. Wei, Z. Wu, Fluorescent gold nanoclusters with interlocked staples and a fully thiolate-bound kernel. Angew. Chem. Int. Ed. 55(38) (2016) 11567–11571, https://doi.org/10.1002/anie.201606661.)*

to achieve enhanced the stability and PL intensity. The solution and crystals of Ag_{29} show inherent emission under light radiation. Compared with the PL in solution, the PL spectrum of crystallization shows an enhanced and broadened features, as well as a red shift, which is attributed to electronic coupling among nanoclusters in the aggregate state (Fig. 5.5) [32]. Besides, doping with heterogeneous metal and promoting the binding of TPP ligands to the surface of Ag_{29} can significantly enhance its PL intensity, which will be discussed in the subsequent section.

Wang group prepared $[Ag_{62}S_{13}(SBut)_{32}](BF_4)_4$ with a core-shell structure in 2010. This nanocluster emits intense red light in both solution and solid state at room temperature with a max emission at 621nm for solid state and 613nm for sample solution. The charge transfer from S(3p) to Ag(5s) is responsible for the bright emission [33]. Recently, Jin

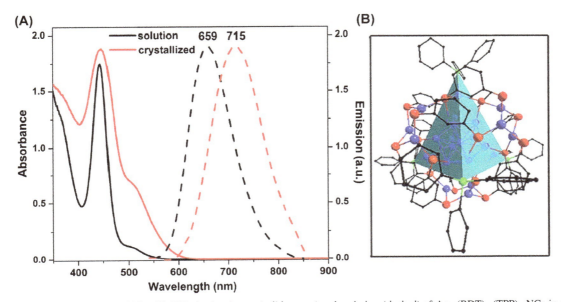

FIG. 5.5 PL of $Ag_{29}(BDT)_{12}(TPP)_4$ NCs. (A) UV-vis absorbance *(solid curves)* and emission *(dashed)* of $Ag_{29}(BDT)_{12}(TPP)_4$ NCs in acetonitrile *(black)* and dried *(red)*. (B) X-ray crystal structure of $Ag_{29}(BDT)_{12}(TPP)_4$. *(From L.G. AbdulHalim, M.S. Bootharaju, Q. Tang, S. Del Gobbo, R.G. Abdul-Halim, M. Eddaoudi, D. Jiang, O.M. Bakr, $Ag_{29}(BDT)_{12}(TPP)_4$: a tetravalent nanocluster, J. Am. Chem. Soc. 137(37) (2015) 11970–11975, https://doi.org/10.1021/jacs.5b04547.)*

et al. determined the structure of $[Ag_{62}S_{12}(SBut)_{32}]^{2+}$ that composed of a face centered cubic Ag_{14} core, a $Ag_{48}(SBut)_{32}$ shell, and 12 S atoms. The nanocluster of $[Ag_{62}S_{12}(SBut)_{32}]^{2+}$ exhibits extremely weak fluorescence emission, which is ~1800 times weaker than $[Ag_{62}S_{13}(SBut)_{32}](BF_4)_4$ [34]. Femtosecond transient absorption spectroscopy shows that free valence electrons significantly change the charge transfer process, resulting in a decrease in fluorescence. Further, the transformation from $[Ag_{62}S_{13}(SBut)_{32}](BF_4)_4$ to $[Ag_{62}S_{12}(SBut)_{32}]^{2+}$ can be achieved by electrochemical reduction due to the unique electrochemical properties of metal nanoclusters [35].

In 2013, Zheng et al. prepared three $Ag_{14}(SC_6H_3F_2)_{12}(PPh_3)_8$, $Ag_{16}(DPPE)_4(SC_6H_3F_2)_{14}$ and $[Ag_{32}(DPPE)_5(SC_6H_4CF_3)_{24}]^{2-}$ nanoclusters protected by phosphine and thiol ligands. $Ag_{14}(SC_6H_3F_2)_{12}(PPh_3)_8$, emits yellow light with the wavelength of 420 and 536 nm after 360 nm excitation [36]. It is found that the 536 nm emission can be excited by 420 nm irradiation, which indicate an energy transfer inside the Ag_{14} nanoclusters. Unlike Ag_{14}, Ag_{16} and Ag_{32} show weak blue emissions with a wavelength of ~440 nm under UV light excitation. A size threshold might be responsible for the weak emission of Ag_{16} and Ag_{32} [37].

5.2.2 Ligand effect on PL of metal nanoclusters

The metal nanocluster is composed of a metal core and a peripheral motif that is composed of metal atoms and protected ligands. The charge transfer between metal and ligand play a significant role for the enhancement of PL, such as LMCT, ligand to metal charge transfer (LMMCT), metal to ligand charge transfer (MLCT). In this context, the modulation of ligand is quite essential to increase the PL emission of metal nanoclusters. Changing the ligand might modify the geometric structure of the nanocluster. Here we will discuss it in two cases: structure maintenance accompanied by ligand substitution; and structural change accompanied ligand substitution.

5.2.2.1 Structure maintenance accompanied by ligand substitution

$Au_{25}(SR)_{18}$ was widely researched due to the advances in synthetic methods. In 2008, T. Pradeep et al. functionalized $Au_{25}(SG)_{18}$ with various ligand, 3-mercapto-2-butanol (MB), N-acetyl-(NAGSH) and N-formyl-glutathione (NFGSH). NMR spectroscopy was used to monitor the ligand exchange process, and a first order kinetics was identified. It is confirmed that the chemical nature of metal structure maintained after ligand exchange. The excitation spectrum of Au_{25}-MB shows one peak at 527 nm, which is different from that of $Au_{25}SG_{18}$, Au_{25}-SGAN, and Au_{25}-SGFN [38]. These nanoclusters emit NIR PL with similar wavelength, which indicates the inherent property of metal core. The emission occurs in the same energy state that can be accessible for different ligands.

To study the ligand effect on PL of nanoclusters, Jin and coworkers prepared four $Au_{25}(SR)_{18}$ caped with various ligands, including $[Au_{25}(SC_2H_4Ph)_{18}]^-$, $[Au_{25}(SC_6H_{13})_{18}]^-$, $[Au_{25}(SC_{12}H_{25})_{18}]^-$, and $[Au_{25}(SG)_{18}]^-$, and found PL intensity of the hydrophobic nanoclusters are much weaker than that of $[Au_{25}(SG)_{18}]^-$. The PL intensity of these nanoclusters follows the order of $[Au_{25}(SG)_{18}]^- > [Au_{25}(SC_2H_4Ph)_{18}]^- > [Au_{25}(SC_{12}H_{25})_{18}]^- > [Au_{25}(SC_6H_{13})_{18}]^-$ [39]. The effect of charge state of nanoclusters on PL was studied based on the $[Au_{25}(SC_2H_4Ph)_{18}]^-$ nanoclusters. It is found that nanocluster with higher charge state have stronger fluorescence intensity.

In 2014, Wang et al. used a direct synthesis method to synthesis the $[Au_{25}(NAPS)_{18}]^-$ protected with 2-(naphthalen-2-yl)-ethanethiolate ligand. The prepared $[Au_{25}(NAPS)_{18}]^-$ shows a similar absorption profile to $[Au_{25}(SC_2H_4Ph)_{18}]^-$, indicating the molecular purity of $[Au_{25}(NAPS)_{18}]^-$. Compared with Au_{25} protected by PET, the fluorescence of $[Au_{25}(NAPS)_{18}]^-$ increases with a 6.5-fold enhancement of QY. In addition, two band emission, 740 and 680 nm, was observed in $[Au_{25}(NAPS)_{18}]^-$, which indicate the important role of ligand for PL property in nanoclusters [40]. The PL of $[Au_{25}(NAPS)_{18}]^q$ ($q=-1, 0, 1$) with various charge state was studied, and the result shows that PL intensity increase with the electropositivity of nanoclusters ($[Au_{25}(NAPS)_{18}]^- < [Au_{25}(NAPS)_{18}]^0 < [Au_{25}(NAPS)_{18}]^+$). Using the ligands with electron-rich atoms and groups can promote the donation of delocalized electron density to metal core, and further enhance the PL intensity.

Zeng et al. reported the thermally stable $Au_{36}(TBBT)_{24}$ nanoclusters. Ligand exchange was performed to study the ligand effect on PL due to the relatively strong PL and stability. 4-fluorothiophenol (FBT), 4-bromobenzenethiol (BBT), and cyclopentanethiol (CPT) were employed as the second ligand, and ESI-MS confirmed the success of ligand substitution. The similar absorption profiles after various ligand exchange indicate the structure of $Au_{36}(SR)_{24}$ was retained. It is found that the PL intensity of these nanoclusters increase with the donating capability of CPT > TBBT > BBT > FBT, which is consistent to the situation of Au_{25}. Besides, in the case of CPT ligand exchange, the PL intensity increase with the exchange time, and the wavelength occur red shift from 735 to 782 nm [41]. The PL increase with the electron donating ability of ligand for Au_{36} and Au_{25} indicate that the LMCT does not be affected with the size of nanoclusters.

Given the interfacial electron transfer between metal and ligands, it can be speculate that chromophore can be ligands to promote resonance energy transfer between the metal and chromophore ligand. In 2010, Dongil Lee and coworkers functionalized $Au_{25}(C_6S)_{18}$ with pyrene chromophore. The functionalized $Au_{25}(C_6S)_{17}PyS$ display manifest PL quenching. Electrochemistry experiments show that the oxidation wave of $Au_{25}(C_6S)_{17}PyS$ shift to higher potential compared with $Au_{25}(C_6S)_{18}$, indicating the electronic polarization effects. Femtosecond transient absorption experiment disclose that the anion radical of pyrene absorb the energy from nanoclusters after excitation, and directional electron transfer from the metal core to pyrene ligand with an ultrafast transfer speed of ~580 fs [42]. Subsequently, Dongil Lee and coworkers functionalized $Au_{22}(SG)_{18}$ nanoclusters with benzyl chloroformate (CBz-Cl) and pyrene (Py). It is found that the PL intensity of Au_{22} increases five times after rigidifying ligand exchange (Fig. 5.6). The energy transfer was observed due to the overlap of emission maximum of Py and absorption of Au_{22}, which is responsible for the enhancement of PL after ligand exchange [43].

$[Ag_{29}(BDT)_{12}(PPh_3)_4]^{3-}$ exhibits inherent fluorescence. Thalappil Pradeep replaced the secondary ligand PPh_3 with different ligands, including the monophosphines of tri(p-tolyl)phosphine (TTP), tris(4-fluorophenyl) phosphine (TFPP) and tris(4-chlorophenyl)phosphine (TCPP) and diphosphines of DPPM, DPPE and DPPP (1,3-bis(diphenylphosphino) propane). UV-vis spectra and mass spectra indicate that the structure of Ag_{29} nanoclusters remains unchanged after ligand replacement. Luminescence spectra show that the PL intensity of $[Ag_{29}(BDT)_{12}(DPPP)_4]$ increases 30 times compared with $[Ag_{29}(BDT)_{12}(PPh_3)_4]$ [44]. A similar enhancement of PL was observed in the case of $[Ag_{51}(BDT)_{19}(PPh_3)_3]^{3-}$ after replace PPh_3 with DPPM. It is demonstrated that the surface rigidity of the nanocluster inducted by noncovalent interaction, LMCT and ligand to ligand charge transfer (LLCT) are responsible for the enhancement of PL.

Kang et al. prepared a strongly emissive Au_2Cu_6 nanocluster by AIE, details is addressed in the section of AIE. In their follow-up work, different ligands were employed to synthesis the Au_2Cu_6 nanoclusters, and the PL properties of these nanoclusters are studied. A series of functionalized Au_2Cu_6 protected by phosphine with electron donating groups (EDGs), including $P(Ph-OMe)_3$ (p-methoxy triphenylphosphine), PPh_3 (triphenylphosphine), $P(Ph-F)_3$ (p-fluorine triphenylphosphine). It is found that methoxy-functionalized Au_2Cu_6 shows the highest fluorescence intensity, and its fluorescence QY is 17.7% [45]. The QY of fluorine group functionalized Au_2Cu_6 shows the lowest PL intensity, and its QY is 5.7%. The QY of Au_2Cu_6 protected by PPh_3 is 12.2%, which is somewhere in between. It can be concluded that the PL intensity of Au_2Cu_6 increases the electron donating ability of the ligands. The LMCT of nanoclusters is responsible for this situation.

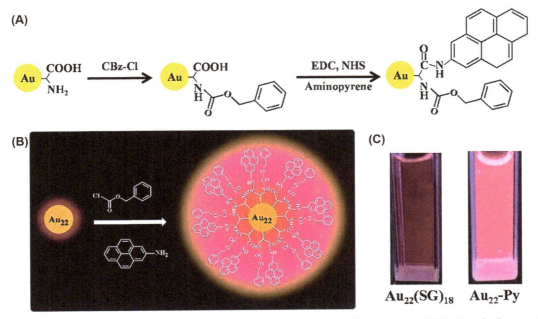

FIG. 5.6 Photoluminescence of Au_{22}-Py. (A) CBz and pyrene functionalization procedure of Au_{22} clusters and (B) schematic diagram of CBz and Py functionalized $Au_{22}(SG)_{18}$ clusters (Au_{22}-Py). (C) The digital photograph of Au_{22} and Au_{22}-Py clusters under long-wavelength UV lamp irradiation (365 nm). The optical densities of both clusters were 0.03 at 520 nm. *(From K. Pyo, V.D. Thanthirige, S.Y. Yoon, G. Ramakrishna, D. Lee, Enhanced luminescence of $Au_{22}(SG)_{18}$ nanoclusters via rational surface engineering, Nanoscale, 8(48) (2016) 20008–20016, https://doi.org/10.1039/C6NR07660B.)*

5.2.2.2 Structural changes inducted by ligand exchange

The protected and coordinating ligands play important roles for the structures and electronic configuration. In 2009, Wang and coworkers replaced the PPh$_3$ (triphenylphosphine) ligand in [(Ph$_3$PAu)$_6$C](BF$_4$)$_2$ (abbreviation as CAu$_6$-PPh$_3$) cluster with dppy (diphenylphoshpino-2-pyridine), and a new [Au$_6$(C)(dppy)$_6$](BF$_4$)$_2$ (abbreviation as CAu$_6$-dppy) was obtained. The Au$_6$-dppy cluster can be installed with Ag atoms due to the multicoordination ability of the dppy ligand. The [Au$_6$Ag$_2$(C)(dppy)$_6$](BF$_4$)$_4$ (abbreviation as CAu$_6$Ag$_2$-dppy) cluster is strongly emissive in solid and solution state [46]. It is observed CAu$_6$Ag$_2$-dppy emit green light in solid state at room temperature. The DCM solution of CAu$_6$Ag$_2$-dppy show red emission with a maximum wavelength of 625 nm. The lifetime of 5.7 ms for CAu$_6$Ag$_2$-dppy indicates the phosphorescence performance. The rigidity of ligand and the coordination of Ag are responsible for the strong emission of CAu$_6$Ag$_2$-dppy clusters. In their later work, [CAu$_6$Cu$_2$(dppy)$_6$](BF$_4$)$_4$ (abbreviation as CAu$_6$Cu$_2$-dppy) was achieved by treating CAu$_6$ with [Cu(MeCN)$_4$]BF$_4$ [47]. The CAu$_6$Cu$_2$-dppy clusters exhibit a trigonal prismatic pattern for six Au atoms with two Cu atoms coordinated with top and down. The obtained CAu$_6$Cu$_2$-dppy clusters are strongly emissive in the solid and solution state, and the maximum wavelength is 619 nm with 4.3% of QY. Similar to CAu$_6$Ag$_2$-dppy, the CAu$_6$Cu$_2$-dppy show a long lifetime of 2.1 ms and large stokes shift, which indicate triplet parentage of emission.

To protect the metal core comprehensively and obtain a high QY for metal clusters, Wang et al. introduced PPhpy$_2$ (bis (2-pyridyl)phenylphosphine) ligand with more coordination site to replace the dppy ligand. The emissive [(C)(AuPPhpy$_2$)$_6$](BF$_4$)$_2$ (abbreviation as CAu$_6$-PPhpy$_2$), [(C)(AuPPhpy$_2$)$_6$Ag$_4$](BF$_4$)$_6$ (abbreviation as CAu$_6$Ag$_4$-PPhpy$_2$) and [(C)-(AuPPhpy$_2$)$_6$Ag$_6$(CF$_3$CO$_2$)$_3$](BF$_4$)$_5$ (abbreviation as CAu$_6$Ag$_6$-PPhpy$_2$) were obtained. The maximum wavelength of these three clusters in solid state are 528, 574, and 569 nm respectively, with the QY of 1.5%, 8.0% and 25% successively (Fig. 5.7) [48]. Significantly, the anions trifluoroacetates and tetrafluoroborates are quite helpful for the emission of clusters. The solution of CAu$_6$Ag$_4$-PPhpy$_2$ and CAu$_6$Ag$_6$-PPhpy$_2$ exhibit orange-red and bright yellow emission centered at 603 and 557 nm with the QY of 4.5% and 92%, respectively.

Further, Wang and coworkers use mdppz (2-(3-methylpyrazinyl)-diphenylphosphine) ligand replace dppy ligand, and a novel NbO topology framework structure material was obtained, which exhibit yellowish green light after irradiation and solvatochromic behavior in solid state due to the rigid framework and channels structures [49]. The details will be discussed in the Assembly Section.

In 2013, Kang et al. etched Pt$_1$Ag$_{24}$(SPhMe$_2$)$_{18}$ with 1-adamantanethiolate (HS-Adm) and triphenyl phosphine (PPh$_3$) ligands, and a new nanocluster of Pt$_1$Ag$_{28}$(SAdm)$_{18}$(PPh$_3$)$_4$ was obtained. The Pt$_1$Ag$_{28}$(SAdm)$_{18}$(PPh$_3$)$_4$ have an FCC Pt$_1$Ag$_{12}$ core and four Ag$_4$(SR)$_6$(PPh$_3$)$_1$ motif. The PL of Pt$_1$Ag$_{28}$ increase extremely with the QY from 0.1% to 4.9% compared with Pt$_1$Ag$_{24}$ [50]. The maximum wavelength of Pt$_1$Ag$_{28}$ is 627 nm. Femto-nanosecond transient absorption spectroscopy (fs-ns TA) and time-correlated single-photon counting (TCSPC) indicated that the suppressed phonon emission of Pt$_1$Ag$_{28}$ is responsible for the enhancement of PL.

In subsequent work, Kang and coworkers prepared a Pt$_1$Ag$_{31}$(SR)$_{16}$(DPPM)$_3$Cl$_3$ nanocluster by treating Pt$_1$Ag$_{28}$(SR)$_{18}$(PPh$_3$)$_4$ with Ag$_2$(DPPM)Cl$_2$ complex. After replacement of PPh$_3$ with DPPM, the obtained Pt$_1$Ag$_{31}$ exhibits a 6 fold enhancement of PL compared with the Pt$_1$Ag$_{28}$ mother nanoclusters (Fig. 5.8) [51]. A bright red emission centered

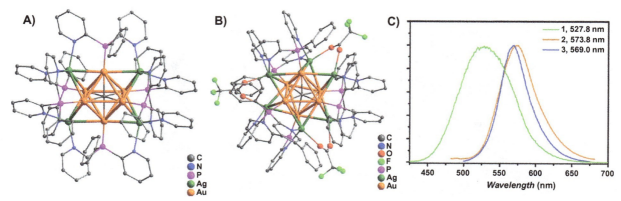

FIG. 5.7 Photoluminescence of [(C)(AuPPhpy$_2$)$_6$Ag$_4$](BF$_4$)$_6$ and [(C)(AuPPhpy$_2$)$_6$Ag$_6$-(CF$_3$CO$_2$)$_3$](BF$_4$)$_5$. (A) Structure of cationic part of [(C)(AuPPhpy$_2$)$_6$Ag$_4$](BF$_4$)$_6$ (2). (B) Structure of cationic part of [(C)(AuPPhpy$_2$)$_6$Ag$_6$-(CF$_3$CO$_2$)$_3$](BF$_4$)$_5$ (3); (C) Emission spectra of 1–3 in the solid state. Right: photos of crystalline 1 (left), 2 (middle), and 3 (right) under 365 nm excitation. *(From Z. Lei, X.-L. Pei, Z.-J. Guan, Q.-M. Wang, Full protection of intensely luminescent gold(I)–silver(I) cluster by phosphine ligands and inorganic anions, Angew. Chem. Int. Ed. 56(25) (2017) 7117–7120, https://doi.org/10.1002/anie.201702522.)*

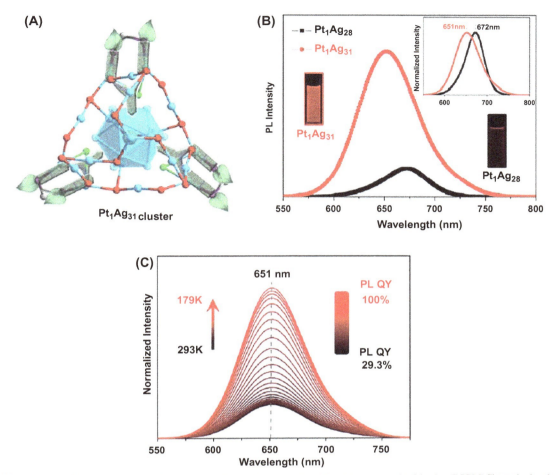

FIG. 5.8 The structure of Pt_1Ag_{31} and its photoluminescence. Schematic illustration of the Y-type growth of the $Ag_2(DPPM)Cl$ terminal and the structure of Pt_1Ag_{31}. In this transformation process, the $Ag-PPh_3$ terminals in Pt_1Ag_{28} are peeled off, and the $Ag_2(DPPM)Cl$ terminals are introduced. Color legends: *dark green sphere*, Pt; *blue sphere*, Ag; *red sphere*, S; *purple sphere*, P; *green sphere*, Cl; *gray sphere*, C. For clarity, all H atoms and some C atoms are omitted. *(From X. Kang, S. Jin, L. Xiong, X. Wei, M. Zhou, C. Qin, Y. Pei, S. Wang, M. Zhu, Nanocluster growth via "graft-onto": effects on geometric structures and optical properties, Chem. Sci. 11(6) (2020) 1691–1697, https://doi.org/10.1039/C9SC05700E.)*

at 672 nm for solution of Pt_1Ag_{31} was observed after UV irradiation. The QY of Pt_1Ag_{31} in DCM solution is 29.3%, significantly higher than that of Pt_1Ag_{28}. Temperature-dependent PL results show that a 4.5 time enhancement of PL as the temperature ranges from 293 to 179 K. The QY of Pt_1Ag_{31} up to 100% after the temperature is below 197 K. The more rigid surface motif of Pt_1Ag_{31} suppresses nonradiative transitions and enhances radiative transitions, and thus enhances the PL of Pt_1Ag_{31} nanoclusters.

Ligand exchange is an effective means to induce structural transformation and property transformation of metal nanoclusters. In the recent work of Kang and coworkers, they achieved the reversible transformation between $Pt_1Ag_{28}(SAdm)_{18}(PPh_3)_4$ (Pt_1Ag_{28}-1, where S-Adm is 1-adamantanethiol) and $Pt_1Ag_{28}(S-c-C_6H_{11})_{18}(PPh_3)_4$ (Pt_1Ag_{28}-2, where $S-c-C_6H_{11}$ is cyclohexanethiol). The FCC core in Pt_1Ag_{28}-1 was changed into icosahedral core in Pt_1Ag_{28}-2. With the transformation of the structure, the PL properties of these two nanoclusters have undergone significant changes. The emission peaks of Pt_1Ag_{28}-1 and Pt_1Ag_{28}-2 are at 672 nm and 720 nm, at room temperature with the QY of 4.9% and 2.7%, respectively [52]. Although the QY of Pt_1Ag_{28}-2 is low at room temperature, the emission increased significantly. The PL intensity increases 63 times as the temperature decreases from 293 to 98 K. The temperature-dependent PL of an intermediate during the ligand exchange process was measured. It is demonstrated that the emission peaks of 680 and 713 nm became sharper and more separated, as well as increased with the decreasement of temperature.

In 2017, Bakr and coworkers reported the bi-metallic $[Pt_2Ag_{23}Cl_7(PPh_3)_{10}]$ nanoclusters that is composed of two icosahedron core [53]. The Pt_2Ag_{23} shows a weak emission, details are discussed in the Alloy and Doping induce PL Section. Based on this work, Kang et al. use a two-step ligand-induced transformation to prepare a brightly emissive Pt_1Ag_{12}

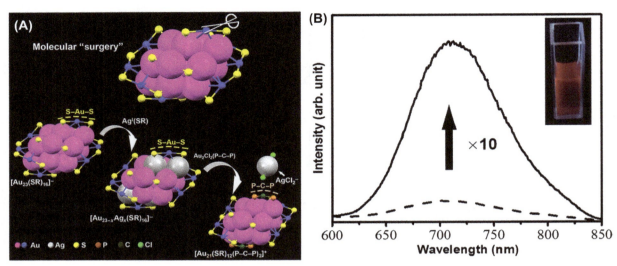

FIG. 5.9 The structure of Au$_{21}$ and its PL. (A) Molecular surgery on the atomically precise 23-gold-atom nanocluster by a two-step metal-exchange method: peeling off two parts of the cluster wrapper and closing the gaps with two P-C-P plasters. (B) PL spectrum of the Au$_{21}$ *(solid line)*; the PL efficiency is enhanced ~10 times compared with Au$_{23}$ *(dashed line)*. Inset shows a photograph of the Au$_{21}$ sample under 365-nm UV light. *(From Q. Li, T.-Y. Luo, M.G. Taylor, W. Shuxin, Z. Xiaofan, S. Yongbo, M. Giannis, N.L. Rosi, J. Rongchao, Molecular "surgery" on a 23-gold-atom nanoparticle, Sci. Adv. 3(5) (2017) e1603193, https://doi.org/10.1126/sciadv.1603193.)*

nanoclusters. In the first step, HS-PhMe$_2$ was used to induce the conversion of Pt$_2$Ag$_{23}$(PPh$_3$)$_{10}$Cl$_7$ to Pt$_1$Ag$_{24}$(SPhMe$_2$)$_{18}$. Next the authors used DPPM ligand to etch the obtained Pt$_1$Ag$_{24}$ nanoclusters. It is speculated that the Ag$_2$(SR)$_3$ motif on the periphery of Pt$_1$Ag$_{24}$ was replaced by DPPM ligands. Various characteristic techniques were employed to verify the composition of Pt$_1$Ag$_{12}$ nanoclusters. The emission peaks of Pt$_2$Ag$_{23}$, Pt$_1$Ag$_{24}$ and Pt$_1$Ag$_{12}$ are centered at 675, 730, and 613 nm, respectively [54]. The deassembled Pt$_1$Ag$_{12}$ nanoclusters display a bright orange red light with a QY of 5.5% that is enhanced 28 times compared with Pt$_2$Ag$_{23}$ nanoclusters.

Wu et al. reported the transformation among Au$_{44}$(TBBT)$_{28}$, Au$_{44}$(2,4-DMBT)$_{26}$, and Au$_{43}$(CHT)$_{25}$ nanoclusters by ligand exchange. A new obtained nanocluster of Au$_{43}$(CHT)$_{25}$ was converted by treat Au$_{44}$(2,4-DMBT)$_{26}$ with cyclohexanethiol ligands, and vice versa. Further, the Au$_{44}$(2,4-DMBT)$_{26}$ and Au$_{44}$(TBBT)$_{28}$ with the same number of metal atoms can also be converted into each other in the presence of a correspond ligands. Particularly, Au$_{43}$(CHT)$_{25}$ can be transferred to Au$_{44}$(TBBT)$_{28}$ after TBBT etching, but Au$_{44}$(TBBT)$_{28}$ cannot be converted to Au$_{43}$(CHT)$_{25}$. The stability of the three clusters from high to low is Au$_{44}$(TBBT)$_{28}$ > Au$_{44}$(2,4-DMBT)$_{26}$ > Au$_{43}$(CHT)$_{25}$. The Au$_{43}$(CHT)$_{25}$ shows the highest PL intensity than the other two gold nanoclusters [55].

In 2017, Li and coworkers used a molecular surgery strategy to modify [Au$_{23}$(SR)$_{16}$]$^-$ nanoclusters into [Au$_{21}$(SR)$_{12}$(P-C-P)$_2$]$^+$ nanocluster. In the first step, Ag(SR) complexes was doped into the [Au$_{23}$(SR)$_{16}$]$^-$ to form a bimetallic [Au$_{23-x}$Ag$_x$(SR)$_{16}$]$^-$ nanocluster. In the second step, Au$_2$Cl$_2$(P-C-P) was added and the surface motif of RS-Au-SR was replaced with P-C-P motifs. A new [Au$_{21}$(SR)$_{12}$(P-C-P)$_2$]$^+$ nanocluster was obtained due to the ligand substitution. Surprisingly, the PL of [Au$_{21}$(SR)$_{12}$(P-C-P)$_2$]$^+$ enhanced 10 times compared with that of [Au$_{23}$(SR)$_{16}$]$^-$ (Fig. 5.9) [56]. The enhancement may be attributed to the enhanced radiative recombination of the electron and hole pair.

5.2.3 Metal doping and alloying

Alloy nanoclusters usually exhibit more excellent and tunable optical properties than monometal nanoclusters, especially fluorescent properties. The alloying process can be achieved in two ways. One is to synthesize a single metal cluster and then dope it, and the other is to directly reduce the bi- or multimetallic salt in the presence of the ligand. For the former, organic M-SR or Au-PR complexes have been added in the solution of as-obtained nanoclusters, and inorganic metallic salts have also been introduced for achieving new doped nanoclusters. For the latter, a series of alloy nanoclusters, such as Au$_{12}$Ag$_{32}$ [57], were prepared by direct reduction of various metal salts in wet chemistry. The intermetallic synergy of alloy nanoclusters usually gives them enhanced PL properties.

Wang et al. doped Au$_{11}$ nanoclusters with Ag-SR complexes and obtained a bi-metallic [Ag$_x$Au$_{25-x}$(PPh$_3$)$_{10}$(SC$_2$H$_4$Ph)$_5$Cl$_2$]$^{2+}$ ($x \leq 13$) nanoclusters. Compared with the Ag$_x$Au$_{25-x}$ species ($x \leq 12$) and Au$_{25}$ nanoclusters,

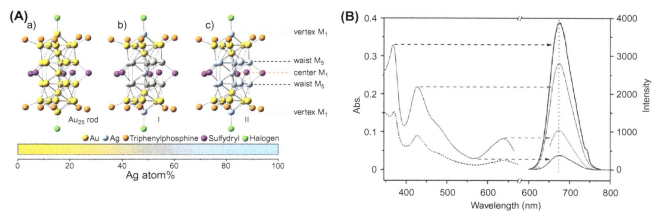

FIG. 5.10 The structure of $[Ag_xAu_{25-x}(PPh_3)_{10}(SC_2H_4Ph)_5Cl_2]^{2+}$ ($x \leq 13$) and its PL. (A) X-ray structures of: (a) $[Au_{25}(PPh_3)_{10}(SC_2H_4Ph)_5Cl_2]^{2+}$, (b) I, and (c) II. M = metal atom. (B) Photoluminescence properties of product II. UV/Vis and excitation spectra (left), and emission spectra (right) at different excitation wavelengths, as indicated by arrows. *(From S. Wang, X. Meng, A. Das, T. Li, Y. Song, T. Cao, X. Zhu, M. Zhu, R. Jin, A 200-fold quantum yield boost in the photoluminescence of silver-doped Ag_xAu_{25-x} nanoclusters: the 13th silver atom matters. Angew. Chem. Int. Ed. 53(9) (2014) 2376–2380, https://doi.org/10.1002/anie.201307480.)*

the Ag_xAu_{25-x} species ($x \leq 13$) show much higher PL intensity, which means the species of $Au_{12}Ag_{13}$ are extremely emissive. The strongest fluorescence emission of $Au_{12}Ag_{13}$ nanocluster is at 680 nm with the QY of 40.1% (Fig. 5.10), which is 400 times that of undoped Au_{25} (QY = 0.1%) [58]. X-ray diffraction results show Ag atoms preferentially occupy the two vertex sites, then the waist site and center position. It is found that the LUMO orbital can be influenced by the 13th Ag atom, higher electron mobility can be observed in $Au_{12}Ag_{13}$.

Alfonso Pedone and coworkers applied the DFT and TD-DFT calculation to investigate the influence of Ag doping into rod Au_{25} on the PL. It is suggested that the first excited state (S1) is affected by the type of metal occupying the central position in the bi-metallic $[Ag_xAu_{25-x}(PPh_3)_{10}(SC_2H_4Ph)_5Cl_2]^{2+}$ ($x \leq 13$) nanoclusters. When the central atom is Ag, the HOMO-LUMO transition is higher, which is responsible for the high PL of $Au_{12}Ag_{13}$ [59]. Femtosecond transient absorption spectroscopy show that $Au_{12}Ag_{13}$ exhibit an ultrafast internal conversion, the nuclear relaxation of $Au_{12}Ag_{13}$ is also much faster than that of Au_{25}. The incorporation of the 13th Ag atom enhances the rigidity of the $Au_{12}Ag_{13}$, thereby enhancing the fluorescence [60].

Unlike the aforementioned adding organometallic complex, Wu and coworkers use antigalvanic reduction (AGR) and treat $Au_{25}(SC_2H_4Ph)_{18}$ with inorganic $AgNO_3$, obtaining new $Au_{25}Ag_2(SC_2H_4Ph)_{18}$ nanoclusters. The composition and structure of $Au_{25}Ag_2$ was determined by UV-vis, MALDI-MS, XRD spectroscopy, and TGA technique. The two Ag atoms adsorb to the motif on the periphery of the Au_{25} nanoclusters. After doping, the PL of $Au_{25}Ag_2$ increase 3.5 times compared with that of $Au_{25-x}Ag_x$ and Au_{25} [61].

In 2015, Bakr and coworkers reported the synthesis and structure of $[Ag_{25}(SR)_{18}]^-$ nanoclusters [62]. Subsequently, Zheng et al. prepared $[PdAg_{24}(SR)_{18}]^{2-}$ and $[PtAg_{24}(SR)_{18}]^{2-}$ nanoclusters that possess a similar structure with Ag_{25} and Au_{25} by direct reduction of metal salts [9].

Further, Bakr et al. synthesis an alloy $[Ag_{25-x}Au_x(SPhMe_2)_{18}]^-$ clusters ($x = 1-8$). ESI-MS was used to determine the number of Au atoms in the alloy nanoclusters. By introducing a postsynthetic galvanic replacement, they obtained pure $AuAg_{24}(SPhMe_2)_{18}$ cluster. The alloy process affects the optical properties of the clusters. The absorption peak of $AuAg_{24}$ was blue-shifted compared with that of Ag_{25}. Accordingly, the wavelength of emission of $AuAg_{24}$ was blue-shifted, and the PL intensity of $AuAg_{24}$ is more than 25 times higher than that of Ag_{25} [63]. Transient absorption results indicate that the excited lifetime of $AuAg_{24}$ is higher after single Au atom doping. In their later work, $[PdAg_{24}(SR)_{18}]^{2-}$ and $[PtAg_{24}(SR)_{18}]^{2-}$ nanoclusters were used as stating material, and Au^+ ions was introduced for alloy. Differently, two bimetallic $[AuAg_{24}(SR)_{18}]^-$, $[Au_2Ag_{23}(SR)_{18}]^-$ nanoclusters were obtained by replacing Pd and Ag with Au, and trimetallic $[Au_xPtAg_{24-x}(SR)_{18}]^{2-}$ ($x = 1-2$) was produced when doping Au into $PtAg_{24}$ nanoclusters [64].

The reason for the increase in fluorescence caused by doping is very worth exploring. Aikens investigated the origin of PL of $Ag_{25}(SR)_{18}^-$, and the ligand and doping effects on the PL of $Ag_{25}(SR)_{18}^-$ nanoclusters by DFT and TD-DFT. It is found that the Stokes shift changed from 0.37 to 0.24 eV when replacing H with aromatic ligands. Doping with heteroatom has an effect on the emission energies and radiative lifetime of S1 states. The PL of Ag_{25} and $AuAg_{24}$ arises from the HOMO-LUMO transition. The extended lifetime of the excited state is the reason for the increase in fluorescence caused by doping [65].

Doping with different metal complexes can result in clusters of different shapes. Kang et al. doped [$Pt_1Ag_{24}(SPhMe_2)_{18}$] with Au-SR complexes, forming a spherical [$Pt_1Au_{6.4}Ag_{17.6}(SPhMe_2)_{18}$] nanocluster, whereas doped with Au-PPh$_3$Br complexes leading to a rod [$Pt_2Au_{10}Ag_{13}(PPh_3)_{10}Br_7$] nanocluster. The PL results reveals that the emission peaks of $Pt_2Au_{10}Ag_{13}$, Pt_1Ag_{24} and Ag_{25} are at 648, 728, and 809 nm, respectively. The QY of $Pt_2Au_{10}Ag_{13}$ is 14.7% that is 150 times higher than 0.1% of Pt_1Ag_{24} [66].

In the recent work of Yuan et al., the fluorescence properties of single metal doped and multimetal doped Ag_{25}, Au_1Ag_{24}, Pt_1Ag_{24}, Au_xAg_{25-x} and $Pt_1Au_xAg_{24-x}$ have been studied. It is found that single Au and Pt atom was doped in the center of metal core, and the extra Au atoms were doped on the surface of core and/or motif. The PL intensities and lifetime of Au_1Ag_{24} and Pt_1Ag_{24} are higher than that of Au_{25}, Au_xAg_{25-x} and $Pt_1Au_xAg_{24-x}$. Temperature-dependent PL results show that the PL intensity of Au_1Ag_{24} increase with the decrease in temperature more obviously than that of Ag_{25} and Au_xAg_{25-x}. It can be concluded that the inward contraction of free valence electrons causes the fluorescence to increase, while the outward expansion causes the fluorescence to decrease [67].

The fluorescence properties of alloy clusters will be affected by the degree of alloying and the number of constituent units. In 2016, Bakr prepared [$Pt_2Ag_{23}Cl_7(PPh_3)_{10}$] nanocluster that is composed of two Pt_1Ag_{12} units [53]. The Pt_2Ag_{23} is emissive. Kang et al. deassembled the cluster to obtain Pt_1Ag_{12} cluster [54]. Recently, Yang et al. prepared a cyclic [$Pt_3Ag_{33}(PPh_3)_{12}Cl_8$]$^+$ nanocluster that is constituted of three Pt_1Ag_{12} building block by sharing a vertex Ag atom. The absorption and emission peak red-shifts as the number of assembled monomers increase. Fluorescence intensity decreases with the increase in assembled units. Interestingly, when doped Au atoms into Pt_3Ag_{33}, the obtained [$Pt_3Au_{12}Ag_{21}(PPh_3)_{12}Cl_8$]$^+$ is unemissive. The synergy between different metals is conducive to obtain strongly emissive clusters [68].

It is reported that $Ag_{29}(BDT)_{12}(TPP)_4$ cluster has weak PL intensity and low QY (0.9%). Bakr et al. used a direct synthesis involving the coreduction of Au and Ag salts in the presence of ligands. As the content of Au increases, the proportion of Au in the clusters increases. When the Au precursor concentration is 40 mmol%, the fluorescence QY of the cluster is as high as 24%. A red shift of PL emission was observed as the increase in Au concentration. Femto-nanosecond transient absorption (fs-ns TA) and time-correlated single-photon counting (TCSPC) experiments show that the heteroatoms doping perturbed the electronic relaxation and dynamics, and further lead to a long lifetime and thus a high PL intensity. The P NMR results indicated that the possible locations of doped Au atoms are at the Ag atoms connected to the TPP ligands [69].

After the experimental results are reported, Hannu Häkkinen et al. studied the stability, electronic structure and PL of Au-doped [$Ag_{29-x}Au_x(BDT)_{12}(TPP)_4$]$^{3-}$ (x = 0–5) nanoclusters. The results of DFT and linear response time-dependent DFT (LR-TDDFT) indicated the coexistence of $Ag_{29-x}Au_x$ nanoclusters with various Au atom dope. The nanoclusters with Au-TPP bond have important roles in increasing the electronic transition at ~450 nm and stability [70].

In 2016, Marte van der Linden identified the [$Ag_{29}(LA)_{12}$]$^{3-}$ clusters by using a serious of characterization technology including MS spectrometry, optical spectroscopy, and analytical ultracentrifugation. These nanoclusters showed a red emission with the QY of 5%. The Ag_{29} clusters display high stability, retaining the red emission more than 18 months in dark [71]. If exposed to light, the degradation rate will increase and the fluorescence will decrease. But if NaBH$_4$ is added, the fluorescence will be restored, indicating a regeneration of Ag_{29} nanoclusters.

de Groot et al. synthesis the bi-metallic $Au_1Ag_{28}(LA)_{12}^{3-}$ nanoclusters and probe the alloy effect on PL of Ag_{29} nanoclusters in 2018. It is found that the QY of clusters is 7.9% when the concentration of doped Au is 2.9%. The lifetime of the cluster prepared with 14% Au is 2.6 ms that is shorter than 4 ms of pure Ag cluster. Low temperature PL experiments showed that the excitation peaks of 400 and 460 nm and emission peak of 660 nm are belonged to Au_yAg_{29-y} clusters. The luminescence decay fitting results suggest a shorter lifetime and an overall distribution of faster decay after doping with 3.6% Au. It is hard to separate the excitation and emission spectra of Ag_{29} and Au_yAg_{29-y}. X-ray absorption spectroscopy XANES and EXAFS indicated that the doped Au atom occupy the central position in Ag_{29} clusters [72].

In 2012, Hedi Mattoussi and coworkers used direct synthesis method to prepare Ag nanoclusters and Ag nanoparticles protected by LA appended with PEG and different terminal group. It is found luminescent Ag nanoclusters can be prepared by direct reduction of Ag salt or size-focusing method [73]. In the later work, Hedi Mattoussi et al. investigated the doping effect on PL for Au nanoclusters (doped with Ag or Cu) and Ag nanoclusters (doped with Au) protected by LA and DHLA. It is found that the doping of Ag or Cu into Au nanoclusters lead to decrease in the PL of alloy nanoclusters. Whereas, when the molar fraction of Au is 8%, the PL of alloy nanocluster is highest, sixfold enhancement, compared with the pure Ag and other AuAg alloy nanoclusters. Similar to the situation of Ag_{29} protected by LA ligand, the alloy AuAg nanoclusters protected by LA-PEG-OMe and DHLA-PEG-OMe show an enhanced PL intensity. By using the phase transfer catalyst, the PL in aqueous can be transferred into organic solution [74].

In addition to the experimental research progress, theoretical calculations also give the fundamental reason for the doping effect on fluorescence enhancement. Michael Dolg et al. employed DFT and TD-DFT methods to study the PL

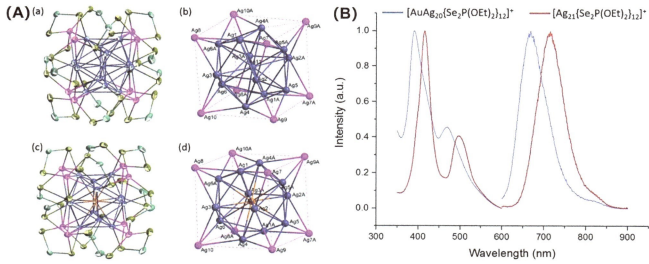

FIG. 5.11 The structure of Ag$_{20}$ and AuAg$_{20}$ and their PL. (A) (a) Total structure of the cationic cluster 4 with ethoxy groups omitted for clarity. (b) The centered icosahedron Ag$_{13}$ inscribed in an Ag$_8$ cube. (c) Total structure of the cationic cluster 5 with ethoxy groups omitted for clarity. (d) The Ag$_8$ cube encapsulating a centered-icosahedral Au@Ag$_{12}$ unit. Au *orange*, Ag *blue/pink*, Se *yellow*, P *green*. (B) UV/Vis absorption (left) and normalized emission spectra (right) of 4 and 5 in chloroform. *(From W.-T. Chang, P.-Y. Lee, J.-H. Liao, K.K. Chakrahari, S. Kahlal, Y.-C. Liu, M.-H. Chiang, J.-Y. Saillard, C. W. Liu, Eight-electron silver and mixed gold/silver nanoclusters stabilized by selenium donor ligands. Angew. Chem. Int. Ed. 56(34) (2017) 10178–10182, https://doi.org/10.1002/anie.201704800.)*

properties of Ag$_{29-x}$Au$_x$ ($x=$0–5) nanoclusters [70]. It is found that the highly doped Ag$_{26}$Au$_3$, Ag$_{25}$Au$_4$, and Ag$_{24}$Au$_5$ are more luminescent. The relativistic effects and heternuclear Au—Ag bonds are responsible for the enhanced PL of Ag$_{29-x}$Au$_x$ ($x=$1–5) nanoclusters.

The alloying process can not only adjust the fluorescence properties of the material, but also the assembly process of the monomer. Zhu group prepared the alloyed Au$_1$Ag$_{22}$(SAdm)$_{12}^{3-}$ and Ag$_{13}$Cu$_{10}$(SAdm)$_{12}^{3-}$ nanocluster monomer. Interestingly, Au$_1$Ag$_{22}$(SAdm)$_{12}^{3-}$ is assembled into framework via SbF$_6^-$ anion, which will be discussed in the following section. [Ag$_{13}$Cu$_{10}$(SAdm)$_{12}$]$^{3-}$ displayed red emission centered at ∼630 nm. The PL increased as the temperature decrease from 300 to 190 K [75]. Beside, Zhu et al. prepared thiolate/phosphine coprotected Ag$_{50}$(Dppm)$_6$(SR)$_{30}$ nanocluster and doped into Au atoms. It is observed the Au$_x$Ag$_{50-x}$(Dppm)$_6$(SR)$_{30}$ alloy nanoclusters exhibited enhanced PL property [76].

In the study of fluorescent alloy nanoclusters, Liu et al. have done novel work. In 2017, Liu et al. prepared Ag$_{20}$[Se$_2$P(OiPr)$_2$]$_{12}$ and Ag$_{21}$[Se$_2$P(OEt)$_2$]$_{12}^+$ by treating Ag$_{20}$[S$_2$P(OiPr)$_2$]$_{12}$ and Ag$_{21}$[S$_2$P(OiPr)$_2$]$_{12}^+$ with Se-donor ligands, respectively. The PL band of Ag$_{20}$[Se$_2$P(OiPr)$_2$]$_{12}$ and Ag$_{20}$[S$_2$P(OiPr)$_2$]$_{12}$ are centered at ∼720 and ∼840 nm. By doping Ag$_{21}$[Se$_2$P(OEt)$_2$]$_{12}^+$ with Au(PPh$_3$)Cl, a bi-metallic AuAg$_{20}$[Se$_2$P(OEt)$_2$]$_{12}^+$ nanocluster was prepared. The absorption spectrum and emission spectrum profiles of AuAg$_{20}$[Se$_2$P(OEt)$_2$]$_{12}^+$ are similar to that of Ag$_{21}$[Se$_2$P(OEt)$_2$]$_{12}^+$. A blue-shift of PL was observed for AuAg$_{20}$ compared with Ag$_{21}$ (Fig. 5.11) [77].

In their later work, Liu et al. doped Au into Ag$_{20}$ template, an alloy nanocluster AuAg$_{19}$[S$_2$P(OnPr)$_2$]$_{12}$ was prepared. The obtained AuAg$_{19}$[S$_2$P(OnPr)$_2$]$_{12}$ can be transformed into AuAg$_{20}$[S$_2$P(OnPr)$_2$]$_{12}^+$ nanoclusters by adding Ag salt. The maximum emissions wavelength of Ag$_{20}$, AuAg$_{19}$ and AuAg$_{20}$ are 835, 782, and 745 nm, respectively. Doping make the PL intensity of AuAg$_{19}$ and AuAg$_{20}$ enhanced 8 times [78]. Besides, they studied the Ag and Au doping effect on [Cu$_{13}$(S$_2$CN^{n}Bu$_2$)$_6$(C/CPh)$_4$][CuCl$_2$] nanoclusters. AuCu$_{12}$ nanoclusters exhibit bright yellow emission at both solid and solution states. The emission peak of 640 nm for AuCu$_{12}$ at 297 K blue-shifted to 616 nm at 77 K. The QY of AuCu$_{12}$ is 59% at 77 K. The maximum PL wavelength of AuCu$_{12}$ solid is at 730 nm [79].

5.2.4 Aggregation-induced emission

Aggregation-induced luminescence refers to a phenomenon in which molecules that do not emit light or emit weak light in a solution have a significant increase in luminescence after aggregation. The concept of AIE was coined by Tang's group in 2001 [80–82]. In a dilute solution, there are active vibrations and rotations inside the AIE molecules. When these molecules absorb energy, various vibrations and rotations dissipate the energy, so the emission is quite weak. When these molecules are gathered together, the interaction between each other restricts the internal motion of the molecules. The dissipation of energy by various vibrations and rotations is hindered due to restriction of intramolecular rotation (RIR), and the radiation transition is enhanced, thus showing the phenomenon of enhanced luminescence.

In 2012, Xie et al. found the mechanism of AIE in Au-thiolate nanoclusters, which is to make nonluminescent oligomeric Au(I)-S complexes aggregate on the surface of Au(0) core [83,84]. The PL intensity and color depend on the degree of aggregation, i.e., the ratio of good solvent and poor solvent or caution. It is found that when the poor solvent ethanol content is 95%, the Au nanoclusters show the strongest fluorescence (Fig. 5.12) [84]. The QY of aggregated Au nanoclusters is up to ~15%. The obtained intra- and intercomplex aurophilic interaction between closed-shell metal centers was responsible for the enhanced emission. On the other hand, the interaction between cautions and ligands lead to the cross links and further aggregation that intern induce the formation of aurophilic bonds and dense aggregates. The exploited procedure is used for preparing luminescent Au nanoclusters capped with various types of thiolate ligands.

In the later work of Xie et al., they used a well-studied $Au_{18}(SG)_{14}$ as a parental Au nanoclusters that possess weak luminescence with the maximum wavelength of 800nm. After adding a certain amount of Ag(I) ions, the characteristic absorption peak of as-prepared $Au_{18}(SG)_{14}$ disappeared, indicating a reaction of $Au_{18}(SG)_{14}$ and added Ag ions. After 15 min of reaction, a bright red emission was observed due to the AIE process. The emission peak of obtained Au@Ag nanoclusters is at 667nm with the QY of ~6.8%. PL lifetime of Au@Ag nanoclusters can reach microsecond level. It is hypothesized that the connection of Au nanocluster and Ag(I) linker and formed large Au-Ag-S motif are responsible for the strong red emission. Besides, the PL can be quenched by adding the thiol ligands with strong bond capability (Cys) [85].

In 2015, Xie et al. further employed the AIE technique for luminescent Ag nanoclusters. They synthesis a core-shell Ag(0)@Ag(I) nanocluster protected by thiol in boiling water, in which high temperature is beneficial for overcoming the reaction barrier and obtaining highly stabile nanoclusters. The as-obtained Ag(0)@Ag(I) nanocluster show strong red emission with an emission peak of 686nm. The QY is ~7.4%. The microsecond PL lifetime hints the AIE luminescence

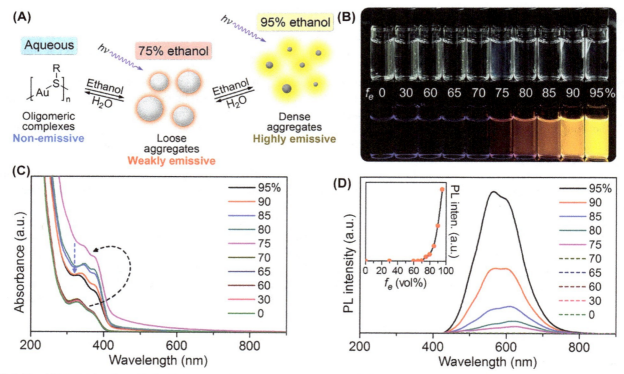

FIG. 5.12 AIE properties of metal NCs. (A) Schematic illustration of solvent-induced AIE properties of oligomeric Au(I)-thiolate complexes. (B) Digital photos of Au(I)-thiolate complexes in mixed solvents of ethanol and water with different Fe under visible (top row) and UV (bottom row) light. (C) UV-vis absorption and (D) photoemission spectra of Au(I)-thiolate complexes in mixed solvents with different Fe. (Inset) Relationship between the luminescence intensity and Fe. The spectra were recorded 30 min after the sample preparation. *(From Z. Luo, X. Yuan, Y. Yu, Q. Zhang, D.T. Leong, J.Y. Lee, J. Xie, From aggregation-induced emission of Au(I)–thiolate complexes to ultrabright Au(0)@Au(I)–Thiolate Core–Shell Nanoclusters, J. Am. Chem. Soc. 134(40) (2012) 16662–16670, https://doi.org/10.1021/ja306199p.)*

type in Ag(0)@Ag(I) nanocluster. The excellent stability was observed in as-obtained Ag nanoclusters due to the high temperature procedure [86].

As mentioned above, the ratio of mixed solvents has an important influence on the AIE properties of the clusters. In this context, Yong Jiang et al. studied the solvent effect on the luminescence of Au nanoclusters. It is found that the obtained Au(I)-$SC_{12}H_{25}$ complexes in toluene exhibit abroad red emission band at 410–440 nm and a band peaked at 620 nm. The PL decreased with the increase in ethanol content. The PL of Au(I)-$SC_{12}H_{25}$ complexes in toluene are attributed to the aggregation of the bilayer Au(I)-SR structure and aurophilic Au...Au interactions [87].

In 2017, Katsuaki Konishi and coworkers reported the aggregation-induced fluorescence to phosphorescence of Au_8 nanoclusters. The Au_8 nanoclusters exhibited an emission band at 596 nm with a lifetime of 55 ps, indicating the fluorescence property of Au_8 in solution state. After adding the poor solvent of MeOH, the clusters aggregate and show a near-IR emission with bands at 708 and 596 nm. The solid of Au_8 showed a 708 nm band emission that corresponds to microsecond-order lifetime of 3.7 (35%) and 0.68 (65%) μs, which indicating the phosphorescence character. The QY of Au_8 in solid state is two orders of magnitude than that in solution. The aggregation process significantly regulates the emission wavelength and improves the PL intensity and QY for Au_8 nanoclusters [88].

Recently, Zhu and coworkers prepared nonfluorescent Au_4Ag_5(dppm)$_2$(SAdm)$_6$(BPh$_4$) nanocluster. The obtained Au_4Ag_5 nanoclusters show no obvious PL in MeOH solution. After introducing the poor solvent of H_2O, the fluorescence Au_4Ag_5 nanocluster gradually increases. When the proportion of water is 95%, (fw=95%) the QY and lifetime of the Au_4Ag_5 nanocluster are 10% and 2.1266 ms, respectively. However, the QY of Au_4Ag_5 is 4% in the solid state. X-ray absorption fine structure spectroscopy (XAFS) and absorption spectra show that Au_4Ag_5 is unchanged in the solution and solid state. The packing structure of Au_4Ag_5 reveals that the restriction of intramolecular motions (RIM) might be responsible for the AIE property of Au_4Ag_5 nanoclusters [89].

Not only Au/Ag nanoclusters can produce AIE, AIE and AIEE also observed in Cu nanoclusters. Wang et al. prepared a weakly luminescent Cu_2 nanocluster by size-focusing process. The as-obtained Cu nanoclusters protected by glutathione showed a faint emission in aqueous solution. After introduction of 80 vol% EtOH, the PL of Cu nanoclusters increase 14.7 times. The QY of this aggregation state is 6.6%, which is much higher than 0.45% in aqueous solution. A strong AIEE effect was observed in Cu nanoclusters protected by GSH or Cys ligands [90].

In 2017, Hao Zhang et al. studied the AIE of Cu nanocluster protected by dodecanethiol. After adding EtOH or MeOH, $Cu_{14}(DT)_{10}$ nanoclusters self-assembly, forming well-defined nanosheets with an approximate width of 50 nm and a length of 200–300 nm. The as-obtained nanosheets showed a bright yellow emission with the maximum wavelength of 550 nm [91]. The absolute QY is 15.4%, much higher than unassembled Cu NCs and the assembled NCs in the absence of ethanol. Low temperature PL experiment indicated that two emission state of 490 (T1 state) and 550 nm (T2 state) are coexisting. It is demonstrated that T1 state at 490 nm belongs to the real LMMCT-determined triplet state, and T2 state at 550 nm belongs to the defects-related emission. Block assemblies of Cu NCs with low specific surface area and minimized metal defects showed emission centered at 490 nm at room and low temperature. Computer simulation further confirms the defects effect on the AIE of Cu nanoclusters [91].

Controlling the aggregation and deaggregation of nanoclusters can control the optical properties related to AIE. Dan Li and coworkers achieved a reversible aggregation process in the sandwich-like Au_3-Ag-Au_3 clusters. A bright yellow emission was observed in Au_3-Ag-Au_3 clusters that are synthesized by a mechanochemical procedure. In addition, the reversible mechanochromic process was captured. It is demonstrated that the formation and breaking of the labile Au—Ag bonding is responsible for the mechanochromism behavior [92].

In the AuCu alloy nanocluster system, Kang et al. assemble render nonluminescent Cu-SR complex into Au(0), forming a red emissive Au_2Cu_6(PPh$_2$Py)$_2$(SAdm)$_6$ nanoclusters. The Au_2Cu_6(PPh$_2$Py)$_2$(SAdm)$_6$ was obtained by reducing the nonemissive Cu-SR complexes in the presence of Au(I) species (Fig. 5.13) [93]. In the contrast experiment, no emission was observed in the absence of Au(I), which indicates the aggregative Au@Cu nanoclusters are highly preferable for the red emission. The QY of obtained Au_2Cu_6(PPh$_2$Py)$_2$(SAdm)$_6$ is 11.7%. Experimental and theoretical results reveal that RIR-based AIE strategy and LMCT process are responsible for the strong fluorescence [93].

Crystal is a special form of aggregation, and the phenomenon of crystal-induced emission (CIE) or crystal-induced emission enhancement (CIEE) has also been widely studied. Chen et al. reported the first metal nanocluster of Au_4Ag_{13}(DPPM)$_3$(SR)$_9$ that exhibits unique CIEE property [94]. The Au_4Ag_{13} nanoclusters were synthesized by a one-pot method. The solution of Au_4Ag_{13} is nonemissive, and the amorphous state of Au_4Ag_{13} is weakly emissive. Interestingly, the crystalline state showed a strong emission centered at 695 nm. The QY of amorphous and the crystal state is 1.14% and 2.68%, respectively. Experiment and calculation results indicated that the triblade fan configuration and multiple weak interactions are responsible for the CIEE process due to the restriction of molecular vibrations and rotations [94].

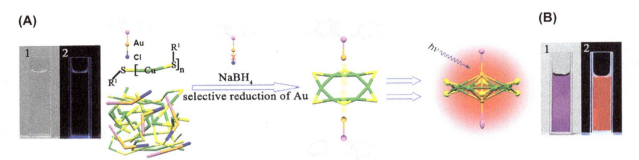

FIG. 5.13 The structure of Au_2Cu_6 and its PL. (A) Illustration of the Au_0-induced aggregation of $CuSR_1$. Au^0 was generated by the selective reduction of $AuPR_2Cl$ with $NaBH_4$. (B) Digital photographs of the corresponding complexes or NCs under visible (1) and UV light (2). Au *gold*, Cl *blue*, Cu *green*, P *violet*, S *yellow*. (From X. Kang, S. Wang, Y. Song, S. Jin, G. Sun, H. Yu, M. Zhu, Bimetallic Au_2Cu_6 Nanoclusters: strong luminescence induced by the aggregation of copper(I) complexes with gold(0) species, Angew. Chem. Int. Ed. 55(11) (2016) 3611–3614, https://doi.org/10.1002/anie.201600241.)

Thalappil Pradeep and coworkers reported a CIEE-feathered $[Ag_{22}(dppe)_4(2,5\text{-DMBT})_{12}Cl_4]^{2+}$ nanoclusters. The crystal of Ag_{22} showed strong red emission centered at ~670 nm, which is 12 times higher than that of the amorphous powder. The extended C–H⋯π and π⋯π interactions are contributed to the highly emissive in the crystalline state [95].

The intermolecular forces in the aggregate state have an important influence on the luminescence of the aggregate state. Thalappil Pradeep changed the crystallization method of $Ag_{29}(BDT)_{12}(TPP)_4^{3-}$ by replacing the slow evaporation of DMF with vapor diffusion of MeOH into DMF solution. A trigonal (T) crystal instead of cubic (C) crystal was obtained. It is found that the PL intensity of the cubic nanoclusters are higher than that of trigonal nanoclusters, which is attributed to the stronger inter C–H⋯π interactions and more rigid structures in trigonal system. A 30 nm of red-shift was observed in trigonal nanoclusters compared with cubic nanoclusters [96].

Recently, Xie et al. employed the aurophilic interactions to constitute nanoribbons based on the self-assembly of ultrasmall $[Au_{25}(p\text{-MBA})_{18}]^-$ nanoclusters by cyclic dialysis. The obtained nanoribbons composed of the building block with longer SR-[Au^I-SR]x motifs ($x > 2$) and a smaller Au(0) core exhibited a red emission at room temperature with the QY of 6.2%. At 77 K, a bright white emission centered at 483 and 620 nm was observed with microsecond-scale lifetimes (19.8 ms for 483 nm, 10.5 ms for 620 nm). The π-π stacking interactions facilitated to the assembly of nanoclusters and further contributed to the optical enhancement [97].

The aggregation and deaggregation of ligands on the surface of the cluster can affect the AIE of the cluster. It is observed the TPP ligands on $Ag_{29}(BDT)_{12}(TPP)_4$ are dynamic aggregation and deaggregation based on the multisignals in the ESI spectrum of $Ag_{29}(BDT)_{12}(TPP)_4$. Kang et al. found that the PL intensity of $Ag_{29}(BDT)_{12}(TPP)_4$ increase with the addition of TPP ligands. $Ag_{29}(BDT)_{12}(TPP)_4$ showed a 13 times QY enhancement with the QY from 0.9% to 11.7%. Low temperature PL experiments showed that two stages enhancement and one stage enhancement of PL as the decreasement of temperature were identified in $Ag_{29}(BDT)_{12}(TPP)_4$ and $Pt_1Ag_{28}(S\text{-Adm})_{18}(TPP)_4$, respectively. The first stage enhancement in the low temperature PL results of $Ag_{29}(BDT)_{12}(TPP)_4$ is associated with the inhibition of nonradiative transitions. The novel mechanism that involves the restriction of ligand dissociation-aggregation process is highly favorable to get enhanced PL property in nanoclusters [98].

5.2.5 Others

The environment of the nanocluster dissolved in or seated in, such as solvent and temperature, has important influences on the PL properties of the nanocluster. In 2017, Wu et al. synthesized and structurally determined $Au_{60}S_6(SCH_2Ph)_{36}$ nanoclusters, which exhibited morphological dependence and solvent polarity dependence characteristics for PL property. The PL intensity of amorphous of $Au_{60}S_6(SCH_2Ph)_{36}$ is 1.7 folds lower than that of crystalline of $Au_{60}S_6(SCH_2Ph)_{36}$, which is attributed to the energy transfer among the 6HLH arranged $Au_{60}S_6(SCH_2Ph)_{36}$ nanoclusters (Fig. 5.14). Besides, the maximum fluorescence emission wavelength blue shifts with the increase in solvent polarity, and the intensity decreased with the increase in solvent polarity [99]. In their following work, $Au_{60}S_7(SCH_2Ph)_{36}$ was identified by treating with $Au_{60}S_6(SCH_2Ph)_{36}$ with excess thiol ligands at 100°C. The obtained $Au_{60}S_7(SCH_2Ph)_{36}$ nanoclusters showed "ABAB" arrangement and an enhanced PL of amorphous [100].

The state of the cluster, the crystal state, and the type of solvent in which the cluster is dissolved, have an important influence on the fluorescence properties of the cluster. Wu and coworkers investigated the crystal PL of $MAg_{24}(SR)_{18}$ (M = Ag/Pd/Pt/Au) nanoclusters. A blue shift of emission wavelength and an enhancement of PL intensity for crystal

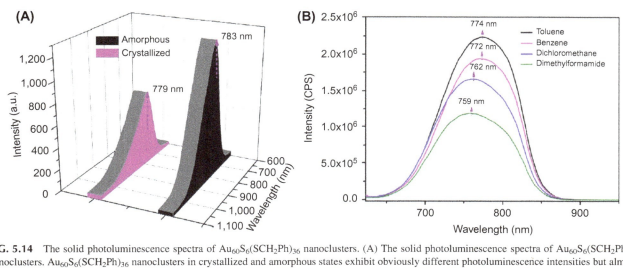

FIG. 5.14 The solid photoluminescence spectra of $Au_{60}S_6(SCH_2Ph)_{36}$ nanoclusters. (A) The solid photoluminescence spectra of $Au_{60}S_6(SCH_2Ph)_{36}$ nanoclusters. $Au_{60}S_6(SCH_2Ph)_{36}$ nanoclusters in crystallized and amorphous states exhibit obviously different photoluminescence intensities but almost identical emission spectrum profiles. Note: the excitation wavelength $l_{ex} = 514$ nm. (B) The solvent-polarity-dependent solution photoluminescence of $Au_{60}S_6(SCH_2Ph)_{36}$ nanoclusters. Not only the maximum emission wavelengths but also the photoluminescence intensities of $Au_{60}S_6(SCH_2Ph)_{36}$ nanoclusters in solution are dependent on the solvent polarity. Note: the excitation wavelength $l_{ex} = 514$ nm. *(From Z. Gan, J. Chen, J. Wang, C. Wang, M.-B. Li, C. Yao, S. Zhuang, A. Xu, L. Li, Z. Wu, The fourth crystallographic closest packing unveiled in the gold nanocluster crystal, Nat. Commun. 8(1) (2017) 14739, https://doi.org/10.1038/ncomms14739.)*

PL was observed in the $AuAg_{24}$, $PtAg_{24}$ and $PdAg_{24}$ compared with the Ag_{25}. Besides, the solvents have an effect on the PL intensity for these nanoclusters. In acetonitrile, the PL intensity follow the sequence of $PtAg_{24} > AuAg_{24} > PdAg_{24} > Ag_{25}$. Surprisingly, the QY of $PtAg_{24}$ is 18.6% in acetonitrile, which is ~100 times higher than that in dichloromethane (0.2%). Experimental and theoretical results indicated a charge-transfer PL mechanism for these four nanoclusters [101].

In 2018, Xing Lu and coworkers constructed a $Ag_{51}(tBuC{\equiv}C)_{32}$ nanocluster, which exhibited a strong solvatochromic effect. In DCM, the Ag_{51} showed a blue emission with the emission peak centered at 418, 436, and 466 nm, and in chloroform, the Ag_{51} displayed similar emission profile with the emission peak at 413, 437, and 469 nm. Whereas, the emission shift to the maximum emission wavelength of 570 and 656 nm in methanol and acetonitrile, respectively [102].

On the other hand, the counterions of metal nanoclusters play an important role for the PL property of nanoclusters. Jean-Marie Basset and coworkers investigated the counterions effect on the PL based on the structurally precise $[PtAg_{28}(BDT)_{12}(PPh_3)_4]^{4-}$ and $[PtAg_{28}(S\text{-}Adm)_{18}(PPh_3)_4]^{2+}$ nanoclusters. It is found that $[PtAg_{28}(BDT)_{12}(PPh_3)_4]^{4-}$ displayed a weak PL with emission peak centered at 750 nm. After combined with TOA counterions, the PL strongly enhanced 17.6 times. A 70 nm blue-shift of emission was observed similar enhancements were also observed in other counterions system, the magnitude of the enhancement was larger for bulkier counterions: TOA > TBA > CTA > TPP. Experimental and theoretical results indicated that the bulky counterions occupy a significant volume and rigidify the ligand shell, leading to the enhancement of PL [103]. On the contrary, ions can quench the PL of metal nanoclusters. Zheng and coworkers reported the $[Au@Ag_8@Au_6(C{\equiv}CtBu)_{12}]^+$ nanoclusters. These alloy nanoclusters can be doped with alkali metal atoms (Na, K, Rb, Cs). The parent cluster exhibited strong emission centered at 818 nm and with a QY of 2%. It is found that the emission can be quenched on substitution with a Naion [104].

Compounds with strong binding ability to metals are expected to induce cluster transformation to form clusters with enhanced optical properties. Recently, Zang and coworkers reported the transformation from Ag_{18} to Ag_{62} nanoclusters by the inducing of hydrogen sulfide, a high-construe PL turn on response was observed accompanied with the transformation [105]. Time-dependent ESI-MS and UV-vis spectra were employed to trace the transformation process by introducing the NaSH. It is observed the red emission centered at 603 nm associated with Ag_{62} nanoclusters enhanced with the increase in the amount of NaSH. It is calculated that the LOD for H_2S gas is 0.13 ppm. The fast response time of 30 s, low detection time and high sensitivity for H_2S is highly desirable for the promising application of metal nanoclusters [105].

Low temperature can suppress the vibration and rotation of the clusters, thereby enhancing the fluorescence performance. Di Sun and Lan-Sun Zheng et al. prepared various Ag and Au nanoclusters with strong thermochromic behavior, including Ag_{18} [106], Ag_{38} [107], Ag_{44} [108], Ag_{45} [109], Ag_{46} [110], Ag_{50} [108], Ag_{52} [111], Ag_{66} [109], Ag_{73} [109], and Ag_{80} [112] nanoclusters. In addition, Zang et al. reported the thermochromic luminescent Ag_{12} [113], Ag_{31} and Ag_{33} nanoclusters [114]. As the temperature decreases, the PL of these nanoclusters increases. For example, Ag_{80} is nonemissive at room temperature. It emits red emission at 77 K [112]. An 18 times enhancement of PL intensity and a blue-shift of emission peak were observed as the temperature decrease. The nanocluster of [S@Ag$_{18}$(tBuC$_6$H$_4$S)$_{16}$(dppp)$_4$]·DMF·5CH$_3$CN·3CH$_3$OH in solid state showed red emission with a emission band centered at 685 nm at room temperature, a new emission peak centered at 590 nm was observed when the temperature was below 148 K [110]. The life time of microsecond was associated with the emission at 590 nm, which is attributed to the LMCT process. The life time of nanosecond was assigned to the emission at 690 nm, which is attributed to a metal centered (d10→d9s1) excited state modified by Ag⋯Ag interactions.

5.3 Magnetism of metal nanoclusters

Single gold, silver, platinum, and copper atoms are paramagnetic because of the existence of unpaired electron. Bulk platinum is paramagnetic, whereas bulk gold, silver, and copper are diamagnetic. Single mercury atom and palladium atom are diamagnetic because there are no unpaired electrons, but bulk palladium is paramagnetic, while liquid mercury is diamagnetic. In addition, iron, cobalt, and nickel are the most common ferromagnetic metals. Besides, rare earth metal atoms have an under filled 4f electron layer and are shielded by the outer 5s and 5d electron layers, which can generate stronger electron spin and orbital magnetic moments. Therefore, rare earth metals are often used to synthesize permanent magnet materials. The evolution of magnetism from single atoms to metal nanoparticles is significant of understanding magnetism. Nanoclusters, a bridge connecting atoms and larger nanoparticles, are considered to be an ideal system for studying the origin and evolution of magnetism. The magnetism of metal nanoclusters comes from the unpaired electrons inside the metal core. The electrons in the rotating state are equivalent to a closed current loop, which generates magnetism. In this part, we will discuss the magnetism of Au$_{25}$-based nanoclusters, Au$_{102}$ and Au$_{133}$ and other metal nanoclusters chronologically.

5.3.1 Magnetic properties of Au$_{25}$-based metal nanoclusters

In 2009, Manzhou Zhu et al. evaluated the magnetic properties of well-defined atomic monodisperse Au$_{25}$(PET)$_{18}$ nanocluster through electron paramagnetic resonance (EPR) spectroscopy. As shown in Fig. 5.15, [Au$_{25}$(SR)$_{18}$]^0 nanocluster samples of microcrystals and frozen solutions showed $S = 1/2$ signal, $g = (2.56, 2.36, 1.82)$. This means that [Au$_{25}$(SR)$_{18}$]^0 each particle has an unpaired spin electron. Furthermore, magnetization measurements using a superconducting quantum interference device (SQUID) shows that [Au$_{25}$(PET)$_{18}$]^0 nanocluster are paramagnetic between 5 K and 300 K, and no hysteresis is observed at 5 K. Studies have shown that the highly delocalized spin density is consistent with the "superatomic" model, and most of the spin density is located in the Au$_{13}$ nucleus. When the nanocluster is oxidized to [Au$_{25}$(SR)$_{18}$]^0, the EPR signal appears and the nanocluster becomes paramagnetic, but when the nanocluster is reduced to a charge state of −1, the EPR

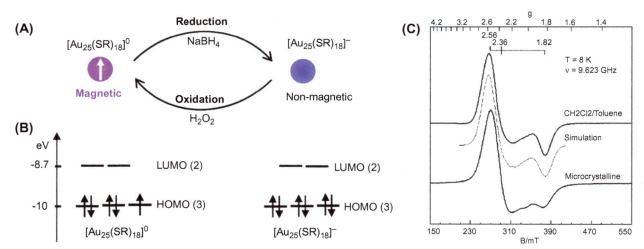

FIG. 5.15 The magnetic properties of Au$_{25}$(PET)$_{18}$ nanocluster. Reversible conversion between the neutral and anionic Au$_{25}$(SR)$_{18}$ nanoparticles. DFT-calculated Kohn-Sham orbital energy-level diagrams for the neutral and anionic nanoparticles, respectively. EPR spectra of [Au$_{25}$(SR)$_{18}$]^0 for the conditions listed. Simulation parameters: $g = (2.556, 2.364, 1.821)$, g strain (σ_g) = 0.03, 13 equivalent $I = 3/2$ nuclei with A = (71, 142, 50) MHz. *(From M. Zhu, C.M. Aikens, M.P. Hendrich, R. Gupta, H. Qian, G.C. Schatz, R. Jin, Reversible switching of magnetism in thiolate-protected Au$_{25}$ superatoms. J. Am. Chem. Soc. 131(7) (2009) 2490–2492, https://doi.org/10.1021/ja809157f.)*

signal can disappear again. Therefore, as shown in Fig. 5.15, the paramagnetism in Au_{25} nanoparticles is turned on or off simply by controlling the charge state of the nanocluster [115].

In 2009, Jiang and Whetten have proved that $Au_{25}(SR)_{18}^-$ is magnetically doped with Fe, Cr and Mn through first-principles DFT studies. The magnetic moment of nanoclusters doped with Cr and Mn can reach $5\mu_B$, while the magnetic moment of nanoclusters doped with Fe can only reach $3\mu_B$. The research results finally show that the magnetic doping of Fe, Cr and Mn in $Au_{25}(SR)_{18}$ is thermodynamically beneficial [116].

In 2009, Arash Akbari-Sharbaf et al. used solid-state EPR to study the unpaired electrons and single-occupied molecular orbital (SOMO) of the positively charged Au_{25} molecular cluster. The EPR spectrum of Au_{25} positively charged (Au_{25}^+) powder. It is found that Au_{25}^+ has the paramagnetism of SOMO, which is mainly located around the central gold atom in the molecular core and has strong p-type atomic properties [117].

In 2010, Jaakko Akola et al. investigated a series of dimer metal nanoclusters, in which [$MnAu_{24}(SMe)_{17}$-BDT-$(SMe)_{17}Au_{25}]^-$ is magnetic, and even $MnAu_{24}(SMe)_{17}$-BDT-$(SMe)_{17}Au_{24}Mn$ is ferromagnetic. Through calculation, the magnetic moment of [$MnAu_{24}(SMe)_{17}$-BDT-$(SMe)_{17}Au_{25}]^-$ is $4.6\mu_B$, and the magnetic moment of $MnAu_{24}(SMe)_{17}$-BDT-$(SMe)_{17}Au_{24}Mn$ is as high as $9.9\mu_B$ [118].

In 2011, Zhou et al. studied several properties of Mn-doped Au_{25} nanoclusters, including electronic, magnetic, spin dependence and so on. The research results show that the most stable doping site of Mn is on the surface rather than in the center. Spin-related transport calculations show that these doped clusters can be used as spin filters [119].

In 2013, Sabrina Antonello et al. used bis(pentafluorobenzoyl) peroxide to oxidize $Au_{25}(SR)_{18}^0$ to $Au_{25}(SR)_{18}^+$. Through EPR testing, it is found that $Au_{25}(SR)_{18}^+$ has no EPR activity in the temperature range of 6–260K, which proves that $Au_{25}(SR)_{18}^+$ is a diamagnetic substance [120].

In 2014, Tiziano Dainese et al. evaluated the interaction between the metal core of the paramagnetic $[Au_{25}(SEt)_{18}]^0$ nanocluster and the capping ligand. The hyperfine interactions between a surface-delocalized unpaired electron and the gold atoms of a nanoclusters was observed by electron nuclear double resonance (ENDOR). The Davies ENDOR spectrum showed that the gold atoms in the cluster is divided into four groups: In addition to the central Au atom and 12 staple Au atoms, there are 12 icosahedral Au atoms divided into 8 atomic groups and 4 atomic groups, the inequality between the icosahedral atoms is due to the slight dissymmetries associated with dissimilar staple motifs. Such a small distortion propagates to the core atom, resulting in a nonequivalent electron density on the icosahedral nucleus, and hyperfine couplings [121].

In 2014, Marco De Nardi et al. synthesized a $[Au_{25}(SBu)_{18}^0]_n$ one-dimensional polymer chain along the (1) lattice plane. As shown in Fig. 5.16, the linear polymer chain is composed of $Au_{25}(SBu)_{18}^0$ units, which are connected to each other by a single Au—Au bond. This polymer is formed by the correct orientation of adjacent clusters and van der Waals interactions between the thiolate ligands. The CW-EPR spectrum of the crystal shows differences compared with the behavior of $Au_{25}(SBu)_{18}$ solid or in frozen solution. With increasing of temperature, the double integrated EPR intensity increases. Since EPR intensity is proportional to the magnetic susceptibility of the sample, it indicates that the crystal exhibits

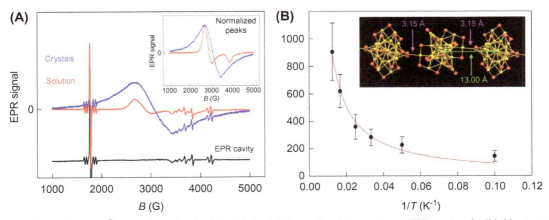

FIG. 5.16 The EPR of $[Au_{25}(SBu)_{18}^0]_n$ one-dimensional polymer chain. (A) Comparison between the cw-EPR spectra of solid (*blue* traces) and frozen toluene solution (*red* traces) of $Au_{25}(SBu)_{18}$ at 20K. The inset shows the same spectra but normalized for peak intensity. The *black* curve corresponds to the EPR cavity signal, subtracted in the inset for clarity. (B) Bonner-Fisher-Hall fit to the experimental data, expressed as rescaled relative double-integrated EPR signal values. *(From M. De Nardi, S. Antonello, D. Jiang, F. Pan, K. Rissanen, M. Ruzzi, A. Venzo, A., Zoleo, F. Maran, Gold nanowired: a linear $(Au_{25})_n$ Polymer from Au_{25} molecular clusters. ACS Nano 8(8) (2014) 8505–8512, https://doi.org/10.1021/nn5031143.)*

antiferromagnetic behavior. At low temperatures, $[Au_{25}(SBu)_{18}^0]_n$ has a nonmagnetic ground state, which is described as a one-dimensional antiferromagnetic system [122]. Since 2015, Zhikun Wu's group has used EPR to study the magnetic properties of single-doped MAu_{24} nanoclusters, where M = Hg, Pd, Pt. the doped atoms can affect the electronic configuration of $[MAu_{24}(SR)_{18}]^0$, transforming paramagnetism to diamagnetism [123,124].

In 2016, Yongbo Song et al. synthesized rod-like $[Au_{25}(PPh_3)_{10}(SePh)_5Cl_2](SbF_6)$ and $[Au_{25}(PPh_3)_{10}(SePh)_5Cl_2]$ $(SbF_6)(BPh_4)$ nanoclusters through kinetic control. They have studied their magnetic properties through EPR and SQUID. The EPR spectrum of $[Au_{25}(PPh_3)_{10}(SePh)_5Cl_2]^+$ nanocluster revealed an S = 1/2 signal, g = (2.40, 2.26, 1.78). However, the EPR spectrum of $[Au_{25}(PPh_3)_{10}(SePh)_5Cl_2]^{2+}$ just revealed a baseline. The result of SQUID also shows that $[Au_{25}(PPh_3)_{10}(SePh)_5Cl_2](SbF_6)$ is paramagnetic, while $[Au_{25}(PPh_3)_{10}(SePh)_5Cl_2](SbF_6)(BPh_4)$ is diamagnetic [125].

In 2016, Mikhail Agrachev et al. used pulsed electron nuclear double resonance (ENDOR) to study the interaction of unpaired electrons in the four structurally related paramagnetic $Au_{25}(SR)_{18}^0$ clusters with the protons of alkanethiolate ligands. The ENDOR signal is successfully correlated with the type of ligand and the distance between the relevant proton and the central gold core. The results show that unpaired electrons can be used as very accurate probes for the main structural features of the interface between the metal core and the capping ligand [126].

In 2016, Tofanelli et al. studied the magnetic behavior of $[Au_{25}(SR)_{18}]^n$, where $n = -1$, 0 and +1. It is concluded that $[Au_{25}(SR)_{18}]^{-1}$ and $[Au_{25}(SR)_{18}]^{+1}$ are diamagnetic, and $[Au_{25}(SR)_{18}]^0$ is typical paramagnetic. The magnetic behavior of $[Au_{25}(SR)_{18}]^0$ can be attributed to the first order Jahn-Teller distortions that leads to the degeneracy of the superatomic P orbital split [127].

In 2017, Mikhail Agrachev et al. used EPR to test the magnetic properties of $[Au_{25}(PET)_{18}]^0$ in different physical states, including: frozen solution, film, single crystal, immobilized single crystal, collection of 10 crystals, immobilized collection of 10 crystals, and microcrystal. The frozen solution is paramagnetic, and the inhomogeneous broadening of the EPR signal in the film is more serious than that in the frozen solution, indicating that there is a weak orientation-dependent interaction in the film. Single crystals have hysteresis and are ferromagnetic. There is no obvious hysteresis for single crystals and single crystal aggregates immobilized in frozen MeCN, but they have magnetic anisotropy in magnetism. However, the microcrystals exhibit obvious paramagnetic, superparamagnetic and ferromagnetic behavior at different conditions. The research results show that the observed magnetic behavior of nanocluster is strongly influenced by the crystalline and physical state of the sample [128].

In 2019, Megumi Suyama et al. reported that the same amount of $[PtAu_{24}(PET)_{18}]^{2-}$ and $[PtAu_{24}(PET)_{18}]^0$ can react to form $[PtAu_{24}(PET)_{18}]^-$. The above reaction is carried out by the spontaneous electron transfer of $[PtAu_{24}(PET)_{18}]^{2-}$. $[PtAu_{24}(PET)_{18}]^0$ and $[PtAu_{24}(PET)_{18}]^{2-}$ are diamagnetic, but $[PtAu_{24}(PET)_{18}]^-$ is paramagnetic [129].

In 2020, Yingwei Li et al. proposed to manipulate the paramagnetism of 7e $[M_{25}(SR)_{18}]^0$ (M = Au/Ag) nanoclusters by doping with Ag. Compared with $[Au_{25}(PET)_{18}]^0$, the EPR spectroscopy showed that the g-value of Ag-doped NCs has changed significantly. The axial split in the g-value linear decreases with the increase in Ag doping because the spin-orbit coupling of Ag is four times lower than that of gold. They also compared the influence of protective ligands on the splitting of g-value. In addition, SCXRD analysis revealed the doping position of Ag, which provides a basis for understanding the source of magnetism [130].

5.3.2 Magnetic properties of other metal nanoclusters

In 2006, Negishi et al. studied a series of $Au_n(SG)_m$ nanoclusters, where $(n, m) = (10,10)$, (15, 13), (18, 14), (22, 16), (25, 18), (29, 20), and (39, 24), in which $Au_{18}(SG)_{14}$ exhibits paramagnetism. Negative and positive peaks were observed at the edges of Au in $Au_{18}(SG)_{14}$ at the L_3 ($2p_{3/2} \rightarrow 5d_{5/2}, 6s_{1/2}$, 11.922 keV) and L_2 ($2p_{1/2} \rightarrow 5d_{3/2}, 6s_{1/2}$, 13.733 keV), respectively. The X-ray magnetic circular dichroism (XMCD) amplitudes are approximately one thousandth of the XAS step height. The XMCD intensities described below are normalized by the corresponding XAS heights, so they are proportional to the magnetic moment of each Au atom. This observation confirmed the inherent spin polarization of the Au atoms in the $Au_{18}(SG)_{14}$ nanocluster. The XMCD peak height of the L_3 edge is 2.5 times that of the L_2 edge. This asymmetry of the spectrum indicates that Au 5d electrons have a considerable orbital magnetic moment. Applying the sum rules, the ratio of the orbital to spin magnetic moment is about 12%. The study also found that in the magnetic field range of −10 to 10 T, the XMCD signal exhibits an approximately linear relationship with the magnetic field, and no obvious hysteresis behavior is observed. The same behavior was observed in the SQUID measurements. The Curie-type behavior is clearly seen in the study of the temperature dependence of the magnetization. The research results show that the magnetic moment of $Au_n(SG)_m$ nanoclusters increases with the increase in the nanocluster core size. The holes produced by Au—S bonding have a greater impact on the spin polarization phenomenon than the quantum size effect of gold nanotubes [131].

In 2007, Magyar et al. studied the magnetic properties of bare uncapped gold nanoclusters through density-functional calculations. The results show that the magnetic properties of gold nanoclusters mainly depend on its size. As the size of the gold nanoclusters increases, the specific surface becomes smaller and the diamagnetic core dominates. In addition, the study has also shown that surface ligands can affect the magnetic properties of gold nanoclusters [132].

In 2010, Muñoz-Márquez et al. recorded hysteresis cycles of the phosphine-capped nanocluster (Au_{11}-TPP) at highest and lowest temperature via SQUID. Au_{11}-TPP exhibits diamagnetism at high and low temperature. The EPR measurement also demonstrates its diamagnetism [133].

In 2010, Beatriz Santiago González et al. synthesized PVP-protected gold clusters, including Au_1, Au_2, and Au_3. According to the Jellium model, Au_3 cluster should be paramagnetic, and Au_2 cluster should be diamagnetic. However, the EPR spectra show that the presence of paramagnetism is observed only for 1 mg/mL. It should contain a proportion of the trimer cluster. On the contrary, 0.5 and 0.1 mg/mL are diamagnetic because of containing mainly dimers [134].

In 2013, McCoy et al. used Evan's NMR method to confirm the paramagnetism of $Au_{102}(pMBA)_{44}$ nanoclusters. In addition, it was found that the solution of $Au_{102}(pMBA)_{44}$ was heated in the oscillating RF magnetic field, but not in the electronic field. Experiments show that the paramagnetic $Au_{102}(pMBA)_{44}$ is heated by the interaction of the spin magnetic moment and the external oscillating magnetic field [135].

In 2013, Katla Sai Krishna et al. synthesized paramagnetic $[Au_{25}(PPh_3)_{10}(SC_{12}H_{25})_5Cl_2]^{2+}$ and $[Au_{25}(SC_2H_4Ph)_{18}]^0$, diamagnetic $Au_{38}(SC_{12}H_{25})_{24}$, and ferromagnetic $Au_{55}(PPh_3)_{12}Cl_6$ nanoclusters. It's seen from the Fig. 5.17, the mixed ligand-stabilized Au_{25} nanocluster (black) exhibits a typical diamagnetic behavior. Fig. 5.17 shows the structures of the four nanoclusters and their respective M-H curves at 5 K. It is observed through experiments that the saturation magnetic moment of the thiol-stabilized Au_{25} nanocluster (red) is $\mu_B = 0.0516$/cluster or $\mu_B = 0.0020$/Au atom. The majority of dodecanethiol stabilized gold nanoparticles, smaller than 2 nm, exhibit room temperature ferromagnetism, but the curve of Au_{38} nanocluster (green) shows surprising diamagnetic behavior. Unexpectedly, the Au_{55} nanoclusters (blue) stabilized by phosphine showed ferromagnetic behavior, a coercive field of $H_c = 1125$ Oe, residual magnetization of 0.0067 emu, and saturation magnetic moment of $\mu_B = 0.0584$/cluster or $\mu_B = 0.0010$/Au atom. Most importantly, although $Au_{55}(PPh_3)_{12}Cl_6$ has been studied for decades, this is the first report of its permanent magnetism at room temperature [136].

In 2014, Jianwei Ji et al. synthesized $Ni_{39}(SC_2H_4Ph)_{24}$ and $Ni_{41}(SC_2H_4Ph)_{25}$ nanoclusters, which exhibit ferromagnetism. Nickel nanoclusters reflect ferromagnetic hysteretic behavior. At 300 K, the saturation magnetization is

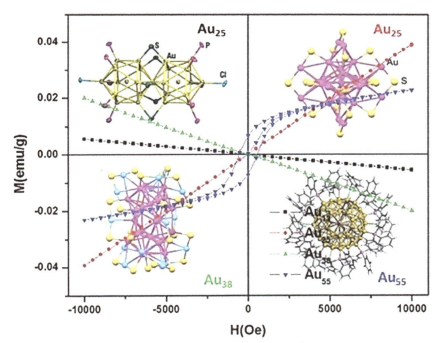

FIG. 5.17 The paramagnetic $[Au_{25}(PPh_3)_{10}(SC_{12}H_{25})_5Cl_2]^{2+}$ and $[Au_{25}(SC_2H_4Ph)_{18}]^0$, diamagnetic $Au_{38}(SC_{12}H_{25})_{24}$, and ferromagnetic $Au_{55}(PPh_3)_{12}Cl_6$ nanoclusters. The structures of $[Au_{25}(PPh_3)_{10}(SC_{12}H_{25})_5Cl_2]^{2+}$, $[Au_{25}(SC_2H_4Ph)_{18}]^0$, $Au_{38}(SC_{12}H_{25})_{24}$, and $Au_{55}(PPh_3)_{12}Cl_6$, and their M vs H curves at 5 K. *(From K.S. Krishna, P. Tarakeshwar, V. Mujica, C.S.S.R. Kumar, Chemically induced magnetism in atomically precise gold clusters, Small 10(5) (2014) 907–911. https://doi.org/10.1002/smll.201302393.)*

0.088 emu/g, remnant magnetization is 0.022 emu/g, and coercivity is 345 Oe. At 1.8 K, saturation magnetization is about 55 emu/g, remnant magnetization is 0.033 emu/g, and coercivity is 0.7 Oe. Since the two clusters cannot be separated further, it is impossible to judge their respective magnetic behavior [137].

In 2014, Shufang Xue et al. prepared a Cu_6Ln_2 (Ln = Dy, Tb, Gd, Y) complex that shows significant antiferromagnetic property. Cu_6Ln_2 nanoclusters show temperature dependence for the $\chi_M T$. At room temperature, the $\chi_M T$ of Cu_6Dy_2, Cu_6Tb_2, and Cu_6Gd_2 are 30.89, 26.67, and 17.88 cm^3 K/mol, respectively. On the basis of Cu_6Y_2, the ferromagnetic contribution is derived from the Gd-O_2-Gd coupling. Due to the Cu...Cu interactions and Cu...Gd coupling, strong magnetic caloric and SMM-like properties were observed [138].

In 2018, Bao-Qian Ji et al. prepared a mononuclear $[Cu_9(bdped)_4(N_3)_4(ba)_3(Ac)_3]\cdot CH_3CH_2OH\cdot H_2O$ nanocluster consist of two Cu_5 ring by sharing a Cu atom. They investigated the magnetization of the sample in the magnetic field range −50 to 50 kOe, and the temperature dependence of the AC susceptibility between 2 and 20 K at four different frequencies. Research results show that it is paramagnetic. The temperature dependence of magnetic susceptibility of per Cu^{2+} ion imply a constant effective magnetic moment of 1.8 m_B, magnetization curve suggest antiferromagnetic interactions between Cu^{2+} ions [139].

In 2019, Chenjie Zeng et al. found paramagnetism in $[Au_{133}(TBBT)_{52}]^0$ nanometer clusters and diamagnetism in $[Au_{133}(TBBT)_{52}]^+$ nanometer clusters through the EPR test. The unpaired spins is removed by oxidizing $[Au_{133}(TBBT)_{52}]^0$ to $[Au_{133}(TBBT)_{52}]^+$, and the nanoclusters correspondingly change from paramagnetism to diamagnetism. Similarly, $[Au_{133}(TBBT)_{52}]^+$ can also be reduced to $[Au_{133}(TBBT)_{52}]^0$, and the magnetism changes accordingly. However, adding more reducing agent will gradually decompose Au_{133} in the presence of a large amount of $NaBH_4$, resulting in a decrease in NMR intensity and peak broadening. It is demonstrated that the unpaired spin is mainly delocalized in the inner icosahedral Au_{13} core. NMR experiments show that the chemical shifts of 52 surface TBBT ligands are affected by the spin of metal core [140].

In 2019, Herbert et al. used magnetic circular dichroism (MCD) spectroscopy to study the monolayer-protective cluster $Au_{102}(pMBA)_{44}$, revealing two distinct magnetic anisotropy behaviors. A low-energy, diamagnetic component is related to the state excitation from the passivating ligand. A high-energy, paramagnetic component is attributed to the excitation of Au core d-band. Au d-band vacancies and magnetic anisotropy are attributable to d^{10}-d^{10} spin-orbit coupling of Au atoms, which ultimately leads to s-p-d orbital rehybridization [141].

In 2020, Window and Ackerson used Evans method NMR measurement to study the magnetic properties of $Au_{102}(SPh)_{44}^{1-/0/1+/2+}$ nanoclusters. According to the χT values of $Au_{102}(SPh)_{44}$ values at different charge states, it is demonstrated that the 1−/1+/2+ charge state is paramagnetic, and the neutral charge state is diamagnetic [142].

In 2020, Qinzhen Li et al. prepared AuCu alloy nanoclusters with various charge states that can be regulated by $H_2O_2/NaBH_4$ or electrochemical redox. Fig. 5.18 is the structures of $[Au_{18}Cu_{32}(SPhCl)_{36}]^{2-}$ (AuCu-I) and $[Au_{18}Cu_{32}(SPhCl)_{36}]^{3-}$ (AuCu-II), and EPR curves of AuCu-I (red) and AuCu-II (black). The EPR spectrum shown that AuCu-II displays a strong signal with S = 1/2 (g = 2.02, 2.35), and AuCu-I only displays a baseline. The test results show that $[Au_{18}Cu_{32}(SPhCl)_{36}]^{3-}$ nanocluster is paramagnetic [143].

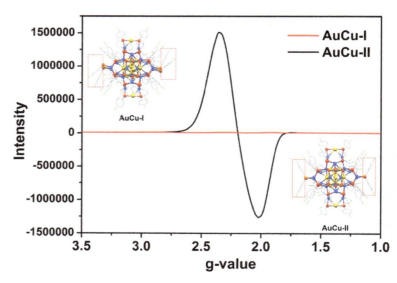

FIG. 5.18 The structure and EPR of AuCu-I and AuCu-II. The structures of $[Au_{18}Cu_{32}(SPhCl)_{36}]^{2-}$ (AuCu-I) and $[Au_{18}Cu_{32}(SPhCl)_{36}]^{3-}$ (AuCu-II), and EPR curves of AuCu-I *(red)* and AuCu-II *(black)*. *(From Q. Li, J. Chai, S. Yang, Y. Song, T. Chen, C. Chen, H. Zhang, H. Yu, M. Zhu, Multiple ways realizing charge-state transform in Au-Cu bimetallic nanoclusters with atomic precision, Small 17(27) (2021) 1907114, https://doi.org/10.1002/smll.201907114.)*

In 2020, Haru Hirai et al. reported the paramagnetic $[IrAu_{12}(dppe)_5Cl_2]^{2+}$ nanoclusters. The EPR spectroscopy shows its paramagnetic single, $g = (2.71, 1.64, 1.42)$ [144].

In 2021, Beiling Liao et al. successfully synthesized two nanoclusters, $[Co_8^{II}Co_4^{III}(L_1)_4(Py)_{12}(CH_3OH)_4(CH_3COO)_4]\cdot(CH_3OH)_{13}$ (1) and $[Co_{18}^{II}(L_2)_6(Py)_{48}]\cdot(DMF)_5\cdot(CH_3OH)_8$ (2), and characterized their structure by single crystal X-ray diffraction studies. The DC magnetic susceptibility of nanoclusters 1 and 2 were studied at a 1000 Oe field from 300 to 2 K. the χMT value of nanocluster 1 is 23.55 cm^3 K/mol at room temperature, and the $\chi_M T$ value of nanocluster 2 is 61.54 cm^3 K/mol. They studied the dynamic magnetic properties of the two nanoclusters via three sets of measurements of the AC susceptibilities. The change of magnetic susceptibility with temperature in a range of frequencies spanning 9–997 Hz shows that in the zero DC field, only the AC magnetic susceptibility of nanocluster two exhibits obvious frequency-dependent behavior [145].

5.4 Chirality of metal nanoclusters

The term chirality means that an object cannot coincide with its mirror image. Chirality is a universal phenomenon in the universe, which is reflected in the emergence and evolution of life. As the initial form of matter, nanoclusters with precise crystal structure are of great value in revealing the origin of chirality. Due to the significant advance of synthesis and structural determination, the studies of chirality of metal nanoclusters have made progress. Various chiral metal nanoclusters, including monometal and alloy metal nanoclusters, have been reported, and the associated literature discussed the chirality of metal nanoclusters. In this chapter, we focus on the origin of clusters chirality, the research method and technique of cluster chirality.

The precise structure of the cluster makes it possible to explore the origin of cluster chirality. The metal cluster protected by the ligand is composed of the metal core, the outer motif and the outermost ligand. Unlike small organic molecules, in which chirality depends on the chiral carbon center. Chirality in metal nanoclusters could have several possible origins due to their more complex structures: (1) Chiral ligand induction, in which the chirality of ligand is transferred into metal nanoclusters; (2) Chiral arrangement on the surface of achiral ligands and Au-S motifs; (3) the asymmetric metal core. Below, we classify according to whether to perform chiral resolution for nanoclusters protected by achiral ligands, that is, identify the chirality through structure and characterize the chirality through chiral high performance liquid chromatography (HPLC) separation and CD spectra.

5.4.1 Identify the chirality through crystal structure

Chenjie Zeng et al. synthesized $Au_{20}(TBBT)_{16}$ nanocluster by ligand exchange method. The obtained $Au_{20}(TBBT)_{16}$ nanoclusters possessed a Au_7 core, a $Au_8(SR)_8$ octameric ring, a trimeric and two monomeric staple motifs. The asymmetric arrangement of surface motif is responsible for the chirality of $Au_{20}(TBBT)_{16}$ [146]. Quan-Ming Wang and co-workers determined structure of the $[Au_{25}(C{\equiv}CAr)_{18}]^-$ nanoclusters (Ar = 3,5-bis(trifluoromethyl)phenyl)), which is different from the previously reported the structure of $Au_{25}(SR)_{18}^-$. The $[Au_{25}(C{\equiv}CAr)_{18}]^-$ displayed a $D3$ symmetry. The chiral arrangement of surface motif due to steric hindrance of alkynyl leaded to the chirality of $[Au_{25}(C{\equiv}CAr)_{18}]^-$ [147]. In 2010, Jin et al. reported the crystal structure of $Au_{38}(SR)_{24}$ (R = C_2H_4Ph) nanocluster, which contained a face-fused biicosahedral Au_{23} core, three monomeric $Au(SR)_2$ staples and six dimeric $Au_2(SR)_3$ staples. The arrangement of $Au_2(SR)_3$ staples resembled a triblade fan with $C3$ axis. The rotation direction of triblade fan in the enantiomers of the left-handed configuration and the right-handed configuration are clockwise and counterclockwise, respectively. The Au_{15} shell leaded to the chirality of $Au_{38}(SR)_{24}$ [148]. Nuclear magnetic resonance (NMR) spectroscopy was employed to identify the chirality of $Au_{38}(SR)_{24}$. It is found that different 1H signals for the two germinal protons in each CH_2 of the ligands on the chiral $Au_{38}(SR)_{24}$ nanocluster were observed, whereas the nonchiral Au_{25} did not display the similar result [149].

In 2015, Zeng et al. reported the crystal structure of $Au_{40}(SR)_{24}$ and $Au_{52}(SR)_{32}$, in which both nanoclusters were chiral. Differently, the chirality of Au_{40} is original from the rotative arrangement of the surface staple motifs, whereas the chirality of Au_{52} is attributing to the chiral metal core (discussed below). The Au_{40} is composed of Au_7 core, a Kekulé ring, six monomeric staples and three trimeric staples. The chiral $D3$ symmetry associated with the rotation direction leaded to the chirality of Au_{40} [150]. Wu et al. reported the crystal structure of $Au_{44}(2,4\text{-}DMBT)_{26}$ that possessed the 18-electron shell closure structure. The arrangement of staples and thiolates on the Au_{29} core and the Au_6 cap atoms leaded to the chirality of $Au_{44}(2,4\text{-}DMBT)_{26}$ [151].

In 2007, Kornberg et al. reported the first crystal structure of Au nanocluster, $Au_{102}(p\text{-}MBA)_{44}$, protected by thiol ligand. The p-MBA ligands interact with Au atoms intercluster and intracluster, in which the sulfur atoms were chiral centers. One enantiomer has 22 sulfur centers with R configuration, 18 with S configuration. The interaction among the ligands and the corresponding surface arrangement are responsible for the chirality of $Au_{102}(p\text{-}MBA)_{44}$ [152]. In terms

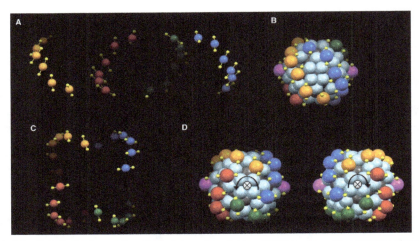

FIG. 5.19 Self-assembled -S-Au-S- helical stripes on the spherical Au_{107} kernel. (A and B) Side views. (C and D) Top views. The two chiral isomers are shown in (D). Each stripe is composed of six monomeric staples stacked into a ladder-like helical structure. *Yellow*: sulfur; *orange/red/blue/green*: gold in the helices; *purple*: gold in the independent monomeric staples. *(From Z. Chenjie, C. Yuxiang, K. Kristin, A. Kannatassen, M.Y. Sfeir, J. Rongchao, Structural patterns at all scales in a nonmetallic chiral $Au_{133}(SR)_{52}$ nanoparticle, Sci. Adv. 1(2) (2015) e1500045, https://doi.org/10.1126/sciadv.1500045.)*

of chiral large-size gold nanoclusters, Zeng et al. synthesized chiral $Au_{133}(SR)_{52}$ and $Au_{246}(p\text{-MBT})_{80}$ nanoclusters, also called nanoparticles. The $Au_{133}(SR)_{52}$ is composed of a Au_{107} kernel and four "helical stripes" on the spherical surface. The clockwise and anticlockwise rotations of the four helices and the self-assembly of carbon tails of the ligands on the surface leaded to chirality in the $Au_{133}(SR)_{52}$ nanoparticle (Fig. 5.19) [153]. The $Au_{246}(p\text{-MBT})_{80}$ possessed a Au_{206} core, $Au_2(SR)_3$ motifs, $Au_1(SR)_2$ motifs and protected thiol ligands, in which 25 of the p-MBTs were rotationally packed into four pentagonal circles. The rotation direction (clockwise and counterclockwise rotational arrangement) induced the chirality of $Au_{246}(p\text{-MBT})_{80}$) [154].

In the chirality research of silver nanoclusters, Zhu et al. reported the structure of $[Ag_{32}(Dppm)_5(SAdm)_{13}Cl_8]^{3+}$ and $[Ag_{45}(Dppm)_4(S\text{-}But)_{16}Br_{12}]^{3+}$. Both nanoclusters contained achiral metal cores and chiral arrangement of the surface motifs and ligands [155]. Sun et al. prepared the chiral $[Ag_6Z_4@Ag_{36}]$ (Z=S or Se) nanoclusters by introducing Ph_3CSH or Ph_3PSe into $[AgS_4@Ag_{36}]$ cluster (Ag_{37}). Interestingly, the obtained chiral $[Ag_6Z_4@Ag_{36}]$ enantiomer crystallizes separately to form homochiral crystallization. The clockwise (CCCC) or anticlockwise (AAAA) patterns are responsible for the chirality of $[Ag_6Z_4@Ag_{36}]$ [156]. In the chirality research of copper nanoclusters, Robinson and coworkers demonstrate the chiral Cu clusters that are constructed to hierarchical architecture. The clockwise and counterclockwise rotation of cap and core and the intramolecular interaction (C-H···π) resulted in the chirality of the whole architecture (Fig. 5.20) [157].

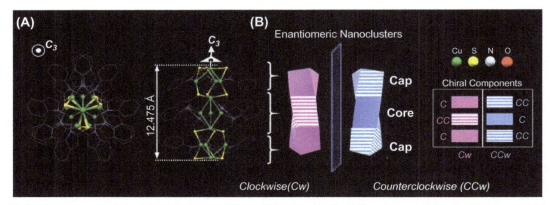

FIG. 5.20 SB-Cu cluster structure derived from single-crystal X-ray diffraction. (A) (Left) Top view, parallel to the C3 axis, and (right) side view, perpendicular to the C3 axis, of the SB-Cu cluster structure. (B) Each of the three components of the SB-Cu clusters (cap, core, cap) possesses either clockwise (C) or counterclockwise (CC) chirality with both caps in a cluster having the same chirality and the core opposite chirality. For the full SB-Cu cluster there are only two configurations, C-CC-C and CC-C-CC (cap-core-cap), resulting in two enantiomeric isomers denoted as Cw and CCw, respectively. *(From H. Han, Y. Yao, A. Bhargava, Z. Wei, Z. Tang, J. Suntivich, O. Voznyy, R.D. Robinson, Tertiary hierarchical complexity in assemblies of sulfur-bridged metal chiral clusters, J. Am. Chem. Soc. 142(34) (2020) 14495–14503, https://doi.org/10.1021/jacs.0c04764.)*

In the chirality research of alloy nanoclusters, Wang et al. determined the structure of [Ag$_{46}$Au$_{24}$(SR)$_{32}$](BPh$_4$)$_2$, which contained an achiral Ag$_2$@Au$_{18}$@Ag$_{20}$ core and a chiral Ag$_{24}$Au$_6$(SR)$_{32}$ shell. The asymmetric arrangement of the two RS groups leaded to the chirality of shell that is further to be transferred to the whole alloy nanoclusters [158]. Zeng et al. reported a chiral [Au$_{80}$Ag$_{30}$(C≡CPh)$_{42}$Cl$_9$]Cl nanocluster. It is found that the M$_{110}$ core patterned with four shell Russian doll architecture, Au$_6$@Au$_{35}$@Ag$_{30}$Au$_{18}$@Au$_{21}$, is responsible for the chirality of Au$_{80}$Ag$_{30}$ [159]. Recently, Liu et al. reported the dimeric assembly of Au$_{25}$(PET)$_{18}$ by two Ag atoms. Interestingly, the two Ag$_1$Au$_{25}$ monomers are mirror-symmetric, meaning the chirality of each monomer. The arrangement of staples in Ag$_1$Au$_{25}$ is responsible for the chirality. The obtained Ag$_2$Au$_{50}$(PET)$_{36}$ molecule is mesomeric due to the assembly of monomer [160].

In the case where the cluster metal core has chirality, Wang et al. prepared the first Au nanocluster containing an intrinsic chiral core. The [Au$_{20}$(PP$_3$)$_4$]Cl$_4$ (PP$_3$ = tris(2-(diphenylphosphino)ethyl)phosphine) was constituted of Au$_{13}$ core and a helical Y-shaped Au$_7$ motif, which possessed $C3$ symmetry. The rotation direction of the Y-shaped spiral structure determines the chirality of Au$_{20}$ [161]. Similarly, Liu and coworkers prepared a chiral Cu$_{20}$H$_{11}$[Se$_2$P-(OiPr)$_2$]$_9$ nanocluster, which contained a chiral Cu$_{19}$ core. The 19 surface Cu atoms can be classified three groups that formed three anticlockwise helices along the $C3$ axis. The helices arrangement of Cu atoms in core is responsible for the chirality of Cu$_{20}$H$_{11}$[Se$_2$P-(OiPr)$_2$]$_9$ nanocluster [162].

As mentioned above, the chirality of Au$_{52}$ originated from the metal core that is composed of 10 tetrahedral units. The tetrahedral are assembled into a double helical superstructure, which is responsible for the chirality of Au$_{52}$ (Fig. 5.21). In 2017, Zheng et al. determined the structure of [Ag$_{141}$X$_{12}$(S-Adm)$_{40}$]$^{3+}$ nanoclusters. The obtained Ag$_{141}$ contained an Ag$_{71}$

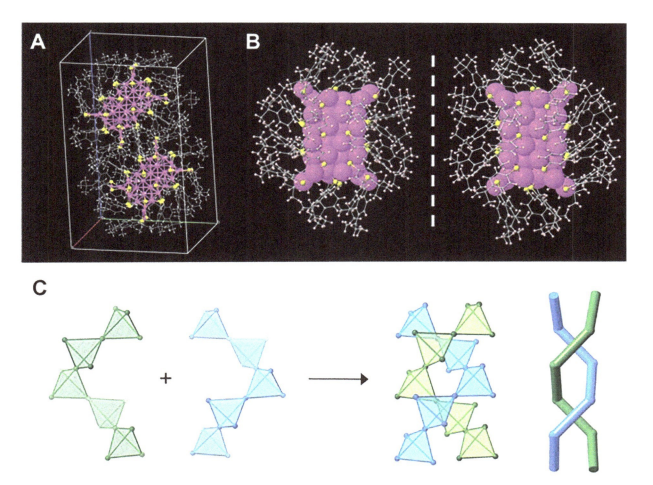

FIG. 5.21 Total structure of the Au$_{52}$(TBBT)$_{32}$ cluster. Two helical pentatetrahedral strands forming the double helical kernel. (A) Unit cell comprising two enantiomers. (B) Mirror symmetry of the enantiomers. (C) Two helical pentatetrahedral strands forming the double helical kernel. *Blue/green*, Au atoms in the kernel. *(From Z. Chenjie, C. Yuxiang, L. Chong, N. Katsuyuki, N.L. Rosi, J. Rongchao, Gold tetrahedra coil up: Kekulé-like and double helical superstructures, Sci. Adv. 1(9) (2015) e1500425, https://doi.org/10.1126/sciadv.1500425.)*

core and an $Ag_{70}(S\text{-}Adm)_{40}$, as well as 12 halides, in which Ag_{71} core is composed of a two shell $Ag_{19}@Ag_{52}$ pattern. The interpenetrating biicosahedra in Ag_{19} slightly twisted, resulting in the chirality of $[Ag_{141}X_{12}(S\text{-}Adm)_{40}]^{3+}$ [163].

5.4.2 Chiral separation of metal nanoclusters

Although the chirality of clusters is identified by structure, the separation of chiral monomers cannot be achieved by structure alone. Chiral high performance liquid chromatography (HPLC) and circular dichroism (CD) are further used to achieve the separation and optical activity characterization of chiral monomers. In 2012, Bürgi et al. for the first time achieved the enantioseparation of Au_{38} nanoclusters. The peak at 8.45 and 17.45 min corresponded to the enantiomer 1 and 2. The UV-Vis spectra of enantiomer 1 and 2 is perfectly similar to the racemic Au_{38}, indicating the two collected fractions are $Au_{38}(SCH_2CH_2Ph)_{24}$. The CD spectra give perfect mirror images and eleven clear signals are observed: 245 (+), 255 (+), 308 (+), 345 (−), 393 (−), 440 (+), 479 (−) 564 (+), 629 (−) and 747 (+) nm. The anisotropy factors of Au_{38} are quite strong with the range of 1×10^{-3}–4×10^{-3} (Fig. 5.22) [164]. Further, Bürgi et al. introduced single Cu atom into $Au_{38}(2\text{-}PET)_{24}$ forming $Au_{38}Cu_1(2\text{-}PET)_{24}$ nanoclusters. MALDI-TOF mass and chiral HPLC were employed to investigate the composition and optical activity [165].

In the later work of Bürgi and coworkers, two enantiomers of the $Au_{40}(2\text{-}PET)_{24}$ cluster were collected using HPLC. In the CD spectra of $Au_{40}(2\text{-}PET)_{24}$, nine pairs of signals are identified that centered at 238, 261, 282, 306, 327, 357, 419, 534, and 642 nm [166]. It is found that racemization temperature of $Au_{40}(2\text{-}PET)_{24}$ is higher than that of $Au_{38}(2\text{-}PET)_{24}$, which is attributed to the negative activation entropy for $Au_{40}(2\text{-}PET)_{24}$ [167].

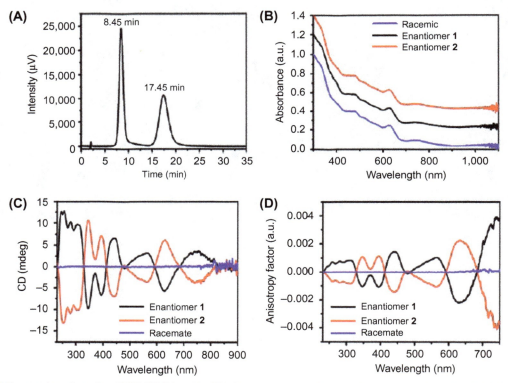

FIG. 5.22 HPLC-separation of rac-$Au_{38}(SCH_2CH_2Ph)_{24}$. (A) HPLC-chromatogram of the enantioseparation of rac-$Au_{38}(SCH_2CH_2Ph)_{24}$ with the ultraviolet-visible detector at 380nm. The peak at 8.45min corresponds to enantiomer 1; the second peak at 17.45 corresponds to enantiomer 2. (B) Ultraviolet-visible spectra of enantiomers 1 *(black)* and 2 *(red)* and of the racemate *(blue)*. The spectra were normalized at 300nm and off-set for clarity. The well-known ultraviolet-visible signature of Au_{38} is perfectly reproduced in all spectra, showing that the two collected fractions are composed of $Au_{38}(SCH_2CH_2Ph)_{24}$. CD spectra and anisotropy factors of $Au_{38}(SCH_2CH_2Ph)_{24}$. (C) CD spectra of isolated enantiomers 1 (black) and 2 (red) and the racemic $Au_{38}(SCH_2CH_2Ph)_{24}$ (blue) before separation; corresponding anisotropy factors of enantiomers 1 and 2 and of the racemate. (D) The spectra exhibit excellent mirror-image relationships and anisotropy factors $g = \Delta A/A$ of up to 4×10^{-3}. *(From I. Dolamic, S. Knoppe, A. Dass, T. Bürgi, First enantioseparation and circular dichroism spectra of Au_{38} clusters protected by achiral ligands, Nat. Commun. 3(1) (2012) 798, https://doi.org/10.1038/ncomms1802.)*

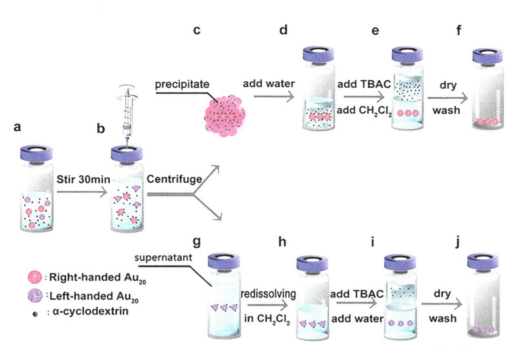

FIG. 5.23 α-CD based supramolecular assembly strategy for chiral separation of Au_{20}. (a) Dissolving Au_{20} and α-CD in DMF. (b) Stirring and adding CH_2Cl_2 as poor solvent. (c) Centrifuging to get precipitate (enriched R-Au_{20}) (d) Dissolving precipitate in water. (e) Transferring R-Au_{20} from water to CH_2Cl_2 with aid of TBAC (removing assembled α-CD). (f) Acquiring bare R-Au_{20} via drying and washing by water to remove residual TBAC. (g) Centrifuging to get supernatant (enriched L-Au_{20}). (h) Redissolving in CH_2Cl_2. (i) Removing residual α-CD in L-Au_{20} with aid of TBAC. (j) Acquiring bare L-Au_{20} via drying and washing by water to remove residual TBAC. Color labels: *blue* solvent, CH_2Cl_2 and DMF; *gray* solvent, H_2O. (From Y. Zhu, H. Wang, K. Wan, J. Guo, C. He, Y. Yu, L. Zhao, Y. Zhang, J. Lv, L. Shi, R. Jin, X. Zhang, X. Shi, Z. Tang, Enantioseparation of $Au_{20}(PP_3)_4Cl_4$ clusters with intrinsically chiral cores, Angew. Chem. Int. Ed. 57(29) (2018) 9059–9063, https://doi.org/10.1002/anie.201805695.)

In 2013, Zeng et al. reported the crystal structure of chiral $Au_{28}(TBBT)_{20}$, which possessed a rotating arrangement of four dimeric staple motifs. Chiral HPLC was used to separate the enantiomers of $Au_{28}(TBBT)_{20}$. The mirror-liked CD spectra showed two strong bands at 300 and 340 nm, weak peaks at 360, 380, 400, 425, and 470 nm, and a broad peak at 600 nm [168]. Jin et al. prepared a pair of structural isomers, one is chiral $[Au_9Ag_{12}(SR)_4(dppm)_6X_6]^{3+}$-C and the other one is $[Au_9Ag_{12}(SR)_4(dppm)_6X_6]^{3+}$-Ac nanoclusters. It is found that the optical isomers of the $[Au_9Ag_{12}(SR)_4(dppm)_6X_6]^{3+}$-C are separated by HPLC. The CD of spectra showed mirror-liked bands at 325, 340, 363, 400, 428, 448, and 483 nm, and a broad peak at ∼590 nm. The arrangement of the extra four Au atoms on kernels leads to the chirality of $[Au_9Ag_{12}(SR)_4(dppm)_6X_6]^{3+}$-C [169].

In view of the structure of $Ag_{29}(BDT)_{12}(TPP)_4$, it's found that the $Ag_{29}(BDT)_{12}(TPP)_4$ is chiral due to the rotation direction of surface motif. Recently, Takuya Nakashima and coworkers separated the enantiomers of $Ag_{29}(BDT)_{12}(TPP)_4$ by HPLC. CD spectra further showed four pairs of mirror signal for the enantiomers. The cluster with single chiral configuration is realized by chiral DHLA induction. DFT calculation showed that the optical activity origin from the electronic transition of molecular orbital in the whole nanoclusters. [170]

Different from the HPLC separation, compounds with inherent chirality is used for chiral resolution. Nature provides excellent examples about selective chiral recognition via multiple weak interactions. Tang et al. used α-cyclodextrin (α-CD) to separate the enantiopure Au_{20} via supramolecular assembly strategy (Fig. 5.23). The favorable configurations of assemblies, in which more O⋯H attraction bonds and thus stronger binding energy between R-Au_{20} and α-CD leaded to the favorable configuration of R-Au_{20}(α-CD)$_n$. The high enantiopurity and large amount separation was attributed to the multiple weak interactions [171].

5.4.3 Chiral ligands/compounds induce chirality for metal nanoclusters

In addition to the inherent chirality of the cluster, the reconstruction of the chirality of the cluster is also an important part of the nanoresearch. Bürgi et al. concluded that the chirality of metal nanoclusters, whether come from metal core or arrangement of surface motif, cannot withstand the driving force from the ligand with opposite absolute configuration

[172]. In their later work, Bürgi et al. made $Au_{25}(SCH_2CH_2Ph)_{18}^-$ react with R/S-BINAS and NILC/NIDC ligands. Although only part of the ligand exchange has occurred, the clusters after the ligand exchange are still optically active [173]. In 2010, they reacted $Au_{38}(2\text{-PET})_{24}$ and $Au_{40}(2\text{-PET})_{24}$ (2-PET: 2-phenylethanethiol) clusters with enantiopure BINAS (BINAS: 1,1'-binaphthyl-2,2'-dithiol) by ligand exchange. It is found that the maximum anisotropy factors are 6.6×10^{-4} [174].

Further, Bürgi et al. used chiral HPLC to monitor the reaction process of racemic $Au_{38}(2\text{-PET})_{24}$ with enantiopure BINAS in situ. It is found that the R-BINAS is preferable to bond the left-handed enantiomer of $Au_{38}(2\text{-PET})_{24}$, leading to a 60% disatereomeric excess [175]. In addition to BINAS, enantiopure planar chiral [2.2]paracyclophane-4-thiol 2-thio [4]helicene are also used for ligand exchange with racemic $Au_{38}(2\text{-PET})_{24}$. The chirality of nanoclusters by chiral ligand exchange was observed at 80°C due to at least three out of four possible symmetry sites [176]. The intensities of CD signals of $Au_{38}(2\text{-PET})_{24-x}(TH_4)_x$ was much lower than that of $Au_{38}(2\text{-PET})_{24}$ [177]. In addition the induction for racemic metal nanocluster, Bürgi et al. functionalized $Ag_{24}Au_1(DMBT)_{18}$ with R/S-BINAS ligand. The obtained $Ag_{24}Au_1(R/SBINAS)_x(DMBT)_{18-2x}$, with $x = 1–7$ showed intense CD signals. The CD signal in the low-energy region was more obviously enhanced with the increase in chiral ligand, which indicated that metal core and surface oligomeric units have effects on the chirality of nanoclusters [178].

In 2010, Zhu et al. prepared the chiral 2-phenylpropane-1-thiol (PET*) ligand, optically active $Au_{25}(pet^*)_{18}$ and $Au_{25}(PPh_3)_{10}(pet^*)_5Cl_2$ were achieved by using this chiral ligand. It is demonstrated that the chirality is original from the mixing of ligand orbitals and the surface metal atoms. The absorption profiles do not affect but the CD spectra of Au_{25} after chiral ligand exchange depend on the ligand type [179]. After this, Zhu and Jin et al. prepared $Au_{38}(R\text{-PET})_{24}$ and $Au_{38}(S\text{-PET})_{24}$ by using the 2-phenylpropane-1-thiol ligand. The CD spectra of $Au_{38}(R\text{-PET})_{24}$ and $Au_{38}(S\text{-PET})_{24}$ are different from that of Au_{38} protected by achiral ligand after HPLC separation, which indicating the ligand effect on the optical activity of metal nanoclusters [180]. The similar chiral ligand effect on the optical activity of metal nanoclusters were also observed in $Au_{25}(Capt)_{18}$, $Au_{25}(SG)_{18}$ and $Au_{25}(PET^*)_{18}$ [181].

In view of inducing effect of chirality, Zhu et al. investigated the chiral influence of chiral solvent and counterion. It was found that chiral solvent and counterion cannot induce chirality for achiral $Au_{25}(SC_2H_4Ph)_{18}^-$ nanocluster [182]. The alloying process is an important means to adjust the properties of metal clusters. Hiroshi Yao et al. investigated the alloy effect on chirality based on the AuAg alloy nanoclusters protected by GSH. The species of $Au_{12.2}Ag_{2.8}(SG)_{13}$, $Au_{14.4}Ag_{3.6}(SG)_{14}$, and $Au_{17.6}Ag_{7.4}(SG)_{18}$ were identified by PAGE. It is found that the alloyed AuAg nanoclusters showed weaker CD signals than that of Au nanoclusters, which is attributed to the increased geometrical isomers and configuration. The averaging of the spectra of the different isomers leads to a decrease in the CD signal [183]. Clusters protected by chiral ligands can generate strong optical activity signals. Rodolphe Antoine synthesis $Au_{10}(SG)_{10}$ nanoclusters that exhibited unique characteristic CD signals at 250–400 nm [184].

In terms of chiral induction by chiral phosphine, extensive progress has been made. In 2006, Tatsuya Tsukuda et al. synthesis chemically pure $[Au_{11}(BINAP)_4X_2]^+$ nanoclusters that exhibited excellent mirror-image CD spectra, which is attributed to the core deformation induced by chiral ligand [185]. Further, Aikens and coworker investigated the origin of intense chiroptical effect in $Au_{11}L_4X_2^+$ nanoclusters (2,2'-bis(diphenylphosphino)-1,1'-binaphtyl (BINAP) or 1,4-diphosphino-1,3-butadiene (dpb)) by TDDFT calculations. The simulated $Au_{11}(dpb)_4X_2^+$ had the similar CD feathers at 480–530 nm and 390–410 nm with the experimental results of $[Au_{11}(BINAP)_4X_2]^+$ and a opposite signal at 300–350 nm. The calculation results demonstrated the intense optical activity of metal nanoclusters is not only original from the metal but also the chiral arrangement and ligands [186].

Recently, inherently chiral metal nanoclusters protected by chiral phosphine or combined with other ligands were extensively studied. Katsuaki Konishi et al. reported the chiral structure and optical activity of $[Au_{24}L_6Cl_4]^{2+}$ nanocluster where L is (R,R)- or (S,S)-diphosphines. The structural chirality was assigned to the two helicene-like hexagold motifs around the achiral Au_8 core. The mirror-liked CD signals were observed at visible and near-IR region [187]. The CD profiles and anisotropy factor is affected by the chiral structure. Tatsuya Tsukuda compared the chirality of $[Au_{11}(R/S\text{-DIOP})_4Cl_2]^+$ (DIOP = 1,4-bis(diphenylphosphino)-2,3-o-isopropylidene-2,3-butanediol) and $[Au_8(R/S\text{-BINAP})_3(PPh_3)_2]^{2+}$. The $[Au_8(R/S\text{-BINAP})_3(PPh_3)_2]^{2+}$ cluster and Au_{11} clusters protected by BINAP ligands showed stronger optical rotator strength than $[Au_{11}(R/S\text{-DIOP})_4Cl_2]^+$. Theoretical results indicated that a chiral arrangement of π-electron system in close vicinity of the Au core can enhance the optical response [188]. Wang et al. reported the structures and chiral properties of $[Au_9(R\text{-/S-BINAP})_4](CF_3COO)_3$ and $[Au_{10}(R\text{-/S-BINAP})_4(p\text{-}CF_3C_6H_4C\equiv C)](CF_3COO)_3$. It is found that the asymmetry of Au_{10} is higher than that of Au_9, and maximum anisotropy factor of Au_{10} is up to 6.6×10^{-3} that is two times higher than that of Au_9 [189].

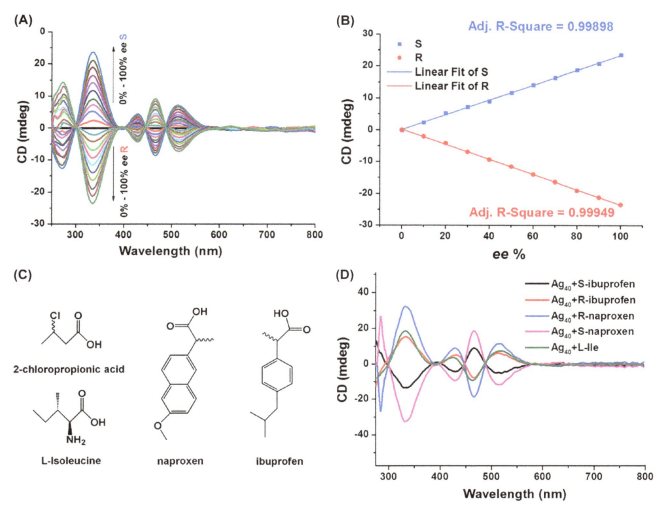

FIG. 5.24 Application of Ag_{40} in chiral amplification. (A) CD spectra of Ag_{40} with R-/S-2-chloropropionic acid from 0% to 100% ee in CH_2Cl_2. (B) Calibration curve by plotting CD readings at 334 nm in panel an against ee values of chiral 2-chloropropionic acid. (C) Seven model chiral carboxylic and amino acids. (D) CD spectra of the reaction mixture of Ag_{40} and the other five model chiral carboxylic acids. *(From W. Du, X. Kang, S. Jin, D. Liu, S. Wang, M. Zhu, Different types of ligand exchange induced by Au substitution in a maintained nanocluster template. Inorg. Chem. 59(3) (2020) 1675–1681, https://doi.org/10.1021/acs.inorgchem.9b02792.)*

Chiral ligands can break or lower the symmetry of metal nanoclusters. Zheng and coworkers synthesized the racemic $[Ag_{78}(DPPP)_6(SR)_{42}]$ (Ag_{78}) where DPPP is the achiral 1,3-bis(diphenyphosphino)propane, chiral BDPP = 2,4-bis-(diphenylphosphino) pentane ligands were used to separate enantiomeric Ag_{78} nanoclusters. The C—C—C bond angles of diphosphines restricted the direction of surface motifs, leading to the chirality of Ag_{78}. The optically pure Ag_{78} exhibited intensive CD signal with maximum anisotropy factor of 2×10^{-3} at 678 nm [190]. In their following work, $[Au_{13}Cu_2(DPPP)_3(SPy)_6]^+$ cluster and optically pure enantiomer $[Au_{13}Cu_2(2r,4r\text{-}BDPP)_3(SPy)_6]^+$ (3-R) and $[Au_{13}Cu_2(2s,4s\text{-}BDPP)_3(SPy)_6]^+$ (3-S) were prepared. The asymmetric arrangement of surface motif and ligands were responsible for the chirality of $Au_{13}Cu_2$. The CD spectra of optically pure 3-R and 3-S showed signals at 279, 288, 297, 319, 334, 360, 395, and 465 nm. Theoretical calculation assigned the CD signals [191].

Carboxylic acid as a new type of ligand has a good protective effect on metal nanoclusters. Zang et al. replaced the weakly coordinating ligands NO_3^- ligands with carboxylic or amino acid in Ag_{20} nanoclusters. It is found that the Ag_{22}, Ag_{23}, and Ag_{24} nanoclusters protected by amino acid showed mirror-liked CD signals. The chiral carbon center in the amino acid induced the chirality of metal nanoclusters [192]. Recently, Du et al. reported the structure and chirality of $Ag_{40}(TBBM)_{22}(CH_3COO)_{10}$ and $AuAg_{39}(TBBM)_{21}(CH_3COO)_{11}$. The CH_3COOH ligand is replaced by chiral carboxylic acid, including 2-chloropropionic acid, ibuprofen, naproxen, and isoleucine (Fig. 5.24). Due to the longer wavelength optical response signal of the cluster, it's used for chiral detection and chiral amplification. The detection range was from 0% to 100% ee, and the detection limit was 3.49% ee (3σ/slope) [193].

On the other hand, a chiral phase transfer or chiral counterions inductions are expected to achieve enantiomers separation. In 2014, Christopher J. Ackerson and coworkers used (−)-1R,2S-N-dodecyl-N-methylephedrinium bromide ((−)-DMEBr) as a phase transfer agent to get enantiomers of $Au_{102}(p$-$MBA)_{44}$ cluster. When low concentration of (−)-DMEBr was used, the transfer is incomplete, the organic phase and water phase exhibited near mirror-liked signals. The different interaction of left- and right-hand enantiomer with (−)-DMEBr was speculate to this phenomenon. When high concentration of (−)-DMEBr was used, the transfer is complete, the organic phase showed optical activity. Theoretical and experiment results showed that the optical activity after fully phase transfer is consistent with one of the diastereomers [194].

In 2016, a facile ion-pairing strategy was employed to achieve asymmetric synthesis of negatively charged chiral $[Ag_{28}Cu_{12}(SR)_{24}]^{4-}$ nanoclusters by Zheng group [195]. The chiral ammonium cations (N-benzylcinchoninium and N-benzylcinchonidinium) were used for separating the racemic enantiomer in postinduction route and asymmetric synthesis in direct synthesis route. The enantiomers showed mirror-imaged CD signals in the visible range, whereas the racemic mixture showed no CD signal. The X-ray crystal diffraction result demonstrated that the arrangement of surface shell is associated with the chirality of $[Ag_{28}Cu_{12}(SR)_{24}]^{4-}$ nanoclusters. Chiral self-recognition and self-discrimination were achieved in $[(R$-$BINAP)_4Au_{10}S_4]Cl_2$ (R-Au_{10}) and $[(S$-$BINAP)_4Au_{10}S_4]Cl_2$ (S-Au_{10}) by adding chiral anions [196]. The 2D regular rhombic nanocrystals were achieved from the racemic mixture of chiral S-Au_{10} and R-Au_{10} nanoclusters. After adding chiral anions, the morphology changes of the NCs from rhombic to strip and quasihexagonal nanocrystals were observed. Reverse and rotational layer-by-layer stacking is responsible for the different morphology of aggregation.

5.4.4 Combination of chirality and other properties

5.4.4.1 Circularly polarized luminescence

Circularly polarized luminescence (CPL) is a combination of chiral and luminescence, which have attract vast interest due to the high sensitivity and spatial resolution. CPL is based on the difference in right and left circular polarized light emission from the excited chiral material. Nanoclusters with PL and chirality can generate CPL. In this section; we will discuss the CPL of metal nanoclusters.

In 2017, Tang and coworkers prepared $Au_3[(R)$-Tol-BINAP$]_3$Cl and $Au_3[(S)$-Tol-BINAP$]_3$Cl clusters, which exhibited strong CD but free of CPL. After aggregation, the aggregated Au_3 nanoclusters showed strong CPL signals in the same wavelength as their PL centered at 583 nm. TEM results displayed the assemblies exhibit a cubic morphology [197]. Recently, Zang et al. reported the racemic anisotropic $Ag_{30}(C_2B_{10}H_9S_3)_8Dppm_6$ (Ag_{30}-rac) nanoclusters. Interestingly, the left-hand and right-hand enantiomer is crystalized separately in dimethylacetamide. In the racemic Ag_{30} nanocluster, the spiral arrangement of the ligands directed by unusual B—H···π and C—H···π bonding interactions among the carborane cages and the benzene rings were found to be responsible for the chirality. In the separate chiral assemblies, various weak interactions including B—H···π, C—H···π, π···π, and vdWs force were found in the helical superstructure, leading to nonclosest packing. The obtained R-Ag_{30} and L-Ag_{30} showed unique CPL property [198].

In addition to CPL of Au and Ag nanoclusters, the research on CPL of Cu nanoclusters have also made great progress. In 2019, Zang et al. synthesized the chiral $[Cu_{14}(R/S$-$DPM)_8](PF_6)_6$ (denoted as R/S-Cu_{14}) nanoclusters by using chiral ligands, which exhibited bright red emission and strong CPL signals in aggregation state. The CIE/AIE and chirality of R/S-Cu_{14} are contribute to the CPL property (Fig. 5.25) [199]. Besides, Zang group replaced the CH_3OH in the racemic mixture of $[K(CH_3OH)_2(18$-crown-6$)]^+[Cu_5(StBu)_6]^-$ with chiral amino alcohols. It was found that impressive CPL signals were observed in $[D/L$-valinol$(18$-crown-6$)]^+[Cu_5(StBu)_6]^-$ [200].

Recently, Chen et al. reported the assembly of chiral $[Au_1Ag_{22}(S$-$Adm)_{12}]^{3+}$ nanoclusters/superatom complex in the presence of SbF_6^- linkers [201]. The obtained 3D superatom complex inorganic framework material SCIF-1 and SICF-2 displayed red emission. Differently, the SCIF-2 showed CPL response centered at ∼660 nm due to the separated aggregation of enantiomers (Fig. 5.26). The direction of rotation of SR-Ag-SR motifs determines the chirality of Au_1Ag_{22} building block. Besides, the SCIF-1 and SCIF-2 framework showed unique sensitive photoluminescence in protic solvents.

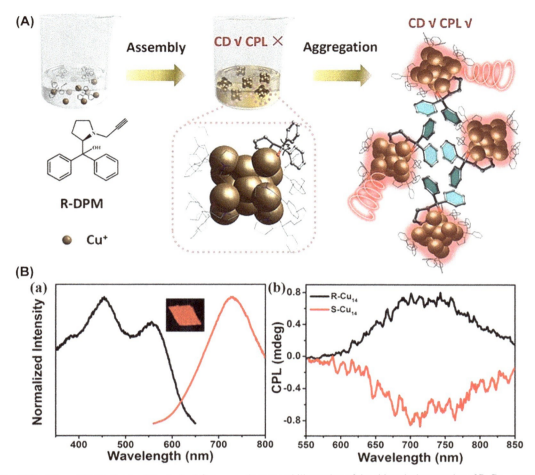

FIG. 5.25 The CPL of Cu_{14}. (A) Synthesis of the desired Cu_{14} nanoclusters and illustration of the chiroptical properties of R-Cu_{14} nanoclusters in the crystalline state and in solution. (B) (a) Normalized excitation (*black* trace, $l_{em} = 726$ nm) and emission (*red* trace, $l_{ex} = 445$ nm) spectra of R-Cu_{14} in the solid state at 290 K. (b) CPL spectra of the R/S-Cu_{14} enantiomers in the solid state. Inset: the corresponding photograph of R-Cu_{14} cluster at room temperature under 365 nm UV irradiation. *(From M.-M. Zhang, X.-Y. Dong, Z.-Y. Wang, H.-Y. Li, S.-J. Li, X. Zhao, S.-Q. Zang, AIE triggers the circularly polarized luminescence of atomically precise enantiomeric copper(I) alkynyl clusters. Angew. Chem. Int. Ed. 59(25) (2020) 10052–10058, https://doi.org/10.1002/anie.201908909.)*

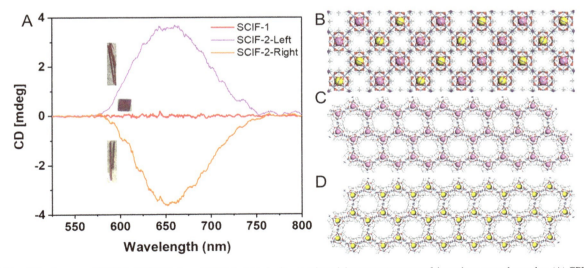

FIG. 5.26 CPL spectrum of SCIF-1, SCIF-2-left, and SCIF-2-right single crystals and the superstructures of these three crystal samples. (A) CPL spectra of SCIF-1, SCIF-2-left, and SCIF-2-right (inserts are photographs of the corresponding crystal). (B) Crystal structure of the SCIF-1 framework. (C) Crystal structure of the SCIF-2-left framework. (D) Crystal structure of the SCIF-2-right framework. *(From S. Chen, W. Du, C. Qin, D. Liu, L. Tang, Y. Liu, S. Wang, M. Zhu, Assembly of the thiolated $[Au_1Ag_{22}(S-Adm)_{12}]^{3+}$ superatom complex into a framework material through direct linkage by SbF_6^- anions, Angew. Chem. Int. Ed. 59(19) (2020) 7542–7547, https://doi.org/10.1002/anie.202000073.)*

5.4.4.2 Vibrational circular dichroism

Vibrational circular dichroism (VCD) is a powerful method for obtaining conformational information, which is more sensitive to conformation than conventional infrared absorption spectroscopy. In 2010, Bürgi et al. prepared small gold nanoparticles capped with (R)-1,1′-binaphthyl-2,2′-dithiol [(R)-BINAS] and (S)-1,1′-binaphthyl-2,2′-dithiol [(S)-BINAS] ligands. The VCD spectra of separated species were collected and compared with DFT simulation results. It is found that the VCD spectra were quite insensitive to the size of particles that is different from CD spectra, meaning that the vibrational characteristic is a local property [202]. Chirality transfer from gold nanocluster to adsorbate was proved in chiral $Au_{38}(2\text{-PET})_{24}$ by Bürgi and coworkers. The solution of chiral $Au_{38}(2\text{-PET})_{24}$ enantiomers showed strong VCD signals. The intense bands in the VCD spectra of the Au_{38} at 1454 and $1418\,cm^{-1}$ were assigned to CH_2 scissoring coupled to the in-plane phenyl ring vibrations and pure CH_2 scissoring modes, respectively. The intense band at $1263\,cm^{-1}$ was correlated to a CH_2 wagging/twisting mode. The intermolecular interactions among the adsorbed ligands on the cluster surface seem to be responsible for this phenomenon [203]. Recently, Li and Jin et al. reported the preparation of $Ag_{23}(PPh_3)_8(SC_2H_4Ph)_{18}$ nanocluster that exhibited chirality. A series of VCD peaks is observed in the range from 1200 to $1000\,cm^{-1}$, which is assigned to the Ag-P-C and Ag-S-C stretching modes. The chirality of Ag_{23} is attributed to the twisted framework of the Ag_{23} cluster. The simulated electronic circular dichroism (ECD) spectra match well with the corresponding absorption bands [204].

5.5 Electrochemical property of metal nanoclusters

Electrochemistry is a science to study the charged interface phenomenon formed by two kinds of conductors and the changes on it. At present, there are many branches of electrochemistry, such as synthetic electrochemistry, quantum electrochemistry, semiconductor electrochemistry, organic conductor electrochemistry, spectroelectrochemistry, bioelectrochemistry and so on. Electrochemistry is widely used in chemical industry, metallurgy, machinery, electronics, aviation, aerospace, light industry, instrument, medicine, materials, energy, metal corrosion and protection, environmental science and other scientific and technological fields. Electrochemistry is closely related to energy, materials, environmental protection and life science.

The battery is composed of two electrodes and the electrolyte between them. The research content of electrochemistry includes two aspects: one is the research of electrolyte, including the conductivity of electrolyte, the transport properties of ions, and the equilibrium properties of ions participating in the reaction; the other is the research of electrode, including the equilibrium properties of electrode and the polarization properties after being electrified, that is, the electrochemical behavior of the electrode at the solute interface. The electrode, electrolyte and other factors will affect the test results. In nanoclusters, we often use open circuit potential, cyclic voltammetry, linear sweep voltammetry and differential pulse voltammetry to characterize the electrochemistry properties of metal nanoclusters.

5.5.1 Nanoelectrochemistry

Most of modern chemistry focuses on small-scale materials, and there is more and more research on nanoscale electrochemistry. Focused on the concept of "nanoelectrochemistry," Murray et al. discussed the electrochemical process of nanoparticles, single nanoelectrodes and nanopores [205]. Recently, a lot of significant research progress has been made in the size range of 10 nm and smaller. In addition, although the preparation of small-size substances is very special and novel, only showing their size, shape, and composition is meaningful for chemical research, and then it will stimulate people's interest in studying their properties.

The optical gap of nanoparticles is the electronic band edge or absorption spectrum of the transition from HOMO (highest occupied molecular orbital) to LUMO (lowest unoccupied molecular orbital). In the near infrared region or higher energy (i.e., >1 eV), the light absorption band edge is detected. Due to the available electrochemical potential window, the energy of the optical gap is greater than that of the electrochemical detection. The electrochemical energy gap is the difference between the first oxidation potential and the first reduction potential (greater than QDL charging energy). The HOMO-LUMO gap can be estimated by reasonably modifying the charging energy [206,207]. In the solvent medium used, the voltage increment generated by producing positive and negative ions from neutral substance or increasing the charge on charged substance is the charging energy. The electronic excitation of optical HOMO-LUMO is not accompanied by the change of the overall charge and does not need charge energy correction.

The electrochemical and optical HOMO-LUMO energy gaps of a series of organothiolate ligand protected Au nanoparticles (MPC) were reported. There are three regions of nanoparticle behavior, range in size from large to small: bulk continuum with no voltammetric features (Au_x), quantized double layer charging for Au_{225} and Au_{140}, molecule-like Au_{75}, Au_{55}, Au_{38}, Au_{25}, and Au_{13}. It is demonstrated that for smaller nanoparticles, HOMO-LUMO gap is forming. With the decrease in MPC size, the number of gold atoms decreases and the gap between these potentials expands faster. Murray et al. also discussed electrochemistry of films of nanoparticles and nanoscopic electrodes. Electrochemistry of nanoscopic electrodes focuses on the fabrication, characterization and performance of nanoscopic electrodes. In addition to the above, we also studied the electrochemistry of single nanopore, including single nanopore in the film, nanopore electrode and so on.

5.5.2 Electrochemical analysis for metal nanoclusters

Ligand-protected metal nanoclusters containing a few to hundreds of core atoms have received much attention in recent years due to their contribution to theoretical science and practical application. In here, we mainly discuss the advances in electrochemistry of atomically precise metal nanoclusters. In recent years, many kinds of clusters have been studied by electrochemistry. The electronic structure of metal nanoclusters is studied and the HOMO-LUMO gap is calculated by electrochemistry. Compared with other metal nanoclusters, the electrochemical research of gold nanoclusters has made great progress.

Recently, gold nanoclusters with various sizes have been synthesized by innovative methods, which help to study the influence of size effect on the properties of clusters. Therefore, great progress has been made in understanding their atomic and electronic structures and quantum size effects, which are manifested in their molecular properties, such as the opening of HOMO-LUMO gap, excision dynamics, and photoluminescence.

Guda Ramakrishna et al. synthesized highly monodisperse n-hexanethiolate-protected gold clusters ranging from Au_{25} to Au_{333} (Fig. 5.27). The size-dependent electrochemical and optical properties were investigated for a series of n-hexanethiolate-protected gold clusters were revealed [208]. Owing to voltammetric studies, the highest occupied molecular orbital-lowest unoccupied molecular orbital (HOMO – LUMO) gaps of these clusters decrease with increasing cluster size. The voltammetry study showed that the electrochemical HOMO – LUMO gap decreases from 1.32 to 0.15 V as the cluster size increases from Au_{25} to Au_{144}.

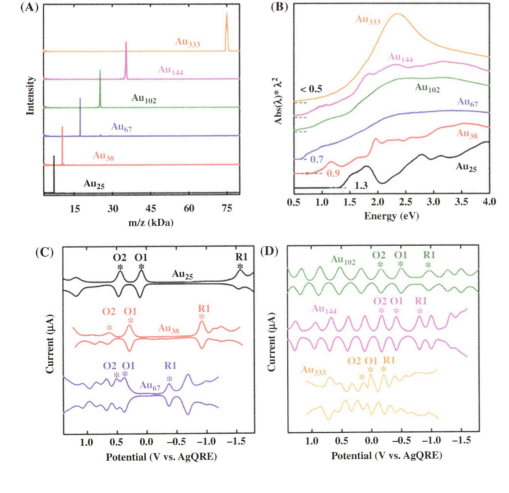

FIG. 5.27 The size-dependent electrochemical and optical properties of Au_{25}, Au_{38}, Au_{67}, Au_{102}, Au_{144}, and Au_{333} clusters. (A) MALDI mass spectra and (B) UV-vis-NIR absorption spectra of Au_{25}, Au_{38}, Au_{67}, Au_{102}, Au_{144}, and Au_{333} clusters. (C) SWVs of Au_{25}, Au_{38}, and Au_{67} clusters and (D) Au_{102}, Au_{144}, and Au_{333} clusters in CH_2Cl_2 containing 0.1 M Bu_4NPF_6. (From K. Kwak, V.D. Thanthirige, K. Pyo, D. Lee, G. Ramakrishna, Energy gap law for exciton dynamics in gold cluster molecules, J. Phys. Chem. Lett. 8(19) (2017) 4898–4905, https://doi.org/10.1021/acs.jpclett.7b01892.)

The absorption and electrochemical spectra of these gold clusters show obvious size effect. In addition, the electrochemical excited state kinetics and excited state kinetics were systematically studied. Cluster molecules have obvious HOMO-LUMO band gap. With the decrease in the band gap, cluster molecules show molecular capacitance and become quantized double-layer charging capacitance. Transient absorption measurements were performed to solve the size dependent exciton dynamics of clusters that have shown pump power independent dynamics (from Au_{25} to Au_{144}). The combined measurement of femtosecond and microsecond transient absorption shows that the exciton decays completely and the lifetime decreases with the increase in cluster size. It is found that the size dependent exciton lifetime is closely related to the HOMO-LUMO energy gap, which indicates that the energy gap law of excitons in these gold clusters is applicable. This study provides a basis for us to further understand the properties and dynamics of excitons in gold clusters, and has practical significance for photovoltaics and photocatalysis.

Voltammogram of $Au_{25}(SCH_2CH_2Ph)_{18}$ show three oxidation peaks at -0.39 V (O_1), -0.04 V (O_2) and 0.69 V (O_3), and the reduction peak at -2.06 V (R_1). According to the potential difference between the first oxidation peak (O_1) and the reduction peak (R_1), the electrochemical gap is 1.67 V. Subtracting the charging energy, which is typically estimated using the gap between O_1 and O_2 (0.35 V), gives a corrected energy of 1.32 eV. For the $Au_{25}(SR)_{18}$ cluster, the open-circuit potential (OCP) was found at -0.49 V, indicating that the $Au_{25}(SR)_{18}$ cluster is in anionic form, i.e., $[Au_{25}(SR)_{18}]^-$ [209,210].

The thiol protected Au_{25} cluster $[Au_{25}(SR)_{18}]^-$ is a closed electron shell of 8 electrons, which leads to its special stability. The bimetallic nanoclusters with superatomic 6-electronic configuration ($1s^2 1p^4$) is obtained by replacing the gold atoms in $[Au_{25}(SR)_{18}]^-$ with Pd or Pt. The voltammetric studies of $[PdAu_{24}(SR)_{18}]^0$ and $[PtAu_{24}(SR)_{18}]^0$ show that the HOMO−LUMO gaps of these clusters are 0.32 eV and 0.29 eV, respectively, indicating that the electronic structure of $[Au_{25}(SR)_{18}]^-$ has undergone dramatic changes when the heterometal is doped.

Compared with $Au_{25}(SR)_{18}$, the SWV of $PdAu_{24}(SR)_{18}$ is quite different. Specifically, there is a doublet of oxidation at -0.03 (O_1) and 0.40 (O_2) and a doublet of reduction at -0.78 (R_1) and -1.10 V (R_2), and the OCP of $PdAu_{24}(SR)_{18}$ is -0.48 V in the middle of the two peaks. The combination of voltammetry and NMR analysis clearly confirmed that the isolated $PdAu_{24}(SR)_{18}$ is electrically neutral, rather than -2 charge state. The oxidation peaks corresponding to $[PdAu_{24}(SR)18]^{1+/0}$ (O_1) and $[PdAu_{24}(SR)_{18}]^{2+/1+}$ (O_2), and the reduction peaks corresponding to $[PdAu_{24}(SR)_{18}]^{0/1-}$ (R_1) and $[PdAu_{24}(SR)_{18}]^{1-/2-}$ (R_2), respectively. The SWV of $PtAu_{24}(SR)_{18}$ is similar to that of $PdAu_{24}(SR)_{18}$, with double oxidation peaks at -0.03 (O_1) and 0.41 (O_2), and reduction peaks at -0.76 (R_1) and -1.10 V (R_2), respectively. The $PtAu_{24}(SR)_{18}$ is also neutral cluster with the OCP was found at -0.49 V. The HOMO-LUMO gaps of $PdAu_{24}(SR)_{18}$ and $PtAu_{24}(SR)_{18}$ deduced from the SWV are determined to be 0.32 eV and 0.29 eV, respectively. The voltammetric results strongly proved that the electronic structure of $Au_{25}(SR)_{18}$ doped with Pd and Pt atoms has changed a lot, and the electronic structures of the resulting $PdAu_{24}(SR)_{18}$ and $PtAu_{24}(SR)_{18}$ clusters are very similar. The optical and electrochemical properties of the doped clusters are completely different from those of the parent clusters. The HOMO-LUMO gap of these clusters measured by voltammetry is reduced, which is consistent with the DFT calculation. Moreover, Dongil Lee and coworkers further successfully proved that the reduction of the HOMO-LUMO gaps in the cluster of 6-electrons $[MAu_{24}(SR)_{18}]^0$ was due to the Jahn-teller like distortion, accompanied by the split of 1p orbit. The results show that metal doping may change their surface binding properties and redox potential, which makes them have high active sites and high selectivity in electrochemical applications.

Heteroatom doping is an effective strategy to adjust the optical and electronic properties of clusters at the atomic level. In order to further explore Jahn-Teller distortion in nanoclusters, Dongil Lee et al. using $Au_{38}(SR)_{24}$ as the parent cluster, Pd-doped nanocluster is $Pd_2Au_{36}(SC_6H_{13})_{24}$, and Pt-doped nanocluster is $Pt_2Au_{36}(SC_6H_{13})_{24}$ [148,211]. They analyzed the atomic and electronic structures of these two clusters, $Pd_2Au_{36}(SC_6H_{13})_{24}$ is electrically neutral and the isolated $Pt_2Au_{36}(SC_6H_{13})_{24}$ was found to be dianionic, i.e., $[Pt_2Au_{36}]^{2-}$. These doped clusters still show different optical and electrochemical properties. Due to the different atoms doped, the redox potential of SWV is obviously changed. The voltammetric pattern changes little after doping Pt, but it changes obviously after doping Pd. The SWV of Pt_2Au_{36} showed that the oxidation peak was at -0.40 (O_1), -0.08 (O_2) and 0.30 (O_3), and the reduction peak was at -1.67 V (R_1), with the OCP was found at -0.58 V. These oxidation peaks correspond to $[Pt_2Au_{36}]^{-/2-}$ (O_1) and $[Pt_2Au_{36}]^{0/-}$ (O_2), respectively, and the reduction peak corresponds to $[Pt_2Au_{36}]^{2-/3-}$ (R_1). It is concluded that the electrochemical gap of $[Pt_2Au_{36}]^{2-}$ is 1.27 V, and the HOMO-LUMO gap is 0.95 V.

On the contrary, the voltammograms of Pd_2Au_{36} cluster are quite different from those of Au_{38} and Pt_2Au_{36}. The double oxidation peaks located at -0.15 (O_1) and 0.17 V (O_2), double reduction occurred at -0.73 (R_1) and -1.02 V (R_2), with the OCP was found at -0.47 V. These redox peaks correspond to $[Pd_2Au_{36}]^{2+/+}$ (O_2), $[Pd_2Au_{36}]^{+/0}$ (O_1), $[Pd_2Au_{36}]^{0/-}$ (R_1), and $[Pd_2Au_{36}]^{-/2-}$ (R_2), respectively. The electrochemical gap and the HOMO-LUMO gap of $[Pd_2Au_{36}]^0$ are dramatically decreased to 0.58 and 0.26 V, respectively.

The HOMO−LUMO gaps of Au_{38}, Pt_2Au_{36} and Pd_2Au_{36} were determined to be 0.86, 0.95, and 0.26 eV by voltammetric analysis, which were consistent with DFT calculation results. The above results show that metal doping can effectively affect the electronic structure of the clusters. The voltammograms of Au_{38} and Pt_2Au_{36} are quite similar, reflecting their similar electronic structure (14 electron configuration). According to the results of DFT calculation, the 12-electron $[Pd_2Au_{36}]^0$ clusters with Jahn-Teller distortion, which leads to the reduced HOMO-LUMO gap of Pd_2Au_{36}. These studies help us to understand the influence of different dopants on the electronic structure of the clusters, so as to regulate the electronic structure reasonably.

At present, the voltammetry of nanoclusters has been proved to be of guiding significance for electronic properties, but it is rarely used to study water-soluble nanoclusters. Because the resolution of the voltage spectrum obtained in aqueous medium is usually very poor, and the electrochemical potential window is usually limited by the oxidation and reduction potentials of water. Dongil Lee et al. transfer water-soluble clusters to organic phase by pairing ions with hydrophobic anion, a water-soluble Au_{25} nanocluster protected with (3-mercaptopropyl)sulfonate (MPS-Au_{25}) was transferred into toluene phase by ion-pairing with tetraoctylammonium cation (TOA^+) [212]. The unique electrochemical properties of phase transfer Au_{25} clusters were revealed by voltammetry in dichloromethane and toluene mixed solvent, that is, the electrochemical HOMO-LUMO gap varies with the polarity of solvent. By transforming water-soluble nanoclusters into organic phase, the feasibility of electrochemical research is proved, and a new possibility for electrochemical characterization of water-soluble nanoclusters is provided, which helps us to explore more properties of water-soluble nanoclusters.

Metal nanoclusters have more prominent quantum effect. The problem of band gap and its dependence on composition and structure remains to be solved. It has an important influence on various electrochemical, optical and other properties. However, in practical experiments, it is difficult to determine how the energy state changes when the energy difference is small, especially in the range of transition from nanoclusters to larger nanomaterials.

$Au_{246}(p\text{-}MBT)_{80}$ is proved to be the largest nonmetallic cluster in quantum size range, while $Au_{329}(PET)_{84}$ is found to have no obvious electrochemical gap and is metallic [154,213,214]. Jin and coworkers reported that $Au_{279}(TBBT)_{84}$ is metallic, but there is still a lack of electrochemical analysis [21]. Dass et al. reported that the electrochemical energy gap of $Au_{133}(TBBT)_{52}$ was not found, and $Au_{144}(PET)_{60}$ was also analyzed, showing a significant ~ 0.37 V peak spacing [215]. Due to the determinacy of the atomic structure and composition of nanoclusters, DFT calculation can provide a very detailed energy diagram; therefore, in an ideal case, experimental exploration and theoretical inference are at least semi quantitative in terms of energy levels and density of states [20,216,217]. The correlation between the energy gap and the structure of the transition region between nanoclusters and nanoparticles needs to be verified.

Wang et al. analyzed the electrochemical properties of atomic accurate $Au_{133}(TBBT)_{52}$, $Au_{144}(BM)_{60}$ and $Au_{279}(TBBT)_{84}$, which are called quantized bilayer charging, namely "continuous" single electron transfer (1e, ETs), to reveal the transition from nonmetallic to metallic transitions (Fig. 5.28) [218]. In voltammetric analysis, the authors use a single ETS as an internal standard for calibration and temperature variation, the energy difference of sub hundred meV is solved from the "usually nearly uniform" peak distance of multiple reversible redox peaks. Cyclic and differential pulse voltammetry experiments show that the energy gaps of Au_{133} and Au_{144} are 0.15 eV and 0.17 eV at 298 K, respectively. It is confirmed that Au_{279} is metallic and shows a "bulk continuous" charge response with no gap. The energy gap and double-layer capacitance of Au_{133} and Au_{144} increase with the decrease in temperature. The temperature dependence of charge energy and HOMO-LUMO gap of Au_{133} and Au_{144} is attributed to the reverse ion permeation and steric hindrance of the ligands and their molecular composition. On the basis of solving the subtle energy difference, the spectroelectrochemical characteristics of Au_{133} and Au_{144} were compared with ultrafast spectra, and a general analysis method was proved to correlate the steady-state and transient energy maps of the process energy. Electrochemiluminescence (ECL) is one of the energy output processes after charge transfer reaction. The ECL intensity of Au_{279} is negligible, while the ECL intensity of Au_{133} and Au_{144} is relatively strong and observable. The results of these atomic precise nanoclusters also show that the combination of voltammetry and spectral analysis, as well as temperature changes, is a powerful tool for revealing subtle differences and gaining insights that other nanomaterials cannot obtain.

5.5.3 Electrochemiluminescence

Electrochemiluminescence (ECL) is a kind of luminescent phenomenon that some special substances are produced by electrochemical method, and then these substances are further reacted with each other or with other substances. It is a combination of chemiluminescence and electrochemical methods. Electrochemiluminescence analysis has the advantages of high sensitivity, simple equipment, easy operation and automation. It is widely used in biology, medicine, pharmacy, clinic, environment and immunity. ECL system mainly includes inorganic system (ruthenium complexes, iridium complexes), organic system and nanoparticle system. Recently, metal complexes, luminol and nanomaterials have been used as

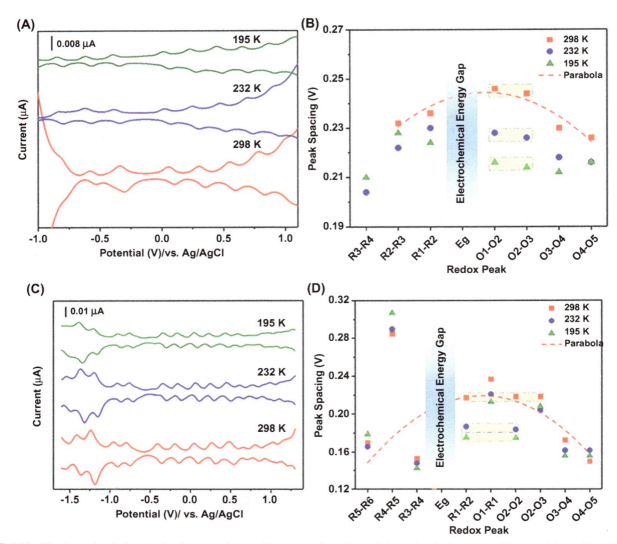

FIG. 5.28 The electrochemical properties of Au_{133} and Au_{144}. Temperature dependence of electrochemical properties of Au_{133} and Au_{144}. The DPVs of Au_{133} and Au_{144} at 298, 232, and 195 K. Peak spacing ΔVs analyzed, respectively. *(From S. Chen, T. Higaki, H. Ma, M. Zhu, R. Jin, G. Wang, Inhomogeneous quantized single-electron charging and electrochemical–optical insights on transition-sized atomically precise gold nanoclusters, ACS Nano 14(12) (2020) 16781–16790, https://doi.org/10.1021/acsnano.0c04914.)*

ECL luminescent materials. ECL has the advantages of high sensitivity, wide dynamic range, fast, simple and stable labeling. In addition, it has achieved great success for in vitro diagnosis by combining immunoassay and DNA probe detection. Metal complexes, such as iridium complexes, provide color tuning for lighting and analytical applications. In addition to molecular luminescent materials, recently discovered nanoluminophores include nanocrystals, metal nanoclusters and carbon nanomaterials. Although many luminescent materials based on nanomaterials have been studied in recent years, they usually have some disadvantages, such as uneven size distribution, easily affected structure, and great changes in luminescent properties after interaction with other materials. Fluorescent metal nanoclusters have the advantages of low toxicity, high stability and convenient preparation. As a new type of ECL emitter, they may have potential applications.

Noble metal nanoclusters have attracted much attention due to their precise structure, ultrasmall size, low toxicity and unique physicochemical properties. At present, there are many studies on their optical properties and few studies on their ECL properties. Ding et al. studied the near-infrared photoluminescence (NIR-PL) and electrochemiluminescence (ECL) of $Au_{25}(SC_2H_4Ph)_{18}^+ C_6F_5CO_2^-$ (Au_{25}^+) clusters (Fig. 5.29) [219]. Generating NIR-ECL light is particularly important in biological imaging applications, but there are few reports about NIR-ECL, and there are no reports about NIR-ECL of Au clusters. This is the first time that NIR-ECL emission has been observed in both annihilation and coreactant paths and explored the mechanisms for the NIR emission. This work clearly shows that NIR ECL of Au_{25} can be observed during

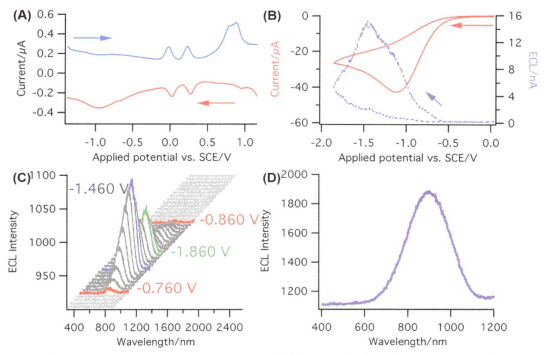

FIG. 5.29 ECL of Au$_{25}^+$ clusters. (A) Differential pulse voltammograms of 0.67 mg/mL Au$_{25}^+$ clusters in 1:1 benzene/acetonitrile solution, with 0.1 M tetra-nbutylammonium perchlorate as supporting electrolyte. (B) Cyclic voltammogram and ECL-voltage curve for an Au$_{25}^+$ cluster solution with 5 mM benzoyl peroxide. (C) Spooled ECL spectra of Au$_{25}^+$ clusters with BPO during one cycle of the applied potential from 0.04 to −1.86 V then back to 0.04 V. Each spectrum was acquired for 1s using an iDUS NIR CCD camera cooled to −75°C. (D) Typical accumulated ECL spectrum of the same coreactant solution collected over 80 s, two cycles of potential scanning between 0.04 and −1.86 V at a scan rate of 0.1 V/s. *(From K.N. Swanick, M. Hesari, M.S. Workentin, Z. Ding, Interrogating near-infrared electrogenerated chemiluminescence of Au$_{25}$(SC$_2$H$_4$Ph)$_{18}^+$ clusters, J. Am. Chem. Soc. 134(37) (2012) 15205–15208, https://doi.org/10.1021/ja306047u.)*

the annihilation of Au$_{25}^{2+}$ and Au$_{25}^{2-}$ species and enhanced in the path of coreaction system with BPO. By means of electrochemistry, photoluminescence and the pseudo-offline spectra the authors clearly show that the luminescence is caused by the electron relaxation of Au$_{25}^-$* excited state to the ground state through the forbidden band of HOMO-LUMO.

Ding et al. also studied the negatively charged Au$_{25}$ cluster, Au$_{25}^-$ have obvious electrochemical and optical properties, which provide an opportunity for the photoelectrochemical study of ECL [220]. Under the self-annihilation condition, Au$_{25}^-$ produces various oxidation and reduction forms under electrochemical conditions, and does not show obvious ECL luminescence, which may be due to the short lifetime of the electrogenerated intermediates. Interestingly, after the addition of tri-n-propylamine (TPrA) or Au$_{25}^-$ benzoyl peroxide (BPO) to form the coreactant system, the corresponding highly reductive and oxidizing intermediates produced by TPrA and BPO emit light at 950 and 890 nm in the near infrared (NIR) region. The effects of different concentrations of TPrA (6.3, 12.5, 25, 50, 100, and 200 mm) and BPO (2.5, 5, 25, and 50 mm) on the strength of ECL were also studied. Using TPrA or BPO as the coreactant, a long-lived strong reducing or oxidizing intermediate was produced, which could react with various Au$_{25}$ charged species to produce NIR-ECL. The Au$_{25}$/TPrA system shows ECL emission at 950 and 900 nm. By spooling ECL spectra, the dual wavelength ECL is attributed to three excited states: Au$_{25}^{-*}$, Au$_{25}^{0*}$ and Au$_{25}^{+*}$. Interestingly, the intensity of ECL is proportional to the local concentration of these excited states and can be adjusted by the applied potential or the concentration of coreactants. The coreactant BPO also enhanced the ECL emission at 950 nm. The effect of TPrA and BPO concentration on the intensity of ECL was studied. It was found that TPrA system could enhance ECL signal more effectively.

The mechanism of near infrared (NIR) ECL of Au$_{38}$(SC$_2$H$_4$Ph)$_{24}$ (Au$_{38}$, SC$_2$H$_4$Ph = 2-phenylethanethiol) nanoclusters in the annihilation and coreaction paths is reported. Due to the short lifetime and/or low reactivity of the intermediate, ECL emission will not occur in the annihilation path. In the range of anode potential, TPrA was used as coreactant to produce strong emission at 930 nm in the near infrared region. The mechanism of ECL was analyzed by ECL potential curve and pseudo ECL spectrum. It was found that Au$_{38}^{+*}$ (and Au$_{38}^{3+*}$) were the main excited states in the process of light emission. In the range of cathode potential, benzoate with high oxidation ability was formed by using BPO as coreactant, which showed a unique peak wavelength of 930 nm in the near infrared region. TPrA and BPO greatly improve the emission of ECL. In the presence of TPrA and BPO, Au_{38}^{+*} (and Au_{38}^{3+*}) and Au_{38}^{-*} participate in the emission process of ECL, respectively. In addition, the cumulative ECL spectra show that the wavelength of ECL emission peak at ~930 nm does not change with the change of TPrA concentration or applied potential. The strength of ECL is adjusted by changing the concentration of coreactant or working electrode potential. The ECL efficiency is 3.5 times higher than that of $Ru(bpy)_3^{2+}$/TPrA system. The NIR-ECL of Au_{38} cluster has potential application in electroanalytical imaging of living cells [221].

Wang et al. investigated the annihilation ECL of Au_8 fluorescent clusters in organic solution and the coreactant ECL of Au_8 fluorescent cluster film in aqueous solution by using the organic soluble Au_8 clusters produced in the process of ligand exchange induced etching of gold nanoparticles. As a new ECL emitter, the obtained Au_8 clusters have potential applications [222]. In view of the ECL enhancement of gold nanoclusters, Wang group established a covalent attachment of coreactants N,N-diethylethylenediamine (DEDA) onto lipoic acid stabilized Au (Au-LA) clusters. The obtained Au-LA-DEDA reduced the mass transport between the Au-LA cluster and coreactants and thus enhances the ECL intensity and efficiency. The ECL intensity of Au-LA-DEDA is multifold higher than the combination of the same amount of Au-LA and DEDA. The ECL intensity of Au-LA-DEDA is dependent on the electrode potential and solution pH, which is beneficial for the further applications [223].

In addition to the electrochemistry of gold nanoclusters, the electrochemistry of alloy nanoclusters has also made some progress. Chen et al. reported the near-infrared ECL of rod-shaped bimetallic $Au_{12}Ag_{13}$ nanoclusters (Fig. 5.30) [224]. Using the standard tris (bipyridine) ruthenium (II) complex ($Ru(bpy)_3^{2+}$) as a reference, the self-annihilation ECL of

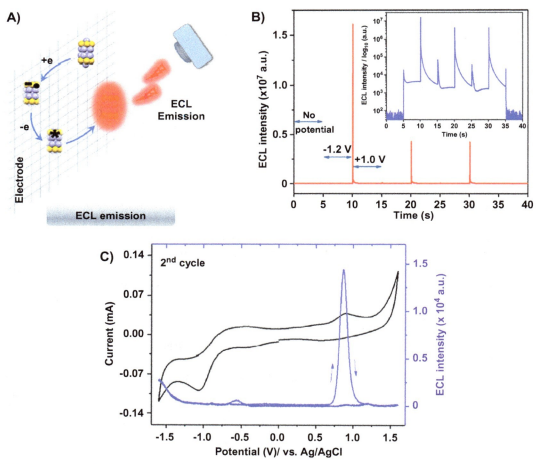

FIG. 5.30 ECL of $Au_{12}Ag_{13}$. (A) Experimental setup. Sketched is the self-annihilation ECL generation from $Au_{12}Ag_{13}$. (B) Potential step self-annihilation ECL of $Au_{12}Ag_{13}$. The $Au_{12}Ag_{13}$ concentration is ca. ~10 μM in 1:1 TOL:ACN with 0.1 M TBAP electrolyte. ECL profile under potential steps between −1.2 and +1.0 V. The electrode potential was held for 5 s at the denoted potentials and stepped cyclically (three cycles shown). No potential was applied in the first and final 5 s. The inset is in log scale for ECL intensity to better illustrate the gradual decay. (C) Cyclic voltammogram (left axis) and ECL-potential (right axis) curves of $Au_{12}Ag_{13}$. *(From S. Chen, H. Ma, J.W. Padelford, W. Qinchen, W. Yu, S. Wang, M. Zhu, G. Wang, Near infrared electrochemiluminescence of rod-shape 25-atom AuAg nanoclusters that is hundreds-fold stronger than that of Ru(bpy)$_3$ standard. J. Am. Chem. Soc. 141(24) (2019) 9603–9609, https://doi.org/10.1021/jacs.9b02547.)*

$Au_{12}Ag_{13}$ nanoclusters is about 10 times higher. The ECL of $Au_{12}Ag_{13}$ is 400 times stronger than that of $Ru(bpy)_3^{2+}$ when 1 mM TPrA is used as the coreactant. Based on the high near-infrared PL selection of bimetallic $Au_{12}Ag_{13}$ nanoclusters, the ECL properties were studied.

Based on the anode and cathode activities of the bimetallic nanoclusters, the mechanism of ECL formation is proposed. The strong ECL of $Au_{12}Ag_{13}$ originates from the 13th Ag atom in the center, which stabilizes the charge on the LUMO orbit, makes the nucleus of rod like $Au_{12}Ag_{13}$ more stable, reduces the nonradiative decay, and enhances the PL and ECL. Especially in the field of fast developing atomic level accurate nanoclusters, the properties related to atomic/electronic structure have been the goal of people. The mechanism insights provided by this basic research can guide the design and synthesis of other nanoclusters or materials for better performance, and further confirm the structure function correlation. The high ECL signal in the near infrared region with less interference provides the comprehensive advantages of high signal, low noise/interference or high contrast, which can be used in a wide range of analytical sensing, immunoassay and other related applications.

5.5.4 Photoelectric conversion

Zang et al. synthesized $Au^{+1/0}$ gold nanocluster Au_{28} with 9-HC≡C-closo-1,2-$C_2B_{10}H_{11}$ as reducing agent and protecting agent. It is worth noting that strong absorption and wide absorption nanoparticles (IBAN) with larger cross-section and wider wavelength range than traditional organic dyes and inorganic quantum dots (QDs) are ideal candidates for light capture applications. Therefore, Au_{28} and Au_{23} may be used in some photocatalytic reactions. In view of this, they tested the photocurrent response of a typical three electrode system (glassy carbon electrode as working electrode, platinum wire as auxiliary electrode, Ag/AgCl as reference electrode) in 0.1mkcl aqueous solution. When LED lights ($\lambda \geq 420$ nm; 50 W; interval 50 s) are used to turn on and off, the photocurrent density increases and decreases accordingly. It is demonstrated that the photocurrent densities of Au_{23} and Au_{28} derived electrodes are 0.83 and 0.91 μ/acm^2, respectively, indicating that both of them have good photoinduced electron/hole pair generation and separation efficiency [225]. At the same time, after several on/off cycles, the photocurrent density remains unchanged, indicating that the response has good reproducibility.

Electrochemical methods can be combined with the advantages of other testing methods to conduct a more comprehensive and in-depth study of nanoclusters, so as to solve the problems that cannot be solved by a single method. Therefore, we have reason to believe that electrochemistry will be a very worthy research direction.

References

[1] M. Zhu, C.M. Aikens, F.J. Hollander, G.C. Schatz, R. Jin, Correlating the crystal structure of a thiol-protected Au_{25} cluster and optical properties, J. Am. Chem. Soc. 130 (18) (2008) 5883–5885, https://doi.org/10.1021/ja801173r.

[2] S. Chen, S. Wang, J. Zhong, et al., The structure and optical properties of the [$Au_{18}(SR)_{14}$] nanocluster, Angew. Chem. Int. Ed. 54 (10) (2015) 3145–3149, https://doi.org/10.1002/anie.201410295.

[3] A. Das, C. Liu, H.Y. Byun, et al., Structure determination of [Au18(SR)14], Angew. Chem. Int. Ed. 54 (10) (2015) 3140–3144, https://doi.org/10.1002/anie.201410161.

[4] S. Chen, L. Xiong, S. Wang, et al., Total structure determination of Au21(S-Adm)15 and geometrical/electronic structure evolution of thiolated gold nanoclusters, J. Am. Chem. Soc. 138 (34) (2016) 10754–10757, https://doi.org/10.1021/jacs.6b06004.

[5] S. Yang, S. Chen, L. Xiong, et al., Total structure determination of Au16(S-Adm)12 and Cd1Au14(StBu)12 and implications for the structure of Au15(SR)13, J. Am. Chem. Soc. 140 (35) (2018) 10988–10994, https://doi.org/10.1021/jacs.8b04257.

[6] E.B. Guidez, V. Mäkinen, H. Häkkinen, C.M. Aikens, Effects of silver doping on the geometric and electronic structure and optical absorption spectra of the Au25−nAgn(SH)18− (n = 1, 2, 4, 6, 8, 10, 12) bimetallic nanoclusters, J. Phys. Chem. C 116 (38) (2012) 20617–20624, https://doi.org/10.1021/jp306885u.

[7] A. Das, T. Li, K. Nobusada, C. Zeng, N.L. Rosi, R. Jin, Nonsuperatomic [Au23(SC6H11)16]− nanocluster featuring bipyramidal Au15 kernel and trimeric Au3(SR)4 motif, J. Am. Chem. Soc. 135 (49) (2013) 18264–18267, https://doi.org/10.1021/ja409177s.

[8] Y. Li, M.J. Cowan, M. Zhou, et al., Atom-by-atom evolution of the same ligand-protected Au21, Au22, Au22Cd1, and Au24 nanocluster series, J. Am. Chem. Soc. 142 (48) (2020) 20426–20433, https://doi.org/10.1021/jacs.0c09110.

[9] J. Yan, H. Su, H. Yang, et al., Total structure and electronic structure analysis of doped thiolated silver [MAg24(SR)18]2− (M = Pd, Pt) clusters, J. Am. Chem. Soc. 137 (37) (2015) 11880–11883, https://doi.org/10.1021/jacs.5b07186.

[10] A. Das, T. Li, K. Nobusada, Q. Zeng, N.L. Rosi, R. Jin, Total structure and optical properties of a phosphine/thiolate-protected Au24 nanocluster, J. Am. Chem. Soc. 134 (50) (2012) 20286–20289, https://doi.org/10.1021/ja3101566.

[11] S. Wang, H. Abroshan, C. Liu, et al., Shuttling single metal atom into and out of a metal nanoparticle, Nat. Commun. 8 (1) (2017) 848, https://doi.org/10.1038/s41467-017-00939-0.

[12] R. Jin, C. Liu, S. Zhao, et al., Tri-icosahedral gold nanocluster [Au37(PPh3)10(SC2H4Ph)10X2]+: linear assembly of icosahedral building blocks, ACS Nano 9 (8) (2015) 8530–8536, https://doi.org/10.1021/acsnano.5b03524.

[13] Y. Song, F. Fu, J. Zhang, et al., The magic Au60 nanocluster: a new cluster-assembled material with five Au13 building blocks, Angew. Chem. Int. Ed. 54 (29) (2015) 8430–8434, https://doi.org/10.1002/anie.201501830.

[14] C. Zeng, Y. Chen, K. Iida, et al., Gold quantum boxes: on the periodicities and the quantum confinement in the Au28, Au36, Au44, and Au52 magic series, J. Am. Chem. Soc. 138 (12) (2016) 3950–3953, https://doi.org/10.1021/jacs.5b12747.

[15] X. Huang, S. Tang, B. Liu, B. Ren, N. Zheng, Enhancing the photothermal stability of plasmonic metal nanoplates by a core-shell architecture, Adv. Mater. 23 (30) (2011) 3420–3425, https://doi.org/10.1002/adma.201100905.

[16] S. Takano, S. Yamazoe, K. Koyasu, T. Tsukuda, Slow-reduction synthesis of a thiolate-protected one-dimensional gold cluster showing an intense near-infrared absorption, J. Am. Chem. Soc. 137 (22) (2015) 7027–7030, https://doi.org/10.1021/jacs.5b03251.

[17] C. Zeng, Y. Chen, G. Li, R. Jin, Magic size Au64(S-c-C6H11)32 nanocluster protected by cyclohexanethiolate, Chem. Mater. 26 (8) (2014) 2635–2641, https://doi.org/10.1021/cm500139t.

[18] Y. Song, K. Lambright, M. Zhou, et al., Large-scale synthesis, crystal structure, and optical properties of the Ag146Br2(SR)80 nanocluster, ACS Nano 12 (9) (2018) 9318–9325, https://doi.org/10.1021/acsnano.8b04233.

[19] C. Yi, M.A. Tofanelli, C.J. Ackerson, K.L. Knappenberger, Optical properties and electronic energy relaxation of metallic Au144(SR)60 nanoclusters, J. Am. Chem. Soc. 135 (48) (2013) 18222–18228, https://doi.org/10.1021/ja409998j.

[20] M. Zhou, C. Zeng, Y. Chen, et al., Evolution from the plasmon to exciton state in ligand-protected atomically precise gold nanoparticles, Nat. Commun. 7 (1) (2016) 13240, https://doi.org/10.1038/ncomms13240.

[21] T. Higaki, M. Zhou, K.J. Lambright, K. Kirschbaum, M.Y. Sfeir, R. Jin, Sharp transition from nonmetallic Au246 to metallic Au279 with nascent surface plasmon resonance, J. Am. Chem. Soc. 140 (17) (2018) 5691–5695, https://doi.org/10.1021/jacs.8b02487.

[22] R. Jin, C. Zeng, M. Zhou, Y. Chen, Atomically precise colloidal metal nanoclusters and nanoparticles: fundamentals and opportunities, Chem. Rev. 116 (18) (2016) 10346–10413, https://doi.org/10.1021/acs.chemrev.5b00703.

[23] H. Yu, B. Rao, W. Jiang, S. Yang, M. Zhu, The photoluminescent metal nanoclusters with atomic precision, in: Special Issue on the 8th Chinese Coordination Chemistry Conference, vol. 378, 2019, pp. 595–617, https://doi.org/10.1016/j.ccr.2017.12.005.

[24] X. Kang, M. Zhu, Tailoring the photoluminescence of atomically precise nanoclusters, Chem. Soc. Rev. 48 (8) (2019) 2422–2457, https://doi.org/10.1039/C8CS00800K.

[25] Y. Huang, L. Fuksman, J. Zheng, Luminescence mechanisms of ultrasmall gold nanoparticles, Dalton Trans. 47 (18) (2018) 6267–6273, https://doi.org/10.1039/C8DT00420J.

[26] D. Li, Z. Chen, X. Mei, Fluorescence enhancement for noble metal nanoclusters, Adv. Colloid Interface Sci. 250 (2017) 25–39, https://doi.org/10.1016/j.cis.2017.11.001.

[27] Y.-P. Xie, Y.-L. Shen, G.-X. Duan, J. Han, L.-P. Zhang, X. Lu, Silver nanoclusters: synthesis, structures and photoluminescence, Mater. Chem. Front. 4 (8) (2020) 2205–2222, https://doi.org/10.1039/D0QM00117A.

[28] S. Link, A. Beeby, S. FitzGerald, M.A. El-Sayed, T.G. Schaaff, R.L. Whetten, Visible to infrared luminescence from a 28-atom gold cluster, J. Phys. Chem. B 106 (13) (2002) 3410–3415, https://doi.org/10.1021/jp014259v.

[29] A. Das, T. Li, G. Li, et al., Crystal structure and electronic properties of a thiolate-protected Au24 nanocluster, Nanoscale 6 (12) (2014) 6458–6462, https://doi.org/10.1039/C4NR01350F.

[30] C. Yao, S. Tian, L. Liao, et al., Synthesis of fluorescent phenylethanethiolated gold nanoclusters via pseudo-AGR method, Nanoscale 7 (39) (2015) 16200–16203, https://doi.org/10.1039/C5NR04760A.

[31] Z. Gan, Y. Lin, L. Luo, et al., Fluorescent gold nanoclusters with interlocked staples and a fully thiolate-bound kernel, Angew. Chem. Int. Ed. 55 (38) (2016) 11567–11571, https://doi.org/10.1002/anie.201606661.

[32] L.G. AbdulHalim, M.S. Bootharaju, Q. Tang, et al., Ag29(BDT)12(TPP)4: a tetravalent nanocluster, J. Am. Chem. Soc. 137 (37) (2015) 11970–11975, https://doi.org/10.1021/jacs.5b04547.

[33] G. Li, Z. Lei, Q.-M. Wang, Luminescent molecular Ag−S nanocluster [Ag62S13(SBut)32](BF4)4, J. Am. Chem. Soc. 132 (50) (2010) 17678–17679, https://doi.org/10.1021/ja108684m.

[34] S. Jin, S. Wang, Y. Song, et al., Crystal structure and optical properties of the [Ag62S12(SBut)32]2+ nanocluster with a complete face-centered cubic kernel, J. Am. Chem. Soc. 136 (44) (2014) 15559–15565, https://doi.org/10.1021/ja506773d.

[35] S. Jin, S. Wang, L. Xiong, et al., Two electron reduction: from quantum dots to metal nanoclusters, Chem. Mater. 28 (21) (2016) 7905–7911, https://doi.org/10.1021/acs.chemmater.6b03472.

[36] H. Yang, J. Lei, B. Wu, et al., Crystal structure of a luminescent thiolated Ag nanocluster with an octahedral Ag64+ core, Chem. Commun. 49 (3) (2013) 300–302, https://doi.org/10.1039/C2CC37347E.

[37] H. Yang, Y. Wang, N. Zheng, Stabilizing subnanometer ag(0) nanoclusters by thiolate and diphosphine ligands and their crystal structures, Nanoscale 5 (7) (2013) 2674–2677, https://doi.org/10.1039/C3NR34328F.

[38] E.S. Shibu, M.A.H. Muhammed, T. Tsukuda, T. Pradeep, Ligand exchange of Au25SG18 leading to functionalized gold clusters: spectroscopy, kinetics, and luminescence, J. Phys. Chem. C 112 (32) (2008) 12168–12176, https://doi.org/10.1021/jp800508d.

[39] Z. Wu, R. Jin, On the Ligand's role in the fluorescence of gold nanoclusters, Nano Lett. 10 (7) (2010) 2568–2573, https://doi.org/10.1021/nl101225f.

[40] S. Wang, X. Zhu, T. Cao, M. Zhu, A simple model for understanding the fluorescence behavior of Au25 nanoclusters, Nanoscale 6 (11) (2014) 5777–5781, https://doi.org/10.1039/C3NR06722J.

[41] A. Kim, C. Zeng, M. Zhou, R. Jin, Surface engineering of Au36(SR)24 nanoclusters for photoluminescence enhancement, Part. Part. Syst. Charact. 34 (8) (2017) 1600388, https://doi.org/10.1002/ppsc.201600388.

[42] M.S. Devadas, K. Kwak, J.-W. Park, et al., Directional electron transfer in chromophore-labeled quantum-sized Au25 clusters: Au25 as an electron donor, J. Phys. Chem. Lett. 1 (9) (2010) 1497–1503, https://doi.org/10.1021/jz100395p.

[43] K. Pyo, V.D. Thanthirige, S.Y. Yoon, G. Ramakrishna, D. Lee, Enhanced luminescence of Au22(SG)18 nanoclusters via rational surface engineering, Nanoscale 8 (48) (2016) 20008–20016, https://doi.org/10.1039/C6NR07660B.

[44] E. Khatun, A. Ghosh, P. Chakraborty, et al., A thirty-fold photoluminescence enhancement induced by secondary ligands in monolayer protected silver clusters, Nanoscale 10 (42) (2018) 20033–20042, https://doi.org/10.1039/C8NR05989F.

[45] X. Kang, X. Li, H. Yu, et al., Modulating photo-luminescence of Au2Cu6 nanoclusters via ligand-engineering, RSC Adv. 7 (46) (2017) 28606–28609, https://doi.org/10.1039/C7RA04743F.

[46] J.-H. Jia, Q.-M. Wang, Intensely luminescent gold(I)−silver(I) cluster with hypercoordinated carbon, J. Am. Chem. Soc. 131 (46) (2009) 16634–16635, https://doi.org/10.1021/ja906695h.

[47] J.-H. Jia, J.-X. Liang, Z. Lei, Z.-X. Cao, Q.-M. Wang, A luminescent gold(i)–copper(i) cluster with unprecedented carbon-centered trigonal prismatic hexagold, Chem. Commun. 47 (16) (2011) 4739–4741, https://doi.org/10.1039/C1CC10497G.

[48] Z. Lei, X.-L. Pei, Z.-J. Guan, Q.-M. Wang, Full protection of intensely luminescent gold(I)–silver(I) cluster by phosphine ligands and inorganic anions, Angew. Chem. Int. Ed. 56 (25) (2017) 7117–7120, https://doi.org/10.1002/anie.201702522.

[49] Z. Lei, X.-L. Pei, Z.-G. Jiang, Q.-M. Wang, Cluster linker approach: preparation of a luminescent porous framework with NbO topology by linking silver ions with gold(I) clusters, Angew. Chem. Int. Ed. 53 (47) (2014) 12771–12775, https://doi.org/10.1002/anie.201406761.

[50] X. Kang, M. Zhou, S. Wang, et al., The tetrahedral structure and luminescence properties of bi-metallic Pt1Ag28(SR)18(PPh3)4 nanocluster, Chem. Sci. 8 (4) (2017) 2581–2587, https://doi.org/10.1039/C6SC05104A.

[51] X. Kang, S. Jin, L. Xiong, et al., Nanocluster growth via "graft-onto": effects on geometric structures and optical properties, Chem. Sci. 11 (6) (2020) 1691–1697, https://doi.org/10.1039/C9SC05700E.

[52] X. Kang, L. Huang, W. Liu, et al., Reversible nanocluster structure transformation between face-centered cubic and icosahedral isomers, Chem. Sci. 10 (37) (2019) 8685–8693, https://doi.org/10.1039/C9SC02667C.

[53] M.S. Bootharaju, S.M. Kozlov, Z. Cao, et al., Doping-induced anisotropic self-assembly of silver icosahedra in [Pt2Ag23Cl7(PPh3)10] nanoclusters, J. Am. Chem. Soc. 139 (3) (2017) 1053–1056, https://doi.org/10.1021/jacs.6b11875.

[54] X. Kang, L. Xiong, S. Wang, Y. Pei, M. Zhu, De-assembly of assembled Pt1Ag12 units: tailoring the photoluminescence of atomically precise nanoclusters, Chem. Commun. 53 (93) (2017) 12564–12567, https://doi.org/10.1039/C7CC05996E.

[55] H. Dong, L. Liao, Z. Wu, Two-way transformation between fcc- and nonfcc-structured gold nanoclusters, J. Phys. Chem. Lett. 8 (21) (2017) 5338–5343, https://doi.org/10.1021/acs.jpclett.7b02459.

[56] L. Qi, T.-Y. Luo, G. Taylor Michael, et al., Molecular "surgery" on a 23-gold-atom nanoparticle, Sci. Adv. 3 (5) (2017), e1603193, https://doi.org/10.1126/sciadv.1603193.

[57] H. Yang, Y. Wang, H. Huang, et al., All-thiol-stabilized Ag44 and Au12Ag32 nanoparticles with single-crystal structures, Nat. Commun. 4 (1) (2013) 2422, https://doi.org/10.1038/ncomms3422.

[58] S. Wang, X. Meng, A. Das, et al., A 200-fold quantum yield boost in the photoluminescence of silver-doped AgxAu25−x nanoclusters: the 13th silver atom matters, Angew. Chem. Int. Ed. 53 (9) (2014) 2376–2380, https://doi.org/10.1002/anie.201307480.

[59] F. Muniz-Miranda, M.C. Menziani, A. Pedone, Influence of silver doping on the photoluminescence of protected AgnAu25−n nanoclusters: a time-dependent density functional theory investigation, J. Phys. Chem. C 119 (19) (2015) 10766–10775, https://doi.org/10.1021/acs.jpcc.5b02655.

[60] M. Zhou, J. Zhong, S. Wang, et al., Ultrafast relaxation dynamics of luminescent rod-shaped, silver-doped AgxAu25−x clusters, J. Phys. Chem. C 119 (32) (2015) 18790–18797, https://doi.org/10.1021/acs.jpcc.5b05376.

[61] C. Yao, J. Chen, M.-B. Li, L. Liu, J. Yang, Z. Wu, Adding two active silver atoms on Au25 nanoparticle, Nano Lett. 15 (2) (2015) 1281–1287, https://doi.org/10.1021/nl504477t.

[62] C.P. Joshi, M.S. Bootharaju, M.J. Alhilaly, O.M. Bakr, [Ag25(SR)18]−: the "golden" silver nanoparticle, J. Am. Chem. Soc. 137 (36) (2015) 11578–11581, https://doi.org/10.1021/jacs.5b07088.

[63] M.S. Bootharaju, C.P. Joshi, M.R. Parida, O.F. Mohammed, O.M. Bakr, Templated atom-precise galvanic synthesis and structure elucidation of a [Ag24Au(SR)18]− nanocluster, Angew. Chem. Int. Ed. 55 (3) (2016) 922–926, https://doi.org/10.1002/anie.201509381.

[64] M.S. Bootharaju, L. Sinatra, O.M. Bakr, Distinct metal-exchange pathways of doped Ag25 nanoclusters, Nanoscale 8 (39) (2016) 17333–17339, https://doi.org/10.1039/C6NR06353E.

[65] K.L.D.M. Weerawardene, C.M. Aikens, Origin of photoluminescence of Ag25(SR)18− nanoparticles: ligand and doping effect, J. Phys. Chem. C 122 (4) (2018) 2440–2447, https://doi.org/10.1021/acs.jpcc.7b11706.

[66] X. Kang, L. Xiong, S. Wang, et al., Shape-controlled synthesis of trimetallic nanoclusters: structure elucidation and properties investigation, Chem. A Eur. J. 22 (48) (2016) 17145–17150, https://doi.org/10.1002/chem.201603893.

[67] Q. Yuan, X. Kang, D. Hu, C. Qin, S. Wang, M. Zhu, Metal synergistic effect on cluster optical properties: based on Ag25 series nanoclusters, Dalton Trans. 48 (35) (2019) 13190–13196, https://doi.org/10.1039/C9DT02493J.

[68] S. Yang, J. Chai, Y. Lv, et al., Cyclic Pt3Ag33 and Pt3Au12Ag21 nanoclusters with M13 icosahedra as building-blocks, Chem. Commun. 54 (85) (2018) 12077–12080, https://doi.org/10.1039/C8CC06900J.

[69] G. Soldan, M.A. Aljuhani, M.S. Bootharaju, et al., Gold doping of silver nanoclusters: a 26-fold enhancement in the luminescence quantum yield, Angew. Chem. Int. Ed. 55 (19) (2016) 5749–5753, https://doi.org/10.1002/anie.201600267.

[70] X.-Y. Xie, P. Xiao, X. Cao, W.-H. Fang, G. Cui, M. Dolg, The origin of the photoluminescence enhancement of gold-doped silver nanoclusters: the importance of relativistic effects and heteronuclear gold–silver bonds, Angew. Chem. Int. Ed. 57 (31) (2018) 9965–9969, https://doi.org/10.1002/anie.201803683.

[71] M. van der Linden, A. Barendregt, A.J. van Bunningen, et al., Characterisation, degradation and regeneration of luminescent Ag29 clusters in solution, Nanoscale 8 (47) (2016) 19901–19909, https://doi.org/10.1039/C6NR04958C.

[72] M. van der Linden, A.J. van Bunningen, L. Amidani, et al., Single Au atom doping of silver nanoclusters, ACS Nano 12 (12) (2018) 12751–12760, https://doi.org/10.1021/acsnano.8b07807.

[73] M.A.H. Muhammed, F. Aldeek, G. Palui, L. Trapiella-Alfonso, H. Mattoussi, Growth of in situ functionalized luminescent silver nanoclusters by direct reduction and size focusing, ACS Nano 6 (10) (2012) 8950–8961, https://doi.org/10.1021/nn302954n.

[74] D. Mishra, V. Lobodin, C. Zhang, F. Aldeek, E. Lochner, H. Mattoussi, Gold-doped silver nanoclusters with enhanced photophysical properties, Phys. Chem. Chem. Phys. 20 (18) (2018) 12992–13007, https://doi.org/10.1039/C7CP08682B.

[75] Y. Bao, X. Wu, H. Gao, et al., The geometric and electronic structures of a Ag13Cu10(SAdm)12X3 nanocluster, Dalton Trans. 49 (47) (2020) 17164–17168, https://doi.org/10.1039/D0DT03638B.

[76] W. Du, S. Jin, L. Xiong, et al., Ag50(Dppm)6(SR)30 and its homologue AuxAg50−x(Dppm)6(SR)30 alloy nanocluster: seeded growth, structure determination, and differences in properties, J. Am. Chem. Soc. 139 (4) (2017) 1618–1624, https://doi.org/10.1021/jacs.6b11681.

[77] W.-T. Chang, P.-Y. Lee, J.-H. Liao, et al., Eight-electron silver and mixed gold/silver nanoclusters stabilized by selenium donor ligands, Angew. Chem. Int. Ed. 56 (34) (2017) 10178–10182, https://doi.org/10.1002/anie.201704800.

[78] Y.-R. Lin, P.V.V.N. Kishore, J.-H. Liao, et al., Synthesis, structural characterization and transformation of an eight-electron superatomic alloy, [Au@Ag19{S2P(OPr)2}12], Nanoscale 10 (15) (2018) 6855–6860, https://doi.org/10.1039/C8NR00172C.

[79] R.P.B. Silalahi, K.K. Chakrahari, J.-H. Liao, et al., Synthesis of two-electron bimetallic Cu–Ag and Cu–Au clusters by using [Cu13(S2CNnBu2)6(C≡CPh)4]+ as a template, Chem. Asian J. 13 (5) (2018) 500–504, https://doi.org/10.1002/asia.201701753.

[80] J. Luo, Z. Xie, J.W.Y. Lam, et al., Aggregation-induced emission of 1-methyl-1,2,3,4,5-pentaphenylsilole, Chem. Commun. 18 (2001) 1740–1741, https://doi.org/10.1039/B105159H.

[81] J. Mei, Y. Hong, J.W.Y. Lam, A. Qin, Y. Tang, B.Z. Tang, Aggregation-induced emission: the whole is more brilliant than the parts, Adv. Mater. 26 (31) (2014) 5429–5479, https://doi.org/10.1002/adma.201401356.

[82] J. Mei, N.L.C. Leung, R.T.K. Kwok, J.W.Y. Lam, B.Z. Tang, Aggregation-induced emission: together we shine, united we soar! Chem. Rev. 115 (21) (2015) 11718–11940, https://doi.org/10.1021/acs.chemrev.5b00263.

[83] N. Goswami, Q. Yao, Z. Luo, J. Li, T. Chen, J. Xie, Luminescent metal nanoclusters with aggregation-induced emission, J. Phys. Chem. Lett. 7 (6) (2016) 962–975, https://doi.org/10.1021/acs.jpclett.5b02765.

[84] Z. Luo, X. Yuan, Y. Yu, et al., From aggregation-induced emission of Au(I)–thiolate complexes to ultrabright Au(0)@Au(I)–thiolate core–shell nanoclusters, J. Am. Chem. Soc. 134 (40) (2012) 16662–16670, https://doi.org/10.1021/ja306199p.

[85] X. Dou, X. Yuan, Y. Yu, et al., Lighting up thiolated Au@Ag nanoclusters via aggregation-induced emission, Nanoscale 6 (1) (2014) 157–161, https://doi.org/10.1039/C3NR04490D.

[86] K. Zheng, X. Yuan, K. Kuah, et al., Boiling water synthesis of ultrastable thiolated silver nanoclusters with aggregation-induced emission, Chem. Commun. 51 (82) (2015) 15165–15168, https://doi.org/10.1039/C5CC04858C.

[87] L. Yang, Y. Cao, J. Chen, et al., Luminescence of Au(I)-thiolate complex affected by solvent, in: Proceedings of the 13th International Symposium on Radiation Physics, vol. 137, 2017, pp. 68–71, https://doi.org/10.1016/j.radphyschem.2016.09.013.

[88] M. Sugiuchi, J. Maeba, N. Okubo, M. Iwamura, K. Nozaki, K. Konishi, Aggregation-induced fluorescence-to-phosphorescence switching of molecular gold clusters, J. Am. Chem. Soc. 139 (49) (2017) 17731–17734, https://doi.org/10.1021/jacs.7b10201.

[89] S. Jin, W. Liu, D. Hu, et al., Aggregation-induced emission (AIE) in Ag–Au bimetallic nanocluster, Chem. A Eur. J. 24 (15) (2018) 3712–3715, https://doi.org/10.1002/chem.201800189.

[90] X. Jia, J. Li, E. Wang, Cu nanoclusters with aggregation induced emission enhancement, Small 9 (22) (2013) 3873–3879, https://doi.org/10.1002/smll.201300896.

[91] Z. Wu, H. Liu, T. Li, et al., Contribution of metal defects in the assembly induced emission of Cu nanoclusters, J. Am. Chem. Soc. 139 (12) (2017) 4318–4321, https://doi.org/10.1021/jacs.7b00773.

[92] W.-X. Ni, Y.-M. Qiu, M. Li, et al., Metallophilicity-driven dynamic aggregation of a phosphorescent gold(I)–silver(I) cluster prepared by solution-based and mechanochemical approaches, J. Am. Chem. Soc. 136 (27) (2014) 9532–9535, https://doi.org/10.1021/ja5025113.

[93] X. Kang, S. Wang, Y. Song, et al., Bimetallic Au2Cu6 nanoclusters: strong luminescence induced by the aggregation of copper(I) complexes with gold(0) species, Angew. Chem. Int. Ed. 55 (11) (2016) 3611–3614, https://doi.org/10.1002/anie.201600241.

[94] C. Tao, S. Yang, C. Jinsong, et al., Crystallization-induced emission enhancement: a novel fluorescent Au-Ag bimetallic nanocluster with precise atomic structure, Sci. Adv. 3 (8) (2017), e1700956, https://doi.org/10.1126/sciadv.1700956.

[95] E. Khatun, M. Bodiuzzaman, K.S. Sugi, et al., Confining an Ag10 core in an Ag12 shell: a four-electron superatom with enhanced photoluminescence upon crystallization, ACS Nano 13 (5) (2019) 5753–5759, https://doi.org/10.1021/acsnano.9b01189.

[96] A. Nag, P. Chakraborty, M. Bodiuzzaman, T. Ahuja, S. Antharjanam, T. Pradeep, Polymorphism of Ag29(BDT)12(TPP)43− cluster: interactions of secondary ligands and their effect on solid state luminescence, Nanoscale 10 (21) (2018) 9851–9855, https://doi.org/10.1039/C8NR02629G.

[97] Z. Wu, Y. Du, J. Liu, et al., Aurophilic interactions in the self-assembly of gold nanoclusters into nanoribbons with enhanced luminescence, Angew. Chem. Int. Ed. 58 (24) (2019) 8139–8144, https://doi.org/10.1002/anie.201903584.

[98] X. Kang, S. Wang, M. Zhu, Observation of a new type of aggregation-induced emission in nanoclusters, Chem. Sci. 9 (11) (2018) 3062–3068, https://doi.org/10.1039/C7SC05317G.

[99] Z. Gan, J. Chen, J. Wang, et al., The fourth crystallographic closest packing unveiled in the gold nanocluster crystal, Nat. Commun. 8 (1) (2017) 14739, https://doi.org/10.1038/ncomms14739.

[100] Z. Gan, J. Chen, L. Liao, H. Zhang, Z. Wu, Surface single-atom tailoring of a gold nanoparticle, J. Phys. Chem. Lett. 9 (1) (2018) 204–208, https://doi.org/10.1021/acs.jpclett.7b02982.

[101] X. Liu, J. Yuan, C. Yao, et al., Crystal and solution photoluminescence of MAg24(SR)18 (M = Ag/Pd/Pt/Au) nanoclusters and some implications for the photoluminescence mechanisms, J. Phys. Chem. C 121 (25) (2017) 13848–13853, https://doi.org/10.1021/acs.jpcc.7b01730.

[102] G.-X. Duan, L. Tian, J.-B. Wen, L.-Y. Li, Y.-P. Xie, X. Lu, An atomically precise all-tert-butylethynide-protected Ag51 superatom nanocluster with color tunability, Nanoscale 10 (40) (2018) 18915–18919, https://doi.org/10.1039/C8NR06399K.

[103] M. Bootharaju, S.M. Kozlov, Z. Cao, et al., Tailoring the crystal structure of nanoclusters unveiled high photoluminescence via ion pairing, Chem. Mater. 30 (8) (2018) 2719–2725, https://doi.org/10.1021/acs.chemmater.8b00328.

[104] Y. Wang, H. Su, L. Ren, et al., Site preference in multimetallic nanoclusters: incorporation of alkali metal ions or copper atoms into the alkynyl-protected body-centered cubic cluster [Au7Ag8(C≡CtBu)12]+, Angew. Chem. Int. Ed. 55 (48) (2016) 15152–15156, https://doi.org/10.1002/anie.201609144.

[105] W.-M. He, Z. Zhou, Z. Han, et al., Ultrafast size expansion and turn-on luminescence of atomically precise silver clusters by hydrogen sulfide, Angew. Chem. Int. Ed. 60 (15) (2021) 8505–8509, https://doi.org/10.1002/anie.202100006.

[106] S.-S. Zhang, H.-F. Su, G.-L. Zhuang, et al., A hexadecanuclear silver alkynyl cluster based NbO framework with triple emissions from the visible to near-infrared II region, Chem. Commun. 54 (84) (2018) 11905–11908, https://doi.org/10.1039/C8CC06683C.

[107] J.-W. Liu, H.-F. Su, Z. Wang, et al., A giant 90-nucleus silver cluster templated by hetero-anions, Chem. Commun. 54 (35) (2018) 4461–4464, https://doi.org/10.1039/C8CC01767K.

[108] Z. Wang, H.-F. Su, C.-H. Tung, D. Sun, L.-S. Zheng, Deciphering synergetic core-shell transformation from [Mo6O22@Ag44] to [Mo8O28@Ag50], Nat. Commun. 9 (1) (2018) 4407, https://doi.org/10.1038/s41467-018-06755-4.

[109] Y.-M. Su, H.-F. Su, Z. Wang, et al., Three silver nests capped by thiolate/phenylphosphonate, Chem. A Eur. J. 24 (56) (2018) 15096–15103, https://doi.org/10.1002/chem.201803203.

[110] X.-Y. Li, Z. Wang, H.-F. Su, et al., Anion-templated nanosized silver clusters protected by mixed thiolate and diphosphine, Nanoscale 9 (10) (2017) 3601–3608, https://doi.org/10.1039/C6NR09632H.

[111] J.-W. Liu, L. Feng, H.-F. Su, et al., Anisotropic assembly of Ag52 and Ag76 nanoclusters, J. Am. Chem. Soc. 140 (5) (2018) 1600–1603, https://doi.org/10.1021/jacs.7b12777.

[112] Y.-M. Su, Z. Wang, G.-L. Zhuang, et al., Unusual fcc-structured Ag10 kernels trapped in Ag70 nanoclusters, Chem. Sci. 10 (2) (2019) 564–568, https://doi.org/10.1039/C8SC03396J.

[113] Q.-Q. Xu, X.-Y. Dong, R.-W. Huang, B. Li, S.-Q. Zang, T.C.W. Mak, A thermochromic silver nanocluster exhibiting dual emission character, Nanoscale 7 (5) (2015) 1650–1654, https://doi.org/10.1039/C4NR05122J.

[114] B. Li, R.-W. Huang, J.-H. Qin, et al., Thermochromic luminescent nest-like silver thiolate cluster, Chem. A Eur. J. 20 (39) (2014) 12416–12420, https://doi.org/10.1002/chem.201404049.

[115] M. Zhu, C.M. Aikens, M.P. Hendrich, et al., Reversible switching of magnetism in thiolate-protected Au25 superatoms, J. Am. Chem. Soc. 131 (7) (2009) 2490–2492, https://doi.org/10.1021/ja809157f.

[116] D. Jiang, R.L. Whetten, Magnetic doping of a thiolated-gold superatom: first-principles density functional theory calculations, Phys. Rev. B 80 (11) (2009), 115402, https://doi.org/10.1103/PhysRevB.80.115402.

[117] A. Akbari-Sharbaf, M. Hesari, M.S. Workentin, G. Fanchini, Electron paramagnetic resonance in positively charged Au25 molecular nanoclusters, J. Chem. Phys. 138 (2) (2013), 024305, https://doi.org/10.1063/1.4773061.

[118] J. Akola, K.A. Kacprzak, O. Lopez-Acevedo, M. Walter, H. Grönbeck, H. Häkkinen, Thiolate-protected Au25 superatoms as building blocks: dimers and crystals, J. Phys. Chem. C 114 (38) (2010) 15986–15994, https://doi.org/10.1021/jp1015438.

[119] M. Zhou, Y.Q. Cai, M.G. Zeng, C. Zhang, Y.P. Feng, Mn-doped thiolated Au25 nanoclusters: atomic configuration, magnetic properties, and a possible high-performance spin filter, Appl. Phys. Lett. 98 (14) (2011), 143103, https://doi.org/10.1063/1.3575203.

[120] S. Antonello, N.V. Perera, M. Ruzzi, J.A. Gascón, F. Maran, Interplay of charge state, lability, and magnetism in the molecule-like Au25(SR)18 cluster, J. Am. Chem. Soc. 135 (41) (2013) 15585–15594, https://doi.org/10.1021/ja407887d.

[121] T. Dainese, S. Antonello, J.A. Gascón, et al., Au25(SEt)18, a nearly naked thiolate-protected Au25 cluster: structural analysis by single crystal X-ray crystallography and electron nuclear double resonance, ACS Nano 8 (4) (2014) 3904–3912, https://doi.org/10.1021/nn500805n.

[122] M. De Nardi, S. Antonello, D. Jiang, et al., Gold nanowired: a linear (Au25)n polymer from Au25 molecular clusters, ACS Nano 8 (8) (2014) 8505–8512, https://doi.org/10.1021/nn5031143.

[123] L. Liao, S. Zhou, Y. Dai, et al., Mono-mercury doping of Au25 and the HOMO/LUMO energies evaluation employing differential pulse voltammetry, J. Am. Chem. Soc. 137 (30) (2015) 9511–9514, https://doi.org/10.1021/jacs.5b03483.

[124] S. Tian, L. Liao, J. Yuan, et al., Structures and magnetism of mono-palladium and mono-platinum doped Au25(PET)18 nanoclusters, Chem. Commun. 52 (64) (2016) 9873–9876, https://doi.org/10.1039/C6CC02698B.

[125] Y. Song, S. Jin, X. Kang, et al., How a single electron affects the properties of the "non-superatom" Au25 nanoclusters, Chem. Mater. 28 (8) (2016) 2609–2617, https://doi.org/10.1021/acs.chemmater.5b04655.

[126] M. Agrachev, S. Antonello, T. Dainese, et al., A magnetic look into the protecting layer of Au25 clusters, Chem. Sci. 7 (12) (2016) 6910–6918, https://doi.org/10.1039/C6SC03691K.

[127] M.A. Tofanelli, K. Salorinne, T.W. Ni, et al., Jahn–Teller effects in Au25(SR)18, Chem. Sci. 7 (3) (2016) 1882–1890, https://doi.org/10.1039/C5SC02134K.

[128] M. Agrachev, S. Antonello, T. Dainese, et al., Magnetic ordering in gold nanoclusters, ACS Omega 2 (6) (2017) 2607–2617, https://doi.org/10.1021/acsomega.7b00472.

[129] M. Suyama, S. Takano, T. Nakamura, T. Tsukuda, Stoichiometric formation of open-shell [PtAu24(SC2H4Ph)18]− via spontaneous electron proportionation between [PtAu24(SC2H4Ph)18]2− and [PtAu24(SC2H4Ph)18]0, J. Am. Chem. Soc. 141 (36) (2019) 14048–14051, https://doi.org/10.1021/jacs.9b06254.

[130] Y. Li, S. Biswas, T.-Y. Luo, et al., Doping effect on the magnetism of thiolate-capped 25-atom alloy nanoclusters, Chem. Mater. 32 (21) (2020) 9238–9244, https://doi.org/10.1021/acs.chemmater.0c02984.

[131] Y. Negishi, H. Tsunoyama, M. Suzuki, et al., X-ray magnetic circular dichroism of size-selected, thiolated gold clusters, J. Am. Chem. Soc. 128 (37) (2006) 12034–12035, https://doi.org/10.1021/ja062815z.

[132] R.J. Magyar, V. Mujica, M. Marquez, C. Gonzalez, Density-functional study of magnetism in bare Au nanoclusters: evidence of permanent size-dependent spin polarization without geometry relaxation, Phys. Rev. B 75 (14) (2007), 144421, https://doi.org/10.1103/PhysRevB.75.144421.

[133] M.A. Muñoz-Márquez, E. Guerrero, A. Fernández, et al., Permanent magnetism in phosphine- and chlorine-capped gold: from clusters to nanoparticles, J. Nanopart. Res. 12 (4) (2010) 1307–1318, https://doi.org/10.1007/s11051-010-9862-0.

[134] B. Santiago González, M.J. Rodríguez, C. Blanco, J. Rivas, M.A. López-Quintela, J.M.G. Martinho, One step synthesis of the smallest photoluminescent and paramagnetic PVP-protected gold atomic clusters, Nano Lett. 10 (10) (2010) 4217–4221, https://doi.org/10.1021/nl1026716.

[135] R.S. McCoy, S. Choi, G. Collins, B.J. Ackerson, C.J. Ackerson, Superatom paramagnetism enables gold nanocluster heating in applied radiofrequency fields, ACS Nano 7 (3) (2013) 2610–2616, https://doi.org/10.1021/nn306015c.

[136] K.S. Krishna, P. Tarakeshwar, V. Mujica, C.S.S.R. Kumar, Chemically induced magnetism in atomically precise gold clusters, Small 10 (5) (2014) 907–911, https://doi.org/10.1002/smll.201302393.

[137] J. Ji, G. Wang, T. Wang, X. You, X. Xu, Thiolate-protected Ni39 and Ni41 nanoclusters: synthesis, self-assembly and magnetic properties, Nanoscale 6 (15) (2014) 9185–9191, https://doi.org/10.1039/C4NR01063A.

[138] S. Xue, Y.-N. Guo, L. Zhao, H. Zhang, J. Tang, Molecular magnetic investigation of a family of octanuclear [Cu6Ln2] nanoclusters, Inorg. Chem. 53 (15) (2014) 8165–8171, https://doi.org/10.1021/ic501226v.

[139] B.-Q. Ji, M. Jagodič, H.-Y. Ma, et al., Solution behavior and magnetic properties of a novel nonanuclear copper(ii) cluster, New J. Chem. 42 (22) (2018) 17884–17888, https://doi.org/10.1039/C8NJ04230F.

[140] C. Zeng, A. Weitz, G. Withers, et al., Controlling magnetism of Au133(TBBT)52 nanoclusters at single electron level and implication for nonmetal to metal transition, Chem. Sci. 10 (42) (2019) 9684–9691, https://doi.org/10.1039/C9SC02736J.

[141] P. Herbert, P. Window, C.J. Ackerson, K.L. Knappenberger, Low-temperature magnetism in nanoscale gold revealed through variable-temperature magnetic circular dichroism spectroscopy, J. Phys. Chem. Lett. 10 (2) (2019) 189–193, https://doi.org/10.1021/acs.jpclett.8b03473.

[142] P.S. Window, C.J. Ackerson, Superatom paramagnetism in Au102(SR)441 −/0/1+/2+ oxidation states, Inorg. Chem. 59 (6) (2020) 3509–3512, https://doi.org/10.1021/acs.inorgchem.9b02787.

[143] Q. Li, J. Chai, S. Yang, et al., Multiple ways realizing charge-state transform in Au□Cu bimetallic nanoclusters with atomic precision, Small 17 (27) (2021) 1907114, https://doi.org/10.1002/smll.201907114.

[144] H. Hirai, S. Takano, T. Nakamura, T. Tsukuda, Understanding doping effects on electronic structures of gold superatoms: a case study of diphosphine-protected M@Au12 (M = Au, Pt, Ir), Inorg. Chem. 59 (24) (2020) 17889–17895, https://doi.org/10.1021/acs.inorgchem.0c00879.

[145] B. Liao, S. Li, G. Yang, Structure types and magnetic behavior of cobalt nanoclusters, Inorg. Chem. 60 (3) (2021) 1839–1845, https://doi.org/10.1021/acs.inorgchem.0c03301.

[146] C. Zeng, C. Liu, Y. Chen, N.L. Rosi, R. Jin, Gold–thiolate ring as a protecting motif in the Au20(SR)16 nanocluster and implications, J. Am. Chem. Soc. 136 (34) (2014) 11922–11925, https://doi.org/10.1021/ja506802n.

[147] J.-J. Li, Z.-J. Guan, Z. Lei, F. Hu, Q.-M. Wang, Same magic number but different arrangement: alkynyl-protected Au25 with D3 symmetry, Angew. Chem. Int. Ed. 58 (4) (2019) 1083–1087, https://doi.org/10.1002/anie.201811859.

[148] H. Qian, W.T. Eckenhoff, Y. Zhu, T. Pintauer, R. Jin, Total structure determination of thiolate-protected Au38 nanoparticles, J. Am. Chem. Soc. 132 (24) (2010) 8280–8281, https://doi.org/10.1021/ja103592z.

[149] H. Qian, M. Zhu, C. Gayathri, R.R. Gil, R. Jin, Chirality in gold nanoclusters probed by NMR spectroscopy, ACS Nano 5 (11) (2011) 8935–8942, https://doi.org/10.1021/nn203113j.

[150] Z. Chenjie, C. Yuxiang, L. Chong, N. Katsuyuki, L. Rosi Nathaniel, J. Rongchao, Gold tetrahedra coil up: Kekulé-like and double helical superstructures, Sci. Adv. 1 (9) (2015), e1500425, https://doi.org/10.1126/sciadv.1500425.

[151] L. Liao, S. Zhuang, C. Yao, et al., Structure of chiral Au44(2,4-DMBT)26 nanocluster with an 18-electron shell closure, J. Am. Chem. Soc. 138 (33) (2016) 10425–10428, https://doi.org/10.1021/jacs.6b07178.

[152] P.D. Jadzinsky, C. Guillermo, C.J. Ackerson, D.A. Bushnell, R.D. Kornberg, Structure of a thiol monolayer-protected gold nanoparticle at 1.1 Å resolution, Science 318 (5849) (2007) 430–433, https://doi.org/10.1126/science.1148624.

[153] Z. Chenjie, C. Yuxiang, K. Kristin, A. Kannatassen, Y. Sfeir Matthew, J. Rongchao, Structural patterns at all scales in a nonmetallic chiral Au133(SR)52 nanoparticle, Sci. Adv. 1 (2) (2015), e1500045, https://doi.org/10.1126/sciadv.1500045.

[154] Z. Chenjie, C. Yuxiang, K. Kristin, K.J. Lambright, J. Rongchao, Emergence of hierarchical structural complexities in nanoparticles and their assembly, Science 354 (6319) (2016) 1580–1584, https://doi.org/10.1126/science.aak9750.

[155] X. Zou, S. Jin, W. Du, et al., Multi-ligand-directed synthesis of chiral silver nanoclusters, Nanoscale 9 (43) (2017) 16800–16805, https://doi.org/10.1039/C7NR06338E.

[156] Z. Wang, J.-W. Liu, H.-F. Su, et al., Chalcogens-induced Ag6Z4@Ag36 (Z = S or Se) core–shell nanoclusters: enlarged tetrahedral core and homochiral crystallization, J. Am. Chem. Soc. 141 (44) (2019) 17884–17890, https://doi.org/10.1021/jacs.9b09460.

[157] H. Han, Y. Yao, A. Bhargava, et al., Tertiary hierarchical complexity in assemblies of sulfur-bridged metal chiral clusters, J. Am. Chem. Soc. 142 (34) (2020) 14495–14503, https://doi.org/10.1021/jacs.0c04764.

[158] S. Wang, S. Jiner, S. Yang, et al., Total structure determination of surface doping [Ag46Au24(SR)32](BPh4)2 nanocluster and its structure-related catalytic property, Sci. Adv. 1 (7) (2015), e1500441, https://doi.org/10.1126/sciadv.1500441.

[159] J.-L. Zeng, Z.-J. Guan, Y. Du, Z.-A. Nan, Y.-M. Lin, Q.-M. Wang, Chloride-promoted formation of a bimetallic nanocluster Au80Ag30 and the total structure determination, J. Am. Chem. Soc. 138 (25) (2016) 7848–7851, https://doi.org/10.1021/jacs.6b04471.

[160] X. Liu, G. Saranya, X. Huang, et al., Ag2Au50(PET)36 nanocluster: dimeric assembly of Au25(PET)18 enabled by silver atoms, Angew. Chem. Int. Ed. 59 (33) (2020) 13941–13946, https://doi.org/10.1002/anie.202005087.

[161] X.-K. Wan, S.-F. Yuan, Z.-W. Lin, Q.-M. Wang, A chiral gold nanocluster Au20 protected by tetradentate phosphine ligands, Angew. Chem. Int. Ed. 53 (11) (2014) 2923–2926, https://doi.org/10.1002/anie.201308599.

[162] R.S. Dhayal, J.-H. Liao, X. Wang, et al., Diselenophosphate-induced conversion of an achiral [Cu20H11{S2P(OiPr)2}9] into a chiral [Cu20H11{Se2P(OiPr)2}9] polyhydrido nanocluster, Angew. Chem. Int. Ed. 54 (46) (2015) 13604–13608, https://doi.org/10.1002/anie.201506736.

[163] L. Ren, P. Yuan, H. Su, et al., Bulky surface ligands promote surface reactivities of [Ag141X12(S-Adm)40]3 + (X = Cl, Br, I) nanoclusters: models for multiple-twinned nanoparticles, J. Am. Chem. Soc. 139 (38) (2017) 13288–13291, https://doi.org/10.1021/jacs.7b07926.

[164] I. Dolamic, S. Knoppe, A. Dass, T. Bürgi, First enantioseparation and circular dichroism spectra of Au38 clusters protected by achiral ligands, Nat. Commun. 3 (1) (2012) 798, https://doi.org/10.1038/ncomms1802.

[165] R. Kazan, B. Zhang, T. Bürgi, Au38Cu1(2-PET)24 nanocluster: synthesis, enantioseparation and luminescence, Dalton Trans. 46 (24) (2017) 7708–7713, https://doi.org/10.1039/C7DT00955K.

[166] S. Knoppe, I. Dolamic, A. Dass, T. Bürgi, Separation of enantiomers and CD spectra of Au40(SCH2CH2Ph)24: spectroscopic evidence for intrinsic chirality, Angew. Chem. Int. Ed. 51 (30) (2012) 7589–7591, https://doi.org/10.1002/anie.201202369.

[167] B. Varnholt, I. Dolamic, S. Knoppe, T. Bürgi, On the flexibility of the gold–thiolate interface: racemization of the Au40(SR)24 cluster, Nanoscale 5 (20) (2013) 9568–9571, https://doi.org/10.1039/C3NR03389A.

[168] C. Zeng, T. Li, A. Das, N.L. Rosi, R. Jin, Chiral structure of thiolate-protected 28-gold-atom nanocluster determined by X-ray crystallography, J. Am. Chem. Soc. 135 (27) (2013) 10011–10013, https://doi.org/10.1021/ja404058q.

[169] S. Jin, F. Xu, W. Du, et al., Isomerism in Au–Ag alloy nanoclusters: structure determination and enantioseparation of [Au9Ag12(SR)4(dppm)6X6]3 +, Inorg. Chem. 57 (9) (2018) 5114–5119, https://doi.org/10.1021/acs.inorgchem.8b00183.

[170] H. Yoshida, M. Ehara, U.D. Priyakumar, T. Kawai, T. Nakashima, Enantioseparation and chiral induction in Ag29 nanoclusters with intrinsic chirality, Chem. Sci. 11 (9) (2020) 2394–2400, https://doi.org/10.1039/C9SC05299B.

[171] Y. Zhu, H. Wang, K. Wan, et al., Enantioseparation of Au20(PP3)4Cl4 clusters with intrinsically chiral cores, Angew. Chem. Int. Ed. 57 (29) (2018) 9059–9063, https://doi.org/10.1002/anie.201805695.

[172] C. Gautier, T. Bürgi, Chiral inversion of gold nanoparticles, J. Am. Chem. Soc. 130 (22) (2008) 7077–7084, https://doi.org/10.1021/ja800256r.

[173] S. Si, C. Gautier, J. Boudon, R. Taras, S. Gladiali, T. Bürgi, Ligand exchange on Au25 cluster with chiral thiols, J. Phys. Chem. C 113 (30) (2009) 12966–12969, https://doi.org/10.1021/jp9044385.

[174] S. Knoppe, A.C. Dharmaratne, E. Schreiner, A. Dass, T. Bürgi, Ligand exchange reactions on Au38 and Au40 clusters: a combined circular dichroism and mass spectrometry study, J. Am. Chem. Soc. 132 (47) (2010) 16783–16789, https://doi.org/10.1021/ja104641x.

[175] S. Knoppe, R. Azoulay, A. Dass, T. Bürgi, In situ reaction monitoring reveals a diastereoselective ligand exchange reaction between the intrinsically chiral Au38(SR)24 and chiral thiols, J. Am. Chem. Soc. 134 (50) (2012) 20302–20305, https://doi.org/10.1021/ja310330m.

[176] L. Beqa, D. Deschamps, S. Perrio, A.-C. Gaumont, S. Knoppe, T. Bürgi, Ligand exchange reaction on Au38(SR)24, separation of Au38(SR)23(SR′)1 regioisomers, and migration of thiolates, J. Phys. Chem. C 117 (41) (2013) 21619–21625, https://doi.org/10.1021/jp408455x.

[177] A. Baghdasaryan, K. Martin, L.M. Lawson Daku, M. Mastropasqua Talamo, N. Avarvari, T. Bürgi, Ligand exchange reactions on the chiral Au38 cluster: CD modulation caused by the modification of the ligand shell composition, Nanoscale 12 (35) (2020) 18160–18170, https://doi.org/10.1039/D0NR03824E.

[178] K.R. Krishnadas, L. Sementa, M. Medves, et al., Chiral functionalization of an atomically precise noble metal cluster: insights into the origin of chirality and photoluminescence, ACS Nano 14 (8) (2020) 9687–9700, https://doi.org/10.1021/acsnano.0c01183.

[179] M. Zhu, H. Qian, X. Meng, S. Jin, Z. Wu, R. Jin, Chiral Au25 nanospheres and nanorods: synthesis and insight into the origin of chirality, Nano Lett. 11 (9) (2011) 3963–3969, https://doi.org/10.1021/nl202288j.

[180] Q. Xu, S. Kumar, S. Jin, H. Qian, M. Zhu, R. Jin, Chiral 38-gold-atom nanoclusters: synthesis and chiroptical properties, Small 10 (5) (2014) 1008–1014, https://doi.org/10.1002/smll.201302279.

[181] S. Kumar, R. Jin, Water-soluble Au25(Capt)18 nanoclusters: synthesis, thermal stability, and optical properties, Nanoscale 4 (14) (2012) 4222–4227, https://doi.org/10.1039/C2NR30833A.

[182] T. Cao, S. Jin, S. Wang, D. Zhang, X. Meng, M. Zhu, A comparison of the chiral counterion, solvent, and ligand used to induce a chiroptical response from Au25− nanoclusters, Nanoscale 5 (16) (2013) 7589–7595, https://doi.org/10.1039/C3NR01782F.

[183] R. Kobayashi, Y. Nonoguchi, A. Sasaki, H. Yao, Chiral monolayer-protected bimetallic Au–Ag nanoclusters: alloying effect on their electronic structure and chiroptical activity, J. Phys. Chem. C 118 (28) (2014) 15506–15515, https://doi.org/10.1021/jp503676b.

[184] F. Bertorelle, I. Russier-Antoine, N. Calin, et al., Au10(SG)10: a chiral gold catenane nanocluster with zero confined electrons. Optical properties and first-principles theoretical analysis, J. Phys. Chem. Lett. 8 (9) (2017) 1979–1985, https://doi.org/10.1021/acs.jpclett.7b00611.

[185] Y. Yanagimoto, Y. Negishi, H. Fujihara, T. Tsukuda, Chiroptical activity of BINAP-stabilized undecagold clusters, J. Phys. Chem. B 110 (24) (2006) 11611–11614, https://doi.org/10.1021/jp061670f.

[186] M.R. Provorse, C.M. Aikens, Origin of intense chiroptical effects in undecagold subnanometer particles, J. Am. Chem. Soc. 132 (4) (2010) 1302–1310, https://doi.org/10.1021/ja906884m.

[187] M. Sugiuchi, Y. Shichibu, K. Konishi, An inherently chiral Au24 framework with double-helical hexagold strands, Angew. Chem. Int. Ed. 57 (26) (2018) 7855–7859, https://doi.org/10.1002/anie.201804087.

[188] S. Takano, T. Tsukuda, Amplification of the optical activity of gold clusters by the proximity of BINAP, J. Phys. Chem. Lett. 7 (22) (2016) 4509–4513, https://doi.org/10.1021/acs.jpclett.6b02294.

[189] J.-Q. Wang, Z.-J. Guan, W.-D. Liu, Y. Yang, Q.-M. Wang, Chiroptical activity enhancement via structural control: the chiral synthesis and reversible interconversion of two intrinsically chiral gold nanoclusters, J. Am. Chem. Soc. 141 (6) (2019) 2384–2390, https://doi.org/10.1021/jacs.8b11096.

[190] H. Yang, J. Yan, Y. Wang, et al., From racemic metal nanoparticles to optically pure enantiomers in one pot, J. Am. Chem. Soc. 139 (45) (2017) 16113–16116, https://doi.org/10.1021/jacs.7b10448.

[191] G. Deng, S. Malola, J. Yan, et al., From symmetry breaking to unraveling the origin of the chirality of ligated Au13Cu2 nanoclusters, Angew. Chem. Int. Ed. 57 (13) (2018) 3421–3425, https://doi.org/10.1002/anie.201800327.

[192] S. Li, X.-S. Du, B. Li, et al., Atom-precise modification of silver(I) thiolate cluster by shell ligand substitution: a new approach to generation of cluster functionality and chirality, J. Am. Chem. Soc. 140 (2) (2018) 594–597, https://doi.org/10.1021/jacs.7b12136.

[193] W. Du, X. Kang, S. Jin, D. Liu, S. Wang, M. Zhu, Different types of ligand exchange induced by Au substitution in a maintained nanocluster template, Inorg. Chem. 59 (3) (2020) 1675–1681, https://doi.org/10.1021/acs.inorgchem.9b02792.

[194] S. Knoppe, O.A. Wong, S. Malola, et al., Chiral phase transfer and enantioenrichment of thiolate-protected Au102 clusters, J. Am. Chem. Soc. 136 (11) (2014) 4129–4132, https://doi.org/10.1021/ja500809p.

[195] J. Yan, H. Su, H. Yang, et al., Asymmetric synthesis of chiral bimetallic [Ag28Cu12(SR)24]4− nanoclusters via ion pairing, J. Am. Chem. Soc. 138 (39) (2016) 12751–12754, https://doi.org/10.1021/jacs.6b08100.

[196] L.-Y. Yao, Z. Chen, K. Zhang, V.W.-W. Yam, Heterochiral self-discrimination-driven supramolecular self-assembly of decanuclear gold(I)-sulfido complexes into 2D nanostructures with chiral anions-tuned morphologies, Angew. Chem. Int. Ed. 59 (47) (2020) 21163–21169, https://doi.org/10.1002/anie.202009728.

[197] L. Shi, L. Zhu, J. Guo, et al., Self-assembly of chiral gold clusters into crystalline nanocubes of exceptional optical activity, Angew. Chem. Int. Ed. 56 (48) (2017) 15397–15401, https://doi.org/10.1002/anie.201709827.

[198] J.-H. Huang, Z.-Y. Wang, S.-Q. Zang, T.C.W. Mak, Spontaneous resolution of chiral multi-thiolate-protected Ag30 nanoclusters, ACS Cent. Sci. 6 (11) (2020) 1971–1976, https://doi.org/10.1021/acscentsci.0c01045.

[199] M.-M. Zhang, X.-Y. Dong, Z.-Y. Wang, et al., AIE triggers the circularly polarized luminescence of atomically precise enantiomeric copper(I) alkynyl clusters, Angew. Chem. Int. Ed. 59 (25) (2020) 10052–10058, https://doi.org/10.1002/anie.201908909.

[200] Y. Jin, S. Li, Z. Han, et al., Cations controlling the chiral assembly of luminescent atomically precise copper(I) clusters, Angew. Chem. Int. Ed. 58 (35) (2019) 12143–12148, https://doi.org/10.1002/anie.201906614.

[201] S. Chen, W. Du, C. Qin, et al., Assembly of the thiolated [Au1Ag22(S-Adm)12]3+ superatom complex into a framework material through direct linkage by SbF6− anions, Angew. Chem. Int. Ed. 59 (19) (2020) 7542–7547, https://doi.org/10.1002/anie.202000073.

[202] C. Gautier, T. Bürgi, Vibrational circular dichroism of adsorbed molecules: BINAS on gold nanoparticles, J. Phys. Chem. C 114 (38) (2010) 15897–15902, https://doi.org/10.1021/jp910800m.

[203] I. Dolamic, B. Varnholt, T. Bürgi, Chirality transfer from gold nanocluster to adsorbate evidenced by vibrational circular dichroism, Nat. Commun. 6 (1) (2015) 7117, https://doi.org/10.1038/ncomms8117.

[204] C. Liu, T. Li, H. Abroshan, et al., Chiral Ag23 nanocluster with open shell electronic structure and helical face-centered cubic framework, Nat. Commun. 9 (1) (2018) 744, https://doi.org/10.1038/s41467-018-03136-9.

[205] R.W. Murray, Nanoelectrochemistry: metal nanoparticles, nanoelectrodes, and nanopores, Chem. Rev. 108 (7) (2008) 2688–2720, https://doi.org/10.1021/cr068077e.

[206] A. Franceschetti, A. Zunger, Pseudopotential calculations of electron and hole addition spectra of InAs, InP, and Si quantum dots, Phys. Rev. B 62 (4) (2000) 2614–2623, https://doi.org/10.1103/PhysRevB.62.2614.

[207] M.J. Weaver, X. Gao, Molecular capacitance: sequential electron-transfer energetics for solution-phase metallic clusters in relation to gas-phase clusters and analogous interfaces, J. Phys. Chem. 97 (2) (1993) 332–338, https://doi.org/10.1021/j100104a012.

[208] K. Kwak, V.D. Thanthirige, K. Pyo, D. Lee, G. Ramakrishna, Energy gap law for exciton dynamics in gold cluster molecules, J. Phys. Chem. Lett. 8 (19) (2017) 4898–4905, https://doi.org/10.1021/acs.jpclett.7b01892.

[209] K. Kwak, Q. Tang, M. Kim, D. Jiang, D. Lee, Interconversion between superatomic 6-electron and 8-electron configurations of M@Au24(SR)18 clusters (M = Pd, Pt), J. Am. Chem. Soc. 137 (33) (2015) 10833–10840, https://doi.org/10.1021/jacs.5b06946.

[210] J.F. Parker, C.A. Fields-Zinna, R.W. Murray, The story of a monodisperse gold nanoparticle: Au25L18, Acc. Chem. Res. 43 (9) (2010) 1289–1296, https://doi.org/10.1021/ar100048c.

[211] M. Kim, Q. Tang, A.V. Narendra Kumar, et al., Dopant-dependent electronic structures observed for M2Au36(SC6H13)24 clusters (M = Pt, Pd), J. Phys. Chem. Lett. 9 (5) (2018) 982–989, https://doi.org/10.1021/acs.jpclett.7b03261.

[212] K. Kwak, D. Lee, Electrochemical characterization of water-soluble Au25 nanoclusters enabled by phase-transfer reaction, J. Phys. Chem. Lett. 3 (17) (2012) 2476–2481, https://doi.org/10.1021/jz301059w.

[213] A. Dass, N.A. Sakthivel, V.R. Jupally, C. Kumara, M. Rambukwella, Plasmonic nanomolecules: electrochemical resolution of 22 electronic states in Au329(SR)84, ACS Energy Lett. 5 (1) (2020) 207–214, https://doi.org/10.1021/acsenergylett.9b02528.

[214] M. Zhou, C. Zeng, Y. Song, et al., On the non-metallicity of 2.2 nm Au246(SR)80 nanoclusters, Angew. Chem. Int. Ed. 56 (51) (2017) 16257–16261, https://doi.org/10.1002/anie.201709095.

[215] A. Dass, S. Theivendran, P.R. Nimmala, et al., Au133(SPh-tBu)52 nanomolecules: X-ray crystallography, optical, electrochemical, and theoretical analysis, J. Am. Chem. Soc. 137 (14) (2015) 4610–4613, https://doi.org/10.1021/ja513152h.

[216] Y. Negishi, T. Nakazaki, S. Malola, et al., A critical size for emergence of nonbulk electronic and geometric structures in dodecanethiolate-protected Au clusters, J. Am. Chem. Soc. 137 (3) (2015) 1206–1212, https://doi.org/10.1021/ja5109968.

[217] M. Shabaninezhad, A. Abuhagr, N.A. Sakthivel, et al., Ultrafast electron dynamics in thiolate-protected plasmonic gold clusters: size and ligand effect, J. Phys. Chem. C 123 (21) (2019) 13344–13353, https://doi.org/10.1021/acs.jpcc.9b01739.

[218] S. Chen, T. Higaki, H. Ma, M. Zhu, R. Jin, G. Wang, Inhomogeneous quantized single-electron charging and electrochemical–optical insights on transition-sized atomically precise gold nanoclusters, ACS Nano 14 (12) (2020) 16781–16790, https://doi.org/10.1021/acsnano.0c04914.

[219] K.N. Swanick, M. Hesari, M.S. Workentin, Z. Ding, Interrogating near-infrared electrogenerated chemiluminescence of Au25(SC2H4Ph)18 + clusters, J. Am. Chem. Soc. 134 (37) (2012) 15205–15208, https://doi.org/10.1021/ja306047u.

[220] M. Hesari, M.S. Workentin, Z. Ding, NIR electrochemiluminescence from Au25 − nanoclusters facilitated by highly oxidizing and reducing co-reactant radicals, Chem. Sci. 5 (10) (2014) 3814–3822, https://doi.org/10.1039/C4SC01086H.

[221] M. Hesari, M.S. Workentin, Z. Ding, Highly efficient electrogenerated chemiluminescence of Au38 nanoclusters, ACS Nano 8 (8) (2014) 8543–8553, https://doi.org/10.1021/nn503176g.

[222] W. Guo, J. Yuan, E. Wang, Organic-soluble fluorescent Au8 clusters generated from heterophase ligand-exchange induced etching of gold nanoparticles and their electrochemiluminescence, Chem. Commun. 48 (25) (2012) 3076–3078, https://doi.org/10.1039/C2CC17155D.

[223] T. Wang, D. Wang, J.W. Padelford, J. Jiang, G. Wang, Near-infrared electrogenerated chemiluminescence from aqueous soluble lipoic acid Au nanoclusters, J. Am. Chem. Soc. 138 (20) (2016) 6380–6383, https://doi.org/10.1021/jacs.6b03037.

[224] S. Chen, H. Ma, J.W. Padelford, et al., Near infrared electrochemiluminescence of rod-shape 25-atom AuAg nanoclusters that is hundreds-fold stronger than that of Ru(bpy)3 standard, J. Am. Chem. Soc. 141 (24) (2019) 9603–9609, https://doi.org/10.1021/jacs.9b02547.

[225] J. Wang, Z.-Y. Wang, S.-J. Li, S.-Q. Zang, T.C.W. Mak, Carboranealkynyl-protected gold nanoclusters: size conversion and UV/Vis–NIR optical properties, Angew. Chem. Int. Ed. 60 (11) (2021) 5959–5964, https://doi.org/10.1002/anie.202013027.

Chapter 6

Theoretical simulations on metal nanocluster systems

Manzhou Zhu and Haizhu Yu

6.1 Foundation of theoretical simulation methods

As the interdisciplinary between computer science and quantum chemistry, computation chemistry (also called quantum chemistry, theoretical chemistry in some other cases) uses computer simulation to assist in solving complex chemical problems (from Boron Nitride Nanotubes in Nanomedicine [1]). Thanks to the rapid progress of computer science in both hardware and softwares, a variety of novel computational platforms have been developed in these years. Specifically, with the continuous progress of network services, and the atomic precision of the large number of metal nanoclusters in the past decade, the application of computational chemistry has been greatly extended to a variety of applications.

To data, theoretical simulation has become an elementary analysis strategy, and has been widely used to settle different types of chemical problems in nanoscience [2], from the structural prediction, electronic structure and optical property analysis in early times, to the thermodynamics, mechanism, and structure-property correlations in these days. Among the various theoretical methods, density functional theory (DFT, see S.M. Blinder, Introduction to Quantum Mechanics (Second Edition), 2021 for a detailed introduction [3]) is undoubtedly the most popular one, predominantly because its balance in computational cost and accuracy.

DFT method has been first developed by Hohenberg and Kohn in the 1960s to simulate the ground state properties of molecules. After the endeavors in these decades, the DFT algorithms (including both the functions and basis sets) have been greatly enriched, making it possible to accurately treat both the ground state and excited state properties of metal nanocluster systems. In addition, the typical DFT methods, such as the traditional B3LYP, PBE, and the newly developed Minnesota density functions (e.g., M06-series of functions) have been extensively available in various quantum chemical platforms.

According to a rough literature screening on the recently reported computational studies, we found that over ten types of computational codes, including Gaussian, SambVca, CP2K, Turbomole, VASP, DMOL^3, ADF, HSE06, LAMMPS, GPAW, PYXAID, ORCA, Multiwfn, Quantum Espresso, and Dalton have been used to analyze the geometric structure, physicochemical property and electronic structure of the atomically precise metal nanoclusters.

In the following section, we will mainly focus on the recent progress of DFT calculations in metal nanocluster systems, and first give a brief introduction to the theoretical basis of the DFT calculations.

6.1.1 Theoretical basis of DFT functionals

Density functional theory (DFT) provides a powerful computational tool for studying the structure and potential energy surface of metal nanoclusters. DFT can describe the microstructure and micro/macro properties based on intermolecular/intramolecular forces, which helps to predict the new structure of metal nanoclusters and analyze their reaction dynamics. In addition, the time-dependent density functional theory (TDDFT) can be used to analyze the excited states and spectral properties, etc., and help to judge the electronic structure-related properties and properties of metal nanoclusters. The calculation of energy and molecular orbital is closely related to the Schrödinger equation.

Conceptionally, Schrödinger equation describes the behavior of a particle in a field of force or the change in a physical quantity. It actually refers to two separate equations, often called the time-dependent and time-independent Schrödinger equations. The time-dependent Schrödinger equation is a partial differential equation that describes how the wavefunction evolves over time, while the time-independent Schrödinger equation [4] is an equation of state for wavefunctions of definite energy. The ground state energy calculation relies on independent form of Schrödinger equation (Eq. 6.1), but without considering the relativistic Born-Oppenheimer approximation.

$$\hat{H}\psi(\vec{r}) = \psi(\vec{r}) \tag{6.1}$$

wherein E is the energy of the particle, and $\hat{H}$ is the Hamiltonian operator, which could be calculated via Eq. (6.2):

$$\hat{H} = \frac{-h^2}{8\pi^2 m}\nabla^2 + V \tag{6.2}$$

Hamiltonian operator $\hat{H}$ is consists of the sum of three items: kinetic energy (T, which is a summation of [2] over all the particles in the molecule), and V is the particle potential including both the interaction with external potential (V_{ext}), and electron interaction (V_{ee}). In the specific material simulation, the external potential of interest is the interaction between electrons and nuclei, which is expressed by Eq. (6.3):

$$\hat{V}_{ext} = -\sum_{\alpha}^{N_{at}} \frac{Z_\alpha}{|r_i - R_\alpha|} \tag{6.3}$$

wherein, r_i is the coordinate of electron i, the charge on the nucleus at R_α is Z_α. Schrödinger equation solves a set of Ψ, with the constraint that Ψ is antisymmetric-if the coordinates of any two electrons are interchanged, they will change their sign. The lowest energy eigenvalue E_0 is the ground state energy, and the probability density of finding an electron with any specific coordinate set of $\{r_i\}$ is $|\Psi_0|^2$.

The average total energy of the specified state from the special Ψ is not necessarily one of the eigenfunctions of the Schrödinger equation, but the expected value of Ψ, i.e.,

$$E[\Psi] = \int \Psi^* \hat{H} \Psi dr \equiv \langle \Psi | \hat{H} | \Psi \rangle \tag{6.4}$$

Symbol $[\Psi]$ shows that energy is a functional of the wavefunction. When $\Psi \neq \Psi^0$, the energy is higher than the energy of the ground state, which is the variational theorem.

The ground state wave function and energy can find the wave function that minimizes the total energy by searching all possible wave functions. As the simplest approximate theories for solving the many-body Hamiltonian, Hartree Fock theory defines Ψ via the composition of the true many-body wavefunction. Assuming the antisymmetric product of function (ϕ_i), each function depends on the coordinates of a single electron, and thus

$$\Psi_{HF} = \frac{1}{\sqrt{N}} det[\varnothing_1, \varnothing_2, \varnothing_3, \cdots \varnothing_N] \tag{6.5}$$

Where det represents the determinant of the matrix. In combination of this equation with the Schrödinger equation in Eq. (6.4), the expression of Hartree Fock energy is obtained:

$$E_{HF} = \int \varnothing_i(r)\left(-\frac{1}{2}\sum_i^N \nabla_i^2 + V_{ext}\right)\varnothing_i(r)dr + \frac{1}{2}\sum_i^N \int \frac{\varnothing_i^*(r_1)\varnothing_i(r_1)\varnothing_j^*(r_2)\varnothing_j(r_2)}{|r_i - r_j|} dr_1 dr_2$$
$$-\frac{1}{2}\sum_{i,j}^N \int \frac{\varnothing_i^*(r_1)\varnothing_i(r_1)\varnothing_j(r_2)\varnothing_j^*(r_2)}{|r_i - r_j|} dr_1 dr_2 \tag{6.6}$$

In the above formula, the second term is the classical Coulomb energy expressed in orbit, and the third term is the exchange energy. Under the constraint of orbit orthogonality, the ground state orbit is determined by applying the variational theorem to the energy expression. Thus, Hartree Fock (or SCF) equation is written as:

$$\left[-\frac{1}{2}\nabla^2 + V_{ext}(r) + \int \frac{\rho(r')}{|r-r'|}dr'\right]\phi_i(r)^\infty + \int V_X(r, r')\phi(r')dr' = \varepsilon_i \phi_i(r) \tag{6.7}$$

Where the nonlocal exchange potential v_X is defined as:

$$\int V_X(rr')\phi_i(r')dr' = -\sum_j^N \int \frac{\phi_j(r)\varnothing_j^*(r')}{|r-r'|}\phi_i(r')dr' \tag{6.8}$$

According to these algorisms, Hartree Fock equation describes the noninteracting electrons under the influence of a mean-field potential composed of classical Coulomb potential and nonlocal exchange potential.

From the Hartree Fock method, the approximations of Ψ and E_0 is obtained by the correlated methods, but the huge computational cost forces chemists to turn their attention to the development of density functional theory with the accurate solutions and flexible descriptions of the wave-function.

6.1.1.1 DFT functionals

For the calculations focusing on the energy surface, DFT provides a practical and potentially precise method to replace the above wave function method. The practicability of the theory depends on the approximate value of $E_{xc}[\rho]$. E_{xc} is the sum of the error generated by the means of noninteraction kinetic energy and the error generated by the classical treatment of electron-electron interaction, that is

$$E_{XC}[\rho] = (T[\rho] - T_S[\rho]) + (V_{ee}[\rho] - V_H[\rho]) \tag{6.9}$$

The use of approximation of E_{xc} has brought a huge and still rapidly expanding research field. In the early applications of DFT, the uniform electron gas system with approximate and accurate results dominates. In these systems, electrons are described by a constant external potential, and the charge density is a constant. Therefore, the system can use a single number to specify the value of the constant electron density $r = N/V$.

According to Thomas and Fermi's research on uniform electron gas in the early 20th century [5–7], the orbit of the system is a plane wave. If the exchange and correlation effects are ignored, the electron-electron interaction can be approximated to the classical Hartree potential, and then the total energy functional can be easily obtained. The dependence of kinetic energy and exchange energy on electron gas density in a homogeneous system is extracted and expressed as a local function of density. Then, for heterogeneous systems, the functional can be approximated as the integral of the local function of charge density. Using the kinetic energy density and exchange energy density of noninteracting uniform electron gas, the kinetic energy and the exchange energy is written as:

$$T[\rho] = 2.87 \int \rho^{5/3}(r) dr \tag{6.10}$$

$$E_X[\rho] = 0.74 \int \rho^{4/3}(r) dr \tag{6.11}$$

These results give the representation of E_{xc} in heterogeneous systems. The local exchange-correlation energy of each electron can be approximated by the simple function of local charge density (i.e., $\varepsilon_{xc}(\rho)$). That is:

$$E_{XC}[\rho] = \int \rho(r) \varepsilon_{XC}(\rho(r)) dr \tag{6.12}$$

in which $\varepsilon_{xc}(\rho)$ is a function consisting of exchange $\varepsilon_x(\rho)$ and correlation density $\varepsilon_c(\rho)$, based on the uniform electron gas model with density ρ. As the local value of electron density is the only function in Eq. (6.13), this treatment is called local density approximation (LDA).

In these days, reliable descriptors have been developed within the LDA method to accurately describe the molecular structures, and properties such as vibrational frequency, elastic modulus, and phase stability. However, LDA may have induced ignorable errors in calculating the energy difference between geometrically different materials. According to reported literatures, the LDA functions might results in the overestimated binding energy (by 20%–30%), or underestimated reaction barriers in some cases.

To improve the exchange-correlation energy of the molecule and solid, generalized gradient approximations (GGA) were developed. GGA adopts a functional form to ensure that the normalization condition and exchange hole are negative definite. In this context, energy functional depends not only on the density and its gradient, but also the analytical properties of exchange-correlation holes inherent in LDA:

$$E_{XC} \approx \int \rho(r) \varepsilon_{XC}(\rho, \nabla \rho) dr \tag{6.13}$$

Compared with LDA methods, GGA generally show significantly improved performance in describing the binding energy of molecules and are widely accepted by chemists since the age of 1990s. After that, the BP, HTBS, MPW, mPBE, OLYP, OPBE, PBE, PBEsol, PW91, revPBE, RPBE [8–13] functionals of GGA series have been developed.

In recent years, quantum chemists have further developed the meta-GGA functional that depends on the semilocal information of spin density or local kinetic energy density in Laplacian. Its functional form is given as follows:

$$E_{XC} \approx \int \rho(r) \varepsilon_{XC}(\rho |\nabla \rho| \nabla^2 \rho, \tau) dr \tag{6.14}$$

Wherein the kinetic energy density τ is given in Eq. (6.15).

$$\tau = \frac{1}{2}\sum |\nabla \varphi_i|^2 \tag{6.15}$$

Up to now, TPSS [14], RTPSS [14], M06L [15], MBJ [16,17], SCAN [18,19], MS0, MS1, MS2 [20,21], RSCAN [22,23], R2SCAN [24], LIBXC [25,26], functional of Meta-GGA serials have been developed.

With the integration of works about electron-electron interactions, an exact adiabatic connection can be constructed between the noninteracting and fully interacting density functionals. Under this adiabatic connection approach [27–29], the functional changes to:

$$E_{XC}[\rho] = \frac{1}{2}\int d\vec{r}d\vec{r}'\int_{\lambda=0}^{1} d\lambda \frac{\lambda e2}{|\vec{r}-\vec{r}'|}\left[\left\langle \rho(\vec{r})\rho(\vec{r}')\right\rangle_{\rho,\lambda} - \rho(\vec{r})(\vec{r}-\vec{r}')\right] \tag{6.16}$$

where $\left\langle \rho(\vec{r})\rho(\vec{r}')\right\rangle$ is the density-density correlation function. Meanwhile, the effective potential at the density $\rho(r)$ could be calculated:

$$V_{eff} = V_{en} + \frac{1}{2}\sum_{i\neq j}\frac{\lambda e^2}{|\vec{r}_i - \vec{r}_j|} \tag{6.17}$$

Given the variation of the density-density correlation function with the coupling constant λ, E_{xc} could be calculated. By replacing the density-density correlation function with the correlation function of the homogeneous electron gas, the description in Eq. (6.17) changes back to LDA.

The adiabatic integration method presents different approximations to the exchange-correlation functional. The noninteracting system is the same as Hartree Fock method when $\lambda = 0$, while LDA and GGA functionals are constructed as fully interacting homogeneous electron gases (i.e., $\lambda = 1$). Therefore, the integral on the coupling constant could be roughly approximated to the weighted sum of the endpoints. In this scenario, E_{xc} is set as:

$$E_{XC} \approx aE_{Fock} + bE_{XC}^{GGA} \tag{6.18}$$

These coefficients will be determined by referencing to a system with accurate results. This approach was adopted by Becke [16] in the definition of a new functional with coefficient determined by fitting atomization energy, ionization potential, proton affinities, and total atomic energy of many small molecules. The energy functional is:

$$E_{XC} = E_{XC}^{LDA} + 0.2\left(E_X^{Fock} - E_X^{LDA}\right) + 0.72\Delta E_X^{B88} + 0.81\Delta E_C^{PW91} \tag{6.19}$$

Where $B88$ exchange energy and $PW91$ correlation energy are widely used in GGA correction of LDA exchange energy and correlation energy [30,31].

Compared with the aforementioned LDA and GGA functionals, hybrid functionals have been widely used in the study of complexes, among which B3LYP functional [32] was suggested reliable in calculating binding energy, geometry and frequency.

6.1.1.2 Basis sets

In computational chemistry, the basis set consists of a set of basic functions, which are used to guess the electron wave function in Hartree Fock method or density functional theory, and to efficiently transform the partial differential equation of the model into an algebraic equation. In the following, we briefly introduce several groups of algorisms that were frequently used in metal nanocluster systems.

In quantum chemistry, Slater type orbitals (STO) are functions used as atomic orbitals [33]. STO is mainly used to calculate the wave functions of atomic and diatomic systems and have been extensively embedded into different commercially-available programs. For example, the different types of STO basis groups are embedded in ADF software [34–37] including SZ, DZ, DZP, TZP, and TZ2P [38–40]

In addition to the STOs type of basis sets, there are Gaussian type basis sets (GTOs). GTOs are used to deal with multicenter integral. The development of GTOs basis group compensates the drawbacks of STOs basis group in treating polyatomic molecular systems.

At present, Gaussian basis sets mainly include the minimal basis sets (STO-nG) [41,42] double-zeta basis sets (DZ, SV, 6-31G) [43–53], Pople basis sets (6-311G*, 6-311+G(2df,2pd), etc.) [54–59], Karlsruhe "def2" basis sets [60,61], correlation-consistent basis sets (cc-pVXZ) [62–66], effective core potentials (e.g., LANL2DZ) [67–69] and auxiliary basis sets (RI-J, RI-JK, "cbas", "cabs") [70].

In Effective core potentials (ECPs), the core electrons of atomic or molecules were replaced with the effective potentials (pseudo-potential), while only the valence electrons were accurately treated in quantum mechanical calculations. This approach greatly reduced the computational cost and could effectively deal with the relativistic effect. For the heavier elements, there are different ways to classify nuclei and valence electrons. In a variety of computational programs, the "small core" and "large core" pseudo-potentials could be designated, while more valence electrons are present in the former case. It should be kept in mind that the pseudo-potential should be used with a consistent basis set.

6.1.1.3 Relativistic effects

Compared with the nonrelativistic theory, the calculations with relativistic effect could show remarkably improved accuracy in describing the systems with a certain number of heavy elements, because the relativistic effect correlates directly with the nuclear charge Z. In this context, Dirac equation [71] is a relativistic wave equation describing electrons and similar kind of particles.

$$\left(\beta mc^2 + c \int_{n=1}^{3} \alpha_n p_n \right) \psi(x,t) = \hbar \frac{\partial \psi(x,t)}{\partial t} \quad (6.20)$$

Where $\psi(x,t)$ is the electron wave function; m is the electronic mass at rest; (x,t) is the space-time coordinate; p_1, p_2, and p_3 are momentum components; c is the speed of light and $\hbar$ is reduced Planck constant. The main purpose behind this formula is to study the relative motion of the electron and to treat the atom as consistent with relativity.

Zero order regular approximation (ZORA) [72–74] is a frequently used relativistic effect in theoretical simulation of metal nanocluster systems. The ZORA equation is:

$$H^{ZORA} \varnothing^{ZORA} = \left(V + \sigma \cdot p \frac{c^2}{2c^2 - V} \sigma \cdot p \right) \varnothing^{ZORA} = \left(H_{SR}^{ZORA} + H_{SO}^{ZORA} \right) \varnothing^{ZORA}$$

$$= \left(V + p \frac{c^2}{2c^2 - V} p + \frac{c^2}{(2c^2 - V)^2} \sigma \cdot (\nabla V \times p) \right) \varnothing^{ZORA} \quad (6.21)$$

Here, the regularized form of the spin-orbit operator has already presented. For comparison, the scalar relativistic ZORA equation is the equation eliminating the spin-orbit operator, that is:

$$H_{SR}^{ZORA} + \varnothing_{SR}^{ZORA} = \left(V + p \frac{c^2}{2c^2 - V} p \right) \varnothing^{ZORA} = E_{SR}^{ZORA} + \varnothing_{SR}^{ZORA} \quad (6.22)$$

6.1.1.4 Time-dependent density functional theory calculations

Time-dependent density functional theory (TDDFT) [75] landed on the Runge–Gross theorem can be considered as an extension of DFT to time-dependent problems. It is an accurate restatement of time-dependent quantum mechanics.

The basic variable of TD-DFT is the electron density of molecules. The advantage is that the multibody wave function is replaced by a density real function which only depends on the three-dimensional vector. Generally, the electron density of molecules is obtained through the auxiliary system of noninteracting electrons, which are affected by the time-dependent Kohn-sham potential.

So far, TD-DFT has become one of the tools to obtain an accurate and reliable prediction of excited state properties in linear and nonlinear states.

6.2 Application of density functional theory simulations

6.2.1 Single point energy calculations

Single point energy calculation-as its name indicates, performs theoretical simulation (on wavefunction, charge distribution, and molecular energy) for a single conformation. As a matter of fact, single point energy calculation is a default task in all types of theoretical calculations. For example, geometry optimization starts with the single point energy calculation on a given starting state, calculates the force on the nuclei with the Hellmann-Feynman theorem, and then adjusts the position of the nuclei toward an equilibrium structure to reduce the forces and finally generate the stationary point (local minimum or transition state) on potential energy surface. Due to the high computational cost of the geometry optimization (compared with that of the single point energy calculations), the combination of a geometry optimization at a lower level of theory and single point energy calculations at a higher level of theory represents an appealing strategy, especially for complex systems with a large number of electrons. Meanwhile, with the aid of the optimized structure, the single point energy calculations could also be conducted to simulate the spectroscopic (such as IR, UV-vis, NMR, and XRD pattern) character of the target cluster.

Of note, the calculated energy cited directly from the computational output file does not necessarily correlate with the relative stability of the concerned species. Herein, we use two groups of examples to support this proposal:

The electronic energy of chloroethane (ClH_2CCH_3) calculated with B3LYP/6-31G(d), B3LYP/6-311++G(2df,2p), M06-2x/6-31G(d), and B3LYP/LANL2DZ (with Gaussian 09, Revision E.01 software [76]) is -539.4262712, -539.4889997, -539.341643 and -94.1612237 Hartree, respectively. The relative energy calculated with B3LYP/LANL2DZ method is remarkably higher energy than that of all other methods, but the results do not mean that the relative stability of ClH_2CCH_3 predicted by B3LYP/LANL2DZ is inferior to that with the other methods. Instead, the higher energy is mainly caused by the use of effective core potential in the LANL2DZ basis set. With this basis set, only the 7 valence electrons of the chlorine are treated explicitly with quantum mechanical calculations, and the 10 inner electrons were treated with the effective potential (such treatment has been widely used on heavy elements to reduce computational cost). Due to the ignorance of the interactions between the nuclear and inner core electrons, the relative energy of the target system is higher than the all-electron systems.

Another set of example relates to the comparison of the electronic energy of ethane with chloroethane. With the same level of calculation method (B3LYP/6-311++G(2df,2p)), the electronic energy of ethane and chloroethane are -79.8617569 and -539.4889997 Hartree. Similar to the aforementioned analysis, the remarkably lower energy of CH_3CH_2Cl than that of CH_3CH_3 is mainly caused by the 16 more electrons in the former case. Both species are metastable molecules. That is, ethane is relatively stable under acid/basic conditions, etc. but could convert to chloroethane upon treatment with chlorine and light irradiation (radical substitution reaction occurs). Meanwhile, CH_3CH_2Cl is the thermodynamic product of the aforementioned substitution reaction but easily undergoes hydrolysis under basic conditions. Therefore, the relative stability of different species could not be evaluated directly via comparing their energy (no matter electronic energy, enthalpy, entropy, or Gibbs free energy was used). Alternatively, the reaction thermodynamics could be determined via constructing reaction equations. For example, the reaction energy of $CH_3CH_2Cl + H_2O \rightarrow CH_3CH_2OH + HCl$ ($\Delta E = E^{CH_3CH_2OH} + E^{HCl} - E^{CH_3CH_2Cl} - E^{H_2O}$) could be used to determine whether the target reaction is thermodynamically feasible (with negative reaction energy) or not (with positive reaction energy).

Herein, the DFT analysis on the relative plausibility of two groups of products during a cluster conversion were given as an example (Of note, geometry optimization is requisite for the energy analysis). In the typical understanding on the mechanism of Brust synthesis, free thiolates are usually suggested as the by-product [77]. But during the ESI-MS monitoring on the size-growth of Au^I(p-MBA) complexes to Au_{25}(p-MBA)$_{18}$ [78], no signal of free thiolates was identified. Instead, the peak intensity of [Au(p-MBA)$_2$]$^-$ increased continuously during the reduction process. To this end, DFT calculations (with Turbomole package) were conducted to determine the thermodynamic facility of the two stoichiometric reactions in Eqs. (6.23) and (6.24). As the reaction energy of Eq. (6.24) is lower than that of Eq. (6.23) by 400 kJ/mol, [Au(p-MBA)$_2$]$^-$ is suggested a more reasonable by-product, which is consistent with the experimental characterizations. On this basis, the thermodynamic facility for formation of [Au(p-MBA)$_2$]$^-$ was suggested the driving force for the target size growth.

$$25/x \left[Au^I(p-MBA)\right]_x + 8e^- \rightarrow \left[Au_{25}(p-MBA)_{18}\right]^- + 7\,p-MBA \tag{6.23}$$

$$32/x \left[Au^I(p-MBA)\right]_x + 8e^- \rightarrow \left[Au_{25}(p-MBA)_{18}\right]^- + 7 \left[Au(p-MBA)_2\right]^- \tag{6.24}$$

6.2.2 Geometry optimization

Geometry optimization is a fundamental function of theoretical simulation. Starting from the single point energy calculation of a given structure, geometry optimization aims to figure out a stationary point on potential energy surface. Herein, the stationary point denotes the points that the gradient of the intramolecular force, i.e., the first derivative of the energy, are zero. Depending on the computational task, the stationary point could be either an equilibrium structure (i.e., local/global minimum) or a saddle point (i.e., transition state, an energy-maximum along the reaction path but a minimum in all other directions). Generally, geometry optimization starts with the energy and energy gradient calculations for an input geometry, and then determines the step width and the direction for structural change (by alternating the position of atoms), according to the most-rapidly decreased direction on potential energy surface. The calculation on energy and gradient, and structural change were conducted alternatively until the calculation is converged, wherein all convergence criteria has been satisfied.

Generally, the convergence criteria include several folds. For example, in Gaussian software, commonly four criteria, including the force, the root-mean-square (RMS) of the forces, the molecular displacement for the next step, and RMS of the displacement should be all lower than the cutoff value (essentially 0 [79]) until the geometry optimization is finished.

Herein, it should be kept in mind that the normal termination of a geometry optimization does not necessarily mean that the calculation is successful. Sometimes, the molecular structure might significantly deviate from the input structure. In this scenario, a careful analysis should be conducted to determine whether the input structure or the calculation outcome is more plausible.

In metal nanocluster systems, geometry optimization is mainly conducted to accomplish two tasks: predicting the atomic structure or generating a plausible atomic structure to prepare for the property calculations. In the following section, we will briefly introduce some recent examples on the utilization of geometry optimization in theoretical simulation of metal cluster systems.

6.2.2.1 Prediction of the experimental structures

With the aid of the experimentally characterized cluster structures (determined by SCXRD, MS, NMR, etc.), geometry optimization is commonly the first step in various theoretical simulations. The most convenient application of geometry optimization in cluster systems is to determine the plausibility of a target atomic packing modes, or in other words, to predict the thermal stability of an input structure.

The atomic structures of the numerous metal nanoclusters have been reported in recent years. Before the development of various synthetic and purification methods, the atomic packing modes of metal nanoclusters have been extensively predicted by theoretical simulations. Most of these studies were conducted with the combination of spectroscopic characterization (such as UV-vis, XRD, etc.) and DFT calculations. The commonly used strategy includes three steps: (i) construct a modeling structure with the aid of certain chemical principals (or obtaining a series of plausible configurations via the molecular dynamic calculations), (ii) verify their structural stability via DFT calculations (the structure maintains after geometry optimization), and (iii) conduct theoretical simulation on the spectra of the obtained structure and compare the theoretical results with the experimental ones. The consistency between experimental and theoretical spectra is requisite to verify the plausibility of theoretical methods and the proposed conclusions. In this scenario, a lot of structures have been theoretically predicted, and some of them have been latter verified by experimental characterizations (such as SCXRD). In the following, we will briefly introduce three typical examples on the theoretical prediction of novel cluster structures.

The first example relates to the theoretical prediction of the thiolate protected $Au_{38}(SR)_{24}$ clusters. Earlier before the atomic precision of the thiolate-protected Au_{102} and Au_{25} clusters, the structure of $Au_{38}(SMe)_{24}$ has been theoretically predicted by DFT calculations. Two plausible structures, one with an Oh-symmetric Au_{14} and six planar, ring-like $Au_4(SR)_4$ motifs [80], and the other one bears a disordered structure with distorted pentagonal bipyramid subunits and Au_mS_n surface structures (**Au_{38}-I1** vs **Au_{38}-I2**, Fig. 6.1) have been theoretically predicted [81]. These two structures were not directly verified by experimental means, but the plausibility of both the six planar, ring-like Au_4S_4 motifs (orientated in an octahedral symmetry), and the pentagonal bipyramid subunits have been verified recently. Using a well-designed thiolate ligand of $HSC(SiMe_3)_3$, Schnept and coworkers recently determined the single crystal structure of $Au_{108}S_{24}(PPh_3)_{16}$ with SCXRD, and found that the octahedral $Au_6@Au_{38}$ kernel is protected by six planar, ring-like Au_4S_4 structures, eight triangular Au_3, and eight $Au(PPh_3)$ blocks [82]. Meanwhile, the pentagonal bipyramid subunits have been extensively observed in thiolate protected nanoclusters, such as the popular $[Au_{25}(SR)_{18}]^-$ (the icosahedral could be viewed as a vertex-sharing assembly of two pentagonal bipyramids) and the multilayered structure of $Au_{102}(p\text{-MBA})_{44}$. Specifically, with the continuous progress of nanocluster science, a variety of cluster isomerism have been identified, demonstrating the variety in M-M/M-L binding modes within a given cluster components (e.g., $Au_{28}(TBBT)_{20}$ vs $Au_{28}(SCy)_{20}$, Fig. 4.39). The lability in binding modes is anticipated to be responsible for the comparable stability of **Au_{38}-I1** and **Au_{38}-I2** predicted by DFT calculations (with PBE/BLYP methods) [80].

With the combination of the "divide and protect" frame (i.e., protecting the metallic core structure with metal thiolate complexes) and the structural characteristics of newly reported Au_{102} and Au_{25} clusters, Zeng and coworkers theoretically constructed the structure of $Au_{38}(SMe)_{24}$ [83]. The central idea is to protect the metallic core with certain number of $Au(SR)_2$ and $Au_2(SR)_3$ staples for the target $Au_n(SR)_m$ cluster. To this end, the formula of a thiolate protected gold cluster $Au_n(SR)_m$ could be described as $(Au)_{a+b}(AuS_2)_c(Au_2S_3)_d$, wherein the $(a+b)$ metallic core atoms are divided into the kernel atoms and core surface atoms, depending on whether the core atom is coordinated by the staples or not. Assuming that the core surface atoms are each coordinated by one thiolate ligand on the AuS_2/Au_2S_3 staples, the number of core surface atoms

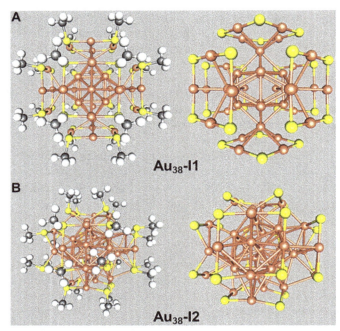

FIG. 6.1 The theoretically predicted atomic structure of $Au_{38}(SR)_{24}$ clusters and the related Au-S framework. *(From H. Häkkinen, M. Walter, H. Grönbeck, J. Phys. Chem. B, 110(20) (2006) 9927–9931, https://doi.org/10.1021/jp0619787, with the permission from American Chemical Society.)*

(b) equals to the sum of 2c and 2d. Besides, $a+b+c+2d=n$, and $2c+3d=m$. In accordance with these principals, the number of a, b, c, and d in $Au_{38}(SR)_{24}$ should follow the equation of $a+b+c+2d=38$, $2c+3d=24$, $b=2c+2d$, and a, b, c, d are all integers. A simple analysis on these three equations demonstrates that c is an integral multiple of 3, while d is an integral multiple of 2. In this context, the number of c could only be 0, 3, 6, 9, and 12 in the concerned $Au_{38}(SR)_{24}$, with d equals to 8, 6, 4, 2, and 0. Accordingly, *a* and *b* values could be possibly (6, 5, 4, 3, 2) and (16, 18, 20, 22, 24), respectively. In other words, the composition of $Au_{38}(SR)_{24}$ could be possibly described as $(Au_{6+16})(Au_2S_3)_8$, $(Au_{5+18})(AuS_2)_3(Au_2S_3)_6$, $(Au_{4+20})(AuS_2)_6(Au_2S_3)_4$, $(Au_{3+22})(AuS_2)_9(Au_2S_3)_2$, or $(Au_{2+24})(AuS_2)_{12}$.

According to Zeng's simulation, the relative energy of $[Au]_{5+18}[Au(SR)_2]_3[Au_2(SR)_3]_6$ (determined by $DMol^3$ program, **Au_{38}-I3**) with a face-sharing bi-icosahedral Au_{23} core is remarkably lower than that of all other configurations. The relative stability of this structure is further supported by Monte Carlo run, the large HOMO-LUMO gap and the large adiabatic ionization/affinity potentials. In addition, the predicted structure (i.e., protection of an Au_{23} core with three AuS_2 and six Au_2S_3 motifs) has been soon (in 2010) verified by the structural elucidation of $[Au_{38}(PET)_{18}]^-$ by Jin and coworkers with SCXRD analysis [84]. Of note, the concept of protecting metallic core structure with Au_nS_{n+1} motifs (not limited to AuS_2 and Au_2S_3) has been found to be widely applicable to comprehend the structural anatomy of a variety of thiolate protected metal nanoclusters. Nevertheless, with the structural elucidation of more metal nanocluster within different size-regime, the core surface metal atoms were found to be possibly ligated by more than one staple motifs, such as the atomic structure of $Au_{28}S_{20}$ in Fig. 4.23.

In the aforementioned example, the atomic structure of $Au_{38}(SR)_{24}$ is mainly deduced from the chemical understanding on the relative stability of core- or motif stability. In addition to such principals, the geometric deduction based on certain structural evolution principal represents another commonly used strategy. In this scenario, the geometry prediction of $[Au_{37}(PH_3)_{10}(SCH_3)_{10}Cl_2]^+$ is a typical example.

In the recent study by Nobusada and coworkers [85], DFT calculations (Turbomole package) were first conducted to simulate the geometric structure and optical absorption spectrum of a theoretically modeled cluster $[Au_{25}(PH_3)_{10}(SCH_3)_5Cl_2]^{2+}$. Similar to the experimental results of $[Au_{25}(PPh_3)_{10}(SC_2H_5)_5Cl_2]^{2+}$, the modeling Au_{25} cluster adopts a vertex-sharing bi-icosahedral metal core structure, two axial Cl, ten shoulder phosphine ligands and five bridging thiolate on the equatorial plane. The theoretical UV-vis of the modeling Au_{25} is close to the experimental one, verifying the used theoretical calculation method and the structural simplification. To this end, Nobusada further predicted the geometric and electronic structure of an experimentally unprecedented vertex-sharing, tri-icosahedral structure of $[Au_{37}(PH_3)_{10}(SCH_3)_{10}Cl_2]^+$ with 10 methanethiolate bridging ligands (Fig. 6.2). The supposed structure was determined to be a local minimum after geometry optimization. Specifically, the simulated UV-vis-NIR (NIR is short for near-infrared) spectrum of Au_{37} is analogous to that of Au_{25} in the range of <600 nm (albeit with slight red-shift). Nevertheless, Au_{37} shows two different characteristic peaks at 761 nm

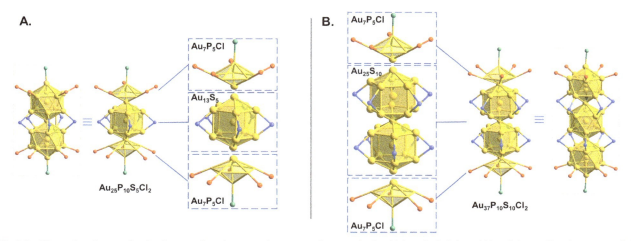

FIG. 6.2 Illustrative diagram for the framework and structural anatomy of vertex-sharing bi-icosahedral Au_{25} (A) and the predicted structure of the vertex-sharing tri-icosahedral Au_{37} clusters (B). *(No permission required.)*

(suggested to correspond to the red-shift of 702 nm peak of Au_{25}, but originated from a different electronic transition) and an additional peak at 1238 nm (which is indistinguishable in optical absorption spectrum of Au_{25}). Both the structural stability of the rod-like tri-icosahedral structure and the appearance of an additional absorption peak at NIR region were recently verified by the experimental characterizations of $[Au_{37}(PPh_3)_{10}(PET)_{10}X_2]^+$ (with SCXRD and UV-vis-NIR tests) reported by Jin et al. [86]

In addition to the structural prediction of an experimentally unprecedented metal nanocluster, DFT (and TD-DFT calculations) have also been widely used to deduce the composition of a cluster system, with the combination of certain experimental characterizations. For example, in the recent study of Jin and coworkers, the size-conversion of $Au_{25}(SePh)_{18}^-$ by $NaBH_4$ was verified by UV-vis spectra [87], which shows a prominent peak at 600 nm and a shoulder peak at 500 nm. The MALDI-MS indicates that the product could be possibly $[Au_{23}(SePh)_{16}]^-$. This proposal is debatable, because fragmentation of cluster samples frequently occurs during laser irradiation in MALDI-MS tests. In this scenario, the composition of the $[Au_{23}(SePh)_{16}]^-$ and its atomic structure were further evidenced by additional spectral analysis and TD-DFT calculations. The overall structural verification includes three steps: (1) The optical absorption spectra of $[Au_{23}(SCy)_{16}]^-$ (i.e., thiolate-counterpart of $[Au_{23}(SePh)_{16}]^-$) were experimentally tested, and two characteristic peaks at 570 and 450 nm were observed. The characteristic peaks of $[Au_{23}(SePh)_{16}]^-$ are systematically red-shift compared with $[Au_{23}(SCy)_{16}]^-$. (2) TD-DFT calculation on the modeling $[Au_{23}(SeH)_{16}]^-$ and $[Au_{23}(SH)_{16}]^-$ shows the red-shift of the prominent absorption peak (622 vs 614 nm), correlating with the experimental results (note: structural simplification is extensively used in theoretical calculations to reduce computational cost); (3) TD-DFT calculation on another group of cluster counterparts, i.e., $[Au_{25}(SeH)_{18}]^-$ and $[Au_{25}(SH)_{18}]^-$, verifies the similar shape of the optical absorption curve, but the slight red-shift of the characteristic peaks after replacing the thiolate ligands with the selenophenols. The agreement of the theoretical and experimental results verifies the composition of $[Au_{23}(SePh)_{16}]^-$ and its structural similarity with that of $[Au_{23}(SCy)_{16}]^-$, which lies the foundation for a more detailed mechanistic analysis of the size-conversion reaction (Please check Section 4.2.2.2 and Fig. 4.21 for the details).

6.2.2.2 Location of the transition state

Compared with the identification of a local minimum, the geometry optimization of a transition state is significantly more challenging. As the transition state features the structural characteristic of both the precursor and the product (of the concerned elementary step), the geometry optimization of the transition state commonly includes several steps (Fig. 6.3). The details for each step are given below:

1. Analyzing the main structural changes in the target elementary step. For example, in the SN2 type of reaction of Cl^- with CH_3Br, the main structural change lies in the lengthening of the C—Br bond in CH_3Br (Fig. 6.4A); while the changes in the R^1-C-C-R^2 dihedral angle are the main structural change in the isomerization process from one staggered conformation to another one (Fig. 6.4B). Although the determination of the most important structural changes in the metal nanocluster systems will be much more complicated, it is generally accepted that the most energy demanding structural changes are the most susceptible parameters (such as the C-Br dissociation compared with that of C-Cl formation in the SN2 step in Fig. 6.4A).

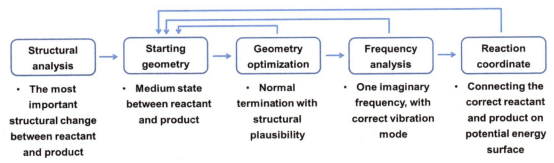

FIG. 6.3 The general procedures for the location of a transition state structure. *(No permission required.)*

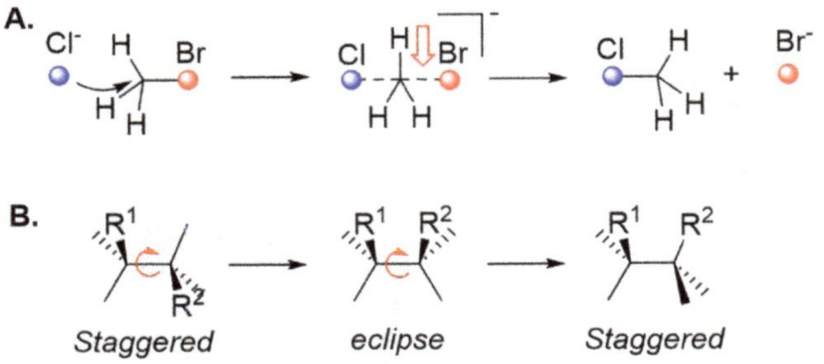

FIG. 6.4 Examples for analysis of the most susceptible structural characteristic in an elementary step. *(No permission required.)*

2. Performing an initial guess on the geometry of the target transition state. This task could be accomplished via different strategies. For example, after inputting the geometric structure of both the precursor and product, the initial geometry could be automatically generated by setting an initial geometry for transition state via certain algorism (such as Synchronous Transit-Guided Quasi-Newton method in Gaussian software). These procedures are commonly easy to implement, but difficult to deal with comprehensive systems with multiple bond formation and dissociation sites or very early/late transition states. Alternatively, based on the figured structural parameter in the first stage, the researchers could construct the initial structure of transition state via changing the key structural parameter of the reactant to the product direction (or via the changing the structure parameter of the product to the reactant direction). For example, in the aforementioned SN2 reaction in Fig. 6.4, we have determined that the most important structural parameter is the C—Br bond distance. The C-Br distance is 1.9 and 4.5 Å in the reactant and product, respectively. In this context, the C—Br bond distance in the transition state should be between 1.9 and 4.5 Å. Meanwhile, the Van der Walls radii of C and Br atoms are 1.70 and 1.85 Å, indicating that the bond lengthening over 3.55 Å might contribute little to the energy increase in Fig. 6.5. Similarly, the distance around 1.9 Å is expected to deviate from the bond distance in the target transition state, as the energy has just begun to increase. To this end, the C—Br bond distance in the starting geometry of the SN2 reaction is expected to be between [2.1–3.3], and the median number (i.e., ~2.7 Å) is suggested as the initially guessed bond distance. Of note, the initial geometry of the transition state could also be generated more reasonably with the aid of partial optimization energy profile, and the details are given in the next section.

3. Geometry optimization of the transition state structure. Different from the common geometry optimization (aiming to locate a local minimum), the optimization of a transition state is commonly necessary to specify certain keywords in the computational input file. For example, the keywords of ts/qst2/qst3 are necessarily stated in the route section of the gaussian input file to specify the job type of transition state calculations. On the basis of the normal termination of the calculation, the researchers are suggested to examine whether the structure fits the expectation, and the key structural parameter lies in between that of the reactant and product.

4. Verification of the transition state structure. This step includes two folds: frequency analysis, and intrinsic reaction coordinate analysis. The frequency calculation is generally conducted at the same level of geometry optimization, and the transition state should contain only one imaginary frequency. Sometimes, we start with a transition state geometry optimization, but the resultant geometry does not contain imaginary frequency. This issue occurs when the initial guess is close to the reactant or product (such as Test-1 and Test-2 in Fig. 6.5). Accordingly, the researchers should adjust the starting geometry to deviate the related intermediate. Otherwise, if the optimized structure is different from both the reactant and the product structures, the researchers are suggested to analyze the plausibility (such as the

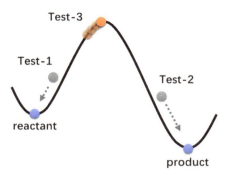

FIG. 6.5 Illustrative diagram for the possible calculation results starting with different initial structure of transition state. *(No permission required.)*

relative energy) of this unexpected structure, and determine whether discarding this new structure or adjust the original mechanistic insights (and adjust the ideas for transition state accordingly). Herein, it is noteworthy that even with only one imaginary frequency; the researchers need to check the imaginary frequency to ensure that it corresponds to the correct vibrational mode.

The last step to verify the transition state structure is the reaction coordination analysis via calculating the intermediate geometries and energies of structures in the reaction path. This step could be implemented spontaneously on different calculation codes via inputting the optimized geometry of the target transition state and designating the job type. As long as the final intermediate structure coincident with the reactant/product of the target step, the entire job for transition state structure location was finished.

6.2.2.3 Partial optimization

Due to the high computational cost of transition state calculations and the current limitation in fast computer hardware and computational programs, the transition state calculations for metal nanoclusters are always very time-consuming. Therefore, partial optimization (and also known as constrained structural optimization in some other cases) was used to roughly determine the energy demands of the target reaction. Commonly in these calculations, certain structural parameter (i.e., the aforementioned parameter that dominates the energy changes) is used as the reaction coordinates. In other words, this parameter is fixed at selected value, while all other parameters are unfrozen and undergoes full structural optimization. In addition, due to the structural bulkiness of the metal nanocluster systems, most structural blocks that are away from the reaction center make little contribution to the reaction energetics, and therefore in some theoretical studies, the atoms beyond certain value from the reaction site was fixed to reduce computational cost. For example, in the recent study on the ligand exchange reaction of $Au_{102}(SH)_{44}$ (i.e., the modeling structure of $Au_{102}(SR)_{44}$, and HSR is para-mercaptobenzoic acid) with methyl thiolate (modeling structure of the para-bromobenzene thiol ligand) reported by Häkkinen, Ackerson and coworkers, the atoms that are over 6 Å away from the reacting center were fixed in some less plausible reaction paths.

As the structures obtained in partial optimization are not stationary points, it is meaningless to conduct frequency calculations. Similarly, the two transition-state structural verification procedures mentioned above are unachievable. Due to the lack of adequate verification procedures, the researchers should be very cautious about the selected structural parameter, and sometimes multiple tests should be conducted to elucidate a more plausible pathway and estimate the energy demands.

The first example relates to the transition states involved in a stepwise ligand exchange process of a $[Au_{25}(SR)_{18}]^-$ cluster, using $[Au_{25}(SH)_{18}]^- + HSMe \rightarrow [Au_{25}(SH)_{17}(SMe)]^- + HSH$ as the modeling system in theoretical calculations. Different reaction sites were considered in the theoretical study by Aikens and coworkers (Fig. 6.6 inset) [88] and herein we mainly focus on the ligand exchange in the core-shell interference (mode A). This process starts with the approaching of the foreign HSMe to the core surface Au atom, and thus the S(HSMe)-Au(core) bond distance was chosen as the constrained structural parameter in the first search for the transition state. According to the calculation results, the S(HSMe)-Au(core) reaches 3.20 Å in the first transition state, which is significantly longer than the Au—S bond distance of the Au_{25} cluster precursor (~2.5 Å). Meanwhile, the original Au(core)—S(staple) bond on the reaction site is lengthened from 2.5 to 3.82 Å, indicating a remarkably weakened bonding interactions therein. The results indicate that the approaching of HSMe results a spontaneous Au(core)—S(staple) bond dissociation, and thus the energy gap of the transition state (note: the most energy demanding species in energy profile) could mainly attributed to two aspects: the energy demand for the Au-S dissociation,

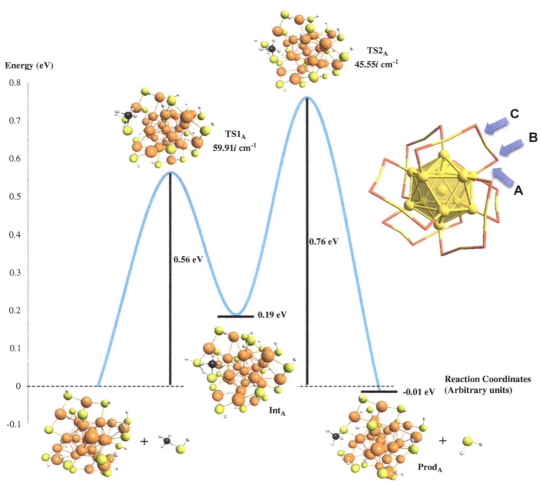

FIG. 6.6 The energy profiles for the ligand exchange of $[Au_{25}(SH)_{18}]^-$ with HSMe by constrained optimization and an illustrative diagram for the possible reaction sites (inset). *(The energy profile is reproduced from A. Fernando, C.M. Aikens, J. Phys. Chem. C 119(34) (2015) 20179–20187, https://doi.org/10.1021/acs.jpcc.5b06833, with the permission from American Chemical Society.)*

and the steric hindrance resulted from the incoming thiol ligands with the original framework. These interactions could be greatly overcome by the Au(core)—S(staple) bonding interactions, and therefore the relative energy of $TS1_A$ is remarkably lower than the Au—S bond dissociation energy (e.g., the average dissociation energy for Au—S bond of $Au_{92}(TBBT)_{44}$/$Au_{102}(TBBT)_{44}$ cluster is ~60 kcal/mol [89]). After the transition state structure, the S(HSMe)-Au(core) bond formation energy overwhelms the energy demands for Au(core)-S(staple) dissociation (note: the bonding has been fairly weak in the transition state, because the covalent radii of Au and S is 1.36 and 1.06 Å [90]) and the increased steric hindrance in the ligand shell. Therefore, the energy decreases after the $TS1_A$, and reaches a local minimum until the formation of intermediate Int_A. In Int_A, the S(HSMe)—Au(core) bond is 2.50 Å, consisting with a common Au—S bond distance in metal nanocluster systems. While the Au(core)-S(staple) distances is slightly lengthened to 4.20 Å, so that the hydrogen from HSMe formally functions as a hydrogen bond and is shared by the HSMe and the staple SH groups. From Int_A, the most prominent structural change lies in the removal of SH group (in the form of H_2S), and thus the related S(staple)—Au(staple) bond distance is used as the constrained structural parameter in the searching for the second transition state. According to the calculation results, the relative energy reaches maximum when the Au(staple)—S(staple) bond distances stretches from 2.38 Å in Int_A to 2.88 Å in $TS2_A$. After that, further lengthening of the Au(staple)—S(staple) bond results in a spontaneous Au(staple)—S(HSMe) bond formation, and the energy decreases dramatically during this process. Finally, the thermodynamically more stable $[Au_{25}(SH)_{17}(SMe)]^-$ and the free H_2S are generated as the products.

6.2.3 Frequency calculation

Namely, frequency calculations compute force constant, vibrational frequencies, and intensities of a target structure. Herein, it is noteworthy that vibrational frequencies are commutated by determining the secondary derivatives of the energy relating to nuclear coordinates and transforming to mass-weighted coordinates. Therefore, the frequency calculations are vivid only for the stationary points, and it is meaningless to conduct frequency calculations on nonstationary points. For example, the geometries obtained by partial optimization (via fixing certain structure parameters) do not correspond to stationary points, and thus it is meaningless to conduct frequency calculation on any of the structures therein. In other words, frequency calculations should be conducted on the basis of the stationary point structure identified by geometry optimization, and at the same level of theory as that of geometry optimization (different theoretical methods results in changes in the structure of stationary points).

Typically, multiple functions could be accomplished via frequency calculations. The most extensively used function is to verify whether the identified structure is a local minimum or a transition state. There should be no imaginary frequency (negative eigenvalue) for a local minimum, and only one imaginary frequency for the transition state. A structure with more than one imaginary frequency demonstrates a higher-order saddle point (note: transition state is a first-order saddle point). This issue could possibly occur in both local-minimum and transition-state optimization and demonstrates the failure of the target calculation. In this case, the person who makes calculation are encouraged to check the vibrational mode of the imaginary points, modify the starting geometry in accordance with the vibrational directions, and then conduct the target geometry optimization again until all unnecessary imaginary frequencies are eliminated.

In addition to determining the character of a certain structure, the vibrational frequencies might also be computed associated with spectroscopic output. Taking Gaussian suit of program as an example, the calculations on vibrational circular dichroism (VCD), Raman optical activity intensities, frequency-dependent polarizabilities and hyperpolarizabilities, and electronic excitation calculations are all available in the frequency calculation procedures (on the premise of specifying the corresponding keywords in the input file). Meanwhile, the thermodynamic data, such as the zero-point energy corrections, thermal correction to enthalpy, thermal correction to Gibbs free energy, and the details on the contribution of entropies (electronic, vibrational, rotational, and transitional entropies) are all available via frequency calculations.

6.2.4 Electronic state elucidation

In early times (since 1992), the "superatom electronic theory" has been successfully utilized to predict the stability and chemistry of the gas-phase uncoordinated metallic clusters, gas-phase metallic clusters with a few simple ligands and Ga-based metalloid clusters. In this context, the aufbau rule of the delocalized superatomic orbitals follows $1S^21P^61D^{10}2S^21F^{14}2P^61G^{18}2D^{10}3S^21H^{22}\ldots$ (S, P, D, F, G, H are angular-momentum character of the target superatomic clusters), and the spherical clusters bearing 2, 8, 18, 34, 58, 92… valence electrons are deduced to be exceptionally stable. Of note, this principal is designated as the superatom electronic theory in some cases, and the clusters such as Au_{11}(8e, $1S^21P^6$), Au_{39}(34e, $1S^21P^61D^{10}2S^21F^{14}$) and Au_{102}(58e, $1S^21P^61D^{10}2S^21F^{14}2P^61G^{18}$) all fit this principal. On the basis of the superatom electronic theory and aufbau rule, a unified electronic model for **solution**-phase ligand protected gold nanoclusters has been proposed by Häkkinen and coworkers, demonstrating that the complete steric shielding of the metallic core with exterior ligands is also a requisite [91]. In their study, $Au_{102}(p\text{-MBA})_{44}$ was first chosen as the modeling cluster, and this structure could be viewed as protecting the Au_{79} core with a $Au_{23}(p\text{-MBA})_{44}$ shell (19*AuS_2 and 2*Au_2S_3, 21 staples in total). Due to the odd number of valence electrons in the Au_{79} core (79e), its angular-momentum-projected local electron density of states (PLDOS, conducted with CP2K and GPAW platforms) shows degenerate HOMO and LUMO states. By contrast, the $Au_{102}(p\text{-MBA})_{44}$ shows discrete HOMO and LUMO states, with a HOMO-LUMO gap of 0.5 eV (Fig. 6.7). The results unambiguously show the effect of the surface ligands shell on the enhanced electronic stability of the cluster structure. On the basis of the PLDOS analysis, and an addition electronic state analysis on a comparative system of $Au_{80}(p\text{-MBA})_2$ (versus Au_{79}, with one AuS_2 block preserving on the core structure), the 79 valence electrons in the Au_{79} core was proposed to be classified into two groups: 58 delocalized electrons in the metallic core structure, and 21 localized one to construct covalent Au—S bond in the core-shell interference. In other words, 21 valence electrons of the metallic core were depleted outside the core structure to make surface covalent bonds with the 21 staples (19*AuS_2 and 2*Au_2S_3 mentioned above). To this end, the number of 58 valence electrons also corresponds to a stable electronic state fitting the superatomic theory. Similarly, the structure of $[Au_{39}(PPh_3)_{14}Cl_6]^{-1}$ could be viewed as extracting 6 valence electrons (with 6 AuCl blocks) from the Au_{39}^{-1} kernel, resulting a 34e metallic core with high electronic stability (fitting superatom model). Herein, it should be noted that the surface phosphine ligands are not included in the electron state analysis, as they are suggested to function mainly as surfactants in thiolated cluster systems. In this context, the binding character of $Au_{11}(PR_3)_7X_3$ (X=X/SR) could be viewed as depleting 3e from the Au_{11} core with the three X groups, while $[Au_{13}(PMe_2Ph)_{10}Cl_2]^{3+}$ adopts an 8e Au_{13} core, with 5 valence electrons being extracted by 2 Cl and 3 counter anions.

According to the aforementioned analysis, the thiolated clusters, in which the metallic core structure is protected by Au_nS_{n+1} ($n=1, 2$) staples or SR/X ligands, could be the viewed to moving certain number of valence electrons (the number

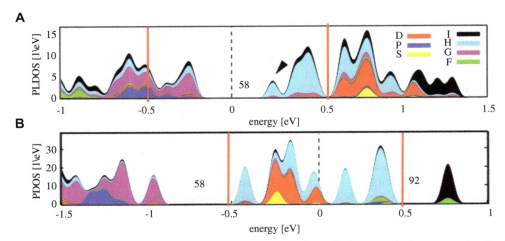

FIG. 6.7 The angular-momentum-projected local density of electron states (PLDOS) (projection up to the I symmetry, i.e., $l = 6$) for the Au_{102}(-p-MBA)$_{44}$ (A) and the Au_{79} core. The zero energy in (A) is the middle of the HOMO-LUMO gap but the HOMO energy in (B). Shell-closing electron numbers are indicated. *(Reproduced from W. Michael, A. Jaakko, L.-A. Olga, P.D. Jadzinsky, C. Guillermo, C.J. Ackerson, R.L. Whetten, G. Henrik, H. Hannu, Proc. Natl. Acad. Sci. 105(27) (2008) 9157–9162, https://doi.org/10.1073/pnas.0801001105.)*

equals to the sum number of Au_nS_{n+1}, SR and X) to construct the core-shell bonding interactions and leaving an extra-stable metallic core structure fitting the superatom electronic theory. Interestingly, this theory could also be used to comprehend the relative stability of the bi-icosahedral, rod-like $[Au_{25}(PR_3)_{10}(SR)_5Cl_2]^{2-}$, which could be viewed as moving $5+2=7e$ from the Au_{25} metallic core structure to construct the Au—S and Au—Cl bond, and thus the remaining 16e could be viewed as 2 equal 8e, icosahedral blocks. The compact geometry, electron shell closing in the metallic core, and the complete steric shielding of the exterior ligands on the metallic core, are 3 fundamental requirements for the formation of a stable metal nanocluster.

The superatom electron theory (and the renewed version) lies the foundation of the deep understanding on the bonding character of the metal nanocluster systems. On the basis of the fundamental ideas of the superatom model, and with the combination of the latter progress in cluster preparation, atomic-structure elucidation, and theoretical simulation, some other widely-applicable electron theories have also been developed in the past decade.

In addition to the superatom model which could successfully explain the extra-stability of gold nanoclusters with magic number of free electrons, Reimers, Ford and coworkers recently developed a novel strategy, i.e., energy fragment decomposition analysis, to shed light on the stability of the thiolate protected gold clusters that bears *un*-magic number of valence electrons (such as the 6e $[Au_{25}(SR)_{18}]^{+1}$, and the 7e $[Au_{25}(SR)_{18}]^{0}$). In this study, both SIESTA and VASP programs have been used for the calculations on test reactions, while the former is used to treat the full structure bearing p-MBA ligands and the latter is used to treat the simplified system with the modeling -SMe ligands. Of note, the two patterns of calculations give the same conclusion (albeit the different reaction energy), and only the results gained from the VASP calculations are mentioned below for clarity reasons. Similar to the aforementioned analysis with superatom model, $Au_{102}(SMe)_{44}$ was used as the target system. Efforts were first made to determine the structure of the bare Au_{79} cluster. The relatively higher energy of the original Au_{79} core structure in the Au_{102} cluster than a reconstructed one demonstrate the critical effect of the ligand shell. In this context, the relative binding affinity of thiolate ligands (in the form of half the dimeric RSSR) with Au_{79} core or the Au(111) surface was examined. Using the relative binding energy of one ligand as a reference, transformation equations could be constructed to correlate any two of the cluster components. For example, the relative stability of $Au_{79}(SR)_{40}$ compared with that of $Au_{102}(SR)_{44}$ could be evaluated by Eq. (6.24), and the exothermicity of 4.7 eV for this reaction (i.e., $\Delta E = -4.7\,eV$) demonstrates that the dissociation of thiolate ligands and the formation of the Au_{102} cluster product is thermodynamically highly feasible. In other words, the formation of $Au_{102}(SMe)_{44}$ is much more favored than that of $Au_{79}(SMe)_{40}$. In this study, $Au_{102}(SMe)_{44}$ has been found a stable cluster product via conducting multiple groups of reaction equations. In combination of these results and the essential bonding character analysis, Reimers and coworkers proposed that the Au—S bond is essentially a covalent bond, and all gold atoms in clusters could be viewed as Au(0). Nevertheless, due to the distinct bonding environment of the core and the staple Au atoms, and the influence of the binding topology by the number of Au—S and Au—Au bonds, the s band contribution of the core and the staple Au atoms are remarkably different. Therefore, the staple Au atoms make little contribution to the frontier orbitals (delocalized ones). Due to the covalent Au—S bonding character and the distinct energy level of the metallic core and the surface staple motifs, the electronic structure of the metallic core could be maintained when removing certain number of electrons (e.g., via single electron oxidation processes). This conclusion well explains the plausibility of $[Au_{25}(SR)_{18}]^n$ structure with different charge states ($n = -1, 0, +1$). Of note, in latter studies, this "energy fragment decomposition analysis" has also been widely applied to analyze the relative stability of different sized metal nanoclusters, to elucidate the origin (such as the stronger core-ligand interactions, or ligand-ligand interactions) of the relative stability of different clusters, and the reversible transformation of a pair of clusters bearing different ligands (such as $Au_{36}(SR)_{24}$ vs $Au_{30}(SR)_{18}$ [92]).

$$\frac{102}{302} Au_{79}(SMe)_{40} \rightarrow \frac{79}{302} Au_{102}(SMe)_{44} + (SMe)_2 \tag{6.25}$$

With the combination of superatom jellium model and valence bond theory, a super valence bond (SVB) model has recently been developed by Yang, Cheng and coworkers [93]. In their study, the SVB model is proposed mainly based on two criteria: (1) The target cluster (namely the supermolecule in the following text) could be divided into several building blocks (i.e., the superatoms mentioned below), and there is no need for clear boundaries among these blocks (i.e., these blocks could possibly share some structural units); (2) The identification of the correct building blocks depends on the Jellium model. In other words, the superatom should be basically "spherical and geometric shell closure."

For clarity reasons, herein we give a brief description on the detailed calculation procedures on the development of the SVB model. First, the geometry and energies of lithium clusters bearing 2–26 Li atoms have been optimized with Gaussian 09 program. The energy difference between the E_{ave} and the Li_N cluster energy (i.e., $E_{ave}-E(Li_N)$, and E_{ave} is obtained via correlation equation of the cluster energy with the number of Li atoms, in the form of $E_{ave} = a + bN^{1/3} + cN^{2/3} + dN$, and a, b, c, d are the fitted parameters, $N = 2-26$) as the index to determine the relative stability of different clusters, the clusters with 8, 10, 12, 14, and 20 Li atoms were found to show superior stability. The relative stability of the Li_{20} cluster has been fully elucidated by the Jellium model, and in Yang's study, efforts are mainly put to provide an alternative insight into the electronic state of Li_8, Li_{10}, Li_{12}, and Li_{14} clusters. Herein, the structure and electronic state analysis of the prolate Li_{14} cluster was used as an example. As shown in Fig. 6.8A, the Li_{14} structure could be viewed as assembly of two Li_{10}^{3+} blocks via sharing the central Li_6^{6+} structure. The electronic structure of Li_{10}^{3+} block (with 7 valence electrons) is close to that of F atom, and thus Li_{10}^{3+} is denoted as a superatom. In this context, the molecular orbital of this Li_{14} cluster shows high similarity to that of the F_2 molecule (Fig. 6.8B), and the bonding/antibonding σ, π-types of orbitals could be easily identified in these two analogous systems. In this context, the Li_{14} was designated as a "supermolecule" with two "superatom" components, and the bonding character between these two super atoms (i.e., Li_{10}^{3+}) resembles that of a super valence bond. This proposal is further evidenced by the adaptive natural density partitioning (AdNDP) analysis (Note: AdNDP is frequently used to determine the nc-2e interactions, while n is the number of bonding atoms, c and e is short for center and electrons. $n=1$, 2 for Lewis bonding interactions, and $n>2$ for delocalized bonding). As shown in Fig. 6.8C, the distribution of the six groups of 10c-2e interactions (in Li_{14}, two groups of lone pair character, and four groups of p-orbital character) is exactly the same as that of the lone pair of the single F atom (in F_2), while the 14c-2e interaction in Li_{14} shows high similarity to that of the 2c-2e σ bonding interaction in F_2. Similarly, the Li_{10} (2*7c-2e), Li_{12} (7c-2e and 10c-2e), and Li_8 (one 4c-4e and 4*4c-1e) could be viewed as supermolecules with the electronic structure resembling to that of N_2, NF, and CH_4. In the following years, the superatomic molecule or supermolecule is frequently used to designate the clusters which could be comprehended as the assembly of several superatoms.

In addition to the structural and electronic state analysis of the gas-phase alkali-metal clusters, the SVB model was also found applicable to the thiolate-protected gold nanoclusters. As shown in Fig. 6.9A, the prolate-shaped $Au_{38}(SR)_{24}$ nanocluster could be viewed as protecting the face-shared bi-icosahedral Au_{23} core with six $Au_2(SR)_3$ and three $Au(SR)_2$ [94]. According to the extended superatom model proposed by Häkkinen and coworkers, each of the oligomeric Au_nS_{n+1} staple extracts one electron from the metallic core, and thus the Au_{23} core formally bears +9 charge states. In this context, the 14e Au_{23}^{9+} core in $Au_{38}(SR)_{24}$ could be viewed as the assembly of two 13c-7e icosahedral Au_{13} blocks, while the electronic state of the 7e Au_{13} superatom is analogous to that of the F atom. As shown in Fig. 6.9B, the frontier orbitals of di-icosahedral Au_{23} core show high similarity with that of F_2 molecule in the presence of the electronic configuration $(\sigma_s)^2(\sigma^*_s)^2(\sigma p_z)^2(\pi_{px,py})^4(\pi^*_{px,py})^4(\sigma^*p_z)^0$. In addition, as shown in Fig. 6.8C, AdNDP analysis on the Au_{23}^{9+} clearly shows two groups of 13c-2e interactions (corresponding to super lone electron pairs), each in one Au_{13} block, four groups of p-type 13c-2e interactions (two on the left Au_{13} block, and two on the right one), and one super σ bond (23c-2e). In this context, the two 7e-Au_{13} blocks in the Au_{23} core shares a pair of electrons in the form of super σ bond to accomplish the 8e shell closure.

Motivated by the discrepancy of the extra-stability (large HOMO-LUMO gap) of the 4e $[Au_{18}(SR)_{14}]$, $[Au_{20}(SR)_{16}]$, and $[Au_{24}(SR)_{20}]$ with the typical superatom model and the earlier proposed SVB model, Yang, Cheng and coworkers proposed the concept of superatom-network (SAN) on the basis of DFT calculations (Gaussian 09 and MOLEKEL 5.4 softwares) [95]. First, a preliminary structural analysis indicates that each of these three clusters comprises two tetrahedral Au_4 blocks, while the intra-block Au-Au distances is remarkably shorter than that of the interblock ones. According to the AdNDP analysis in Fig. 6.10, the Au_8^{4+} core of each of these 4e cluster adopt a pair of 4c-2e bonds. In other words, the Au_8^{4+} core could be viewed as the assembly of two discrete 2e, tetrahedral Au_4^{2+} core. Specifically, incorporating the ligand shell

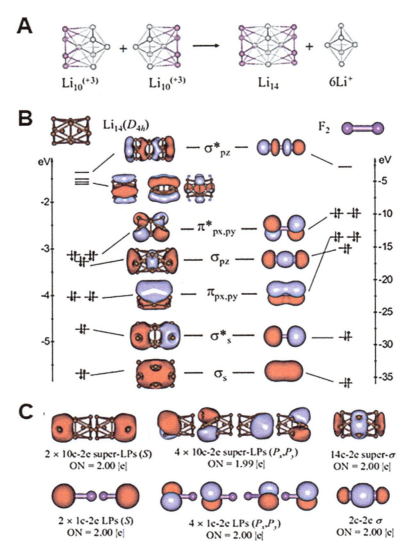

FIG. 6.8 The structural anatomy of the Li$_{14}$ cluster (A), Kohn-Shan molecular orbital (MO) diagrams of Li$_{14}$ (B: left) and F$_2$ (B: right), and the ADNDP analysis on Li$_{14}$ (C: upper) and F$_2$ (C: lower). *(Reproduced from L. Cheng, J. Yang, J. Chem. Phys. 138(14) (2013) 141101, https://doi.org/10.1063/1.4801860, with the copyright permission from AIP Publishing.)*

into AdNDP analysis gives the same conclusion, verifying that the four valence electrons locate exclusively on the two Au$_4^{2+}$ structures, while the core-shell interactions, or the interactions within the ligand shell makes little perturbation to the distribution of valence electrons. Of note, the SAN model based on the concept of combining n*2e blocks has been found applicable to a variety of thiolated gold nanoclusters, including [Au$_8$(SR)$_6$] (1*2e), [Au$_{12}$(SR)$_8$] (2*2e), [Au$_{18}$(SR)$_{12}$] (3*2e), [Au$_{19}$(SR)$_{13}$] (3*2e), [Au$_{20}$(SR)$_{12}$] (4*2e). Meanwhile, the structure of [Au$_{40}$(SR)$_{24}$] could be viewed as protecting the Au$_{26}$ core (i.e., the combination of two 8e, Au$_{13}$ blocks) with six AuS$_2$ and four Au$_2$S$_3$ staples.

6.2.5 Reaction mechanism analysis

6.2.5.1 Thermodynamic analysis

With the aid of first-principal DFT calculations (Turbomole package: BP-86/def2-SV(P) method), Mpourmpakis and coworkers recently proposed a thermodynamic stability theory, which is widely applicable to determine the thermal stabilization of a variety of thiolate protected metal nanolcusters (Fig. 6.11 the sizes range from Au$_{18}$(SR)$_{14}$ to Au$_{102}$(SR)$_{44}$) [96].

This theory first categorizes the metal atoms in a certain cluster into core and shell metal atoms, according to the natural bond orbital (NBO) charge analysis [97] or S-bonding methods. For example, the metal atoms with less than 0.2 charge or

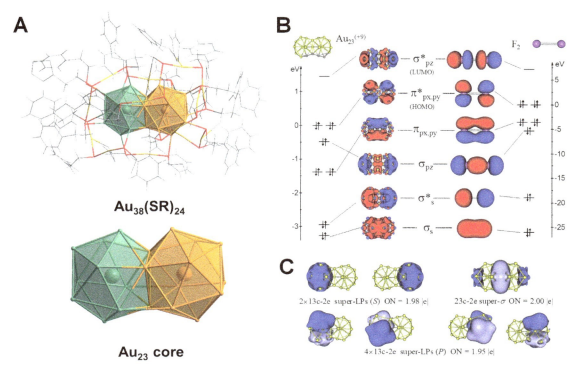

FIG. 6.9 Illustrative diagram for the complete structure of $Au_{38}(SR)_{24}$ structure and the bi-icosahedral Au_{23} core (A), the frontier orbitals of Au_{23}^{9+} core (B: left) and F_2 molecule (B: right), and the localized natural bonding orbitals of Au_{23}^{9+} core by ADNDP analysis (C). *((B and C) Reproduced from L. Cheng, C. Ren, X. Zhang, J. Yang, Nanoscale 5(4) (2013) 1475–1478, https://doi.org/10.1039/C2NR32888G, with the copyright permission from Royal Society of Chemistry.)*

FIG. 6.10 AdNDP localized chemical bonding of the Au_8^{4+} core of $[Au_{20}(SR)_{16}]$ (left), $[Au_{18}(SR)_{14}]$ (middle), and $[Au_{24}(SR)_{20}]$ (right). *(Reproduced from L. Cheng, Y. Yuan, X. Zhang, J. Yang, Angew. Chem. Int. Ed. 52(34) (2013) 9035–9039, https://doi.org/10.1002/anie.201302926, with the copyright permission from Wiley-VCH Verlag GmbH & Co. KGaA, Weinheim.)*

with less than two Au—S bonds are categorized into core atoms. By contrast, the metal atoms with more positive charge (> +0.2e), or with two or more Au—S bonds are categorized into shell metal atoms. These two methods produce identical results for all the examined clusters. Meanwhile, for anionic clusters, such as $[Au_{25}(SR)_{18}]^-$, the negative charge was set on the region of ligand shell in view of the vertical electron affinity calculations.

In this context, the shell-to-core binding energy (BE^{SC}) is calculated according to Eq. (6.26)

$$BE^{SC} = \frac{E^{full} - E^{core} - E^{shell}}{n^{core-shell}} \qquad (6.26)$$

In Eq. (6.26), E^{full}, E^{core}, and E^{shell} denotes the electronic energy of the entire cluster, the metallic core, and the shell section (E^{core} and E^{shell} are obtained from single point energy calculations on the core and shell moieties), respectively. Different multiplicities of each core structures were tested, and the electronic energy of the lowest-energy one was used as E^{core}. Meanwhile, $n^{core-shell}$ denotes the number of the core-shell interactions. Mainly two types of interactions are count in, the first is the interaction between the core metal and shell metal atoms (core-shell metallic interaction, wherein the distance between the core and shell metal atoms are within 4 Å in the optimized geometry[a]), and the second is the interaction between the core metal atoms with the thiolate ligand directly bound the core metal atoms (and is not bound to shell

---

a. The cutoff of 4.0 Å for Au—Au bond is approximately 2.5 times the van der Walls radii of Au metal, which is 1.66 Å in A. Bondi, van der Waals Volumes and Radii, J. Phys. Chem. 68 (3) (1964) 441–451. https://doi.org/10.1021/j100785a001.

FIG. 6.11 Correlation of the core cohesive energy and shell-to-core binding energy. The shell metal atoms are given in *blue* color (ball and stick model), while S, C, and H are given in capped sticks, with *yellow*, *grey*, and *white* colors respectively. The core silver and copper atoms in [Ag$_{25}$(SR)$_{18}$]$^-$ and [Cu$_{25}$(SR)$_{18}$]$^-$ (hypothetical, deduced from the optimized geometry of its Au analogue) are given in *light green* and *red*, respectively. *(Reproduced from M.G. Taylor, G. Mpourmpakis, Thermodynamic stability of ligand-protected metal nanoclusters, Nat. Commun. 8(1) (2017) 15988, https://doi.org/10.1038/ncomms15988, with the permission from Authors.)*

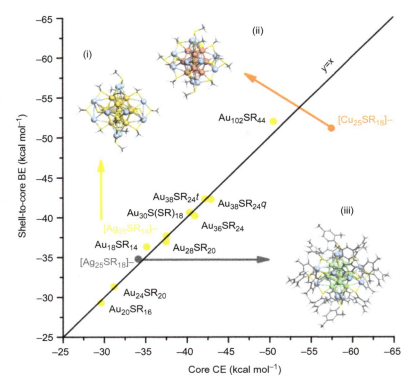

Au atoms, with Au—S bond distance within 3.01 Å). To this end, BE^{SC} means the averaged core-shell interaction of the two types of interactions, i.e., Au^{core}-Au^{shell} and Au^{core}-S interactions.

Another set of energy involved in this theory is the core cohesive energy (CE^{core}), and this energy is calculated according to Eq. (6.27):

$$CE^{core} = \frac{E^{full} - n^{core}E^{metal} - E^{shell}}{n^{core} + n^{core-shell}} \tag{6.27}$$

Herein, n^{core} and E^{metal} denote the number of core atoms, and the electronic energy of a discrete metal atom. To this end, the cohesive energy of the gas phase minimum energy clusters could be simply calculated with Eq. (6.28) (due to the lack of ligand shell):

$$CE^{core} = \frac{E^{full} - n^{core}E^{metal}}{n^{core}} \tag{6.28}$$

According to this theory, almost perfect match of BE^{SC} and CE^{core} was observed for the 10 reported Au$_n$(SR)$_m$ clusters. The deviation of [Cu$_{25}$(SR)$_{18}$]$^-$ from the linear correlation of BE^{SC} and CE^{core} indicates an unbalanced CE^{core} by shell-to-core interfacial interactions therein.

Specifically, BE^{SC} and CE^{core} shows negative linear correlation with $[n^{core}]^{-1/3}$ (due to the decreased fraction of low-coordinated surface sites), and positive linear correlation with the average coordination number (CN) of core metal atoms.[b] The linear correlation of the core cohesive energy/core-to-shell binding energy with these simple structural characteristics makes it easy to deduce the relative stability of a target Au$_n$(SR)$_m$ cluster, eliminating the necessity to carry out comprehensive density functional theory calculations.

---

b. The same covalent radii + buffer method was used to determine the coordination number of all atoms. For example, each Au—Au bond with distance of ≤3.32 Å and each Au—S bond with distances of ≤3.01 Å contributes one coordination number. Meanwhile, the cutoff of 3.01 Å for Au—S bond is determined by the covalent radii + buffer method. The covalent radii of Au and S is 1.36 and 1.06 Å (cited from Ref. [90]), and the buffer is 0.6 Å.

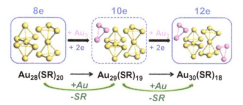

FIG. 6.12 Core structure of the three structurally highly correlated Au$_{28}$, Au$_{29}$, and Au$_{30}$ clusters. *(Reproduced from L. Xiong, B. Peng, Z. Ma, P. Wang, Y. Pei, Nanoscale 9(8) (2017) 2895–2902, https://doi.org/10.1039/C6NR09612C, with the permission of copyright from Royal Society of Chemistry.)*

6.2.5.2 "Gold atom insertion-thiolate group elimination" mechanism

The atomic precision of the various metal nanoclusters has elucidated fantastic structural correlation between different clusters and lies the foundation for assessing the thermodynamic/kinetic (depending on whether the transition states were obtained) facility of a target transformation. For example, in the recent study of Pei and coworkers, a theoretical principal designated as "gold-atom insertion, thiolate-group elimination" [98] has been proposed to predict the atomic structure of Au$_{29}$(SR)$_{19}$, and to probe the energy changes via the conversion from Au$_{28}$(SR)$_{20}$ to Au$_{29}$(SR)$_{19}$ and then to the Au$_{30}$(SR)$_{18}$. This study is enlightened by the presence of 2e, 4e, 6e, 8e, 12e, 14e ... clusters but the lack of atomic structure of the 10e cluster. Meanwhile, the cluster peak of a 10e cluster Au$_{29}$(PET)$_{19}$ has been observed on MALDI-TOF-MS in an earlier study [99], demonstrating the plausibility for the existence of 10e clusters. Interestingly, the formula of Au$_{29}$(PET)$_{19}$ is essentially differentiated from those of the reported Au$_{28}$(SR)$_{20}$ and Au$_{30}$(SR)$_{18}$ (both structures have been elucidated by SCXRD) by one {+Au − SR} moiety. That is, Au$_{28}$(SR)$_{20}$ + 2Au − 2SR → Au$_{29}$(SR)$_{19}$ + Au − SR → Au$_{30}$(SR)$_{18}$. Specifically, the Au$_{28}$ and Au$_{30}$ clusters show high structural similarity. As shown in Fig. 6.12, Au$_{28}$(SR)$_{20}$ (cited from the single crystal structure of Au$_{28}$(SC$_6$H$_{11}$)$_{20}$) comprises a Au$_{14}$ core, which could be viewed as a pair of vertex-sharing bi-tetrahedral Au$_7$ structures. The Au$_{14}$ structure has also been observed in the Au$_{30}$(SR)$_{18}$ (cited from the single crystal structure of Au$_{30}$(StBu)$_{18}$), wherein the Au atoms in the two triangular Au$_3$ structures were also categorized to the core atoms. In this context, the Au$_{14}$ core in Au$_{28}$ cluster is further protected by two AuS$_2$ and four Au$_3$S$_4$ motifs, while Au$_{20}$ core of the Au$_{30}$ cluster is protected four AuS$_2$, two Au$_2$S$_3$ and two bridging thiolate ligands. In view of the similarity in the molecular formula and the structural similarity of the series of Au$_{28}$(SR)$_{20}$, Au$_{29}$(SR)$_{19}$, and Au$_{30}$(SR)$_{19}$, Pei and coworkers suggest Au$_{29}$ to be an intermediate state between Au$_{28}$ and Au$_{30}$ clusters. To this end, a theoretical model of Au$_{29}$(SR)$_{19}$ cluster is constructed by combining half Au$_{28}$ and half Au$_{30}$ structure. Geometry optimization and frequency calculations (at PBE/DND:DSPP level with Dmol^3 software) demonstrate the theoretical structure to be a local minimum, and its energy is slightly lower than that of 1/2[Au$_{28}$(SR)$_{20}$ + Au$_{30}$(SR)$_{19}$] (R = Me in DFT calculations) by 0.06 eV.

With the atomic structure of Au$_{29}$(SR)$_{19}$ and a careful analysis, the evolution principal from Au$_{28}$ to Au$_{29}$ and then to Au$_{30}$ was deduced to occur via a "gold-atom insertion, thiolate-group elimination" mechanism. The mechanism is proposed on the basis of the following concerns. First, the single crystal structure of both Au$_{30}$(SR)$_{18}$ and Au$_{30}$S(SR)$_{18}$ has been reported [100,101]. The structures of these two clusters are differentiated only in one surface μ^3-S, while the cluster peak of Au$_{30}$(SR)$_{18}$ was also observed in the MALDI-MS of Au$_{30}$S(SR)$_{18}$. The results indicate that the release of one μ^3-S from a cluster precursor is plausible. Second, the single crystal structures of Au$_{23}$(Adm)$_{16}$, Au$_{24}$(Adm)$_{16}$ and Au$_{25}$(SAdm)$_{16}$ indicate that the same core structure might be tolerated to different shell structures, as the Au$_{24}$ and Au$_{25}$ clusters could be viewed as adding one or two bridging Au atoms to connect two adjacent Au$_n$S$_{n+1}$ motifs [102]. To this end, adding single Au atom between the adjacent Au$_n$S$_{n+1}$ motifs on the surface motif (i.e., gold atom insertion step) is a plausible elementary step. On the basis of the structural similarity between the Au$_{28}$(SR)$_{20}$, Au$_{29}$(SR)$_{19}$, and Au$_{30}$(SR)$_{19}$ clusters, and the deduction of the gold atom insertion and the thiolate group steps, Pei and coworkers proposed an "gold atom insertion-thiolate group elimination" mechanism to account to the size-evolution from Au$_{28}$ to Au$_{30}$ clusters. In this mechanism, the Au addition on ligand shell, removal of alkynyl ligands, and elimination of μ^3-S on Au$_{28}$ cluster occur successively to generate Au$_{29}$ and then to Au$_{30}$ clusters. DFT calculations indicates that the formation of the Au$_{29}$ and Au$_{30}$ clusters is thermodynamic feasible, while the preference for the site of Au addition on both Au$_{28}$ and Au$_{29}$ precursors correlates with the thermodynamic facility.

In recent years, the aforementioned "gold atom insertion-thiolate group elimination" mechanism has also been utilized to explain the conversion mechanism of other cluster systems, such as the conversion of Cd$_1$Au$_{14}$(StBu)$_{12}$ to Au$_{16}$(SAdm)$_{12}$ [103] (see Section 4.2.2.5 for the details).

6.2.5.3 Reaction mechanism of metal nanocluster catalysis and conversion

In the past decade, the atomically precise metal nanoclusters have been successfully developed as a novel category of nanocatalyst, and have shown highly promising applications in a variety of reactions, such as the electrocatalytic CO$_2$ reduction, photocatalytic water splitting, alcohol oxidation, etc. [104]. Nevertheless, the mechanistic understanding on the nanocluster catalysis is remains limited, and the main challenge lies in identifying the active catalytic center/site and elucidating the catalytic mechanism. In this context, the "single atom effect" in catalytic reactions (i.e., remarkably changed catalytic efficiency after incorporating an additional metal atom or extruding one metal atom on the original cluster catalyst) has been recently elucidated in a series of metal nanocluster catalyzed reactions. For example, both the rod-like, vertex-sharing bi-icosahedral $Au_{25}(PPh_3)_{10}(PET)_5Cl_2$ cluster and its analog of $Au_{24}(PPh_3)_{10}(PET)_5Cl_2$ (with one central Au atom missing) show catalytic efficiency on methane oxidation reactions [105], while the catalytic efficiency of the Au_{24} cluster is relatively higher. Interestingly, the interconversion of Au_{24} and Au_{25} clusters could be easily adjusted via the addition of H_2O_2 ($Au_{25} + H_2O_2 \rightarrow Au_{24}$) or PPh_3 ($Au_{24} + PPh_3 \rightarrow Au_{25}$), resulting in the reversible switch of the catalytic activity. On the basis of the characterization of hydroxyl (HO) during the reaction processes, DFT calculations (with Gaussian 09 software) were conducted to elucidate the detailed reaction mechanism of both Au_{24} and Au_{25} catalyzed systems, and to shed light on the origin for the remarkably enhanced catalytic activity of the Au_{24} clusters. Herein, it is noteworthy that two groups of structural simplification were conducted in their study to reduce the computational cost. First, the PPh_3 and PET groups on the cluster catalysts were modeled with PH_3 and SMe, and the calculations were conducted with PBE/aug-cc-pVDZ(-pp) method. This simplified model is mainly used to elucidate the binding-preference for the hydroxyl groups on different sites of the Au_{24} and Au_{25} catalysts, and to perform an initial guess on the configuration of the concerned intermediates and transition states. On the basis of the potential methane activation pathway, a relatively more comprehensive model, by changing the simplified PH_3 and SMe around the reaction center into the original ones (i.e., PPh_3 and PET) were used to accurately evaluate the effect of the side groups (with PBE/LANL2DZ calculation method).

According to the DFT calculations, the methane oxidation catalyzed by Au_{25} starts with the binding of two hydroxyl groups on the neighboring two waist Au atoms, associating with the cleavage of one group of the bridging Au—S—Au bonds (Fig. 6.13A). After that the methane activation occurs favorably at the open Au site, and via a formally triangular transition state of **25d**. From the formed intermediate **25e**, the hydroxyl migration occurs to generate the intermediate **25f**, in which both the hydroxyl and the methyl groups ligand on the open Au site. After that, the C—O bond formation occurs easily to generate the intermediate **25h**, bearing a weakly coordinated MeOH on the open site. Finally, regeneration of the catalyst **25a** could be achieved via the dissociation of MeOH from **25h**. Throughout the catalytic cycle, the most energy-demanding step is the hydroxyl migration step (i.e., **25e** → **25f**).

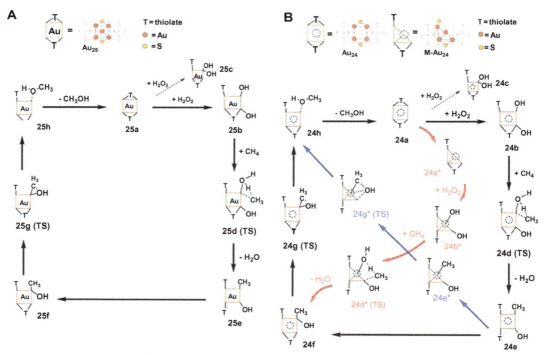

FIG. 6.13 The proposed mechanism for the Au_{25} (A) and Au_{24} (B) catalyzed methane oxidation reactions. *(Reproduced from X. Cai, G. Saranya, K. Shen, et al., Angew. Chem. Int. Ed. 58(29) (2019) 9964–9968, https://doi.org/10.1002/anie.201903853, with the copyright permission from Wiley-VCH Verlag GmbH & Co. KGaA, Weinheim.)*

Of note, the Au_{24} cluster catalyst could possibly undergo the same catalytic pathway comprising the double hydroxyl coordination (**24a** → **24b**, Fig. 6.13B), methane activation (**24b** → **24d** → **24e**), hydroxyl migration (**24e** → **24f**), C—O bond formation (**24f** → **24g** → **24h**), and methoxy cleavage steps (i.e., **24g** → **24h**). Nevertheless, the energy demands are very close to that of the Au_{25}-mediated pathway and are unlikely to explain the significantly higher catalytic activity of Au_{24} compared with that of Au_{25}. Alternatively, due to the presence of the central Au vacancy in Au_{24} clusters (compared with the compact bi-icosahedral Au_{25} cluster structure), two shortcut pathways could possibly occur to eliminate the energy-demanding hydroxyl migration step). In the first shortcut pathway (labeled with red arrows in Fig. 6.13B), a waist Au atom migration first occurs to occupy the central position at the rod framework. To this end, the methyl and hydroxyl group locate on the same Au site after the methane activation step (**24b*** → **24d*** → **24f**). In the second shortcut pathway (labeled with blue arrows in Fig. 6.13B), the rod-like structure of the Au_{24} catalyst maintains until the formation of the intermediate **24e** (formed after the methane activation step). After that, waist Au migration occurs to occupy the central position, and then a two-site CH_3—OH bond formation occurs to generate the intermediate **24h**. The energy demands for these two shortcut pathways are both remarkably lower than that of the hydroxyl-migration involved pathway, and the reason is mainly attributed to isomerized metallic framework induced increased mobility the absorbed hydroxyl groups.

The aforementioned example refers to the utilization of DFT calculations in elucidating the mechanism of a catalytic reaction. That is, the metal nanocluster catalysts react with the substrate, mediate the reaction of between different species, and are finally regenerated after one catalytic cycle. Commonly in these reaction systems, the active metal site(s) on the cluster undergoes a formally "substitution" (i.e., the incoming ligand replaces the original ligand) or an "addition" (i.e., coordination with an extra-substance) reaction to initiate the reaction. Such elementary steps are also widely involved in the conversion reaction of metal nanoclusters. Therefore, DFT calculations play a positive role in elucidating the origin, the structural and energy changes of the nanocluster conversion reactions. In this context, the mechanism of size-maintained ligand exchange reaction is an easy-handling system, as the maintained framework could remarkably reduce the mechanistic complexity and thus the computation costs. Herein, the DFT calculations on the mechanism of ligand exchange of $Au_{102}(p$-$MBA)_{44}$ with p-BBTH to form $Au_{102}(p$-$MBA)_{40}(p$-$BBT)_4$ was used as an example [106]. A brief introduction of this reaction has been provided in Section 4.2.2.3, and thus we mainly focus on the details of DFT calculations in this section. In Ackerson's study, the experimental p-MBA ligand and the incoming p-BBT ligands are modeled with -SH and -SMe groups, respectively. On the basis of a typical associative mechanism (a stepwise pathway in which the incoming thiol ligand first coordinates to one surface Au atom to form an additive, and then the dissociation of the outgoing ligand occurs), the plausibility of the different surface Au atoms acting as the active reaction site was first examined via the solvent accessibility calculation (performed with GPAW program). The results indicate two solvent accessible atoms, i.e., Au_{23} and Au_{24} in Fig. 6.14 inset. In this context, constrained optimization with DFT calculations on modeling system was conducted to shed light on kinetic profile of the ligand exchange processes. Using the substitution of the modeling cluster ligand (i.e., -SH) with the modeling incoming ligand (i.e., -SMe) on Au_{23} site as an example, the mechanistic details are given in Fig. 6.14. It can be seen that the overall reaction involves two stages: the formation of intermediate (iii) via the approaching of the incoming HSMe ligand, and this process need to overcome an energy barrier of ~15 kcal/mol. In the intermediate (iii), the incoming HSMe ligates on the Au_{23} site, while the Au_{23}-SH partially dissociates, with the H atom (of HSMe) locating in between the incoming S (of -SMe) and the outgoing S (of -SH) atoms. The relative energy of intermediate (iii) is slightly higher than the starting state of (i) by 4.3 kcal/mol. After that, an H transfer (from the incoming HSMe to the outgoing -SH) occurs easily due to the geometric facility, and then the desorption of the outgoing ligand occurs via the transition point (iv) to form the ligand-exchanged structure of (v). The energy barrier for the second stage is 18.7 kcal/mol, and the total effective activation barrier throughout this pathway is ~23 kcal/mol. Due to the thermoneutral character of the target ligand exchange reaction (i.e., the relative energy of the final state (v) is comparable to that of (i)), the entropic mixing of the chemical entities in the ligand shell was proposed the main driving force for the ligand exchange reactions.

6.2.6 Theoretical simulation on optical properties

Due to the discrete distribution of energy levels, the optical absorption of the ultrasmall metal nanoclusters with several to hundreds of metal atoms is always highly structure and size-dependent. In this scenario, the UV-vis spectra have been frequently used as a fingerprint spectra for clusters, to provide preliminarily guess on the molecular component/structure of a prepared nanocluster. Meanwhile, as mentioned above, the theoretical simulation of UV-vis spectra (i.e., optical absorption curve) is one of the most extensively implemented calculation tasks. Generally, the agreement of the theoretical and experimental UV-vis spectra is a requisite to deduce the atomic structure of a novel cluster (whose structure was not available by experimental means) or to analyze the electronic states.

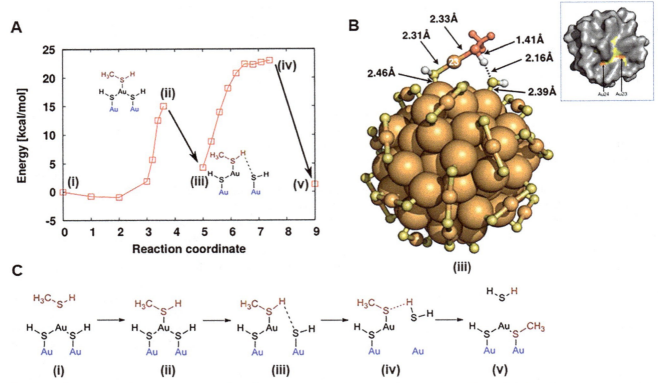

FIG. 6.14 The energy profile for the associative ligand-change in the modeling system between $Au_{102}(SH)_{44}$ and HSMe (A), the framework of the optimized geometry of the intermediate (iii) (B) and the deduced feasible reaction sites by solvent accessibility calculation (B inset), and the illustrative diagram around the reaction center in states (i)–(v) (C). *(Reproduced from C.L. Heinecke, T.W. Ni, S. Malola, V. Mäkinen, O.A. Wong, H. Häkkinen, C.J. Ackerson, J. Am. Chem. Soc. 134(32) (2012) 13316–13322, https://doi.org/10.1021/ja3032339, with permission, Copyright American Chemical Society.)*

As to the theory of UV-vis spectroscopy, the exposure of analyst to an incident optical irradiation result in the electronic transition of one or more electrons from a low-energy ground state to a series of excited states with relatively higher energy. Therefore, theoretical simulation of UV-vis spectroscopy involves the excited state calculations, and TD-DFT (short for time-dependent density functional theory) represents the most popular method for both the low computational cost and the high accuracy in treating electron correlation effects. Of note, TD-DFT is suggested to carried out to analyze the low-lying excitation states with lower energy than the Rydberg state [107]. Otherwise, the first ionization of the target molecule and some charge transfer excited states might be involved, and results in a less-accurate absorption spectra.

Despite the relatively lower cost of the TD-DFT methods than the other excitation-state calculation methods (such as CASSCF method-Complete Active Space Multiconfiguration, based on a combination of an SCF computation with a full configuration interaction calculation on a subset of orbitals), the TD-DFT calculations remain highly computational expensive than the ground state calculations, especially for the relatively large nanocluster systems bearing hundreds/ thousands of electrons. There, structural simplification (via modeling the bulk organic ligands on the experimentally characterized cluster with a relatively smaller one) has been frequently used in TD-DFT calculations. For example, the experimentally used PR_3, SR groups in metal nanoclusters have been frequently modeled with PH_3/PMe_3 and SH/SMe in theoretical calculations. The plausibility of such treatment mainly lies on the weak contribution of the hydrocarbon groups in the frontier orbitals of metal clusters (and this point could be supported by Kohn-Sham orbital analysis). Nevertheless, for some aryl substituted phosphate/thiolate ligands, the hydrocarbon groups might show nonnegligible contribute to frontier orbitals, and especially the degenerated HOMOs and LUMOs. In these clusters, the structural simplification should be very cautious, and a rough simplification of the aromatic ligands into -H or -Me might results in low accuracy.

Herein, the theoretical simulation of the UV-vis spectra of $Au_{25}(PET)_{18}^-$ and the related electronic state analysis was used as an example [108]. First, geometry optimization was carried out with ADF program to locate the local minimum, using the single crystal structure of $Au_{25}(PET)_{18}^-$ (all PET groups are simplified with -SH) as the starting point. With the optimized geometry in hand (i.e., a local minimum), the Kohn-sham (KS) energy level could be gained via a single point

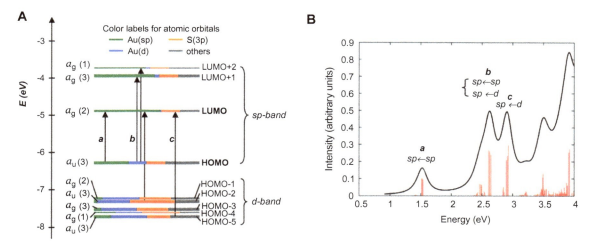

FIG. 6.15 Kohn-sham orbital analysis (A), and simulated UV-vis (B) for the modeling $Au_{25}(SH)_{18}^{-}$ cluster. The length of the colored bar correlates with the orbital contribution (in percentage) of the related components to the molecular orbital. *(Reproduced from M. Zhu, C.M. Aikens, F.J. Hollander, G.C. Schatz, R. Jin, J. Am. Chem. Soc. 130(18) (2008) 5883–5885, https://doi.org/10.1021/ja801173r, with permission. Copyright American Chemical Society.)*

energy and energy level calculations (the calculation on energy levels is a default task of geometry optimization procedures in many suit of programs). Herein, it is noteworthy that all molecular orbitals are available in the output file, while only the ones nearby HOMO and LUMO are shown for clarity reasons. As shown in Fig. 6.15A, Au(sp) (i.e., the sum of Au(6s) and Au(6p) components), Au(d) (i.e., Au(5d)), and S(3p) orbitals make the major contribution to the concerned frontier orbitals (HOMO-5 → LUMO+2). In addition, due to the dominance of Au(sp) components in HOMO and the three lowest unoccupied orbitals (LUMO → LUMO+2), the energy band comprising these orbitals are denoted as the "sp band." For the same reason, the major contribution of HOMO-5 → HOMO-1 originates from the Au(d) orbitals, and the bands comprising these orbitals are denoted as the "d-band." Aside from the energy level analysis of the ground state structure with DFT calculations, TD-DFT calculations were also conducted on the optimized geometry to gain the simulated UV-vis spectra. In metal nanocluster systems, the number of excited states is necessarily defined in calculation route section, and commonly >50 excited states are need to be calculated for a better correlation with the experimental spectra. In Jin's study, 200 allowed excited states were calculated, and the theoretical spectra shows high similarity with the experimental one (albeit with systematic red-shift of the characteristic peaks, vide infra). In this context, the contribution to each characterized absorption peak was analyzed (the detailed electronic transition of the involved molecular orbitals is given in the output files). As shown in Fig. 6.15B, the lowest energy-lying transition appears at 1.52 eV, and it corresponds to the HOMO → LUMO transition. As mentioned above, the KS orbital analysis indicates both states belong to sp band, and thus the transition basically corresponds to a sp → sp transition. Meanwhile, the peak at 2.63 eV involves the contribution of multiple transitions, while the energy levels in both d-band (lower-lying HOMOs, such as HOMO-2) and sp-band (HOMO) are involved. To this end, the peak at 2.63 eV was proposed to arise from both intra-band sp → sp and interband d → sp transitions. In view of the systematic red-shift of the calculated characteristic peaks compared with the related one on the experimental spectra, Jin and coworkers proposed that the structural simplification might result in the overestimated energy levels of HOMOs: the HOMO-1 → HOMO-5 in Fig. 6.15 show clearly the contribution of S(3p) orbitals, and thus using the complete PET ligands in experiments could further stabilize the related orbitals (lower down the orbital energy). To this end, the actual energy gaps of the experimental clusters are expected to be larger, and thus the peaks are blue-shift compared with the theoretical ones.

6.2.7 Calculations on NMR spectra

Compared with the extensively investigated UV-vis spectra with the DFT and TD-DFT calculations, theoretical simulation on the NMR spectra is not seen very often in metal nanocluster system. The main reason relates to the limited experimental NMR spectra and the high computational cost. For example, with the same group of computers (32 core, 96 GB memory) and with the same PBE/TZ2P (using spin-orbit relativistic) level of theory, it takes about 270 and 37 min to accomplish the NMR spectra and single point energy calculations of $Au_2(PPh_3)_2SMe$. Of note, the computational time will be remarkably

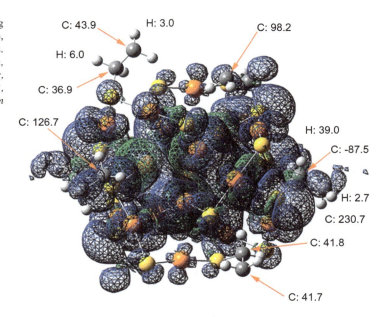

FIG. 6.16 Isosurface of the spin distribution of the modeling $[Au_{25}(PET)_{18}]^0$ cluster (*blue* and *green* for the α and β spin surface, isovalue = 0.0001) and the chemical shifts of selected atoms. *(Reproduced from A. Venzo, S. Antonello, J.A. Gascón, I. Guryanov, R.D. Leapman, N.V. Perera, A. Sousa, M. Zamuner, A. Zanella, F. Maran, Anal. Chem. 83(16) (2011) 6355–6362, https://doi.org/10.1021/ac2012653, with the permission from 2011, American Chemical Society.)*

differentiated, and the NMR calculations (so does the calculations on optical absorption/emission spectra) will be especially expensive when larger systems are concerned. For this reason, structural simplification by modeling the comprehensive hydrocarbon groups with -Me or H atoms has been frequently used in theoretical calculations, unless a highly accurate analysis on the ligand effect is desired.

An excellent example on theoretical simulation of the NMR spectra of a group of $[Au_{25}(PET)_{18}]^n$ ($n=+1, 0, -1$) clusters were recently reported by Maran and coworkers [109]. The calculations were conducted with the Gaussian 09 software [76], using the B3LYP functional and the mixed basis functions (i.e., the LANL2DZ effective core potential for Au and the total electron basis set of 6-31 g** for all other atoms). Gauge-including atomic orbitals (GIAOs) formalism has been used to calculate the NMR shielding tensors. Herein, the gas-phase calculation was conducted to simulate the low-polarity solvation environments, and such strategy has been frequently used in theoretical calculations. According to the calculation results, the spin distribution of neural $[Au_{25}(PET)_{18}]^0$ significantly affect the ^1H and ^{13}C chemical shift of the neighboring atoms (Fig. 6.16). For example, the unevenly distributed α and β spin results in the remarkably different ^{13}C chemical shift of the methene group nearby the metallic core structure (most spin density spread in the metallic core and the core-surface), resulting in the greatly broadened peak on NMR spectra. By contrast, due to the relatively long distance of the second methane group away from the core-shell interference, the chemical shift of remote carbon atoms varies only in a very narrow range. Briefly, the calculation results agree with the experimental outcome mainly in the following aspects: (1) The chemical shift of the two groups of methene protons on the exterior thiolates (i.e., the thiolates away from the core-shell surface) has been well reproduced by theoretical calculations; (2) the temperature-dependency (i.e., lower proton chemical shift at higher temperature) of the calculated chemical shift correlates with the experimental results; and (3) The chemical shift of the core-shell interference neighboring ligands is relatively more sensitive to the changed temperatures. To this end, the calculation on NMR spectra could be used verify the purity of the prepared cluster samples. More specifically, for some comprehensive structures, the NMR spectra, and especially the specific chemical shift of certain types of H/C/P atoms, could be practically used to correlate with the experimental results, and to deduce the origin of the different types of resonances.

6.2.8 Structure-property correlation

6.2.8.1 Bonding interaction analysis

The bonding character within the metal nanocluster systems has long been an intriguing question, and a series of beautiful theories have been proposed in these decades. Nevertheless, most of the reported studies focus on the homometallic metal nanoclusters, wherein the bonding interactions within the metallic core and in the core-shell interference have been extensively studied [110]. Due to the relatively short history of the atomically precise alloy metal nanoclusters, the bonding characteristics within the alloy metal clusters systems are remain limited. Herein, the DFT calculations on the rod-like, bi-icosahedral $[Au_xAg_{25-x}(PR_3)(SR)_5Cl_2]^{2+}$ clusters were used as an example to illustrate the theoretical insights into the origin of enhanced fluorescence by the alloying effect. In Jin, Zhu and coworkers' study, two groups of alloy

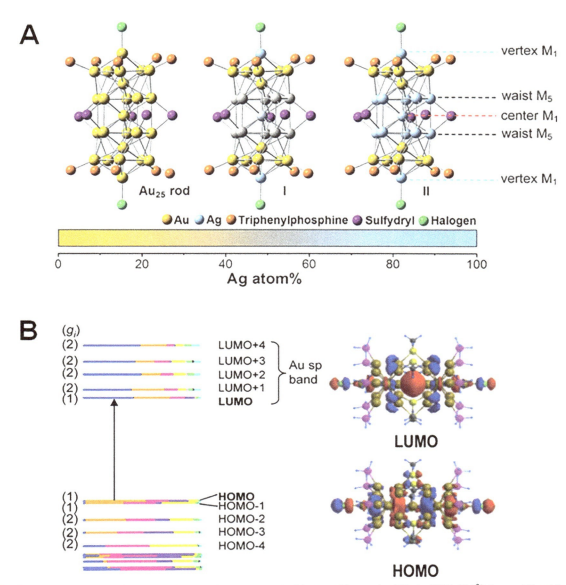

FIG. 6.17 Illustrative diagram for the metal distribution and the framework of the target $[Au_{25-x}Ag_x(PPh_3)_{10}(PET)_5Cl_2]^{2+}$ clusters (A) and the KS orbital of the homometallic $[Au_{25}(PPh_3)_{10}(PET)_5Cl_2]^{2+}$ (B). ((A and B) Reproduced from S. Wang, X. Meng, A, Das, T. Li, Y. Song, T. Cao, X. Zhu, M. Zhu, R. Jin, Angew. Chem. Int. Ed. 53(9) (2014) 2376–2380. https://doi.org/10.1002/anie.201307480, Copyright Wiley-VCH Verlag GmbH & Co. KGaA, Weinheim and M.Y. Sfeir, H. Qian, K. Nobusada, R. Jin, J. Phys. Chem. C 115(14) (2011) 6200–6207, https://doi.org/10.1021/jp110703e, Copyright American Chemical Society.)

Au-Ag clusters, i.e., $[Au_{25-x}Ag_x(PPh_3)_{10}(PET)_5Cl_2]^{2+}$ (group I: $x=0$–12; group II: $x=0$–13) with the same framework but different Au:Ag components were prepared via two independent synthetic procedures [111]. The differences in alloy cluster components mainly lies in the absence or presence of the $[Au_{12}Ag_{13}(PPh_3)_{13}(PET)_5Cl_2]^{2+}$, which is determined mainly by ESI-MS. Interestingly, these two clusters show remarkably different optical emission activity. That is, the emission quantum yield of group II clusters is about ~400 folds stronger than that of the group I clusters (40.1% vs 0.21%). In this context, partial occupancy analysis was first used to analyze the possible distribution site of Ag atoms in the target clusters, and it was found that the Ag in group II clusters show significantly higher occupancy ration in the waist positions of the framework than that in group I clusters (Fig. 6.17A). According to the combination of UV-vis and DPV analysis, the photo emission was found to originate from the LUMO-to-HOMO transition, and therefore the difference in the HOMO, LUMO orbitals of different cluster components will be responsible for the distinct fluorescence. Nevertheless, due to the statistic occupation of Ag in the target alloy clusters and the mixture composition in both cluster systems, it is difficult to construct a modeling system of either $[Au_{13}Ag_{12}(PPh_3)_{10}(PET)_5Cl_2]^{2+}$ or $[Au_{12}Ag_{13}(PPh_3)_{10}(PET)_5Cl_2]^{2+}$ to distinguish the effect of the pivotal 13th Ag atom. To this end, the energy levels of the homometallic $[Au_{25}(PPh_3)_{10}(PET)_5Cl_2]^{2+}$ cluster [112] was used as a reference for electronic state analysis. As shown in Fig. 6.17B, the HOMO of the rod-like Au_{25} cluster mainly distributes on the waist Au atoms, originating mainly from the Au(6sp) orbitals. Meanwhile, the LUMO of Au_{25} cluster mainly locates on the central Au atom (i.e., the shared vertex between the two icosahedral structures). Due to the high susceptibility of the 13 Ag atoms in occupying the waist positions (the occupancy preference follows the order of vertex > waist > center), the electronic state of both HOMO and LUMO are significantly changed compared with that of Au_{25}. Meanwhile, the energy of Ag(5s) is relatively higher than that of Au(6s), and Ag shows less sp hybridization tendency than that of Au (due to the weaker relativistic effect). Therefore, the HOMO-LUMO energy gap of the group II component is relatively larger than that of the homometallic Au_{25} and group I clusters. Meanwhile, due to the less sd hybridization in Ag (than that of Au), the closer masses of the excited electron and the hole in $Au_{12}Ag_{13}$ cluster results in higher recombination rate, and thus stronger fluorescence.

In addition to the aforementioned metallic interactions and the typical M—L (L=S, P, Cl, etc.) bonds, the bonding character of Au-H has been recently elucidated by Jiang and coworkers [113]. According to the reaction energetics analysis using VASP: PBE/PAW method, the absorption of 1–6 H (in the form of half H_2, as shown in Eq. 6.29) on the modeling diphosphate (-H_2P-$(CH_2)_8$-PH_2-, abbreviated as L^8) protected Au_{22} clusters (i.e., $Au_{22}(L^8)_6$) is thermodynamically feasible (with $\Delta E^0 < 0$ and $\Delta G^0 \sim 0$ eV for $n = 0$–6).

$$Au_{22}H_n(L^8)_6 + \frac{1}{2}H_2 \rightarrow Au_{22}H_{n+1}(L^8)_6 \ (n = 0 - 6) \quad (6.29)$$

The DFT calculation results indicate that the foreign H atoms favorably absorb on the long-bridge cus Au sites. After all the long-bridge sites were occupied, the H tend to absorb on the noncub Au sites with low coordination number (Fig. 6.18A and B). With the aid of Bader charge analysis, the H atoms in all these absorbed structures were found to carry partial negative charges (ranging from -0.17 to -0.10 $|e|$), in strong contrast to the metallic H states (with ~ 0 $|e|$) in the thiolated Au nanoclusters. The results demonstrate the "hydride" character of these absorbed structures, and charge transfer from the cus Au atoms to the hydrides were further evidenced by the charge density difference analysis, which clearly shows the charge depletion in the cus Au atoms, and charge accumulation on the hydrides (the thermodynamic stable structure of $Au_{22}H_2(L^8)_6$ was chosen as an example, Fig. 6.18C).

6.2.8.2 Determination on the role of noncovalent interactions

In the recently study of Jin and coworkers, DFT calculations were conducted to determine the relative thermostability of different doping structures of the alloy MAu_{24} clusters (i.e., the modeling $[AgAu_{24}(PH_3)_{10}(SH)_5Cl_2]^{2+}$ and $[CuAu_{24}(PH_3)_{10}(SH)_5Cl_2]^{2+}$ [114]. The calculation results (PAW method implemented in the Quantum Espresso Package) indicate that the Ag atoms favorably occupy the central position, while the most plausible position for Cu doping is the vertex and central positions. Interestingly, these results do not agree with the experimental outcome, because SCXRD indicates the Ag atom favorably locates in the vertex and Cu is orientated preferentially at the vertex/waist positions. Of note, the results do not dependent on the used theoretical methods, as the Grimme-D2 [115] and the exchange hole dipole moment [116–118] methods with the van der Waals (vdW) interactions gives very similar results with those of the DFT calculations. In other words, the incorporation of the vdW interactions or not does not change the main conclusions in the target alloying system. To this end, Jin and coworkers proposed the possible influence of the reaction dynamics and the metal statics, which was further evidenced by the DFT calculations on the kinetic profiles (see Section 4.2.2.5 for the details).

6.3 Application of quantum mechanics-molecular mechanics calculations

Compared with DFT calculations, the molecular mechanics (MM) and quantum mechanics-molecular mechanics (i.e., QM-MM) calculations have been relatively less used in treating metal nanocluster systems. The QM functional is good at describing the changes in electronic structures, such as electron transition, bond formation/dissociation processes,

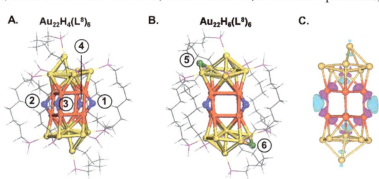

FIG. 6.18 Selected optimized structures of the H-absorbed Au_{22} clusters bearing four and six H atoms (A, B, note: the numbers incircles labels the sequential sites for H absorption), and the charge density difference for $Au_{22}(L^8)_6$ with absorbing two H (C). (Reproduced from G. Hu, Z. Wu, D. Jiang, J. Mater. Chem. A 6(17) (2018) 7532–7537, https://doi.org/10.1039/C8TA00461G, with the copyright formation from Royal Society of Chemistry.)

but the computationally cost is too high when treating large and comprehensive systems. As the physicochemical properties of metal nanocluster are mainly determined by the metallic core, and core-shell interference interactions, ligand simplification have been extensively used in DFT calculation of metal nanocluster systems to reduce computational cost in the premise of adequate accuracy. By contrast, the MM force field is unable to describe the changes in the electronic structure, but good at describing the cluster systems wherein the small-amplitude vibrations, torsions, van der Waals interactions, and electrostatic interactions play an important role. Therefore, in metal nanocluster systems, the MM calculations have been frequently implemented associating with the DFT calculations, so as to achieve well balance the accurate description on dynamic effect/electronic state calculations and the computational cost.

In a recent study by Yarovsky, Stevens and coworkers, the photoluminescence origin of a series of oligopeptide protected gold nanoclusters ($Au_{25}(SR)_{18}$, and SR is the oligopeptide containing N-terminal cysteine fragment) have been systematically studied by means of combined experimental and theoretical strategies [119]. In this study, MD calculations (with GROMACS 4.6.5 software) are conducted to shed light on the influence of the peptide sequence (or more exactly, the substituent effect on the oligopeptide ligands) on the ligand dynamics and the ligand-metal core interactions. Some key structural parameters, such as the $Au_{13}@Au_{12}(SR)_{18}$ framework, and the Au-S-C-C dihedral angles are cited from the reported single crystal structure and the DFT calculations (with Gaussian09 software). According to the experimental observation, the N-terminal acetyl capping on the oligopeptide ligands significantly affect the luminescence intensity, as the gold clusters with N-capped peptide ligands showed significantly stronger emission than those with N-uncapped ones. To this end, the molecular dynamic (MD) calculations are mainly conducted to assess the influence of peptide structure on the core-shell interactions, which is mainly responsible for the photoluminescence of Au nanoclusters. The calculation results indicate that the N-capping considerably increases the number of electron-rich atoms (e.g., N, O) around the core-shell interference, diminishing the water solvent from the ligand shell (and thus the hydrogen bonds). Both effects contribute to the enhanced luminescence in the N-capped peptide protected cluster systems. Meanwhile, the presence of the capping ligands results in the prohibited electrostatic interaction between the positively charged N-terminal residues with the negatively charged N-terminal fragments, and thus the less compressed ligand layer and the relatively larger size-hydrodynamic cluster volume (Fig. 6.19). The higher flexibility of the N-capped peptide ligands partially smears out the features of the scattering function.

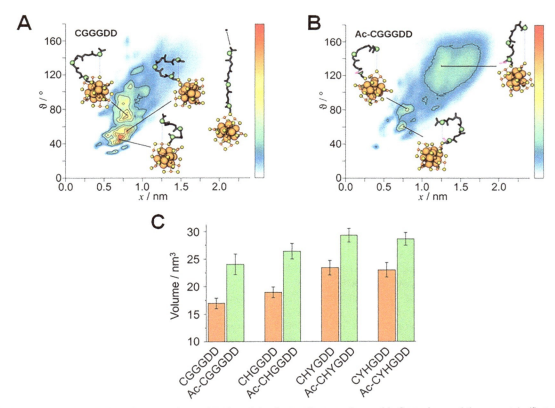

FIG. 6.19 Density maps of peptide backbone angle as a function of the distance between the peptide C-terminus and the nearest Au/S atom, using the CGGGDD peptide (A) and the N-capped Ac-CGGDD (B) ligands coated cluster systems as examples, and the average $Au_{25}(SR)_{18}$ volumes in different peptide or N-capped peptide protected cluster systems (C). *(Reproduced from Y. Lin, P. Charchar, A.J. Christofferson, et al., J. Am. Chem. Soc. 140(51) (2018) 18217–18226, https://doi.org/10.1021/jacs.8b04436, with the permission of Copyright from American Chemical Society.)*

With the equilibrated MD simulation snapshots in hand, QM calculations were performed, using Dmol3 program for geometry optimization and Gaussian09 for electrostatic potential atomic partial charge calculations. The calculation results indicate the increased electron density of the Au_{13} kernel of the modeling $Au_{25}(Cys)_{18}$ after the N-capping with acetyl. In other words, the N-capping indeed results in the higher electron donating effect of the ligand shell to the metallic core structure, associating with the reduction in the kernel dipole moment. Both effects contribute to the enhanced photoluminescence in the N-capped peptide ligand protected cluster systems.

6.4 Conclusion and prospect

According to the review of the above chapters, it can be seen that theoretical calculation plays an important role in the study of metal nanoclusters. With the rapid development of theoretical calculation and simulation methods, the structural evolution/transformation, electronic structure and structural properties of metal nanoclusters have been gradually elucidated by theoretical calculations. In the evolution/transformation of collective structure, people put forward the mechanism of structural transformation through theoretical simulation to verify the phenomena observed in the experiment. In the evolution of electronic structure, researchers put forward the evolution model of cluster electronic structure, such as superatom electron theory, super valence bond model and superatom-network model. In addition, for the stability factors of metal nanoclusters, the core cohesive energy and shell-to-core binding energy were decomposed and analyzed by theoretical calculation. However, it is worth noting that the simulation of experimentally unprecedented cluster structure is remains highly challenging. For example, the predicted conformation of clusters varies when different calculation methods was used, and therefore multiple strategies are necessary to check the reliability of the theoretical conclusion. In addition, because the size of metal nano-clusters ranges from several to hundreds of atoms, a single method in theoretical simulation might not adapt to both large and small clusters, and the computational cost of large clusters is fairly high in this days. The development of more powerful simulation methods and software for metal nanoclusters is an ongoing research issue.

References

[1] S. Adhikari, Y. Bando, S. Bhandari, et al., Micro and Nano Technologies, William Andrew Publishing, Boston, MA, 2016, pp. xi–xiii, https://doi.org/10.1016/B978-0-323-38945-7.00017-1.

[2] G.C. Schatz, Proc. Natl. Acad. Sci. U. S. A. 104 (17) (2007) 6885–6892, https://doi.org/10.1073/pnas.0702187104.

[3] S.M. Blinder, Density functional theory, in: Introduction to Quantum Mechanics (Second edition), Academic Press, San Diego, CA, 2021, pp. 235–244, https://doi.org/10.1016/B978-0-12-822310-9.00022-7.

[4] S.M. Blinder, Waves and particles, in: Introduction to Quantum Mechanics, Academic Press, San Diego, CA, 2021, pp. 21–34, https://doi.org/10.1016/B978-0-12-822310-9.00010-0.

[5] L.H. Thomas, The calculation of atomic fields, Math. Proc. Camb. Philos. Soc. 23 (5) (1927) 542–548, https://doi.org/10.1017/S0305004100011683.

[6] E. Fermi, Eine statistische Methode zur Bestimmung einiger Eigenschaften des Atoms und ihre Anwendung auf die Theorie des periodischen Systems der Elemente, Z. Phys. 48 (1) (1928) 73–79, https://doi.org/10.1007/BF01351576.

[7] M.L. Cohen, M. Schlüter, J.R. Chelikowsky, S.G. Louie, Self-consistent pseudopotential method for localized configurations: molecules, Phys. Rev. B 12 (12) (1975) 5575–5579, https://doi.org/10.1103/PhysRevB.12.5575.

[8] J.P. Perdew, W. Yue, Accurate and simple density functional for the electronic exchange energy: generalized gradient approximation, Phys. Rev. B 33 (12) (1986) 8800–8802, https://doi.org/10.1103/PhysRevB.33.8800.

[9] J.P. Perdew, Erratum: density-functional approximation for the correlation energy of the inhomogeneous electron gas, Phys. Rev. B 34 (10) (1986) 7406, https://doi.org/10.1103/PhysRevB.34.7406.

[10] D.C. Langreth, M.J. Mehl, Beyond the local-density approximation in calculations of ground-state electronic properties, Phys. Rev. B 28 (4) (1983) 1809–1834, https://doi.org/10.1103/PhysRevB.28.1809.

[11] A.D. Becke, Density-functional exchange-energy approximation with correct asymptotic behavior, Phys. Rev. A 38 (6) (1988) 3098–3100, https://doi.org/10.1103/PhysRevA.38.3098.

[12] J.P. Perdew, K. Burke, M. Ernzerhof, Generalized gradient approximation made simple, Phys. Rev. Lett. 77 (18) (1996) 3865–3868, https://doi.org/10.1103/PhysRevLett.77.3865.

[13] J.P. Perdew, K. Burke, M. Ernzerhof, Generalized gradient approximation made simple [Phys. Rev. Lett. 77, 3865 (1996)], Phys. Rev. Lett. 78 (7) (1997) 1396, https://doi.org/10.1103/PhysRevLett.78.1396.

[14] J. Sun, M. Marsman, G.I. Csonka, et al., Self-consistent meta-generalized gradient approximation within the projector-augmented-wave method, Phys. Rev. B 84 (3) (2011) 035117, https://doi.org/10.1103/PhysRevB.84.035117.

[15] Y. Zhao, D.G. Truhlar, J. Chem. Phys. 125 (19) (2006), 194101, https://doi.org/10.1063/1.2370993.

[16] A.D. Becke, E.R. Johnson, J. Chem. Phys. 124 (22) (2006), 221101, https://doi.org/10.1063/1.2213970.

[17] F. Tran, P. Blaha, Accurate band gaps of semiconductors and insulators with a semilocal exchange-correlation potential, Phys. Rev. Lett. 102 (22) (2009), 226401, https://doi.org/10.1103/PhysRevLett.102.226401.

[18] J. Sun, R.C. Remsing, Y. Zhang, et al., Accurate first-principles structures and energies of diversely bonded systems from an efficient density functional, Nat. Chem. 8 (9) (2016) 831–836, https://doi.org/10.1038/nchem.2535.

[19] J. Sun, A. Ruzsinszky, J.P. Perdew, Strongly constrained and appropriately normed semilocal density functional, Phys. Rev. Lett. 115 (3) (2015), 036402, https://doi.org/10.1103/PhysRevLett.115.036402.

[20] J. Sun, R. Haunschild, B. Xiao, I.W. Bulik, G.E. Scuseria, J.P. Perdew, J. Chem. Phys. 138 (4) (2013), 044113, https://doi.org/10.1063/1.4789414.

[21] J. Sun, B. Xiao, A. Ruzsinszky, J. Chem. Phys. 137 (5) (2012), 051101, https://doi.org/10.1063/1.4742312.

[22] A.P. Bartók, J.R. Yates, J. Chem. Phys. 150 (16) (2019), 161101, https://doi.org/10.1063/1.5094646.

[23] D. Mejía-Rodríguez, S.B. Trickey, J. Chem. Phys. 151 (20) (2019), 207101, https://doi.org/10.1063/1.5120408.

[24] J.W. Furness, A.D. Kaplan, J. Ning, J.P. Perdew, J. Sun, J. Phys. Chem. Lett. 11 (19) (2020) 8208–8215, https://doi.org/10.1021/acs.jpclett.0c02405.

[25] S. Lehtola, C. Steigemann, M.J.T. Oliveira, M.A.L. Marques, SoftwareX 7 (2018) 1–5, https://doi.org/10.1016/j.softx.2017.11.002.

[26] M.A.L. Marques, M.J.T. Oliveira, T. Burnus, Comput. Phys. Commun. 183 (10) (2012) 2272–2281, https://doi.org/10.1016/j.cpc.2012.05.007.

[27] O. Gunnarsson, B.I. Lundqvist, Exchange and correlation in atoms, molecules, and solids by the spin-density-functional formalism, Phys. Rev. B 13 (10) (1976) 4274–4298, https://doi.org/10.1103/PhysRevB.13.4274.

[28] J. Harris, R.O. Jones, The surface energy of a bounded electron gas, J. Phys. F Metal Phys. 4 (8) (1974) 1170–1186, https://doi.org/10.1088/0305-4608/4/8/013.

[29] D. Pines, P. Nozières, Response and Correlation in Neutral Systems, in: The Theory of Quantum Liquids, W.A. Benjamin Inc., 1989, pp. 82–146, http://doi.org/10.1201/9780429495717-3.

[30] A.D. Becke, J. Chem. Phys. 88 (2) (1988) 1053–1062, https://doi.org/10.1063/1.454274.

[31] J.P. Perdew, Y. Wang, Accurate and simple analytic representation of the electron-gas correlation energy, Phys. Rev. B 45 (23) (1992) 13244–13249, https://doi.org/10.1103/PhysRevB.45.13244.

[32] C. Lee, W. Yang, R.G. Parr, Development of the Colle-Salvetti correlation-energy formula into a functional of the electron density, Phys. Rev. B 37 (2) (1988) 785–789, https://doi.org/10.1103/PhysRevB.37.785.

[33] J.C. Slater, Atomic shielding constants, Phys. Rev. 36 (1) (1930) 57–64, https://doi.org/10.1103/PhysRev.36.57.

[34] G. te Velde, F.M. Bickelhaupt, E.J. Baerends, C. Fonseca Guerra, S.J.A. van Gisbergen, J.G. Snijders, T. Ziegler, Chemistry with ADF, J. Comput. Chem. 22 (9) (2001) 931–967, https://doi.org/10.1002/jcc.1056.

[35] D.P. Chong, Augmenting basis set for time-dependent density functional theory calculation of excitation energies: slater-type orbitals for hydrogen to krypton, Mol. Phys. 103 (6-8) (2005) 749–761, https://doi.org/10.1080/00268970412331333618.

[36] D.P. Chong, E. Van Lenthe, S. Van Gisbergen, E.J. Baerends, J. Comput. Chem. 25 (8) (2004) 1030–1036, https://doi.org/10.1002/jcc.20030.

[37] E. Van Lenthe, E.J. Baerends, J. Comput. Chem. 24 (9) (2003) 1142–1156, https://doi.org/10.1002/jcc.10255.

[38] E. Clementi, C. Roetti, At. Data Nucl. Data Tables 14 (3) (1974) 177–478, https://doi.org/10.1016/S0092-640X(74)80016-1.

[39] A.D. McLean, R.S. McLean, At. Data Nucl. Data Tables 26 (3) (1981) 197–381, https://doi.org/10.1016/0092-640X(81)90012-7.

[40] J.G. Snijders, P. Vernooijs, E.J. Baerends, At. Data Nucl. Data Tables 26 (6) (1981) 483–509, https://doi.org/10.1016/0092-640X(81)90004-8.

[41] J.B. Collins, P.v.R. Schleyer, J.S. Binkley, J.A. Pople, J. Chem. Phys. 64 (12) (1976) 5142–5151, https://doi.org/10.1063/1.432189.

[42] W.J. Hehre, R.F. Stewart, J.A. Pople, J. Chem. Phys. 51 (6) (1969) 2657–2664, https://doi.org/10.1063/1.1672392.

[43] J.S. Binkley, J.A. Pople, W.J. Hehre, J. Am. Chem. Soc. 102 (3) (1980) 939–947, https://doi.org/10.1021/ja00523a008.

[44] R. Ditchfield, W.J. Hehre, J.A. Pople, J. Chem. Phys. 54 (2) (1971) 724–728, https://doi.org/10.1063/1.1674902.

[45] K.D. Dobbs, W.J. Hehre, J. Comput. Chem. 7 (3) (1986) 359–378, https://doi.org/10.1002/jcc.540070313.

[46] M.M. Francl, W.J. Pietro, W.J. Hehre, et al., J. Chem. Phys. 77 (7) (1982) 3654–3665, https://doi.org/10.1063/1.444267.

[47] M.S. Gordon, Chem. Phys. Lett. 76 (1) (1980) 163–168, https://doi.org/10.1016/0009-2614(80)80628-2.

[48] M.S. Gordon, J.S. Binkley, J.A. Pople, W.J. Pietro, W.J. Hehre, J. Am. Chem. Soc. 104 (10) (1982) 2797–2803, https://doi.org/10.1021/ja00374a017.

[49] P.C. Hariharan, J.A. Pople, The influence of polarization functions on molecular orbital hydrogenation energies, Theor. Chim. Acta 28 (3) (1973) 213–222, https://doi.org/10.1007/BF00533485.

[50] P.C. Hariharan, J.A. Pople, Accuracy of AH_n equilibrium geometries by single determinant molecular orbital theory, Mol. Phys. 27 (1) (1974) 209–214, https://doi.org/10.1080/00268977400100171.

[51] W.J. Hehre, R. Ditchfield, J.A. Pople, J. Chem. Phys. 56 (5) (1972) 2257–2261, https://doi.org/10.1063/1.1677527.

[52] W.J. Pietro, M.M. Francl, W.J. Hehre, D.J. DeFrees, J.A. Pople, J.S. Binkley, J. Am. Chem. Soc. 104 (19) (1982) 5039–5048, https://doi.org/10.1021/ja00383a007.

[53] V.A. Rassolov, M.A. Ratner, J.A. Pople, P.C. Redfern, L.A. Curtiss, J. Comput. Chem. 22 (9) (2001) 976–984, https://doi.org/10.1002/jcc.1058.

[54] R.C. Binning Jr., L.A. Curtiss, J. Comput. Chem. 11 (10) (1990) 1206–1216, https://doi.org/10.1002/jcc.540111013.

[55] J.-P. Blaudeau, M.P. McGrath, L.A. Curtiss, L. Radom, J. Chem. Phys. 107 (13) (1997) 5016–5021, https://doi.org/10.1063/1.474865.

[56] L.A. Curtiss, M.P. McGrath, J. Blaudeau, N.E. Davis, R.C. Binning, L. Radom, J. Chem. Phys. 103 (14) (1995) 6104–6113, https://doi.org/10.1063/1.470438.

[57] P.J. Hay, J. Chem. Phys. 66 (10) (1977) 4377–4384, https://doi.org/10.1063/1.433731.

[58] M.P. McGrath, L. Radom, J. Chem. Phys. 94 (1) (1991) 511–516, https://doi.org/10.1063/1.460367.

[59] A.D. McLean, G.S. Chandler, J. Chem. Phys. 72 (10) (1980) 5639–5648, https://doi.org/10.1063/1.438980.

[60] F. Weigend, Phys. Chem. Chem. Phys. 8 (9) (2006) 1057–1065, https://doi.org/10.1039/B515623H.
[61] F. Weigend, R. Ahlrichs, Phys. Chem. Chem. Phys. 7 (18) (2005) 3297–3305, https://doi.org/10.1039/B508541A.
[62] T.H. Dunning, J. Chem. Phys. 90 (2) (1989) 1007–1023, https://doi.org/10.1063/1.456153.
[63] R.A. Kendall, T.H. Dunning, R.J. Harrison, J. Chem. Phys. 96 (9) (1992) 6796–6806, https://doi.org/10.1063/1.462569.
[64] K.A. Peterson, D.E. Woon, T.H. Dunning, J. Chem. Phys. 100 (10) (1994) 7410–7415, https://doi.org/10.1063/1.466884.
[65] A.K. Wilson, T. van Mourik, T.H. Dunning, J. Mol. Struct. Theochem. 388 (1996) 339–349, https://doi.org/10.1016/S0166-1280(96)80048-0.
[66] D.E. Woon, T.H. Dunning, J. Chem. Phys. 98 (2) (1993) 1358–1371, https://doi.org/10.1063/1.464303.
[67] P.J. Hay, W.R. Wadt, J. Chem. Phys. 82 (1) (1985) 270–283, https://doi.org/10.1063/1.448799.
[68] P.J. Hay, W.R. Wadt, J. Chem. Phys. 82 (1) (1985) 299–310, https://doi.org/10.1063/1.448975.
[69] W.R. Wadt, P.J. Hay, J. Chem. Phys. 82 (1) (1985) 284–298, https://doi.org/10.1063/1.448800.
[70] J.A. Montgomery, M.J. Frisch, J.W. Ochterski, G.A. Petersson, J. Chem. Phys. 110 (6) (1999) 2822–2827, https://doi.org/10.1063/1.477924.
[71] P.A.M. Dirac, R.H. Fowler, Proc. R. Soc. Lond., Ser. A 117 (778) (1928) 610–624, https://doi.org/10.1098/rspa.1928.0023.
[72] C. Chang, M. Pelissier, P. Durand, Regular two-component Pauli-like effective Hamiltonians in Dirac theory, Phys. Scr. 34 (5) (1986) 394–404, https://doi.org/10.1088/0031-8949/34/5/007.
[73] J.-L. Heully, I. Lindgren, E. Lindroth, S. Lundqvist, A.-M. Martensson-Pendrill, Diagonalisation of the Dirac Hamiltonian as a basis for a relativistic many-body procedure, J. Phys. B Atomic Mole. Phys. 19 (18) (1986) 2799–2815, https://doi.org/10.1088/0022-3700/19/18/011.
[74] E. van Lenthe, E.J. Baerends, J.G. Snijders, J. Chem. Phys. 99 (6) (1993) 4597–4610, https://doi.org/10.1063/1.466059.
[75] M.A.L. Marques, E.K.U. Gross, Annu. Rev. Phys. Chem. 55 (1) (2004) 427–455, https://doi.org/10.1146/annurev.physchem.55.091602.094449.
[76] M.J. Frisch, G.W. Trucks, H.B. Schlegel, et al., Gaussian 09 Revision E.01, Gaussian Inc., Wallingford, 2016.
[77] I. Chakraborty, T. Pradeep, Chem. Rev. 117 (12) (2017) 8208–8271, https://doi.org/10.1021/acs.chemrev.6b00769.
[78] T. Chen, V. Fung, Q. Yao, Z. Luo, D. Jiang, J. Xie, J. Am. Chem. Soc. 140 (36) (2018) 11370–11377, https://doi.org/10.1021/jacs.8b05689.
[79] J.B. Foresman, Æ. Frisch, Exploring Chemistry with Electronic Structure Methods, second ed., Gaussian Inc., Pittsburgh, PA, 1996.
[80] H. Häkkinen, M. Walter, H. Grönbeck, J. Phys. Chem. B 110 (20) (2006) 9927–9931, https://doi.org/10.1021/jp0619787.
[81] I.L. Garzón, C. Rovira, K. Michaelian, et al., Do thiols merely passivate gold nanoclusters? Phys. Rev. Lett. 85 (24) (2000) 5250–5251, https://doi.org/10.1103/PhysRevLett.85.5250.
[82] S. Kenzler, C. Schrenk, A. Schnepf, Angew. Chem. Int. Ed. 56 (1) (2017) 393–396, https://doi.org/10.1002/anie.201609000.
[83] Y. Pei, Y. Gao, X.C. Zeng, J. Am. Chem. Soc. 130 (25) (2008) 7830–7832, https://doi.org/10.1021/ja802975b.
[84] H. Qian, W.T. Eckenhoff, Y. Zhu, T. Pintauer, R. Jin, J. Am. Chem. Soc. 132 (24) (2010) 8280–8281, https://doi.org/10.1021/ja103592z.
[85] K. Nobusada, T. Iwasa, J. Phys. Chem. C 111 (39) (2007) 14279–14282, https://doi.org/10.1021/jp075509w.
[86] R. Jin, C. Liu, S. Zhao, et al., ACS Nano 9 (8) (2015) 8530–8536, https://doi.org/10.1021/acsnano.5b03524.
[87] Y. Song, H. Abroshan, J. Chai, et al., Chem. Mater. 29 (7) (2017) 3055–3061, https://doi.org/10.1021/acs.chemmater.7b00058.
[88] A. Fernando, C.M. Aikens, J. Phys. Chem. C 119 (34) (2015) 20179–20187, https://doi.org/10.1021/acs.jpcc.5b06833.
[89] N.A. Sakthivel, L. Sementa, B. Yoon, U. Landman, A. Fortunelli, A. Dass, J. Phys. Chem. C 124 (2) (2020) 1655–1666, https://doi.org/10.1021/acs.jpcc.9b08846.
[90] B. Cordero, V. Gómez, A.E. Platero-Prats, et al., *Dalton Trans.* (21) (2008) 2832–2838, https://doi.org/10.1039/B801115J.
[91] W. Michael, A. Jaakko, L.-A. Olga, et al., Proc. Natl. Acad. Sci. 105 (27) (2008) 9157–9162, https://doi.org/10.1073/pnas.0801001105.
[92] A. Dass, T.C. Jones, S. Theivendran, L. Sementa, A. Fortunelli, J. Phys. Chem. C 121 (27) (2017) 14914–14919, https://doi.org/10.1021/acs.jpcc.7b03860.
[93] L. Cheng, J. Yang, J. Chem. Phys. 138 (14) (2013) 141101, https://doi.org/10.1063/1.4801860.
[94] L. Cheng, C. Ren, X. Zhang, J. Yang, Nanoscale 5 (4) (2013) 1475–1478, https://doi.org/10.1039/C2NR32888G.
[95] L. Cheng, Y. Yuan, X. Zhang, J. Yang, Angew. Chem. Int. Ed. 52 (34) (2013) 9035–9039, https://doi.org/10.1002/anie.201302926.
[96] M.G. Taylor, G. Mpourmpakis, Thermodynamic stability of ligand-protected metal nanoclusters, Nat. Commun. 8 (1) (2017) 15988, https://doi.org/10.1038/ncomms15988.
[97] A.E. Reed, R.B. Weinstock, F. Weinhold, J. Chem. Phys. 83 (2) (1985) 735–746, https://doi.org/10.1063/1.449486.
[98] L. Xiong, B. Peng, Z. Ma, P. Wang, Y. Pei, Nanoscale 9 (8) (2017) 2895–2902, https://doi.org/10.1039/C6NR09612C.
[99] S.M. Reilly, T. Krick, A. Dass, J. Phys. Chem. C 114 (2) (2010) 741–745, https://doi.org/10.1021/jp9067944.
[100] D. Crasto, S. Malola, G. Brosofsky, A. Dass, H. Häkkinen, J. Am. Chem. Soc. 136 (13) (2014) 5000–5005, https://doi.org/10.1021/ja412141j.
[101] A. Dass, T. Jones, M. Rambukwella, et al., J. Phys. Chem. C 120 (11) (2016) 6256–6261, https://doi.org/10.1021/acs.jpcc.6b00062.
[102] D. Crasto, G. Barcaro, M. Stener, L. Sementa, A. Fortunelli, A. Dass, J. Am. Chem. Soc. 136 (42) (2014) 14933–14940, https://doi.org/10.1021/ja507738e.
[103] S. Yang, S. Chen, L. Xiong, et al., J. Am. Chem. Soc. 140 (35) (2018) 10988–10994, https://doi.org/10.1021/jacs.8b04257.
[104] Y. Du, H. Sheng, D. Astruc, M. Zhu, Chem. Rev. 120 (2) (2020) 526–622, https://doi.org/10.1021/acs.chemrev.8b00726.
[105] X. Cai, G. Saranya, K. Shen, et al., Angew. Chem. Int. Ed. 58 (29) (2019) 9964–9968, https://doi.org/10.1002/anie.201903853.
[106] C.L. Heinecke, T.W. Ni, S. Malola, et al., J. Am. Chem. Soc. 134 (32) (2012) 13316–13322, https://doi.org/10.1021/ja3032339.
[107] M.E. Casida, D. Chong, Recent advances in density functional methods, in: Recent Advances in Computational Chemistry, 1995, pp. 53–78, doi:10.1142/9789812830586_0002.
[108] M. Zhu, C.M. Aikens, F.J. Hollander, G.C. Schatz, R. Jin, J. Am. Chem. Soc. 130 (18) (2008) 5883–5885, https://doi.org/10.1021/ja801173r.
[109] A. Venzo, S. Antonello, J.A. Gascón, et al., Anal. Chem. 83 (16) (2011) 6355–6362, https://doi.org/10.1021/ac2012653.

[110] S. Wang, Y. Song, S. Jin, et al., J. Am. Chem. Soc. 137 (12) (2015) 4018–4021, https://doi.org/10.1021/ja511635g.
[111] S. Wang, X. Meng, A. Das, et al., Angew. Chem. Int. Ed. 53 (9) (2014) 2376–2380, https://doi.org/10.1002/anie.201307480.
[112] M.Y. Sfeir, H. Qian, K. Nobusada, R. Jin, J. Phys. Chem. C 115 (14) (2011) 6200–6207, https://doi.org/10.1021/jp110703e.
[113] G. Hu, Z. Wu, D. Jiang, J. Mater. Chem. A 6 (17) (2018) 7532–7537, https://doi.org/10.1039/C8TA00461G.
[114] S. Wang, H. Abroshan, C. Liu, et al., Shuttling single metal atom into and out of a metal nanoparticle, Nat. Commun. 8 (1) (2017) 848, https://doi.org/10.1038/s41467-017-00939-0.
[115] S. Grimme, J. Comput. Chem. 27 (15) (2006) 1787–1799, https://doi.org/10.1002/jcc.20495.
[116] A.D. Becke, E.R. Johnson, J. Chem. Phys. 122 (15) (2005), 154104, https://doi.org/10.1063/1.1884601.
[117] A.D. Becke, E.R. Johnson, J. Chem. Phys. 124 (1) (2006), 014104, https://doi.org/10.1063/1.2139668.
[118] A.D. Becke, E.R. Johnson, J. Chem. Phys. 127 (15) (2007), 154108, https://doi.org/10.1063/1.2795701.
[119] Y. Lin, P. Charchar, A.J. Christofferson, et al., J. Am. Chem. Soc. 140 (51) (2018) 18217–18226, https://doi.org/10.1021/jacs.8b04436.

Chapter 7

Assembly of metal nanoclusters

Manzhou Zhu and Shan Jin

7.1 Introduction—A brief introduction to the assembly of nanomaterials

The process of self-assembly is very common in nature and science technology [1]. People are interested in the self-assembly for the following reasons: (i) People are naturally interested in turning disorder into order; (ii) Cells have assembling behavior, to better understand life, it is necessary to understand self-assembly behavior first; (iii) Self-assembly is an effective strategy for constructing nanostructures, which is very important for nanoscience; (iv) From a functional point of view, self-assembly can stabilize, optimize and amplify the properties of the building element, and even generate new functions through the aggregation effect, which is of great significance to its application [2–5].

The scale of self-assembly ranges from the microscopic molecular level to the macroscopic planetary level. The interaction forces between the building blocks of self-assembly at different scales are different. The concept of self-assembly is increasingly applied in a variety of disciplines, while the focus of researchers in the field of chemistry and materials is mainly on self-assembly in the microscopic field. Self-assembly reflects the basic information of constructing primitives, such as morphology, size, surface properties, chargeability, polarizability, magnetic dipole, and mass. These basic information determine the interaction between them. The self-assembled structure is very important. Designing the building element to assemble the desired structure and function is the key to self-assembly research, and some self-assembly processes cannot be fully resolved because it is difficult to design the building element to have all the desired characteristics. The guess of the relationship between structural primitives and assembly structure, so it is of great significance to continuously introduce new assembly systems to gradually understand the self-assembly process from different angles. Since the self-assembly requires that the building element is movable, it generally occurs in the liquid phase or on a smooth surface.

For microscopic self-assembly, the specific size of the building elements will also bring about different interactions between the building elements. For self-assembly at the molecular level, the interactions are mainly noncovalent or weak covalent interactions, such as van der Waals forces, electrostatic forces, hydrophobic interactions, host-guest interactions, hydrogen bonds and coordination bonds, etc. For larger-scale nanoparticles or even colloidal particles, the interaction forces generally include van der Waals forces, electrostatic forces, magnetic forces, hydrophobic interactions, spatial repulsive interference and confinement, solvent interactions, and capillary interactions, convection, lubrication and friction, surface force and entropy effect, etc. Supramolecular assembly at the molecular scale is indispensable for biological functions, and in many aspects, supramolecular chemistry is deeply inspired by biology. There are many examples of complex assembly from living cells, such as double-layer membranes, ribosomes, Nucleic acid transcription machinery, etc. In some cases, thermodynamically driven molecular assembly processes are effective for biological functions, such as lipid self-assembly to produce cell membranes or the formation of stable host-guest complexes. However, for many more advanced biological machines, only thermodynamic self-assembly is ineffective, because the required functions only appear in the dynamics and continuous energy dissipation process. Sustained free energy activates complex supramolecular machines, thereby enabling chemical reactions to proceed and functionalizing the work of cells. The building blocks at the molecular scale include small molecules, polymers, and some biomolecules. The study of self-assembly at the molecular scale helps to enhance people's understanding of the human system and is also used in drug transport and transmission, sensing or catalysis and other fields [6–9].

The rise of nanoparticle-scale self-assembly is due to the rapid development of nanoscience in the past two to three decades. Numerous kinds of nanoparticles of different sizes, shapes, material properties and surface properties have been obtained, and these independent building elements have been assembled into larger-scale materials have received extensive attention. Nanoscale spheres, polyhedrons, rods, ellipsoids, sheets, triangular or quadrangular pyramids, core-shell, and hollow particles, nanocages and dumbbell-shaped particles can be synthesized in large quantities and have high monodispersity. Nanoscale synthesis methods can be applied to a variety of metals, semiconductors, oxides, inorganic salts and polymers, etc. [10–24]. These nanoscale materials can be flexibly functionalized by using organic ligands according to the desired surface properties, such as solubility in specific solvents, specificity for small molecules or larger biological agents, nonspecific adsorption resistance, net charge and electrochemical activity, etc. [25–28]. In many cases, it is difficult to apply independent nanomaterials. It is necessary to organize the individual nanomaterials into assemblies through interactions in a specific way. This is the actual requirement of self-assembly. There are several examples of successful application of self-assembly at the nanoscale: from the system of ultrasensitive biosensors to the uniform width of highly conductive nanowires, and the ordered two-dimensional nanoparticle arrays with unique electrical properties, and then the overall microdimensional size of the material exhibits unusual overall performance. These examples prove the significance of the self-assembly work, and also show that the self-assembly process is very dependent on our understanding and regulation of the interaction between the building elements, so that the building elements are assembled into the designed structure. Without such capabilities, it is impossible to design structures, and to customize the functional properties of nanomaterials [29–32].

In recent years, self-assembly based on metal nanoclusters has received extensive attention. Nanoclusters are the bridge between atoms and nanoscience. Due to the volatility of their physical and chemical properties, they not only provide a functionally rich material library for the field of materials but also provide theoretical research on the changes of matter from atoms to macroscopic properties. It is a functional unit with a unique structure, which is of great significance for studying and regulating their interaction and realizing specific assembly designs. Various diversified and useful materials based on nanoclusters are obtained through assembly design, which is an important basis for the application of nanoclusters. Many devices and luminescent materials in actual use are based on the assembled structure of nanoclusters. Using nanocluster interactions to construct materials with novel structures and practical functions will be one of the most common and important research contents of cluster science. Recently, multiple techniques have been developed to form assembly material of metal nanoclusters through self-assembly. Further progress of these techniques will promote the development of nanomaterials that take advantage of the characteristics of metal nanoclusters. On the one hand, the size of metal nanoclusters is between small molecules and nanoparticles. For example, the self-assembly of metal nanoclusters will help to strengthen the understanding of subnanoscale; on the other hand, the self-assembly of metal nanoclusters has indeed brought many performance improvements. And most of researches are based on assembly-induced fluorescence enhancement. In addition, self-assembly can also improve the stability of metal nanoclusters, which promotes its application. Due to the size of the metal core and the types of surface ligands, various combinations can be carried out, especially alloys and multimatches exist. The bulk coating of metal clusters makes it more selective as a building element.

Early research was to assemble gold clusters on the surface of the substrate [33,34]. Afterwards, in 2013, Wu et al. assembled alkyl mercaptan-protected gold clusters into self-supporting two-dimensional nanosheets in a binary high-boiling solvent. The solvent microphase separation provided layered soft templates for assembly. The hydrophobic interaction between the gold cluster ligands is stimulated to complete the assembly of the gold cluster. At the same time, it is also found that the concentration of the gold cluster and the solvent ratio have a great influence on the assembly morphology [35]. Subsequently, Wu et al. found through experiments and computer simulations that the self-assembly of the two-dimensional orientation of the gold cluster originated from the one-dimensional dipole force of the gold cluster itself. The one-dimensional preassembly of the gold cluster caused the redistribution of the ligands on the surface, resulting in asymmetrical Van der Waals force finally completed the two-dimensional assembly. By adjusting the reaction temperature, reaction time, heating rate, and gold cluster concentration, the gold clusters are assembled into different morphologies [36]. Yao et al. used metal halide, thiourea as a reducing agent, and CTAB as a surfactant to assemble water-soluble gold, platinum and palladium nanoclusters into assemblies with clear mesoscale boundaries [37]. Lin et al. assembled chiral gold clusters into crystalline nanocubes. After assembly, the circular dichroism intensity of gold clusters was significantly enhanced, and circularly polarized light was generated [38]. By controlling the solvent conditions, Nonappa et al. assembled atomically precise gold clusters into a hexagonal monolayer of two-dimensional colloidal crystals by hydrogen bonding and folded them into spherical hollow shells [39]. Metal clusters can be assembled into larger-scale self-assembled structures, or they can be assembled in pairs to produce larger-scale metal clusters. This is of great significance for studying the formation process of metal clusters and the essential driving force of assembly, and it is also possible to bring new properties to metal clusters [40–42]. The metal clusters are coassembled with other functional or template materials to form a composite to improve the diversity of its functionalization and assembly, and begin to enter people's field of vision, various polymers, nanoparticles, DNA, etc. have been used to coassemble with metal clusters. Zhang et al. assembled fluorescent gold clusters in peptide nanofibers, and their fluorescence quantum yield was increased by 70 times and then used for temperature sensing and cell imaging [43]. Xue et al. also assembled the amino-modified blue nanoparticles and glutathione-coated gold clusters into spherical nanoparticles through electrostatic interaction. The fluorescence of the gold clusters was enhanced by the assembly induced fluorescence. The dual emission nanocomposite can be used to detect protamine and trypsin [44]. Wang et al. assembled gold clusters and polymers into uniform-sized nanocomposites in the aqueous phase through the modification of host and guest molecules. After assembly, the gold clusters exhibited assembly induced fluorescence enhancement [45]. Shen et al. assembled silver clusters and polyethyleneimine into nanospheres and nanovesicles through electrostatic interaction, and the assembly enhanced the fluorescence of silver clusters. The assembled nanocomposite can detect aluminum ions with high sensitivity [46].

We hope this chapter focusing coin metal nanoclusters allow readers to obtain a general understanding of the formation and functions of nanoclusters and that the obtained knowledge will help to establish clear design guidelines for fabricating new nanoclusters with desired functions in the future.

7.2 Cluster-based host-gust nanosystem for assemble nanomaterial

7.2.1 Supramolecular interactions of NCs as hosts with cyclodextrins (CDs)

The size of nanoclusters is between molecular atoms and nanocrystals, so many of their properties will be like both. So, the interactions of a cluster as the host with other molecules may be tuned to precisely functionalize the surface of the NCs. CDs are water-soluble cyclic oligosaccharides having hydrophobic cavities that form inclusion complexes with specific molecules. Now, three types of cyclodextrins have been most thoroughly researched: α-CD (joining 6 glucose subunits), β-CD (joining 7 glucose subunits), and γ-CD (joining 8 glucose subunits). The inherent hydrophobic cavities and hydrophilic external surface endow the supramolecular interactions of the cyclodextrins with different species such as the transition metal clusters, POMs, boron clusters [47–65], as well as the inorganic complexes containing thiol groups like BBSH [66] and adamantanethiol [67–69], which promoted the development of supramolecular interactions of coin metal nanoclusters as hosts with cyclodextrins (CDs; Fig. 7.1).

In 2014, Pradeep and coworkers had synthesized $Au_{25}SBB_{18} \cap CD_n$ ($n=1–4$), where $X \cap Y$ denotes an inclusion complex between the host $Au_{25}SBB_{18}$ and receptor β-CD, using interactions between β-cyclodextrin (CD) and the ligand anchored on the cluster [70]. The synthesized $Au_{25}SBB_{18} \cap CD_n$ ($n=1–4$) was characterized such as NMR spectroscopy and electrospray ionization mass spectrometry (ESI-MS). Via carefully control of the ratio of SBB/CD molar, the amount of CD_n increase. When at 1:1.2, $Au_{25}SBB_{18} \cap CD_4$ can be observed clearly (Fig. 7.1A), maybe the geometric stability of the $Au_{25}SBB_{18} \cap CD_4$ cluster adducts. Schematic structures of $Au_{25}SBB_{18} \cap CD_n$ ($n=1–4$) are presented in Fig. 7.1B. Computational studies were performed to clarify the nature of the interaction. The optimized structure of $Au_{25}SBB_{18} \cap CD_4$ (Fig. 7.1C) represented when the four CDs located at tetrahedral locations, the interactions between CDs be minimal, contributing to the high stability of the overall structure, whose spicule model is shown in Fig. 7.1D.

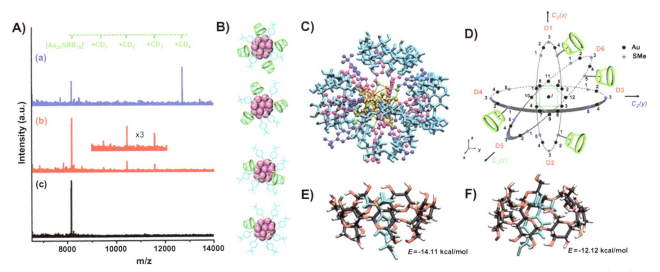

FIG. 7.1 The assemble of $Au_{25}SBB_{18}$ with cyclodextrins (CDs). (A) ESI MS spectra for the $Au_{25}SBB_{18}$ cluster and its CD adducts. Spectra a, b, and c are at SBB/CD molar ratios of 1:1.2, 1:1, and 1:0, respectively. (B) Schematic representations of $Au_{25}SBB_{18} \cap CD_n$ ($n=1–4$). (C) Computed structure of $Au_{25}SBB_{18} \cap CD_4$ with four CDs (shown in cyan). H atoms have been omitted. (D) A spicule representation of $Au_{25}SBB_{18} \cap CD_4$. (E, F) BBSH∩CD inclusion complexes where the tert-butyl group of BBSH undergoes complexation through the (E) narrower and (F) wider rims of β-CD. In (E) and (F), all the atoms of BBSH are colored cyan for clarity, whereas atoms of CD are colored differently. *(From A. Mathew, G. Natarajan, L. Lehtovaara, H. Häkkinen, R.M. Kumar, V. Subramanian, A. Jaleel, T. Pradeep, Supramolecular functionalization and concomitant enhancement in properties of Au_{25} clusters, ACS Nano 8(1) (2014) 139–152, https://doi.org/10.1021/nn406219x.)*

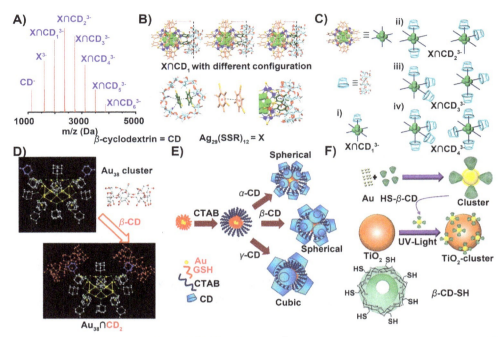

FIG. 7.2 Cluster-based cluster-CD hybrid materials. (A) ESI-MS of $[X\cap(CD)_n]^{3-}$ ($n=1–6$) supramolecular complexes, where X and CD represent $Ag_{29}(SSR)_{12}$ and β-CD, respectively. (B) (top) DFT-optimized lowest energy structures of the $[X\cap(CD)_1]^{3-}$ where CD stands for α, β, and γ-CD. (bottom) The C–H...π interactions between SSR ligands and β-CD. The π...π interaction between two neighboring SSR ligands, and the Ag...O interactions with $Ag_{29}\cap(β-CD)_1$. (C) Simple representation of isomers of $[X\cap(CD)n]^{3-}$, $n=1–4$. (D) Interactions between $Au_{38}S_2(SR)_{20}$ and CD, and $\cap(CD)_2$ hybrid material. (E) Proposed arrangements of the three types of CDs around the GSH stabilized Au nanoclusters. (F) Schematic illustration of the fabrication of the hybrid TiO_2-Au clusters $\cap(β-CD)$ photocatalyst. *(From A. Nag, P. Chakraborty, G. Paramasivam, M. Bodiuzzaman, G. Natarajan, T. Pradeep, Isomerism in supramolecular adducts of atomically precise nanoparticles, J. Am. Chem. Soc. 140 (42) (2018) 13590–13593, https://doi.org/10.1021/jacs.8b08767; C. Yan, C. Liu, H. Abroshan, Z. Li, R. Qiu, G. Li, Surface modification of adamantane-terminated gold nanoclusters using cyclodextrins, Phys. Chem. Chem. Phys. 18 (33) (2016) 23358–23364, https://doi.org/10.1039/c6cp04569c; S. Bhunia, S. Kumar, P. Purkayastha, Gold nanocluster-grafted cyclodextrin suprastructures: formation of nanospheres to nanocubes with intriguing photophysics, ACS Omega 3 (2) (2018) 1492–1497, https://doi.org/10.1021/acsomega.7b01914; H. Zhu, N. Goswami, Q. Yao, T.K. Chen, Y.B. Liu, Q.F. Chen, J.M. Lu, J.P. Xie, Cyclodextrin-gold nanocluster decorated TiO_2 enhances photocatalytic decomposition of organic pollutants, J. Mater. Chem. A 6 (3) (2018) 1102–1108, https://doi.org/10.1039/c7ta09443d.)*

The H-H interactions between CDs and SR ligand are responsible for the anchoring the CDs onto the surface of nanoclusters. And the complexation through the narrower rim of CD facing the cluster core was more favorable than complexation through the wider rim (Fig. 7.1E, F). In contrast, the interaction between $Au_{25}(PET)_{18}$ and CDs has not been reported, the possible reason maybe the precise orientation of the SBB, which pointed outward form the core [71,72].

Further, Pradeep and coworkers again investigated the supramolecular adducts of atomically precise nanoclusters, $[Ag_{29}(BDT)_{12}\cap(CD)_n]^{3-}$ ($n=1–6$), which $Ag_{29}(BDT)_{12}$ as host and CDs as receptor (Fig. 7.2) [73]. The $[Ag_{29}(BDT)_{12}\cap(CD)_{1-6}]^{3-}$ clusters was synthesized via mixing $[Ag_{29}(BDT)_{12}]^{3-}$ and CDs (α, β, and γ) in various proportions, which further confirmed via electrospray ionization mass spectrometry (ESI-MS). ESI-MS results shows that there are up to six β-CDs that can be adsorbed onto the surface of the $[Ag_{29}(BDT)_{12}]^{3-}$ nanocluster (Fig. 7.2A). HNMR spectra of $[Ag_{29}(BDT)_{12}]^{3-}\cap CD$ was also performed. Combining the HNMR spectra and ESI-MS data, each CD encapsulating a pair of BDT ligands can be rational considered, and the six CDs on the cluster surface may be arranged in an octahedron. To further understand the global structure of $[Ag_{29}(BDT)_{12}]^{3-}\cap CD$, molecular docking and DFT calculations was performed. Firstly, the lowest energy docked structures of $[Ag_{29}(BDT)_{12}]^{3-}\cap CD_1$ (α, β, and γ) were used for DFT optimization. DFT optimized lowest energy structures of $[Ag_{29}(BDT)_{12}\cap(CD)_1]^{3-}$ suggested that the CDs (α, β, and γ) encapsulate a pair of SSR ligands protecting the Ag_{29} kernel (Fig. 7.2B) and the diameter of CDs can affect the binding energy (BE) values of CDs with $Ag_{29}(BDT)_{12}$, which in turn affect the penetration of SSR ligands in the cavity. For example, the C–H···π interactions between BDTs and β-CD and the π···π interaction between two BDTs ligand as well as the weak ionic Ag···O, hydrogen bonding and van der Waals interactions led to the such complexation. More importantly, due to the different ways of addition, the unique geometry (octahedral geometry with six vertices) endowed the supramolecular adducts $[Ag_{29}(BDT)_{12}]^{3-}\cap CD$ various isomers (Fig. 7.2C). In detail, $Ag_{29}(BDT)_{12}]^{3-}\cap(CD)_1$, $Ag_{29}(BDT)_{12}]^{3-}\cap(CD)_5$, and $Ag_{29}(BDT)_{12}]^{3-}\cap(CD)_6$ have only one structural isomer, while $Ag_{29}(BDT)_{12}]^{3-}\cap(CD)_2$, $Ag_{29}(BDT)_{12}]^{3-}\cap(CD)_3$, or $Ag_{29}(BDT)_{12}]^{3-}\cap(CD)_4$ have two possible isomers. Ion mobility mass spectrometry

(IM-MS) can effectively separate structural isomers in the gas phase. The discovery of new clusters with amenable ligands would enhance better stabilization of such supramolecular adducts.

Due to the inclusion complexes of adamantanethiol with CD were also reported, Yan et al. investigated the supramolecular chemistry of atomically precise nanoclusters $Au_{38}S_2(SAdm)_{20}$ nanocluster (SAdm = 1-adamantanethiolate) with α, β, and γ-CDs [74]. Via UV-vis and NMR spectroscopies, MALDI mass spectrometry, only β-CDs were found to be able to chemisorb onto the nanocluster surface to form $Au_{38}S_2(SAdm)_{20}\cap(\beta\text{-CD})_1$ and $Au_{38}S_2(SAdm)_{20}\cap(\beta\text{-CD})_2$. Molecular dynamics (MD) simulations were performed to analyze the fragmentation results of MALDI-MS of the nanoclusters with CDs. The results demonstrated both the cavity size and ligand affecting the binding energy, like that of the α, β and γ-CDs with $Ag_{29}(BDT)_{12}$. And due to the steric hindrance arising from nearby ligands, the two ligands at the surface corner can completely fit into the cavity of β-CDs shown in Fig. 7.2D, minimizing their mutual steric hindrance. Furthermore, the introduction of β-CD into the surface of the nanocluster enhances the stability of the $Au_{38}S_2(SAdm)_{20}$ nanocluster, making it unaffected by oxidation or reduction-possibly due to the hindrance of charge transfer. This unfolds the possibility of functionalizing other Adm-protected clusters like $Au_{30}(SAdm)_{18}$, $Pt_1Ag_{28}(SAdm)_{18}(PPh_3)_4$, $Au_{24}(SAdm)_{16}$, $Au_{22}(SAdm)_{16}$, etc. with CDs in a similar fashion [75–78].

Similarly, Wu and coworkers reported a novel adamantane thiolate-protected $Au_{40}(S\text{-Adm})_{22}$ nanocluster bound with γ-CD-MOF, which was later used for the HRP-mimicking reaction system. Via simple mixing γ-CD and KOH, followed by filtering and recrystallizing, the γ-CD-MOF was synthesized. The Au_{40}/γ-CD-MOF was obtained by mixing $Au_{40}(S\text{-Adm})_{22}$ with γ-CD-MOF. The introduction of the γ-CD-MOF not only endows water solubility but also endows the catalytic activity to the nanoclusters, different from the virgin $Au_{40}(S\text{-Adm})_{22}$ nanoclusters. The Au_{40}/γ-CD-MOF has a higher affinity toward TMB, and the detailed HRP-mimicking catalysis mechanism was proposed and supported by DFT calculation. The success of water-phase in the $Au_{40}(S\text{-Adm})_{22}@$ γ-CD-MOF system provide a chance to use host-guest chemistry to activate the gold nanocluster and endow the liposoluble nanocluster with water solubility, which will be possible to promote the synthesis of composites of cyclodextrins (CDs) and cluster compounds [79].

The nonemissive, GSH-protected Au nanoclusters reacting with α, β and γ-CDs were reported by Bhunia and coworkers. Firstly, protonated GSH-Au nanoclusters are deprotonated under alkaline conditions, accompanied with -COOH groups to -COO$^-$. When adding cetyltrimethylammonium bromide (CTAB) (a cationic surfactant) into the alkaline solution, the electrostatic interaction between CTAB and -COO$^-$ groups resulted in Au NCs became hydrophobic, decreasing the fluorescence of the Au nanoclusters. Interestingly, the PL of the quenched GSH-CTAB-coated Au nanoclusters can be revived as the addition of α, β-CDs. The addition of α, β-CDs could enhance the fluorescence, in contrast to the addition of γ-CDs further decrease the fluorescence. Unlike the CDs reacting with $Au_{25}SBB_{18}$, $Ag_{29}(SSR)_{12}$ and $Au_{38}S_2(SR)_{20}$ nanoclusters, the reacting of GSH-CTAB-coated Au nanoclusters with CDs could induce the nanoclusters self-assemble into suprastructures. And due to the difference in the dimension of the CDs, α- and β-CDs induce cluster $\cap$(α-CD)$_n$ and cluster $\cap$(β-CD)$_n$ assembled into spherical suprastructures, whereas cluster $\cap$(γ-CD)$_n$ aggregate into cubic suprastructures (Fig. 7.2E) [80]. The results provided important information on guest-host interaction between CDs and Au nanoclusters.

Novelty, the β-CD could be modified with an SH-group and the resulting β-CD-SH could be used to stabilize Au nanoclusters [81]. As usual, such β-CD guests could affect the optical properties of the cluster hosts. After, the β-CD-S capped Au nanoclusters were loaded onto the TiO_2 nanoparticles (Fig. 7.2F), and the obtained hybrid material showed improved photocatalytic performance over TiO_2 nanoparticles: for the photocatalytic decomposition of methyl orange, the degradation efficiency of pure TiO_2 nanoparticles was 47%, while for the hybrid materials, this value was 98%.

7.2.2 Supramolecular interactions of NCs as hosts with fullerene

Like the CDs as the guest to assemble nanoclusters, the fullerenes could also serve as guest to construct the nanocluster@Host-guest nanosystems [82]. Because of its large number of lowest empty molecular orbitals, fullerenes can act as electron acceptor, which could serve as an attractive candidate to be included into cluster-based host-guest compounds (Fig. 7.3).

Simply mixing $[Ag_{29}(BDT)_{12}]^{3-}$ nanoclusters and C_{60} molecules, a host-guest supracompounds $\{[Ag_{29}(BDT)_{12}]\cap(C_{60})_n\}^{3-}$ ($n=1$–9) was obtained by Pradeep and coworkers [83]. Specifically, the concentration of C_{60} will influence the number of C_{60} additions. A lesser amount of C_{60} addition generated the supracompounds $\{[Ag_{29}(BDT)_{12}]\cap(C_{60})_n\}^{3-}$ ($n=1$–4). The excess concentration produced a higher number of C_{60} anchoring on the surface of the cluster $[Ag_{29}(BDT)_{12}]^{3-}$, forming $\{[Ag_{29}(BDT)_{12}]\cap(C_{60})_n\}^{3-}$ ($n>4$) (Fig. 7.3A). The structural details of $\{[Ag_{29}(BDT)_{12}]\cap(C_{60})_n\}^{3-}$ ($n=1$–9) was studied by UV-vis, high-resolution electrospray ionization mass spectrometry (ESI-MS), collision-induced dissociation (CID), and nuclear magnetic resonance spectrometry (NMR). Firstly, the parent $[Ag_{29}(BDT)_{12}]^{3-}$ clusters without apex-encapsulated TPP ligands can only maintain stable only for a few hours, the

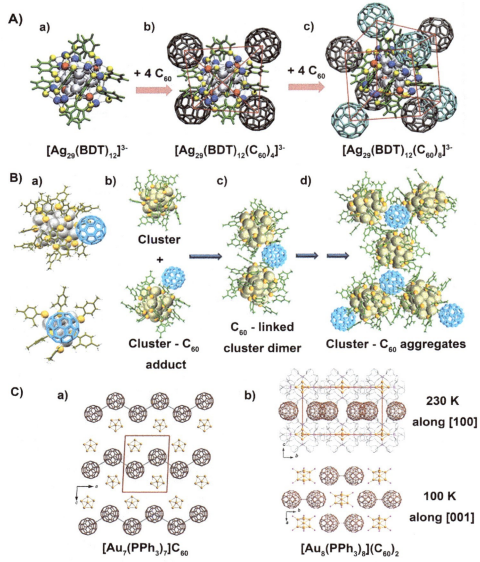

FIG. 7.3 Cluster-based hybrid materials. Cluster-based hybrid materials. (A) Anchoring four or eight C_{60} molecules onto the $[Ag_{29}(BDT)_{12}]^{3-}$ nanocluster surface; (B) Lowest energy structure of the adduct, $[Ag_{25}(DMBT)_{18}(C_{60})]$ as well as the aggregates, obtained from DFT; (C) Projection of the crystal structure of $[Au_7(PPh_3)_7]C_{60} \cdot THF$ and $Au_8(PPh_3)_8 \cap (C_{60})_2$ (view along [100] at 230 K and along [001] at 100 K). *(From P. Chakraborty, A. Nag, G. Paramasivam, G. Natarajan, T. Pradeep, Fullerene-Functionalized Monolayer-Protected Silver Clusters: [Ag29(BDT)12(C60)n]3- (n = 1–9). ACS Nano. 12 (3) (2018) 2415–2425, https://doi.org/10.1021/acsnano.7b07759; P. Chakraborty, A. Nag, B. Mondal, E. Khatun, G. Paramasivam, T. Pradeep. Fullerene-Mediated Aggregation of M25(SR)18- (M = Ag, Au) Nanoclusters. J. Phys. Chem. C. 124 (27) (2020) 14891–14900, https://doi.org/10.1021/acs.jpcc.0c03383; M. Schulz-Dobrick, M. Jansen, Intercluster Compounds Consisting of Gold Clusters and Fullerides: [Au7(PPh3)7] C60·THF and [Au8(PPh3)8](C60)2. Angew. Chem. Int. Ed. 47 (12) (2008) 2256–2259, https://doi.org/10.1002/anie.200705373.)*

C_{60} guests linked with the $[Ag_{29}(BDT)_{12}]^{3-}$ host enhanced the thermal-stability of the cluster host. Like predicting supracompound's structure, the DFT calculations and molecular docking simulations was also performed to the rational the location of C_{60} molecule on the Ag_{29} nanoclusters surface and the interaction between $[Ag_{29}(BDT)_{12}]^{3-}$ and C_{60}. For the $\{[Ag_{29}(BDT)_{12}] \cap (C_{60})_4\}^{3-}$, the most plausible structure of it was that the four C_{60} located at corners of the $[Ag_{29}(BDT)_{12}]^{3-}$, forming a tetrahedral configuration. The π⋯π interaction as well as van der Waals interactions between C_{60} and the benzene rings of the BDT ligands of the $[Ag_{29}(BDT)_{12}]^{3-}$ led to this attachment. Furthermore, for the $[Ag_{29}(BDT)_{12}(TPP)_4]^{3-}$, the addition of the C_{60} failed to induce C_{60} to replace PPh_3, while the PPh_3 ligands were added to the solution of $\{[Ag_{29}(BDT)_{12}] \cap (C_{60})_n\}^{3-}$ (n=1–4), the anchored C_{60} molecules were replaced by the PPh_3, giving rise to the $Ag_{29}(SSR)_{12}(PPh_3)_4$. These combined results reaffirmed the binding sites (vertex sites) of the fullerenes to the

$[Ag_{29}(BDT)_{12}]^{3-}$. As the number of fullerene attachments onto the cluster surface increased, the structures of the conjugates ($\{[Ag_{29}(BDT)_{12}]\cap(C_{60})_n\}^{3-}$) has been investigated via the DFT calculations. The $\{[Ag_{29}(BDT)_{12}]\cap(C_{60})_8\}^{3-}$ was a distorted cube, where all the tetrahedral vertex and face positions are occupied by C_{60} molecules. When another C_{60} located between two C_{60} in the vertex and face, the $\{[Ag_{29}(BDT)_{12}]\cap(C_{60})_9\}^{3-}$ could be obtained. Except the C_{60}, another C_{70} also exhibited similar behavior, evidenced by ESI-MS spectrometry.

Following, another host-guest supracompound composed of $M_{25}(SR)_{18}^-$ (M=Ag, Au and -SR is a thiolate ligand) clusters nanoclusters and C_{60}/C_{70} molecules was reported [84]. Via supramolecular interaction with fullerenes, supracompounds such as $[\{M_{25}(SR)_{18}\}_n(C_{60})]^{n-}$ ($n=1$–5), $[\{M_{25}(SR)_{18}\}_n(C_{60})_{n-1}]^{n-}$ ($n=2$–5), and $[\{M_{25}(SR)_{18}\}_n(C_{60})_n]^{n-}$ ($n=1, 2, 3, ...,$ etc.) were formed, which were studied by electrospray ionization mass spectrometry. Similar supracompounds with C_{70} were also observed. DFT calculations and molecular docking simulations were used to predict the structure of the supracompounds. In the early reported fullerene adducts of $[Ag_{29}(BDT)_{12}]^{3-}$, multiple fullerenes were attached on the surface of a single cluster. While for the supracompounds of $[M_{25}(SR)_{18}]^-$ (M=Ag, Au), multiple fullerenes were not observed to be attached to a single cluster, but it is observed that fullerenes were connected to multiple $[M_{25}(SR)_{18}]^-$, causing aggregation, which can be clearly observed in the ESI-MS spectra. Similarity, the fullerene functionalization aggregates also enhance the stability of $[M_{25}(SR)_{18}]^-$, which resulting the aggregates can be stored stably at 4°C, more than 7 days. The lowest energy structure of $[Ag_{25}(DMBT)_{18}(C_{60})]^-$ (Fig. 7.3B) showed that C_{60} was trapped in a cavity on the surface of the cluster, which was surrounded by six 2,4-DMBT ligands. Fig. 7.3B shows an expanded view of the interaction between the six DMBT ligands attached to the cluster surface and C_{60}. DFT calculations show that the main force to stabilize the $[Ag_{25}(DMBT)_{18}(C_{60})]^-$ is the vdW interaction and the C-H$\cdots\pi$ contact between the -H of the ligand -CH_3 group and the π system of C_{60}. The lowest energy geometry of $[(Ag_{25}(DMBT)_{18})_2(C_{60})]^{2-}$ shows that C_{60} interacts with the DMBT ligands of the two $Ag_{25}(DMBT)_{18}^-$. For the $[(Ag_{25}(DMBT)_{18})_3(C_{60})]^{3-}$, the lowest energy level structure shows that C_{60} was indeed encapsulated by three $Ag_{25}(DMBT)_{18}^-$. The structure of the heavier adduct $[(Ag_{25}(DMBT)_{18}^-)_n(C_{60})_{n-1}]^{n-}$ ($n=3$–4) obtained by molecular docking showed two isomer possibilities in each case. For the structures of $[(Ag_{25}(DMBT)_{18}^-)_n(C_{60})]^{n-}$ ($n=1$–3) and $[(Ag_{25}(DMBT)_{18}^-)_n(C_{60})_{n-1}]^{n-}$ ($n=2$–4), the C_{60}s were surrounded by clusters. However, the equivalent sites on the cluster surface are free and can be used to capture more fullerenes, thereby further binding more clusters. Therefore, these assemblies can be extended further, leading to the formation of larger aggregates of such cluster-fullerene complexes.

Intercluster compounds consisting of $Ag_7(PPh_3)_7/Au_8(PPh_3)_8$ nanoclusters and C_{60} molecules was reported by Schulz-Dobrick and Jansen [85]. For the $[Au_7(PPh_3)_7]C_{60}\cdot THF$, the crystal structure of gold nanocluster showed double layer and C_{60}s formed zigzag chains along [100]. The short-range attractive interactions such as the C-H-$\pi\cdots\pi$ contacts between the phenyl rings in the ligand periphery and the $\pi\cdots\pi$ interactions between fullerides contributed the arrangement of $[Au_7(PPh_3)_7]$ and C_{60} in the crystal structure. The crystal structure of $Au_8(PPh_3)_8\cap(C_{60})_2$ demonstrated a layer-by-layer configuration along the [100] axis, and a staggered configuration along the [001] axis (Fig. 7.3C). Differently, at 100K, it was found that the fullerene cage dimerizes through covalent bonds across short intermolecular contacts in the $Au_8(PPh_3)_8\cap(C_{60})_2$, which resulted from the electron-transport in such a cluster-fullerene nanosystem. However, in $[Au_7(PPh_3)_7]C_{60}\cdot THF$ structure, fullerenes were always separated. The long-range Coulomb interaction between clusters and fullerenes as one force to drive the $[Au_7(PPh_3)_7]C_{60}\cdot THF/Au_8(PPh_3)_8\cap(C_{60})_2$ formation, and the short-range intermolecular interactions between clusters and fullerenes as an important driving force to direct the arrangement of cocrystallization.

7.2.3 Supramolecular interactions of NCs as hosts with other molecules

In addition to these CDs and C_{60}s, several other types of molecules have been used as guests to modulate the structure or properties of host nanoclusters. The SR modified polyhedral oligomeric silsesquioxane (POSS) could serve as protected ligands to prepared two atomically precise Ag_{12} nanoclusters, $Ag_{12}(POSS)_6(CF_3COO)_6(C_4H_8O)_6$ (Ag_{12}-POSS-1, C_4H_8O=tetrahydrofuran) and $Ag_{12}(POSS)_6(CF_3COO)_6(C_3H_6O)_2$($Ag_{12}$-POSS-2, C_3H_6O=acetone), which were reported by Zang and coworkers. The configurations of Ag_{12}-POSS-1 and Ag_{12}-POSS-2 could be reversibly transformed by adding C_4H_8O or C_3H_6O solvents. The different capped ligands resulting in the Ag_{12} kernel in Ag_{12}-POSS-1showing a flattened cubooctahedral, whereas the Ag_{12} kernel in Ag_{12}-POSS-2 a normal cubo-octahedral. The POSS shell in Ag_{12}-POSS-1 was pseudooctahedral, which arranged to quasioctahedral when the cluster was transformed to Ag_{12}-POSS-2. A film matrix modified by Ag_{12}-POSS-1 or Ag_{12}-POSS-2 showed different hydrophobicities [86].

An $Ag_2Au_6(C-Ag_2Au_6(L)_6)$ nanocluster capped by an innermost C atom and x 2-(diphenylphosphino)-5-pyridinecarboxaldehyde ligands had been served as a host to absorb mono-amines guests based on the reacting of amines with the outward aldehyde groups on phosphine ligands. The interaction between the chiral amine- and cluster transferring

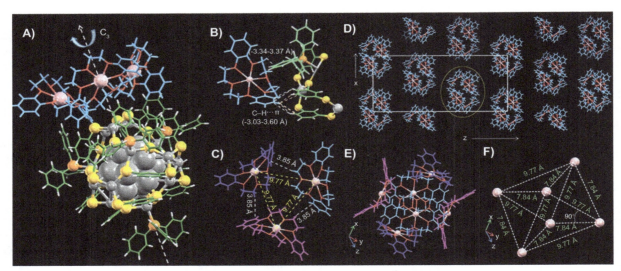

FIG. 7.4 Supramolecular host-guest interaction-based coassembly of [Ag$_{29}$(BDT)$_{12}$(TPP)$_4$]$^{3-}$ nanoclusters with crown ether (dibenzo-18-crown-6; DB$_{18}$C$_6$). (A) Structure of [Ag$_{29}$(BDT)$_{12}$(TPP)$_4$][(DB$_{18}$C$_6$Na)$_3$]. (B) Expanded view showing the interaction between one of the DB$_{18}$C$_6$Na$^+$ molecules and BDT/TPP ligands of the NC, and (C) expanded view of the three DB$_{18}$C$_6$Na$^+$ molecules attached on the NC surface. (D) Packing of DB$_{18}$C$_6$Na$^+$ molecules into hexameric units throughout the crystal lattice view from the y-axis. (E) Expanded view of one of the hexameric units showing the formation of cage-like structures. Opposite crown ethers, shown in similar colors, are related by a center of inversion. (F) The orientation of Na$^+$ of the crown ether hexamer in a rectangular bipyramidal geometry. *(From P. Chakraborty, A. Nag, K.S. Sugi, T. Ahuja, B. Varghese, T. Pradeep, Crystallization of a supramolecular coassembly of an atomically precise nanoparticle with a crown ether, ACS Mater. Lett. 1 (5) (2019) 534–540, https://doi.org/10.1021/acsmaterialslett.9b00352.)*

the chirality from the amine to the cluster can be observed, which resulted in the overall structure of the Ag$_2$Au$_6$ was chiral. The finding was further confirmed by X-ray structural determination and the circular dichroism (CD) signals of S-Ag$_2$Au$_6$ and R-Ag$_2$Au$_6$ nanoclusters. More significantly, the ee (enantiomeric excess) values of the introduced amines will affect the intensity of CD signals, making the C-Ag$_2$Au$_6$(L)$_6$ serving as a sensor for detecting the existence and the ee values of amines in solution. Three mono-amines including 1-phenylethylamine, 1-cyclohexylethylamine, and 2-aminooctane were used for checking the practicability of this Ag$_2$Au$_6$-based chirality detector. So, the Ag$_2$Au$_6$-based host-guest nanosystem constitutes an efficient approach for detecting the ee values of amine [87].

Experiments showed that [Ag$_{29}$(BDT)$_{12}$]$^{3-}$ was really a good host molecule, except that [Ag$_{29}$(BDT)$_{12}$]$^{3-}$ can assemble with CDs and C$_{60}$ to form a supramolecular host-guest nanosystem, another supramolecular host-guest interaction-based coassembly of [Ag$_{29}$(BDT)$_{12}$(TPP)$_4$]$^{3-}$ nanoclusters with crown ether (dibenzo-18-crown-6; DB$_{18}$C$_6$) was again reported by Pradeep group [88]. Coassembled [Ag$_{29}$(BDT)$_{12}$(TPP)$_4$][(DB$_{18}$C$_6$Na)$_3$] crystals was obtained via vapor-diffusion of methanol into the DMF solution of [Ag$_{29}$(BDT)$_{12}$(TPP)$_4$]$^{3-}$ and DB$_{18}$C$_6$ (Fig. 7.4). The electrostatic interaction between the cationic DB$_{18}$C$_6$Na$^+$ units and anionic [Ag$_{29}$(BDT)$_{12}$(TPP)$_4$]$^{3-}$ promoted the assembly. Further intra- and intermolecular C···H···π interactions between the crown ether and TPP/BDT resulted in long-range assembled NC crystals composed of hexameric DB$_{18}$C$_6$Na$^+$ units. Similar to other assembly changes the performance of the clusters, the assembled [Ag$_{29}$(BDT)$_{12}$(TPP)$_4$][(DB$_{18}$C$_6$Na)$_3$] crystals also showed enhanced PL (3.5-fold) compared with the [Ag$_{29}$(BDT)$_{12}$(TPP)$_4$]$^{3-}$. Considering the experimental results of the reaction of Ag$_{29}$ as the host and guest molecules (such as CD, C$_{60}$/C$_{70}$ and DB$_{18}$C$_6$), we can predict that there will be a great possibility of observing Ag$_{29}$ supramolecular assemblies in future studies.

7.2.4 Clusters as guests

As the host molecule, cluster compound can construct supramolecular assembly with other guest molecules, so cluster compound can also be used as guest molecule to bond with other large host molecules to construct supramolecular cluster compound. So, when nanoclusters serve as guests, they can be arranged onto the surface or into the cavity of nanowires, nanoparticles, and MOFs. Like the intermolecular interaction observed in cluster-host supramolecular assembly, the interactions between hosts and cluster-guests also induced by the hydrogen bonding, π-π/C-H...π interactions, chemical bonding, and soon.

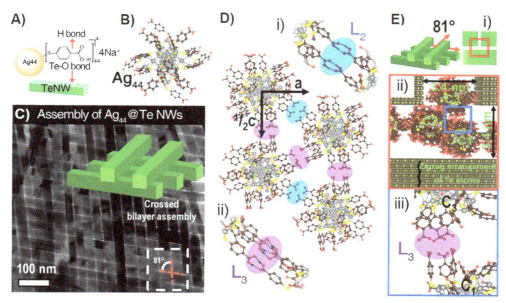

FIG. 7.5 Assembly of Ag$_{44}$@Te NWs. (A) Schematic of arranging Ag$_{44}$(SR)$_{30}$ nanoclusters onto Te NWs. (B) Crystal structure of the Ag$_{44}$(SR)$_{30}$ nanocluster. (C) TEM image of Ag$_{44}$@Te NWs. (D) H-bonding induced interactions between adjacent Ag$_{44}$(SR)$_{30}$ nanoclusters. (E) (i) Schematic of the crossed bilayer assembly of NWs. (ii) Expanded view of the region modeled in (i). (iii) Expanded view of the region modeled in (ii).C1, C2, and C3 are different Ag$_{44}$(SR)$_{30}$ clusters. *(From A. Som, I. Chakraborty, T.A. Maark, S. Bhat, T. Pradeep, Cluster-mediated crossed bilayer precision assemblies of 1D nanowires, Adv. Mater. 28 (14) (2016) 2827–2833, https://doi.org/10.1002/adma.201505775.)*

7.2.4.1 Nanowire@nanocluster

The water-soluble Ag$_{32}$(SG)$_{19}$ nanocluster as the guest clusters assembled onto the 1D tellurium nanowires (Te NWs) was reported by Pradeep and coworkers. The reaction of Ag$_{32}$(SG)$_{19}$ with Te NWs generated the formation of nodular growth at the surface of the Te NWs, giving rise to the final Ag nodule-decorated Te NWs. In contrast, the interactions between Ag$^+$ ions and Te NWs resulted in the formation of Ag$_2$Te NWs, showing that clusters are different from ions. Te NWs modified with hybrid silver nodules are sensitive to temperature, and their controlled solid solution annealing causes their morphology to change, and dumbbell-shaped Ag-Te-Ag NWs are formed. There was a difference in chemical reactivity between Te NWs decorated with silver nodules and dumbbell-shaped Ag-Te-Ag NWs [89].

In subsequent work, the assembled of the Te NWs with atomically precise Ag$_{44}$(S-PhCOOH)$_{30}$ nanoclusters was again reported by Pradeep and coworkers (Fig. 7.5) [90]. The dispersion of Ag$_{44}$@Te NWs in 1-butanol was spread over water, taken in a petri dish. As butanol evaporated, a freestanding assembly was formed on water. The Ag$_{44}$(S-PhCOOH)$_{30}$@ Te NWs showed a crossed-bilayer structure, whereas a monolayer assembly which Te NWs are parallel to each other was obtained without Ag$_{44}$(SR)$_{30}$ guests. TEM represented the crossed-bilayer Ag$_{44}$@Te NWs followed a woven-fabric-like configuration, wherein the Te NWs were parallel in the same layer but were arranged with an unusual angle of 81 degrees in adjacent layers. The Te-O interactions induced the Ag$_{44}$(SR)$_{30}$ nanoclusters to adsorb on the surface of the Te NWs. Of note, during the assembly process, changes in the concentration of Ag$_{44}$(S-PhCOOH)$_{30}$ concentration will cause changes in the assembly geometry. Based on these results and modeling using the crystal structure of the Ag$_{44}$(S-PhCOOH)$_{30}$, the different states of the aggregated Ag$_{44}$ nanoclusters were responsible for stabilizing the above woodpile-like structure. In the crystal lattice of Ag$_{44}$(S-PhCOOH)$_{30}$, two types of H-bonding between neighboring clusters could be observed due to the *p*-MBA ligands are present either in bundles of two (L$_2$) or three (L$_3$) units. Computational modeling revealed that the arrangement of the Te NWs between two layers with an angle of 81 degrees led to the strongest hydrogen bonding because of three pairs (L$_3$) of H-bonding between the Ag$_{44}$(SR)$_{30}$ nanoclusters coated over Te NWs. However, angles of 90 and 0 degrees allowed only either two or one pairs of H-bonds, respectively, which was detrimental to the thermal-stability of Ag$_{44}$@TeNWs. Further, another Au$_{102}$(S-PhCOOH)$_{44}$ nanoclusters were used to coat the Te NWs. similar crossed-bilayer structure with an angle of 77 degrees between the two adjacent layers was observed. The difference in geometrical arrangements of the ligands in two nanoclusters led to different angles of two layers arrangement. Overall, precise angular control between highly ordered Te NWs could be achieved by using atomically precise nanoclusters with directional H-bonding surface ligands.

7.2.4.2 Nanoparticle@nanocluster

In addition to Ag_{44}(S-PhCOOH)$_{30}$ can be adsorbed on the surface by Te tellurium nanowires, Ag_{44}(SPhCOOH)$_{30}$ nanoclusters being adsorbed onto the surface of gold nanorods has also been exploited which H-bonding affected the arrangement (Fig. 7.6) [91]. The nanoclusters served as guests for the encapsulation of Au nanorod hosts. The gold nanorods functionalized with HS-PhCOOH were incubated together with Ag_{44}(SPhCOOH)$_{30}$ nanoclusters in DMF. During, H-bonding were formed between the H atoms of adjacent S-PhCOOH ligands and both the Ag_{44} nanoclusters and gold nanorods, resulting in a multilayer shell encapsulating the individual nanorods, further causing the Au nanorods to transform into cage-like nanostructures. The morphology of the hybrid materials was octahedral, confirmed by TEM, scanning transmission electron microscopy (STEM), and 3D tomographic reconstructions. The morphology resulted from the preferable anchoring of the Ag_{44}(SR)$_{30}$ nanoclusters onto the four alternative <110> facets over the <100> facets of gold nanorods. The absorption spectrum of the Au nanorod@Ag_{44} hybrid material contains the peaks generated by the gold nanorods and Ag_{44}(SR)$_{30}$ nanometer clusters. This can be seen from the peak fitting, indicating that the constituent materials (Au nanorods and Ag_{44} nanoclusters). Clusters retained in the final hybrid nanostructure.

The Au_{102}(S-PhCOOH)$_{44}$ and Au_{250}(S-PhCOOH)$_n$ nanoclusters were also used as guests to enwrap Au nanorod-hosts [92]. Similar encapsulation of the gold nanorods by the introduced nanoclusters was also observed. It has been proposed that some of these S-PhCOOH stabilized Au nanoclusters would be deprotonated (i.e., into S-PhCOONa) in the assembly process, and thus not all the ligands of these two nanoclusters are involved in H-bonding, giving rise to the final hybrid structures with a less compact surface.

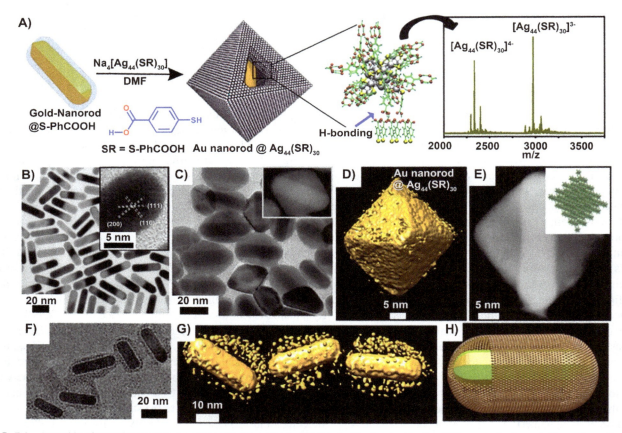

FIG. 7.6 Assembly of nanoclusters onto the plasmonic Au nanorods. (A) Schematic of arranging Ag_{44}(SR)$_{30}$ nanoclusters onto the surface of an Au nanorods, induced by the H bonding interactions. Right: crystal structure and ESI-MS spectrum of the Ag_{44}(SR)$_{30}$. (B) TEM image of S-PhCOOH stabilized Au nanorods. (C) TEM image of Au nanorod@Ag_{44}(SR)$_{30}$. (D and E) TEM tomographic and dark-field STEM images of Au nanorod@Ag_{44}(SR)$_{30}$. Inset: Theoretical model of the assembly of Ag_{44}(SR)$_{30}$ nanoclusters forming an octahedral shape. (F) TEM image of Au nanorod@Au_{250}. (G and H) 3D reconstructed structures and 3D graphical representation of Au nanorod@Au_{250}, respectively. *(From A. Chakraborty, A.C. Fernandez, A. Som, et al., Atomically precise nanocluster assemblies encapsulating plasmonic gold nanorods, Angew. Chem. Int. Ed. 57 (22) (2018) 6522–6526, https://doi.org/10.1002/anie.201802420.)*

7.2.4.3 MOFs@nanocluster

In MOFs, the arrangement of organic ligands and metal ions or clusters has obvious directionality, and can form different framework pore structures, thereby exhibiting different adsorption properties, optical properties, and electromagnetic properties. It has the advantages of high porosity, low density, large specific surface area, regular pores, adjustable pore size, diversity of topological structure and tailorability [93–97]. These properties can be used in fields including coal gas storage, chemical separation, light harvesting, and chemical sensing. MOFs contain numerous junction sites on the surface, or in the channels, making them potential hosts to anchor nanocluster guests.

A general strategy to integrate atomically precise $Au_{25}(SG)_{18}$ (SG=glutathione) nanoclusters with ZIF-8 was developed by Luo et al. (Fig. 7.7) [98]. Through the coordination-assisted self-assembly, $Au_{25}(SG)_{18}$ nanoclusters were uniformly encapsulated into a ZIF-8 framework, forming $Au_{25}(SG)_{18}$@ZIF-8 (Fig. 7.7A). In contrast, simple impregnation will only promote the $Au_{25}(SG)_{18}$ nanoclusters loading along the outer surface of ZIF-8 (termed as $Au_{25}(SG)_{18}$/ZIF-8, shown in Fig. 7.7B). The distribution of $Au_{25}(SG)_{18}$ nanoclusters in the two hybrid nanocomposites can be observed via STEM. N_2 absorption-desorption isothermal indicated the porous structure of ZIF-8 matrix remains intact after encapsulating $Au_{25}(SG)_{18}$ and $Au_{25}(SG)_{18}$ are not embedded into the pores of ZIF-8. Furthermore, accessibility of $Au_{25}(SG)_{18}$ in ZIF-8 matrix was evaluated by the 4-nitrophenol reduction reaction. The results indicated $Au_{25}(SG)_{18}$-ZIF-8 nanocomposites possess the chemical stability under the reaction condition. Further, the fluorescence of $Au_{25}(SG)_{18}$ retained in both nanocomposites. So, the hybrid nanocomposites retain the high porosity and thermal stability of the ZIF-8, while also exhibited the desired catalytic and optical properties derived from the integrated $Au_{25}(SG)_{18}$ nanoclusters. $Au_{25}(SG)_{18}$@ZIF-8 also contained inward-facing catalytic sites (from Au_{25}) as well as pores with a specific size (from ZIF-8), and thus is a promising, size selective heterogenous catalyst.

Recently, a new sandwich composites ZIF-8@Au_{25}@ZIF-67[tkn] and ZIF-8@Au_{25}@ZIF-8[tkn] [tkn=thickness of shell] was reported by Zhu and coworkers, which were also achieved by coordination-assisted self-assembly (Fig. 7.8) [99] Via physical impregnation, the nanomaterial Au_{25}/ZIF-8 was achieved by Au_{25}(L-Cys)$_{18}$ slowly pouring into the ZIF-8 solution, which was characterized by TEM and IR spectroscopy. With the aid of $Co(NO_3)_2$ and 2-MeIm, the

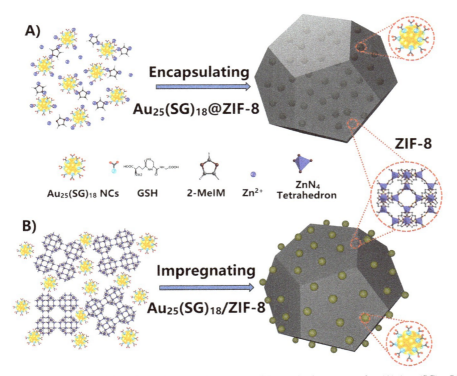

FIG. 7.7 Assembly of nanoclusters into/onto ZIF-8 MOFs. Schematic illustration of the synthesis processes for: (A) $Au_{25}(SG)_{18}$@ZIF-8 (i.e., assembling of $Au_{25}(SG)_{18}$ into ZIF-8 matrix) and (B) $Au_{25}(SG)18$@ZIF-8 (i.e., impregnating $Au_{25}(SG)_{18}$ on outer surface of ZIF-8). *(From Y. Luo, S. Fan, W. Yu, Z. Wu, D.A. Cullen, C. Liang, J. Shi, C. Su, Fabrication of $Au_{25}(SG)_{18}$-ZIF-8 nanocomposites: a facile strategy to position $Au_{25}(SG)_{18}$ nanoclusters inside and outside ZIF-8Adv, Adv. Mater. 30 (2018) 1704576.)*

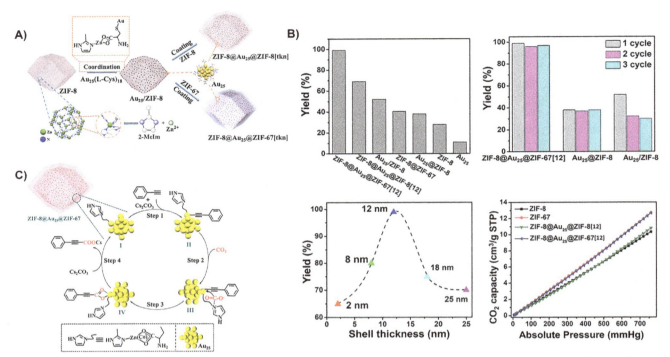

FIG. 7.8 A new sandwich assembly of ZIF-8@Au$_{25}$@ZIF-67[tkn] and ZIF-8@Au$_{25}$@ZIF-8[tkn]. (A) Synthetic route for the sandwich structures of ZIF-8@Au$_{25}$@ZIF-67[tkn] and ZIF-8@Au$_{25}$@ZIF-8[tkn] [tkn = thickness of shell]; (B) Catalytic activity of various catalysts for the carboxylation of phenylacetylene; (C) Proposed catalytic mechanism of the reaction between terminal alkynes and CO$_2$ over ZIF-8@Au$_{25}$@ZIF-67[tkn]. *(From Y. Yun, H. Sheng, K. Bao, L. Xu, Y. Zhang, D. Astruc, M. Zhu, Design and remarkable efficiency of the robust sandwich cluster composite nanocatalysts ZIF-8@Au$_{25}$@ZIF-67, J. Am. Chem. Soc. 142(9) (2020) 4126–4130, https://doi.org/10.1021/jacs.0c00378, with permission from the American Chemical Society, copyright 2020.)*

sandwich composite ZIF-8@Au$_{25}$@ZIF-67 was obtained by gentle agitation at room temperature for 24 h. Interestingly, the shell thickness of ZIF-8@Au$_{25}$@ZIF-67[tkn] was controlled at will by precisely adjusting the concentration of 2-MeImA and Co^{2+} solution. The ZIF-8@Au$_{25}$@ZIF-8[tkn] were also fabricated using the same strategy. ZIF-8@Au$_{25}$@ZIF-67[12] showed superior catalytic performance on the carboxylation using CO$_2$ of the terminal alkyne i.e., remarkable maintenance of 99% yield even after 3 cycles. Based on the experimental data and literature reports, a carboxylation mechanism is proposed. First, phenylene is deprotonated, forming intermediate II under the combined action of Cs$_2$CO$_3$ and Au$_{25}$ (confirmed by IR, Raman, and NMR), which is consistent with previous reports. Then CO$_2$ is captured (confirmed by TPD-CO$_2$) and activated by 2-MIem of the ZIF-67 shell to give intermediate III. Next, activated CO$_2$ inserts into the C≡C—Au bond to form intermediate IV. Finally, the products are released with regeneration of catalyst I. Further, compared with Au$_{25}$@ZIF-8 and Au$_{25}$/ZIF-8, ZIF-8@Au$_{25}$@ZIF-67[12] also exhibited superior catalytic performance on 4-nitrophenol(4-NP) reduction.

Encapsulation of the nanocluster guest into the MOF host can also be achieved through electrostatic attraction. Negatively charged [Au$_{12}$Ag$_{32}$(SR)$_{30}$]$^{4-}$, [Ag$_{44}$(SR)$_{30}$]$^{4-}$, and [Ag$_{12}$Cu$_{28}$(SR)$_{30}$]$^{4-}$ nanoclusters can form hybrid material with ZIF-8, ZIF-67 (ZIF-8 = Zn(mlm)$_2$, and ZIF-67 = Co(mlm), where mlm is 2-methylimidazole), and MHCF(Mn$_4$[Fe(CN)$_6$]$_{2.667}$·15.84H$_2$O) frameworks, while the positively charged [Au$_{24}$Ag$_{46}$(SR)$_{32}$]$^{2+}$ nanoclusters cannot, which indicated the matrix of the MOFs was prone to absorb negatively charged nanoclusters. The obtained cluster@MOFs hybrid materials were all structurally characterized by the FT-IR, UV-vis, XRD, STEM and TEM analyses. Among, the obtained Au$_{12}$Ag$_{32}$(SR)$_{30}$@ZIF-8 hybrid material showed excellent catalytic performance in capturing CO$_2$ and converting phenylacetylene into phenylpropiolate under mild conditions, encompassing the benefits of porous and molecular sieving behavior characterized by the MOF matrix together with the functional behavior characteristic of nanoclusters [100].

Mono-disperse gold nanoclusters in the pores of the MOFs with an in situ synthetic method was obtained by Zhu and coworkers. The obtained Au$_{11}$(PPh$_3$)$_8$Cl$_2$@ZIF-8 and Au$_{13}$Ag$_{12}$(PPh$_3$)$_{10}$Cl$_8$@MIL-101(Cr)(MIL-101(Cr) = [Cr$_3$F

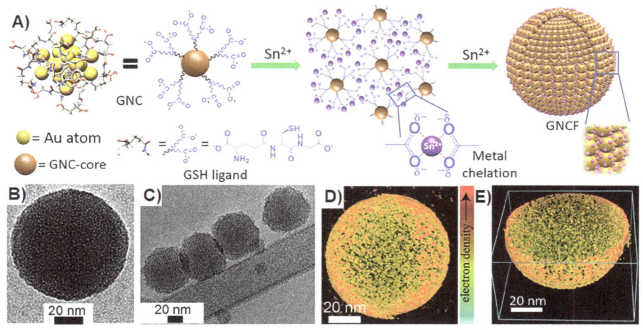

FIG. 7.9 Luminescent Au NC frameworks (GNCFs) with carboxylate anions on the NC surface. (A) Representation of a luminescent [Au@SG] NC, the possible interactions between the metal ions and COO- groups of the ligand over NCs through chelation, and the schematic illustration for the formation of GNCFs. (B) HR-TEM micrograph of GNCF-100 dispersed in water, (C) cryo-TEM micrograph of GNCF-100, (D) ET reconstruction of a GNCF-100, and (E) its cross-sectional view. *(From S. Chandra, G. Nonappa, Beaune, A. Som, S. Zhou, J. Lahtinen, H. Jiang, J.V.I. Timonen, O. Ikkala, R.H.A. Ras, Highly luminescent gold nanocluster frameworks, Adv. Opt. Mater. 7(20) (2019), https://doi.org/10.1002/adom.201900620, with permission.)*

($H_2O)_2O(bdc)_3$], where bdc is 1,4 benzenedicarboxylate) hybrid composites showed excellent performance in capturing the CO_2 and converting phenylacetylene into phenylpropiolate under mild conditions [101].

Liu et al. reported a cation exchange method, which successfully encapsulated the presynthesized $Au_{133}(SR)_{52}$ nanoclusters into a bMOF102/106 crystal(bMOF-102/106 = $Zn_8(ad)_4(linker)_6O_{24}$ cation, ad = adeninate, cation = dimethylammonium, linker = azobenzene 4,4'-dicarboxylate in bMOF-102, or = 20-nitro-1,1':4',1''-terphenyl 4,4''-dicarboxylate in bMOF-106). Of note, the cation exchange method required the H_2O_2 oxidant to pretreat Au nanoclusters in order to promote cation exchange-driven diffusion [102].

Chandra et al. reported the fabrication of highly luminescent Au NC frameworks (GNCFs) by coordinating divalent cations with carboxylate anions present on the NC surface (Fig. 7.9) [103]. [Au@SG] NCs were prepared by heating the presynthesized Au-thiolates at 70°C for 24h. Surprisingly, the addition of Sn^{2+} ions into the above prepared NCs had shown a sevenfold enhancement in their PL due to the coordination-assisted assembly. Interaction of negatively charged carboxylate group of SG with divalent Sn^{2+} cations resulted in the formation of GNCFs via inter-NC crosslinking, which was confirmed by observing 3D spherical colloidal structures in STEM and ET.

In general, this section reviews cluster-based host-guest nanosystems. In these nanosystems, nanoclusters can serve as hosts for absorbing small, functionalized guests, or as guests arranged on/in large hosts. Remarkably, conjugates tend to include the benefit of two components, which provides an effective method for producing cluster-based hybrid materials with desired functions.

7.3 Assembly in crystal lattice for assemble nanomaterial via supramolecular interactions

The determination of the crystal structure of the cluster is helpful to our understanding of the assembly method of the inter-crystal cluster. And the intercluster interactions play a crucial role in the orderly arrangement of nanoclusters in their crystal lattices. Similar to the assembled supramolecule constructed by the cluster compound and the above-mentioned summary

246 Metal nanocluster chemistry

of molecules, showing intermolecular forces like C-H⋯π, π⋯π, van der Waals, hydrogen bonding, and electrostatic interactions to strong covalent bonding, the forces guiding the organization of the nanoclusters may vary from weak intermolecular forces like them. We focus on the real combination of nanoclusters, for simply revealing (i) assembly modes of nanoclusters in crystal lattices, and (ii) atomically precise interactions between assembled nanoclusters from large-size nanoclusters to small-size nanoclusters.

Entire self-assembled structure of the $Au_{246}(SR)_{80}$ nanocluster (SR=S-Ph-tBu) at the atomic (packing of Au atoms), molecular (packing of surface ligands), and nanoscale (packing of nanoclusters) levels was successed by Jin group (Fig. 7.10) [104]. Surprisingly, $Au_{246}(SR)_{80}$ nanocluster are stacked into a lattice in the order of orientation, rotation and translation, which can be guided by the symmetry and density of the complex patterns constructed on the surface of the cluster by the peripheral ligands. A square configuration of $Au_{246}(SR)_{80}$ nanoclusters was observed, akin to the packing mode in the {100} plane of the fcc lattice (Fig. 7.10A). The interlocking (interaction) of surrounding ligands

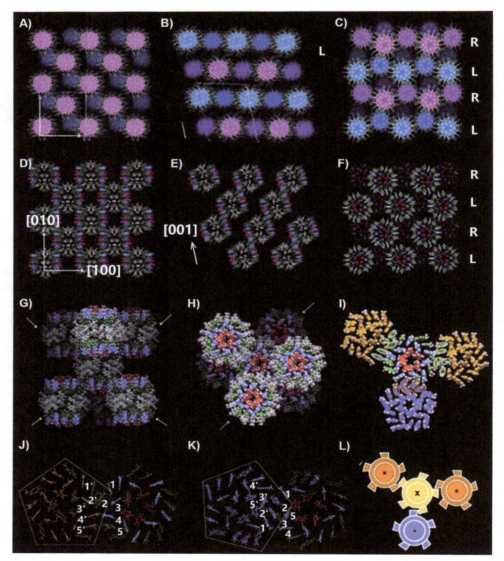

FIG. 7.10 Intercluster self-assembly and the symmetry of surface patterns of the $Au_{246}(SR)_{80}$ nanoclusters in crystal lattices. (A–C) Packing structure of the $Au_{246}(SR)_{80}$ in single crystals from different view. (D–F) Alignment of surface ligands among nanoclusters. (G–I) Coordination geometry of nanoclusters in the crystal lattice. (I) Contacting environment among the intercluster ligands. (J) Side-by-side stacking of the ligands in the nanoclusters with the same chirality. (K) Point-to-point stacking of the ligands in the nanoclusters with opposite chirality. (L) Scheme showing the directional packing of nanoclusters achieved through matching the symmetry of surface patterns. Au kernels in (D–L) are omitted for clarify. *(From C. Zeng, Y. Chen, K. Kirschbaum, K.J. Lambright, R. Jin, Emergence of hierarchical structural complexities in nanoparticles and their assembly, Science 354(6319) (2016) 1580–1584, https://doi.org/10.1126/science.aak9750.)*

on adjacent clusters results in a distance between clusters of 3.1 nm, which is less than 3.3 nm of the size of $Au_{246}(SR)_{80}$. Obliquely stacked was observed when along the [001] direction or [100] direction (Fig. 7.10B and C). The alignment of the surface ligands on the nanocluster affected inter cluster interactions, thereby affecting the accumulation mode (Fig. 7.10D-F). Of note, the $Au_{246}(SR)_{80}$ is organized with an ABAB pattern in crystal lattice, with A(R-isomer)#B(L-isomer)#A(R-isomer)#B(L-isomer). The intercluster assembly of the $Au_{246}(SR)_{80}$ nanocluster was further understand by insight into the coordination environment of the surface ligands (Fig. 7.10G-L). Each cluster has six adjacent clusters-four of them are arranged in the same square layer and have the same chirality as the center, while the other two in different layers have the opposite chirality (Fig. 7.10G and H). In the packing pattern, the van der Waals interaction is maximized, in favor of the compact packing in crystal lattices. From another perspective, surface symmetry is very important for the assembly of nanoclusters in the lattice. In order to stack these nanoclusters in an orderly manner, the peripheral ligands of the cluster interact with other ligands from three adjacent clusters, and five pairs of interacting ligands contribute to each interaction (Fig. 7.10I). In addition, for adjacent nanoclusters with the same chirality, the above-mentioned interacting ligands are arranged along the plane of symmetry (Fig. 7.10J, from 1–1' to 5–5'); in contrast, for the opposite chirality in adjacent nanoclusters, the interacting ligands are arranged along the center of symmetry and have opposite chirality (Fig. 7.10K, from 1–1' to 5–5'). This symmetrical matching strategy is reminiscent of mechanical gears, and each contact area is similar to a tooth of a gear (Fig. 7.10L).

Like the packing pattern of $Au_{246}(SR)_{80}$, the $Au_{133}(SR)_{52}$ (SR = S-C_2H_4Ph) and $Au_{144}(SR)_{60}$ (SR = S-CH_2Ph) nanoclusters were also assembled with an ABAB packing pattern in crystal lattices with A(R-isomer)#B(L-isomer)#A(R-isomer)#B(L-isomer) packing mode. The asymmetric arrangement of surface structures resulted in nanocluster isomerism [105,106]. Interactions are found at the interface of neighboring $Au_{144}(SR)_{60}$ nanoclusters. The R- and L-enantiomers are organized layer by layer with a square configuration or a hexagonal pattern from different view. In crystal lattices, each $Au_{144}(SR)_{60}$ nanocluster is wrapped by six $Au_{144}(SR)_{60}$ nanoclusters with the opposite chirality. The distance between clusters is smaller than the size of $Au_{144}(SR)_{60}$ nm clusters, indicating that adjacent nanoclusters are closely packed. This close packing affects the interaction between the ligands on adjacent nanoclusters -C-H...p and H...H interactions wrap a total of six thiolates, which act as "clamps" to fix adjacent nanoclusters, leading to the formation of ordered and compact lattice.

The self-assembled structure of the $[Au_{52}Cu_{72}(SR)_{55}]^+$ (SR = S-Ph-tBu) nanoalloy nanoclusters was resolved by Song et al. (Fig. 7.11) [107]. The $[Au_{52}Cu_{72}(SR)_{55}]^+$ exhibited a pentagonal shape (Fig. 7.11A), and consisted of the first Au_5Cu_2 core, the second Au_{47} core and the third shell- a Cu-thiolate cage. Similarly, the assembly pattern can be observed in the crystal lattices, shown in Fig. 7.11B. Fig. 7.11C and D shows the four nanoclusters from different views. Such interweaving interactions remind us of the "quadruple-gear" meshing mechanism, in which the rotation of any of the gears would lead to an integrated movement. However, the two nanoclusters at diagonal positions (green and light orange nanoclusters) also showed C-H···π interactions between their ligands, making the four gears interlock with each other. Specifically, two close nanoclusters in the edge-to-edge assembly show significant interactions between the ligands within and between the nanoclusters. Interestingly, the interactions between the South clusters (represented by the blue dashed line) all range from methyl H to the benzene ring, while the interactions between the South clusters (represented by the yellow dashed line) all originate from the benzene ring H to the benzene ring. In general, the six ligands belong to two nanoclusters, forming a triangular mosaic pattern, which is more compact than the herringbone pattern. The intercluster interaction between the benzene ring H and the benzene ring of two nanoclusters close to the tip is shown in Fig. 7.11E. As for the two nanoclusters approaching each other in a tip-to-edge manner, the intercluster and intra-cluster interactions can also be observed, and uniformly, the intra-cluster interaction is still from the benzene ring H to the benzene ring interaction and the clusters interact effect is due to the methyl H near the benzene ring. In addition, this interlocking pattern can be repeated for a long distance along the x-axis, but along the y-axis or z-axis, there is basically no interaction between nanoclusters belonging to different sets.

The structure of the $Au_{103}S_2(SR)_{41}$ (SR = 2-naphthalene-thiol) nanocluster displays complex surface patterns through intracluster ligand interactions, including both C-H···π and H···H interactions (Fig. 7.12) [108]. Intercluster ligand-ligand interactions give rise to the compact assembly in crystal lattice of this nanocluster (Fig. 7.12A). The peripheral ligands of neighboring nanoclusters are crosslinked with T-shaped C-H···π interactions that account for the intercluster assembly. Such an orderly T-shape arrangement gives rise to a "herringbone pattern," ubiquitous in crystal structures of polycyclic aromatic hydrocarbons (Fig. 7.12B). Fig. 7.12C depicts a full picture of the cluster assembly in crystal lattices, and the intercluster interactions are highlighted by white dotted circle. When the interface of two adjacent nanoclusters is magnified, four parts of thiolates are found enwound by both C-H···π interactions, which further fix the interface ligands and the neighboring nanoparticles (Fig. 7.12D). The $Au_{103}S_2(SR)_{41}$ was quite similar with that of $Au_{102}(p$-MBA$)_{44}$, however, the assemble mode was different [109].

Osman M. Bakr and coworkers reported a high-nuclearity copper nanocluster formulated as $[Cu_{81}(PhS)_{46}(tBuNH_2)_{10}(H)_{32}]^{3+}$, which has an irregular shape (Fig. 7.13) [110]. Structurally, the Cu_{81} can be divided

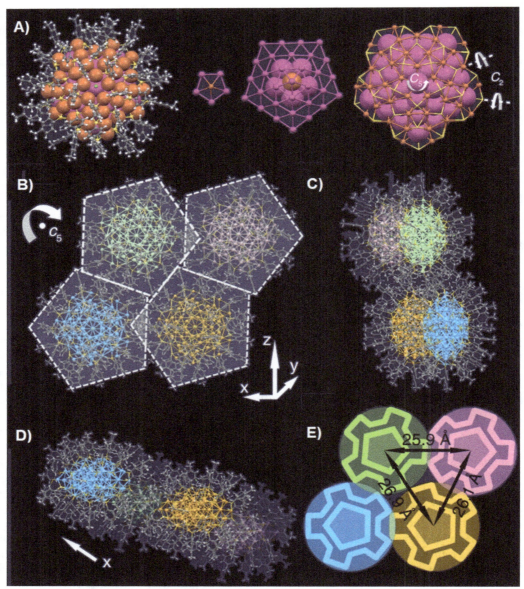

FIG. 7.11 The self-assembled structure of the $[Au_{52}Cu_{72}(SR)_{55}]^+$ (A) The structure of $[Cu_{72}Au_{52}(p\text{-}MBT)_{55}]^+$. (B–E) The assembly of $[Cu_{72}Au_{52}(SR)_{55}]+$. (B) Four nearest pentagonal $[Cu_{72}Au_{52}(SR)_{55}]^+$ particles with tip-to-edge, edge-to-edge, and tip-to-tip interactions; (C, D) top and side views of the four interacting particles; (E) the "quadruple-gear" interlock diagram demonstrating the interactions among the four nearest particles. *(From Y. Song, Y. Li, H. Li, F. Ke, J. Xiang, C. Zhou, P. Li, M. Zhu, R. Jin, Atomically resolved $Au_{52}Cu_{72}(SR)_{55}$ nanoalloy reveals marks decahedron truncation and penrose tiling surface, Nat. Commun. 11(1) (2020), https://doi.org/10.1038/s41467-020-14400-2.)*

two parts: the main body and the peripheral motifs. The six motifs are divided into three types: linear monomers, curved monomers, and cyclic trimmers, which are connected to the main structure at different angles. The main skeleton has three layers, a total of 71 Cu atoms, 32 PhS⁻ ligands and two tBuNH$_2$ ligands, the 30 Cu atoms and 20 PhS⁻ ligands in the top layer are regarded as a "moon bridge" constructed by one curved "bridge deck" and two M-shaped "piers." The 17 copper atoms in the middle layer interact through Cu⋯Cu bonding to form 20 triangles; the remaining 24 copper atoms and 12 ligands are tiled on the bottom layer. It is worth noting that although the top and bottom layers have similar compositions, the patterns they construct are obviously different. The research on the patterning rules of surface protection ligands in nanoclusters has attracted more and more attention, because it not only helps to determine the driving force and rules driving the self-organization of surface ligands, but also is important for understanding the structure and stability. The top layer is covered by 16 PhS⁻ ligands. The benzene ring on the ligand

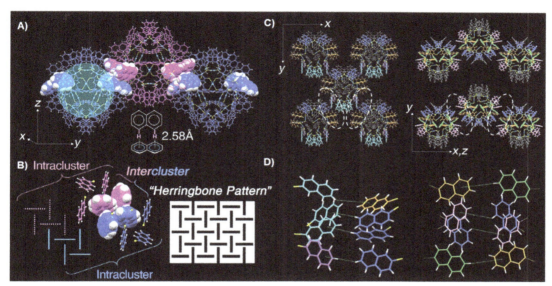

FIG. 7.12 Intercluster self-assembly of the $Au_{103}S_2(SR)_{41}$ nanocluster in crystal lattices. (A) Intercluster ligand-ligand interactions via C-H⋯p (average distance = 2.58 Å) in the crystal packing structure. (B) Combination of the intercluster and the intra-cluster interactions to form a "herringbone pattern." (C) Intercluster interactions between neighboring nanoclusters. (D) C-H⋯p interactions between ligands on adjacent nanoclusters. Only peripheral ligands are presented, and the Au kernels are omitted. *(From T. Higaki, C. Liu, M. Zhou, T.Y. Luo, N.L. Rosi, R. Jin, Tailoring the structure of 58-electron gold nanoclusters: $Au_{103}S_2(S-Nap)_{41}$ and its implications, J. Am. Chem. Soc. 139(29) (2017) 9994–10001, https://doi.org/10.1021/jacs.7b04678.)*

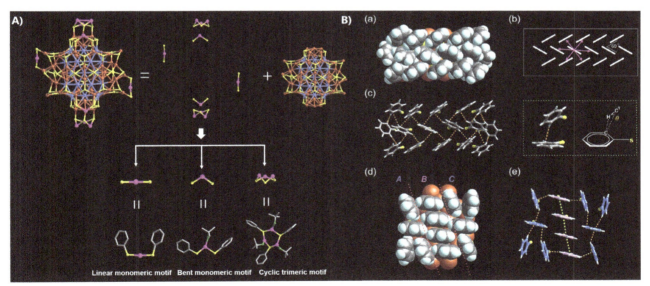

FIG. 7.13 The structure and intercluster self-assembly of Cu_{81}. (A) the total skeleton of Cu_{81} composed of the main body and epitaxial surface-protecting motifs. (B) intercluster self-assembly in crystallization state: (a) the arrangement of aromatic rings on the "bridge-deck" of the upper layer displayed in the space-filling model. (b) A schematic of the aromatic rings on the "bridge-deck" of the upper layer showing a herringbone like pattern. (c) The C-H⋯π interaction network *(orange dashed lines)* in the upper layer. The H-C_1-C angle is defined as θ, and θ should be smaller than 60 degrees. *Brown*: Cu; *yellow*: S; *light turquoise*: H. (d) The arrangement of aromatic rings on the bottom layer displayed in the space-filling model. (e) The C-H⋯π *(orange dashed lines)* and π⋯π *(yellow dashed lines)* interactions on the bottom layer. *Brown*: Cu; *yellow*: S; *blue/lavender*: C; *light turquoise*: H. *(From R.-W. Huang, J. Yin, C. Dong, A. Ghosh, M.J. Alhilaly, X. Dong, M.N. Hedhili, E. Abou-Hamad, B. Alamer, S. Nematulloev, Y. Han, O.F. Mohammed, O.M. Bakr, $[Cu_{81}(PhS)_{46}(tBuNH_2)_{10}(H)_{32}]^{3+}$ reveals the coexistence of large planar cores and hemispherical shells in high-nuclearity copper nanoclusters, J. Am. Chem. Soc. 142(19) (2020) 8696–8705, https://doi.org/10.1021/jacs.0c00541.)*

presents a chevron-like pattern, resulting in an average dihedral angle and centroid distance (between adjacent molecules) of 60 degrees and 4.92 Å, respectively. According to our observations, noncovalent interactions between clusters seem to play a crucial role in stabilizing the Cu 81 structure. The PhS$^-$ coverage on the top layer is relatively high, and they are connected to each other through weak C-H$\cdots\pi$ interactions on the limited Cu surface, and the whole surface is arched through self-assembly at a certain angle. Owing to the different weak interaction forces, the 12 benzene rings at the bottom are divided into three groups: ABC, A and C are arranged in opposite directions, adjacent benzene rings are assembled and connected to each other through C-H$\cdots\pi$, while the layer B four benzene rings are in a parallel displacement configuration. There are $\pi\cdots\pi$ interactions between adjacent aromatic rings, and they are connected to each other through the C-H$\cdots\pi$ and AB layer. The above results indicate that the self-organization behavior of the ligand on the surface of the nanocluster is not only affected by the shape of the ligand structure, but also by weak molecular interactions. In addition, in order to stabilize the structure, regardless of the shape of the surface layer, the protective ligand should cover as much surface as possible through various connection methods as much as possible to finally achieve equilibrium.

Sun and coworkers reported the atomic precisely silver nanocluster [Ag$_{78}$(^{i}PrPhS)$_{30}$(dppm)$_{10}$Cl$_{10}$]$^{4+}$ (dppm = bis-(diphenylphosphino)methane) synthesized through a one-pot reaction (Fig. 7.14) [111]. Importantly, the rhombic

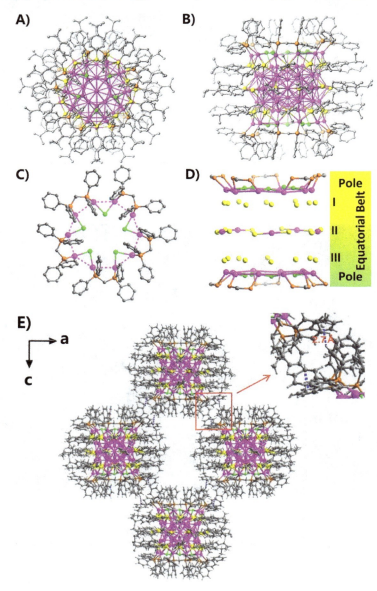

FIG. 7.14 The total molecular structure of [Ag$_{78}$(^{i}PrPhS)$_{30}$(dppm)$_{10}$Cl$_{10}$]$^{4+}$ and intercluster self-assembly (A–D) The total molecular structure of [Ag$_{78}$(^{i}PrPhS)$_{30}$(dppm)$_{10}$Cl$_{10}$]$^{4+}$. (E) The C-H$\cdots\pi$ interactions *(blue dashed lines)* between dppm ligands on adjacent clusters. *(From W.J. Zhang, Z. Liu, K.P. Song, C.M. Aikens, S.S. Zhang, Z. Wang, C.H. Tung, D. Sun, A 34-electron superatom Ag$_{78}$ cluster with regioselective ternary ligands shells and its 2D rhombic superlattice assembly, Angew. Chem. Int. Ed. 60(8) (2021) 4231–4237, https://doi.org/10.1002/anie.202013681.)*

superlattice assembled from [$Ag_{78}(^iPrPhS)_{30}(dppm)_{10}Cl_{10}$]$^{4+}$ through intercluster C-H$\cdots\pi$ interactions can be formed by a simple drop-casting treatment. Because of used ternary ligands, there are diverse metal-ligand interfacial reciprocity, such as S-Ag-S, Cl-Ag, and S-Ag. The cluster can be split into a $Ag_{13}@Ag_{40}$ core in the center and capped with a [$Ag_{25}(^iPrPhS)_{30}(dppm)_{10}Cl_{10}$] shell. The outermost shell has a $Ag_{10}(dppm)_5Cl_5$ ring and filled of CH_3CN molecule in the cavity, stabilized by C-H$\cdots$Cl hydrogen bonds. Furthermore, the cluster has two $dppmO_2$ molecules in the cell, and if no silver salt in the synthesis process, then no $dppmO_2$. Which determined in the presence of Ag (I) ion, the dppm can be situ oxidized by O_2. Ag_{78} shows high stability in dichloromethane solution, this was demonstrated by high-angle annular dark-field scanning transmission electron microscopy, the cluster-size (≈ 1.16 nm) are similar to the metallic core size (1.1 nm), this tight arrangement makes the clusters very stable. Besides, the cluster can form a rhombic superlattice by solvent-drying-induced assembly at room temperature, according to the results of SCXRD, there are C-H$\cdots\pi$ interactions between the dppm ligands and another neighboring cluster generated the 2D rhombic superlattice assembly. This 2D superlattice assembly rarely found in nanoclusters but commonly in the colloidal nanoparticles.

The two basic, well-known nanocluster stacking modes were found: ABAB stacking mode in $Au_{246}(SR)_{80}$, $Au_{144}(SR)_{60}$, $Au_{133}(SR)_{52}$); or ABCD-4H packaging mode in $Au_{92}(SR)_{44}$) [112,113]. When studying the crystalline sample of the $Au_{60}S_6(SR)_{36}$ cluster (SR = S-CH_2-Ph) [114], another different ABCDEF-6H packing mode was found by Gan et al. Along the (001) plane of the nanocluster, the clusters are evenly arranged, and each cluster is surrounded by six identical clusters with the same tropism. From another perspective, these six nanoclusters are arranged in a spiral; therefore, the collection of $Au_{60}S_6(SR)_{36}$ nanoclusters is called a 6H left-handed helix (6HLH) arrangement. In subsequent work, Gan et al. tailored the surface structure of $Au_{60}S_6(SR)_{36}$ and transformed it into $Au_{60}S_7(SR)_{36}$. Structurally, the 6HLH arrangement in the $Au_{60}S_6(SR)_{36}$ lattice was converted into the ABAB stacking mode in $Au_{60}S_7(SR)_{36}$, which both clusters are crystallized under the same conditions [115]. The contrast stacking of these lattices is reflected in their different optical properties: the PL intensity of $Au_{60}S_6(SR)_{36}$ in the 6HLH lattice is weaker than that of the amorphous state, while the relationship between $Au_{60}S_7(SR)_{36}$ nm clusters is opposite.

After the ABCDEF-6H packing mode, ACB#ACB packing mode was observed in the crystal lattices of $Ag_{46}S_7(SR)_{24}$ (SR = S-$PhMe_2$) [116]. Each Ag_{46} cluster contacts four adjacent nanoclusters. The intra-cluster C-H $\cdots \pi$ interaction occurs between adjacent ligands, resulting in an orderly arrangement of the ligands on the surface of the nanocluster. For the arrangement between clusters, hydrogen bonding determines the crystal packing of nanoclusters. These interactions (intra-cluster C-H $\cdots \pi$ and intercluster HH interactions) cause the $Ag_4S_1(SR)_4$ quadrilaterals on adjacent nanoclusters to be arranged in a mirror image, although all nanoclusters in the lattice are the same.

The well determined structure of the $Ag_{44}(S-PhCOOH)_{30}$ nanocluster by X-ray crystallography showed that the lattice contains two types of $Ag_{44}(SR)_{30}$ nm clusters (a and b), which are stacked in the fcc lattice by specular reflection symmetry (Fig. 7.15). Due to the mirror symmetry of a-b, the ACB layers are stacked in a sequence of 6 layers: Aa-Cb-Ba-Ab-Ca-Bb (Fig. 7.15A and B). Adjacent $Ag_{44}(SR)_{30}$ nanoclusters interact through a complete H-bond network formed between the carboxyl groups of the interface ligands where the thiol is bonded to the adjacent nanoclusters. The S-PhCOOH ligands are linked in two patterns: double-bundles and triple-bundles (Fig. 7.15C and D), which affected the assemble of host $Ag_{44}(S-PhCOOH)_{30}$ with other gust nanomaterial. The n-layer bonding between adjacent nanoclusters (aa or bb) involves the interface of double bond ligands, while the interaction between adjacent ab nanoclusters involves the interface of triple bond ligands (Fig. 7.15D). By using DFT calculations, the changes in the lattice arrangement of $Ag_{44}(S-PhCOOH)_{30}$ nm clusters in response to hydrostatic compression are also predicted. Fig. 7.15E and G describes the arrangement of nanoclusters under different hydrostatic pressure levels, where V_0 is the initial equilibrium volume and $V/V_0 = 0.71$ represents the compressed volume of the unit cell. As shown in the left Fig. 7.15E ($V/V_0 = 1$), the interaction of the double bundling is linear in both layers a and b. In contrast, when $V/V_0 = 0.71$, adjacent nanoclusters are close to each other; this disrupts the hydrogen bonds between adjacent nanoclusters: the double bond interaction still exists, but the linear cluster-SPhH-HPhS-cluster units rotate relative to each other (Fig. 7.15F and G). Both the linear mode and the rotation mode are described in Fig. 7.15F and G. The DFT calculation clearly shows that the rotation of the interaction between the clusters is driven by the compression-induced buckling of the ligand [117,118].

Ligands also affect the assembly mode of nanoclusters in the lattice. For example, the stacking of $Au_{12}Ag_{32}(SR)_{30}$ nanometer clusters depends on the nature of the thiol ligand: when SR = S-PhF or S-PhF_2, the space group is P-1; when SR = S-PhF or S-PhF_2 When, the steric group is P-1. However, when SR = S-$PhCF_3$, the space group is $P2_1/n$. It has been suggested that different stabilizing ligands will lead to different interactions between nanoclusters, leading to different aggregation methods of $Ag_{44}(SR)_{30}$ nanoclusters in the crystal lattice [119].

A racemic anisotropic [$Ag_{30}(C_2B_{10}H_9S_3)_8(Dppm)_6$] NCs, ($Ag_{30}$-rac) protected with a mixture of achiral carborane trithiolate and phosphine ligands ($C_2B_{10}H_9S_3$ = 8, 9,12-trimercapto-1,2-closo-carborane and Dppm = bis(diphenylphosphino)methane) was reported by Huang and coworkers (Fig. 7.16) [120]. It can be observed that the entire metal structure of [$Ag_{30}(C_2B_{10}H_9S_3)_8(Dppm)_6$] was a regular triangle Ag_3 surrounded by a helically arranged irregular Ag_{27} crown like

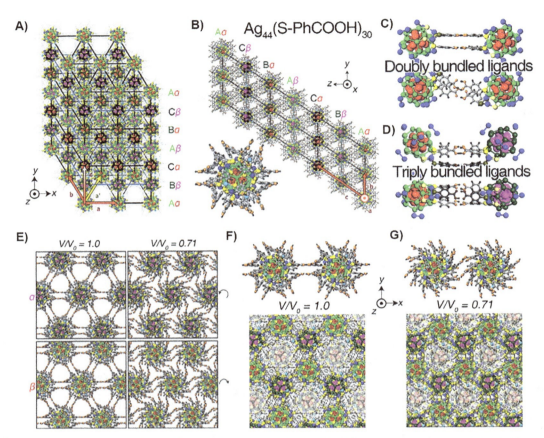

FIG. 7.15 Intercluster self-assembly of the Ag$_{44}$(SR)$_{30}$ nanocluster in crystal lattices (A and B) Interlayer and intra-layer H-bonded structures within 3D and 2D views. (C and D) Intercluster interactions with doubly bundled or triply bundled mode. (E) Arrangement of the nanoclusters in crystal lattices, viewed in the *a* and *b* plane; a and b are two different neighboring layers. Left: the configurations correspond to the $P = 0$ equilibrium state ($V/V_0 = 1$) (F); (G) right: the configurations correspond to the end state of the volume compression process ($V/V_0 = 0.71$). Left: intercluster H-bonding mode in crystal lattices with $V/V_0 = 1$; right: intercluster H-bonding mode in crystal lattices with $V/V_0 = 0.71$. Color codes: *red/green/blue*, Ag in different cluster sites; *yellow*, S; *gray*, C. (*From B. Yoon, W.D. Luedtke, R.N. Barnett, et al., Hydrogen-bonded structure and mechanical chiral response of a silver nanoparticle superlattice, Nat. Mater. 13 (8) (2014) 807–811, https://doi.org/10.1038/nmat3923.*)

motif. In the crystal lattices, it could be found that the B-H···π and C-H···π bonding interactions between the carborane cages and benzene rings directed the spiral arrangement of the ligands, inducing chirality for Ag$_{30}$-rac. During the crystallization of Ag$_{30}$-rac in dimethylacetamide, the race mates (R-Ag$_{30}$ and L-Ag$_{30}$) were self-resolved into the racemic conglomerates and the enantiomeric nanoclusters self-assembled into helical structures. The helical superstructures can be ascribe to the noncovalent interactions, including B-H···π, C-H···π, π···π, and van der Waals interactions.

C-H···π interaction driven crystallization of [Ag$_{29}$(BDT)$_{12}$(TPP)$_4$]$^{3-}$ NCs using the solvent evaporation technique was reported by AbdulHalim et al. (Fig. 7.17) [121]. The self-assembled macroscopic cubic lattice (C) crystals can be formed by evaporation of [Ag$_{29}$(BDT)$_{12}$(TPP)$_4$]$^{3-}$ DMF solution on a glass slide at room temperature (Fig. 7.17A–C). Later, the self-assembled [Ag$_{29}$(BDT)$_{12}$(TPP)$_4$]$^{3-}$ NC crystals with trigonal (T) lattices was observed by Nag et al. because of the solvent-dependent CH···π interaction of TPP molecules (Fig. 7.17D, E) [122]. Interestingly, the crystal packing of [Ag$_{29}$(BDT)$_{12}$(TPP)$_4$]$^{3-}$ NCs in trigonal (T) and cubic (C) lattices can be tuned by tailoring crystallization methods. Solvent evaporation of [Ag$_{29}$(BDT)$_{12}$(TPP)$_4$]$^{3-}$ NCs resulted in the formation of C crystals, vapor-diffusion of methanol into DMF solution of NCs produced T crystals. The intra- and intermolecular interactions between BDT and TPP molecules bound on the NC surface drive the self-assembly. In the NC internal interaction, the C-H bond of the TPP molecule interacts with the benzene ring of the BDT molecule bound to the same nanocluster. However, in the interaction between NCs, the C—H bond of BDT interacts with the benzene ring of the TPP ligand connected in the nearby NC. The main C-H···π interaction of TPP molecules during solvent evaporation leads to the formation of C crystals. Due to restricted molecular rotation and vibration, C lattice crystals show enhanced PL (Fig. 7.17F). The electronic coupling between nearby NCs in the C crystal causes the red shift of its PL spectrum [122].

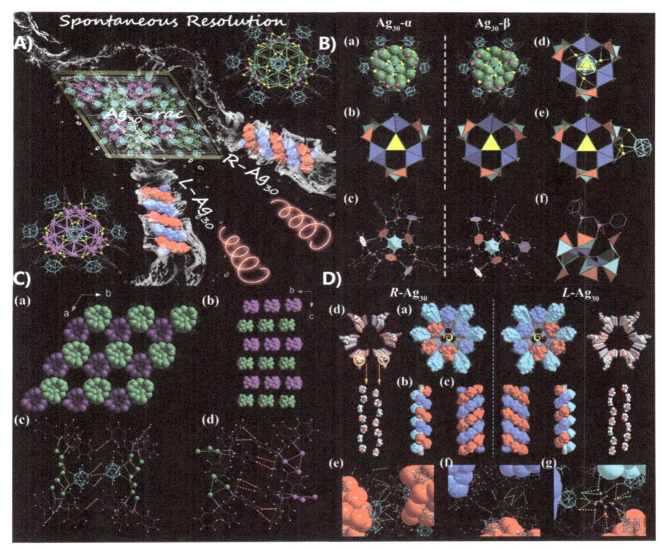

FIG. 7.16 The racemic anisotropic [Ag$_{30}$(C$_2$B$_{10}$H$_9$S$_3$)$_8$(Dppm)$_6$] and its assembly. (A) crystallization of Ag$_{30}$-rac in dimethylacetamide inducing enantiomeric nanoclusters self-assembled into helical structures; (B) Atomic structure of the cluster in Ag$_{30}$-rac; (C) Packing of enantiomers in the lattice of Ag$_{30}$-rac; (D) Packing mode analysis of the R-Ag$_{30}$ and L-Ag$_{30}$. *(From J.H. Huang, Z.Y. Wang, S.Q. Zang, T.C.W. Mak, Spontaneous resolution of chiral multi-thiolate-protected Ag$_{30}$ nanoclusters, ACS Cent. Sci. 6 (11) (2020) 1971–1976, https://doi.org/10.1021/acscentsci.0c01045.)*

The self-assembly of cationic and anionic Au-Ag bimetallic NCs was reported by the Wu and coworkers with the formula with [Ag$_{26}$Au(2-EBT)$_{18}$(TPP)$_6$]$^+$[Ag$_{24}$Au(2-EBT)$_{18}$]$^-$ NCs, which was called as a double nanocluster ion compound (DNIC) (Fig. 7.18). A single nanocluster ion compound (SNIC) [PPh$_4$]$^+$[Ag$_{24}$Au(2-EBT)$_{18}$]$^-$ was later synthesized [123]. Interestingly, strong electrostatic interactions and weak C-H···π/π···π interactions can be observed in DNIC, leading to cationic [Ag$_{26}$Au(2-EBT)$_{18}$(TPP)$_6$]$^+$ and anionic [Ag$_{24}$Au(2-EBT)$_{18}$]$^-$ NCs assemble in an alternating array of cationic and anionic stacking structure. SNIC displayed a k-vector-differential crystallographic arrangement (Fig. 7.18A–C). The significant difference in crystal arrangement was determined by the excitation effects of ligand and counterion. Later, another atomically precise nanoclusters double nanocluster ion compound [AuAg$_{24}$(S-c-C$_6$H$_{11}$)$_{18}$]$^-$ [Au$_2$Ag$_{41}$(S-c-C$_6$H$_{11}$)$_{26}$(Dppm)$_2$]$^+$ was reported by Zhu and coworkers. X-ray single crystal structure analysis revealed the cocrystallization of [AuAg$_{24}$(S-c-C$_6$H$_{11}$)$_{18}$]$^-$ [Au$_2$Ag$_{41}$(S-c-C$_6$H$_{11}$)$_{26}$(Dppm)$_2$]$^+$ also in a 1:1 M ratio, and their hierarchical assembly in the triclinic space group. Different the [Ag$_{26}$Au(2-EBT)$_{18}$(TPP)$_6$]$^+$[Ag$_{24}$Au(2-EBT)$_{18}$]$^-$, the firstly obtained [AuAg$_{24}$(S-c-C$_6$H$_{11}$)$_{18}$]$^-$ was partial converted into [Au$_2$Ag$_{41}$(S-c-C$_6$H$_{11}$)$_{26}$(Dppm)$_2$]$^+$, then form a eutectic assembly via strong electrostatic interactions [124].

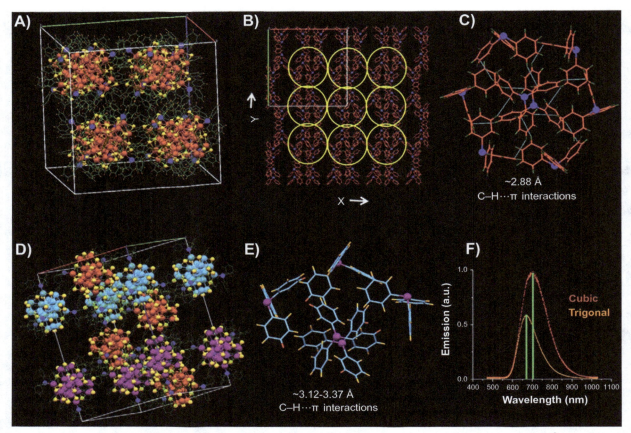

FIG. 7.17 C–H···π interaction driven crystallization of [Ag$_{29}$(BDT)$_{12}$(TPP)$_4$]$^{3-}$ NCs. (A) Cubic unit cell of [Ag$_{29}$(BDT)$_{12}$(TPP)$_4$]$^{3-}$ NCs. (B) Packing of TPP subunits viewed from the z-axis in the C system. (C) Strong C···H···π interactions (T-shaped) between the eight TPP subunits forming a hexagonal shape in the C system. (D) The trigonal unit cell of NCs. (E) C···H···π interactions between the TPP subunits in the T system. (F) Emission spectra obtained from single crystals of the C and T lattices. The C structure exhibits stronger and slightly red-shifted (≈30 nm) PL compared with the T structure. *(From A. Nag, P. Chakraborty, M. Bodiuzzaman, T. Ahuja, S. Antharjanam, T. Pradeep, Polymorphism of Ag$_{29}$(BDT)$_{12}$(TPP)$_4^{3-}$ cluster: interactions of secondary ligands and their effect on solid state luminescence. Nanoscale, 10.)*

Nanocluster assembly in crystal lattices can also be affected by the cocrystallized counter-ions (Fig. 7.19) [125]. [Au$_{23}$(S-c-C$_6$H$_{11}$R)$_{16}$]$^-$[TOA]$^+$ assembles with a layer-by-layer packing mode A(cluster)#B(TOA)#A(cluster)#B(TOA)) in crystal lattices (Fig. 7.19A) [126]. Ligand-exchanging this [Au$_{23}$(S-c-C$_6$H$_{11}$R)$_{16}$]$^-$[TOA]$^+$ nanocluster with Dppm ligands gives rise to an [Au$_{21}$(SR)$_{12}$(Dppm)$_2$]$^+$[AgCl$_2$]$^-$, representing a linear packing mode in crystal lattice (Fig. 7.19B). Furthermore, changing the [AgCl$_2$]-counter-ion of this Au$_{21}$ nanocluster into [Cl]$^-$, results in an [Au$_{21}$(SR)$_{12}$(Dppm)$_2$]$^+$[Cl]$^-$-nanocluster, which still exhibits a linear packing mode (Fig. 7.19C). The different packing modes of the [Au$_{23}$(S-c-C$_6$H$_{11}$R)$_{16}$]$^-$[TOA]$^+$, [Au$_{21}$(SR)$_{12}$(Dppm)$_2$]$^+$[AgCl$_2$]$^-$ and [Au$_{21}$(SR)$_{12}$(Dppm)$_2$]$^+$[Cl]$^-$ nanoclusters are proposed to result from their different surface structures and different counter-ions. The π···Cl···π interactions are observed on surfaces of both [Au$_{21}$(SR)$_{12}$(Dppm)$_2$]$^+$[AgCl$_2$]$^-$ and [Au$_{21}$(SR)$_{12}$(Dppm)$_2$]$^+$[Cl]$^-$ nanoclusters (Fig. 7.19D–F). Such interactions act as "hooks" to fix the adjacent nanoclusters. So, the linear arrangement of these Au$_{21}$ nanoclusters is driven by intercluster interactions induced by interacting the phenyl ligands on nanocluster surface and the surface hooks (i.e., [AgCl$_2$]$^{-1}$ or [Cl]$^{-1}$). However, the π···Cl···π interactions are different between [Au$_{21}$(SR)$_{12}$(Dppm)$_2$]$^+$[AgCl$_2$]$^-$ and [Au$_{21}$(SR)$_{12}$(Dppm)$_2$]$^+$[Cl]$^-$. The [AgCl$_2$]$^{-1}$ hook in [Au$_{21}$(SR)$_{12}$(Dppm)$_2$]$^+$[AgCl$_2$]$^-$ is larger and contains two Cl, resulting in a larger gap between two adjacent nanoclusters. Accordingly, the cluster packing in [Au$_{21}$(SR)$_{12}$(Dppm)$_2$]$^+$[Cl]$^-$ is more compact relative to that in [Au$_{21}$(SR)$_{12}$(Dppm)$_2$]$^+$[AgCl$_2$]$^-$ (Fig. 7.19F). Of note, modulation of the packing modes of Au$_{21}$ nanocluster crystals by tailoring the associated counter-ions dramatically influences the electrical transport properties of the corresponding self-assembled solids by two orders of magnitude. The assembly of the [Au$_{21}$(SR)$_{12}$(Dppm)$_2$]$^+$[Cl]$^-$ nanoclusters (with hooks) from 1D to 3D takes place in two stages (Fig. 7.19G): (i) from individual nanoclusters into 1D nanofibrils; and (ii) from 1D nanofibrils to 3D hierarchical crystals. In contrast, the Au$_{23}$ nanoclusters that contain no surface hooks show a 2D to 3D assembly behavior; in other words, no 1D nanofibrils are observed for in the assembly of the Au$_{23}$ nanocluster [126].

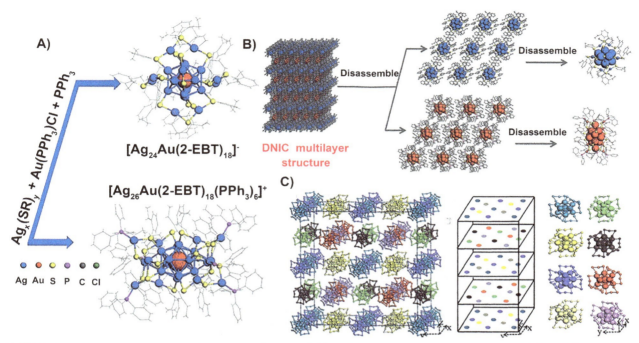

FIG. 7.18 The self-assembly of cationic and anionic Au-Ag bimetallic NCs. (A) X-ray crystal structure of $[Ag_{26}Au(2-EBT)_{18}(TPP)_6]^+[Ag_{24}Au(2-EBT)_{18}]^-$. (B) Illustration of the self-assembly of $[Ag_{26}Au(2-EBT)_{18}(TPP)_6]^+[Ag_{24}Au(2-EBT)_{18}]^-$ NCs. (C) Crystallographic arrangement of $[PPh_4]^+[Ag_{24}Au(2-EBT)_{18}]^-$ NCs. *(From L. He, Z. Gan, N. Xia, L. Liao, Z. Wu, Alternating array stacking of $Ag_{26}Au$ and $Ag_{24}Au$ nanoclusters, Angew. Chem. Int. Ed. 58 (29) (2019) 9897–9901, https://doi.org/10.1002/anie.201900831.)*

Li et al. observed that the symmetry of the kernel of the nanocluster can affect the symmetry of its crystal lattice [127]. Recently, the aurophilic interaction-induced colloidal (3D) self-assembly in $[Au_{25}(p\text{-MBA})_{18}]^-$ nanoclusters was reported by Xie and coworkers, which promoted to understand the structure-directional assembly and their enhanced PL (Fig. 7.20) [128]. $[Au_{25}(p\text{-MBA})_{18}]^-$ NCs was initially synthesized by chemical reduction of gold thioglycolate by adjusting the pH in the presence of $NaBH_4$. The cyclic dialysis process has been used to lower the pH of the NC solution, and finally the parent $[Au_{25}(p\text{-MBA})_{18}]^-$ is converted into a surface primitive with Au_{10} center and d^{10} electronic configuration to reconstruct nanoclusters. These surface reconstructed NCs consist of long $SR\text{-}[Au^I\text{-}SR]_x$ patterns ($x>2$) and small Au^0 cores. The overnight incubation of the above-mentioned NC solution resulted in the formation of self-assembled nanoribbons. Self-assembly is triggered by the conversion of the NC surface motif from short $SR\text{-}[Au^I\text{-}SR]_2$ units to long $SR\text{-}[Au^I\text{-}SR]_x$ ($x>2$) staples, accompanied by the internal structure of the Au_{13} core variety. The enhancement in the Au-S and Au_{staple}-Au_{staple} character revealing w the formation of long $SR\text{-}[Au^I\text{-}SR]_x$ was confirmed by the Fourier transform extended X-ray absorption fine-structure (FT-EXAFS). In the main stage of assembly, the rich Au^I content and the short aurophilic $Au^I \cdots Au^I$ interaction promoted the entry of nanoclusters into one-dimensional nanowires. The $\pi \cdots \pi$ stacking between p-MBA ligands from the adjacent nanowires, evidenced from the reduced internanowire spacing are responded to the formation of nanoribbons from nanowires. The assembled nanoribbons showed enhanced PL with a 6.2% quantum yield (at 77 K) and two independent emission bands (at 77 K) and temperature-responsive luminescence. A similar strategy to fabricate the self-assembled nanoribbons from $[Au_{38}(p\text{-MBA})_{24}]$ and $[Au_{44}(p\text{-MBA})_{26}]^{2-}$ NCs could also be observed.

$Au_{25}(SR)_{18}$ as the captain of the great nanocluster ship, whose geometrical/electronic structures and physical/chemical properties studied extensively can be used to construct 1D assembly nanomaterial through Au—Au bonds (Fig. 7.21). In 2014, Maran et al. fabricated a 1D CS composed of $[Au_{25}(SR)_{18}]^0$ nanoclusters. In the study, $[Au_{25}(S\text{-Bu})_{18}]^-$ was first synthesized, followed being oxidized into neutral $[Au_{25}(S\text{-Bu})_{18}]^0$ in open column packed with silica gel, and single crystals were grown by slow evaporation. The determined structure of $[Au_{25}(S\text{-Bu})_{18}]^0$ had almost the same framework structure as that of $[Au_{25}(PET)_{18}]^0$ or $[Au_{25}(S\text{-Et})_{18}]^0$. However, Au—Au bonds between adjacent $[Au_{25}(S\text{-Bu})_{18}]^0$ can be observed, while cannot be observed in other $[Au_{25}(PET)_{18}]^0$ or $[Au_{25}(S\text{-Et})_{18}]^0$ crystals. The difference indicated that $[Au_{25}(S\text{-Bu})_{18}]^0$ is a suitable structural unit to form 1D assemble nanomaterial. The Au-Au distance between adjacent

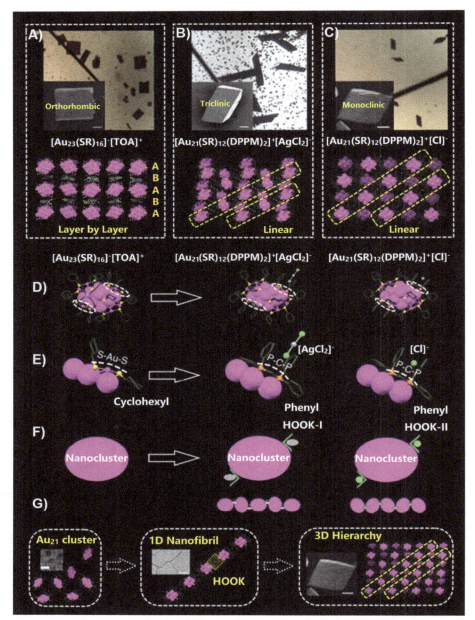

FIG. 7.19 Counter-ion effects on stacking nanoclusters in crystal lattices. Hierarchical 3D crystals of the (A) $[Au_{23}(SR)_{16}]^-[TOA]^+$, (B) $[Au_{21}(SR)_{12}(Dppm)_2]^+[AgCl_2]^-$, and (C) $[Au_{21}(SR)_{12}(Dppm)_2]^+[Cl]^-$ nanoclusters. Insets: SEM (scanning electron microscopic) micrographs. $[Au_{23}(SR)_{16}]^-[TOA]^+$ grows with a layer-by-layer pattern-A(cluster)#B(TOA)#A(cluster)#B(TOA). In comparison, linear packing (highlighted with *yellow* rectangles) is observed in $[Au_{21}(SR)_{12}(Dppm)_2]^+[AgCl_2]^-$ and $[Au_{21}(SR)_{12}(Dppm)_2]^+[Cl]^-$ nanoclusters. (D) Crystal structure of the three nanoclusters. (E) Different surface structures and counter-ions of the three nanoclusters. (F) Surface hooks on these nanoclusters, and the corresponding cluster-based line. (G) Schematic diagram of the assembly of the $[Au_{21}(SR)_{12}(Dppm)_2]^+[Cl]^-$-nanocluster from 1D to 3D. Color codes: *purple sphere*, Au; *yellow sphere*, S; *orange sphere*, P; *green sphere*, C. For clarity, the H atoms are omitted. *(From Q. Li, J.C. Russell, T.-Y. Luo, et al., Modulating the hierarchical fibrous assembly of Au nanoparticles with atomic precision, Nat. Commun. 9 (2018) 3871, https://doi.org/10.1038/s41467-018-06395-8.)*

$Au_{25}(S\text{-}Bu)_{18}]^0$ of 3.15 Å was within the range of aurophilic interactions (2.9–3.5 Å) and shorter than the nonbonding Au-Au distance (3.80 Å) estimated from the van der Waals radius of Au. To form such a 1D CS, it was considered that the repulsion between the ligands was suppressed and an attractive force between the ligands was induced because the adjacent NCs twisted and approached each other (twist-and-lock mechanism). Because S-Et has a short alkyl group that leads to a weak attractive force between ligands and PET with a bulky functional group has large steric repulsion between ligands 1D assemble nanomaterial cannot be observed in other $[Au_{25}(PET)_{18}]^0$ or $[Au_{25}(S\text{-}Et)_{18}]^0$ crystals. Another 1D

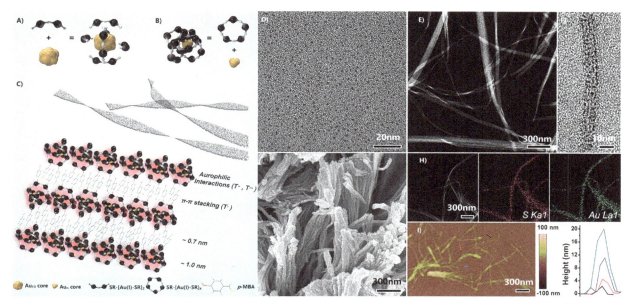

FIG. 7.20 Schematic shows the evolution of $[Au_{25}(p\text{-}MBA)_{18}]^-$ NCs into well-defined self-assembled nanoribbons. The structure of (A) $[Au_{25}(p\text{-}MBA)_{18}]^-$ NCs and (B) the as evoked smaller NCs, and (C) their spatial arrangement within nanoribbons. (D) TEM image of $[Au_{25}(p\text{-}MBA)_{18}]^-$ NCs, (E) high-angle dark-field scanning TEM, (F) high-resolution TEM, and (G) SEM images of the nanoribbons. (H) High-angle dark field scanning TEM (left), corresponding energy-dispersive X-ray (EDX) element maps (middle and right), and (I) AFM image of the representative nanoribbons are shown. *(From Z. Wu, Y. Du, J. Liu, Q.F. Yao, T.K. Chen, Y.T. Cao, H. Zhang, J.P. Xie, Aurophilic interactions in the self-assembly of gold nanoclusters into nanoribbons with enhanced luminescence, Angew. Chem. Int. Ed. 58 (24) (2019) 8139–8144, https://doi.org/10.1002/anie.201903584.)*

assemble nanomaterial of $[Au_{25}(S\text{-}Pen)_{18}]^0$ (S-Pen = pentanethiolate) was obtained in 2017. The distance between adjacent $[Au_{25}(S\text{-}Pen)_{18}]^0$ was shorter than that in the 1D $[Au_{25}(S\text{-}Bu)_{18}]^0$. In addition, in 2019, 1D CSs of $[Au_{24}Hg(S\text{-}Bu)_{18}]^0$ and $[Au_{24}Cd(S\text{-}Bu)_{18}]^0$ were also reported again [129–131].

Furthermore, electronic structures and physical properties of $[Au_{25}(S\text{-}Bu)_{18}]^0$ was changed when individual nanoclusters formed 1D. The $[Au_{25}(S\text{-}Bu)_{18}]^0$ exhibited paramagnetism in solution due to unpaired electrons. Conversely, the 1D assemble nanomaterial of $[Au_{25}(S\text{-}Bu)_{18}]^0$ was nonmagnetic. This change is mainly attributed to the formation of 1D assemble nanomaterial, resulting in close proximity of nanoclusters, allowing unpaired electrons of adjacent $[Au_{25}(S\text{-}Bu)_{18}]^0$ to form electron pairs. Due to the formation of such an electron pair, the conduction band of the 1D assemble nanomaterial was full and its valence band is empty. Therefore, the predicted 1D assemble nanomaterial has the characteristics of a semiconductor.

Via using $[Au_4Pt_2(SR)_8]^0$ as a template cluster, Negishi group conducted a detailed study on the factors responsible for the formation of 1D CSs via Au—Au bonds. A similar $[Au_4Pd_2(PET)_8]^0$, was reported by Wu and colleagues (Fig. 7.22). The SC-XRD of the series of $[Au_4Pt_2(SR)_8]^0$ crystals revealed the following three points for $[Au_4Pt_2(SR)_8]^0$: (i) $[Au_4Pt_2(SR)_8]^0$ is a metal NC that can become a structural unit of 1D assemble nanomaterial via Au—Au bond formation (Fig. 7.22A); (ii) although all $[Au_4Pt_2(SR)_8]^0$ NCs have similar structures, the intra-cluster ligand interactions vary depending on the ligand structure. As a result, the distribution of the ligands in $[Au_4Pt_2(SR)_8]^0$ changes depending on the ligand structure; (iii) the differences in the ligand distributions influence the intercluster ligand interactions, which in turn affect the formation of 1D assemble nanomaterial and change their structure (Fig. 7.22B). This study also explored the effects of 1D assemble nanomaterial formation on the electronic structure of NCs. The results revealed that the formation of the 1D assemble nanomaterial caused the band gap of the NCs to decrease (Fig. 7.22C and D) [132–135].

The 1D assembled nanomaterials based on the $(AuAg)_{34}(A\text{-}Adm)_{20}$ alloy NCs (A-Adm = 1-ethynyladamantane) and formed via Ag—Au—Ag bond was reported by Zheng and coworkers (Fig. 7.23) [136]. Solvent effect can control whether the cluster compound was presented in the form of monomer or assembly structure of $(AuAg)_{34}(A\text{-}Adm)_{20}$. Monomeric $(AuAg)_{34}(A\text{-}Adm)_{20}$ could be converted to 1D assembled nanomaterials $[(AuAg)_{34}(A\text{-}Adm)_{20}]_n$ by dissolving in an appropriate solvent. (Fig. 7.23A and B). Density functional theory (DFT) calculations was performed to study the electronic structure of $[(AuAg)_{34}(A\text{-}Adm)_{20}]_n$, showing $[(AuAg)_{34}(A\text{-}Adm)_{20}]_n$ have a band gap of about 1.3 eV (Fig. 7.23C). The field effect transistor (FET) manufactured with $[(AuAg)_{34}(A\text{-}Adm)_{20}]_n$ (Fig. 7.23D) shows highly anisotropic p-type semiconductor characteristics, and the conductivity in the polymer direction is about 1800 times than that of the cross

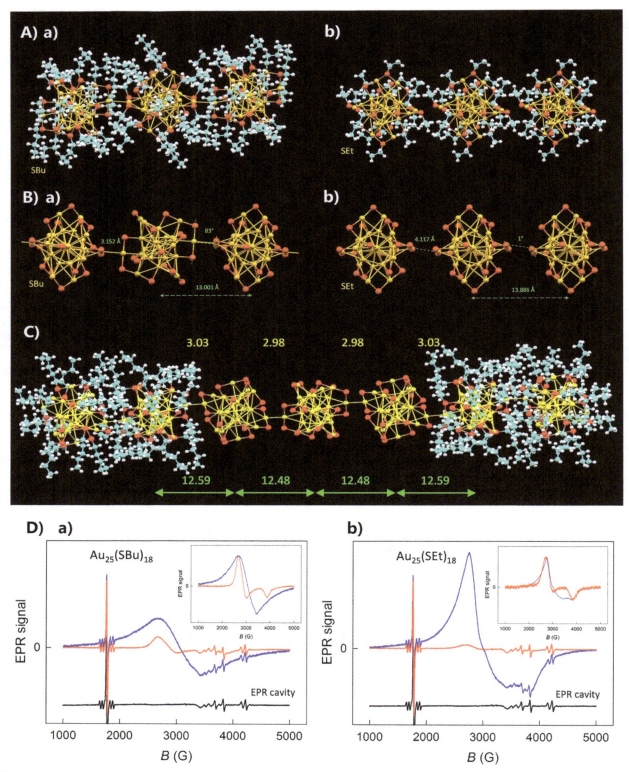

FIG. 7.21 Structure of $[Au_{25}(SR)_{18}]^0$ and interclusters assembly. (A and B) Crystal structures of (a) $[Au_{25}(S\text{-}tBu)_{18}]^0$ and (b) $[Au_{25}(S\text{-}Et)_{18}]^0$. In (B), R groups are omitted. (C) Crystal structure of $[Au_{25}(S\text{-}Pen)_{18}]^0$. (D) Comparison of the continuous wave-electron paramagnetic resonance (EPR) spectra of solid (*blue* traces) and frozen toluene solution (*red* traces) for (a) $[Au_{25}(S\text{-}tBu)_{18}]^0$ and (b) $[Au_{25}(S\text{-}Et)_{18}]^0$ at $-253°C$. The inset shows the same spectra with normalized peak intensity. The *black* curve corresponds to the EPR cavity signal, which is subtracted in the inset for clarity. All spectra were obtained by using the following parameters: microwave frequency = 9.733 GHz; microwave power = 150 μW; amplitude modulation = 1 G. (*From M. De Nardi, S. Antonello, D.E. Jiang, F.F. Pan, K. Rissanen, M. Ruzzi, A. Venzo, A. Zoleo, F. Maran, Gold nanowired: a linear $(Au_{25})_n$ polymer from Au_{25} molecular clusters, ACS Nano 8 (8) (2014) 8505–8512, doi:10.1021/nn5031143; S. Antonello, T. Dainese, F. Pan, K. Rissanen, F. Maran, Electrocrystallization of monolayer-protected gold clusters: opening the door to quality, quantity, and new structures, J. Am. Chem. Soc. 139 (11) (2017) 4168–4174, https://doi.org/10.1021/jacs.7b00568.*)

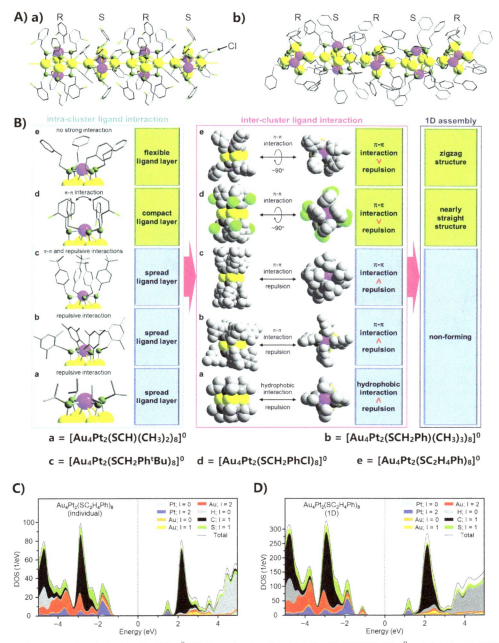

FIG. 7.22 1D CSs via Au—Au bonds for [Au$_4$Pt$_2$(SR)$_8$]^0 (A) Crystal unit cells of (a) [Au$_4$Pt$_2$(SCH$_2$PhCl)$_8$]^0 and (b) [Au$_4$Pt$_2$(PET)$_8$]$_0$. Au = *yellow*, Pt = *magenta*, S = *green*, Cl = *light green*, C = *gray*. R and S indicate two enantiomers in each NC. (B) Relationships between intra-cluster ligand interactions, which are related to the distribution of the ligands within each cluster, intercluster ligand interactions, and 1D assembly. Projected density of states of (C) an individual [Au$_4$Pt$_2$(PET)$_8$]^0 NC and (D) the 1D CS of [Au$_4$Pt$_2$(PET)$_8$]^0. *(From S. Hossain, Y. Imai, Y. Motohashi, Z.H. Chen, D. Suzuki, T. Suzuki, Y. Kataoka, M. Hirata, T. Ono, W. Kurashige, T. Kawawaki, T. Yamamoto, Y. Negishi, Understanding and designing one-dimensional assemblies of ligand-protected metal nanoclusters, Mater. Horiz. 7 (3) (2020) 796–803, https://doi.org/10.1039/c9mh01691k.)*

direction (Fig. 7.23E). It was interpreted that the conductivity and charge carrier mobility was increased by several orders of magnitude in [(AuAg)$_{34}$(A-Adm)$_{20}$]$_n$ via direct linking of the metal NCs by the -Ag-Au-Ag-chains in the crystal. This result provides hope for the further design of functional cluster-based materials with highly anisotropic semiconductor characteristics.

When silver nanoclusters consisted acetic acid ions (CH$_3$COO$^-$), trifluoroacetic acid ions (CF$_3$COO$^-$), or nitrate ions (NO$_3^-$) as one of ligand, Ag—O bonds was possible to connect Ag nanocluster forming nanomaterial (Fig. 7.24)

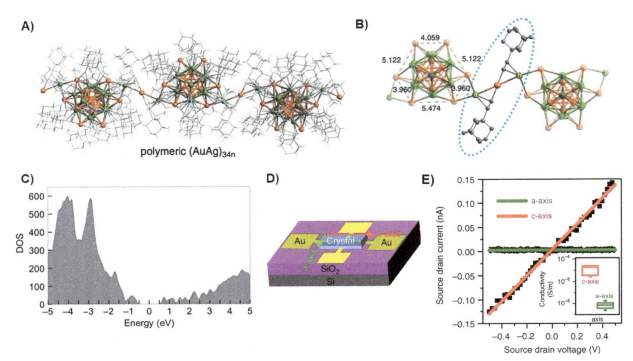

FIG. 7.23 The 1D assembled nanomaterials based on the $(AuAg)_{34}$(A-Adm)$_{20}$ alloy NCs (A) Structures of the cluster polymer (approximately orthogonal to the c-axis). (B)Au-Au distances in the distorted Au_6 hexagon and Ag-Ag distance in the "Ag-Au-Ag" unit of between alloy NCs. Au/Ag = *golden* and *green*, C = *gray*. All hydrogen atoms are omitted for clarity. (C) DFT-computed electronic density of states (DOS) of the cluster polymer crystal. Cluster model was used to build the periodic crystal, and the integration over the Brilloinzone was done in a $4 \times 4 \times 4$ Monkhorst-Packk-point mesh. The band gap is centered around zero. (D and E) Electrical transport properties of the cluster polymer crystals; (D) structure of the polymer crystal FET; (E) I-V plot of the polymer crystal along a-axis and c-axis, respectively, with the range of corresponding conductivity values shown in the inset. *(From P. Yuan, R. Zhang, E. Selenius, P. Ruan, Y. Yao, Y. Zhou, S. Malola, H. Häkkinen, B.K. Teo, Y. Cao, N. Zheng, Solvent-mediated assembly of atom-precise gold–silver nanoclusters to semiconducting one-dimensional materials, Nat. Commun. 11(1) (2020), https://doi.org/10.1038/s41467-020-16062-6.)*

[137–142]. A 1D assembled nanomaterials via Ag—O bonds was reported by Su et al., which based on the $Ag_{20}(CO_3)$(S-tBu)$_{10}$(CH$_3$COO)$_8$(DMF)$_2$ (CO_3^{2-} = carbonate anion; S-tBu = tert-butylthiolate, DMF = N,N-dimethylformamide). The 1D assembled nanomaterials was obtained by crystallization. In the crystal lattice, the Ag NCs were connected in one dimension via two Ag—O—Ag bonds (Fig. 7.24A and B). The obtained 1D assembled nanomaterials enhance the stability of $Ag_{20}(CO_3)$(S-tBu)$_{10}$(CH$_3$COO)$_8$(DMF)$_2$ in both solid and solution states and exhibited reversible thermochromic emission. Furthermore, the 2D assembled nanomaterials by changing the SR structure was formed by this group again. Formation of 1D assembled nanomaterials based on the similar method was reported by Mak and coworkers. The $Ag_{18}(CO_3)$(S-tBu)$_{10}$(NO$_3$)$_6$(DMF)$_4$ was linked by the formation of Ag—O bonds (Fig. 7.24C and D). Recently, Sun et al. synthesized an $Ag_{44}(V_{10}O_{28})$(S-Et)$_{20}$(PhSO$_3$)$_{18}$(H$_2$O)$_2$ ($V_{10}O_{28}^{6-}$, PhSO$_3^-$ = benzene sulfonic acid ion) which (POM)$V_{10}O_{28}^{6-}$ as an anion template. This 1D CS assembled nanomaterials was obtained due to two Ag—O bonds were formed between two PhSO$_3^-$ in the ligand layer and one $Ag_{44}(V_{10}O_{28})$(S-Et)$_{20}$ nanocluster (Fig. 7.24E and F). The 1D nanomaterial of $[Au_7Ag_9(Dppf)_3(CF_3COO)_7BF_4]_n$ (Dppf = 1,10-bis(diphenylphosphino)ferrocene, BF$_4$ = tetrafluoroboric acid) were reported by Wang and coworkers. Via Ag—O bond, each NC was also connected into 1D nanomaterial.

Crystallization induced emission enhancement (CIEE) was rarely reported in nanoclusters. Zhu and coworkers reported a novelty nanocluster with a formula of Au_4Ag_{13}(Dppm)$_3$(SR)$_9$ exhibiting CIEE (Dppm = bis(diphenylphosphino)methane, HSR = 2,5-dimethylbenzenethiol) (Fig. 7.25) [143]. The four Au-doped icosahedral Au_4Ag_9 kernel was surrounded by three AgS_2P, and one AgS_3 staple motifs (Fig. 7.25A). In the crystal lattices, each unit cell contained four pair of enantiomers, which can be classified into one clockwise and one anticlockwise structure according to rotation direction of the surface ligand. Due to the crystallization induced emission enhancement (CIEE) of the Au_4Ag_{13}, strong luminescence in crystalline state was observed, compared with the weak emission in the amorphous state and hardly any emission in solution (Fig. 7.25C). The triblade fan configuration benefits to restrict the vibrational/rotational movement during the electronic transitions and thus altering the optical properties. In the crystal lattice, the weak interactions C-H⋯π

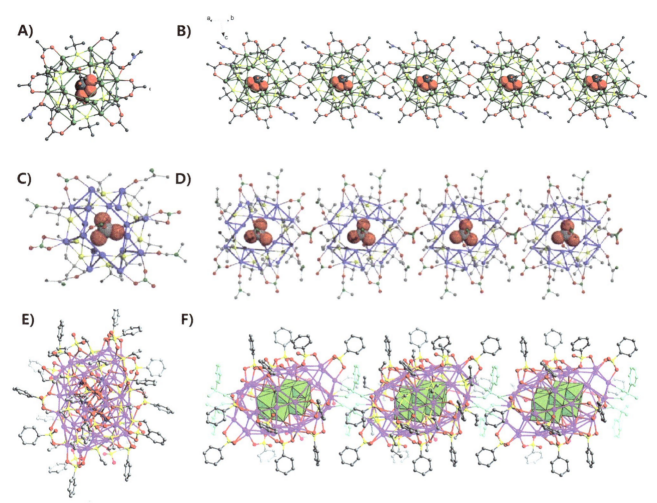

FIG. 7.24 A 1D assembled nanomaterials via Ag—O bonds (A) Structure of $Ag_{20}(CO_3)(S\text{-}tBu)_{10}(CH_3COO)_8(DMF)_2$, (B) Ball-and-stick view of the 1D chain of $Ag_{20}(CO_3)(S\text{-}tBu)_{10}(CH_3COO)_8(DMF)_2$, Ag = *green*, S = *yellow*, N = *blue*, O = *red*, C = *gray*.; (C) Structure of $Ag_{18}(CO_3)(S\text{-}tBu)_{10}(NO_3)_6(DMF)_4$. (D) Ball-and-stick view of the 1D chain of $Ag_{18}(CO_3)(S\text{-}tBu)_{10}(NO_3)_6(DMF)_4$. Ag = *blue*, S = *yellow*, O = *red*, C = *gray*, N = *green*; (E) Structure of $Ag_{44}(V_{10}O_{28})(S\text{-}Et)_{20}(PhSO_3)_{18}(H_2O)_2$; (F) 1D chain structure of $Ag_{44}(V_{10}O_{28})(S\text{-}Et)_{20}(PhSO_3)_{18}(H_2O)_2$ with all bridging $PhSO_3$-ligands highlighted in cyan and $V_{10}O_{28}^{6-}$ anions shown as *green* polyhedra. Ag = *purple*, V = *dark blue*, S = *yellow*, C = *gray*, O = *red*. All H atoms are omitted. *(From K. Zhou, C. Qin, X.L. Wang, K.Z. Shao, L.K. Yan, Z.M. Su, Unexpected 1D self-assembly of carbonate-templated sandwich-like macrocycle-based $Ag_{20}S_{10}$ luminescent nanoclusters, CrstEngComm 16 (34) (2014) 7860–7864, doi:10.1039/ c4ce00867g; Z.Y. Chen, D.Y.S. Tam, L.L.M. Zhang, T.C.W. Mak, Silver thiolate nano-sized molecular clusters and their supramolecular covalent frameworks: an approach toward pre-templated synthesis, Chem. Asian J. 12 (20) (2017) 2763–2769, doi:10.1002/asia.201701150; Z. Wang, Y.M. Sun, Q.P. Qu, et al., Enclosing classical polyoxometallates in silver nanoclusters, Nanoscale 11 (22) (2019) 10927–10931, https://doi.org/10.1039/c9nr04045e.)*

interactions between the hydrogen atom (para to the S atom) of the thiolate ligand and the centroid of the nearby phenyl ring of the dppm ligand are widely spread in the entire crystal structure (Fig. 7.25B). Structural analysis and density functional theory calculations show that the C-H...π interaction significantly limits the rotation and vibration in the molecule, thus greatly enhancing the radiation transition in the crystalline state. Due to the noncovalent interactions can be easily modulated via varying the chemical environments, the CIEE phenomenon might represent a general strategy to amplify the fluorescence from weakly (or even non-) emissive nanoclusters.

Following, crystallization-induced emission enhancement (CIEE) phenomenon was observed in $[Cu_{15}(PPh_3)_6(PET)_{13}]^{2+}$ (Fig. 7.26) [144]. The $[Cu_{15}(PPh_3)_6(PET)_{13}]^{2+}$ was composed of unconventional copper framework and twisted triangular antiprismatic Cu_6 core. In the crystal lattice, a chiral structure resembling a "tri-blade fan" configuration with clockwise and anticlockwise enantiomers was observed. Upon excitation at 473 nm, the crystalline state of $[Cu_{15}(PPh_3)_6(PET)_{13}]^{2+}$ displays intense PL in the NIR region with an emission at 720 nm, while the solution states of it showed weaker PL. Due to both enantiomers crystallize as a racemic mixture in the packing model, the location of the aromatic ligands on the cluster's surface effect

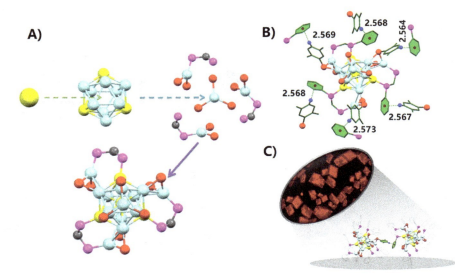

FIG. 7.25 The structure of Au$_4$Ag$_{13}$ and its C-H$\cdots\pi$ interactions. (A) The framework of the Au$_4$Ag$_{13}$ NC. Color labels: Ag, *light blue*; Au, *yellow*; P, *magenta*; S, *red*; C, *gray*; all other carbon (except the bridging carbon atoms between two P atoms in DPPM) and hydrogen atoms are omitted. (B) Distances between the hydrogen atom of one thiolate ligand and the centroid of nearby phenyl group of the DPPM ligand. (C) Optical microscopy image of the fluorescent Au$_4$Ag$_{13}$(Dppm)$_3$(SR)$_9$ NCs and the key C-H$\cdots\pi$ interactions. *(From T. Chen, S. Yang, J. Chai, Y. Song, J. Fan, B. Rao, H. Sheng, H. Yu, M. Zhu, Crystallization-induced emission enhancement: a novel fluorescent Au-Ag bimetallic nanocluster with precise atomic structure, Sci. Adv. 3(8) (2017), https://doi.org/10.1126/sciadv.1700956.)*

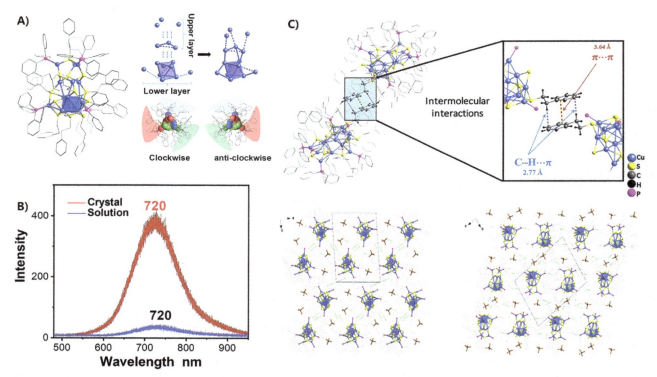

FIG. 7.26 The structure of [Cu$_{15}$(PPh$_3$)$_6$(PET)$_{13}$]$^{2+}$ and its intermolecular ligand interactions. (A) the structure of the [Cu$_{15}$(PPh$_3$)$_6$(PET)$_{13}$]$^{2+}$ containing one pair Cu$_{15}$ enantiomers; (B) PL spectrum of [Cu$_{15}$(PPh$_3$)$_6$(PET)$_{13}$]$^{2+}$ in crystalline and solution forms; (C) Intercluster π-π stacking between the benzene rings of PET ligands enhanced by C-H$\cdots\pi$ interactions and the packing of the Cu$_{15}$ along the b-direction (left in C) and along the a-direction (right in C). *(From S. Nematulloev, R. Huang, J. Yin, A. Shkurenko, C. Dong, A. Ghosh, B. Alamer, R. Naphade, M. Nejib, H.P. Maity, Cu$_{15}$(PPh3)$_6$(PET)$_{13}$]$^{2+}$: a copper Nanocluster with crystallization enhanced photoluminescence, Small, 17 (2021), https://doi.org/10.1002/smll.202006839.)*

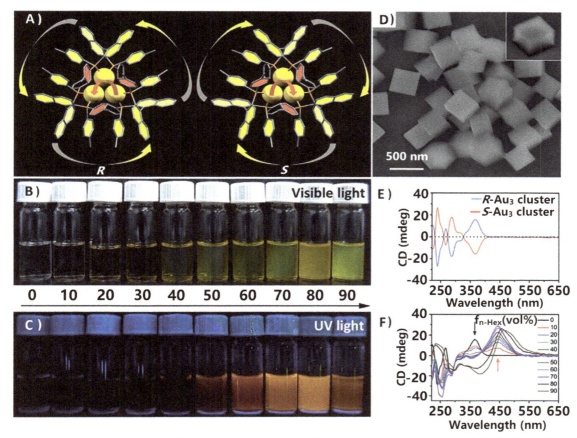

FIG. 7.27 The intermolecular interactions amplified the circularly polarized luminescence (CPL). (A) Chiral [Au$_3${(R)-Tol-BINAP}$_3$Cl] and [Au$_3${(S)-Tol-BINAP}$_3$Cl] NCs. Photographs of [Au$_3${(R)-Tol-BINAP}$_3$Cl] NCs in DCM with a different fraction of n-hexane under (B) visible light and (C) UV light. (D) SEM image of [Au$_3${(R)-Tol- BINAP}$_3$Cl] NC assembly fabricated in DCM with 70% n-hexane. (E) CD spectra of [Au$_3${(R)-Tol-BINAP}$_3$Cl] and [Au$_3${(S)-Tol-BINAP}$_3$Cl] NCs in DCM. (F) CD spectra of [Au$_3${(R)-Tol-BINAP}$_3$Cl] NCs in DCM with various n-hexane contents. *Arrows* show the direction of change with increasing n-hexane content. *(From L. Shi, L. Zhu, J. Guo, L.J. Zhang, Y.N. Shi, Y. Zhang, K. Hou, Y.L. Zheng, Y.F. Zhu, J.W. Lv, S. Q. Liu, Z.Y. Tang, Self-assembly of chiral gold clusters into crystalline nanocubes of exceptional optical activity, Angew. Chem. Int. Ed. 56 (48) (2017) 15397–15401, https://doi.org/10.1002/anie.201709827.)*

the optical properties. Weak intermolecular ligand interactions including C-H···π and π-π stacking resulted in centrosymmetric supramolecular dimer formation and limit the vibrational and rotational movement of ligands, enhancing the emission properties of [Cu$_{15}$(PPh$_3$)$_6$(PET)$_{13}$]$^{2+}$ [145].

The smallest chiral [Au$_3${(R)-Tol-BINAP}$_3$Cl] and [Au$_3${(S)-TolBINAP}$_3$Cl] NCs (BINAP=2,2′-bis(di-p-tolylphosphino)-1,1′-binaphthyl) were obtained firstly (Fig. 7.27). The drop-wise addition of antisolvent n-hexane into the nanocluster dichloromethane solution induced a strong intermolecular interaction among the ligands. An orange emission was observed from the NC solution when the concentration of n-hexane reaches 40%, indicates the formation of assembled body-centered cubic (BCC) nanocubes via strong C-H···π interactions. Due to the restricted movements of ligands, the intermolecular interactions amplified the circularly polarized luminescence (CPL) and chiral responses from the assembled superstructures [38,146].

7.4 Cluster-based metal-organic framework via linker

Metal-Organic Frameworks (MOFs) materials are porous materials with infinite network structure formed by coordination with metal ions or metal clusters as nodes and organic ligands as connectors. As a new type of functional molecular material, MOFs material has many advantages: large specific surface area, diverse pore structure, modifiable pore surface, adjustable structure and function, and numerous unsaturated metal sites. Therefore, it has good application prospects in the storage and separation of gases, the screening of mixed gases, luminescent materials and drug carriers, and heterogeneous catalysis [147–162]. The introduction of intercluster linker such as the counterion or organic molecular can benefit to assemble nanoclusters building blocks into cluster-based metal-organic-frameworks. These metal clusters have their own properties and these cluster-based assembled porous materials often exhibit properties different from assembly nanoclusters building blocks, showing enhanced properties, such as the PL, thermal stability, etc. In addition, the porosity of these nanomaterials can be used for molecular detecting, gas sensing and storage.

With the aid of different solvents and Cs^+ ions and with $Ag_{29}(BDT)_{12}$ as the node, Zhu and coworkers constructed a hierarchical assembly of $Ag_{29}(BDT)_{12}$ clusters into linear chains $[Cs_3Ag_{29}(BDT)_{12}(DMF)_x]_{2n}$ (Ag_{29}-1D, DMF = dimethylformamide), 2D grid networks $[Cs_3Ag_{29}(BDT)_{12}(NMP)_x]_n$, ($Ag_{29}$-2D; NMP = N-methyl-2-pyrrolidone), and 3D super structures $[Cs_2Ag_{29}(BDT)_{12}(TMS)_x]_{n-}$ (Ag_{29}-3D; TMS = tetramethylene sulfone) in which the Cs^+ cations are surrounded by solvent ligand molecules (Fig. 7.28) [147,148]. The interactions of cluster-Cs^+-solvent interaction (solvent = DMF,

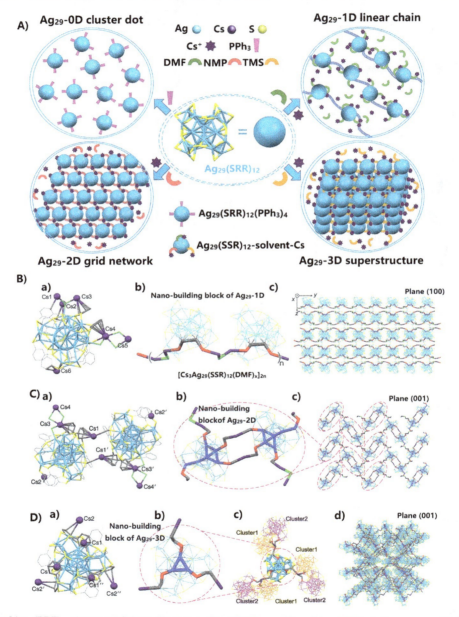

FIG. 7.28 Assembly of $Ag_{29}(BDT)_{12}$ clusters (A) Scheme illustration of the 1D-3D assemblies of $Ag_{29}(SSR)_{12}$ in the presence of Cs^+ and different solvents; (B) Crystal structures and crystalline packing modes of Ag_{29}-1D: (a) Crystal structure of Ag_{29}-1D. (b) Nanobuilding block of Ag_{29}-1D. (c) Packing of Ag_{29}-1D in the crystal lattice; (C); Crystal structure and crystalline packing mode of Ag_{29}-2D. (a) Crystal structure (nanobuilding block) of Ag_{29}-2D. (b) Is the enlargement of the circled section in (c). (c) Packing of Ag_{29}-2D grid network in the crystal lattice; (D) Crystal structure and crystalline packing mode of Ag_{29}-3D. (a) Crystal structure (nanobuilding block) of Ag_{29}-3D. (b) Is the enlargement of the circled section in (c). (c) Each Ag_{29}-3D cluster is surrounded by six adjacent Ag_{29}-3D nanoclusters. (d) Packing of the Ag_{29}-3D superstructure in the crystal lattice. Color codes: *light blue/blue/orange/magenta sphere/ stick*, Ag; *yellow/red sphere/stick*, S; *gray sphere/stick*, C; *dark purple sphere/stick*, Cs. For clarity, all H, O, N atoms, some C, Cs^+ atoms and TMS molecules are omitted. Each green atom (O) represents a TMS molecule. *(From X. Wei, X. Kang, Z. Zuo, F. Song, S. Wang, M. Zhu, Hierarchical structural complexity in atomically precise nanocluster frameworks, Natl. Sci. Rev. 8 (3) (2021), https://doi.org/10.1093/nsr/nwaa077.)*

NMP and TMS) assemble the anionic clusters into silver cluster-assembled material. In Ag_{29}-1D, three Cs^+ ions are located on the surface of $Ag_{29}(SSR)_{12}$ in five positions. Driven by a Cs—S bond, a Cs-DMF-Ag and a Cs···π interaction, the nanocluster molecules are assembled into one-dimensional cluster-based lines $[Cs_3Ag_{29}(BDT)_{12}(DMF)_x]_{2n}$ in the crystal lattice. For Ag_{29}-2D, two $Ag_{29}(SSR)_{12}$ compounds are connected by two Cs^+ cations through Cs-C. The Cs-π interactions, and the outward interactions from other four Cs^+ conjunction sites driven the intercluster assembly. For Ag_{29}-3D, each Ag_{29} cluster is surrounded by six adjacent cluster molecules, and each Cs^+ cation links two clusters. The different cluster-Cs^+-solvent coordination modes in Ag_{29}-1D, Ag_{29}-2D, Ag_{29}-3D as well as the $(Ag_{29}(SSR)_{12})(PPh_3)_4$ resulted in the difference of the surface structures and crystalline packing modes, influencing the optical absorptions and emissions of these Ag_{29}-based assemblies. The optical absorptions of these Ag_{29} nanoclusters in the solution (Ag_{29}-0D, Ag_{29}-1D in DMF; Ag_{29}-2D in NMP; Ag_{29}-3D in TMS) were very similar, but the PL intensities of Ag_{29}-0D, Ag_{29}-1D, and Ag_{29}-3D showed 1.7-, 2.1-, and 2.3-fold enhancement, respectively, relative to that of Ag_{29}-2D with the lowest PL intensity, reflecting both the structural effect and the solvent effect on nanocluster emissions. Furthermore, Ag_{29}-0D, Ag_{29}-1D, Ag_{29}-2D, and Ag_{29}-3D nanomaterials showed distinct surface area, exhibiting difference in the nitrogen adsorption-desorption tests. Due to the Ag_{29}-2D crystal possessing larger pore sizes, the values of the specific surface areas can reach $19 m^2/g$, more than that of Ag_{29}-0D, Ag_{29}-1D, and Ag_{29}-3D (about 6, 4, and $8 m^2/g$, respectively). The hierarchical assembly of $Ag_{29}(BDT)_{12}$ clusters will promote the superstructure with higher capacity of interactions.

Like using Cs^+ ions and solvent molecules to build assembly materials, Zhu and coworkers synthesized two types 3D assembled nanomaterials in which $[Au_1Ag_{22}(S-Adm)_{12}]^{3+}$ NCs (S-Adm = 1-adamantanethiolate) were used as nanocluster nodes and the linker was hexafluoroantimonateions (SbF_6^-) (Fig. 7.29) [149]. The $[Au_1Ag_{22}(S-Adm)_{12}]^{3+}$ consisted of an icosahedral Au_1Ag_{12} alloy core (Fig. 7.29A(a)) and a cage shell of $Ag_{10}(S-Adm)_{12}$. (Fig. 7.29A(b),(c)). The direction of rotation of $Ag_{10}(S-Adm)_{12}$ led to the $[Au_1Ag_{22}(S-Adm)_{12}]^{3+}$ NCs to form a pair of optical isomers. And the bonding method of SbF_6^- on the surface of the cluster determines the three-dimensional structure of the assembly material. For the 3D assembled nanomaterials based on the $[Au_1Ag_{22}(S-Adm)_{12}](SbF_6)_2Cl$ node, two SbF_6^- were connected to $[Au_1Ag_{22}(S-Adm)_{12}]^{3+}$ via an Ag_{u3}-F bond (Ag_{u3} atoms are connected to three SR group) to construct the 3D assembled nanomaterials (Fig. 7.29A(d–h)). And in $\{[Au_1Ag_{22}(S-Adm)_{12}](SbF_6)_2Cl\}_n$, same chirality $Au_1Ag_{22}(S-Adm)_{12}(SbF_6)_2Cl$ was connected, forming two separate assembly networks. Through interpenetrate of two assembly networks, 3D framework with channels of approximately 6.2 was achieved. For the 3D assembled nanomaterials based on the $[Au_1Ag_{22}(S-Adm)_{12}]^{3+}$, three SbF_6^- was connected to $[Au_1Ag_{22}(S-Adm)_{12}]^{3+}$ via Ag_{u2}-F bonds (Fig. 7.29B(a–c)). This structure only contained clockwise or counterclockwise optical isomers, different from that in $\{[Au_1Ag_{22}(S-Adm)_{12}](SbF_6)_2Cl\}_n$ assembled nanocluster (Fig. 7.29B(d)). As a result, the $\{[Au_1Ag_{22}(S-Adm)_{12}](SbF_6)_3\}_n$ had a larger pore diameter (15 Å, Fig. 7.29B(e)) than that of the $\{[Au_1Ag_{22}(S-Adm)_{12}](SbF_6)_2Cl\}_n$ assembled nanomaterials (6.2 Å). Both two types 3D assembled nanomaterials exhibit protic-solvent-sensitive photoluminescence (PL) in the solid state. The red PL in the presence of polar solvents such as CH_3OH, ethanol, and water will be disappeared after the solvent was evaporated. When assembled nanomaterials with solvent removed was added to the polar solvents again, the fluorescence will reappear, indicating two types 3D assembled nanomaterials can serving as sensors for polar solvents. Further, $\{[Au_1Ag_{22}(S-Adm)_{12}](SbF_6)_3\}_n$ composed of only the right-or left-handed enantiomer exhibited circularly polarized luminescence (CPL). The new approach to create new open frame materials with superatomic complexes will facilitate the further development of 3D frame materials for sensing and other applications.

Silver cluster-assembled material $[Ag_{27}S_2(tBuS)_{14}(CF_3COO)_8(TPyP-H_2)](CF_3COO)\}_n$ determined by single crystal X-ray crystallography was reported by Sun and coworkers (Fig. 7.30). The saddle-shaped Ag_{27} cluster was protected by 14 $tBuS^-$ and eight CF_3COO^- ligands, as well as templated by two enclosed S^{2-} ions. $TPyP-H_2$ ligand was used as the linker to construct the robust 2D metal-organic framework. Due to the stericity of the $TPyP-H_2$ ligand, four Ag atoms in the corner were used in conjunction with the pyridyl nitrogen atom, thereby reducing the stereochemical reaction and providing a beneficial function to amplify the interaction between the silver atom and the linker molecule. The obtained silver cluster-assembled material $[Ag_{27}S_2(tBuS)_{14}(CF_3COO)_8(TPyP-H_2)](CF_3COO)\}_n$ showed highly efficient heterogeneous catalytic actively for the cyclization of both terminal and internal propargylamine with CO_2 under atmospheric pressure. Density functional theory (DFT) calculations illustrated that the high catalytic activity and broad substrate scope were attributable to the saddle-shaped metallic node in Ag_{27}-MOF, which features an accessible platform with high-density silver atoms as π-Lewis acid sites for activating CC triple bonds [150].

In 2019, Zang and colleagues formed a 1D silver cluster-assembled material $\{[Ag_{18}(PhPO_3)(StBu)_{10}(CF3COO)_2(Ph-PO_3H)_4(bpy-NH_2)_2]\cdot(PhPO_3H_2)\}_n$ ($PhPO_3^{2-}$ = phenylphosphinic diion; $PhPO_3H^-$ = phenylphosphinic acid ion) NC nodes with bipyridine(3-amino-4,4'-bipyridine (bpy-NH_2, (Fig. 7.31A). The node $Ag_{18}(PhPO_3)(S-tBu)_{10}$NCs contained $PhPO_3^{2-}$ as an anion template in their center. Seven Ag_4S square pyramids and three Ag_3S tetrahedrons construct the $Ag_{18}S_{10}$ core featuring a sandwich-like structure via corner-sharing mode. The adjacent $Ag_{18}S_{10}$ nodes are connected by two bpy-NH_2 linkers through Ag—N coordination bonds. (Fig. 7.31B). The obtained 1D silver cluster-assembled material was stable up to 110°C in a nitrogen (N_2) atmosphere. Interestingly, mechanical stimulation will influence the luminescent of 1D silver cluster-assembled material, accompany with the emission color changing from blue to cyan (Fig. 7.31C).

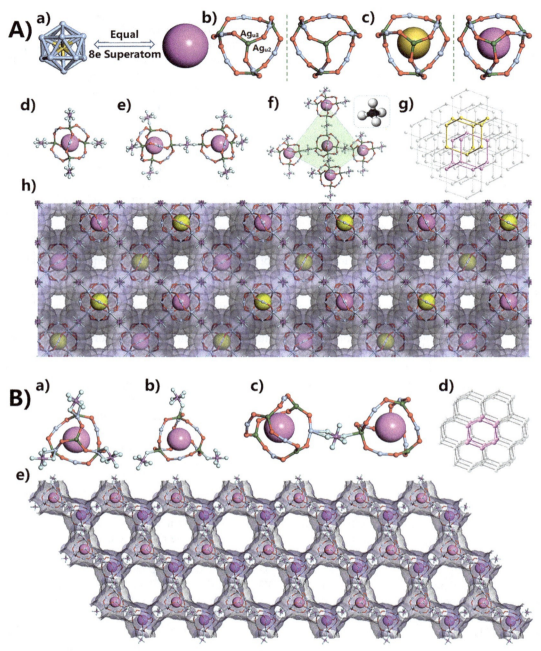

FIG. 7.29 3D assembled nanomaterials Based on [Au$_1$Ag$_{22}$(S-Adm)$_{12}$]$^{3+}$ nanoclusters (A) Structure of the [Au$_1$Ag$_{22}$(S-Adm)$_{12}$](SbF$_6$)$_2$Cl superatom complex and interpenetrating 3D channel framework assembled (a) Icosahedral Au$_1$Ag$_{12}$core, (b) cage-like Ag$_{10}$(SR)$_{12}$ complex shell, (c) a pair of Au$_1$Ag$_{22}$ isomers, (d) the connection of SbF$_6^-$ and alloy NCs, (e) two alloy NCs connected by SbF$_6^-$, (f) tetrahedral structure of NC monomers (the inset shows methane), (g) topology of the diamond-like structure, and (h) interconnected channels of Au$_1$Ag$_{22}$ along the z-axis. The left-and right-handed enantiomers in (c), (g), and (h) are highlighted in pink and yellow, respectively. (B) Crystal and channel structure of left-handed chiral 3D channel framework (C and H atoms are omitted for clarity). (a) The connection of Ag and SbF$_6^-$, (b) the connection of Ag and SbF$_6^-$, (c) two alloy NCs linked by SbF$_6^-$, (d) illustration of the hexagonal network structure, and (e) schematic of the large hexagonal channel structure. Note that the packing pattern of the right-handed chiral 3D channel framework was the same as that of the left-handed chiral 3D channel framework. Atoms are denoted in conventional colors: Au = *gold*, Ag in core and the Ag$_{\mu 2}$ motif = *pale blue*, Ag in the Ag$_{\mu 3}$ motif = *green*, S = *red*, F = *light turquoise*, and Sb = *purple*. (*From S. Chen, W. Du, C. Qin, et al., Assembly of the thiolated [Au$_1$Ag$_{22}$(SAdm)$_{12}$]$^{3+}$ superatom complex into a framework material through direct linkage by SbF$_6^-$ anions, Angew. Chem. Int. Ed. 59 (19) (2020) 7542–7547, https://doi.org/10.1002/anie.202000073.*)

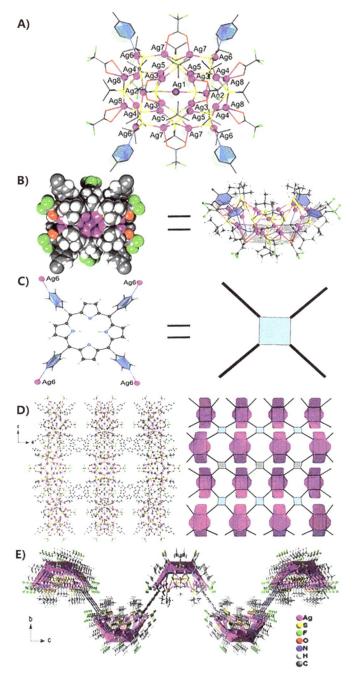

FIG. 7.30 Silver cluster-assembled material [Ag$_{27}$S$_2$(tBuS)$_{14}$(CF$_3$COO)$_8$(TPyP-H$_2$)](CF$_3$COO)}$_n$ (A) Structure of [Ag$_{27}$S$_2$(tBuS)$_{14}$(CF$_3$COO)$_8$(TPyP-H$_2$)] silver cluster. (B) Space-filling and ball-and-stick representations of silver cluster; (C) Coordination mode of the TPyP-H$_2$ ligand, acting as a 4-connected linker. (D and E) 2D structure along b and a directions, respectively, and the sql topology for [Ag$_{27}$S$_2$(t-BuS)$_{14}$(CF$_3$COO)$_8$(TPyP-H$_2$)](CF$_3$COO)}$_n$. The polyhedron *(purple)* represents the saddle-shaped silver cluster unit, and square node *(sky-blue)* represents the TPyP-H$_2$ ligand. *(From M. Zhao, S. Huang, Q. Fu, W.F. Li, R. Guo, Q.X. Yao, F.L. Wang, P. Cui, C.-H. Tung, D. Sun, Ambient chemical fixation of CO2 using a robust Ag$_{27}$ cluster-based two-dimensional metal-organic framework, Angew. Chem. Int. Ed. 59 (45) (2020) 20031–20036, https://doi.org/10.1002/anie.202007122.)*

And the PL can recovered when a mechanically stimulated one-dimensional CS sample is recrystallized (Fig. 7.31C), suggesting the PL of{[Ag$_{18}$(PhPO$_3$)(StBu)$_{10}$(CF$_3$COO)$_2$(PhPO$_3$H)$_4$(bpy-NH$_2$)$_2$]·(PhPO$_3$H$_2$)}$_n$ exhibited reversible mechanochromism. Because this assembled material emitted light at two wavelengths and its PL intensity ratio changed with temperature (thermochromism; Fig. 7.31D), suggesting that this assembled materials [Ag$_{18}$(PhPO$_3$)(StBu)$_{10}$(CF$_3$COO)$_2$(PhPO$_3$H)$_4$(bpy-NH$_2$)$_2$]·(PhPO$_3$H$_2$)}$_n$ could be applied as a thermometer [151].

In 2019, Bakr et al. reported a single Ag nanocluster and two silver cluster-assembled material via an organic linker, 4,4′-bipyridine (bpy) linkers (Fig. 7.32) [152]. With one-pot method, the single Ag$_{16}$Cl(S-tBu)$_8$(CF$_3$COO)$_7$(DMF)$_4$(H$_2$O) cluster is attained without organic linker (no bpy) in the synthesis process, which crystallizes as a discrete zero-dimensional structure. And when different ratio bpy linker was added during the synthesis, the 1-D silver cluster-assembled material {[Ag$_{15}$Cl(StBu)$_8$(CF$_3$COO)$_{5.67}$(NO$_3$)$_{0.33}$(bpy)$_2$(DMF)$_2$]·4.3-(DMF)·H$_2$O}$_n$ and 2-D assembled material {[Ag$_{14}$Cl(StBu)$_8$(CF$_3$COO)$_5$(bpy)$_2$(DMF)]$_2$(DMF)}$_n$ were obtained (Fig. 7.32A). As shown in Fig. 7.32B, distorted square

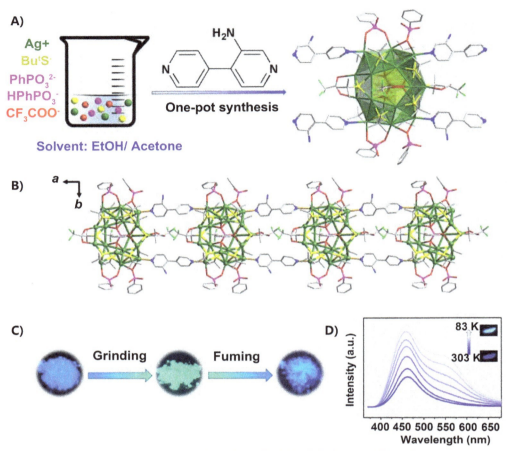

FIG. 7.31 1D silver cluster-assembled material{[Ag$_{18}$(PhPO$_3$)(StBu)$_{10}$(CF$_3$COO)$_2$(PhPO$_3$H)$_4$(bpy-NH$_2$)$_2$]·(PhPO$_3$H$_2$)}$_n$ (A) One-pot synthesis of [Ag$_{18}$(PhPO$_3$)(StBu)$_{10}$(CF$_3$COO)$_2$(PhPO$_3$H)$_4$(bpy-NH$_2$)$_2$]·(PhPO$_3$H$_2$). (B) 1D structure of {[Ag$_{18}$(PhPO$_3$)(StBu)$_{10}$(CF$_3$COO)$_2$(PhPO$_3$H)$_4$(bpy-NH$_2$)$_2$]·(PhPO$_3$H$_2$)}$_n$. Ag = *green*, S = *yellow*, C = *gray*, N = *blue*, O = *red*, F = *light green*, P = *purple*. H atoms are omitted for clarity. (C) Luminescent images of the as-synthesized, ground, and fumed [Ag$_{18}$(PhPO$_3$)(S-tBu)$_{10}$(CF$_3$COO)$_2$(PhPO$_3$H)$_4$(bpy-NH$_2$)$_2$] under ultraviolet light irradiation. (D) Temperature-dependent luminescence spectra of assembled material in the solid state. The inset is photo graphs. *(From X.H. Ma, J.Y. Wang, J.J. Guo, Z.Y. Wang, S.Q. Zang, Reversible wide-range tuneable luminescence of a dual-stimuli responsive silver cluster-assembled material, Chin. J. Chem. 37 (11) (2019) 1120–1124, https://doi.org/10.1002/cjoc201900314.)*

gyrobicupola Ag$_{16}$Cl core of Ag$_{16}$Cl(S-tBu)$_8$(CF$_3$COO)$_7$(DMF)$_4$(H$_2$O) nanocluster consisted of 8 triangular and 10 distorted square faces, in which each square face was capped by a tBuS-ligand (except for the top and bottom faces). When one Ag was lost of Ag$_{16}$Cl of Ag$_{16}$Cl(S-tBu)$_8$(CF$_3$COO)$_7$(DMF)$_4$(H$_2$O), the Ag$_{15}$Cl core was obtained. And the Ag$_{14}$Cl core can be achieved by one Ag was lost from the Ag$_{15}$Cl core of the Ag$_{15}$Cl(S-tBu)$_8$(CF$_3$COO)$_{5.67}$(NO$_3$)$_{0.33}$(DMF)$_2$ nanocluster again. When Ag$_{15}$Cl mode was connected with three adjacent Ag$_{15}$Cl nodes via four bpy molecules, the ladder-structure 1D {[Ag$_{15}$Cl(StBu)$_8$(CF$_3$COO)$_{5.67}$(NO$_3$)$_{0.33}$(bpy)$_2$(DMF)$_2$]·4.3-(DMF)·H$_2$O}$_n$ was obtained (Fig. 7.32C). Furthermore, the 2D assembled material with Ag$_{14}$Cl(S-tBu)$_8$(CF$_3$COO)$_5$(DMF) as a node was formed by changing the concentration of bpy during synthesis. Each Ag$_{14}$Cl(S-tBu)$_8$(CF$_3$COO)$_5$(DMF) cluster was also coordinated to four bpy linkers. Firstly, two clusters and two parallel bpy construct a double-cluster unit, which like the double-cluster unit in 1D assemble material. Further differences in cross-linking methods lead to different dimensions of assembly materials. When each dual cluster unit connected two other units in the same rows by a single bpy bridge, 1D assemble material was formed. While when each dual cluster unit was connected to four other units in the upper and lower rows by a single bpy bridge, 2D frame was obtained (Fig. 7.32D). Distinct characteristics in photoluminescence (PL) and thermal stability could be observed for one silver NC and two cluster-based assembled networks. 2D cluster-based assembled networks showed enhanced PL and higher thermal stability. Based on the results of a DFT calculation, it was interpreted that the enhancement of PL intensity was caused by a linker-to-cluster charge transfer excitation. This study will also pave the way for synthesizing new functional silver cluster-assembled material.

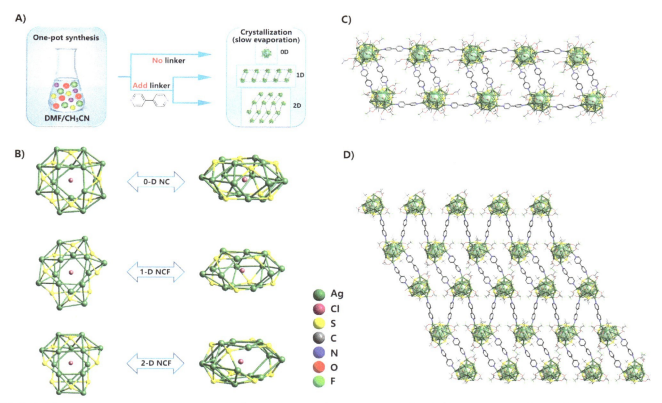

FIG. 7.32 The single Ag nanocluster and two silver cluster-assembled material via an organic linker, 4,4′-bipyridine(bpy) linkers (A) Synthesis of Ag NCs and NC-based frameworks. (B) Top views of the core structures of (a) Ag$_{16}$Cl(S-tBu)$_8$(CF$_3$COO)$_7$(DMF)$_4$(H$_2$O), (b) Ag$_{15}$Cl(S-tBu)$_8$(CF$_3$COO)$_{5.67}$(NO$_3$)$_{0.33}$(DMF)$_2$ and (c) Ag$_{14}$Cl(S-tBu)$_8$(CF$_3$COO)$_5$(DMF)(bpy)$_2$. (C) Crystal structure of the corresponding 1D CS. (D) Crystal structure of the corresponding 2D CS. Free (cocrystallized) DMF molecules are not shown. The green semitransparent spheres in the Ag clusters are shown as a visual guide. H atoms were omitted for clarity. *(From M.J. Alhilaly, R.W. Huang, R. Naphade, B. Alamer, M.N. Hedhili, A.-H. Emwas, P. Maity, J. Yin, A. Shkurenko, O. F. Mohammed, M. Eddaoudi, O.M. Bakr, Assembly of atomically precise silver nanoclusters into nanocluster-based frameworks, J. Am. Chem. Soc. 141 (24) (2019) 9585–9592, https://doi.org/10.1021/jacs.9b02486.)*

When functional 1,2-dithiolate-o-carborane was used as a ligand to capped the silver nanoclusters in the CH$_3$CN solution, the Ag$_{14}$(C$_2$B$_{10}$H$_{10}$S$_2$)$_6$(CH$_3$CN)$_8$·4CH$_3$CN was obtained, which the structure also had a face-centered cubic structure (octahedral Ag$_6^{4+}$ inner core and outer cubic Ag$_8^{8+}$ shell) similar to other Ag$_{14}$ clusters (Fig. 7.33). Due to easier dissociation of coordination solvent molecules (CH$_3$CN) molecules, Ag$_{14}$(C$_2$B$_{10}$H$_{10}$S$_2$)$_6$(CH$_3$CN)$_8$·4CH$_3$CN are not much stable (**NC-1**). Replacing CH$_3$CN with monodendate N-heteroaromatic ligands, ultra-stable clusters [Ag$_{14}$(C$_2$B$_{10}$H$_{10}$S$_2$)$_6$(pyridine/p-methylpyridine)$_8$] (**NC-2/NC-3**) with high thermal stability, up to 150°C were obtained, whose stability was confirmed by variable-temperature PXRD patterns. Moreover, when using variable-length bidentate N-heteroaromatic ligands (pyrazine(py)), 4,4′-Bipyridine(bpy), di-pyridin-4-yl-diazene (dpz), 1,4-bis(4-pyridyl)benzene(dpbz) as linkers, 1D- to 3D-superatomic silver cluster-assembled materials (**SCAM**) including {[Ag$_{14}$(C$_2$B$_{10}$H$_{10}$S$_2$)$_6$(pyrazine)$_{6.5}$(DMAc)(CH$_3$CN)$_{0.5}$]·2DMAc}$_n$ (Ag$_{14}$-py), {[Ag$_{14}$(C$_2$B$_{10}$H$_{10}$S$_2$)$_6$(bpy)$_4$]}$_n$ (Ag$_{14}$-bpy), {[Ag$_{14}$(C$_2$B$_{10}$H$_{10}$S$_2$)$_6$(dpz)$_2$(CH$_3$CN)$_4$]·DMAc}$_n$ (Ag$_{14}$-dpz), {[Ag$_{14}$(C$_2$B$_{10}$H$_{10}$S$_2$)$_6$(dpbz)$_4$]}$_n$ (Ag$_{14}$-dpbz) are synthesized. For the 1D silver cluster-assembled materials Ag$_{14}$-py, due to the small size of the ligand, the pyrazines were connected to each Ag$_{14}$(C$_2$B$_{10}$H$_{10}$S$_2$)$_6$ only in the diagonal direction, leading to a right-handed helix along the c axis. Further, a 2D silver cluster-assembled materials were obtained with di-pyridin-4-yl-diazene (dpz) linker, in which the Ag$_{14}$(C$_2$B$_{10}$H$_{10}$S$_2$)$_6$ are coprotected by four CH$_3$CN and four dipyridin-4-yl-diazeneligands. Although Ag$_{14}$-py and Ag$_{14}$-dpz were silver cluster-assembled materials, the stability of them was quite low probably due to their easily dislodged coordinating solvent molecules. When 4,4′-bipyridines was applied, the 3D silver cluster-assembled materials Ag$_{14}$-bpy were obtained. Eight vertices of each [Ag$_{14}$(C$_2$B$_{10}$H$_{10}$S$_2$)$_6$] cluster in Ag$_{14}$-bpy related to linear 4,4′-bipyridines. However, 3D Ag$_{14}$-bpy still exhibits poor stability, maybe the escape of solvent molecules leads to the lack of rigidity of the overall frame. So, a new stable 3D silver cluster-assembled materials Ag$_{14}$-dpbz with an interpenetrating framework by using 1,4-bis(4-pyridyl)benzene(dpbz) was synthesized. The obtained (Ag$_{14}$-dpbz)) showed high thermal stability, remaining

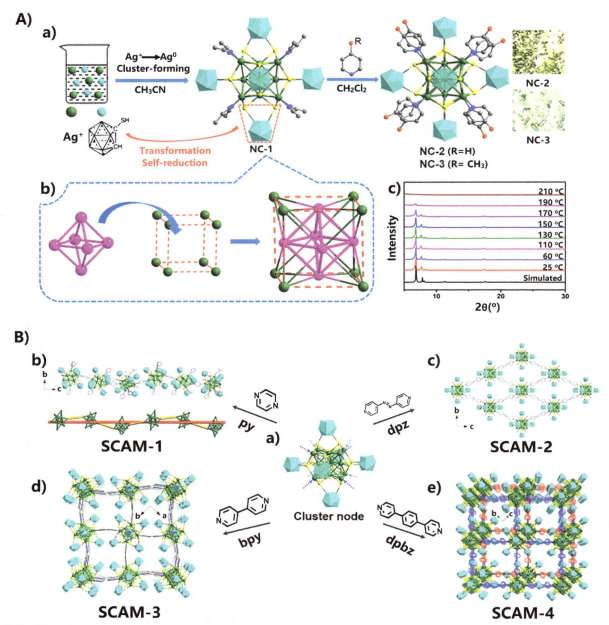

FIG. 7.33 Silver cluster-assembled materials based on [Ag$_{14}$(C$_2$B$_{10}$H$_{10}$S$_2$)$_6$(CH$_3$CN)$_8$]·4CH$_3$CN (A) (a) Synthesis of desired Ag$_{14}$ NCs; note that the (fcc) array (represented by *green spheres*) of Ag$_{14}$ superatoms are stabilized by face-capping 1,2-dithiolate-o-carborane(C$_2$B$_{10}$H$_{10}$S$_2$) ligands. (b) Structural dissection of NC-1. (c) Variable-temperature PXRD patterns of NC-2. Color codes: *green and pink* = silver; *yellow* = sulfur; *gray* = carbon; *blue* = nitrogen; *turquoise* = carborane. (B) Bidentate N-containing ligands bridge mixed-valence (0/1+) Ag$_{14}$ building blocks to form 1D-to-3D Silver cluster-assembled materials (SCAM). *(From Z.Y. Wang, M.Q. Wang, Y.L. Li, R. Luo, T.-T. Jia, R.-W. Huang, S.-Q. Zang, T.C.W. Mak, Atomically precise site-specific tailoring and directional assembly of superatomic silver nanoclusters, J. Am. Chem. Soc. 140 (3) (2018) 1069–1076, https://doi.org/10.1021/jacs.7b11338.)*

intact up to 220°C, and possessed small size pores. The study constitutes a major step toward the development of ligand-modulation of the structure, stability, assembly, and functionality of superatomic silver nanoclusters [153].

Huang et al. synthesized silver cluster assembled materials [Ag$_{12}$(StBu)$_8$(CF$_3$COO)$_4$(bpy)$_4$]$_n$, bpy = 4,4′-bipyridine) by using bidentate linkers, in which silver cluster Ag$_{12}$(StBu)$_8$(CF$_3$COO)$_4$ are embedded as nodes (Fig. 7.34). In the framework of [Ag$_{12}$(StBu)$_8$(CF$_3$COO)$_4$(bpy)$_4$]$_n$, each [Ag$_{12}$(StBu)$_8$(CF$_3$COO)$_4$] node was linked to four adjacent nodes via eight bpy bridges. Followed, the effect of modification of the substituents on the [Ag$_{12}$(StBu)$_8$(CF$_3$COO)$_4$(bpy)$_4$]$_n$ on room-temperature emission was explored, such as [Ag$_{12}$(SBut)$_8$(CF$_3$COO)$_4$(bpy-NH$_2$)$_4$]$_n$ having 3-amino-4,

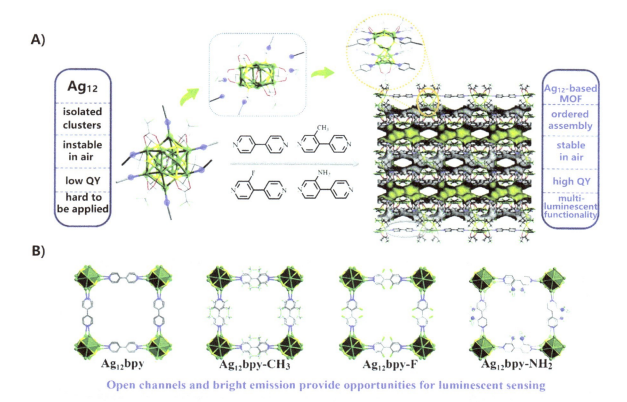
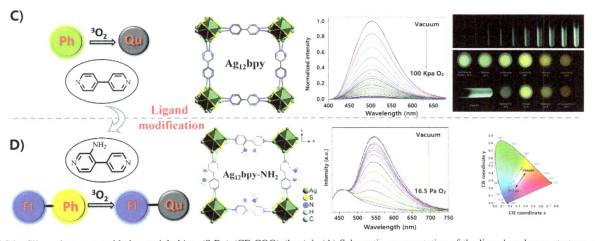

FIG. 7.34 Silver cluster assembled materials [Ag$_{12}$(StBu)$_8$(CF$_3$COO)$_4$(bpy)$_4$]$_n$ (A) Schematic representation of the ligand-exchange strategy used to obtain Ag$_{12}$-bpy crystals. (B) Perspective views of open channels of the iso structures of Ag$_{12}$bpy, Ag$_{12}$bpy-NH$_2$, Ag$_{12}$bpy-CH$_3$ and Ag$_{12}$bpy-F. (C) Luminescence switching of Ag$_{12}$bpy responding to O$_2$ and VOCs. From left to right, the mechanism of O$_2$ quenching phosphorescence, structure of Ag$_{12}$bpy, O$_2$ partial pressure-dependent emission spectra, images of a glass tube containing Ag$_{12}$bpy powder samples from vacuum to air excited by UV light and the image of Ag$_{12}$bpy when responding to different VOCs under a UV light. (D) From left to right, the mechanism of ratiometric O$_2$ sensing based on dual-emission fluorescence and phosphorescence; structure of Ag$_{12}$bpy-NH$_2$; ratiometric emission spectra of Ag$_{12}$bpy-NH$_2$; the CIE coordinate diagram illustrating the color changing with O$_2$ pressure. Fl = fluorescence, Ph = phosphorescence and Qu = quenched. *(From R.W. Huang, Y.S. Wei, X.Y. Dong, et al., Hypersensitive dual-function luminescence switching of a silver-chalcogenolate cluster-based metal-organic framework, Nat. Chem. 9 (7) (2017) 689–697, doi:10.1038/nchem.2718; X.Y. Dong, Y. Si, J.S. Yang, et al., Ligand engineering to achieve enhanced ratiometric oxygen sensing in a silver cluster-based metal-organic framework, Nat. Commun. 11 (1) (2020), https://doi.org/10.1038/s41467-020-17200-w.)*

4′-bipyridine (bpy-NH$_2$) bearing an electron-donating -NH$_2$ group, [Ag$_{12}$(SBut)$_8$(CF$_3$COO)$_4$(bpy-CH$_3$)$_4$]$_n$, possessing 3-methyl-4,4′-bipyridine(bpy-CH$_3$) with a weakly electron-donating methyl group, [Ag$_{12}$(SBut)$_8$(CF$_3$-COO)$_4$(bpy-F)$_4$]$_n$ containing 3-fluorine-4,4′-bipyridine (bpy-F) (Fig. 7.34A,B). In detail, [Ag$_{12}$(StBu)$_8$(CF$_3$COO)$_4$(bpy)$_4$]$_n$ showed significantly enhanced stability more than 1 year, a enhance QY (12.1%) and a lifetime of 200 ns at room temperature than that of [Ag$_{12}$(StBu)$_8$(CF$_3$COO)$_4$] node. And this [Ag$_{12}$(StBu)$_8$(CF$_3$COO)$_4$(bpy)$_4$]$_n$ materials can function in ultrafast dual-function luminescences witching (<1 s), that is, turn-off by O$_2$ and multicolored turn-on by volatile organic compounds (VOCs) (Fig. 7.34C). The switching mechanism was distinctly resolved by SCXRD structural analysis of O$_2$/VOCs-included crystals, highlighting the importance of structural precision. Fluorescence (Fl) and phosphorescence (Ph) are cases of luminescence. Fl originates from the first excited singlet state S1, while Ph originates from the triplet state T1. The amino-modified [Ag$_{12}$(SBut)$_8$(CF$_3$COO)$_4$(bpy-NH$_2$)$_4$]$_n$ displayed two emissive peaks, one of which is a blue subnanosecond Fl component (0.37 ns under vacuum) at approximately 456 nm and the other Ph at 556 nm (3.12 ms under vacuum). Compared with the 200 ns Ph of [Ag$_{12}$(SBut)$_8$(CF$_3$COO)$_4$(bpy)$_4$]$_n$, the lifetime of the Ph component of [Ag$_{12}$(SBut)$_8$(CF$_3$COO)$_4$(bpy-NH$_2$)$_4$]$_n$ is prolonged by approximately 15,000-fold, with an approximate 2% increase in QY. Thus, [Ag$_{12}$(SBut)$_8$(CF$_3$COO)$_4$(bpy-NH$_2$)$_4$]$_n$ could detect trace oxygen gas in 0.3 s with a limit of detection (LOD) as low as 0.1 ppm in a ratiometric method, in which blue insensitive Fl functioned as a reference signal (Fig. 7.34D). When the O$_2$ concentration varied in the range of zero to 20 ppm, the lighting color of [Ag$_{12}$(SBut)$_8$(CF$_3$COO)$_4$(bpy-NH$_2$)$_4$]$_n$ contemporaneously varied, enabling the visible color response (Fig. 7.34D). When the -CH$_3$ groups are anchored on the bpy linkers, they can interfere with the process by which the oxygen molecules interact with the emissive framework, and greatly decease the response sensitivity to molecular oxygen. The combined luminescence characteristics of [Ag$_{12}$(SBut)$_8$(CF$_3$COO)$_4$(bpy-NH$_2$)$_4$]$_n$ and [Ag$_{12}$(SBut)$_8$(CF$_3$COO)$_4$(bpy-CH$_3$)$_4$]$_n$ and prepared the mixed-linker [Ag$_{12}$(SBut)$_8$(CF$_3$COO)$_4$(bpy-NH$_2$/CH$_3$)$_4$]$_n$, will extend the ratiometric sensing range. The variable substitutes on likers significantly tuned the luminescence sensing performance, which promises more novel and unprecedented phenomena worthy of exploring [154,155].

By tuning the synthesis methods, the cuboctahedra of Ag$_{12}$(SBut)$_6$(CF$_3$COO)$_6$ nodes could be further assembled in to 2D and 3D network structures (Fig. 7.35) [156]. In detail, the reported 3D network structures Ag$_{12}$(S-tBu)$_8$(CF$_3$COO)$_4$(bpy)$_4$ can be converted into a new trigonal structure [Ag$_{12}$(SBut)$_6$(CF$_3$COO)$_6$-(bpy)$_3$·(DMAc$_x$·toluene$_y$)]$_n$ when the Ag$_{12}$(S-tBu)$_8$(CF$_3$COO)$_4$(bpy)$_4$ crystals immersed in DMAc/toluene(3:1) for a sufficient duration at room temperature. And the [Ag$_{12}$(SBut)$_6$(CF$_3$COO)$_6$-(bpy)$_3$·(DMAc$_x$·toluene$_y$)]$_n$ can be converted into Ag$_{12}$(S-tBu)$_8$(CF$_3$COO)$_4$(bpy)$_4$.solvent again when [Ag$_{12}$(SBu^t)$_6$(CF$_3$COO)$_6$-(bpy)$_3$·(DMAc$_x$·toluene$_y$)]$_n$ immersed in CH$_3$CN/EtOH(1,1) for 6 h. As shown in Fig. 7.35A and B, the core structures are changed in the process of reversible transformation of [Ag$_{12}$(SBut)$_6$(CF$_3$COO)$_6$-(bpy)$_3$·(DMAc$_x$·toluene$_y$)]$_n$ and Ag$_{12}$(S-tBu)$_8$(CF$_3$COO)$_4$(bpy)$_4$, resulting in the geometrical structure of 3D Ag$_{12}$(S-tBu)$_8$(CF$_3$COO)$_4$(bpy)$_4$ structure changed to 2D [Ag$_{12}$(SBut)$_6$(CF$_3$COO)$_6$-(bpy)$_3$·(DMAc$_x$·toluene$_y$)]$_n$ structure. This suggested the solvent molecular could tailor the structure of metal NCs and their assembled structure. For the new highly symmetric 2D [Ag$_{12}$(SBut)$_6$(CF$_3$COO)$_6$-(bpy)$_3$·(DMAc$_x$·toluene$_y$)]$_n$ network, each Ag$_{12}$(S-tBu)$_6$(CF$_3$COO)$_6$ nanocluster was linked to six adjacent Ag$_{12}$(S-tBu)$_6$(CF$_3$COO)$_6$ nanoclusters via linkers. The adjacent lays were separated by 7.23 Å, with weak van der Waal's interactions and hydrogen bonds (H—F and H—O) (Fig. 7.35C). Furthermore, the structural deformation leaded to changes in the electronic structure and PL properties. 3D Ag$_{12}$(S-tBu)$_8$(CF$_3$COO)$_4$(bpy)$_4$ displays PL at a single wavelength, independent of temperature and excitation wavelength, while 2D CS displays PL of two colors (blue and red) depending on to the excitation wavelength (Fig. 7.35D). In order to enhance the blue emission of [Ag$_{12}$(SBut)$_6$(CF$_3$COO)$_6$-(bpy)$_3$·(DMAc$_x$·toluene$_y$)]$_n$ network, the researchers used bpy-NH$_2$ as a linker with emit blue light to create a 2D CS containing two types of linkers, bpy and bpy-NH$_2$. The intensity ratio of the red and blue PL signals depends on the mixing ratio of the linker molecules. At the optimal joint mixing ratio, the PL intensity ratio of the red and blue peaks depends on the temperature. Therefore, this 2D [Ag$_{12}$(SBut)$_6$(CF$_3$COO)$_6$(bpy)$_3$/(bpy-NH$_2$)$_3$·(DMAc$_x$·toluene$_y$)]$_n$ containing two types of connectors can be used as a temperature sensor (Fig. 7.35E).

By changing the linker molecular, Zang group and coworkers synthesized an 2D assembled nanomaterials [Ag$_{12}$(StBu)$_6$(CF$_3$COO)$_6$(TPPA)$_6$(DMAc)$_x$]$_n$-AA (TPPA = tris(4-pyridylphenyl)-amine, AA = AA packing pattern) through a one-pot method (Fig. 7.36) [157]. [Ag$_{12}$(StBu)$_6$(CF$_3$COO)$_6$(TPPA)$_6$(DMAc)$_x$]$_n$-AA crystallizes in trigonal space group P-3. The Ag$_{12}$(StBu)$_6$(CF$_3$COO)$_6$-AAnodes are surrounded by six TPPA in each layer, and the guest dimethylacetamide (DMAc) molecules existed between layers after synthesis. In AA packing model, C-H...F and C-H...O weak interactions could be observed. Interestingly, like the DMAc molecular affecting the structure of [Ag$_{12}$(SBut)$_6$(CF$_3$COO)$_6$(bpy)$_3$·(DMAc$_x$·toluene$_y$)]$_n$, the DMAc molecules in [Ag$_{12}$(StBu)$_6$(CF$_3$COO)$_6$(TPPA)$_6$(DMAc)$_x$]$_n$ could also influence the packing pattern. In AA model, the guest molecules were highly disordered in the channels of 2D layer structure. When the DMAC was partially removed from 2D layer structure, the overlap of the 2D layers changed, forming [Ag$_{12}$(StBu)$_6$(CF$_3$COO)$_6$(TPPA)$_6$(DMAc)$_{1.7}$]$_n$-AB with an ABAB packing pattern. Structure shrinkage can be noticed along the c-axis, where node-to-node and N center to N center stackings have been changed to node-to-N center stacking

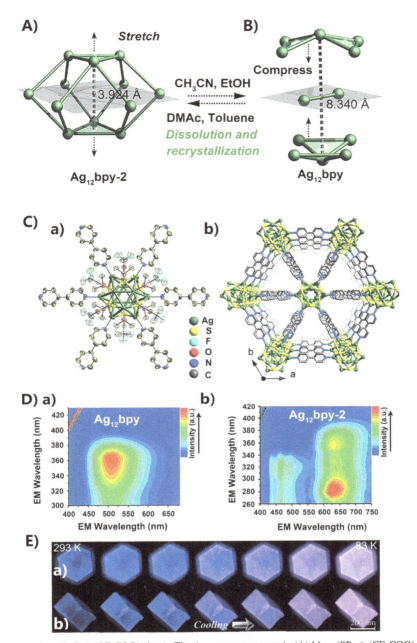

FIG. 7.35 3D network structures Ag$_{12}$(S-tBu)$_8$(CF$_3$COO)$_4$(bpy)$_4$ The Ag$_{12}$ core structures in (A) [Ag$_{12}$(SBut)$_6$(CF$_3$COO)$_6$(bpy)$_3$·(DMAc$_x$·toluene$_y$)]$_n$ and (B) Ag$_{12}$(S-tBu)$_8$(CF$_3$COO)$_4$(bpy)$_4$. (C) (a) the view of [Ag$_{12}$(SBut)$_6$(CF$_3$COO)$_6$ with six pendent bpy linkers and (b) stacking of 2D network structure of [Ag$_{12}$(SBut)$_6$(CF$_3$COO)$_6$(bpy)$_3$·(DMAc$_x$·toluene$_y$)]$_n$. (D) The 3D-EEM PL spectra of (a) Ag$_{12}$(S-tBu)$_8$(CF$_3$COO)$_4$(bpy)$_4$ and (b) [Ag$_{12}$(SBut)$_6$(CF$_3$COO)$_6$(bpy)$_3$·(DMAc$_x$·toluene$_y$)]$_n$ measured at 83 K. (E) Thermochromic images of the exterior {001} surfaces (a) and the exposed interior {010}/{100} planes (b) of Ag$_{12}$bpy-$_2$/NH$_2$·solvents(20:1) single crystals under UV-light irradiation. *(From R.W. Huang, X.Y. Dong, B.J. Yan, et al., Tandem silver cluster isomerism and mixed linkers to modulate the photoluminescence of cluster-assembled materials, Angew. Chem. Int. Ed. 57 (28) (2018) 8560–8566, https://doi.org/10.1002/anie.201804059.)*

from AA packing model to AB model. Furthermore, when [Ag$_{12}$(StBu)$_6$(CF$_3$COO)$_6$(TPPA)$_6$(DMAc)$_{1.7}$]$_n$-AA structure was immersed in the mother liquor, [Ag$_{12}$(StBu)$_6$(CF$_3$COO)$_6$(TPPA)$_6$(DMAc)$_{0.9}$]$_n$-ABC adopting an ABC stacking mode was observed. From [Ag$_{12}$(StBu)$_6$(CF$_3$COO)$_6$(TPPA)$_6$(DMAc)$_{1.7}$]$_n$-AA transformed to [Ag$_{12}$(StBu)$_6$(CF$_3$COO)$_6$(TPPA)$_6$(DMAc)$_{1.7}$]$_n$-AB or to [Ag$_{12}$(StBu)$_6$(CF$_3$COO)$_6$(TPPA)$_6$(DMAc)$_{0.9}$]$_n$-ABC, morphology distinction could be found and the a significant shift have also been observed for the emission of Ag$_{12}$TPPA-AA, Ag$_{12}$TPPA-AB and Ag$_{12}$TPPA-ABC.

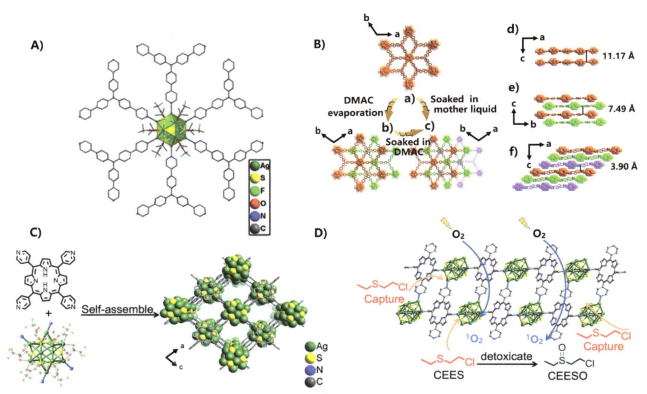

FIG. 7.36 An 2D assembled nanomaterials [Ag$_{12}$(S*t*Bu)$_6$(CF$_3$COO)$_6$(TPPA)$_6$(DMAc)$_x$]$_n$-AA (A) Perspective view of an Ag$_{12}$(SBu*t*)$_6$(CF$_3$COO)$_6$ with six TPPA linkers in each layer. (B) Stacking of the 2D network structure of Ag$_{12}$(SBu*t*)$_6$(CF$_3$COO) with TPPA from different views. (C) The synthesis of Ag$_{12}$(S-*t*Bu)$_6$(CF$_3$COO)$_3$(TPyP). (D) photodetoxification of CEES by Ag$_{12}$(S-*t*Bu)$_6$(CF$_3$COO)$_3$(TPyP). *(From X.S. Du, B.J. Yan, J.Y. Wang, X.J. Xi, Z.Y. Wang, S.Q. Zang, Layer-sliding-driven crystal size and photoluminescence change in a novel SCC-MOF, Chem. Commun. 54 (43) (2018) 5361–5364, doi:10.1039/C8CC01559G; M. Cao, R. Pang, Q.Y. Wang, H.Y. Li, S.Q. Zang, Porphyrinic silver cluster assembled material for simultaneous capture and photocatalysis of mustard-gas simulant, J. Am. Chem. Soc. 141 (37) (2019) 14505–14509, https://doi.org/10.1021/jacs.9b05952.)*

Changing fluorescent molecules (Bpy/TPPA) into molecules 5,10,15,20-tetra(4-pyridyl)porphyrin(TPyP) with photosensitizing effect, Zang and coworkers successfully synthesized a silver cluster-assembled materials [Ag$_{12}$(S*t*Bu)$_6$(CF$_3$COO)$_3$(TPyP)]$_n$ (Fig. 7.36C,D) [158]. The Ag$_{12}$(S*t*Bu)$_6$(CF$_3$COO)$_3$ cluster served as a four-connected node in assembly with the μ$_4$-TPyP linkers to form the 2D silver cluster-assembled materials with the AB stacking mode, showing remarkable structural stability. Due to the photosensitizing effect, the ability of the 2D silver cluster-assembled materials to degrade the toxic substance 2-chloroethyl ethyl sulfide (CEES) was explored. When compared with other reported assembled materials, the obtained 2D silver cluster-assembled materials showed higher photocatalytic activity, which could be ascribed to the synergistic effect of the porphyrin photosensitizer units (TpyP) and the silver clusters Ag$_{12}$(S*t*Bu)$_6$(CF$_3$COO)$_3$. The 2D silver cluster-assembled materials maintained its crystallinity after the photocatalytic reaction and were able to be used repeatedly. When selecting appropriate AgNCs and organic molecular linkers, photocatalytic activity could be further increased in the future work.

Using 2,5-bis(4-cyanophenyl)-1,4-bis(4-(pyridine-4-yl)-phenyl)-1,4-dihydropyrrolo[3,2-*b*]pyrrole(CPPP) with a nitrile group (-CN) as the connecting ligand, Zang group obtained a 3D silver cluster-assembled material of [Ag$_{12}$(S*t*Bu)$_6$(CF$_3$COO)$_6$(CPPP)$_2$(DMAc)$_{12}$]$_n$ (Fig. 7.37) [159]. The Ag$_{12}$(S*t*Bu)$_6$(CF$_3$COO)$_6$ nanocluster was surrounded by eight CPPP ligands; four of them through pyridine groups, and the other four ligands through nitrile groups. This coordination pattern is special and is rarely mentioned in reported silver cluster-assembled material. Twelve DMAc molecules are distributed in the pores of the framework. The 3D silver cluster-assembled material [Ag$_{12}$(S*t*Bu)$_6$(CF$_3$COO)$_6$(CPPP)$_2$(DMAc)$_{12}$]$_n$ showed an intense emission at 492 nm with a high quantum yield of 61%. Because of the aggregation-induced quenching, the emission intensity of CPPP is much weaker in solid state or in solution. While in [Ag$_{12}$(S*t*Bu)$_6$(CF$_3$COO)$_6$(CPPP)$_2$(DMAc)$_{12}$]$_n$, the aggregation-induced quenching of CPPP was suppressed, leading to a high quantum yield. And the good photo stability of [Ag$_{12}$(S*t*Bu)$_6$(CF$_3$COO)$_6$(CPPP)$_2$(DMAc)$_{12}$]$_n$ was also observed, which will promote the application in the future work.

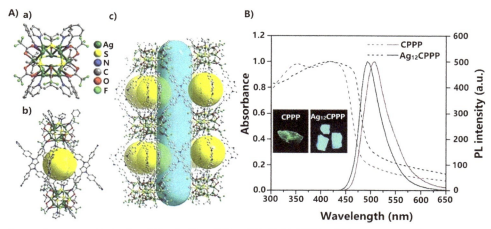

FIG. 7.37 3D silver cluster-assembled material of $[Ag_{12}(StBu)_6(CF_3COO)_6(CPPP)_2(DMAc)_{12}]_n$. (A) Structures of $[Ag_{12}(StBu)_6(CF_3COO)_6(CPPP)_2(DMAc)_{12}]_n$. (a) $Ag_{12}(StBu)_6(CF_3COO)_6(CPPP)_2$ nodes, (b) cage in $Ag_{12}(S-tBu)_6(CF_3COO)_6(CPPP)_2$, and (c) distribution of the cages in $[Ag_{12}(StBu)_6(CF_3COO)_6(CPPP)_2(DMAc)_{12}]_n$. (B) Solid-state absorption *(dashed lines)* and emission *(solid lines)* spectra of $[Ag_{12}(StBu)_6(CF_3COO)_6(CPPP)_2(DMAc)_{12}]_n$ and CPPP at room temperature. Inset are photographs of the crystals of CPPP and $[Ag_{12}(StBu)_6(CF_3COO)_6(CPPP)_2(DMAc)_{12}]_n$ under 365 nm ultraviolet light irradiation. *(From Z. We, X.H. Wu, P. Luo, J.Y. Wang, K. Li, S.Q. Zang, Matrix coordination induced emission in a three-dimensional silver cluster-assembled material, Chem. Eur. J. 25 (11) (2019) 2750–2756, https://doi.org/10.1002/chem.201805381.)*

Using 1,1,2,2-tetrakis(4-(pyridin-4-yl)phenyl)ethene(TPPE) as a molecular linker, in dimethylacetamide (DMAC) solvent, another 3D silver cluster-assembled material $\{[Ag_{12}(StBu)_6(CF_3CO_2)_6(tppe)_{1.5}](DMAC)_{39}\}$ was obtained by Zang group (Fig. 7.38) [160]. One $Ag_{12}(StBu)_6(CF_3CO_2)_6$ silver cluster are surrounded by nine other $Ag_{12}(StBu)_6(CF_3CO_2)_6$ silver clusters through Ag—N bonds by TPPE linkers to form 3D network. Because of the aggregation-induced emission (AIE) properties of the TPPE ligand $\{[Ag_{12}(StBu)_6(CF_3CO_2)_6(tppe)_{1.5}](DMAC)_{39}\}$ exhibited adjustable intense luminescence. The presence or absence of solvent molecules DMAC can influence the fluorescence properties. Novelty, the diameter of the cubic cage in $\{[Ag_{12}(StBu)_6(CF_3CO_2)_6(tppe)_{1.5}](DMAC)_{39}\}$ was as high as 32 Å, endowing it with satisfactory encapsulation capacity for different guest molecules to achieve tailored luminescence properties, including D-menthol (D-MT), L-menthol (L-MT), 4,4′-dimethoxybenzil (DMB), and 3,6-di(2-thienyl)diketopyrrolopyrrole(DPP). By adjusting functionalized guest molecules, circularly polarized luminescence, white-light-emitting, and room-temperature phosphorescence were readily realized. $\{[Ag_{12}(StBu)_6(CF_3CO_2)_6(tppe)_{1.5}](DMAC)_{39}\}$ could enrich limited kinds of mesoporous silver cluster-assembled material and offers the prospect of application in host guest chemistry. More importantly, this work provides a strategy, versatile for tailoring luminescence in crystalline porous material.

For better application, Zang and coworkers developed a facile strategy to engineer discrete silver cluster-assembled material into a flexible membrane (Fig. 7.39A) [161]. Firstly, the $Ag_{12}(StBu)_6(CF_3COO)_6$ with an amino group functionalized organic linker (1,4-bis(pyrid-4-yl)benzenenamine (NH_2-bpz) assembled into an silver cluster-assembled material $[Ag_{12}(StBu)_6(CF_3COO)_6(NH_2\text{-bpz})_3]_n$ with enhanced luminescence (Fig. 7.39B). The $Ag_{12}(StBu)_6(CF_3COO)_6$ bound to NH_2-bpz linkers via Ag—N bonds, resulting in extensive 2D networks with an AA stacking mode The amino group provided a chance to further polymerize. The decorated $[Ag_{12}(StBu)_6(CF_3COO)_6(NH_2\text{-bpz})_3]_n$ was subsequently covalently cross-linked with acrylate monomers through a convenient photopolymerization approach to produce a silver cluster-assembled material membrane (Fig. 7.39C, D). In detail, the 2D $[Ag_{12}(StBu)_6(CF_3COO)_6(MA\text{-bpz})_3]_n$ film was synthesized by reacting the $[Ag_{12}(StBu)_6(CF_3COO)_6(MA\text{-bpz})_3]_n$ with methacrylic anhydride (MA), which MA bound to the amino group of bpz-NH_2 (Fig. 7.39C) and was then polymerized with acrylate monomers butyl methacrylate (BMA) and triethylene glycol dimethacrylate (TEGDMA) (Fig. 7.39D). The resulting membrane significantly alters the physical and chemical properties of the silver cluster-assembled material, with improved chemical stability as well as enhanced quantum yield. The study provided an efficient route to developing high-performance silver cluster-assembled material. The rich researches based on $Ag_{12}(StBu)_6(CF_3COO)_6$ as a node promoted the development of the silver cluster-assembled material and guides the subsequent research on the assembly.

Like the tailoring of luminescent properties of nanoclusters via ligand engineering, aggregation-induced emission luminogens (AIEgens) as ligands has proven to be an effective strategy for constructing metal-organic frameworks (MOFs) with intense luminescent properties. Successfully, a typical AIE ligand, tppe (1,1,2,2-tetrakis(4-(pyridin-4-yl)phenyl)-ethene) ligands combining two silver chalcogenolate cluster nodes was assembled to a dual-node 3D silver chalcogenolate cluster

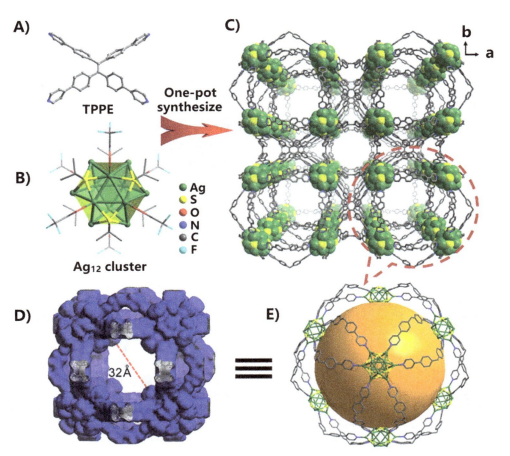

FIG. 7.38 3D silver cluster-assembled material {[Ag$_{12}$(StBu)$_6$(CF$_3$CO$_2$)$_6$(tppe)$_{1.5}$](DMAC)$_{39}$} (A and B) Molecular structure of TPPE ligand and Ag$_{12}$(StBu)$_6$(CF$_3$CO$_2$)$_6$ cluster. (C) [Ag$_{12}$(StBu)$_6$(CF$_3$CO$_2$)$_6$(tppe)$_{1.5}$] ⊃ DMAC framework viewed along the c axis. (D and E) Enlargement of the circled section in (B): a cubic cage surrounded by 24 TPPE ligands and 8 Ag$_{12}$(StBu)$_6$(CF$_3$CO$_2$)$_6$ clusters. *(From X.H. Wu, Z. Wei, B.J. Yan, R.W. Huang, Y.Y. Liu, K. Li, S.Q. Zang, T.C.W. Mak, Mesoporous crystalline silver-chalcogenolate cluster-assembled material with tailored photoluminescence properties, CCS Chem. 1 (5) (2019) 553–560, https://doi.org/10.31635/ccschem.019.20190024.)*

{[Ag$_{12}$(StBu)$_6$(CF$_3$CO$_2$)$_6$]$_{0.5}$[Ag$_8$(StBu)$_4$(CF$_3$CO$_2$)$_4$](tppe)$_2$(DMAC)$_{10}$}$_n$ in the dimethylacetamide solution (Fig. 7.40) [162]. The Ag$_{12}$Ag$_8$tppe is not able because it incorporates two types of Ag$_8$ and Ag$_{12}$ cluster nodes. The structural diversity of SCAMs and the introduction of attractive AIE linkers could bring about a broader application range. As shown in Fig. 7.40A, the Ag NC-based MOF was composed of tppe, Ag$_{12}$(S-tBu)$_6$(CF$_3$COO)$_6$, and Ag$_8$(S-tBu)$_4$(CF$_3$COO)$_4$. One Ag$_{12}$(S-tBu)$_6$(CF$_3$COO)$_6$NC and three Ag$_8$(S-tBu)$_4$(CF$_3$COO)$_4$ NCs were bound to the four N atoms of tppe, forming the 3D net (Fig. 7.40B). Single-crystal X-ray diffraction analysis confirmed the solvent molecular DMAC presenting in the pores of the 3D net immediately after synthesis (Fig. 7.40C). And the solvent molecular DMAC play an important role in tailoring the PL wavelength of the 3D net. Firstly, infrared spectra and TGA data showed that the framework of the 3D net still maintain combining the morphology of crystal remained unchanged after DMAC removing when the obtained 3D net was exposed to the atmosphere (Fig. 7.40D). While, the PL wavelength of the 3D net gradually changed from blue to green and the maximum emission wavelength shifted to 532 nm when exposed to the atmosphere. Interestingly, this fluorescence change was reversible when treated with DMAC again. Possible fluorescence decay paths for the light-emitting molecule tppe in the 3D net are proposed in Fig. 7.40Ea. When DMAC was present in the pores of the framework, the steric hindrance between tppe and DMAC suppressed the intramolecular rotation of tppe, resulting in a high excited-state dynamic of the 3D net. The change in the emission behavior of the Ag NC-based MOF induced by DMAC is attributed to this change in its excited state characteristics (Fig. 7.40Ea). For exploring the influence of guest molecules on the fluorescence of [Ag$_{12}$(St-Bu)$_6$(CF$_3$CO$_2$)$_6$]$_{0.5}$[Ag$_8$(StBu)$_4$(CF$_3$CO$_2$)$_4$](tppe)$_2$, three solvents (THF, toluene, and DMF) were used. The obtained 3D nets named 1 ⊃ THF, 1 ⊃ Toluene, and 1 ⊃ DMF showed highly analogous emission spectra and fluorescence lifetimes

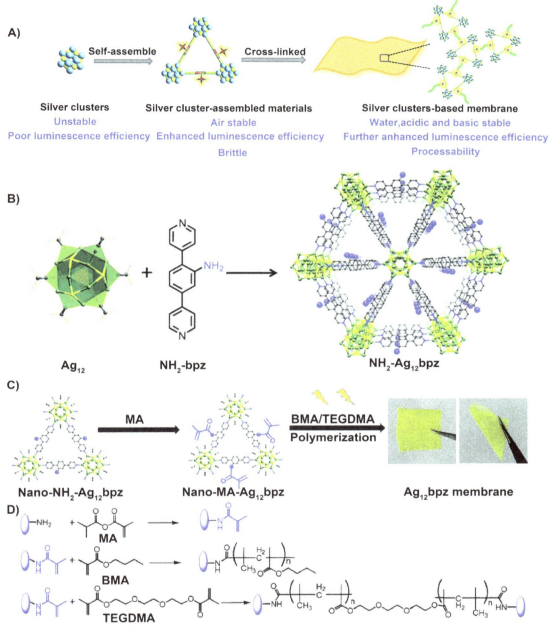

FIG. 7.39 A facile strategy to engineer discrete silver cluster-assembled material into a flexible membrane (A) Schematic illustration of the fabrication process of an AgNC-based membrane. (B) Structure views of $Ag_{12}(S\text{-}tBu)_6(CF_3COO)_6(bpz\text{-}NH_2)_3$. (C) Fabrication process of the membrane. (D) Chemical reactions in the postmodification and cross-linking steps. *(From Y.M. Wang, J.W. Zhang, Q.Y. Wang, H.Y. Li, X.Y. Dong, S. Wang, S. Q. Zang, Fabrication of silver chalcogenolate cluster hybrid membranes with enhanced structural stability and luminescence efficiency, Chem. Commun. 55 (97) (2019) 14677–14680, https://doi.org/10.1039/C9CC07797A.)*

(Fig. 7.40Eb), demonstrating that the polarity of the guest molecules in $[Ag_{12}(StBu)_6(CF_3CO_2)_6]_{0.5}[Ag_8(StBu)_4(CF_3CO_2)_4](tppe)_2$ have little effect on its fluorescence properties and further confirmed that the steric hindrance between tppe and solvents was the cause of blue fluorescence.

Zang et al. have also synthesized another smaller functional Ag NC-based MOFs (Fig. 7.41). A flexible Ag NC-based MOF $[Ag_{10}(StBu)_6(CF_3COO)_2(PhPO_3H)_2(bpy)_2]_n$ was reported [163]. The structure first form 2D layers via $Ag_{10}(S\text{-}tBu)_6(CF_3COO)_2(PhPO_3H)_2$NCs coordinating with bpy. After, stacking through hydrogen bond (O—H···O) between the intercluster $PhPO_3H$ linkers and C-H···O interactions between interlayer pyridine and CF_3COO^- groups, the 3D silver

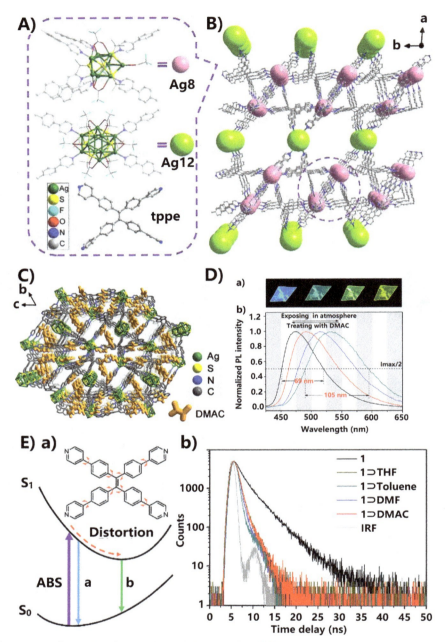

FIG. 7.40 An effective strategy for constructing metal-organic frameworks (MOFs) with intense luminescent properties (A) the structure of [Ag_8($StBu$)$_4$(CF_3CO_2)$_4$], [Ag_{12}($StBu$)$_6$(CF_3CO_2)$_6$] cluster and tppe ligand. (B) Single net of {[Ag_{12}($StBu$)$_6$(CF_3CO_2)$_6$]$_{0.5}$[Ag_8($StBu$)$_4$(CF_3CO_2)$_4$](tppe)$_2$(DMAC)$_{10}$}$_n$ framework. (C) Distribution of DMAC molecule in {[Ag_{12}($StBu$)$_6$(CF_3CO_2)$_6$]$_{0.5}$[Ag_8($StBu$)$_4$(CF_3CO_2)$_4$](tppe)$_2$(DMAC)$_{10}$}$_n$ framework. (D) (a) gradual fluorescence changes of the {[Ag_{12}($StBu$)$_6$(CF_3CO_2)$_6$]$_{0.5}$[Ag_8($StBu$)$_4$(CF_3CO_2)$_4$](tppe)$_2$(DMAC)$_{10}$}$_n$ (1 ⊃ DMAC) crystal under atmospheric exposure. (b) Normalized fluorescence spectra of 1 ⊃ DMAC. (E) proposed fluorescence decay paths in {[Ag_{12}($StBu$)$_6$(CF_3CO_2)$_6$]$_{0.5}$[Ag_8($StBu$)$_4$(CF_3CO_2)$_4$](tppe)$_2$(DMAC)$_{10}$}$_n$ (path a) and [Ag_{12}($StBu$)$_6$(CF_3CO_2)$_6$]$_{0.5}$[Ag_8($StBu$)$_4$(CF_3CO_2)$_4$](tppe)$_2$ (path b). And fluorescence-decay profiles [Ag_{12}($StBu$)$_6$(CF_3CO_2)$_6$]$_{0.5}$[Ag_8($StBu$)$_4$(CF_3CO_2)$_4$](tppe)$_2$ in DMAC, THF, toluene, and DMF (1 ⊃ DMAC, 1 ⊃ THF, 1 ⊃ Toluene, 1 ⊃ DMF, and 1 for short). *(From X.H. Wu, P. Luo, Z. Wei, Y.Y. Liu, R.W. Huang, X.Y. Dong, K. Li, S.Q. Zang, B.Z. Tang, Guest-triggered aggregation-induced emission in silver chalcogenolate cluster metal-organic frameworks, Adv. Sci. 6 (2) (2019) 1801304, https://doi.org/10.1002/advs.201801304.)*

cluster-assembled material was achieved (Fig. 7.41A and B). Due to the weak interactions between the 2D layers, facilitating the sliding of the layers, the 3D silver cluster-assembled material can undergo structural deformation from closed silver cluster-assembled material to open silver cluster-assembled material in response to guest organic molecules (Fig. 7.41C). The free solvent [Ag_{10}($StBu$)$_6$(CF_3COO)$_2$($PhPO_3H$)$_2$(bpy)$_2$]$_n$ exhibited green PL in air. After the guest organic molecule is included, the PL color of [Ag_{10}($StBu$)$_6$(CF_3COO)$_2$($PhPO_3H$)$_2$(bpy)$_2$]$_n$ solvent depends on the guest

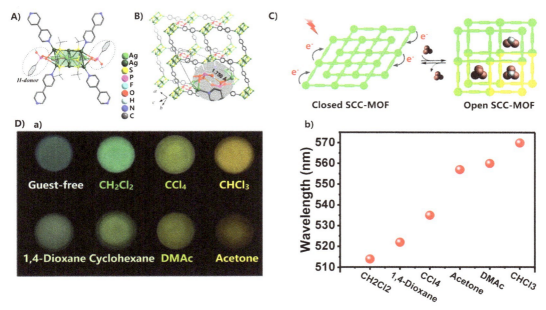

FIG. 7.41 Smaller functional Ag NC-based MOFs (A) the $Ag_{10}(S\text{-}tBu)_6$ core in $Ag_{10}(S\text{-}tBu)_6(CF_3COO)_2(PhPO_3H)_2(bpy)_2$. (B) Two-layer stack of the host framework of $Ag_{10}(S\text{-}tBu)_6(CF_3COO)_2(PhPO_3H)_2(bpy)_2$ with complementary hydrogen bonding (O—H···O; the H···O distance is 1.750 Å) between inter layer-PO_2OH moieties. (C) Illustration of reversible pore open/closed structural transformation induced by CH_2Cl_2, $CHCl_3$, and CCl_4 (represented as space-filling models) and switchable solvatochromism. (D) (a) Luminescence images of $Ag_{10}(S\text{-}tBu)_6(CF_3COO)_2(PhPO_3H)_2(bpy)_2$/solvent (guest free, CH_2Cl_2, $CHCl_3$, CCl_4, 1,4-dioxane, cyclohexane, DMAC, and acetone) combinations under 365 nm ultraviolet light irradiation. (b) Emission maxima of various $Ag_{10}(S\text{-}tBu)_6(CF_3COO)_2(PhPO_3H)_2(bpy)_2$/solvent combinations at room temperature. *(From X.Y. Dong, H.L. Huang, J.Y. Wang, H.Y. Li, S.Q. Zang, A flexible fluorescent SCC-MOF for switchable molecule identification and temperature, Chem. Mater. 30 (6) (2018) 2160–2167, https://doi.org/10.1021/acs.chemmater.8b00611.)*

organic molecule (Fig. 7.41D). So, 3D silver cluster-assembled material $[Ag_{10}(StBu)_6(CF_3COO)_2(PhPO_3H)_2(bpy)_2]_n$ could be served as a sensor for distinguishing volatile organic compounds by its PL color.

Like $Ag_{12}(StBu)_6(CF_3CO_2)_6$ as a node to build silver cluster-assembled material with a variety of linker molecules, Xu and coworkers reported the 1D, 2D and 3D silver-based cluster-assembled material, which using $Cd_6Ag_4(SPh)_{16}(DMF)_3(CH_3OH)$ as a node (Fig. 7.42). Less bpe linker (1 dose bpe) (bpe = trans-1,2-bis(4-pyridyl)ethylene) in the synthesis will promote the formation of a zigzag-type one-dimensional assembled material $[Cd_6Ag_4(SPh)_{16}(DMF)(H_2O)(bpe)]_n$ (Fig. 7.42A). In the obtained 1D assembled material, $Cd_6Ag_4(SPh)_{16}$ are two-connected by bpe linkers through the coordination bonds of Cd—N, not Ag—N. While, the use of a lot of bpe in the synthesis resulted in a 3D diamond-assembled material $[Cd_6Ag_4(SPh)_{16}(DMF)_4(bpe)_2]_n$. Each cluster in $Cd_6Ag_4(SPh)_{16}(DMF)_4(bpe)_2$ was four-connected by bpe linkers through the coordination bonds of Cd(cluster)-N(bpe) (Fig. 7.42C). When changed the linker molecular form bpe to a flexible ligand 4,4′-trimethylenedipiperidine(tmdp), 2D coordination polymers $Cd_6Ag_4(SPh)_{16}(DMF)_4(tmdp)_2$ was obtained (Fig. 7.42B). The 1D, 2D and 3D silver-based cluster-assembled material can acted as a visible light (>420 nm)-responsive photocatalyst to decompose the organic dye Rhodamine B in water, and all showed higher photocatalytic activity than that of the $Cd_6Ag_4(S\text{-}Ph)_{16}$ NCs and high stability during the photocatalytic reaction [164–166].

In the early research, Wang and colleagues reported the formation of a 3D assembled materials, using metal NCs as linkers, not as nodes. Via using Ag ions as nodes and $[C(Au\text{-}mdppz)_6](BF_4)_2]$(mdppz = 2-(3-methylpyrazinyl) diphenylphosphine,) NCs as linkers, an NbO-type MOF was obtained. The luminescent $[C(Au\text{-}mdppz)_6](BF_4)_2$ has a framework with C in the center (Fig. 7.43), whose core was capped by N donors serving as an anchors for Ag atoms (Fig. 7.43A). The 3D assembled materials were formed by the outer N atom of mdppz binding to an Ag ion (Fig. 7.43B and C). The obtained 3D assembled materials consisted of two interpenetrating frameworks (Fig. 7.43D and E) with a 1D channel in the c-axis direction (Fig. 7.43C). Because a luminescent NC was used as the linker, the obtained MOF also showed green PL. The 3D CS displayed a PL QY of 25.6%, which was much higher than that of the luminescent NCs (1.5%). This increase of QY was caused by the strengthening of the framework of the linker NCs upon MOF formation and the excited-state perturbation induced by the coordination of Ag ions [167].

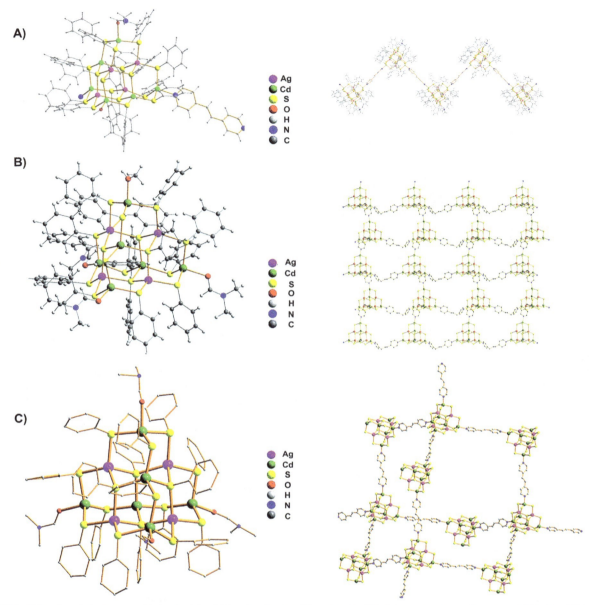

FIG. 7.42 Silver-based cluster-assembled material, which using $Cd_6Ag_4(SPh)_{16}(DMF)_3(CH_3OH)$ as a node (A) Structures of $Cd_6Ag_4(SPh)_{16}(DMF)(H_2O)$ and 1D assembled material $Cd_6Ag_4(SPh)_{16}(DMF)(H_2O)(bpe)$; (B) Crystal structures of cluster $Cd_6Ag_4(SPh)_{16}(DMF)_3(CH_3OH)$, the 2D assembled material $Cd_6Ag_4(SPh)_{16}(DMF)_4(tmdp)_2$; (C) Molecular structures of cluster $Cd_6Ag_4(SPh)_{16}(DMF)_4$ and the 2D assembled material $\{[Cd_6Ag_4(SPh)_{16}](bpe)_2\}_n$. (From C. Xu, N. Hedin, H.T. Shi, Q.F. Zhang, A semiconducting microporous framework of $Cd_6Ag_4(Sph)_{16}$ clusters interlinked using rigid and conjugated bipyridines, Chem. Commun. 50 (28) (2014) 3710–3712, doi:10.1039/c3cc49660k; Z.K. Wang, M.M. Sheng, S.S. Qin, et al., Assembly of discrete chalcogenolate clusters into a one-dimensional coordination polymer with enhanced photocatalytic activity and stability, Inorg. Chem. 59 (4) (2020) 2121–2126, doi:10.1021/ acs.inorgchem.9b03578; C. Xu, M.M. Sheng, H.T. Shi, M. Strømme, Q.F. Zhang, Interlinking supertetrahedral chalcogenolate clusters with bipyridines to form two-dimensional coordination polymers for photocatalytic degradation of organic dye, Dalton Trans. 48 (17) (2019) 5505–5510, https://doi.org/10.1039/c9dt00480g.)

7.5 Perspective

In this chapter, we mainly summarize the assembly phenomenon of clusters, mainly on the Au, Ag, Cu and their alloy nanoclusters. When atomical precise coin metal nanoclusters encounter assembly, cluster science blossoms. The internanoclusters assembly can bloom in supramolecular chemistry, such as the crystal lattice, cluster-based MOFs, cluster-based AIE material as well as cluster-based host-guest nanosystems. Crystallization is an important means of self-assembly of

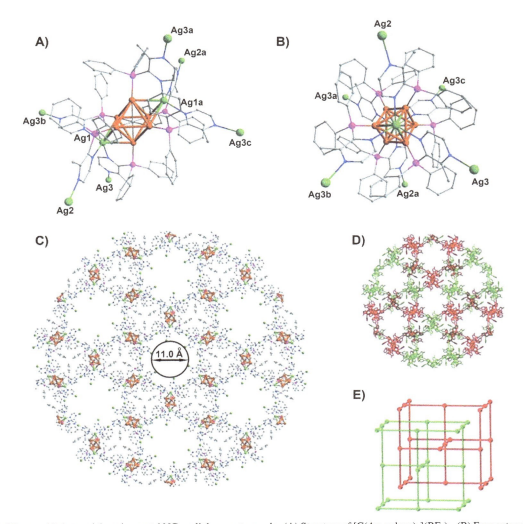

FIG. 7.43 A 3D assembled materials, using metal NCs as linkers, not as nodes (A) Structure of [C(Au-mdppz)$_6$](BF$_4$)$_2$. (B) Four extensions: Ag$_3$, Ag$_{3a}$, Ag$_{3b}$, and Ag$_{3c}$. (C) Perspective view of the 3D CS along the c direction. Au = *orange*, Ag = *green*, P = *purple*, N = *blue*, C = *gray*. (D) Two interpenetrating nets shown in different colors; anions and solvent molecules are omitted for clarity. (E) Schematic representation of NbO topology in the 3D CS. *(From Z. Lei, X.L. Pei, Z.G. Jiang, Q.M. Wang, Cluster linker approach: preparation of a luminescent porous framework with NbO topology by linking silver ions with gold(I) clusters, Angew. Chem. Int. Ed. 53 (47) (2014) 12771–12775, https://doi.org/10.1002/anie.201406761.)*

nanoclusters. Adjust the stacking mode of nanocluster building blocks in the unit cell has been exploring and developing. Linking the nanocluster building group to the cluster-based MOF is an effective method that can adjust the chemical and physical properties of a single nanocluster. Some suitable inorganic salt counterions and N-containing multidentate ligand binders had been found to be effective in constructing MOF. Compared with nonemission and weak emission nanoclusters, cluster-based AIE materials show greatly improved optical properties. The interaction between the host and the guest can adjust the geometric/electronic structure and chemical/physical properties of a single nanocluster. At the same time, the correlation between the structure and properties of the nanoclusters was studied, and the property changes of the nanoclusters caused by the assembly of the clusters were also studied. Supramolecular chemistry (intercluster assemblies and cluster-based hybrid materials) of nanocluster opens more opportunities for preparing cluster-based nanomaterials bearing desirable functionalities.

References

[1] G.M. Whitesides, B. Grzybowski, Self-assembly at all scales, Science 295 (5564) (2002) 2418–2421, https://doi.org/10.1126/science.1070821.
[2] Y. Min, M. Akbulut, K. Kristiansen, Y. Golan, J. Israelachvili, The role of interparticle and external forces in nanoparticle assembly, Nat. Mater. 7 (7) (2008) 527–538, https://doi.org/10.1038/nmat2206.

[3] L.L.K. Taylor, I.A. Riddell, M.M.J. Smulders, Self-assembly of functional discrete three-dimensional architectures in water, Angew. Chem. Int. Ed. 58 (5) (2019) 1280–1307, https://doi.org/10.1002/anie.201806297.

[4] F. Yu, V.M. Cangelosi, M.L. Zastrow, et al., Protein design: toward functional metalloenzymes, Chem. Rev. 114 (7) (2014) 3495–3578, https://doi.org/10.1021/cr400458x.

[5] L.A. Churchfield, F.A. Tezcan, Design and construction of functional supramolecular metalloprotein assemblies, Acc. Chem. Res. 52 (2) (2019) 345–355, https://doi.org/10.1021/acs.accounts.8b00617.

[6] S. Turega, W. Cullen, M. Whitehead, C.A. Hunter, M.D. Ward, Mapping the internal recognition surface of an octanuclear coordination cage using guest libraries, J. Am. Chem. Soc. 136 (23) (2014) 8475–8483, https://doi.org/10.1021/ja504269m.

[7] C. Lin, Y. Liu, H. Yan, Designer DNA nanoarchitectures, Biochemistry 48 (8) (2009) 1663–1674, https://doi.org/10.1021/bi802324w.

[8] T.R. Cook, P.J. Stang, Recent developments in the preparation and chemistry of metallacycles and metallacages via coordination, Chem. Rev. 115 (15) (2015) 7001–7045, https://doi.org/10.1021/cr5005666.

[9] Y. Li, H. Zhou, J. Chen, S. Anjum Shahzad, C. Yu, Controlled self-assembly of small molecule probes and the related applications in bioanalysis, Biosens. Bioelectron. 76 (2016) 38–53, https://doi.org/10.1016/j.bios.2015.06.067.

[10] K.J.M. Bishop, C.E. Wilmer, S. Soh, B.A. Grzybowski, Nanoscale forces and their uses in self-assembly, Small 5 (14) (2009) 1600–1630, https://doi.org/10.1002/smll.200900358.

[11] J. Park, J. Joo, S.G. Kwon, Y. Jang, T. Hyeon, Synthesis of monodisperse spherical nanocrystals, Angew. Chem. Int. Ed. 46 (25) (2007) 4630–4660, https://doi.org/10.1002/anie.200603148.

[12] H. Dai, E.W. Wong, Y.Z. Lu, S. Fan, C.M. Lieber, Synthesis and characterization of carbide nanorods, Nature 375 (6534) (1995) 769–772, https://doi.org/10.1038/375769a0.

[13] B. Nikoobakht, M.A. El-Sayed, Preparation and growth mechanism of gold nanorods (NRs) using seed-mediated growth method, Chem. Mater. 15 (10) (2003) 1957–1962, https://doi.org/10.1021/cm020732l.

[14] B.J. Wiley, Y. Chen, J.M. McLellan, et al., Synthesis and optical properties of silver nanobars and nanorice, Nano Lett. 7 (4) (2007) 1032–1036, https://doi.org/10.1021/nl070214f.

[15] R. Klajn, A.O. Pinchuk, G.C. Schatz, B.A. Grzybowski, Synthesis of heterodimeric sphere-prism nanostructures via metastable gold supraspheres, Angew. Chem. Int. Ed. 46 (44) (2007) 8363–8367, https://doi.org/10.1002/anie.200702570.

[16] L. Manna, E.C. Scher, A.P. Alivisatos, Synthesis of soluble and processable rod-, arrow-, teardrop-, and tetrapod-shaped CdSe nanocrystals, J. Am. Chem. Soc. 122 (51) (2000) 12700–12706, https://doi.org/10.1021/ja003055+.

[17] M.C. Daniel, D. Astruc, Gold nanoparticles: assembly, supramolecular chemistry, quantum-size-related properties, and applications toward biology, catalysis, and nanotechnology, Chem. Rev. 104 (1) (2004) 293–346, https://doi.org/10.1021/cr030698+.

[18] Y. Yin, R.M. Rioux, C.K. Erdonmez, S. Hughes, G.A. Somorjai, A.P. Alivisatos, Formation of hollow nanocrystals through the nanoscale Kirkendall effect, Science 304 (5671) (2004) 711–714, https://doi.org/10.1126/science.1096566.

[19] J. Chen, B. Wiley, Z.Y. Li, et al., Gold nanocages: engineering their structure for biomedical applications, Adv. Mater. 17 (18) (2005) 2255–2261, https://doi.org/10.1002/adma.200500833.

[20] Y. Wei, R. Klajn, A.O. Pinchuk, B.A. Grzybowski, Synthesis, shape control, and optical properties of hybrid Au/Fe$_3$O$_4$ "nanoflowers", Small 4 (10) (2008) 1635–1639, https://doi.org/10.1002/smll.200800511.

[21] T. Trindade, P. O'Brien, N.L. Pickett, Nanocrystalline semiconductors: synthesis, properties, and perspectives, Chem. Mater. 13 (11) (2001) 3843–3858, https://doi.org/10.1021/cm000843p.

[22] S. Sun, H. Zeng, Size-controlled synthesis of magnetite nanoparticles, J. Am. Chem. Soc. 124 (28) (2002) 8204–8205, https://doi.org/10.1021/ja026501x.

[23] S. Vaucher, M. Li, S. Mann, Synthesis of Prussian blue nanoparticles and nanocrystal superlattices in reverse microemulsions, Angew. Chem. Int. Ed. 39 (10) (2000) 1793–1796, https://doi.org/10.1002/(SICI)1521-3773(20000515)39:10<1793::AID-ANIE1793>3.0.CO;2-Y.

[24] R. Gref, Y. Minamitake, M.T. Peracchia, V. Trubetskoy, V. Torchilin, R. Langer, Biodegradable long-circulating polymeric nanospheres, Science 263 (5153) (1994) 1600–1603, https://doi.org/10.1126/science.8128245.

[25] E. Katz, I. Willner, Integrated nanoparticle-biomolecule hybrid systems: synthesis, properties, and applications, Angew. Chem. Int. Ed. 43 (45) (2004) 6042–6108, https://doi.org/10.1002/anie.200400651.

[26] M. Zheng, F. Davidson, X. Huang, Ethylene glycol monolayer protected nanoparticles for eliminating nonspecific binding with biological molecules, J. Am. Chem. Soc. 125 (26) (2003) 7790–7791, https://doi.org/10.1021/ja0350278.

[27] A.M. Kalsin, M. Fialkowski, M. Paszewski, S.K. Smoukov, K.J.M. Bishop, B.A. Grzybowski, Electrostatic self-assembly of binary nanoparticle crystals with a diamond-like lattice, Science 312 (5772) (2006) 420–424, https://doi.org/10.1126/science.1125124.

[28] Y. Jin, W. Lu, J. Hu, X. Yao, J. Li, Site-specific DNA cleavage of EcoRI endounclease probed by electrochemical analysis using ferrocene capped gold nanoparticles as reporter, Electrochem. Commun. 9 (5) (2007) 1086–1090, https://doi.org/10.1016/j.elecom.2006.12.028.

[29] J.M. Nam, C.S. Thaxton, C.A. Mirkin, Nanoparticle-based bio-bar codes for the ultrasensitive detection of proteins, Science 301 (5641) (2003) 1884–1886, https://doi.org/10.1126/science.1088755.

[30] H. Yan, S.H. Park, G. Finkelstein, J.H. Reif, T.H. LaBean, DNA-templated self-assembly of protein arrays and highly conductive nanowires, Science 301 (5641) (2003) 1882–1884, https://doi.org/10.1126/science.1089389.

[31] E.V. Shevchenko, D.V. Talapin, N.A. Kotov, S. O'Brien, C.B. Murray, Structural diversity in binary nanoparticle superlattices, Nature 439 (7072) (2006) 55–59, https://doi.org/10.1038/nature04414.

[32] R. Klajn, K.J.M. Bishop, M. Fialkowski, et al., Plastic and moldable metals by self-assembly of sticky nanoparticle aggregates, Science 316 (5822) (2007) 261–264, https://doi.org/10.1126/science.1139131.

[33] A.W. Snow, E.E. Foos, M.M. Coble, G.G. Jernigan, M.G. Ancona, Fluorine-labeling as a diagnostic for thiol-ligand and gold nanocluster self-assembly, Analyst 134 (9) (2009) 1790–1801, https://doi.org/10.1039/b906510p.

[34] S. Ren, S.K. Lim, S. Gradeak, Synthesis and thermal responsiveness of self-assembled gold nanoclusters, Chem. Commun. 46 (34) (2010) 6246–6248, https://doi.org/10.1039/c0cc01829e.

[35] Z. Wu, C. Dong, Y. Li, et al., Self-assembly of Au_{15} into single-cluster-thick sheets at the interface of two miscible high-boiling solvents, Angew. Chem. Int. Ed. 52 (38) (2013) 9952–9955, https://doi.org/10.1002/anie.201304122.

[36] Z. Wu, J. Liu, Y. Li, et al., Self-assembly of nanoclusters into mono-, few-, and multilayered sheets via dipole-induced asymmetric van der Waals attraction, ACS Nano 9 (6) (2015) 6315–6323, https://doi.org/10.1021/acsnano.5b01823.

[37] Y. Zhou, H.C. Zeng, Simultaneous synthesis and assembly of noble metal nanoclusters with variable micellar templates, J. Am. Chem. Soc. 136 (39) (2014) 13805–13817, https://doi.org/10.1021/ja506905j.

[38] L. Shi, L. Zhu, J. Guo, et al., Self-assembly of chiral gold clusters into crystalline nanocubes of exceptional optical activity, Angew. Chem. Int. Ed. 56 (48) (2017) 15397–15401, https://doi.org/10.1002/anie.201709827.

[39] Nonappa, T. Lahtinen, J.S. Haataja, T.R. Tero, H. Häkkinen, O. Ikkala, Template-free supracolloidal self-assembly of atomically precise gold nanoclusters: from 2D colloidal crystals to spherical capsids, Angew. Chem. Int. Ed. 55 (52) (2016) 16035–16038, https://doi.org/10.1002/anie.201609036.

[40] X. Kang, J. Xiang, Y. Lv, et al., Synthesis and structure of self-assembled $Pd_2Au_{23}(PPh_3)_{10}Br_7$ nanocluster: exploiting factors that promote assembly of icosahedral nano-building-blocks, Chem. Mater. 29 (16) (2017) 6856–6862, https://doi.org/10.1021/acs.chemmater.7b02015.

[41] R. Jin, C. Liu, S. Zhao, et al., Tri-icosahedral gold nanocluster $[Au_{37}(PPh_3)_{10}(SC_2H_4Ph)_{10}X_2]^+$: linear assembly of icosahedral building blocks, ACS Nano 9 (8) (2015) 8530–8536, https://doi.org/10.1021/acsnano.5b03524.

[42] J.W. Liu, L. Feng, H.F. Su, et al., Anisotropic assembly of Ag_{52} and Ag_{76} nanoclusters, J. Am. Chem. Soc. 140 (5) (2018) 1600–1603, https://doi.org/10.1021/jacs.7b12777.

[43] W. Zhang, D. Lin, H. Wang, et al., Supramolecular self-assembly bioinspired synthesis of luminescent gold nanocluster-embedded peptide nanofibers for temperature sensing and cellular imaging, Bioconjug. Chem. 28 (9) (2017) 2224–2229, https://doi.org/10.1021/acs.bioconjchem.7b00312.

[44] F. Xue, F. Qu, W. Han, L. Xia, J. You, Aggregation-induced emission enhancement of gold nanoclusters triggered by silicon nanoparticles for ratiometric detection of protamine and trypsin, Anal. Chim. Acta 1046 (2019) 170–178, https://doi.org/10.1016/j.aca.2018.09.033.

[45] X. Wang, C. Wang, N. Yang, J. Xia, L. Li, Preparation of fluorescent nanocomposites based on gold nanoclusters self-assembly, Colloids Surf. A Physicochem. Eng. Asp. 548 (2018) 27–31, https://doi.org/10.1016/j.colsurfa.2018.03.059.

[46] J. Shen, Z. Wang, D. Sun, et al., PH-responsive nanovesicles with enhanced emission co-assembled by Ag(I) nanoclusters and polyethyleneimine as a superior sensor for Al^{3+}, ACS Appl. Mater. Interfaces 10 (4) (2018) 3955–3963, https://doi.org/10.1021/acsami.7b16316.

[47] D. Prochowicz, A. Kornowicz, J. Lewiński, Interactions of native cyclodextrins with metal ions and inorganic nanoparticles: fertile landscape for chemistry and materials science, Chem. Rev. 117 (22) (2017) 13461–13501, https://doi.org/10.1021/acs.chemrev.7b00231.

[48] K.I. Assaf, W.M. Nau, The chaotropic effect as an assembly motif in chemistry, Angew. Chem. Int. Ed. 57 (43) (2018) 13968–13981, https://doi.org/10.1002/anie.201804597.

[49] J. Szejtli, Introduction and general overview of cyclodextrin chemistry, Chem. Rev. 98 (5) (1998) 1743–1753, https://doi.org/10.1021/cr970022c.

[50] W. Saenger, J. Jacob, K. Gessler, et al., Structures of the common cyclodextrins and their larger analogues - beyond the doughnut, Chem. Rev. 98 (5) (1998) 1787–1802, https://doi.org/10.1021/cr9700181.

[51] G. Chen, M. Jiang, Cyclodextrin-based inclusion complexation bridging supramolecular chemistry and macromolecular self-assembly, Chem. Soc. Rev. 40 (5) (2011) 2254–2266, https://doi.org/10.1039/c0cs00153h.

[52] P. Klüfers, H. Piotrowski, J. Uhlendorf, Homoleptic cuprates(II) with multiply deprotonated α-cyclodextrin ligands, Chem. A Eur. J. 3 (4) (1997) 601–608, https://doi.org/10.1002/chem.19970030416.

[53] H. Xu, S. Rodríguez-Hermida, J. Pérez-Carvajal, J. Juanhuix, I. Imaz, D. Maspoch, A first cyclodextrin-transition metal coordination polymer, Cryst. Growth Des. 16 (10) (2016) 5598–5602, https://doi.org/10.1021/acs.cgd.6b01115.

[54] A.A. Ivanov, C. Falaise, P.A. Abramov, et al., Host–guest binding hierarchy within redox- and luminescence-responsive supramolecular self-assembly based on chalcogenide clusters and γ-cyclodextrin, Chem. A Eur. J. 24 (51) (2018) 13467–13478, https://doi.org/10.1002/chem.201802102.

[55] P.A. Abramov, A.A. Ivanov, M.A. Shestopalov, et al., Supramolecular adduct of γ-cyclodextrin and $[\{Re_6Q_8\}(H_2O)_6]^{2+}$ (Q = S, Se), J. Clust. Sci. 29 (1) (2018) 9–13, https://doi.org/10.1007/s10876-017-1312-z.

[56] M.A. Moussawi, M. Haouas, S. Floquet, et al., Nonconventional three-component hierarchical host-guest assembly based on Mo-blue ring-shaped giant anion, γ-cyclodextrin, and dawson-type polyoxometalate, J. Am. Chem. Soc. 139 (41) (2017) 14376–14379, https://doi.org/10.1021/jacs.7b08058.

[57] K. Kirakci, V. Šícha, J. Holub, P. Kubát, K. Lang, Luminescent hydrogel particles prepared by self-assembly of β-cyclodextrin polymer and octahedral molybdenum cluster complexes, Inorg. Chem. 53 (24) (2014) 13012–13018, https://doi.org/10.1021/ic502144z.

[58] M.A. Moussawi, N. Leclerc-Laronze, S. Floquet, et al., Polyoxometalate, cationic cluster, and γ-cyclodextrin: from primary interactions to supramolecular hybrid materials, J. Am. Chem. Soc. 139 (36) (2017) 12793–12803, https://doi.org/10.1021/jacs.7b07317.

[59] M. Stuckart, N.V. Izarova, J. van Leusen, et al., Host–guest-induced environment tuning of 3d ions in a polyoxopalladate matrix, Chem. A Eur. J. 24 (67) (2018) 17767–17778, https://doi.org/10.1002/chem.201803531.

[60] Y. Wu, R. Shi, Y.L. Wu, et al., Complexation of polyoxometalates with cyclodextrins, J. Am. Chem. Soc. 137 (12) (2015) 4111–4118, https://doi.org/10.1021/ja511713c.

[61] C. Falaise, M.A. Moussawi, S. Floquet, et al., Probing dynamic library of metal-oxo building blocks with γ-cyclodextrin, J. Am. Chem. Soc. 140 (36) (2018) 11198–11201, https://doi.org/10.1021/jacs.8b07525.

[62] B. Zhang, W. Guan, F. Yin, J. Wang, B. Li, L. Wu, Induced chirality and reversal of phosphomolybdate cluster: via modulating its interaction with cyclodextrins, Dalton Trans. 47 (5) (2018) 1388–1392, https://doi.org/10.1039/c7dt03669h.

[63] K.I. Assaf, M.S. Ural, F. Pan, et al., Water structure recovery in chaotropic anion recognition: high-affinity binding of dodecaborate clusters to γ-cyclodextrin, Angew. Chem. Int. Ed. 54 (23) (2015) 6852–6856, https://doi.org/10.1002/anie.201412485.

[64] W. Wang, X. Wang, C. Xiang, et al., Orthogonal molecular recognition of chaotropic and hydrophobic guests enables supramolecular architectures, ChemNanoMat 5 (1) (2019) 124–129, https://doi.org/10.1002/cnma.201800377.

[65] W. Wang, X. Wang, J. Cao, et al., The chaotropic effect as an orthogonal assembly motif for multi-responsive dodecaborate-cucurbituril supramolecular networks, Chem. Commun. 54 (17) (2018) 2098–2101, https://doi.org/10.1039/c7cc08078f.

[66] B.L. May, J. Gerber, P. Clements, et al., Cyclodextrin and modified cyclodextrin complexes of E-4-tert-butylphenyl- 4′-oxyazobenzene: UV-visible,1H NMR and ab initio studies, Org. Biomol. Chem. 3 (8) (2005) 1481–1488, https://doi.org/10.1039/b415594g.

[67] J.H. Park, S. Hwang, J. Kwak, Nanosieving of anions and cavity-size-dependent association of cyclodextrins on a 1-adamantanethiol self-assembled monolayer, ACS Nano 4 (7) (2010) 3949–3958, https://doi.org/10.1021/nn1008484.

[68] I. Böhm, K. Isenbügel, H. Ritter, R. Branscheid, U. Kolb, Cyclodextrin and adamantane host−guest interactions of modified hyperbranched poly(ethyleneimine) as mimetics for biological membranes, Angew. Chem. Int. Ed. 50 (2011) 7896–7899, https://doi.org/10.1002/anie.201101604.

[69] D. Harries, D.C. Rau, V.A. Parsegian, Solutes probe hydration in specific association of cyclodextrin and adamantane, J. Am. Chem. Soc. 127 (7) (2005) 2184–2190, https://doi.org/10.1021/ja045541t.

[70] A. Mathew, G. Natarajan, L. Lehtovaara, et al., Supramolecular functionalization and concomitant enhancement in properties of Au25 clusters, ACS Nano 8 (1) (2014) 139–152, https://doi.org/10.1021/nn406219x.

[71] M.W. Heaven, A. Dass, P.S. White, K.M. Holt, R.W. Murray, Crystal structure of the gold nanoparticle $[N(C_8H_{17})_4][Au_{25}(SCH_2CH_2Ph)_{18}]$, J. Am. Chem. Soc. 130 (12) (2008) 3754–3755, https://doi.org/10.1021/ja800561b.

[72] M. Zhu, C.M. Aikens, F.J. Hollander, G.C. Schatz, R. Jin, Correlating the crystal structure of a thiol-protected Au25 cluster and optical properties, J. Am. Chem. Soc. 130 (18) (2008) 5883–5885, https://doi.org/10.1021/ja801173r.

[73] A. Nag, P. Chakraborty, G. Paramasivam, M. Bodiuzzaman, G. Natarajan, T. Pradeep, Isomerism in supramolecular adducts of atomically precise nanoparticles, J. Am. Chem. Soc. 140 (42) (2018) 13590–13593, https://doi.org/10.1021/jacs.8b08767.

[74] C. Yan, C. Liu, H. Abroshan, Z. Li, R. Qiu, G. Li, Surface modification of adamantane-terminated gold nanoclusters using cyclodextrins, Phys. Chem. Chem. Phys. 18 (33) (2016) 23358–23364, https://doi.org/10.1039/c6cp04569c.

[75] T. Higaki, C. Liu, C. Zeng, et al., Controlling the atomic structure of Au_{30} nanoclusters by a ligand-based strategy, Angew. Chem. Int. Ed. 55 (23) (2016) 6694–6697, https://doi.org/10.1002/anie.201601947.

[76] X. Kang, M. Zhou, S. Wang, et al., The tetrahedral structure and luminescence properties of Bi-metallic Pt 1 Ag 28 (SR) 18 (PPh 3) 4 nanocluster, Chem. Sci. 8 (4) (2017) 2581–2587, https://doi.org/10.1039/c6sc05104a.

[77] D. Crasto, G. Barcaro, M. Stener, L. Sementa, A. Fortunelli, A. Dass, $Au_{24}(SAdm)_{16}$ nanomolecules: x-ray crystal structure, theoretical analysis, adaptability of adamantane ligands to form $Au_{23}(SAdm)_{16}$ and $Au_{25}(SAdm)_{16}$, and its relation to $Au_{25}(SR)_{18}$, J. Am. Chem. Soc. 136 (42) (2014) 14933–14940, https://doi.org/10.1021/ja507738e.

[78] Q. Li, S. Yang, T. Chen, et al., Structure determination of a metastable $Au_{22}(SAdm)_{16}$ nanocluster and its spontaneous transformation into $Au_{21}(SAdm)_{15}$, Nanoscale 12 (46) (2020) 23694–23699, https://doi.org/10.1039/d0nr07124b.

[79] Y. Zhao, S. Zhuang, L. Liao, et al., A dual purpose strategy to endow gold nanoclusters with both catalysis activity and water solubility, J. Am. Chem. Soc. 142 (2) (2020) 973–977, https://doi.org/10.1021/jacs.9b11017.

[80] S. Bhunia, S. Kumar, P. Purkayastha, Gold nanocluster-grafted cyclodextrin suprastructures: formation of nanospheres to nanocubes with intriguing photophysics, ACS Omega 3 (2) (2018) 1492–1497, https://doi.org/10.1021/acsomega.7b01914.

[81] H. Zhu, N. Goswami, Q. Yao, et al., Cyclodextrin-gold nanocluster decorated TiO_2 enhances photocatalytic decomposition of organic pollutants, J. Mater. Chem. A 6 (3) (2018) 1102–1108, https://doi.org/10.1039/c7ta09443d.

[82] F. Diederich, M. Gómez-López, Supramolecular fullerene chemistry, Chem. Soc. Rev. 28 (5) (1999) 263–277, https://doi.org/10.1039/a804248i.

[83] P. Chakraborty, A. Nag, G. Paramasivam, G. Natarajan, T. Pradeep, Fullerene-functionalized monolayer-protected silver clusters: $[Ag_{29}(BDT)_{12}(C_{60})_n]^{3-}$ (n = 1–9), ACS Nano 12 (3) (2018) 2415–2425, https://doi.org/10.1021/acsnano.7b07759.

[84] P. Chakraborty, A. Nag, B. Mondal, E. Khatun, G. Paramasivam, T. Pradeep, Fullerene-mediated aggregation of $M_{25}(SR)_{18}^-$ (M = Ag, Au) nanoclusters, J. Phys. Chem. C 124 (27) (2020) 14891–14900, https://doi.org/10.1021/acs.jpcc.0c03383.

[85] M. Schulz-Dobrick, M. Jansen, Intercluster compounds consisting of gold clusters and fullerides: $[Au_7(PPh_3)_7]C_{60}\cdot THF$ and $[Au_8(PPh_3)_8](C_{60})_2$, Angew. Chem. Int. Ed. 47 (12) (2008) 2256–2259, https://doi.org/10.1002/anie.200705373.

[86] S. Li, Z.Y. Wang, G.G. Gao, et al., Smart transformation of a polyhedral oligomeric silsesquioxane shell controlled by thiolate silver(I) nanocluster core in cluster@clusters dendrimers, Angew. Chem. Int. Ed. 57 (39) (2018) 12775–12779, https://doi.org/10.1002/anie.201807548.

[87] Y. Yang, X.L. Pei, Q.M. Wang, Postclustering dynamic covalent modification for chirality control and chiral sensing, J. Am. Chem. Soc. 135 (43) (2013) 16184–16191, https://doi.org/10.1021/ja4075419.

[88] P. Chakraborty, A. Nag, K.S. Sugi, T. Ahuja, B. Varghese, T. Pradeep, Crystallization of a supramolecular coassembly of an atomically precise nanoparticle with a crown ether, ACS Mater. Lett. 1 (5) (2019) 534–540, https://doi.org/10.1021/acsmaterialslett.9b00352.

[89] A. Som, A.K. Samal, T. Udayabhaskararao, M.S. Bootharaju, T. Pradeep, Manifestation of the difference in reactivity of silver clusters in contrast to its ions and nanoparticles: the growth of metal tipped Te nanowires, Chem. Mater. 26 (10) (2014) 3049–3056, https://doi.org/10.1021/cm403288w.

[90] A. Som, I. Chakraborty, T.A. Maark, S. Bhat, T. Pradeep, Cluster-mediated crossed bilayer precision assemblies of 1D nanowires, Adv. Mater. 28 (14) (2016) 2827–2833, https://doi.org/10.1002/adma.201505775.

[91] A. Chakraborty, A.C. Fernandez, A. Som, et al., Atomically precise nanocluster assemblies encapsulating plasmonic gold nanorods, Angew. Chem. Int. Ed. 57 (22) (2018) 6522–6526, https://doi.org/10.1002/anie.201802420.

[92] P. Chakraborty, A. Nag, A. Chakraborty, T. Pradeep, Approaching materials with atomic precision using supramolecular cluster assemblies, Acc. Chem. Res. 52 (1) (2019) 2–11, https://doi.org/10.1021/acs.accounts.8b00369.

[93] O.M. Yaghi, M. O'Keeffe, N.W. Ockwig, H.K. Chae, M. Eddaoudi, J. Kim, Reticular synthesis and the design of new materials, Nature 423 (6941) (2003) 705–714, https://doi.org/10.1038/nature01650.

[94] M. O'Keeffe, O.M. Yaghi, Deconstructing the crystal structures of metal-organic frameworks and related materials into their underlying nets, Chem. Rev. 112 (2) (2012) 675–702, https://doi.org/10.1021/cr200205j.

[95] J. Jiang, Y. Zhao, O.M. Yaghi, Covalent chemistry beyond molecules, J. Am. Chem. Soc. 138 (10) (2016) 3255–3265, https://doi.org/10.1021/jacs.5b10666.

[96] S. Wang, C.M. McGuirk, A. Aquino, J.A. Mason, C.A. Mirkin, Metal–organic framework nanoparticles, Adv. Mater. 30 (6) (2018) 1800202, https://doi.org/10.1002/adma.201800202.

[97] A. Kirchon, L. Feng, H.F. Drake, E.A. Joseph, H.C. Zhou, From fundamentals to applications: a toolbox for robust and multifunctional MOF materials, Chem. Soc. Rev. 47 (23) (2018) 8611–8638, https://doi.org/10.1039/c8cs00688a.

[98] Y. Luo, S. Fan, W. Yu, et al., Fabrication of $Au_{25}(SG)_{18}$–ZIF-8 nanocomposites: a facile strategy to position $Au_{25}(SG)_{18}$ nanoclusters inside and outside ZIF-8, Adv. Mater. 30 (6) (2018) 1704576, https://doi.org/10.1002/adma.201704576.

[99] Y. Yun, H. Sheng, K. Bao, et al., Design and remarkable efficiency of the robust sandwich cluster composite nanocatalysts ZIF-8@Au_{25}@ZIF-67, J. Am. Chem. Soc. 142 (9) (2020) 4126–4130, https://doi.org/10.1021/jacs.0c00378.

[100] L. Sun, Y. Yun, H. Sheng, et al., Rational encapsulation of atomically precise nanoclusters into metal-organic frameworks by electrostatic attraction for CO_2 conversion, J. Mater. Chem. A 6 (31) (2018) 15371–15376, https://doi.org/10.1039/c8ta04667k.

[101] L. Liu, Y. Song, H. Chong, et al., Size-confined growth of atom-precise nanoclusters in metal-organic frameworks and their catalytic applications, Nanoscale 8 (3) (2016) 1407–1412, https://doi.org/10.1039/c5nr06930k.

[102] C. Liu, C. Zeng, T.Y. Luo, A.D. Merg, R. Jin, N.L. Rosi, Establishing porosity gradients within metal-organic frameworks using partial postsynthetic ligand exchange, J. Am. Chem. Soc. 138 (37) (2016) 12045–12048, https://doi.org/10.1021/jacs.6b07445.

[103] S. Chandra, Nonappa, G. Beaune, et al., Highly luminescent gold nanocluster frameworks, Adv. Opt. Mater. 7 (20) (2019), https://doi.org/10.1002/adom.201900620.

[104] C. Zeng, Y. Chen, K. Kirschbaum, K.J. Lambright, R. Jin, Emergence of hierarchical structural complexities in nanoparticles and their assembly, Science 354 (6319) (2016) 1580–1584, https://doi.org/10.1126/science.aak9750.

[105] C. Zeng, Y. Chen, K. Kirschbaum, K. Appavoo, M.Y. Sfeir, R. Jin, Structural patterns at all scales in a nonmetallic chiral $Au_{133}(SR)_{52}$ nanoparticle, Sci. Adv. 1 (2) (2015), https://doi.org/10.1126/sciadv.1500045.

[106] A. Dass, S. Theivendran, P.R. Nimmala, et al., $Au_{133}(SPh-^tBu)_{52}$ nanomolecules: x-ray crystallography, optical, electrochemical, and theoretical analysis, J. Am. Chem. Soc. 137 (14) (2015) 4610–4613, https://doi.org/10.1021/ja513152h.

[107] Y. Song, Y. Li, H. Li, et al., Atomically resolved $Au_{52}Cu_{72}(SR)_{55}$ nanoalloy reveals Marks decahedron truncation and Penrose tiling surface, Nat. Commun. 11 (1) (2020), https://doi.org/10.1038/s41467-020-14400-2.

[108] T. Higaki, C. Liu, M. Zhou, T.Y. Luo, N.L. Rosi, R. Jin, Tailoring the structure of 58-electron gold nanoclusters: $Au_{103}S_2(S-nap)_{41}$ and its implications, J. Am. Chem. Soc. 139 (29) (2017) 9994–10001, https://doi.org/10.1021/jacs.7b04678.

[109] P.D. Jadzinsky, G. Calero, C.J. Ackerson, D.A. Bushnell, R.D. Kornberg, Structure of a thiol monolayer-protected gold nanoparticle at 1.1 Å resolution, Science 318 (5849) (2007) 430–433, https://doi.org/10.1126/science.1148624.

[110] R.-W. Huang, J. Yin, C. Dong, et al., $[Cu_{81}(PhS)_{46}(tBuNH_2)_{10}(H)_{32}]^{3+}$ reveals the coexistence of large planar cores and hemispherical shells in high-nuclearity copper nanoclusters, J. Am. Chem. Soc. 142 (19) (2020) 8696–8705, https://doi.org/10.1021/jacs.0c00541.

[111] W.J. Zhang, Z. Liu, K.P. Song, et al., A 34-electron superatom Ag_{78} cluster with regioselective ternary ligands shells and its 2D rhombic superlattice assembly, Angew. Chem. Int. Ed. 60 (8) (2021) 4231–4237, https://doi.org/10.1002/anie.202013681.

[112] C. Zeng, C. Liu, Y. Chen, N.L. Rosi, R. Jin, Atomic structure of self-assembled monolayer of thiolates on a tetragonal Au92 nanocrystal, J. Am. Chem. Soc. 138 (28) (2016) 8710–8713, https://doi.org/10.1021/jacs.6b04835.

[113] L. Liao, J. Chen, C. Wang, et al., Transition-sized Au_{92} nanoparticle bridging non-fcc-structured gold nanoclusters and fcc-structured gold nanocrystals, Chem. Commun. 52 (81) (2016) 12036–12039, https://doi.org/10.1039/c6cc06108g.

[114] Z. Gan, J. Chen, J. Wang, et al., The fourth crystallographic closest packing unveiled in the gold nanocluster crystal, Nat. Commun. 8 (2017), https://doi.org/10.1038/ncomms14739.

[115] Z. Gan, J. Chen, L. Liao, H. Zhang, Z. Wu, Surface single-atom tailoring of a gold nanoparticle, J. Phys. Chem. Lett. 9 (1) (2018) 204–208, https://doi.org/10.1021/acs.jpclett.7b02982.

[116] X. Liu, J. Chen, J. Yuan, et al., A silver nanocluster containing interstitial sulfur and unprecedented chemical bonds, Angew. Chem. Int. Ed. 57 (35) (2018) 11273–11277, https://doi.org/10.1002/anie.201805594.

[117] B. Yoon, W.D. Luedtke, R.N. Barnett, et al., Hydrogen-bonded structure and mechanical chiral response of a silver nanoparticle superlattice, Nat. Mater. 13 (8) (2014) 807–811, https://doi.org/10.1038/nmat3923.

[118] A. Desireddy, B.E. Conn, J. Guo, et al., Ultrastable silver nanoparticles, Nature 501 (7467) (2013) 399–402, https://doi.org/10.1038/nature12523.

[119] H. Yang, Y. Wang, H. Huang, et al., All-thiol-stabilized Ag_{44} and $Au_{12}Ag_{32}$ nanoparticles with single-crystal structures, Nat. Commun. (2013), https://doi.org/10.1038/ncomms3422.

[120] J.H. Huang, Z.Y. Wang, S.Q. Zang, T.C.W. Mak, Spontaneous resolution of chiral multi-thiolate-protected Ag_{30} nanoclusters, ACS Cent. Sci. 6 (11) (2020) 1971–1976, https://doi.org/10.1021/acscentsci.0c01045.

[121] A. Nag, P. Chakraborty, M. Bodiuzzaman, T. Ahuja, S. Antharjanam, T. Pradeep, Polymorphism of $Ag_{29}(BDT)_{12}(TPP)_4^{3-}$ cluster: interactions of secondary ligands and their effect on solid state luminescence, Nanoscale 10 (2018) 9851–9855, https://doi.org/10.1039/C8NR02629G.

[122] L.G. AbdulHalim, M.S. Bootharaju, Q. Tang, et al., $Ag_{29}(BDT)_{12}(TPP)_4$: a tetravalent nanocluster, J. Am. Chem. Soc. 137 (37) (2015) 11970–11975, https://doi.org/10.1021/jacs.5b04547.

[123] L. He, Z. Gan, N. Xia, L. Liao, Z. Wu, Alternating array stacking of $Ag_{26}Au$ and $Ag_{24}Au$ nanoclusters, Angew. Chem. Int. Ed. 58 (29) (2019) 9897–9901, https://doi.org/10.1002/anie.201900831.

[124] D. Liu, W. Du, S. Chen, et al., Interdependence between nanoclusters $AuAg_{24}$ and Au_2Ag_{41}, Nat. Commun. 12 (2021) 778, https://doi.org/10.1038/s41467-021-21131-5.

[125] Q. Li, J.C. Russell, T.-Y. Luo, et al., Modulating the hierarchical fibrous assembly of Au nanoparticles with atomic precision, Nat. Commun. 9 (2018) 3871, https://doi.org/10.1038/s41467-018-06395-8.

[126] A. Das, T. Li, K. Nobusada, C. Zeng, N.L. Rosi, R. Jin, Nonsuperatomic $[Au_{23}(SC_6H_{11})_{16}]^-$ nanocluster featuring bipyramidal Au_{15} kernel and trimeric $Au_3(SR)_4$ motif, J. Am. Chem. Soc. 135 (49) (2013) 18264–18267, https://doi.org/10.1021/ja409177s.

[127] Y. Li, T.Y. Luo, M. Zhou, Y. Song, N.L. Rosi, R. Jin, A correlated series of Au/Ag nanoclusters revealing the evolutionary patterns of asymmetric Ag doping, J. Am. Chem. Soc. 140 (43) (2018) 14235–14243, https://doi.org/10.1021/jacs.8b08335.

[128] Z. Wu, Y. Du, J. Liu, et al., Aurophilic interactions in the self-assembly of gold nanoclusters into nanoribbons with enhanced luminescence, Angew. Chem. Int. Ed. 58 (24) (2019) 8139–8144, https://doi.org/10.1002/anie.201903584.

[129] M. De Nardi, S. Antonello, D.E. Jiang, et al., Gold nanowired: a linear $(Au_{25})_n$ polymer from Au_{25} molecular clusters, ACS Nano 8 (8) (2014) 8505–8512, https://doi.org/10.1021/nn5031143.

[130] S. Antonello, T. Dainese, F. Pan, K. Rissanen, F. Maran, Electrocrystallization of monolayer-protected gold clusters: opening the door to quality, quantity, and new structures, J. Am. Chem. Soc. 139 (11) (2017) 4168–4174, https://doi.org/10.1021/jacs.7b00568.

[131] W. Fei, S. Antonello, T. Dainese, et al., Metal doping of $Au_{25}(SR)_{18}^-$ clusters: insights and hindsights, J. Am. Chem. Soc. 141 (40) (2019) 16033–16045, https://doi.org/10.1021/jacs.9b08228.

[132] S. Hossain, Y. Imai, Y. Motohashi, et al., Understanding and designing one-dimensional assemblies of ligand-protected metal nanoclusters, Mater. Horiz. 7 (3) (2020) 796–803, https://doi.org/10.1039/c9mh01691k.

[133] J. Chen, L. Liu, X. Liu, et al., Gold-doping of double-crown Pd nanoclusters, Chem. A Eur. J. 23 (72) (2017) 18187–18192, https://doi.org/10.1002/chem.201704413.

[134] D.E. Jiang, S. Dai, From superatomic $Au_{25}(SR)_{18}^-$ to superatomic $M@Au_{24}(SR)_{18}^q$ core-shell clusters, Inorg. Chem. 48 (7) (2009) 2720–2722, https://doi.org/10.1021/ic8024588.

[135] H. Qian, D.E. Jiang, G. Li, et al., Monoplatinum doping of gold nanoclusters and catalytic application, J. Am. Chem. Soc. 134 (39) (2012) 16159–16162, https://doi.org/10.1021/ja307657a.

[136] P. Yuan, R. Zhang, E. Selenius, et al., Solvent-mediated assembly of atom-precise gold–silver nanoclusters to semiconducting one-dimensional materials, Nat. Commun. 11 (1) (2020), https://doi.org/10.1038/s41467-020-16062-6.

[137] K. Zhou, C. Qin, X.L. Wang, K.Z. Shao, L.K. Yan, Z.M. Su, Unexpected 1D self-assembly of carbonate-templated sandwich-like macrocycle-based $Ag_{20}S_{10}$ luminescent nanoclusters, CrstEngComm 16 (34) (2014) 7860–7864, https://doi.org/10.1039/c4ce00867g.

[138] Z.Y. Chen, D.Y.S. Tam, L.L.M. Zhang, T.C.W. Mak, Silver thiolate nano-sized molecular clusters and their supramolecular covalent frameworks: an approach toward pre-templated synthesis, Chem. Asian J. 12 (20) (2017) 2763–2769, https://doi.org/10.1002/asia.201701150.

[139] Z. Wang, Y.M. Sun, Q.P. Qu, et al., Enclosing classical polyoxometallates in silver nanoclusters, Nanoscale 11 (22) (2019) 10927–10931, https://doi.org/10.1039/c9nr04045e.

[140] Z.R. Wen, Z.J. Guan, Y. Zhang, Y.M. Lin, Q.M. Wang, $[Au_7Ag_9(dppf)_3(CF_3CO_2)_7BF_4]_n$: a linear nanocluster polymer from molecular Au_7Ag_8 clusters covalently linked by silver atoms, Chem. Commun. 55 (86) (2019) 12992–12995, https://doi.org/10.1039/c9cc05924e.

[141] X.-Y. Li, H.-F. Su, J. Xu, A 2D layer network assembled from an open dendritic silver cluster $Cl@Ag_{11}N_{24}$ and an N-donor ligand, Inorg. Chem. Front. (2019) 3539–3544, https://doi.org/10.1039/c9qi01100e.

[142] T. Wu, D. Yin, X. Hu, et al., A disulfur ligand stabilization approach to construct a silver(i)-cluster-based porous framework as a sensitive SERS substrate, Nanoscale 11 (35) (2019) 16293–16298, https://doi.org/10.1039/c9nr05301h.

[143] T. Chen, S. Yang, J. Chai, et al., Crystallization-induced emission enhancement: a novel fluorescent Au-Ag bimetallic nanocluster with precise atomic structure, Sci. Adv. 3 (8) (2017), https://doi.org/10.1126/sciadv.1700956.

[144] S. Nematulloev, R. Huang, J. Yin, et al., $[Cu_{15}(PPh_3)_6(PET)_{13}]^{2+}$: a copper nanocluster with crystallization enhanced photoluminescence, Small 17 (27) (2021) 2006839, https://doi.org/10.1002/smll.202006839.

[145] S. Kolay, S. Maity, D. Bain, S. Chakraborty, A. Patra, Self-assembly of copper nanoclusters: isomeric ligand effect on morphological evolution, Nanoscale Adv. 3 (19) (2021) 5570–5575, https://doi.org/10.1039/D1NA00446H.

[146] H. Wu, X. He, B. Yang, C.-C. Li, L. Zhao, Assembly-induced strong circularly polarized luminescence of spirocyclic chiral silver(I) clusters, Angew. Chem. Int. Ed. 60 (3) (2021) 1535–1539, https://doi.org/10.1002/anie.202008765.

[147] X. Wei, X. Kang, Q. Yuan, et al., Capture of cesium ions with nanoclusters: effects on inter-and intramolecular assembly, Chem. Mater. 31 (13) (2019) 4945–4952, https://doi.org/10.1021/acs.chemmater.9b01890.

[148] X. Wei, X. Kang, Z. Zuo, F. Song, S. Wang, M. Zhu, Hierarchical structural complexity in atomically precise nanocluster frameworks, Natl. Sci. Rev. 8 (3) (2021), https://doi.org/10.1093/nsr/nwaa077.

[149] S. Chen, W. Du, C. Qin, et al., Assembly of the thiolated [Au$_1$Ag$_{22}$(SAdm)$_{12}$]$^{3+}$ superatom complex into a framework material through direct linkage by SbF$_6^-$ anions, Angew. Chem. Int. Ed. 59 (19) (2020) 7542–7547, https://doi.org/10.1002/anie.202000073.

[150] M. Zhao, S. Huang, Q. Fu, et al., Ambient chemical fixation of CO$_2$ using a robust Ag$_{27}$ cluster-based two-dimensional metal-organic framework, Angew. Chem. Int. Ed. 59 (45) (2020) 20031–20036, https://doi.org/10.1002/anie.202007122.

[151] X.H. Ma, J.Y. Wang, J.J. Guo, Z.Y. Wang, S.Q. Zang, Reversible wide-range tuneable luminescence of a dual-stimuli- responsive silver cluster-assembled material, Chin. J. Chem. 37 (11) (2019) 1120–1124, https://doi.org/10.1002/cjoc.201900314.

[152] M.J. Alhilaly, R.W. Huang, R. Naphade, et al., Assembly of atomically precise silver nanoclusters into nanocluster-based frameworks, J. Am. Chem. Soc. 141 (24) (2019) 9585–9592, https://doi.org/10.1021/jacs.9b02486.

[153] Z.Y. Wang, M.Q. Wang, Y.L. Li, et al., Atomically precise site-specific tailoring and directional assembly of superatomic silver nanoclusters, J. Am. Chem. Soc. 140 (3) (2018) 1069–1076, https://doi.org/10.1021/jacs.7b11338.

[154] R.W. Huang, Y.S. Wei, X.Y. Dong, et al., Hypersensitive dual-function luminescence switching of a silver-chalcogenolate cluster-based metal-organic framework, Nat. Chem. 9 (7) (2017) 689–697, https://doi.org/10.1038/nchem.2718.

[155] X.Y. Dong, Y. Si, J.S. Yang, et al., Ligand engineering to achieve enhanced ratiometric oxygen sensing in a silver cluster-based metal-organic framework, Nat. Commun. 11 (1) (2020), https://doi.org/10.1038/s41467-020-17200-w.

[156] R.W. Huang, X.Y. Dong, B.J. Yan, et al., Tandem silver cluster isomerism and mixed linkers to modulate the photoluminescence of cluster-assembled materials, Angew. Chem. Int. Ed. 57 (28) (2018) 8560–8566, https://doi.org/10.1002/anie.201804059.

[157] X.S. Du, B.J. Yan, J.Y. Wang, X.J. Xi, Z.Y. Wang, S.Q. Zang, Layer-sliding-driven crystal size and photoluminescence change in a novel SCC-MOF, Chem. Commun. 54 (43) (2018) 5361–5364, https://doi.org/10.1039/C8CC01559G.

[158] M. Cao, R. Pang, Q.Y. Wang, H.Y. Li, S.Q. Zang, Porphyrinic silver cluster assembled material for simultaneous capture and photocatalysis of mustard-gas simulant, J. Am. Chem. Soc. 141 (37) (2019) 14505–14509, https://doi.org/10.1021/jacs.9b05952.

[159] Z. We, X.H. Wu, P. Luo, J.Y. Wang, K. Li, S.Q. Zang, Matrix coordination induced emission in a three-dimensional silver cluster-assembled material, Chem. Eur. J. 25 (11) (2019) 2750–2756, https://doi.org/10.1002/chem.201805381.

[160] X.H. Wu, Z. Wei, B.J. Yan, et al., Mesoporous crystalline silver-chalcogenolate cluster-assembled material with tailored photoluminescence properties, CCS Chem. 1 (5) (2019) 553–560, https://doi.org/10.31635/ccschem.019.20190024.

[161] Y.M. Wang, J.W. Zhang, Q.Y. Wang, H.Y. Li, X.Y. Dong, S. Wang, S.Q. Zang, Fabrication of silver chalcogenolate cluster hybrid membranes with enhanced structural stability and luminescence efficiency, Chem. Commun. 55 (97) (2019) 14677–14680, https://doi.org/10.1039/C9CC07797A.

[162] X.H. Wu, P. Luo, Z. Wei, et al., Guest-triggered aggregation-induced emission in silver chalcogenolate cluster metal-organic frameworks, Adv. Sci. 6 (2) (2019) 1801304, https://doi.org/10.1002/advs.201801304.

[163] X.Y. Dong, H.L. Huang, J.Y. Wang, et al., A flexible fluorescent SCC-MOF for switchable molecule identification and temperature, Chem. Mater. 30 (6) (2018) 2160–2167, https://doi.org/10.1021/acs.chemmater.8b00611.

[164] C. Xu, N. Hedin, H.T. Shi, Q.F. Zhang, A semiconducting microporous framework of Cd$_6$Ag$_4$(Sph)$_{16}$ clusters interlinked using rigid and conjugated bipyridines, Chem. Commun. 50 (28) (2014) 3710–3712, https://doi.org/10.1039/c3cc49660k.

[165] Z.K. Wang, M.M. Sheng, S.S Qin, et al., Assembly of discrete chalcogenolate clusters into a one-dimensional coordination polymer with enhanced photocatalytic activity and stability, Inorg. Chem. 59 (4) (2020) 2121–2126, https://doi.org/10.1021/acs.inorgchem.9b03578.

[166] C. Xu, M.M. Sheng, H.T. Shi, M. Strømme, Q.F. Zhang, Interlinking supertetrahedral chalcogenolate clusters with bipyridines to form two-dimensional coordination polymers for photocatalytic degradation of organic dye, Dalton Trans. 48 (17) (2019) 5505–5510, https://doi.org/10.1039/c9dt00480g.

[167] Z. Lei, X.L. Pei, Z.G. Jiang, Q.M. Wang, Cluster linker approach: preparation of a luminescent porous framework with NbO topology by linking silver ions with gold(I) clusters, Angew. Chem. Int. Ed. 53 (47) (2014) 12771–12775, https://doi.org/10.1002/anie.201406761.

Chapter 8

Practical applications of metal nanoclusters

Manzhou Zhu and Yuanxin Du

8.1 Introduction

Metal nanoclusters (NCs) (core size below 2 nm) are composed of several to a few hundred metal atoms, and exhibits discrete electronic states and molecule-like characteristics because its size is comparable with the Fermi wavelength of electrons. It has greatly extended the horizon of nanomaterials science by filling the gap between isolated metal atoms and plasmonic metal nanoparticles. Metal NCs possess dramatically unique physical and chemical properties different from its larger counterparts-nanoparticles (core sizes above 2 nm), such as well-defined molecular structures, HOMO-LUMO transitions, strong photoluminescence, and high catalytic properties. Such uniqueness endows it significant roles in practical applications. Metal NCs have attracted a widespread interest in biological applications, such as biochemical sensing, biolabeling/imaging, and disease diagnostics and therapy, benefiting from its strong luminescence, high quantum yields, tunable fluorescence emission, large Stokes shift, excellent photostability and high biocompatibility. Moreover, metal NCs with ultrasmall size, abundant active sites, large specific surface area and unique electronic structure, establishing it as a new kind of highly efficient nanocatalyst has shown high catalytic activity and selectivity in many reactions. Particularly, for those with precise compositions and structures, it can be utilized as model catalysts to shed light of the structure-performance relationship, which has crucial effect on rational designing high efficiency catalysts. In this chapter, we will focus on the practical application of metal NCs such as sensors, biological application, and catalysis. We will review the recent progress in corresponding field and provide prospects regarding potential opportunities and challenges of metal NCs in application.

8.2 Sensors

As a new type of optical materials, metal NCs have a series of advantages including small size, high optical stability, long lifetime, large Stokes shift, adjustable optical properties, nontoxic, and so on. Such intriguing benefits make up for some drawbacks in traditional optical materials such as organic fluorescent dyes, fluorescent protein, and fluorescence quantum dots. Furthermore, metal NCs exhibit highly sensitive and selective response to external analytes, showing on the obvious change in optical absorption and luminescence properties, which makes metal NCs a promising optical sensor. Therefore, various metal NC-optical sensors based on luminescence quenching, luminescence enhancement, luminescence color change, fluorescence resonance energy transfer (FRET), ratiometric fluorescence, electrochemical luminescence (ECL) have been constructed and widely used in the analysis of sensing field. According to the types of analytes, we will summarize the metal NC in sensing application in the following three aspects.

8.2.1 Chemical sensor

8.2.1.1 Heavy metal ion sensing

With the progress of science and the improvement of human health awareness, heavy metal ions in the environment, such as mercury ion, lead ion, arsenic ion, chromium ion, and copper ion, have been paid more and more attention. It is well known that heavy metal ions have serious adverse effects on the human body, especially on the central nervous system, kidneys, liver, skin, bones, teeth, etc. These heavy metal ions are found to be nondegradable and can enter the body, for example through the food chain. After entering, they will react with proteins and enzymes, causing protein denaturation and changing the function of enzymes, resulting in cell metabolism disorders. Besides, they can also accumulate in the organs and lead to chronic poisoning. Because of their high toxicity, the tolerance limits for Hg^{2+}, Pb^{2+}, As^+, Cr^{6+}, and Cu^{2+} in drinking water recommended by the environmental protection agency (EPA) of the United States are 0.002, 0.015, 0.01, 0.05, and 1.3 ppm, respectively.

These heavy metal ions are difficult to detect in complex environmental and biological samples. At present, the detection methods of metal ions include atomic absorption spectrometry, inductively coupled plasma emission spectrometry, mass spectrometry, etc. Although inductively coupled plasma emission spectrometry has the advantages of high sensitivity, small interference and wide linear range, the instrument is expensive and the detection cost is high. Although atomic absorption spectrometry has the advantages of high accuracy, fast analysis speed and many kinds of analysis, it cannot be used for simultaneous determination of multiple analytes. For mass spectrometry detection method, the sample preparation and the preconcentration steps are complicated, and the instrument is complex, which requires skilled professionals. Generally speaking, although the above methods have good selectivity and sensitivity to a certain extent, they either need expensive and complex equipment or require complicated procedures in the detection process, so they are not suitable for large-scale detection and real-time detection. Therefore, it is necessary to design an easy to operate, low cost, high sensitivity and real-time sensor system to identify and detect metal ions.

Hg^{2+} sensing

In the early days, DNA-labeled Au nanoparticles (NPs) were used to detect mercury ions based on the specific T-Hg^{2+}-T coordination [1–3]. The presence of Hg^{2+} induces the aggregation of DNA-labeled Au NPs and thus causes the change of surface plasmon resonance (SPR) optical properties of Au NPs. However, the cost of label is relatively high and impractical for routine detection. Different from large metal NPs detecting metal ions by using remarkable SPR optical properties [3], metal NCs has been acted as metal ions sensors based on the fluorescence properties [4]. Ying et al. developed fluorescent bovine serum albumin (BSA)-Au NCs [5] as a label-free sensor to detect Hg^{2+} based on the fluorescence quenching caused by the high-affinity metallophilic Hg^{2+}-Au^+ interactions [6]. The synthesized Au NCs was comprised of 25 Au atoms, which had about 17% Au^+ on the surface of the cluster could provide a strong and specific interaction with Hg^{2+}. The limit of the detection (LOD) for Hg^{2+} was estimated to be 0.5 nM and the liner detection range was from 1 to 20 nM. The Au NCs had a high fluorescence emission at 640 nm and showed a bright red color under UV light. Therefore, they extended the Au NCs sensor to a paper test strip to realize the visualized detection with the naked eye. Furthermore, the metallophilic bonding between $3d^{10}$ Cu^+ with $4d^{10}$ Ag^+ had similar interactions as the bonding between $5d^{10}$ Hg^{2+} and $5d^{10}$ Au^+. Based on this concept, they further synthesized Au@Ag NCs to realize the selectively detection for Cu^{2+}. Cai et al. also synthesized BSA-Au NCs and applied it into Hg^{2+} sensing by fluorescence quenching effect [7]. Yu and Zhu et al. found that Ag^+ could amplify the fluorescent signal of BSA-Au NCs. They used the amplification effect of Ag^+ to realize the high sensitive detection for Hg^{2+} with a lower LOD of 6 nM, compared with the system without Ag^+ with a LOD of 8 nM. The detection upper limit was extended from 20 μM to 0.3 mM [8].

Besides BSA, there are many other proteins that can be used as reduction and stabilized agents to synthesize fluorescent Au NCs. For example, ARAKAWA et al. synthesized trypsin-stabilized Au NCs with a red emission at 645 nm, which was about 1 nm and had a good photostability which was similar to that of CdSe quantum dots (QDs). This Au NC was sensitive and selective to Hg^{2+}, accompanied by an obvious fluorescence quenching with the addition of Hg^{2+}. The trypsin-stabilized Au NCs sensor for the quantitative detection of Hg^{2+} had a wide range and low concentration, ranging from 50 to 600 nM [9]. Lu et al. prepared lysozyme (Lys)-stabilized Au NCs and applied to the detection of Hg^{2+}. The Lys-Au NCs had two fluorescence emission peaks which were located at 660 nm and 445 nm, respectively. Interestingly, with the increase of Hg^{2+}, the peak at 660 nm gradually decreased while the 445 nm peak had a negligible change. Based on this phenomenon, it can be used as a ratiometric fluorescent sensor [10]. However, they did not realize this at that time. In the next years, there have been a lot of sensors based on ratiometric fluorescence mechanism, which we will illustrate in detail in the later.

Chang et al. used a series of alkanethiol with different chain length such as 2-mercaptoethanol (2-ME), 6-mercaptohexanol (6-MH), and 11-mercaptoundecanoic acid (11-MUA) to synthesize water soluble Au NCs and realize the control of the fluorescent properties. Among them, 11-MUA-Au NCs showed the highest quantum yield (QY). Due to the Hg^{2+} induced the aggregation of Au NCs, the fluorescence intensity of Au NC was decreased. Thanks to the chelating ligand 2,6-pyridinedicarboxylic acid (PDCA), the 11-MUA-Au NCs sensor can realize selectively detection for Hg^{2+}. The presence of PDCA eradicated the interference of Pb^{2+} and Cd^{2+}. Even the concentration of Pb^{2+} and Cd^{2+} was 400-fold higher than that of Hg^{2+}, the 11-MUA-Au NCs sensor still displayed a single sensitivity to Hg^{2+} [4].

Although these above sensors showed high sensitive and selective response to Hg^{2+}, most of them detected Hg^{2+} only in simple aqueous medium. It is necessary to explore a sensor which is suitable for all kinds of complex matrix. Chang et al. incorporated poly(N-isopropylacrylamide) microgels (PNIPAM MGs) with fluorescent Au NCs (namely Au NC-PNIPAM MGs). The hybrid was more stable than Au NCs and had a high tolerance to pH and salt concentration variation. The Au NC-PNIPAM MGs had the LOD for Hg^{2+} is 1.7 nM in phosphate solution pH = 7 (Fig. 8.1A). When the solution was contained with 500 mM NaCl, the LOD was 1.9 nM. Furthermore, they used the Au NC-PNIPAM MGs sensor to detect Hg^{2+} in

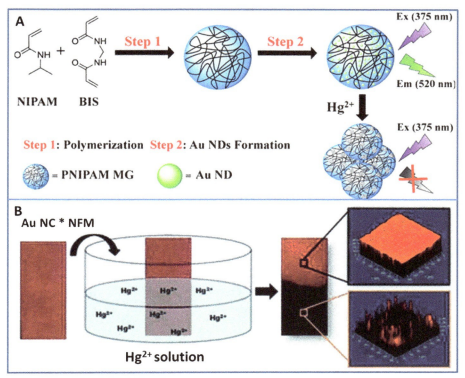

FIG. 8.1 Schematic diagram of Au NC for detection of mercury ions (A) Schematic diagram of the synthesis of fluorescent Au NC-PNIPAM MGs and its detection of mercury ions; (B) self-standing Au NCs membrane sensor for visual colorimetric detection of Hg^{2+}.

fish samples. The detected result showed a high consistency with the true value, indicating the practical application potential of this sensor [11].

Nienhaus et al. utilized microwave-assisted method to rapidly synthesize dihydrolipoic acid (DHLA)-protected Au NCs with high near-infrared (NIR) fluorescence emission (fivefold enhancement of QY). Due to the specific interaction between Hg^{2+} and Au^+ on the surface of Au NCs, the fluorescence quenching of DHLA-Au NCs occurred upon the treatment of Hg^{2+}. The DHLA-Au NCs was sensitive to Hg^{2+} with the LOD as low as 0.5 nM. In addition, the DHLA-Au NCs had a good stability in the pH range of 5–10 and had a low cytotoxicity toward HeLa cells. Therefore, they used the DHLA-Au NCs to detect Hg^{2+} in living cells. [12] Firstly, the HeLa cells needed to be incubated with Au NCs for 2 h, and then Au NCs was localized inside the cells. Under spinning disc laser microscopy, bright luminescence cells could be observed. Secondly, adding a series different concentration of Hg^{2+}, an obvious quenching of fluorescence could be seen within 10 min. The results indicated the Au NCs could detect Hg^{2+} not only in simple environment but also in complex system like biological sample.

Except the detection environment, the detected specie of Hg is still limited. Tseng et al. prepared lysozyme type VI-stabilized Au NCs (Lys VI-Au NCs) and realized the ultrasensitive detection of Hg^{2+} and CH_3Hg^+ in seawater [13]. The fluorescence properties and particle sizes of the Au NCs were highly depended on the concentration of Lys VI. With the increasing of Lys VI, the Au NCs had a higher QY. In the combination of cost, the optimal concentration of Lys VI was determined at 25 mg/mL, the formed Lys VI-Au NCs (namely Au-631) had a fluorescence emission peak at 631 nm with a high QY ~9%. The Au-631 had more Au^+ on the surface of Au core than BSA-Au NCs. Due to the stronger interaction with Hg^{2+}/CH_3Hg^+, the Au-631 sensor exhibited a more sensitive detection to Hg^{2+} and CH_3Hg^+. The LOD for Hg^{2+} and CH_3Hg^+ in deionized water were 3 pM and 4 nM, respectively. In addition to the low detection limit, the selectivity of Au-631 sensor had an attractive selectivity which was more than 500-fold for Hg^{2+} over any other metal ions. The Au-631 could be stabilized in high concentration of NaCl. Therefore, they used it to detect mercury ion in complex matrix, such as seawater. The LODs for Hg^{2+} and CH_3Hg^+ in seawater were 0.51 and 5.90 nM, respectively.

Above Au NCs sensors are mostly solution-based, limiting the potential effectiveness and practical applications. Senthamizhan and Uyar et al. designed a self-standing and easy handling membrane-based Au NCs sensor by using a flexible fluorescent electrospun nanofibrous membrane (Fig. 8.1B) [14]. They have incorporated Au NCs with polyvinyl alcohol (PVA) nanofibers, the integrated Au NCs membrane exhibited red fluorescence under UV light with high stability (at least 6 months). The Au NCs membrane could be still usable even at the high temperature up to 100°C. They further

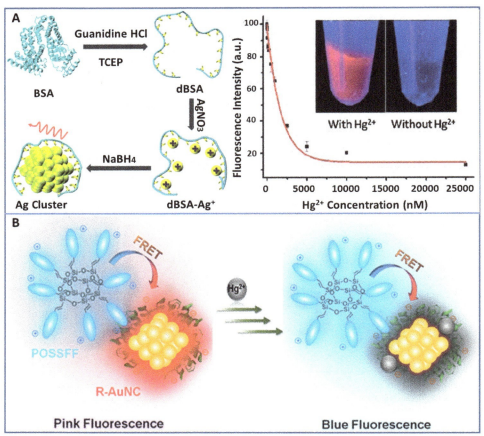

FIG. 8.2 Schematic diagram of Ag NCs and Au NCs FRET sensor for Hg^{2+} detection (A) Schematic diagram of the synthesis of dBSA-Ag NCs and its detection of Hg^{2+}; (B) POSSFF and R-Au NCs FRET sensor used for Hg^{2+} sensing.

extended the water-soluble Au NCs membrane to water-insoluble one by cross-linking with glutaraldehyde vapor. The Au NCs membrane could be used as naked-eye Hg^{2+} sensor with a low LOD of 1 ppb and high selectivity and good reproducibility.

Besides fluorescent Au NCs, Ag NCs also can be used as probes to fabricate sensors. Irudayaraj et al. used denatured bovine serum albumin (dBSA) as stabilizing agents to synthesize water-soluble Ag NCs with a high fluorescence emission at ~637 nm. The Ag NCs had high stability in high concentrated NaCl. Due to the $5d^{10}(Hg^{2+})$-$4d^{10}(Ag^{+})$ metallophilic interaction, there was a specific fluorescence quenching of Ag clusters when exposed to Hg^{2+} (Fig. 8.2A). Therefore, the Ag NCs was used as a sensitive Hg^{2+} sensor with the LOD of 10 nM [15]. Wang et al. synthesized oligonucleotide-stabilized Ag NCs with a fluorescence emission at 650 nm and a high QY of $28 \pm 1\%$. This Ag NCs showed a rapid and sensitive response to Hg^{2+}. After 3 min of adding Hg^{2+}, the fluorescence intensity of Ag NCs has decreased to its maximum. The LOD of this Ag NCs sensor for Hg^{2+} was 5 nM, and the detection widely ranged from 5 nM to 1.5 μM [16]. Banerjee et al. synthesized water-soluble DHLA-Ag NCs and realized ultrasensitive Hg^{2+} sensing with a low LOD down to 0.1 nM [17]. They supposed the fluorescence quenching was caused by the Hg^{2+}-induced aggregation. The free carboxylic acid which was present in reduced lipoic acid on the surface of Ag NCs could interact with Hg^{2+}. In the presence of Hg^{2+}, the Ag NCs aggregated and became a large mass which can be confirmed by the HRTEM image, there was no particle size <7 nm.

Among all the metal ions, there is a specific attraction between mercury and sulfur because of soft-soft interaction, therefore, lots of Hg^{2+} sensors based on such chemical interactions have been developed. However, due to lack of well-defined molecular composition and structure, all of these sensing systems are not able to give detailed mechanism of the specific interaction. Pradeep et al. took Ag_{25} NC as an example to detect Hg^{2+} [18]. The UV-vis absorption spectrum of Ag_{25} has shown a dramatic change exposure under 5d block ions (the decrease of 478 nm peak and the disappearance of 640 nm peak). When the concentration of metal ions reduced to 10 ppm, only Hg^{2+} caused an obvious changed in UV-vis absorption property of Ag_{25}, while other ions showed little effect on it. Due to the formation of Ag-Hg alloy [19,20], the 478 nm peak of Ag_{25} was blue-shifted accompanied with a new hump at 420 nm. With the sensitive change of optical absorption, the Ag_{25} sensor could detect Hg^{2+} at lower limit of 1 ppb. Besides, the fluorescence intensity of Ag_{25} decreased with the addition of Hg^{2+}, the sensing of Hg^{2+} also could be realized by this phenomenon.

It is common to detect Hg^{2+} by using fluorescence quenching. Nevertheless, few studies have discussed the detailed sensing mechanism. Pradeep et al. used Fourier transform infrared spectroscopy (FT-IR), scanning electron microscopy (SEM), high resolution transmission electron microscopy (HRTEM), X-ray diffraction (XRD), and X-ray photoelectron spectroscopy (XPS) to investigate the specific interaction between Hg^{2+} and Ag_{25}. In FT-IR spectra analysis, it was found that the band at $1658\,cm^{-1}$ corresponding to the C=O stretching has shifted to $1643\,cm^{-1}$ with Hg^{2+} (in both situations, 10 ppm and 10 ppb). It was supposed that the Hg^{2+} has bound with the carbonyl moiety of the glutathione (the ligand of Ag_{25}). When adding 10 ppm Hg^{2+} into Ag_{25}, the cluster became aggregated, meanwhile, the high energy electronic beam caused aggregation of NC. Therefore, there were large particles in TEM image. The large particle showed a lattice corresponding to the (021) plane of Ag_3Hg_2, which was also confirmed by the XRD pattern. That was because high concentration Hg^{2+} could bind with the sulfur of the ligand and interact with the Ag surface resulting in the reduction of Hg^{2+} to Hg and the formation of AgHg alloy. At low concentration of Hg^{2+}, the XPS spectra of Ag 3d showed a peak at 367.8 eV which was attributed to the Ag^0, it was indicated the low concentration Hg^{2+} could not affect the Ag_{25} core and the oxidation could not happen. At high concentration of Hg^{2+}, the XPS spectra of Hg 4f showed a peak at 99.9 eV which was assigned to metallic mercury, and the XPS spectra of Ag 3d showed a peak at 367.2 eV corresponding to the oxidation state of Ag. These features indicated the occurrence of oxidation-reduction process. Based on the above experimental evidence, they proposed a mechanism for Ag_{25} sensing Hg^{2+}. With the low addition of Hg^{2+}, the S atom and carbonyl group (on the ligand of Ag_{25}) bound with Hg^{2+} through weakly interaction. With the increase of Hg^{2+}, it could gradually approach the Ag atom (on the surface of Ag_{25}), thus, the redox reaction happened and the AgHg alloy formed. It is briefly summarized that Hg^{2+} can interact with the metal core as well as the functional groups of the capping ligand and the interaction is concentration-dependent. It is worth mentioned that Hg^{2+} cannot be reduced by Ag (in bulk state), while the abnormal phenomenon can be achieved when Ag is present in the NC state.

Metal NCs with single metal component has been widely used as probes to detect Hg^{2+}, however, the limit of detection is not satisfied. By utilizing the synergetic effect, the bimetallic NCs have a more sensitivity to metal ions. Xiao and Choi et al. synthesized BSA-protected AuAg NC via microwave irradiation method [21]. The AuAg NC exhibited a strong yellow fluorescent. Upon treatment with Hg^{2+}, the fluorescence intensity of AuAg NC displayed a significant decline. Based on the quenching effect, the AuAg NC could be used to sense the concentration of Hg^{2+}. The limit of the detection of Hg^{2+} was determined as 13 nM for BSA-AuAg sensor. The Hg^{2+} could cause the similar fluorescence quenching for monometallic NC (BSA-Au NC), but the quenching degree was different from the BSA-AuAg NCs. With adding the same concentration Hg^{2+} (9.5 μm), the fluorescence intensity of BSA-AuAg was decreased much higher (83.99%) than that of BSA-Au (19.95%). The sensitivity of BSA-AuAg sensor was about 6.3-fold higher than that of BSA-Au. The coeffect of Au and Ag greatly improved the sensitivity of the fluorescent sensor.

All above Au NCs, Ag NCs, and AuAg NCs Hg^{2+} sensors are mostly relied on fluorescence quenching. Wu et al. developed Ag NCs sensor to detect Hg^{2+} based on the change of absorbance of NCs. Different from the majority of water-soluble NCs sensors, Wu et al. has reported amphiphilic $Ag_{30}(Capt)_{18}$ NCs which can dissolve in water and other common organic solvents, such as methanol, ethanol, acetone, acetonitrile, dichloromethane and ethyl acetate. Its good solubility in various solvents makes it a promising sensor for Hg^{2+} detection in different kinds of environmental samples, such as lake water, and soil solution. This Ag_{30} NCs sensor identified Hg^{2+} via the change of absorption spectra rather than others based on the fluorescence changing. Besides, the related mechanism was also different. They ruled out the possibility of Hg^{2+}-induced aggregation or discomposing of Ag_{30} through experimental evidence. Based on the supposed antigalvanic reduction mechanism, Hg^{2+} could oxidize Ag_{30} leading to a notable decrease in the absorbance of Ag_{30}, thus Ag_{30} NCs could be acted as a colorimetric sensor Hg^{2+} [22].

Moreover, Liu et al. fabricated an Au NCs sensor for Hg^{2+} sensing based on fluorescence resonance energy transfer (FRET) (Fig. 8.2B) [23]. The blue fluorescent cationic-oligofluorene-substituted polyhedral oligomeric silsesquioxane (POSSFF) was chosen as energy donor and combined with acceptor-red fluorescent Au NCs (R-Au NCs) to construct efficient FRET pairs. POSSFF and R-Au NCs had opposite charge induced the formation of the hybrid complex via the electrostatic attraction between them. The distance between them was short enough to provide efficient FRET and emitted pink fluorescence. In the presence of Hg^{2+}, the R-Au NCs was connected with Hg^{2+} due to the strong metallophilic Hg^{2+}/Au^+ interaction, and thus the FRET was destroyed, the fluorescence emission of the hybrid system turned from pink to blue. Therefore, the hybrid system could realize visual detection for Hg^{2+} with a low LOD of 0.1 nM. Benefiting from the whole-cell permeability and ion-selective FRET in cells, the hybrid system further realized multicolor intracellular sensing of Hg^{2+}. This signal readout based on the variation of fluorescence emission was more suitable and practical for visual sensing with naked eye than that based on fluorescence quenching.

Cu^{2+} sensing

Besides Hg^{2+} sensing, fluorescent metal NCs can be used as Cu^{2+} sensor based on the fluorescence quenching effect. Burda and Zhu et al. synthesized highly fluorescent (8% QY) and water-soluble Au NCs with NIR emission at 670 nm. The fluorescence of Au NCs decreased with the increase of Cu^{2+}. The linear detection range was wide from 1 nM to 1250 μM and the LOD was 0.3 nM [24]. Zhang and Dong et al. used glutathione (GS)-protected Au NCs (GS-Au NCs) to detect Cu^{2+}. Due to the coordination between the amino, carboxyle group of GSH molecule and Cu^{2+}, the formed complex could cause the fluorescence quenching of GS-Au NCs. However, they found Hg^{2+} and Pb^{2+} were capable to quench the fluorescent of GS-Au NCs as well besides Cu^{2+}. They utilized the ethylenediaminetetraacetate (EDTA) as a strong chelator to help distinguish the difference of Hg^{2+}, Pb^{2+}, and Cu^{2+} by decomplexing the metal-GS-Au NCs. With the addition of EDTA, the fluorescence intensity of GS-Au NCs containing Cu^{2+} was recovered to ~80% of its initial value. While, for the solution containing Hg^{2+}, the presence of EDTA could not change the fluorescent. The results also confirmed the quenching mechanism was different between Cu^{2+} (Cu^{2+}-GS-ligand interaction) and Hg^{2+} (Hg^{2+}-Au^+ interaction). Moreover, the addition of EDTA not only recovered the fluorescent of GS-Au NCs containing Pb^{2+} but also turned the turbid solution to clear. This phenomenon could be used to distinguish Cu^{2+} from Pb^{2+} [25]. Lin and Yu et al. used cost-efficient papain as a capping and reducing agent to synthesize Au NCs with a red emission at 660 nm. The Au NCs was quite stable at a buffer pH range from 6 to 12 and its fluorescent properties remained unchanged at room temperature for more than 3 months. Based on fluorescence quenching, the Au NCs could act as a label-free Cu^{2+} sensor with a LOD of 3 nM [26]. Besides Au NCs sensors, fluorescent Ag NCs was also applied to Cu^{2+} sensing. Dong et al. synthesized poly(methacrylic acid) (PMAA)-templated Ag NCs. The free carboxylic groups of PMAA polymers could bind with Cu^{2+} and then quench the luminescence of Ag NCs. The PMAA-Ag NCs could detect Cu^{2+} with a LOD of 8 nM [27]. Zhu et al. synthesized $[Ag_{62}S_{13}(SBut)_{32}]^{4+}$ NCs with orange emission. Cu^{2+} could be reduced by the Ag_{62} NCs and then induced the fluorescent quenching with a quantitative relationship. Therefore, the Ag_{62} NCs served as a sensor to detect Cu^{2+} (Fig. 8.3A) [28].

In addition to detecting the Cu^{2+} in aqueous sample, Chen et al. reported dithiothreitol (DTT)-protected Au NCs and applied it into detecting Cu^{2+} in serum. The DTT-Au NCs showed single response to Cu^{2+} without the interference from other metal ions. Besides, the detecting results from DTT-Au NCs sensor were consistent with that obtained from traditional colorimetric method [29]. In addition to expanding the complexity of test samples, we should also broaden the forms of sensors. Pradeep et al. made a freestanding Cu^{2+} sensor film similar to pH test paper, which was composed of highly luminescent Au_{15} NCs. The sensing mechanism was confirmed by XPS analysis, which was the reduction of Cu^{2+} to Cu^{1+}/Cu^0 either by the GSH ligand or Au_{15} metal core. The reduction caused the obvious fluorescence quenching of Au_{15} film sensor which could be visually detected by naked-eye (Fig. 8.3B). The luminescent changes were proportional to with the concentration of Cu^{2+} and the film sensor could identify the Cu^{2+} as low as 1 ppm at visual sensitivity. Besides, the film

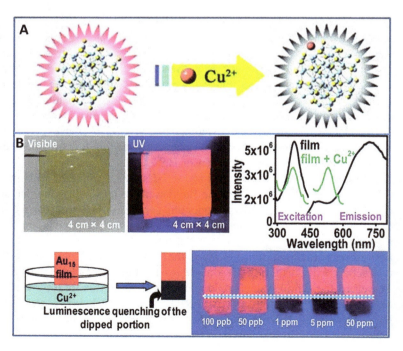

FIG. 8.3 Schematic diagram of metal NCs for Cu^{2+} sensing (A) Schematic diagram of Ag_{62} NCs for Cu^{2+} sensing; (B) freestanding Au_{15} NCs luminescent sensor for Cu^{2+} detection.

sensor was selective to Cu^{2+}, the fluorescent properties were unchanged with the exposure of other metal ions such as Hg^{2+}, As^{3+}, and As^{5+} [30].

There are some other fluorescent NCs sensors which can not only detect Cu^{2+} but also has response to other analytes. These NCs sensors can be used as dual-functional sensors. Pradeep et al. synthesized lactoferrin (Lf)-stabilized Au NCs and applied it into detection of metal ions. The Cu^{2+} caused the quenching of luminescence intensity of Lf-Au NCs, while the luminescent was increased about two times in terms of Ag^+ [31]. Zhang and Yu et al. also found the similar phenomenon. They synthesized GS-Au NCs with tunable fluorescence emissions and utilized it as metal ions sensors. The Cu^{2+} induced the quenching of fluorescent while Ag^+ enhanced the photoluminescent of Ag NCs [32]. Sreenivasan et al. used BSA-Au NCs to detect Cu^{2+} based on the fluorescence quenching which was caused by the interaction between Cu^{2+} and BSA. The BSA-Au NCs exhibited high selectivity to Cu^{2+} at various pH range from 2 to 12. The BSA-Au NCs sensors were also effective in determination of Cu^{2+} in living cells with a visual LOD of 100 μM. The fluorescent could be retrieved up to 72% of its initial intensity with the addition of glycine, a chelator, which had specific interaction with Cu^{2+}. Therefore, the BSA-Au NCs could be acted as "turn off" sensor for Cu^{2+} as well as "turn on" sensor for glycine [33]. Chi et al. fabricated Au NCs membrane sensors to detect Cu^{2+} by utilizing the fluorescence quenching mechanism. The histidine could recover the quenched fluorescent by Cu^{2+} to its initial state. The fluorescence quenching-recovery process could be repeated many times, and thus the Au NCs membrane could be acted as a reversible and recyclable Cu^{2+}-histidine sensor (Fig. 8.4A–E). This membrane sensor was further used in detecting the Cu^{2+} in local drinking tap water, and the results were consistent with that measured by atomic absorption spectrophotometric (AAS) method [34].

Mostly the sensitivity of fluorescence quenching sensors cannot meet the higher detection requirement. Therefore, lots of highly sensitive sensing modes are needed to be designed. Yang et al. synthesized lysine-stabilized Au NCs (Au NCs@Lys) and cooperated with Au NCs@BSA. The mixed Au NCs exhibited much higher fluorescence intensity. When detecting Cu^{2+}, the luminescent quenching was much more dramatically and thus the detection limit was much lower (0.8×10^{-12} mol/L) than that of only one type Au NCs [35].

Compared with the fluorescence quenching single-channel detection, ratiometric fluorescent sensing mode realized by measuring fluorescence signals of two different wavelengths simultaneously and calculating their intensity ratio is more accurate and can eliminate the ambiguities of fluorescent signals. Ye et al. synthesized vitamin B_2 (riboflavin)-stabilized Au NCs (Ri-Au NCs) which has two fluorescence emissions at 530 and 840 nm. In the presence of Cu^{2+}, the fluorescence intensity at 840 nm was decreased while the fluorescence emission peak at 530 nm was enhanced, therefore the Ri-Au NCs could be used as a ratiometric sensor for Cu^{2+}. The Ri-Au NCs was further applied to detecting Cu^{2+} in tap water and river

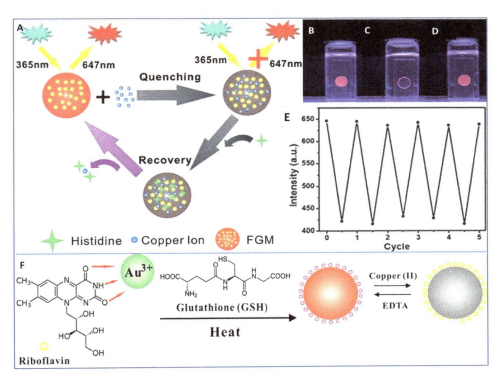

FIG. 8.4 Schematic diagram of ratiometric fluorescent probe for Cu^{2+} detection (A) Schematic diagram of the fluorescence of FGM quenched by copper ions and recovered by histidine; (B–D) the fluorescence of FGM under UV light excitation (365 nm) in sequence in water, in 1 mM Cu^{2+} ion, and in 10 mM histidine; (F) the Ri-AuNCs ratiometric fluorescent probe for detection of Cu^{2+}.

water, the detection results were similar to those measured by ICP-MS method, which indicated the Ri-Au NCs ratiometric sensor had a great accuracy and reliability in practical application. In addition, the Ri-Au NCs Cu^{2+} sensor could be reversible and recycled by using EDTA as a switching agent (Fig. 8.4F) [36]. Li et al. integrated the virtues of quantum dots (QDs) and metal NCs by covalent linking BSA-Au NCs and amino functionalized $CdTe@SiO_2$ spheres. In the CdTe/Silica/Au NCs hybrid spheres, the CdTe QDs were used as reference signal by providing a built-in correction for environmental effect and the Au NCs acted as response signal by selectively binding with Cu^{2+} and quenching the fluorescence intensity. Therefore, the CdTe/Silica/Au NCs hybrid sphere with two emissions could be served as a ratiometric fluorescent sensor for Cu^{2+}. This ratiometric fluorescent sensor provided an effective, accurate, and reliable platform for Cu^{2+} sensing in complex real samples such as tomato and cucumber. The LOD was 4.1×10^{-7} mol/L. [37]

Besides ratiometric fluorescent, fluorescent enhancement and fluorescent variation also can be used as a sensing mode, which is relatively precise compared with fluorescent quenching. Chang et al. constructed a fluorescent enhancement Cu^{2+} sensor by combining 3-mercaptopropionic acid (MPA) with DNA-Cu/Ag NCs. The DNA-Cu/Ag NCs exhibited a high fluorescent emission with a QY as high as 47.8%. The MPA could change the conformation of DNA and thus induce the fluorescent quenching of NCs. Because the Cu^{2+} can result in the oxidation of MPA, when the solution was contained with Cu^{2+}, the quenched fluorescent of MPA-DNA-Cu/Ag NCs was recovered. On this basis, the MPA-DNA-Cu/Ag NCs can detect Cu^{2+} as low as 2.7 nM with remarkable selectivity (2300-fold over other tested metal ions). The sensor was further used in the Cu^{2+} detection in Montana soil and pond water samples to demonstrate good practicality [38]. Pradeep et al. obtained stabilized-Au_{15} NCs in cyclodextrin (αCD, βCD, and γCD) cavities. The NCs could identify Cu^{2+} by fluorescence emission shifting. With the addition of 1 μM Cu^{2+}, there was no obvious change in UV-vis absorbance spectra, indicating that the NCs were stable in the presence of Cu^{2+}. However, in the case of photoluminescent, the red emission of Au_{15} NCs was disappeared, accompanied by the emergence of yellow fluorescence emission. The fluorescence emission peak had a ∼100 nm blue shift. This is the first time detecting Cu^{2+} by using fluorescent emission variation [39].

Ag^+ sensing

Chang et al. utilized Al_2O_3 NPs to support Au NCs, the obtained Al_2O_3 NP@Au NCs displayed different fluorescence emissions from 510 nm (dark cyan) to 630 nm (red) via simply controlling the reaction time (Fig. 8.5A–E). Moreover, the Al_2O_3 NP@Au NCs are stable at ambient temperature for more than 3 months. In the presence of PA as a masking agent, the Al_2O_3 NP@Au NCs are used to detect Ag^+ based on the fluorescent quenching due to the d^{10}-d^{10} interaction. The linearly detection range is between 3 and 300 nM and the LOD is 1.5 nM [40].

Different from Hg^{2+} or Cu^{2+} causing fluorescent quenching of metal NCs in most cases, Ag^+ usually induces fluorescent enhancement. Most metal NCs sensors for Ag^+ are based on fluorescent enhancement. Li and Wu et al. synthesized BSA-Au_{16} NCs via one-step microwave-assisted method. In the presence of Ag^+, the fluorescence emission of Au_{16} NCs is blue shifted from 604 to 567 nm accompanied by an obvious enhancement of intensity [41]. They further revealed the sensing mechanism by means of MALDI-TOF mass spectra and XPS analyses. The evidence confirmed that it is the Au core in BSA-Au_{16} NCs that reduces Ag^+ to Ag^0, and then forming the hybrid Au@Ag NCs induces the enhancement of fluorescent (Fig. 8.6A–D) [42]. It seems confused and has broken the common sense in chemical. But, it is worth noticing that when the size of Au is decreased down to 2 nm, the reactivity is different from that in the bulk state. Decreasing the core sizes will enhance the reductive ability of Au NCs. Bulk Au cannot reduce Ag^+ because Au is less reactive than Ag, while Au NCs can. This is so-called antigalvanic theory. Wu and Jin et al. utilized $Au_{25}(SG)_{18}$ to realize Ag^+ sensing by fluorescent enhancement mode. They deeply investigated the factors in the unique fluorescence enhancement, which are attributed to the oxidation of Au_{25}^- to higher oxidation states (contributing 126% fluorescent increase), the interaction between Ag^+ and Au_{25} (contributing 55% fluorescent increase) and the interaction between Ag^0 and Au_{25} (contributing 62% fluorescent increase). Moreover, the type of ligand has greatly influence on the replacement between Ag^+ and Ag^0 in Au_{25}. In $Au_{25}(SC_2H_4Ph)_{18}$, the replacement between it and Ag^+ occurs quickly, while in $Au_{25}(SG)_{18}$, the replacement is prevented [43].

Other metal ions sensing

Chen et al. prepared MSA-AgAu NCs with red fluorescent by etching the core of Ag NPs and a galvanic exchange reaction. By covalently linking to m-PEG-NH_2, the PEGylated MSA-AgAu NCs exhibit high stability in high ionic conditions (up to 1 M NaCl). Due to the deposition of Al^{3+} on the surface of the metal core, the fluorescence intensity of PEGylated MSA-AgAu NCs is increased with the increasing of Al^{3+}. Therefore, the PEGylated MSA-AgAu NCs is used as a Al^{3+} quantitative sensor (Fig. 8.7A) [44]. Banerjee et al. synthesized dicysteine capped Au NCs with a high QY of 41.3% via core etching method. This Au NCs is used to sensitively detect As^{3+} through fluorescent enhancement. When adding 115 μM As^{3+}, the fluorescence intensity can increase 66%. The LOD of this sensor is 53.7 nM. Moreover, this Au NCs

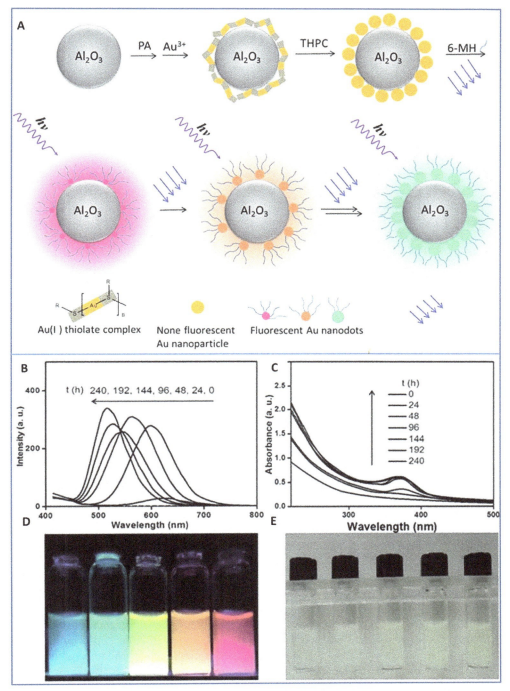

FIG. 8.5 Schematic illustration of the synthesis and spectrum information of Al$_2$O$_3$ NP@Au NCs (A) Schematic illustration of the formation of Al$_2$O$_3$ NP@Au NCs; (B–E) PL, UV-vis absorption spectra, and corresponding photographs of Al$_2$O$_3$ NP@Au NCs.

sensor can be recycled used by adding succinic acid to chelate with As^{3+} to realize the reversibility of the fluorescent (Fig. 8.7B–D) [45]. Ho et al. prepared L-3,4-Dihydroxyphenylalanine (L-DOPA)-capped Au NCs via a rapid one-pot method. The DOPA-Au NCs can be used as Fe^{3+} sensor based on the fluorescent quenching caused by the Fe^{3+}-induced aggregation. The LOD of this sensor is 3.5 μM. This sensor has a highly feasibility which is confirmed by applying into detecting Fe^{3+} in tap water, lake water, and iron supplements [46]. Huang and Zheng et al. synthesized tannic acid (TA) stabilized Cu NCs with a QY of 14%. The TA-Cu NCs can detect Fe^{3+} as low as 10 nM based on the fluorescent quenching due to the electron transfer mechanism. Due to the good biocompatibility and low cytotoxicity of Cu NCs, the sensor was further used for imaging Fe^{3+} in living cells [47].

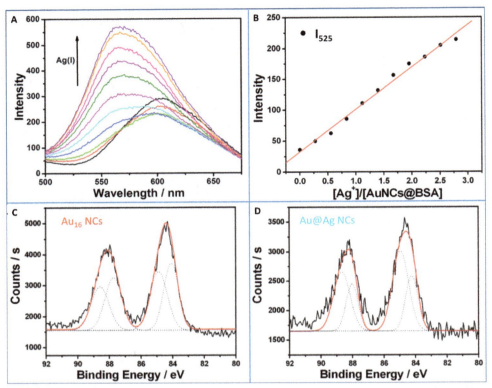

FIG. 8.6 Schematic diagram of Au NCs for Ag ion detection (A) Fluorescence spectra of $Au_{16}NCs@BSA$ after adding different amount Ag^+; (B) Plot of PL intensity at 525 nm vs the molar ratio of Ag^+ to $Au_{16}NCs@BSA$; (C and D) Au 4f XPS spectra of $Au_{16}NCs@BSA$ before and after Ag^+ addition.

Liu et al. found that Cr^{3+} and Cr^{6+} have different fluorescent quenching capabilities of GSH-Au NCs in different pH environment (Fig. 8.8). 25 μM Cr^{6+} can induce 99% and 92% fluorescent quenching of GSH-Au NCs at pH of 1 and 3.5, respectively. But it only caused 10% and 5% fluorescent decrease at pH of 5 and 6.5, respectively. 25 μM Cr^{3+} can lead to 32%–45% fluorescence quenching at the pH range of 3.5–6.5. Therefore, they used pH assisted stepwise detection strategy to determine Cr^{3+} and Cr^{6+}. Cr^{3+} can be detected at pH of 6.5 because Cr^{6+} basically does not cause fluorescent quenching. Detecting Cr^{6+} is based on the difference of its relative fluorescence intensities between pH 3.5 and 5.0, because in this range, the quenching caused by Cr^{3+} is similar. This sensor for Cr^{3+} and Cr^{6+} has a wide detection range and low LOD. For Cr^{3+} and Cr^{6+}, the linearly detection range is 25–3800 μg/L and 5–500 μg/L, the LOD is 2.5 μg/L and 0.5 μg/L, respectively [48]. Zhu et al. synthesized highly fluorescent Ag NCs by microassisted method. The Ag NCs is applied to Cr^{3+} sensing based on the selectively fluorescent quenching effect. The LOD of this sensor for Cr^{3+} is 28 nM [49].

Pradeep and Pal et al. used BSA to synthesize Cu NCs with blue fluorescent. The Cu NCs is highly stable. It displays similar emission spectra after 2 months stored at room temperature. Besides, the emission property is not changed at pH 7–12. The Cu NCs can be served as Pb^{2+} sensor based on the fluorescent quenching induced by the NCs aggregation (Fig. 8.9A). The aggregation of NCs is caused by the complexation between BSA and Pb^{2+}, which is confirmed by the dynamic light scattering (DLS) measurement [50]. Kawasaki et al. utilized pH-dependent synthesis method to prepare pepsin-stabilized Au NCs with different fluorescent emissions. Based on the MS analysis, the red emission NCs is Au_{25}, the green emission NCs is Au_{13}, and the blue emission NCs is Au_5 and Au_8 mixture. The pepsin-Au_{25} NCs is more stable than others, therefore it is used to detect Pb^{2+} based on fluorescent enhancement sensing mode [51]. He et al. synthesized GSH/MUA protected Au NCs with green luminescent by ligand exchange method. The GSH has high affinity toward Pb^{2+}, therefore, in the presence of Pb^{2+}, the GSH/MUA-Au NCs become aggregated and the fluorescent is quenched (Fig. 8.9B). On this basis, the GSH/MUA-Au NCs can detect Pb^{2+} with a LOD of 2 nM. In the GSH/MUA-Au NCs dual ligand NCs, MUA is responsible for the luminescent and GSH is sensitive to Pb^{2+}. The MUA-Au NCs can be extended to detect other analytes by replacing GSH with other target-sensitive sensing molecules [52].

8.2.1.2 Anion sensing

Among various anions, cyanide is a highly toxic pollutant that binds to the heme cofactor and inhibits the end of the respiratory chain enzyme cytochrome *c* oxidase, causing death in humans and other organisms. Lu et al. used BSA-Au NCs to achieve

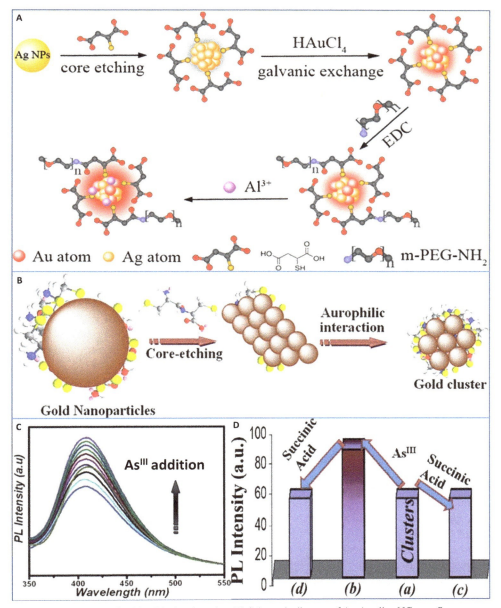

FIG. 8.7 Schematic diagram of metal NCs for Al and As ion detection (A) Schematic diagram of Au-Ag alloy NCs as a fluorescence-enhanced probe for Al^{3+} sensing; (B) Schematic diagram of the synthesis of Au NCs through a top-down approach; (C) PL spectra of the dicysteine capped Au NCs with the addition of As^{3+}; (D) fluorescence recovery after the complexation of As^{3+} with succinic acid.

highly sensitively and selectively CN^- detection based on the CN^- etching-triggered fluorescence quenching (Fig. 8.10A). The high selectivity is attributed to the unique Elsner reaction between CN^- and Au: $4Au + 8CN^- + 2H_2O + O_2 \rightarrow 4Au(CN)^{2-} + 4$-$OH^-$. To realize the optimization of sensing, a series of factors are considered. The reaction is completed at 20 min; therefore, the optimized detection time is 20 min. Because CN^- is inclined to combine with the available protons in the solution, the increase in pH of the solution can decrease the competition for CN^- between the H^+ and Au. The fluorescent quenching effect reaches maximum at pH of 12. Therefore, the optimized pH is 12. Under the optimized conditions, the LOD of this sensor for CN^- is 200×10^{-9} M which is ~14 times lower than the maximum permission required by the World Health Organization (WHO). The linear correlation range is 200×10^{-9} M to 9.5×10^{-6} M. The sensing system is highly selective to CN^- over other anions, the tolerance is at least 20 times. The sensor is further directly used in real water samples such as local groundwater, tap water, pond water, and lake water with good recoveries (>93%). This CN^- sensor holds promising potential without the need of complex organic synthesis or complicated instruments [53]. Dong et al. also synthesized lysozyme-stabilized Au NCs (Lys-Au NCs) and applied it as a fluorescent sensor to detect CN^-. The Lys-Au NCs exhibit a red emission at 650 nm. When adding the

FIG. 8.8 Schematic diagram of Au NCs for Cr^{3+} and Cr^{6+} ion detection Scheme for preparation of GSH-Au NCs and application in selective determination of Cr(III) and Cr(VI).

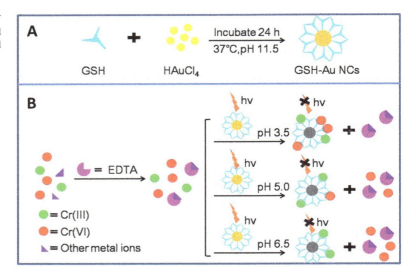

CN^-, the fluorescence intensity of Lys-Au NCs decreases due to the formation of $[Au(CN)_2]^-$ caused by the strong coordination between Au* and CN^-. The linear detection range is from 5.00×10^{-6} to 1.20×10^{-4} M and the LOD is 1.9×10^{-7} M. The Lys-Au NCs sensor can be used to visualize identify CN^- with naked eye. Under visible light, the Lys-Au NCs is yellow, after adding CN^-, the color changes to transparent. Correspondingly, under UV light, the red fluorescent changes to blue in the presence of CN^-. The blue fluorescent is generated by lysozyme. In the addition of CN^-, the Lys-Au NCs gradually etched and the covered blue fluorescent has appeared [54].

Lu et al. used molecular Au(I) NCs to fabricate solid-state microporous sensing film for colorimetric detecting CN^-. Different from cyanide etching-induced fluorescence quenching of Au NCs, this sensing system is based upon the phosphorescent of Au(I) NCs. The ligand of Au(I) NCs is 3-mercapto-1,2,4-triazole (MTA), and the precise composition is determined as $[Au(I)MTA]_3$ confirmed by matrix-assisted laser desorption/ionization time-of-flight mass spectrometry (MALDI-TOF-MS). The $[Au(I)MTA]_3$ NCs exhibit a strong red emission at 640 nm with a high QY of 38%, but the lifetime is only 10 μs. This microsecond-scaled lifetime and the large Stokes shift reveal that the emission is a spin forbidden transition and is phosphorescent. In the presence of CN^-, the phosphorescence intensity of $[Au(I)MTA]_3$ NCs is decreased. On this basis, a rapid, sensitive CN^- sensor is constructed. The detection only needs 8 min and the LOD is 80 nM. Besides, the $[Au(I)MTA]_3$ NCs sensor is not limited to pH value and can be used in a broad pH range. The sensing mechanism is investigated by X-ray absorption near edge structure (XANES) and extended X-ray absorption fine structure (EXAFS) analyses. With the increasing of CN^-, the peak at 11.928 keV in XANES spectra is enhanced, indicating the coordination between CN^- and Au(I) species. There are new peaks appeared at 1.55 and 2.63 Å in the Fourier transforms (FTs) of the EXAFS spectra, which is corresponded to Au-C and Au-N, respectively. Besides, the peak at 1.93 Å related to the Au—S bonding is diminished. They speculated the coordination between CN^- and Au(I) induce the dissociation of MTA ligand, break the ligand-to-metal charge transfer (LMCT) emission process and thus cause the quench of the phosphorescence of the $[Au(I)MTA]_3$ NCs.

FIG. 8.9 Schematic diagram of metal NCs for detection of lead ions (A) Schematic diagram of detection of lead ions by copper NCs; (B) Synthetic strategy of Au NCs and Pb^{2+} sensing principle.

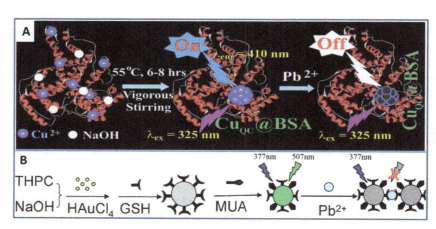

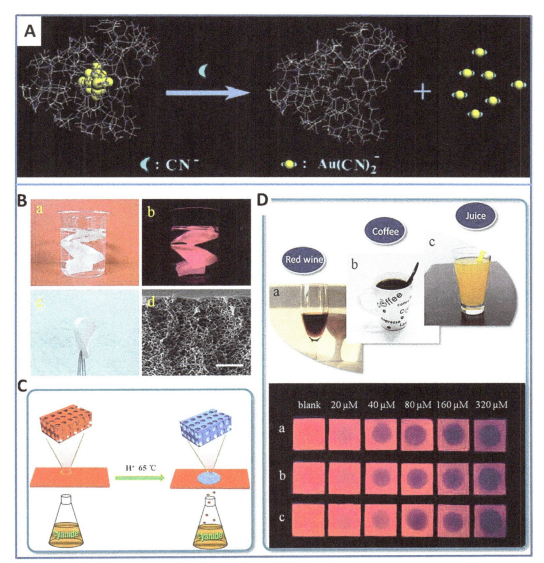

FIG. 8.10 Schematic diagram of detection of CN⁻ in real samples by Au NCs (A) Schematic diagram of detection of CN⁻ by Au NCs; (B) Photographs of the Au NCs porous films in water under day light (a) and under UV illumination (b) after drying; (d) SEM image of the porous film; (C) Scheme illustration of the experiment setup for the porous film toward cyanide detection; (D) Photographs of the selected model systems including (a) red wine, (b) coffee, and (c) juice, and the corresponding color changes of the film after the above detection procedure.

They further made a solid-state film sensor by in situ forming [Au(I)MTA]$_3$ NCs within blue fluorescent poly(acrylonitrile) (PAN) matrix to detect CN⁻ in colored samples and complex mixtures, such as red wine, coffee, juice, and soil. The free-standing film is comprised of uniform [Au(I)MTA]$_3$ NCs with pink luminescent, exhibiting high flexibility and stability (Fig. 8.10B–D). The luminescent of this film can remain after 5 months stored in ethanol. The film has several advantages: (1) Due to the macroporous structure, the film can enhance the interaction sites and facilitate the diffusion of the analyte, which benefits the rapid detection; (2) The macropores are beneficial to capture and concentrate the volatilized CN⁻, which favors the amplification of signal; (3) The blue luminescent PAN matrix can be used as internal calibration, which enables the reliable detection. The film sensors were successfully applied to the CN⁻ detection in real samples, with the increase of CN⁻, the film color changes from pink to blue. Furthermore, the film sensor can be served as a fast and portable monitor to determine time-dependent CN⁻ release process in cassava manufacturing [55].

Nitrite is a common food additive, and its excessive use will do a lot of harm to human body, and even cause cancer in serious cases. Therefore, it is very important to quantitatively detect nitrite in food industry. Zhu et al. applied BSA-Au NCs as fluorescent sensor to detect NO_2^-. Due to the NO_2^-/BSA interaction, the Au NCs became aggregated in the presence of

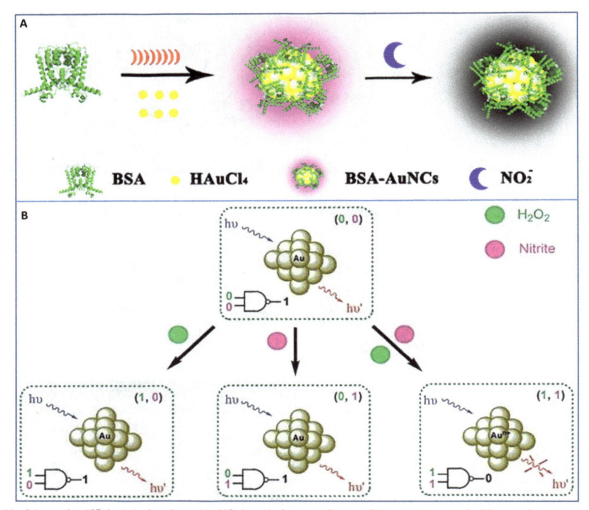

FIG. 8.11 Scheme of Au NCs in nitrite detection and Au NCs-based logic gate (A) Scheme of the synthetic strategy for BSA-AuNCs and the principle of nitrite sensing; (B) The binary Boolean NAND logic gate based on the BSA-protected gold nanoclusters.

NO_2^-, and thus causing the fluorescent quenching of Au NCs (Fig. 8.11A). The fluorescent intensity decreased linearly with the increase of NO_2^- from 2.0×10^{-8} to 5.0×10^{-5} M. Under the optimized conditions (pH=7.4, 20 min, 40 nM BSA-AuNCs), the sensor has a LOD of 1 nM. Benefiting from the high sensitivity and selectivity, the sensor was further used to detect NO_2^- in natural water samples, such as local groundwater, tap water, pond water, and wastewater. The results from this sensor have no obvious difference from the standard addition [56]. Different from the NO_2^- induced aggregation triggered fluorescent quenching, Hsu and Huang et al. proposed a distinct mechanism, they attributed the fluorescent quenching to the oxidation of Au(0) atoms to Au(I) atoms in the core of BSA-Au NCs mediated by nitrite ions, based on the results obtained from (UV-vis) absorption spectroscopy, X-ray photoelectron spectroscopy (XPS), fluorescence measurements, circular dichroism (CD) spectroscopy, zeta potential and dynamic light scattering (DLS) studies. They further constructed BSA-Au NC-modified nitrocellulose membrane sensor and realized NO_2^- detection in human urine samples with a LOD of 250 nM [57]. Huang and Chang et al. utilized the nonspecific hydrophobic interactions between amphiphilic ligands (ALs) and 11-mercaptoundecanol (11-MU) to prepare a series of hybridized ligands-protected Au NCs and realize tunable optical properties as well. The ALs include fatty acids (FAs) and quaternary ammonium surfactants (QASs), such as decanoic acid (DA), dodecanoic acid (DDA), tetradecanoic (TA), dodecyltrimethylammonium bromide (DTAB), myristyltrimethylammonium bromide (MTAB), and hecadecyl trimethylammonium bromide (HTAB). Among these bilayered Au NCs, the TA/11-MU-Au NCs exhibit the strongest fluorescent and highest dispersibility, therefore it was used to detect NO_2^- with high sensitivity (LOD: 40 nM) and selectivity (>100-fold tolerance against other anions) [58]. Yang et al. used BSA-Au NCs as fluorescent sensor and designed a Boolean NAND logic gate with NO_2^- and

H_2O_2 as inputs and apply it into NO_2^- detection in real samples (Fig. 8.11B). Firstly, the concentrations of NO_2^- and H_2O_2 were defined as inputs and the fluorescent intensity ratio of Au NCs was outputs. The presences of NO_2^- above and below 21.7 µM (according to the maximum content allowed in drinking water permitted by U.S. EPA) were defined as "1" and "0," respectively. The presences of H_2O_2 more than and lower than 0.2 nM were defined as "1" and "0," respectively. For the output, the fluorescence intensity ratio over and below the threshold value 0.2 is considered as "1" and "0" states, respectively. Only when both inputs are present, the output is inactivated. The Boolean NAND logic gate was successfully applied to monitoring NO_2^- in multiple samples, such as tap water, commercial mineral water, milk powder, ham sausage, and human urine [59].

Sulfide ions are widely distributed in natural water and wastewater, and the concentration of S^{2-} ions is an important environmental index. Long-term exposure to low concentrations of S^{2-} causes chronic diseases. Therefore, it is necessary to detect S^{2-}. Liu et al. applied BSA-Au NCs into S^{2-} detection based on the fluorescent quenching effect. The proposed mechanism is that Ag_2S could be formed due to the reaction between S^{2-} and Au, resulting in the structure degradation of BSA-Au NCs, and thus the fluorescent is decreased. To achieve the highest sensitivity, selectivity and accuracy detection, a series of effect factors were optimized. Under the best detection condition: 1.00 mL BSA-AuNCs, pH=7.0 NaAc-HAc buffer solution, 20 min at 60°C incubation, the LOD is 0.029 µM [60]. Chang et al. synthesized DNA-templated gold/silver nanoclusters (DNA-Au/Ag NCs). The DNA-Au/Ag NCs showed stronger fluorescence intensity and higher QY (4.5%) than DNA-Ag NCs (3.9%), and are more stable than DNA-Ag NCs in high ionic media (200 mM NaCl). In the presence of S^{2-}, the fluorescent of DNA-Au/Ag NCs is decreased, which is because the interaction between S^{2-} and Au/Ag atoms/ions leading to the DNA conformation change from packed hairpin to random coil structures. The sensitivity of DNA-Au/Ag NCs sensor is higher (LOD=0.83 nM) than DNA-Ag NCs (110 nM). The sensor exhibits high selectivity toward S^{2-} than other majority anions, except I^-. But $S_2O_8^{2-}$ could be used as masking agent to minimize the interference from I^-. The DNA-Au/Ag NCs sensor was further used in S^{2-} detection in hot spring and seawater samples and the results were in good consistent with those detected by conventional methylene blue (MB) method [61]. Different from the S^{2-} induced fluorescent quenching, He et al. utilized fluorescent recovery to detect S^{2-}. They synthesized 1-(10-mercaptodecyl)-5-methylpyrimidine-2,4-dione (TSH) and 11-mercaptoundecanoic acid (MUA) coprotected Au NCs. The TSH has poor solubility, therefore, after 2 days aging, the TSH/MUA-Au NCs became aggregated and thus the fluorescent is diminished. However, the aggregates disassembled and the fluorescent recovered after adding S^{2-} due to the increased surface charge and static repulsion. HRTEM images, DLS and zeta potential data all confirmed the disaggregation induced fluorescent recovery phenomenon. TSH ligand is the crucial factor to construct fluorescence turn-on S^{2-} sensor. Besides its poor solubility leading to the aggregation of Au NCs, its large steric hindrance effect can prevent MUA ligands overly covering on Au NCs surface and thus provide more adsorption sites for S^{2-}. This aspect can be confirmed by the comparison counterparts such as MUA-Au NCs or GSH/MUA-Au NCs [62].

Halide ions play essential roles in industrial, medical, and environmental processes. The specific determination of halide concentrations in natural resources is important. Cai and Wu et al. utilized $Au_{25}(SG)_{18}$ to detect I^-. Different from the usual ions induced fluorescence quenching, I^- can lead to the red-shift and enhance of emission peak due to its stronger affinity toward Au than other anions. The $Au_{25}(SG)_{18}$ sensor can detect I^- as low as 400 nm and exhibit excellent selectivity against 12 types of anions, especially F^-, Br^-, Cl^- [63]. Luo and Li et al. synthesized hyperbranched polyethyleneimine (PEI)-capped Ag NCs and realized three kinds of halide ions (Cl^-, Br^-, I^-) sensing simultaneously. The halide ions could induce the fluorescent quenching of Ag NCs, but the quenching degree is different for the three halide ions because of the different solubility constants of corresponding silver compounds, and thus it can be used to discriminate the three halide ions. The fluorescent quenching is resulted from the halide-induced oxidative etching and aggregation of Ag NCs which is confirmed by the DLS and Zeta potential measurements. The PEI-Ag NCs sensor exhibits high selectivity to halide ions against most of anions and cations (except S^{2-} and Hg^{2+}). The PEI-Ag NCs sensor shows wide linear ranges: 0.5–80 µM for Cl^-, 0.1–14 µM for Br^-, and 0.05–6 µM for I^-, respectively, and low LOD: 200, 65, and 40 nM for Cl^-, Br^-, and I^-, respectively [64].

8.2.1.3 Chemical sensing

Besides heavy metal ions and anions, metal NCs are also used in the detection of various chemicals. Pradeep et al. anchored luminescent Ag_{15} NCs on silica-coated Au mesoflowers with a size of 4 µm (namely $Au@SiO_2@Ag_{15}$ MFs) to realize ultrasensitive detection of trinitrotoluene (TNT) based on the surface-enhancement luminescent. Utilizing Au MFs not only enhances the luminescent of Ag_{15} NCs, but also takes its distinct shape as an advantage to achieve immediate identification of analyte by optical microscopy. The luminescent of $Au@SiO_2@Ag_{15}$ MFs quickly decreased within 1 min under the exposure of TNT even at less than 1 zeptomole. That fluorescent quenching is caused by the Meisenheimer complex formed due to the chemical interaction between TNT and the free amino groups in BSA. The specific Meisenheimer complex makes the Au@SiO$_2$@Ag$_{15}$ MFs highly selective toward TNT over other related molecules, such as 2,4-dinitrotoluene (2,4-DNT), cyclotrimethylenetrinitramine (RDX), and 4-nitrotoluene. To achieve high resolution visualized detection, it is better to apply luminescent change method rather than luminescent quenching strategy. Therefore, they first functionalize a TNT-insensitive fluorophore-fluorescein isothiocyanate (FITC) on the Au@SiO$_2$ MFs. The FITC exhibits different green fluorescent when it is excited at the same energy with Ag$_{15}$ NCs (red emission). Secondly, the Au@SiO$_2$-FTIC MFs were coated by Ag$_{15}$ NCs to construct Au@SiO$_2$-FTIC@Ag$_{15}$ MFs, which have a red emission originating from Ag$_{15}$ luminescent and the FTIC luminescent was suppressed. In the presence of TNT, the Au@SiO$_2$-FTIC@Ag$_{15}$ MFs fluorescent color change from red to green, because Ag$_{15}$ NCs are sensitive to TNT, easily forming complex and decreasing the luminescent while FTIC is inert to TNT (Fig. 8.12A and B). This strategy can be extended to detect other analytes by choosing specific fluorophore or ligand of NCs [65].

Methotrexate is an antimetabolite used in the treatment of certain types of cancer. However, its over dosage may cause irreversible damage to normal tissues. Chen et al. utilized BSA-Au NCs to detect methotrexate. Due to the interaction between methotrexate and BSA-Au NCs, the addition of methotrexate induced the strong fluorescence intensity of the Au NCs decreased (Fig. 8.12C). The LOD is 0.9 ng/mL, and the linear detection range is from 0.0016 to 24 mg/mL. Compared with other assays for the methotrexate quantification, the BSA-Au NCs sensor shows a lower LOD and wider linear range. This sensor was further used into methotrexate detection in real samples including injections, human urine, and human serum, the results were satisfied [66].

Theophylline is a potent bronchodilator and respiratory stimulator used in the treatment of asthma and chronic obstructive pulmonary disease. However, if the dosage is over the acceptable therapeutic range, it may promote permanent neurological damage at concentrations. Park et al. fabricated a label-free basic (AP) site-incorporated duplex DNA-Ag NCs fluorescent sensor to detect theophylline by utilizing the strong and selective binding between AP site-incorporated duplex DNA and the target. The kind of nucleobase opposite to the AP site in the duplex DNA was investigated, only positioned opposite the AP site is cytosine can efficiently induce the high fluorescent Ag NCs. In addition, only duplex DNA containing cytosines positioned opposite the AP site can selectively bind theophylline. Without theophylline, Ag NCs in AP site-incorporated duplex DNA can be synthesized due to the interactions between silver ions and cytosine moieties and the obtained Ag NCs can emit red fluorescent. In the presence of theophylline, due to the specific binding between theophylline and AP site, the competitive binding inhibits the silver ions binding to the cytosine nucleobases, therefore, suppressing the formation of fluorescent Ag NCs. On this basis, the fluorescent decrease can be used to detect theophylline. The linear detection range is from 0 to 40 μM, and the LOD is 1.8 μM. The sensor shows high selectivity to theophylline over other methylxanthine derivatives such as theobromine, caffeine, D-glucose, and creatinine. The sensor was further successfully used to detect theophylline in human serum [67].

Organophosphorus compounds (OPs) include nerve agents and organophosphate pesticides, are highly toxic and can severely threaten public safety and the environment. Besides, OPs can inhibit the catalytic activity of acetycholinesterase (AChE) by binding to it at certain active sites of amine acids (i.e., serine). Phosphorylation-induced inactivation of AChE in the body can cause serious clinical complications and even death. Therefore, it is necessary to monitor OPs and evaluate the catalysis and phosphorylation of AChE simultaneously. Wang et al. reported a "lab-on-a-drop" protocol with biocompatible and high fluorescent Au NCs to realize label-free evaluation of the catalysis and OP-induced inhibition of AChE and quantitatively determination of OPs at the same time. BSA-Au NCs and acetylthiocholine (ATC) were mixed via electrostatic interactions to serve as "a drop" of fluorimetric reaction substrate. With the addition of AChE, the thiocholine was produced by the AChE catalyzing the hydrolysis of ATC. The generated thiocholine cap onto the BSA-Au NCs, then causes the aggregation of them, and thus induces the fluorescent decrease (Fig. 8.12D). Based on this fluorescent quenching, the system can be used to evaluate the catalysis progress of AChE promoting ATC hydrolysis. The system for AChE sensing has a linear relationship in the range of 5.0×10^{-6} to 5.0×10^{-2} U/mL with a LOD of 2.0×10^{-6} U/mL. On the contrary, after introducing OPs (i.e., dimethyl-dichlorovinyl phosphate DDVP), the aggregation and decrease in the fluorescence of BSA-AuNCs could be curbed to some degree due to the inhibition of Ops to AChE activity. Therefore, the system can be used to determine OPs. The system can detect DDVP as low as 13.67 pM, and the linear detection range is from 3.2×10^{-5} to 2.0×10^{-2} mM. The sensor was further used for the detection of DDVP residues in cabbage, the results were highly consistent with those obtained by classic high-performance liquid chromatography (HPLC) method. The sensor has promising potential in rapid evaluating physiologic catalytic activity of various enzymes and early warning and accurate diagnosis of OPs [68].

Zhang and Han et al. synthesized high-level incorporation of Ag in Au NCs (BSA-Ag$_{28}$Au$_{10}$ NCs) by controlling the molar ratio of Ag$^+$ and Au^{3+}. The NCs was used for the sensitive detection of H$_2$O$_2$ and herbicide propazine. Different from fluorescent quenching of other metal NCs, the BSA-Ag$_{28}$Au$_{10}$ NCs displayed a fluorescence redshift induced by H$_2$O$_2$. On

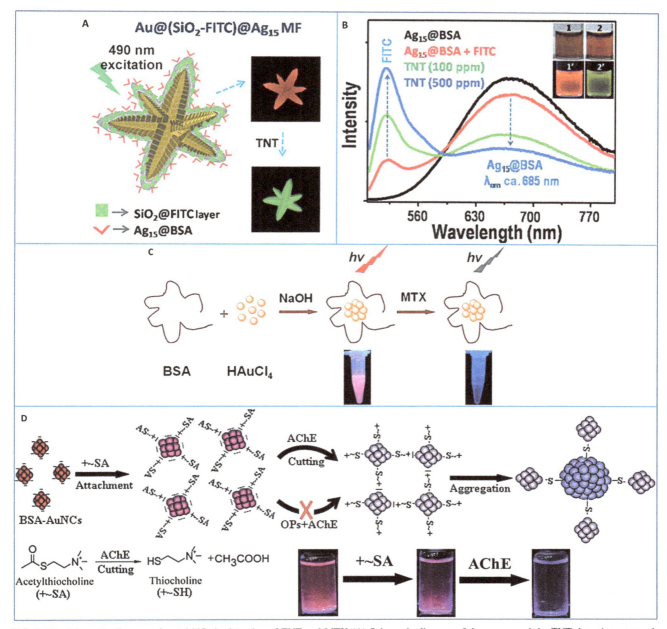

FIG. 8.12 Schematic diagram of metal NCs in detection of TNT and MTX (A) Schematic diagram of the sensor and the TNT detection approach; (B) Effect on the emission spectra of the bare cluster upon mixing with FITC dye and subsequent exposure to TNT solutions of varying concentrations; (C and D) Schematic illustration of the fluorescent BSA-AuNCs for MTX detection and the assay procedure with fluorescent changes of BSA-stabilized AuNCs (BSA-AuNCs), including the AChE-catalyzed hydrolysis of ATC and the phosphorylation-induced inhibition of AChE catalyzed activities by OPs.

the basis of obvious fluorescence peak shift, the NCs sensor can be used to identify H_2O_2. Besides H_2O_2 detection, the NCs can be used for the detection of herbicide propazine. The fluorescence intensity of NCs is enhanced in the presence of propazine due to strong coordination between propazine and surface Ag^+ of NCs. The LOD is 0.1 nM which is far below the permission regulated by U.S. EPA [69].

Hou et al. used BSA-Au NCs to detect glutaraldehyde (GA). Due to the cross-linking nature of BSA with GA, the fluorescent of NCs is quenched. The sensor can be used for quantitative determination of GA in tap water and rive water with a LOD 0.2 μM and a linear relationship in the concentration of 0.8–6 μM [70].

Yong et al. utilized density functional theory (DFT) calculations to prove the Ag_7Au_6 NCs has the potential to be used as gas sensor to detect CO, HCN, and NO. They found the electronic properties such as electric conductivity of the

Ag_7Au_6 NCs might have dramatic changes after these gas molecules adsorption by simulation. Therefore, the Ag_7Au_6 NCs is expected to sense these gases [71].

8.2.2 Biological sensor

Metal NCs have attracted considerable attention because of ease of synthesis, bright photoluminescence, low toxicity, good chemical stability, and high biocompatibility. Owing to these advantages, metal NCs have become promising candidate as biological sensor, especially for in vivo biological analysis. Here, we will discuss metal NCs-based biosensor from biomolecule sensing, protein sensing, DNA sensing, and immunoassay sensing.

8.2.2.1 Biomolecule sensing

Phosphate-containing metabolites sensing

Adenosine-5′-triphosphate (ATP) is the main energy carrier in cells, and pyrophosphate (PPi) is the generated product when ATP is hydrolyzed into adenosine monophosphate (AMP). Both of them can be used as indicators for health conditions of individuals. Chen et al. used GSH-Au NCs and Fe^{3+} to construct "turn on" fluorescent sensor for phosphate-containing metabolites detection (Fig. 8.13A). In the presence of ATP or PPi, the quenched fluorescent of GSH-Au NCs induced by Fe^{3+} is recovered, which can be obvious observed by naked eye. Therefore, the GSH-Au NCs-Fe^{3+} sensor can realize quantitative analysis of ATP and PPi with the LOD of ∼43 and ∼28 μM, respectively. The linear response range is 50–500 μM for ATP, 50–100 μM for PPi, respectively. The detection is rapid (15 min reaching the equilibrium of competitive chelating process) and inexpensive (without the requirement of expensive chemicals). The sensor can also be used to detect phosphate-containing metabolites in the cell lysates and human blood samples [72]. Similarly, Chen et al. synthesized chicken egg white proteins capped Au NCs (Au NCs@ew) with a red emission. The Cu^{2+} can induce the fluorescent quenching of Au NCs@ew. The phosphate functional groups have strong affinity toward Cu^{2+}. Therefore, in the presence

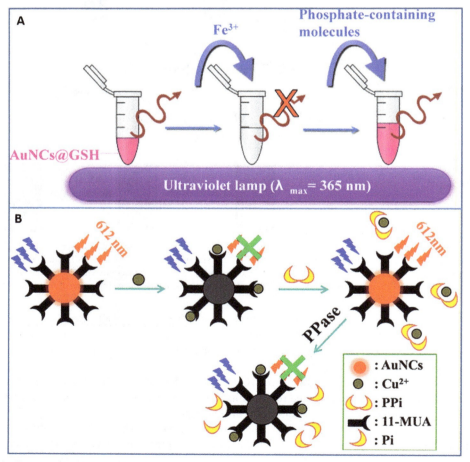

FIG. 8.13 Schematic diagram of Au NCs for phosphate-containing metabolites detection (A and B) Schematic diagram of AuNCs@GSH-Fe^{3+} for phosphate-containing metabolites detection and the assay of inorganic pyrophosphatase activity based on the fluorescent Au NCs.

of ATP and PPi, the quenched fluorescent can be restored, which can be used to detect ATP and PPi. The LOD of the Au NCs@ew-Cu^{2+} sensor for ATP and PPi are $\sim$19 and $\sim$5 μM, respectively. Besides, because there are abundant glycan moieties in ew ligand of Au NCs, Con A with glycan binding sites can specifically bind with Au NCs@ew. Therefore, the Au NCs@ew was used as label-free sensor to detect Con A in serum samples. The LOD was estimated to be $\sim$600 nM [73]. Yang et al. utilized competition assay method to realize the determination of pyrophosphatase (PPase) activity (Fig. 8.13B). They used 11-MUA-Au NCs as fluorescent probe, in the presence of Cu^{2+}, the fluorescent is decreased due to the interaction between 11-MUA and Cu^{2+}. However, the binding affinity between PPi and Cu^{2+} is stronger than that between 11-MUA and Cu^{2+}. Therefore, after adding PPi, the quenched fluorescent of 11-MUA-Au NCs-Cu^{2+} is recovered. PPase can promote the hydrolysis of PPi, therefore, after adding PPase into the 11-MUA-Au NCs-Cu^{2+}-PPi system, the Cu^{2+} can be released lead to the fluorescent quenching again. On the above concept, the system can be used as a monitor to detect the PPase activity with a LOD of <1 mU [74].

Guanosine 3′-diphosphate-5′-di(*tri*)phosphate (ppGpp) is a bacterial alarmone which is generated when bacteria face stress circumstances such as nutritional deprivation. Except for this, ppGpp also play significant roles in bacterial physiology affecting many biological functions of microorganisms. Huang et al. established a "turn on" fluorescent sensor-Cu^{2+} mediated DNA-Ag NCs to detect ppGpp. Cu^{2+} can quench the fluorescent of DNA-Ag NCs via electron or energy transfer, however, the ppGpp with rich electrons can strongly bind with Cu^{2+}, therefore the Cu^{2+} are disassociated from Ag NCs-Cu^{2+} complex and the fluorescent is recovered. Under the optimal conditions (800 nM Cu^{2+}, pH=7.4), the linearly proportional detection is in the range 2–200 μM and the LOD is 0.75 μM [75].

H_2O_2 sensing

H_2O_2 is an important intermediate species in many biological processes. Aimed at detecting H_2O_2, Zhang et al. rationally designed horseradish peroxidase (HRP) functionalized fluorescent Au NCs. The HRP-Au NCs exhibit dual functions including fluorescent and catalytic abilities. HRP could catalyze the reaction between Au NCs and H_2O_2, resulting in the fluorescent quenching (Fig. 8.14B). Based on this, the HRP-Au NCs can be used to detect H_2O_2. Under the optimal conditions (7.8 μM HRP-Au NCs in 50 mM pH=9.0 glycine buffer at 25°C), the system can detect H_2O_2 as low as 30 nM. Besides H_2O_2, the HRP-Au NCs are also sensitive to other reactive oxygen species (ROS), such as O_2^-, tert-butyl hydroperoxide (TBPH), OCl^-, and •OH. The HRP-Au NCs can be used as ROS sensor in biological samples [76]. Wang et al. prepared highly fluorescent Cu NCs. In the presence of H_2O_2, the Cu centers of NCs were oxidized to Cu^{2+}, leading to the fluorescent quenching. The Cu NCs were used to detect H_2O_2 with a LOD of 0.01 mM [77].

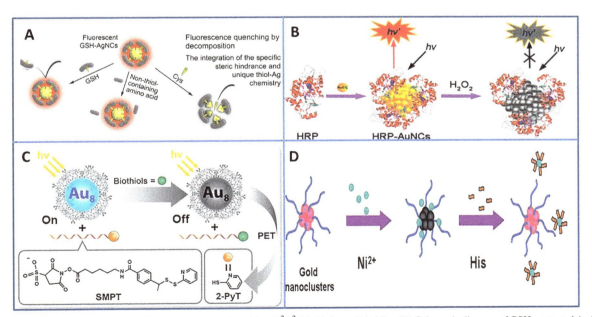

FIG. 8.14 Schematic diagram of metal NCs for detection of cysteine, H_2O_2, biothiols, and histidine (A) Schematic diagram of GSH-protected Ag NCs as fluorometric and colorimetric probe for cysteine sensing; (B) schematic of the H_2O_2 directed quenching of HRP-Au NCs; (C) Schematic illustrations of the nonconjugated sensor (Au$_8$ NCs + SMPT) for detection of biothiols; (D) Schematic diagram of Ni^{2+}-modified Au NCs as fluorescence turn-on probe for histidine detection.

Biothiol sensing

Biothiols such as cysteine (Cys), homocysteine (Hcy) and glutathione (GSH) play a critical role in the process of important cellular functions including detoxification and metabolism. Analyzing their levels in the human plasma or urine has important reference value for the early diagnosis of a variety of diseases. Dong et al. synthesized poly(methacrylic acid) (PMAA)-templated Ag NCs and used it to detect Cys based on the fluorescent quenching effect caused by the thiol-adsorption-accelerated oxidation of the emissive Ag NCs. The sensor allows for Cys detection in the range of 2.5×10^{-8}–6.0×10^{-6} M with a LOD of 20 nM [78]. Xie et al. utilized GSH-Ag NCs to identify and detect Cys by combining two advantages of the unique thiol-Ag chemistry and specific steric hindrance (Fig. 8.14A). By the ESI-MS analysis, it is confirmed that the fluorescent quenching is induced by the decomposing of the NCs, when Cys penetrate the GSH protecting layer. The GSH-Ag NCs show high selectivity to Cys against other 19 natural amino acids (nonthiol-containing) owing to the specific thiol-Ag interaction. The GSH-Ag NCs can also discriminate the usual interference biothiol-GSH or other bulky thiol-containing biomolecules (i.e., protein) and the target analyte-Cys because of the steric hindrance. The sensor provides a sensitive Cys detection with a LOD of <3 nM. Besides the change in fluorescent, the color of NC solution also has an obvious change from brown to transparent in the presence of Cys, therefore, the GSH-Ag NCs can be used as a dual functional sensor for Cys based on fluorometrically and colorimetrically [79]. Lin et al. utilized a thiol/disulfide exchange to trigger the fluorescence quenching through a photoinduced electron transfer (PET) process between the Au_8-cluster (an electron donor) and 2-pyridinethiol (2-PyT) (as an electron acceptor) for biothiols detection (Fig. 8.14C). In the presence of biothiols, the disulfide bond is cleavage and 2-PyT is released, PET process occurs from Au_8 NCs to 2-PyT, leading to the fluorescent quenching. The LOD is 15.4 µM and the linear detection range is 0–1500 µM. Thanks to the less steric hindrance of Au_8 NCs, it can detect not only less molecular weight thiols (i.e., GSH and Cys) but also high molecular weight thiols (i.e., protein thiols). Therefore, the Au_8 NCs sensor was further used for the imaging of protein thiols in living cells [80].

Different from the fluorescent quenching sensing, Liu et al. realized the Cys detection based on the fluorescent enhancement of BSA-Au NCs. The reason of fluorescent increase is that Cys could decrease the surface defects of Au NCs. The sensor can detect Cys in a wide linear range from 2.0 to 800 nmol/mL with a LOD of 1.2 nmol/mL. The sensor was further used in the Cys detection in human serum samples with satisfied results [81]. Besides fluorescent sensor, metal NCs-based electrochemical sensor was also designed to detect L-Cys by fabricating graphene oxide/Au nanocluster (GO-Au NCs) composite. Yu et al. utilized electrostatic interactions to realize linker-free connection between GO and Au NCs. The GO-Au NCs exhibit electrocatalytic active to the oxidation of L-Cys. Therefore, the anodic peak current is increased with the increasing of L-Cys. Based on the relationship between peak current intensity and concentration of L-Cys, the GO-Au NCs can be used to determine L-Cys within a linear range of 0.05–20.0 µmol/L and a LOD of 0.02 µmol/L. The sensor was further applied to detect L-Cys in human urine. The results show a good agreement with those obtained by chemiluminescence method [82].

Wang et al. reported single-stranded DNA stabilized Ag NCs (DNA-Ag NCs) to detect cysteine (Cys), homocysteine (Hcy) and glutathione (GSH) at the same time. The fluorescent of NCs is static quenched due to the formation of nonfluorescent coordination complex between DNA-Ag NCs and biothiols, which is confirmed by the shift in the absorption spectrum and unchanged lifetime of DNA-Ag NCs. The sensor shows high selectivity (10-fold tolerance) to biothiols over other amino acids. The LOD is 4.0 nmol/L, 4.0 nmol/L, and 0.2 µmol/L, for Cys, GSH, and Hcy, respectively. The practicability of this sensor was presented by successfully application in human plasma samples detection [83]. Qu and Ren et al. chosen appropriate DNA templates to synthesize biothiol targeted DNA-Ag NCs and used it as a "turn on" fluorescent sensor to detect Cys, Hcy, and GSH simultaneously. The sensor shows high specificity to biothiol over other amino acids, and it can be used in quantitative analysis of biothiol in human plasma samples. The LOD for Cys, Hcy, and GSH is 2.1, 1.5, and 6.2 nM, respectively [84].

Song et al. fabricated a fluorescent switch by combining histidine with Ni^{2+}-BSA Au NCs. The fluorescent of Ni^{2+}-BSA Au NCs is quenched due to the formation of nonfluorescence ground state complex between the surface of BSA-Au NCs and Ni^{2+}. In the addition of histidine, the fluorescent is restored because the histidine can bind with Ni^{2+} and remove it from the BSA-Au NCs surface. On the basis, the Ni^{2+}-BSA Au NCs can be used as a turn on fluorescent sensor to detect histidine with a LOD of 30 nM (Fig. 8.14D). Its practicality was validated by applying to the detection of histidine in human urine [85].

Bilirubin sensing

Free bilirubin concentration in blood serum is a better parameter to assess the risk caused by hyperbilirubemia in case of neonatal jaundice. Goswami et al. synthesized human serum albumin (HSA) stabilized gold nanoclusters (HSA-AuNCs) to detect bilirubin based on both fluorometric and colorimetric methods. On the one hand, the fluorescent of HSA-AuNCs is decreased due to the inherent specific interaction between bilirubin and HAS. Therefore, the HSA-AuNCs can be used as fluorescent probe to detect bilirubin with a LOD of 248 ± 12 nM and a linear detection concentration range of 1–50 μM. On the other hand, the HSA-Au NCs exhibit intrinsic peroxidase-like activity. In the presence of HSA-Au NCs, the free bilirubin can be oxidized to a colorless compound by H_2O_2. Therefore, the HSA-Au NCs can be used a naked eye sensor to detect free bilirubin with a LOD of 200 ± 19 nM. The feasibility of HSA-Au NCs was further confirmed by taking the application in detection of bilirubin in human serum [86].

Dopamine sensing

Dopamine (DA) is well known as an important neurotransmitter, which plays a vital role in the function of the central nervous, renal, hormonal and cardiovascular systems. Sony et al. used Cu^{2+}-BSA Au NCs as a "turn on" fluorescent sensor to detect DA. In the presence of DA, the quenched fluorescent of Cu^{2+}-BSA Au NCs is restored due to the removal of Cu^{2+} from BSA Au NCs surface induced by the interaction between DA and Cu^{2+}. The Cu^{2+}-BSA Au NCs demonstrated high selectivity and sensitivity to DA over other catecholamines and coexisting substances with a LOD of 0.01 μM. The sensor was further used to detect DA in human serum and urine samples with satisfied results [87]. Qu and Ren et al. utilized BSA-Au NCs to realize fluorometric and colorimetric dual channel detect DA. In the presence of DA, due to the photo-induced electron transfer process from the electrostatically attached DA to the BSA-AuNCs, the fluorescent of NCs is dramatically decreased. The fluorometric sensing method has a LOD of 10 nM. Besides, the intrinsic peroxidase-like activity of BSA-Au NCs was efficiently restrained after the addition of DA. The BSA-AuNCs could catalyze the oxidation of peroxidase substrates 3,3′,5,5′-tetramethylbenzidine (TMB) by H_2O_2 to produce a bright blue color. Therefore, AuNCs-TMB-H_2O_2 system could be used as the colorimetric indicator to detect DA (Fig. 8.15A and B). The great diagnostic potential of this sensor was demonstrated by the application in hydrochloride injection sample, human serum sample and PC12 cells [88]. Wang and Li et al. designed a ratiometric fluorescence GSH-Au NCs sensor to detect tyrosinase (TYR) activity and DA by utilizing the special reaction between them (Fig. 8.15C and D). TYP is a typical polyphenol oxidase, which can catalyze the oxidation of DA to o-quinone and therefore quenched the red fluorescent of Au NCs at 610 nm. Besides, the reaction between them can generate a new emission at 400 nm. The two-signal change can be used to construct a ratiometric fluorescent sensing mode. The sensor can detect TYP activity and DA as low as 0.006 unit/mL and 1.0 nM, respectively. Moreover, the sensor can be extended to detect TYP inhibitor, such as kojic acid (KA) [89].

Besides fluorescent sensing mode, metal NCs-based electrochemical sensing mode are also developed. Lee et al. utilized GSH-Au_{25} modified electrode to realize DA detection based on the excellent electrocatalytic activity to DA [90]. Zhu et al. utilized the electrogenerated chemiluminescence (ECL) behavior of BSA-Au NCs-ITO to detect DA (Fig. 8.15E and F). The cathodic ECL of BSA-Au NCs-ITO showed an obvious increase with the increasing of DA in the electrolyte. The ECL enhancement is due to acceleration of the electron injection into the conduction-band of ITO induced by the formation of a charge transfer complex between DA and ITO. The linear detection range is from 2.5 to 47.5 μM [91].

Heparin sensing

Monitoring of heparin levels is of crucial significance to avoid the risk such as hemorrhage and thrombocytopenia because it is widely used as the anticoagulant during numerous surgical procedures involving extracorporeal blood circulation. Yan et al. combined trypsin stabilized gold nanoclusters (try-AuNCs) with cysteamine modified gold nanoparticles (cyst-AuNPs) to construct a surface plasmon enhanced energy transfer (SPEET) sensor to detect heparin (Fig. 8.16A). The SPEET was formed because of the close distance between try-AuNCs and cyst-AuNPs due to the electrostatic interaction between them. The other indispensable factor is the large overlap between absorption spectrum of cyst-AuNPs SPR and excitation spectrum of try-AuNCs. Therefore, the fluorescent of try-AuNCs was dramatically quenched by cyst-AuNPs. In the presence of heparin, the cyst-Au NPs became aggregated, leading to a redshift of SPR absorption peak. Besides, the distance between try-AuNCs and cyst-AuNPs was increased because of the stronger electrostatic interaction between heparin and cyst-AuNPs. Therefore, the SPEET was weakened and the quenching degree of try-AuNCs fluorescent was less. Under the optimal conditions (20 min incubation, 10 mM pH 5.0 BR buffer), the sensor can detect as low as 0.05 μg/mL heparin and the linear detection range is 0.1–4.0 μg/mL. Its applicability was validated by applying to the determination of heparin in human serum samples, the results were in good agreement with those obtained by the medical diagnostic method in the hospital [92].

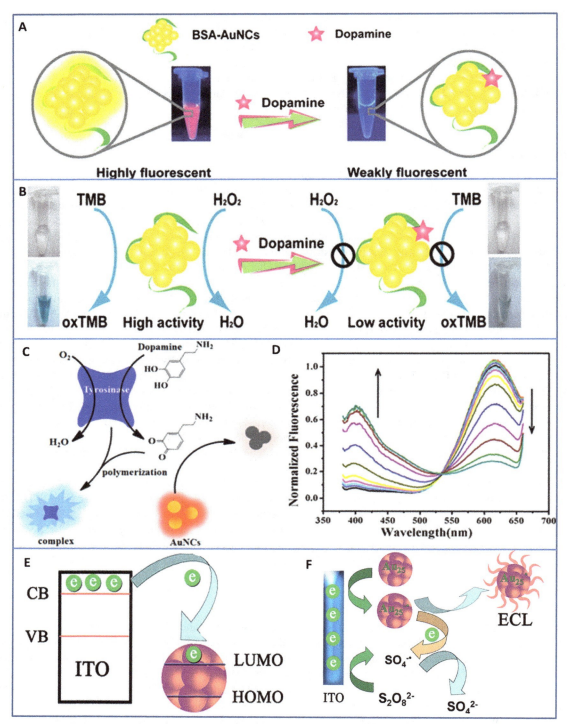

FIG. 8.15 Schematic illustration of (A) the fluorescence response of the BSA-Au NCs to DA and (B) the peroxidase-like catalytic color reaction for sensitive sensing of DA; (C and D) Schematic illustration of thiolate-protected Au NCs as ratiometric fluorescence probe for tyrosinase activity and DA detection; Schematic illustration of (E) electron transfer between ITO and Au NCs and (F) the ECL mechanisms of Au NCs.

Reactive oxygen species (ROS) sensing

Reactive oxygen species (ROS) are important signaling molecules that are generated in metabolic processes and play key roles in regulating a wide range of physiological functions. Due to the strong oxidant properties that can directly oxidize critical components of cells, leading to potentially serious damage in living cells, therefore, it is necessary to detect ROS in

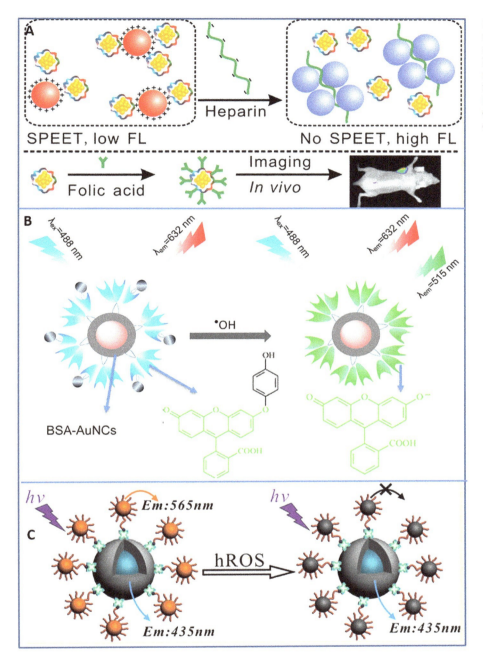

FIG. 8.16 (A) Schematic illustration for selective detection of heparin based on surface plasmon enhanced energy transfer between cyst-AuNPs and try-AuNCs; (B) working principle of the Au NC@HPF fluorescent probe for •OH detection; (C) Schematic illustration of a dual-emission fluorescent probe (Au NCs-decorated silica NPs) for hROS detection.

biological environment. Tian et al. used BSA-Au NCs and organic molecule 2-[6-(4′-hydroxy)phenoxy-3H-xanthen-3-on-9-yl]benzoic acid (HPF), respectively, acted as reference fluorophore and response signal origin to construct a ratiometric fluorescence sensor for •OH (Fig. 8.16B). Without •OH, the system exhibit only a red emission at 637 nm generated by NCs due to the nonfluorescent of HPF. However, in the presence of •OH, the HPF can specifically react with •OH to generate the product-dianionic fluorescein, which emitted at 515 nm. With the increase of •OH, the green emission is gradually increased, while the red fluorescent remains. Therefore, the ratiometric fluorescent sensor, Au NC@HPF, can determine •OH with a LOD of ∼0.68 μM, and the linear range is from 1 to 150 μM. The sensor showed high selectivity for •OH against other ROS, reactive nitrogen species (RNS), metal ions, and other biological species. Benefiting from the long-term stability against light illumination and pH, good cell permeability, low cytotoxicity, and high biocompatibility, the sensor was further used to monitoring of •OH changes in live cells upon oxidative stress [93]. Chu et al. assembled different luminescent dye-encapsulated silica particles and Au NCs to construct dual-emission fluorescence probe to detect highly ROS (hROS) including •OH, ONOO$^-$, and ClO$^-$ (Fig. 8.16C). The fluorescent of Au NCs at 565 nm is quenched due to the special response to hROS. On the other hand, the dye (i.e., CF 405S succinimidyl ester)-encapsulated silica particle is inert to hROS with a stable fluorescent at 435 nm and thus it can be served as an internal reference. Besides this, the silica particle has several advantages such as high biocompatibility and stability against photobleaching, which makes it a good matrix in intracellular sensing. The dye on the silica particle and its concentration can be carefully selected. Such a combination offers a well-resolved, intensity-comparable dual-emission signal, thus affording a high contrast and long track of hROS detection in living cells such as HeLa cells, HL-60 cells, and RAW 264.7 cells [94]. Qu and Ren et al. also fabricated a dual emission ratiometric fluorescent sensor by combining Au NCs and C-dots to detect hROS. The combination not only enhances the fluorescent of Au NCs but also stabilizes the fluorescent of C-dots. Similar to the silica particles and Au NCs system, C-dots and Au NCs are acted as internal reference and responsive signal, respectively. The C-dots-Au NCs are also used in monitoring hROS in living cells and further extended into the detection microenvironment of local ear inflammation [95].

Glucose sensing

Level of body fluid glucose is used for diagnosis of diabetes or hypoglycemia. Thus, accurate detection of glucose is very important. Wang et al. synthesized glucose oxidase-functionalized fluorescent gold nanoclusters (GOD-AuNCs) to detect glucose. The GOD can catalyze the glucose and O_2 to produce H_2O_2, and thus quench the fluorescent of Au NCs. The sensor has a LOD of 0.7×10^{-6} M and a linear range of 2.0×10^{-6}–140×10^{-6} M [96]. Sahu et al. utilized the fluorescent quenching triggered by the interaction between β-D-glucose with L-cys Au NCs to detect β-D-glucose. The sensor was used to detect the glucose level in human serum and the results were comparable to the pathological data obtained from the local hospital [97]. Qi and Li et al. used ovalbumin-protected Au NCs and the Alizarin Red S-3-aminophenyl boronic acid (ARS-APBA) to serve as reference and response signal, respectively. The Au NCs and ARS-APBA were connected by the linker, homopolymer of N-acryloxysuccinimide (PNAS). The 610 nm emission was used as built in correction due to the stability of Au NCs. The 567 nm emission of PNAS-APBA-ARS was used to identify the glucose, therefore, a ratiometric fluorescent sensor was fabricated (Fig. 8.17A). The sensor was further applied to the continuously determination of glucose in the global cerebral calm/ischemia surgeries of the rat brain [98].

Urea sensing

Urea is an important marker for evaluating uremic toxin levels and kidney and hepatocellular functions. Jayasree and Ajayaghosh et al. synthesized urease functionalized Au NCs with a NIR emission around 750 nm and a high QY of $25 \pm 6\%$. Due to the aggregation of Au NCs induced by urea, the fluorescent is quenched (Fig. 8.17B). On the basis, the sensor was used to detect urea in whole blood and blood serum samples [99]. Chen et al. synthesized N-acetyl-L-cysteine (NAC) stabilized Au NCs and applied it to determination of urea, urease, and urease inhibitors. The fluorescent of Au NCs is decreased with the increase in the concentration of urea or the enzyme activity of urease. The linear detection range is 0.055–0.55 mM and 2.2–55 U/L, and the LOD is 0.055 mM and 0.55 U/L for urea and urease, respectively. To validate the practicality, the sensor was further used to detect urea in human urine sample and detect *Helicobacter pylori* in human gastric tissue. The results were in line with the conventional diacetyl monoxime and bromothymol blue method, respectively [100].

Folic acid sensing

Folic acid (FA) deficiency in pregnant women leads to foetal development defects. Besides, the level of FA is a risk factor for coronary artery disease and stroke. Hemmateenejad et al. used BSA-Au NCs to detect FA based on the fluorescent quenching. Under the optimal conditions (0.1 M pH = 7.4 phosphate buffer solution), the sensor showed a linear detection range is 120.0 ng/mL–33.12 μg/mL with a LOD of 18.3 ng/mL. The sensor was further applied to the determination of FA in the tablet formulations, the results were in satisfactory agreement with the labeled values [101].

Protein sensing

Human serum proteins are useful diagnostic tools, and the alteration of the expression of some serum proteins is an early sign of altered physiology. Ouyang et al. utilized low-temperature oxygen plasma (LTP) treated BSA-Au NCs as a fluorescent sensor to realize highly sensitive detection for human serum proteins after native polyacrylamide gel electrophoresis (PAGE). The interaction between BSA-Au NCs and the proteins in human serum is induced by van der Waals force and hydrogen-bonding interactions, which is studied by performing isothermal titration calorimetry. By a series of optimization experiments, the oxygen LTP-treated BSA-Au NCs fluorescence imaging method (LOD = 15 ng) is 14 times more

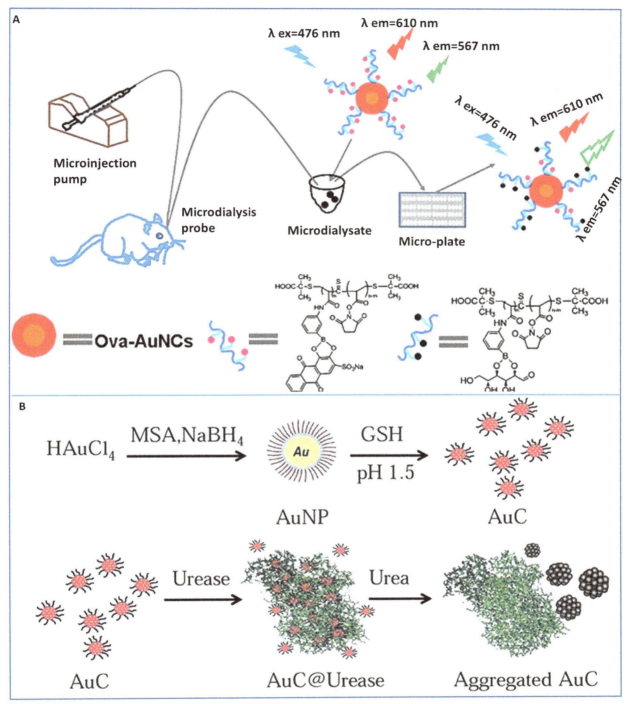

FIG. 8.17 (A) Schematic illustration of ratiometric fluorescent probe based on Au NCs and alizarin red-boronic acid for monitoring glucose in brain microdialysate; (B) Schematic illustration of the AuC@Urease as fluorescent quenching probe for urea detection.

sensitive than the CBB-R250 staining (LOD=248 ng). The sensor was successfully applied to distinguish the serum samples of patients with liver diseases from normal serum [102].

Hemoglobin (Hb) is an iron-containing oxygen-transport metalloprotein found in red blood cells, the variations of Hb level is associated with several diseases, including anemia (low Hb), erythrocytosis (high Hb), and thalassemia (abnormal chain synthesis). Chang et al. used 11-MUA-Au NCs to detect Hb based on the fluorescent quenching effect (Fig. 8.18A). In the presence of Hb, the redox reaction happened between the Fe^{2+} in Hb and the Au(I) of NCs, leading to the fluorescent

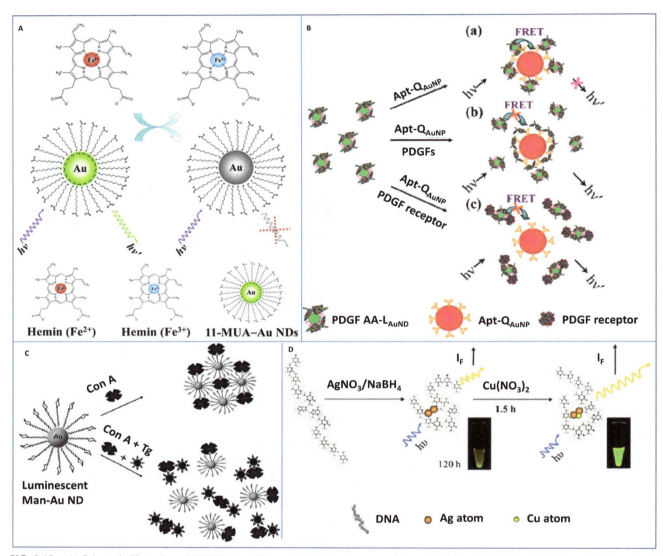

FIG. 8.18 (A) Schematic illustration of 11-MUA-Au NCs as photoluminescence quenching probe for hemin detection; (B) Schematic illustration of PDGF and PDGF receptor nanosensors based on the modulation of the photoluminescence quenching between PDGF AA-L$_{AuND}$ and Apt-Q$_{AuNP}$; (C) Schematic illustration of a Tg nanosensor based on Tg-mediated modulation of the interaction between Con A and G-Man-Au NCs; (D) Schematic illustration of the preparation of DNA-Ag and DNA-Cu/Ag NCs.

quenching, which can be confirmed by a decrease in the oxidation state of the Au$^+$ in XPS data and the appearance of high-spin Fe^{3+} signal in electron paramagnetic resonance (EPR) analysis. The sensor showed a linear detection range for Hb from 1.0 to 10 nM, a LOD of 0.5 nM in biological buffer, and a good selectivity over other nonheme-containing, such as proteins human serum albumin, b-casein, and carbonic anhydrase. The sensor was further used in the detection of Hb in diluted human blood samples [103].

Platelet-derived growth factor (PDGF) is an important protein for cell transformation and tumor growth and progression. Chang et al. fabricated a bioconjugate by combining photoluminescent Au nanodots (L$_{AuND}$) as donors and spherical Au NPs(Q$_{AuNP}$) as acceptors. L$_{AuND}$ and Q$_{AuNP}$ were modified with a breast cancer marker protein, platelet-derived growth factor AA (PDGF AA) and high affinity PDGF binding aptamers (Apt), respectively. Due to the FRET between Apt-Q$_{AuNP}$ and PDGF AA-L$_{AuND}$, the photoluminescent of Apt-Q$_{AuNP}$/PDGF AA-L$_{AuND}$ is quenched. In the presence of PDGFs, the interaction between Apt-Q$_{AuNP}$ and PDGF AA-L$_{AuND}$ were suppressed due to the competitive reactions between Apt-Q$_{AuNP}$ and PDGFs, leading to a recovery of luminescent. Similarly, in the presence of PDGFα-receptor, due to the competitive reactions between PDGFα-receptor and PDGF AA-L$_{AuND}$, the FRET between Apt-Q$_{AuNP}$ and PDGF AA-L$_{AuND}$ decreased, resulting in the less quenching in luminescent. Based on the competitive homogeneous photoluminescence quenching assays, the PDGF AA-L$_{AuND}$/Apt-Q$_{AuNP}$ realized proteins analysis (Fig. 8.18B). The LOD for PDGF AA and PDGFα-receptor were 80 pM and 0.25 nM, respectively. Moreover, the Apt-Q$_{AuNP}$ was demonstrated to be an effective selector for enriching PDGF AA from large-volume cell media and urine samples. Therefore, this sensor can remove most of the background fluorescence and thus detect PDGF AA as low as 10 pM, which is more sensitive than enzyme linked immunosorbent assay (ELISA) PDGF kits [104].

Chang et al. synthesized α-D-mannose conjugated Au NCs with good water solubility and high luminescent to detect concanavalin A (Con A) and *Escherichia coli* (*E. coli*). Due to the aggregation of Man-Au NCs induced by the Con A, the fluorescent is decreased in a linear relationship with the increase in Con A in the range of 1.0–10.0 nM. The LOD is 0.7 nM. Moreover, compared with the long detection time of standard methods for pathogen detection (2 days), the Man-Au NCs can be used to rapidly detect *E. coli* in 3 h. A brightly green fluorescent was yield due to the binding between Man-Au NCs and bacteria after incubation with *E. coli*. The linear detection range is from 1.00×10^6 to 5.00×10^7 cells/mL and the LOD is 7.20×10^5 cells/mL [105].

Thyroglobulin (Tg) is a representative glycoprotein biomarkers and therapeutic targets for thyroid cancer. Chang et al. synthesized Man-Au NCs by the irradiation of a light emitting diode (LED). The irradiation method enhanced the QY, altered the fluorescent properties and shortened the preparation time. The Man-Au NCs was used to detect Tg based on the competition between Tg and Man-Au NCs for the interaction with the Con A. As mentioned above, Con A can decrease the fluorescent of Man-Au NCs. However, the Tg inhibited the association between Con A and Man-Au NCs, leading to a recovery of fluorescent (Fig. 8.18C). The LOD is 48 pM. The practicality of this sensor was further validated in the detection in the serum samples [106].

C-reactive protein (CRP) is a prototypic acute-phase protein in humans and an important biomarker. Kawasaki et al. synthesized 2-methacryloyloxyethyl phosphorylcholine (MPC)-protected Au$_4$ NCs by using the size focusing approach. The Au$_4$(MPC)$_4$ NCs exhibits a yellow emission with a quantum yield (3.6%) and an average lifetime of 1.5 μs. CRP can induce the aggregation of Au$_4$(MPC)$_4$ NCs via specific interactions between the CRP and the MPC ligand and thus cause the fluorescent quenching. The LOD is 5 nM, which is far enough for the clinical diagnosis of inflammation [107].

Chen et al. utilized the selective binding between glutathione *S*-transferase (GST) and GSH to realize GST sensing by using fluorescent GSH-Au NCs. The fluorescent is decreased upon the addition of GST. The sensing approach is low cost without need of expensive antibodies and rapid (15 min) [108].

Change et al. synthesized DNA-Cu/Ag NCs with a higher QY of 51.2% and a shorter preparation time (1.5 h), compared with the DNA-Ag NCs with 11.5% QY and 120 h reaction time needed. The fluorescent of DNA-Cu/Ag NCs is quenched in the presence of single-stranded DNA binding protein (SSB), due to the specific interaction between DNA and SSB (Fig. 8.18D). The DNA-Cu/Ag NCs can detect SSB as low as 0.2 nM and the linear detection range is 1–50 nM [109].

To improve the practicality, a fluorescent solid array sensor was designed. Chang et al. constructed a fluorescent sensor array by using eight dual-ligand functionalized Au NCs with similar fluorescent properties but different surface characteristics. These Au NCs acted not only as efficient protein receptors but also as competent signal transducers. The sensor array can be used to discriminate eight proteins (i.e., BSA, human serum albumin (HSA), lysozyme (Lys), trypsin (Try), myoglobin (Myo), cytochrome C (CytC), histone (His), streptavidin (SA)) with a variety of sizes and charges at a low concentration (A$_{280}$ = 0.005), even can be extended to 48 unknown proteins [110]. To improve the sensitivity of metal NCs fluorescent sensor, Zhang et al. decorated five kinds of protein-protected Au NCs on plasmonic Ag substrates to obtain about 20-fold enhancement in the fluorescence intensity. Targeted protein analytes can interact with these Au NCs and induce the fluorescence alternation, leading to a distinct fluorescent image pattern, which can be used to detect protein analytes [111].

Enzyme sensing

Proteases can catalyze the hydrolytic cleavage of specific peptide bonds in target proteins and breaking up proteins into smaller fragments, which are involved in the control of a variety of essential physiological and pathological processes. Xu et al. utilized BSA-Au NCs to detect trypsin, the biomarker for pancreatitis. Trypsin can enzymatically hydrolyze the BSA templates, leading to the breakdown of the BSA-Au NCs, and thus the fluorescent of NCs is decreased. The sensor showed a linear detection range of 10 ng/mL–100 μg/mL with a LOD of 2 ng/mL. The sensor displayed a good selectivity to trypsin over other enzymes such as lysozyme, glucose oxidase, thrombin, papain, pepsin, or Nα-Tosyl-Lys-chloromethylketone (TLCK) inhibited trypsin. The potential application of this sensor was demonstrated in detection of trypsin in the human urine samples [112]. Fang and Chen et al. also utilized the enzymatic hydrolysis of template protein trigged fluorescent quenching to detect trypsin and chymotrypsin. The LOD are 1.9 and 1.4 ng/mL and the linear ranges are $8.5–1 \times 10^5$ ng/mL and $5.7–1 \times 10^5$ ng/mL for trypsin and chymotrypsin, respectively [113].

Human α-thrombin is a coagulation protein in the bloodstream, which converts soluble fibrinogen into insoluble strands of fibrin as well as catalyzing many other coagulation related reactions. Martinez et al. synthesized highly fluorescent DNA aptamer-templated Ag NCs which can be acted as both a fluorescent label and a specific recognition part to human α-thrombin. Therefore, in the presence of human α-thrombin, the fluorescent of DNA aptamer-templated Ag NCs is quenched due to the specific binding between it and human α-thrombin. The LOD is 1 nM [114]. Compared with the fluorescent quenching sensing, Zhu and Le et al. reported a binding-induced fluorescent enhancement strategy to detect human α-thrombin. The sensing system includes three parts, which are affinity recognition part, binding-induced DNA hybridization part, and the fluorescent enhancement of Ag NCs part. Two aptamers, Apt 29 and Apt 15 bind to heparin and fibrinogen-binding site which are located at opposite sides of human α-thrombin molecule, respectively. Apt 15 is modified with a 12-nucleotide (nt) sequence and a nanocluster nucleation sequence, while, Apt 29 is modified with a complementary sequence (12-nt) and a G-rich overhang. Martinez and Werner et al. found the fluorescent of DNA-Ag NCs can be enhanced 500-fold when it is close to a G-rich DNA segment [115]. In the presence of human α-thrombin, the two aptamers are linked together and thus promote the DNA hybridization between the complementary linker sequences attached to each aptamer. Therefore, the G-rich overhang is getting close proximity to Ag NCs, leading to a fluorescent enhancement. The sensor can detect human α-thrombin as low as 1 nM and was further used in the human serum samples. Wang and Dong et al. also used the similar target induced DNA hybridization promoting G-rich sequence close to NCs strategy to detect thrombin and target DNA [116]. Only if appropriate recognition parts which can specifically link with target are available, the sensing strategy can be extended to other analytes detection [117]. Besides fluorescent sensing, He and Yang et al. fabricated a colorimetric sensor by using DNA-Ag/Pt NCs. Apt 29 was first introduced into the streptavidin-coated 96-well microplates via biotin-streptavidin interaction. The DNA sequence in the Ag/Pt NCs contains Apt 15. In the presence of human α-thrombin, the DNA-Ag/Pt NCs is immobilized on the plates due to the specific binding of the two aptamers to the target protein. The increase in absorbance intensity at 452 nm is linear with the increase of human α-thrombin in the range of 1–50 nM and the LOD is 2.6 nM [118].

Similar to the binding-induced fluorescent turn on detection strategy, Zhang et al. designed and synthesized DNA-Ag NCs to detect EcoRI activity and inhibition. EcoRI, a type II restriction endonuclease, can protect living cells against foreign DNA by recognizing and cleaving a defined DNA sequence GAATTC. Its activity and inhibition play important roles in the fields of modern molecular biology. The DNA template was composed of a cytosine-rich single-stranded DNA with the sequence of EcoRI recognition site (GAATTC). The synthesized DNA-Ag NCs exhibit a dark fluorescent. After introducing the other partly complementary DNA sequence which has been rationally modified by adding a sequence of GAATTC and a G-rich overhang sequence, the G-rich overhang was close to Ag NCs due to the DNA hybridization, resulting in the fluorescent enhancement of Ag NCs. In the presence of EcoRI, the G-rich sequence and Ag NCs were separated by the specific DNA cleavage induced by EcoRI. Therefore, the fluorescent is decreased linearly with the increase of EcoRI in the range of 5.0×10^{-4}–3.0×10^{-3} U/μL. The LOD is 3.5×10^{-4} U/μL. Due to the endonuclease cleavage of DNA can be restrained by the presence of inhibitors the sensor was further used to monitor endonuclease inhibitors. Here, 5-fluorouracil was tested as the inhibitor of EcoRI endonuclease. The activity of EcoRI decreased with the increase of 5-fluorouracil. In the presence of 0.1 mM 5-fluorouracil, the activity of EcoRI was reduced to 50%. The preparation of DNA-metal NCs is simple only including reduction and hybridization step and the detection is easy without the need for separation, precipitation, and washing. Moreover, the sensing system can be extended to measure the activity and inhibition of other endonucleases by rationally designing appropriate DNA substrate sequences. The sensing system has great promising potential in practical application [119].

Adenosine deaminase (ADA), an important enzyme in all human tissues, can catalyze the conversion of adenosine to inosine. ADA deficiency or overexpression will cause the severe combined immunodeficiency disease (SCID) or hemolytic anemia. Ye et al. designed a DNA sequence which is composed of three regions, a cytimidine-rich (C-rich) sequence used to synthesize Ag NCs, an aptamer region used to assemble adenosine, a 15G overhang region used to enhance the fluorescent (Fig. 8.19A). They synthesized the specific DNA sequence templated-Ag NCs. In the presence of adenosine, due to its induced allosteric effect, the aptamer region assembles into a hairpin structure, the 15G overhang region is close to the Ag NCs, leading to the fluorescent increase. By correlating the fluorescent increase and the adenosine concentration, the system can realize the detection of adenosine. However, the ADA can transform adenosine to inosine and thus release the aptamer region, and then the 15G region is back and far away from Ag NCs, resulting in the fluorescent decrease. Therefore, the system can detect ADA by the fluorescent quenching [120]. Similarly, Zhu et al. utilized DNA-Ag NCs to detect ADA and its inhibition. The DNA stand is designed of three parts, including an aptamer region for adenosine assembly, a sequence complementary to the region of the adenosine aptamer, and an inserted six bases cytosine-loop. In the presence of adenosine, the DNA is a close-packed tight structure. While, after adding ADA, the adenosine is converted to inosine, leading to a flexible structure and leaving more abundant vacant sites for the complementary sequence to

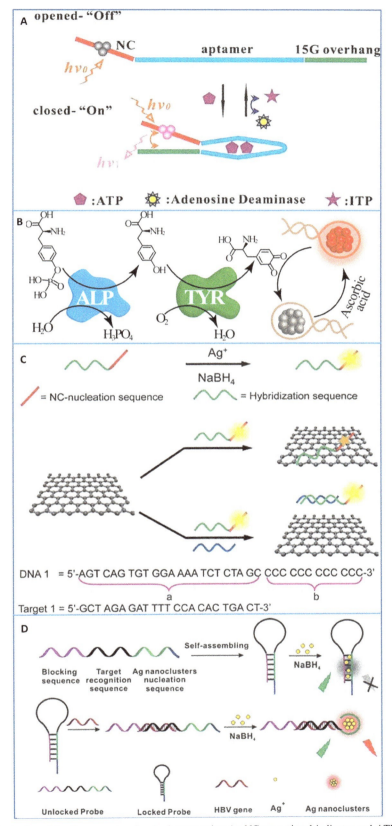

FIG. 8.19 (A) Schematic illustration of the fluorescent molecular beacon using Ag NCs as a signal indicator and ATP and adenosine deaminase as mechanical activators; (B) Schematic illustration of Luminescent DNA/Ag NCs to probe biocatalytic transformations; (C) Schematic illustration of Ag NCs-GO as probe for label-free DNA detection; (D) Schematic illustration of the assay for HBV gene detection using the hairpin DNA probe.

hybridize with the regions of adenosine aptamer, which plays a key role to generate a red-emitting Ag NCs. Therefore, the fluorescent enhancement can be used to indicate the activity of ADA. The LOD is 0.05 U/L. The sensor was further used to detect ADA in human serum with accurate results. Furthermore, the sensor can monitor the inhibition of ADA, taking erythro-9-(2-hydroxy-3-nonyl) adenine hydrochloride (EHNA) as the inhibitor. With 107 nM EHNA, the activity of ADA is inhibited by half. The sensor can detect EHNA as low as 100 fM and the linear response range is from 30 to 500 nM [121].

Protein kinase CK2 (formerly casein kinase II) can catalyze protein phosphorylation and plays a vital role in the early diagnosis of cancers and the discovery of new drugs in treat cancer-related phosphorylation. Qiu et al. reported Zr^{4+}-mediated peptide-Au NCs to detect CK2 activity. The peptide was phosphorylated by CK2 firstly, after being treated with Zr^{4+}, the NCs became aggregated due to the coordination between Zr^{4+} and phosphate groups, leading to a significant fluorescent quenching. Under the optimal conditions (7.5 pH, 0.25 mM Zr^{4+}, 60 min incubation), the LOD is 0.027 U/mL and the linear range is from 0.08 to 2.0 U/mL. The sensor was further applied to detect the inhibition of CK2 in human serum samples by measuring the content of ellagic acid—the inhibitor for CK2. Compared with other inhibitors like DRB, emodin, and quercetin, the ellagic acid shows the best enhancing degree in fluorescent. It is 0.045 μM ellagic acid that can inhibit 50% activity of CK2 [122].

Protein posttranslational modifications (PTMs), which are chemical modifications and most often regulated by enzymes, play key roles in functional proteomics. Jiang and Tang et al. synthesized fluorescent Au NCs with different peptide templates. The peptide was carefully designed with the ability to interact with target enzyme. The peptide 1 CCIHK(Ac) as a compact coating suppressing the O_2 induced quenching was used to synthesize Au NCs with an emission at 455 nm. In the presence of histone deacetylase 1 (HDAC 1), due to the deacetylation of peptide 1 by HDAC 1 destroying the compactness of the coating and the O_2 can diffuse into contact with Au NCs core and thus quench the fluorescent. They further extend the sensing strategy to another PTM-protein kinase A (PKA) detection. The peptide 3 CCLRRASLG, was designed for PKA detection. The fluorescent of peptide 3 protected Au NCs is decreased with the increase of PKA. The specific peptide protected Au NCs was used to quantitatively analyze HDAC 1 and PKA with a LOD of 5 pM and 6 pM, respectively [123]. Xia et al. also utilized PKA destroying peptide template promoting O_2^- induced the fluorescent quenching to realize PKA detection [124]. Wang and Yang et al. used 12 polycytosine-templated silver nanoclusters (dC_{12}-Ag NCs) to monitor the PKA activity and inhibition. In the presence of ATP, the fluorescent of dC_{12}-Ag NCs is enhanced. However, PKA can catalyze the hydrolysis of ATP to ADP, therefore, the fluorescent intensity change can be used to indicate the content of PKA. The sensor was further used to detect the PKA inhibitor, H-89, and the screen the drug-induced PKA activation in HeLa cells [125].

Similarly to the enzyme catalysis cleavage of template peptide promoting the O_2^- diffuse induced fluorescent quenching, Jiang and Kuang et al. used this sensing strategy to detect esterase. The designed template peptide is CCAAA, in which the CC is used to assemble to Au NCs and the AAA is used for the elastase. The CCAAA-Au NCs exhibit fluorescent due to the peptide coating suppressing the O_2^- mediated fluorescent quenching. However, in the presence of esterase, the peptide is destroyed and the O_2 diffuse into the Au NCs core, leading to the fluorescent quenching. The sensor provides a linear range of 50 pM–100 nM and a LOD of 30 pM [126]. Lin and Yu et al. utilized enzyme catalyzing substrate hydrolysis triggered top-down etching process for the fluorescent NCs generation to detect enzyme activity. For example, esterase and alkaline phosphatase (ALP) can catalyze the hydrolysis of substrate 1 and substrate 2 to generate 6-mercapto-1-hexanol (MCH), an alkanethiol ligand, which can be used to etch Au NPs to generate fluorescent Au NCs. Therefore, the fluorescent intensity is linearly positive related to the concentration of esterase or ALP. The LODs are 0.04 and 0.005 mU/mL and the linear range is 0.1–10 and 0.01–10 mU/mL, for esterase and ALP, respectively. The sensor can also be used to detect enzyme activity in biological samples such as calf serum and A549 cell lysate. Besides, the sensor can detect the inhibition of the enzyme as well. Na_3VO_4, a commonly used ALP inhibitor was tested. The presence of 10 mm Na_3VO_4 can inhibit the activity of 50 mU/mL ALP [127].

Phospholipase C (PLC), an enzyme of the phospholipase superfamily, plays important roles in a range of biological processes, including metabolism, digestion, inflammation response, membrane trafficking, and intercellular signaling. Chang and Huang et al. synthesized 11-MUA-Au NCs and liposome hybrids (11-MUA-Au NC/Lip). Its fluorescence intensity is quenched due to the O_2. PLC can catalyze the hydrolysis of phosphatidylcholine units from Lip to yield diacylglycerol (DAG) and phosphocholine (PC). The DAG can interact with 11-MUA-Au NCs by hydrophobic interactions, resulting in the inhibition of O_2^- induced fluorescent quenching. The sensor realizes the PLC detection with a LOD of 0.21 nM. The sensor was used to detect PLC activity in breast cancer cells (MCF-7 and MDA-MB-231 cell lines) and nontumor cells (MCF-10A cell line). The concentrations of PLC in MCF-10A, MCF-7 and MDA-MB-231 are 111.5 ± 9.3, 165.8 ± 13.8, and 376.4 ± 33.9 nM, respectively. The sensor was further used to probe the interaction of PLC and its inhibitor D609. $3.81 \pm 0.22\,\mu M$ D609 can inhibit 50% activity of PLC; the results are in good agreement with that detected by the Amplex Red PC-PLC assay [128].

Aetylcholinesterase (AChE) has pivotal functions in Alzheimer's disease, inflammatory processes, and nerve-agent poisoning. Zhang et al. utilized enzyme hydrolysis triggered fluorescent enhancement of NCs to detect AChE. AChE can convert catalytic hydrolysis of Acetylthiocholine (ATCh) chloride to generate thiocholine (TCh). The fluorescence of 12 polycytosine-templated silver nanoclusters (dC_{12}-Ag NCs) is increased in the presence of TCh due to the formation of Ag-S complex. Therefore, the fluorescent enhancement can be used as an indicator to detect AChE. The LOD is $0.5 \times 10^{-4}\,U/mL$. The sensor was further used to detect AChE in human blood red cell membranes with accurate results comparable to those obtained by Ellman's method. The sensor was also used to screen the AChE inhibition. $4.7 \pm 0.7\,nM$ tacrine can inhibit 50% activity of AChE [129].

Willner et al. utilized enzyme biocatalytic transformation to realize specific enzyme detection by using DNA-Ag NCs (Fig. 8.19B). For example, H_2O_2 can quench the fluorescent of DNA-Ag NCs, meanwhile, glucose oxidase can oxidize glucose and produce H_2O_2 and gluconic acid. Therefore, with the increase of glucose oxidase, the fluorescent of NCs is decreased (in the presence of glucose), the DNA-Ag NCs can be used to detect glucose oxidase. Similar to the biocatalytic oxidase-stimulated H_2O_2-generating biotransformations, the sensor can be extended to detect another enzyme such as tyrosinase. Because tyrosinase can catalyze the oxidation of tyrosine, dopamine, or tyramine and produce quinone derivatives, it can quench the fluorescent of NCs. Furthermore, the sensing strategy was evolved to probe the bienzyme biocatalytic cascades, such as the alkaline phosphatase/tyrosinase coupled hydrolysis and oxidation of o-phospho-L-tyrosine and the acetylcholine esterase/choline oxidase hydrolysis of acetylcholine and subsequent oxidation of choline [130].

DNA sensing

The sensitive and selective detection of nucleic acids plays important roles in gene expression profiling, clinical disease diagnostics, and the drug industry. Ren et al. combined single-stranded DNA (ssDNA)-Ag NCs and graphene oxide (GO) to construct a label-free fluorescent "turn on" sensor to detect DNA (Fig. 8.19C). GO plays a role of fluorescent quencher, the ssDNA-Ag NCs-GO exhibits weak fluorescent due to the energy transfer from Ag NCs to GO by π-π stacking interaction. In the presence of target DNA, due to the different adsorption affinity for ssDNA and double-stranded DNA (dsDNA) to GO, the fluorescent is increased. Therefore, the fluorescent intensity enhancement can be quantitatively related to the concentration of target DNA. The LOD is estimated to be 1 nM. Besides, the sensor shows high selectivity to target DNA, even single-nucleotide difference can be detected. The large surface area of GO allows the accommodation of many ssDNA-Ag NCs with different fluorescent emissions. Therefore, it can be used as a multiple DNA sensor to detect different target DNA at the same time [131]. Willner et al. also used nucleic-acid functionalized Ag NCs and GO hybrids to detect a series of genes of infectious pathogens, like hepatitis B virus gene (HBV), the immunodeficiency virus gene (HIV), and the syphilis (*Treponema pallidum*) gene, only specific designed nucleic acid required [132]. Wang et al. constructed DNA-Ag NCs and G-quadruplex/hemin complexes and utilized the photoinduced electron transfer between them induced fluorescent quenching to detect target DNA [133].

West et al. used enzyme-DNase1 as template to synthesize Au NCs. The as-prepared $DNase1:Au_8$ NCs exhibit blue fluorescence whereas the $DNase1:Au_{25}$ NCs are red emitting. The DNase1:Au NCs remain the endodeoxyribonuclease activity. Therefore, in the presence of dsDNA, the DNase1: Au NCs can digest the dsDNA and cause fluorescent change. The sensor can detect dsDNA as low as $2\,\mu g/mL$ [134].

Petty et al. used bifunctional oligonucleotide-stabilized Ag NCs to realize the target DNA sensing. The bifunctional oligonucleotide includes two parts, one is used to synthesize Ag NCs and the other is used to bind with target DNA. First, the bridging quencher stand is introduced to quench the fluorescent emission. Second, in the presence of target DNA, the competition binding occurred and the fluorescent is enhanced. On this basis, the sensor can detect target DNA [135]. Liu et al. designed a specific DNA sequence including three parts, one is used for the Ag NCs nucleation, one is used for target recognition, and one is used for blocking nucleation. Without target DNA (here using HBV gene as a model), the blocking part can hybridize with the nucleation part and form a hairpin DNA template which prohibits Ag NCs growth, therefore, the fluorescent is "off". In the presence of HBV gene, the recognition part specific binding with it and the locked hairpin DNA is unlocked and releases the nucleation part, thus, the fluorescent Ag NCs is generated. Therefore, the fluorescent enhancement of Ag NCs can be used to detect HBV gene (Fig. 8.19D). The LOD is 3.0 nM and the linear detection range is 10–200 nM. The sensor can be used to detect diverse target DNA only by embedding different recognition sequences [136]. Zhang and Wang et al. also used the specific designed hairpin DNA probe to in suit generate fluorescent DNA-Ag NCs to detect the Influenza A virus subtype H1N1 gene DNA and subtype H5N1 gene DNA [137]. Lei et al. designed specific DNA sequence which has two parts can generate two kinds of Ag NCs with different fluorescent. In the presence of target DNA, the one part of DNA sequence turns into loop and enhance the fluorescent of Ag NCs 1, while, the fluorescent of Ag NCs 2 is decreased. The ratio of two fluorescent emission intensities is used to detect the content of the target [138].

Single nucleotide polymorphisms (SNPs) are the most common inherited types of sequence variations in the human genome and important to various medical and physiological features of human. Wang et al. expanded ssDNA to hybridized dsDNA to synthesize Ag NCs. The as-prepared NCs has a highly sequence dependent fluorescent property, and it can be used to discriminate single nucleotide acid difference [139]. Besides, Wang et al. used DNA-Cu NCs to distinguish single nucleotide mismatch and identify the mismatch type in specific DNA sequence at room temperature [140]. Martinez, Werner and Yeh et al. reported a chameleon DNA-Ag NCs sensor to identify single-nucleotide polymorphisms (SNPs). The used DNA template is composed of an ssDNA for synthesizing Ag NCs and a particular DNA sequence with G-rich for fluorescent enhancement and recognition target. Different enhancer sequence can bind with particular target DNA and light up distinct colors which can be easily identified by naked eye. With only single nucleotide mismatch, the luminescent of Ag NCs is significantly different due to the different alignment between the Ag NCs and the enhancer. The sensor was successfully used to discriminate single-nucleotide substitution scenarios in three synthetic DNA targets, six disease-related SNP targets, and two clinical samples [141].

miRNA sensing

MicroRNAs (miRNAs) are regulatory small RNAs that play vital roles in numerous developmental, metabolic, and disease processes. The levels of miRNAs can be useful biomarkers for cellular events or disease diagnosis. Gao et al. utilized hybridization chain reaction (HCR) to detect miRNA. Two designed DNA sequences are stable in solution. One is a hairpin DNA sequence (MB1) with a ploy-cytosine nucleotide loop which is used to synthesize highly fluorescent Ag NCs. The other hairpin DNA sequence (MB2) contains a poly-guanine nucleotide sticky end. In the presence of let-7a, due to the hybridization between it and MB1, the hairpin structure is opened, inducing MB2 unlocked and exposing a new single strand of MB2 which is identical in sequence to let-7a. The process is a cascade of hybridization events and leads to a nicked double helix. Thanks to the HCR, a small amount of let-7a can cause a significant conformational change of large amount of MB1, resulting in an obvious fluorescent quenching of NCs. Therefore, the NCs can be used to detect let-7a by the decrease in fluorescent intensity. The sensor shows high selectivity to let-7a among SNP in the let-7a miRNA family [142]. Yang and Vosch et al. designed a DNA sequence which contains a specific complementary region toward a target miRNA and a 12 nucleotide scaffold for NCs synthesis. The DNA template Ag NCs emit red fluorescent, however, in the presence of target miRNA, the fluorescent is decreased. Based on the fluorescent quenching, the sensor can be used to detect target miRNA (Fig. 8.20A) [143].

Ye et al. utilized target-assisted isothermal exponential amplification (TAIEA) coupled with fluorescent DNA-scaffolded AgNCs to detect miRNA with attomolar sensitivity. The TAIEA includes two parts, one is unimolecular DNA as mechanical activator, and the other one is target miRNA as the trigger. The unimolecular DNA was designed of three functional regions, the amplification template, polymerases, and nicking enzymes. The TAIEA strategy can realize large amount (10^9-fold) of reporter oligonucleotides converted from small amount of miRNA within minutes. The reporter oligonucleotides were used as scaffolds to synthesize fluorescent Ag NCs as signal indicators. Owing to the TAIEA strategy, the sensing system can detect 2 aM miRNA and discriminate the difference between miRNA family members. The sensing system can be extended to other kinds of miRNA detection, only if designing particular unimolecule DNA with complementary sequence with target. The fluorescent emission can also be tuned by changing the reporter oligonucleotides sequence, which can be used to realize multiple targets analysis [144]. Similarly, Wang et al. used a rolling circle replication (RCR) technique to synthesize concatemeric dsDNA-templated fluorescent Cu NCs to realize ultrasensitive miRNA detection. In this sensing strategy, the miRNA is triggered to initiate the RCR and a long concatemeric dsDNA scaffold is obtained via hybridization. The concatemeric dsDNA-template was used to synthesize Cu NCs which can emit highly red fluorescent. The sensitivity for miRNA detection of the concatemeric dsDNA-templated Cu NCs is highly improved by ~10,000 fold compared with the monomeric dsDNA-Cu NCs and the LOD is 10 pM for let-7d [145]. Besides, Qu and Ren et al. also utilized target miRNA triggered HCR to obtain large amount of dsDNA to synthesize high fluorescent Cu NCs, thereby realizing target miRNA detection [146].

Zhang et al. utilized oligonucleotide encapsulated Ag NCs as electrochemical label to detect miRNA. The oligonucleotide sequence includes two regions, one is recognition region for hybridization with the target miRNA, and the other is template region for the synthesis of Ag NCs. The gold electrode was preimmobilized with molecular beacon (MB) probe. The hybridization between MB and the oligonucleotide sequence brings Ag NCs to the electrode surface. Because the Ag NCs has metal mimic enzyme properties for catalyzing H_2O_2 reduction, the response to H_2O_2 reduction causing an

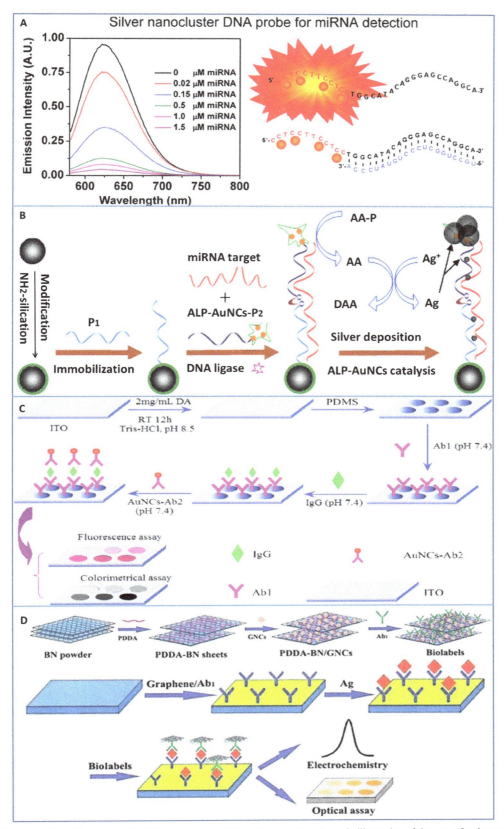

FIG. 8.20 (A) Schematic illustration of Ag NCs DNA probe for miRNA detection; (B) schematic illustration of the assay for the sandwich-type electroanalysis of miRNAs using Fe_3O_4-P_1 as magnetic particles-loaded DNA capture probe and ALP-Au NCs-P_2 as ALP-Au NCs-labeled DNA detection probe; (C) schematic illustration of the optical immunosensor using luminescent Au NCs as labels; (D) schematic illustration of the fabrication and measurement process of the sandwich-type immunosensor.

electrochemical signal change can be used to detect the content of miRNA. The sensitivity of the sensor is high with a LOD of 67 fM [147]. Wang et al. reported a sandwich-type analysis method to detect miRNA by using electrochemical signal amplification of catalytic silver deposition (Fig. 8.20B). The alkaline phosphatase (ALP)-stabilized Au NCs was synthesized with unexpected, enhanced catalysis activities. Besides, the Au NCs can catalyze silver deposition. The magnetic particles (i.e., Fe_3O_4 NPs) and the ALP-Au NCs were bound to DNA capture probes (P1) and DNA detection probes (P2), respectively. The DNA ligase catalyzed the linking of two DNA probes and the combination hybridized with miRNA. The ALP of ALP-Au NCs hydrolyzes AA-P to AA which can be used to reduce Ag^+ to Ag. The Ag deposited magnetic electrode shows signal amplification, thereby for the miRNA ultrasensitive detection. The LOD for the detection of free miRNAs in blood is 21.5 aM [148].

8.2.2.2 Immunoassays

Chang and Huang et al. synthesized protein A (a *Staphylococcus aureus* protein)-modified Au NCs (PA-Au NCs) and utilized the specific interaction between protein A and human immunoglobulin G (hIgG) to realize the detection of hIgG. Based on the hIgG-induced turning on of the fluorescent of the PA-Au NCs, the sensor can detect hIgG as low as 10 nM. The sensor was further used to detect hIgG in human plasma samples with a good consistent result with that obtained by ELISA [149]. Zhu et al. utilized Au NCs as labels for signal amplification to construct an optical immunosensor. The indium tin oxide (ITO) chip immobilizes a stable and robust polydopamine film and then modifies goat antihuman IgG (Ab1). Au NCs were conjugated with rabbit antihuman IgG (Ab2). hIgG could be qualitatively and quantitatively detected by colorimetry, fluorescence method, as well as eye observation through sandwiched immuno reaction (Fig. 8.20C). Based on the dual signal amplification of Au NCs and silver enhancement, the sensor can detect hIgG as low as 5 pg/mL [150]. Fernández and Pereiro et al. synthesized lipoic acid stabilized Au NCs and conjugated the desired antibody to realize quantitatively immunoassay for immunoglobulin E (IgE) detection in human serum. They compared two typical immunoassay configurations, competitive and sandwich. With the sandwich format, the LOD is 10 ng/mL and the competitive format provided a LOD of 0.2 ng/mL [151]. Zhu et al. fabricated boron nitride-Au NCs composite to immunosense interleukin-6 (IL-6) based on fluorescent or electrochemical signal change (Fig. 8.20D). Fluorescent Au NCs were decorated on poly-diallyldimethylammonium chloride (PDDA)-BN sheets by electrostatic layer-by-layer assembly. Then, antibody conjugates (Ab2) were immobilized on Au NCs. The PDDA-BN-Au NC-Ab2 acted as a fluorescent and electrochemical probe to detect IL-6 by sandwich bioaffinity immunoassay. The LOD is 0.03 ng/mL and 1.3 pg/mL for fluorescent and electrochemical method, respectively [152]. They also synthesized calcium carbonate-Au NCs hybrid spheres and used it as an immunosensor to detect the cancer biomarker neuron-specific enolase (NSE) by fluorescent and electrochemical method. The LODs were 2.0 and 0.1 pg/mL for fluorescent and electrochemical detection, respectively [153].

8.2.3 Other sensor

8.2.3.1 pH meters

Wang et al. developed highly luminescent and stable Cu NCs which exhibits the aggregation-induced emission (AIE) feature. Due to the AIE feature, the Cu NCs can be used as a pH meter (Fig. 8.21A and B). When the pH is below 5.0, the Cu NCs become aggregated and emit red luminescent under UV light. When the pH is above 6.2, the Cu NCs become separated and exhibit no fluorescent. When the pH is from 5.0 to 6.2, the fluorescent intensity is linearly decreased. The luminescent of Cu NCs were on and off according to the pH cycling between pH 3.1 and 7.1 using acid and base as modulators [77]. Chen et al. synthesized *N*-acetyl-L-cysteine (NAC)-Au NCs exhibits ultrasensitive (ΔpH of 0.35 between ON/OFF states) pH-responsive properties in the range of 6.05–6.40 [100]. Chang et al. synthesized BSA-Ce/Au NCs with dual fluorescent emissions at 410 and 650 nm which are pH dependent and independent, respectively. Therefore, the ratio of the two emission peak intensities can be used as pH indicator with a linear relationship between 6.0 and 9.0. Furthermore, due to the stability and biocompatibility of BSA-Ce/Au NCs, it can be further used to monitor internal pH of HeLa cells [154].

8.2.3.2 Thermometers

Nienhaus et al. used fluorescent Au NCs to sense temperature variation in the living cells (Fig. 8.21C and D). Au NCs can enter the cells and localize mainly in endosomes/lysosomes through the endocytosis. The fluorescent intensity and lifetime of Au NCs are both dependent on the temperature. On the basis of time-correlated single photon counting (TCSPC)-based fluorescence lifetime imaging microscopy (FLIM), the Au NCs realized the spatially resolved temperature measurements in HeLa human cancer cells [155]. Baker et al. employed protein-passivated Au NCs as optical sensors for accurate thermometers. By halide doping, sol-gel incarceration and thermal denaturation of the starting protein, the pronounced

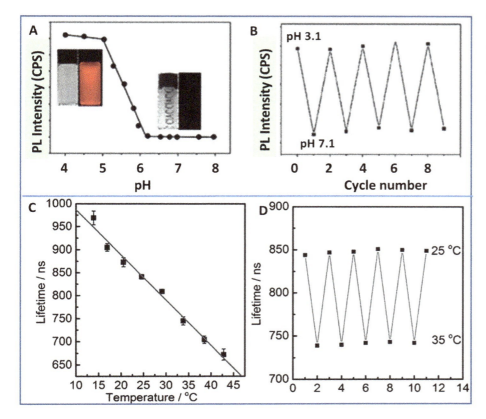

FIG. 8.21 (A) Performance assessment of Cu NC-based fluorescence pH meter; (B) Performance assessment of Au NC-based fluorescence thermometry on HeLa cells. (C) Average fluorescence lifetime of intracellular Au NCs versus temperature; (D) Lifetime of intracellular Au NCs upon cycling the temperature five times between 25°C and 35°C.

hysteresis in the luminescence intensity of Au NCs upon thermal cycling has been eliminated and a rigorously precise optical thermometer with a temperature resolution of 0.2°C has been obtained [156]. Zhu and Yu et al. designed pentapeptide-Au NCs with fluorescent intensity linearly and reversibly changes with the change of solution temperature, making it a potential temperature sensor [157].

8.3 Biological application

Thanks to the ultrasmall size, good biocompatibility, low cytotoxicity, excellent photo and chemical stability, the metal NCs has shown promising potential in various biological applications. Especially, the water-soluble fluorescent metal NCs exhibits good aqueous stability, sufficient brightness, large Stoke shift, long lifetime, and cell permeability, it can be used as a perfect probe to mark cells and living animals and realize in vitro and in vivo labeling and imaging. In addition, due to its ease in bioconjugation, it can be modified with specific molecules to locate at special sites such as cancer cells or tumor tissue to realize targeted imaging, therapy, and drug delivery. Furthermore, metal NCs have also been extensively developed to inhibit and reduce the growth of detrimental bacteria owing to its antibacterial ability. Therefore, in this part, we summarize the biological applications of metal NCs in three aspects: biolabeling/imaging, disease diagnostics and therapy, and antimicrobial agents.

8.3.1 Biolabeling/imaging

Metal NCs is an ideal fluorescent probe with many unique properties and is suitable for biolabeling/imaging. For example, the basic requirement—the good water solubility, biocompatibility and low cytotoxicity; the ultrasmall particle size would not disturb the normal biological functions of cells; the good cell membrane permeability can directly deliver the probes into the cells without the need for additional agents; the NIR fluorescent can avoid the interference from the autofluorescent from biological media or scattering light; the excellent luminescent properties such as large Stokes shift, minimal photobleaching and high QY, all of these contribute to the high signal-noise ratio (S/N) ratio imaging. Besides the passive endocytosis induced untargeted imaging, the NCs can be bioconjugated with recognition molecules like protein, small biomolecule, enzyme, and nucleotide sequence, to realize targeted imaging. Moreover, it can combine with other functional materials to integrate merits to realize multimodal imaging.

8.3.1.1 Untargeted imaging

Chou and Ho et al. investigated the two-photon absorption of 11-MUA-Au NCs and applied it in the human mesenchymal stem cells (hMSCs) imaging in vitro [158]. Xu et al. investigated one- and two-photon emission properties of GSH-Au NCs and utilized its good photostability and low toxicity to realize live cell fluorescence imaging in both one- and two-photon excitation [159]. Nienhaus et al. synthesized D-penicillamine (DPA) capped Au NCs with good stability over the pH range of 5–9. The DPA-Au NCs show low cytotoxicity which is confirmed by a trypan blue exclusion assay. The DPA-Au NCs are applied in HeLa cells imaging by internalization. The DPA has carboxylic and amino groups which can be easily functionalized with other target molecules; therefore, it may be used as specific tagging in biological applications [160]. Wang et al. synthesized chiral L, D, DL-penicillamine (PA) protected Au NCs with different optical properties. Among these PA-Au NCs, L-PA-Au NCs and D-PA-Au NCs have a strong fluorescence emission at 630 nm, while the DL-PA-Au NCs has no fluorescent. Due to the good photostability and low cytotoxicity, the L-PA-Au NCs and D-PA-Au NCs can be used for imaging living HeLa cells [161]. Guevel et al. loaded BSA-Au NCs in a 100 nm SiO_2 NP with an improved monodispersity and stability over more than 5 months. The BSA-Au NCs@SiO_2 composite displayed potential in tumor lung cells imaging [162]. Zhang et al. utilized the good biocompatible, water-dispersible, red fluorescent of BSA-Cu NCs to realize CAL-27 cells imaging [163]. Schneider et al. synthesized a series of distinct fluorescent properties Ag NCs with different emissions and lifetimes by ligand-etching method. Among them, the Ag NCs with yellow emission has a high QY of over 65%. The Ag NCs has no cytotoxicity even at 1 mg/mL high concentration and exhibits good stability for a wide range of pH from 4 to 8. Therefore, the Ag NCs is used for epithelial lung cancer cells imaging through the endocytotic process and the NCs are accumulated mainly in the cytoplasm and the vesicles sites [164]. Zhu et al. synthesized Ag_xAu_{25-x} ($x = 1$–13) NCs with 400 times enhanced quantum yield compared with rod-like Au_{25} NCs. The cell viability of the alloy clusters was investigated by MTT assay, the clusters showed no cytotoxicity when its concentration is in the range from 0 to 80 μg/mL. The alloy NCs has a broad exciting spectrum and the emission peak is almost unchanged. The fluorescence imaging of the NCs in HCC cells was clear and with high intensity regardless of the excitation wavelength (Fig. 8.22A and B) [165].

Wang and He et al. utilized NIR fluorescent Au NCs as imaging agents for tumor imaging in vivo. The Au NCs display high contrast images and can be easily distinguished from the background. Because of ultrasmall size of Au NCs, it can effectively avoid suffering from the extremely high reticuloendothelial system uptake. The Au NCs had no potential toxicity to the body which is verified by the no obvious change in the weight of mice body. The Au NCs can be passively accumulated into the tumor site based on the enhanced permeability and retention effect, as demonstrated by in vivo and ex vivo imaging studies by using MDA-MB-45 and Hela tumor xenograft models [166].

Besides using the fluorescent intensity of NCs to act as an imaging mode, the fluorescent lifetime and time gated intensity of NCs can also be used as imaging modes. Gryczynski and Raut et al. synthesized long fluorescence lifetime (>1 μs) of BSA-Au NCs as a cellular/tissue, time gated intensity imaging probe. This long lifetime is several hundred fold longer than the autofluorescence lifetime (~7 ns) and can be effectively used to off-gate cellular autofluorescence background and enhance the clarity and specificity by time gated imaging [167]. Zhang and Wang et al. synthesized two water-soluble fluorescent Au NCs with mercaptosuccinic acid (MSA) and tiopronin thiolate as ligands, respectively. The Au NCs exhibits strong NIR fluorescent emission and long lifetimes. After conjugated polyethylene glycol (PEG) moieties, the NCs can be more efficiently taken into HeLa cells. Owing to the long lifetime, the autofluorescent from the cellular can be easily distinguished from the emission fluorescent from Au NCs. Therefore, the NCs can be used for cell imaging not only by fluorescent intensity but also by fluorescent lifetime [168]. Besides, the dihydrolipoic acid (DHLA) stabilized Au NCs synthesized by Nienhaus et al. has an attractive long fluorescent lifetime, therefore, the DHLA-Au NCs is used to realize HeLa cells imaging via fluorescence lifetime imaging technique [169].

8.3.1.2 Organelle-targeted imaging

Liu et al. modified chitosan coated Au NCs with triphenylphosphonium (TPP) cations by utilizing the covalent linking, the obtained Au NCs@CS-TPP exhibits a special mitochondria-targeted functional. The Au NCs@CS-TPP displays a blue emission at 440 nm with a QY of 8.5% and shows good photostability (no obvious fluorescent decrease after 8 min irradiation) and low cytotoxicity in cells (even at a concentration of 60 μg/mL). The Au NCs@CS-TPP can precisely target and accumulate into mitochondria in living cells such as HeLa cells and HepG2 cells [170]. Gao et al. modified fluorescent Ag

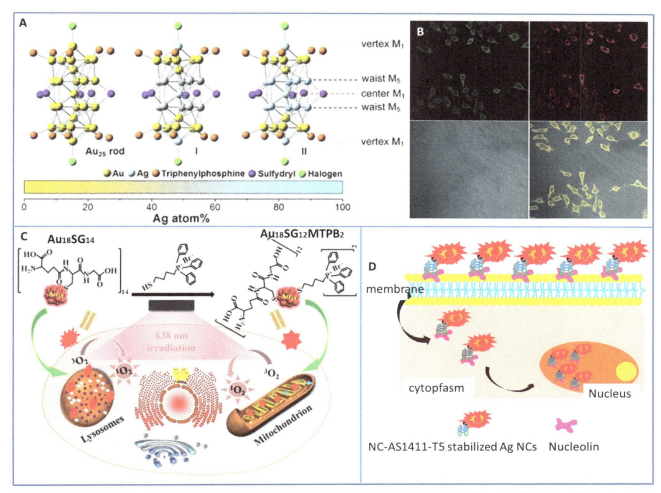

FIG. 8.22 (A and B) Schematic illustration of Ag_xAu_{25-x} ($x=1$–13) NCs with enhanced fluorescent for cell imaging; (C) Schematic illustration of Au_{18} NCs targeted organelle switching fluorescence imaging through ligand exchange; (D) Schematic illustration of modified Ag NCs targeted cancer cell nuclear imaging.

NCs with sgc8c aptamer, a specific aptamer for the endosomes of CCRF-CEM cells. After incubation with CCRF-CEM living cells, the nucleus exhibits significant red fluorescent because the Au NCs specifically target and localize in the nucleus [171]. They designed a bifunctional peptide to synthesize fluorescent Au NCs and realize nuclei targeting and imaging. The peptide contains two parts, one is for Au NCs synthesis and the other one is for nucleus targeting and localization. The prepared peptide-Au NCs displays a red emission at 677 nm and can specifically target and stain the nuclei of three cell lines [172]. Similarly, they designed another dual functional peptide to synthesize Cu NCs and applied it to cell nuclei imaging via two-photon fluorescence. The peptide includes two parts, one is for Cu NCs synthesis sequence, and the other one is derived from the simian virus 40 [SV40] large T antigen used for cell nuclei targeting. The as prepared Cu NCs is mainly composed of Cu_{14} and exhibits blue two-photon fluorescence. Therefore, the as prepared Cu NCs can specifically mark the nuclei of both HeLa and A549 cell lines with a good S/N ratio [173]. Sivakumar and Verma et al. synthesized 8-mercapto-9-propyladenine capped Au NCs with green fluorescent. The purine-stabilized Au NCs can mark cell nuclei of various cell lines via a macropinocytosis pathway [174]. Yang et al. synthesized water-soluble $Au_{18}SG_{14}$ with high fluorescence properties by one-step method and then synthesized Au NCs with mitochondrial targeting ligand molecules (4-mercaptobutane triphenyl phosphine bromide, MTPB) $Au_{18}SG_{12}MTPB_2$ by ligand exchange method (Fig. 8.22C). Cell fluorescence imaging experiments showed that $Au_{18}SG_{14}$ NCs tended to accumulate in lysosome sites, while $Au_{18}SG_{12}MTPB_2$ NCs tended to accumulate in mitochondrial sites, indicating that NCs could achieve targeted organelle switching fluorescence imaging through ligand exchange [175].

8.3.1.3 Targeted imaging in vitro and in vivo

Schneider et al. biolabeled Au NCs with human transferrin (Tf). The biolabeled-Au NCs are stable in a wide range of pH and its fluorescent does not vary whether it is in the presence of iron or conjugation to antibodies. The biolabeled-Au NCs preserve the protein activity and the receptor target ability which is confirmed by colorimetric assay and antibody-induced aggregation. As verified by cell viability tests, the biolabeled-Au NCs are nontoxic and can be up-taken into cellular. Those results suggest that the biolabeled-Au NCs are a promising biolabel for biological applications [176]. Yan et al. combined Tf-functionalized Au NCs with GO to fabricate a turn-on NIR imaging probe for cancer cells and small animals. Tf not only acts as a stabilizer for Au NCs synthesis but also serve as a recognition site for transferrin receptor (TfR) targeting. The fluorescent of Tf-Au NCs/GO is "off" due to the quenching effect of GO. In the presence of TfR, due to the specific interaction between Tf and TfR being stronger than that between Tf and GO, the PRET system is broken up and the fluorescent is restored. Therefore, the Tf-Au NCs/GO can be used for turn-on NIR fluorescent imaging TfR overexpressed cancer cells and Hela tumor sites in mice [177].

Zhu and Yu et al. synthesized Met-Au NCs with high fluorescent in a wide pH range from 4–12 and good stability without fluorescent decrease after 3 months. Due to the L-type amino acid transporters overexpressed in cancer cells can specifically recognize Met ligand, the Met-Au NCs showed high specificity in the recognition of cancer cells: after 1 h incubation with Met-Au NCs, all cancer cells (including A549, Hela, MCF-7, HepG2) showed fluorescence, while normal cells (Wi-38 and Cho) showed no fluorescence response [178].

Zhou et al. utilized bovine pancreatic ribonuclease A (RNase-A) as template to synthesize Au NCs. The as-prepared RNase-A-Au NCs exhibit a NIR fluorescent emission with a 12% enhanced QY, a 210 nm large Stokes shift, and a 1.5 μs long fluorescent lifetime. These advantages make it a good candidate for bioimaging. The RNase-A-Au NCs was further conjugated with vitamin B_{12} (VB_{12}) to construct VB_{12}-R-AuNC nanoplatform for specific in vitro tumor targeting and imaging by efficiently internalized into Caco-2 cells via VB_{12} receptor-mediated endocytosis [179].

Koyakutty et al. conjugated folic acid (FA) onto the BSA-Au NCs through amide linkage with BSA. The FA-BSA-Au NCs show good stability in a wide pH range of 4–14 and display a NIR emission at ∼674 nm. The FA-BSA-Au NCs show no toxicity toward cells even at a high concentration of 500 μg/mL. The FA-BSA-Au NCs are internalized into folate receptor positive oral carcinoma cells with significantly higher concentrations than that in the negative control cell lines, suggesting FA-BSA-Au NCs successfully used in molecular-receptor-based targeted imaging of cancer [180]. Pradeep et al. also conjugated FA with BSA-Au NCs and used it for oral carcinoma KB cells imaging through FA-receptor mediated endocytosis [181]. Qi et al. modified FA onto ovalbumin (Ova) protected-Au NCs by using the homopolymer of N-acryloxysuccinimide (PNAS) as the linker. The FA-Ova-Au NCs can be used to target and image cancer cells by the endocytosible by the overexpressed FA receptor molecule [182]. Yan et al. synthesized thiol-terminated polyethyleneimine (SH-PEI) stabilized Au NCs with NIR fluorescent and then conjugated FA to realize FA receptor positive MCF-7 cells and human breast tumor-bearing mouse targeting and imaging. The SH-PEI can facilitate postsurface modification with other functional biomolecules, if a special biomolecule is modified on SH-PEI-Au NCs; it can realize specific target imaging [183]. Yan et al. prepared trypsin stabilized Au NCs (try-AuNCs) and modified them with FA. The FA-try-Au NCs has specific affinity for tumors, therefore, it is used for in vivo NIR fluorescent imaging of high FA receptor expressing Hela tumor [92]. Huang et al. synthesized protein protected-Au_{20} NCs and modified with FA and hyaluronic acid (HA) to recognize target tumor. Hep-2 cancer cells can overexpress the FA receptor and A549 cells can overexpress CD44 and thus can selectively bind to HA. They found the uptake of Au_{20} NCs by both cancer cells and tumor-bearing nude mice via receptor-mediated endocytosis are more efficient than that via endocytosis. The Au_{20} NCs are selectively accumulated at tumor sites and realize tumor-targeted imaging in vitro and in vivo [184].

Pradeep et al. synthesized bright-NIR-emitting Au_{23} NCs by core etching Au_{25} NCs. The Au_{23} NCs is further modified with streptavidin and used for human hepatoma cells (HepG2) imaging based on the avidin-biotin interaction [185].

Lo and Yang et al. synthesized a peptide functionalized Au NCs (namely NES-linker-DEVD-linker-NLS-Au NCs). The NES, NLS, DEVD are represented as nuclear export signal, nuclear localization signal, and capsase-3 recognition sequence. Upon the induction of apoptosis, the activated caspase-3 makes the functional peptide moiety on Au NCs separated, resulting in changes of subcellular distribution of NCs. The change can be quantified by the ratios of Au NCs photoluminescence in nucleus to that in cytoplasm. Therefore, the functionalized Au NCs can be used to real-time monitor the apoptosis of cell [186].

Nie and Zhao et al. utilized horse spleen ferritin as a nanoreactor to synthesize various paired Au NCs with tunable fluorescent from green to far-red. Compared with single Au NCs, the obtained Au-Ft complex showed enhanced fluorescent and a red shift in emission. The far-red Au-Ft complex was used for ferritin receptor-mediated targeting and imaging in vitro and in vivo. Particularly, the Au-Ft complex tends to accumulate in the kidney and liver [187].

Medintz et al. directly synthesized bidentate-poly(ethylene glycol) (PEG) stabilized Au NCs with NIR fluorescent emission at 820 nm. Due to the different terminated groups of PEG dithiolane ligands (amine, carboxyl, azide, and methoxy); the PEG-Au NCs can be modified with DNA, dye and peptide through EDC, NHS-amine, and Azide-alkyne chemistry. Despite the presence of different terminal groups on the ligands, the Au NCs still displays minimal cytotoxicity. The PEG-Au NCs is further microinjected or bioconjugated to cell penetrating peptides (CPPs) and then its utility is verified by COS-1 and HeLa cells imaging with standard fluorescence or multiphoton microscopy [188].

In addition to proteins, enzymes, and specific small molecules that can target cancer cells, nucleotides and aptamers can also be carefully engineered to target cancer cells. Zhu and Jiang et al. utilized AS1411, an antiproliferative G-rich phosphodiester oligonucleotide, an aptamer to nucleolin and an anticancer agent, to functionalize Ag NCs, realizing targeted cancer cell nuclear imaging (Fig. 8.22D). The AS1411 is connected with poly(cytosine) by a T5 loop and used to synthesize Ag NCs. The AS1411-Ag NCs exhibits high fluorescent with a QY of 40.1%, and shows an enhanced efficiency of growth inhibition of tumor. Therefore, by receptor-mediated endocytosis, the AS1411-Ag NCs is used for nucleus imaging of MCF-7 human breast cancer cells and anticancer as well. The design of AS1411 for the synthesis of Ag NCs can be extended to other aptamers, such as Sgc8c and mucin 1 aptamer, which can specifically bind to acute lymphoblastic leukemia and epithelial malignancies, respectively. The Sgc8c and MUC1 aptamer are linked to poly(cytosine) also via a T5 loop and then used to synthesize fluorescent Ag NCs. The as prepared Ag NCs has potential in targeting corresponding cancer cells [189]. Similarly, Wang et al. designed a special oligonucleotide structure with a cancer-targeted DNA aptamer and cytosine-rich sequence for biotarget recognition and Ag NCs synthesis at the same time. By assembling specific aptamer, the sgc8c-Ag NCs and TD05-Ag NCs exhibited specific binding to target tumor cells CCRF-CEM cells and Ramos cells, respectively. The aptamer-NCs assembly can be extended to recognize other target tumor cells by only modifying specific selected aptamer [190]. Zhu and Xu et al. designed a multifunctional DNA scaffold for the preparation of fluorescent Ag NCs to realize intracellular tumor-related mRNA imaging. The DNA scaffold contains three regions, one is sgc8c aptamer used for specific internalization; one is used for Ag NCs nucleation; another one is used for hybridization with target TK1 mRNA, a marker for tumor growth. The NCs can be used as a switchable fluorescence excitation and emission wavelength imaging agent to target tumor-related mRNA. Moreover, the designed NCs exhibit high signal-to-background (S/B) ratio even in relatively low expressed mRNA imaging [191].

In addition to cell imaging by uptake fluorescent NCs through endocytosis, fluorescence NCs can also be generated in situ in cells to achieve imaging function simultaneously. Dickson et al. used nucleolin to serve as a guide template to produce fluorescent Ag NCs in the nucleoli of cells by amibient-temperature photoactivation and realize in situ high S/N ratio cell nuclear imaging [192]. Wang et al. designed a self-bioimaging fluorescent Au NCs which can be spontaneously biosynthesized by cancerous cell (i.e., human hepatocarcinoma cell line-HepG2, leukemia cell line-K562). The Au NCs is synthesized in the cells cytoplasmas and mainly accumulated in the nucleoli. The Au NCs cannot be synthesized in noncancerous cells such as human embryo liver cells (L02). Therefore, the Au NCs can be used as a self-bioimaging probe for tumor targeting and marking in vivo [193]. Besides in situ synthesized fluorescent Au NCs, Wang et al. further reported in situ self-imaging of cancer cells and tumors in vivo by their special spontaneous ability to biosynthesize NIR fluorescent Ag NCs upon the [Ag(GSH)]$^+$ treatment while it cannot happen as fast in normal cells and tissues. The precise and selective self-imaging of cells and tumors by in situ generated Ag NCs are demonstrated by both ex vivo experiments comparing cancer cell models to normal cells and in vivo imaging of subcutaneous xenografted tumor (cervical carcinoma model) in nude mice. Besides rapid and specific bioimaging, the in situ generated Ag NCs can drastically reduce the tumor size and lead to complete remission. That makes it a valuable potential agent for noninvasive treating cancer tissues without significant damage to healthy tissues [194].

8.3.1.4 Multimodal imaging

Early recognition and tracking of cancer cells is necessary to lower the death rate. Even though there are lots of early cancer diagnosis techniques, they have their own weaknesses and strengths. For example, computed X-ray tomography (CT) is able to present high-resolution 3D structure details of tissues, but it relies on CT contrast agents which are predominantly iodinated compounds, with short imaging time (<10 min) and even potential renal toxicity. Magnetic resonance imaging (MRI) has excellent spatial resolution and deep tissue penetration but low sensitivity. In contrast, fluorescence imaging has much higher detection sensitivity but with limited tissue penetration depth. Therefore, multimodal imaging is highly desirable for accurate diagnosis because it can provide complementary information from each imaging modality. Wu and Chen et al. designed nontoxic self-illuminating ^{64}Cu-doped Au NCs based on Cerenkov resonance energy transfer (CRET) to realize positron emission tomography (PET) and near-infrared (NIR) fluorescence dual-modality imaging in both vitro and vivo (Fig. 8.23A). The radioactive ^{64}Cu acts as positron emitting radionuclide (tracer) for PET imaging.

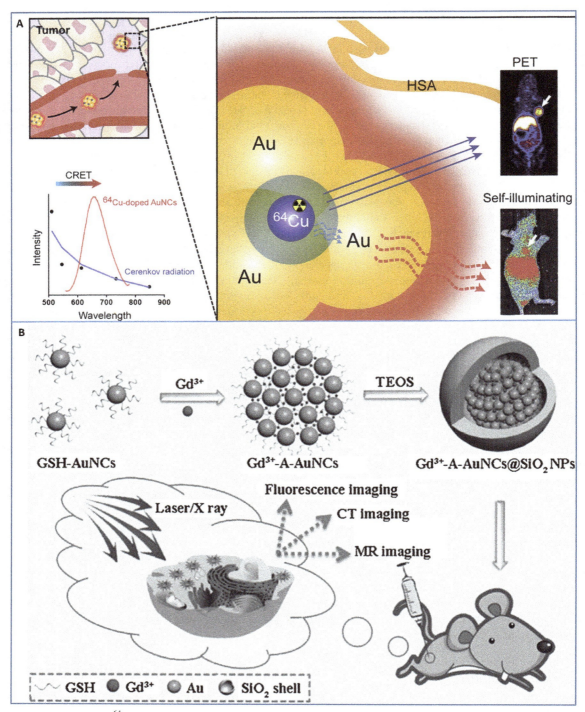

FIG. 8.23 (A) Self-illuminating ^{64}Cu-doped Au NCs for in vivo synergistic dual-modality PET and self-illuminating NIR imaging; (B) Schematic illustration of the synthetic procedure of Gd^{3+}-A-Au NCs@SiO$_2$ NPs with AEF property for in vitro and in vivo multimodal imaging.

Besides, the ^{64}Cu and Au NCs act as energy donor and acceptor, respectively, for self-illuminating NIR fluorescent imaging. The ^{64}Cu-doped Au NCs displayed high tumor uptake (14.9% ID/g at 18 h) and generated satisfactory tumor self-illuminating NIR images without external excitation in a U87MG glioblastoma xenograft model [195]. Zhao and Sun et al. utilized covalent linking to combine BSA-Au NCs with Gd-DTPA, the composite exhibits bright red fluorescent and high relaxivity and it is used for fluorescent and MRI dual mode imaging [196]. Yan et al. synthesized BSA-Gd$_2$O$_3$/Au hybrid with good stability, intense NIR fluorescent and excellent MRI ability. The BSA-Gd$_2$O$_3$/Au is further modified with arginine-glycine-aspartic acid peptide c (RGDyK) (RGD) and then successfully used for targeted tumor imaging in vivo by NIR fluorescent and MRI dual modes [197].

Cai et al. synthesized hybrid Au-Gd NCs and used it into NIR fluorescent (NIRF)/CT/MRI triple-modal imaging in vivo. The Au-Gd NCs can penetrate into the solid tumor with efficient tumor targeting capacity and low body residues. By using MCF-7 tumor-bearing mice as models, after tail-vein injection of the hybrid Au-Gd NCs, it exhibited clear and consistent NIRF/CT/MRI signals [198]. Lin and Wang et al. synthesized silica-encapsulated Gd^{3+}-aggregated Au NCs (Gd^{3+}-A-Au NCs@SiO$_2$ NPs) with 3.8 times improved fluorescent intensity and used it as nanoprobes for in vitro and in vivo multimodal (NIRF/CT/MRI) cancer cell imaging (Fig. 8.23B) [199].

Biju et al. anchored Au NCs onto the streptavidin-functionalized Fe$_3$O$_4$ NPs and then conjugated the epidermal growth factor (EGF) to convert it to a multimodal bioimaging probe. Interestingly, the system can sensitize dissolved oxygen and produce ^{1}O$_2$ due to the long-living excited state. Therefore, the produced ^{1}O$_2$ can be monitored in solutions or human lung epithelial adenocarcinoma cells by using a singlet oxygen sensor green (SOSG) dye, which generates green-fluorescent endoperoxide upon reaction with ^{1}O$_2$. Besides the NIR fluorescent and MRI imaging modes, the system can provide a third modality for live cell imaging by the green fluorescence of the endoperoxide of SOSG triggered by ^{1}O$_2$ [200].

8.3.2 Disease diagnostics and therapy

8.3.2.1 Cancer diagnosis and detection

Sensitive and specific detection of cancer cells is of great significance not only for the early diagnosis of cancer, but also for the study of tumor metastasis. In cancer cells detection, the recognition molecules are important, it is needed to choose based on the specific cancer cells. Some cancer cells overexpress specific protein or biomolecule receptor, some cancer cells has particular DNA sequence. Therefore, we can modify the NCs with specific protein, molecule, or DNA sequence to target cancer cells.

Koyakutty et al. developed a nano-bioprobe based on monoclonal antibody conjugated red-NIR emitting Au NCs to detect acute myeloid leukemia (AML) cells using flow cytometry. The Au NCs was bioconjugated with the antibody against CD33 myeloid antigen, and thus can be recognized by AML stem/progenitor cells. After 2 h of incubation, the CD33-Au NCs can be taken up by ~95.4% of leukemia cells that is much higher compared with the nonspecific uptake in normal blood cells (~ 8.2%). Based on this, the conjugated NCs can be used for flow-cytometric detection of leukemia [201].

Wang et al. designed two specific DNA sequences. One is a hairpin-shaped recognition probe consisting of a specific aptamer for target recognition and a G-rich part for fluorescent enhancement, and the other one is a signal probe consisting of a region for Ag NCs synthesis and a link region for the complementary recognition probe. They utilized the recognition-induced conformation alteration of aptamer and hybridization-induced fluorescence enhancement effect to detect cancer cells. For example, the sgc8c aptamer was used to detect CCRF-CEM cancer cell. In the presence of CCRF-CEM cancer cell, the conformation of sgc8c aptamer was changed, leading to the hybridization of the two DNA probes and bringing the Ag NCs close to the G-rich part, resulting in the fluorescent enhancement. This phenomenon not only can be used to recognize and image the target cell but also can be used to detect cancer cells (Fig. 8.24A). The system can detect as low as 150 CCRF-CEM cells in the 200 μL binding buffer by flow cytometry. The universality and practicality were verified by extended other cancer cells detection such as Ramos cancer cells (changing the sgc8c aptamer to TD05) [202].

Qu and Ren et al. synthesized Au NCs and GO composite via electrostatic interactions. The composite exhibits peroxidase-like activity. Benefiting from the rich variety of functional groups and larger surface areas, GO can easily modify with other molecules. The FA was used to recognize the cancer cells because cancer cells can overexpress FA receptors. Therefore, the GO-Au NCs system was conjugated with FA to detect cancer cells. GO-Au NCs-FA showed much stronger binding affinity to cancer cells such as HeLa and MCF-7 cells than that of NIH-3T3 normal cells. Based on the peroxidase-like activity of the GO-Au NCs-FA system, it can be used to quantitatively identify the cancer cells. In the presence of TMB and H$_2$O$_2$, the GO-Au NCs-FA can specifically target cancer cells and catalyze a reaction with an obvious color change easily judged by naked eye and quantitatively detected by the absorbance change at 652 nm as well. The system can detect as low as 1000 MCF-7 cells [203]. Sheng and Hu et al. also utilized the NIR fluorescent and peroxidase activity of FA-Au NCs to realize cancer diagnosis by microscopic imaging with bright field and fluorescent images. The two imaging are mutually complementary and exhibit high precision with free false-positive and false-negative results [204].

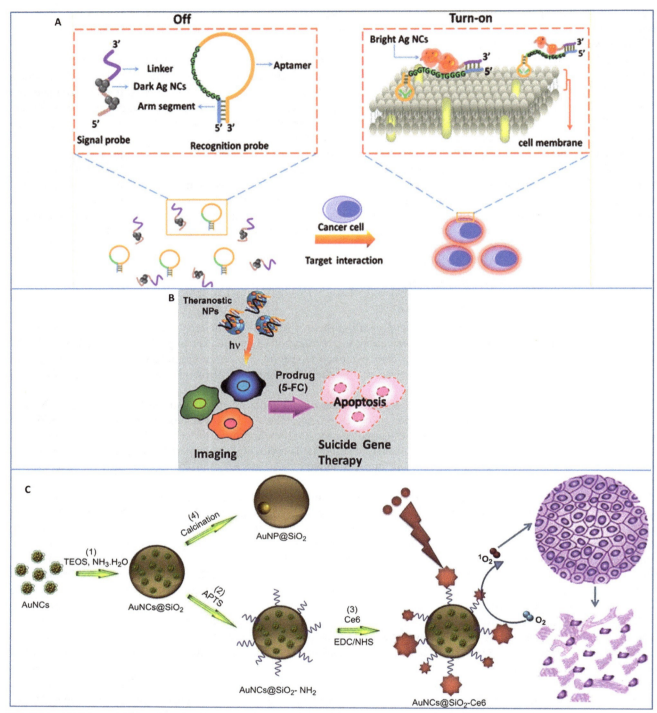

FIG. 8.24 (A) Schematic illustration of the label-free and turn-on aptamer strategy for cancer cell detection based on DNA-Ag NC fluorescence upon recognition-induced hybridization; (B) Scheme for the fluorescent Au NCs for anticancer gene theranostics; (C) Schematic illustration of the synthetic procedure of Au NCs@SiO$_2$-Ce6 for photodynamic therapy.

In healthy cells, more than 90% of the total glutathione exists in the reduced form (GSH). An increased GSSG-to-GSH ratio is an indicator of oxidative stress. In cancer cells, the ROS stress is much greater than that in normal cells. Therefore, more reduced glutathione (GSH) than oxidized glutathione (GSSG) is produced, which can have ligand exchange with the Au NCs, leading to enhanced fluorescent within the cancer cells. Chen and Wu et al. reported Au NCs can selectively distinguish cancer cells (e.g., lung A549 and liver Hep G2 cancer cells) from normal cells (e.g., lung ATII and liver L02 cells) based on the different GSH concentrations in normal cells and cancer cells [205].

8.3.2.2 Cancer therapy

At present, the mainstream cancer treatment plan is divided into chemical drug therapy, radiation therapy, targeted drug therapy, and light therapy. Gold cores in Au NCs act as radiosensitizers and can be used in radiotherapy. Metal NCs can generate singlet oxygen and also combine with photosensitizers to reduce their toxicity, which can be used as photodynamic therapy for cancer. Metal NCs can also functionalize a variety of drugs and recognition sites and have the potential to target cancer therapy. Due to the small size of metal NCs, they are easy to be cleaned out through the kidney and will not stay in the organism for a long time to cause side effects. Therefore, metal NCs have a good application prospect in the treatment of cancer. This part has summarized the application of metal NCs in cancer therapy.

One of the thorny problems with oncology chemotherapeutic drugs is that they are highly cytotoxic to both normal and cancerous cells. In addition, their short life span in the living body limits their ability to effectively reach tumor sites. Many of them may be metabolized and cleared by the liver and kidneys during the first circulation. Therefore, there is a need to develop targeted drug vectors to enhance the selective cytotoxicity of chemotherapy drugs. Metal NCs are ideal carriers for binding anticancer drug to form prodrug because they circulate in the blood for a long time and accumulate preferentially in the tumor site due to enhanced permeability and retention effect.

Gu et al. functionalized Au NCs with doxorubicin (DOX), a widely used clinical anticancer drug, to form a prodrug for tumor therapy. The Au NCs was first conjugated with methionine (Met) to obtain tumor targeting capability which is confirmed by in vitro and in vivo tumor imaging. The cytotoxicity and biodistribution of the Au NCs-Met probe were also examined in different tumor bearing mouse models. Then, the Au NCs-Met was immobilized with DOX. By a series of experiments such as cellular uptake, apoptosis study, tissue distribution and tumor therapy evaluation, the therapeutic efficacy of the Au NCs-Met-DOX was evaluated. The Au NCs-Met-DOX showed higher tumor inhibition capability in cells and living subjects than DOX and Au-DOX [206]. Similarly, Zhu and Yu et al. also modified Au NCs to carry DOX to achieve antitumor property [207]. Besides, Wang et al. combined RGO with Au NCs and then carry DOX for oncotherapy. Due to the synergy effect inducing karyopyknosis, the composite can be used to inhibit cancer cells [208].

Chattopadhyay and Ghosh et al. utilized chitosan and mercapto propionic acid (MPA) to synthesize Au NCs with red, green, and blue light emitting. The composite is favorable for cells uptake due to the chitosan. The composite delivers the suicide gene, corresponding to *E. coli* cytosine deaminase uracil phosphoribosyltransferase (CD-UPRT) enzyme to HeLa cells. The CD-UPRT enzyme converts nontoxic prodrug 5-fluorocytosine (5-FC) to 5-fluorouracil (5-FU) and other toxic metabolites, inducing apoptosis and thus killing the cancer cells. The composite can be used as a theranostic device for induction of apoptosis in mammalian cancer cells, based on suicide gene therapy (Fig. 8.24B) [209]. Similarly, Irudayaraj et al. conjugated Herceptin with BSA-Au NCs (namely Au NCs-Her). The Au NCs-Her inherits the fluorescent property and enhances the therapeutic efficacy of Herceptin. Taking ErbB2 overexpressing breast cancer cells and tumor tissue as models, the Au NCs-Her can escape the endolysosomal pathway and specifically target and localize in the nuclear area. The diffusion time and the number of diffusers and concentration of the NCs in the nucleus of live cells are investigated. Due to the nuclear localization effect, the Au NCs-Her enhances the anticancer therapeutic efficacy of Herceptin by inducing DNA damage [210]. Nie and Wu et al. integrated multifunctional element to construct enhanced antitumor drug delivery and efficacy. The composite contains poly(N-isopropyl acrylamide-co-acrylic acid) nanogels (NGs) with thermo- and pH-responsive properties, chemotherapeutic drug DOX, BSA-Au NCs with fluorescent, and tumor targeting peptide iRGD. The composite can achieve a controlled drug release in tumor tissues and track the process simultaneously. Due to the iRGD, the composite can specific target to tumor and enhance the cellular uptake, which is confirmed by flow cytometry and fluorescent imaging. The enhanced antitumor efficacy and controlled drug release are proved by the in vitro cytotoxicity study [211]. Wan et al. embedded Au NCs and DOX into a photosensitive liposomeby by using a supercritical CO_2 method to construct a liposome-based drug carrier. The Au NCs/DOX dual-loaded liposome can controllably release drug in a short time and achieve tumor therapy under light irradiation [212]. Zhao and Zhang et al. synthesized multifunctional nanocarriers (Au NCs-PLA-GPPS-FA) by using Au NCs as core and (FA)-conjugated amphiphilic hyperbranched block copolymer as shell based on poly(L-lactide) (PLA) inner arm and FA-conjugated GPPS (GPPS — FA) outer arm. The composite carries hydrophobic anticancer drug, camptothecin (CPT). Amphiphilic hyperbranched block copolymer can improve nanocarriers stability and drug loading ability. FA and Au NCs are used for targeting cancer cells and their imaging. The composite realizes tumor-targeted drug delivery and release and tracks at the cellular level for cancer therapy [213].

Lu and Xie et al. construct a theranostic composite by combining fluorescent Au NCs, FA, paclitaxel (PTX), and pH-responsive amphiphilic polymeric nanocarrier. The composite firstly selectively accumulates in tumor tissues by FA-receptor mediated endocytosis, which is confirmed by in vivo fluorescent imaging study. The composite then following release the drug payload in mildly acidic endosomal/lysosomal compartments within the targeted cells by the action of the pH-labile linkages in the polymer. The composite integrates multifunctions such as cancer cell imaging, targeted drug delivery and controlled drug release [214].

Light stimulus cancer therapy such as photodynamic therapy (PDT), photothermal therapy (PTT) and phototriggered chemotherapy has been extensively applied owing to its specific spatial and temporal controllable ability. Among these, PDT is an emerging external photoactivation therapy used to treat a variety of diseases that involves three key components: photosensitizers, light (usually lasers), and tissue oxygen. Under the light, photosensitizer is able to transfer the absorbed photon energy to surrounding oxygen molecules to produce reactive oxygen species (ROS), such as singlet oxygen or free radicals, which induce cell death and tissue destruction in the dark without damaging the surrounding healthy tissue. Chen and Yang et al. synthesized chlorin e6 (Ce6) photosensitizer-conjugated silica-coated Au NCs (AuNCs@SiO$_2$-Ce6) for fluorescence imaging-guided in vitro and in vivo PDT (Fig. 8.24C). The AuNCs@SiO$_2$-Ce6 exhibits high Ce6 loading ability, and it does not cause nonspecific release of Ce6 during circulation. In addition, compared with free Ce6, the AuNCs@SiO$_2$-Ce6 shows significantly enhanced cellular uptake efficiency and an improved photodynamic therapeutic efficacy [215]. Ajayaghosh and Jayasree et al. utilized lipoic acid as ligand to synthesize NIR fluorescent Au$_{18}$L$_{14}$ NCs. By incorporating with a tumor-targeting agent-FA and a photosensitizer-protoporphyrin IX (PPIX), the Au NCs system exhibits an excellent PDT tumor reduction property. After being combined with Au NCs, the PPIX showed significantly higher singlet oxygen generation efficiency. Due to the better localization of the Au NCs system, it can facilitate tumor cells death even under a low dose of laser irradiation. Besides, the Au NCs system can realize real-time monitor the PDT progress [216].

Zhu and Shen et al. constructed an integrated system (Au NCs@GTMS-FA) composed of water-soluble fluorescent Au NCs, two-level mesoporous canal silica, gelatin and FA. The composite shows 13-fold enhanced fluorescent in comparison to Au NCs. Cytotoxicity tests showed that Au NCs@GTMS-FA complex had good biocompatibility and could specifically target cancer cells. Confocal fluorescence imaging showed that the fluorescence stability of Au NCs@GTMS-FA in cells was much better than that of Au NCs. Under a certain light condition, the composite not only has a certain photodynamic effect, but also has certain photothermal conversion ability. After 9 min of light exposure with 808 nm laser, the temperature of Au NCs@GTMS-FA dispersion can rise from the initial temperature of 22.6°C to 43.7°C. Therefore, it can not only realize cancer cell-specific targeting and aggregation-enhanced fluorescence imaging, but also achieve synergistic photothermal/photodynamic therapies [217]. They further synthesized two different Au NCs, one is Au$_{18}$SG$_{14}$ NCs tended to accumulate in lysosome sites, the other is Au$_{18}$SG$_{12}$MTPB$_2$ NCs (MTPB, (4-mercaptobuty l)triphenylphosphonium bromide) tended to accumulate in mitochondrial sites. Singlet oxygen detection experiments showed that both NCs could obviously produce a certain amount of ^{1}O$_2$ under the illumination of 638 nm, but the ^{1}O$_2$ generation efficiency of the Au$_{18}$SG$_{12}$MTPB$_2$ NCs was slightly lower than that of Au$_{18}$SG$_{14}$, possibly due to the replacement of two-SG molecules. Interestingly, although the photodynamic effect of Au$_{18}$SG$_{12}$MTPB$_2$ NCs was not as significant as that of Au$_{18}$SG$_{14}$ NCs, their killing effects on cells were similar through cytotoxicity test, indicating that the apoptosis of mitochondria inside cells had a greater effect on cells than the apoptosis of lysosomes [175].

Gao et al. modified Au$_{25}$ NCs with positively charged tridecapeptides. The peptides help Au$_{25}$ NCs to penetrate the membrane and specifically bind to the TrxR1 in the cytoplasm by electrostatic attraction. TrxR1 plays important roles in controlling the intracellular redox status and tumor cell apoptosis. They found that the peptide modified Au$_{25}$ can be bound to the active site of TrxR1 and form coordination bonds, leading to the suppression of TrxR1 and induce the apoptosis of tumor cells due to the increase of ROS [218].

Metal NCs can be used to treat cancer through chemotherapy and photodynamic methods by carrying anticancer drugs and photosensitizers, in addition to that, they can also be used as radiosensitizers for radiotherapy. With high radiotherapy enhancement, good tumor targeting capability, good biocompatibility, efficient renal clearance, and low toxicity, the metal NCs is considered as an ideal radiosensitizer. Xie et al. reported GSH protected-Au$_{25}$ NCs and GSH-Au$_{10-12}$ NCs with good biocompatibility and strong radiosensitizing effect [219,220]. The GSH shell not only makes the NCs good biocompatible but also makes it good tumor deposition by affecting its in vivo pharmacokinetics such as escaping the reticuloendothelial system (RES), and activating the GSH transporters inside the body. The NCs can preferentially accumulate in tumor via the improved enhanced permeability and retention effect resulting in sufficient enhancement for cancer radiotherapy, which is detailed investigated by the analyses of the cell response, DNA damage, and changes in tumor volume and weight before and after the treatment. Due to its ultrasmall size lower than the threshold of kidney filtration (~5.5 nm), the NCs has effective renal clearance, and can minimize toxic side effects after the treatment, which is confirmed by the experimental evidence from the pathology, biochemistry, organ index, and biodistribution studies. The GSH-Au NCs are demonstrated to be ideal radiotherapy sensitizers with enhanced the safety and efficacy of radiotherapy.

8.3.2.3 Gene delivery

Gene therapy offers a promising paradigm for the treatment of a variety of acquired or congenital diseases. However, safe and effective gene delivery remains a considerable obstacle due to rapid enzyme digestion, limited cell membrane translocations, and inefficient endosomal release. One of the major challenges of gene therapy is to find suitable gene vehicle with low toxicity and high transfection efficiency. Polyethyleneimine (PEI), a nonviral delivery system, has been considered as a promising vector for gene delivery due to its synthetic maneuverability and high DNA-binding ability. But the cytotoxicity limits its practical application. Qu and Ren et al. utilized ligand-induced etching strategy to synthesize PEI-templated Au NCs. The PEI-Au NCs inherits the advantages of PEI and Au NCs and compensates the disadvantages of each other. The PEI-Au NCs has decreased cytotoxicity and enhanced gene transfection efficiency compared with PEI and Au NCs. In addition, the fluorescent property of Au NCs enables us to track the transfection behavior [221]. In clinical application, the noninvasive monitoring gene delivery and therapy process is also important. Zhu and Min et al. coupled sgc8c aptamer-functionalized Ag NCs with biotinylated small interfering RNA (siRNA) by modular streptavidin bridge. The sgc8c aptamer has specific internalization ability, the Ag NCs has fluorescent property, and the siRNA has the capability to silence targeted gene expression (Fig. 8.25A). Therefore, the composite is used to specifically deliver the therapeutic siRNA targeting VEGF mRNA into HeLa cells and visualize the transport process simultaneously [222].

8.3.3 Self-vaccine

Gold nanomaterials, as a novel platform for vaccine development, have been broadly applied for vaccine antigens or adjuvants delivery. Ren et al. utilized ovalbumin (OVA)-cytosine-phosphate-guanine (CpG) conjugates as the template to synthesize Au NCs with self-vaccine ability (Fig. 8.25B). The Au NCs can induce specific immunological response via dual-delivery of protein antigen and CpG oligodeoxynucleotides (ODNs) into the same antigen presenting cells (APCs) [223]. They further designed a special peptide, H_2N-CCYSIINFEKL-COOH (abbreviated as CCYSIINFEKL), in which the CCY part is responsible for fluorescent Au NCs synthesis, and the OVA peptide SIINFEKL is used to stimulate the immune response. The synthesized peptide-Au NCs exhibits high fluorescent and displays enhanced immunostimulatory ability. Moreover, the thiolated CpG ODNs is further involved in the synthesis of AuNCs (peptide-Au NCs-CpG). The conjugates are able to effectively deliver the antigen and the adjuvant CpG ODNs to the same APCs, inducing strong and long-term immune response, which is confirmed by both in vitro and in vivo studies [224].

8.3.4 Antimicrobial agents

It is well known that ionic silver is an excellent antibacterial agent, and many research have demonstrated Ag NP can release silver ions, possess high antibacterial activity against both Gram-positive and Gram-negative bacteria and can be used as effective growth inhibitors in various microorganisms. Smaller Ag NCs have higher silver ions releasing speed due to the larger specific surface area and thus the effectiveness of smaller Ag NCs in killing bacteria will be higher than that of larger Ag NPs. Wang and Lu et al. utilized SiO_2 nanospheres to encapsulate Ag NCs and prevent the active silver from aggregation. The Ag NCs-SiO_2 can continually release Ag ions and exhibit durable antimicrobial activity against both Gram-negative *E. coli* and Gram-positive *S. aureus*, which is confirmed by minimum inhibitory concentration (MIC), minimal bactericidal concentration (MBC), and the modified Kirby-Bauer method [225]. Pradeep et al. utilized polyacrylamide gel as cavity to control the growth of GSH-Ag NCs. They synthesized GSH-Ag NCs with large yield by using sunlight without using chemical reducing agent. The as-synthesized GSH-Ag NCs shows enhanced antibacterial properties against a Gram negative and Gram positive organism, *E. coli* and *S. aureus*, in comparison to large Ag NPs protected with glutathione (Fig. 8.25C) [226].

Besides the conventional Ag ions release antimicrobial mechanism, researchers have found a novel antimicrobial mechanism. They think when the size of Ag NPs decreases to the sub-2 nm range to form Ag NCs, the dissociation of Ag^+ ions from the NPs inducing the broad-spectrum antimicrobial properties may not be entirely applicable. Xie et al. synthesized well-defined size and structure Ag NCs ($Ag_{16}(SG)_9$ and $Ag_9(SG)_6$) by a reduction-decomposition-reduction cycle method. The obtained Ag NCs can generate a high concentration of intracellular ROSs to kill bacteria such as the multidrug-resistant bacteria *Pseudomonas aeruginosa* [227]. Meanwhile, they synthesized two GSH-Ag NCs with same size and surface ligand, but different oxidation states of the core silver, GSH-protected Ag^+-rich NCs and GSH-Ag^0-R NCs. The GSH-protected Ag^+-rich NCs exhibits higher antimicrobial activities toward both gram-negative and gram-positive bacteria than GSH-Ag^0-R NCs. They supposed the antimicrobial property originated from undissociated Ag^+-R NCs armed with abundant Ag^+ ions on the surface (Fig. 8.25D) [228].

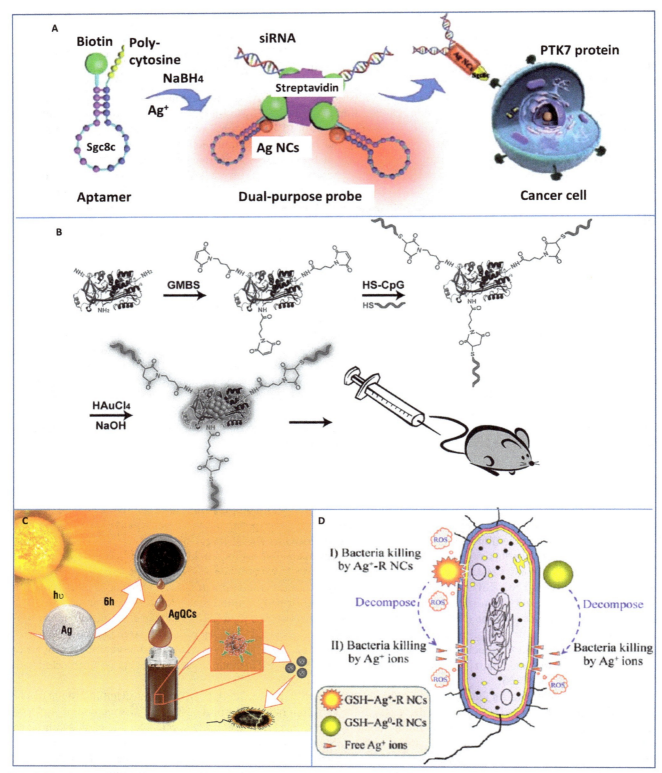

FIG. 8.25 (A) Schematic illustration of aptamer-functionalized Ag NCs-mediated cell type-specific siRNA delivery and tracking; (B) Schematic illustration of the synthesis of the OVA-AuNCs-CpG conjugates to induce immune response; (C) Schematic illustration of the synthesis of Ag NCs under sunlight and its antibacterial properties; (D) Schematic of possible antimicrobial mechanisms of GSH-Ag$^+$-R NCs and GSH-Ag^0-R NCs.

In addition to the Ag NCs killing bacteria, Au NCs can also have antimicrobial properties with the help of functional ligand. Lysozyme should be capable of hydrolyzing the cell walls of all pathogenic bacteria, including troublesome antibiotic-resistant bacteria. Chen et al. synthesized bioactive and fluorescent lysozyme-Au NCs and used it as antimicrobial agents for antibiotic-resistant bacteria such as pan-drug-resistant *Acinetobacter baumannii* and vancomycin-resistant *Enterococcus faecalis* and broad-band labeling agents for pathogenic bacteria as well [229]. Zhu and Yu et al. designed pentapeptide-protected Au NCs and loaded vancomycin (Van) via a spontaneous process based on the specific binding between Van and the custom-designed peptide to obtain self-regulated drug loading and release capabilities. The Au NCs loaded with Van could effectively inhibit the growth of gram-positive bacteria, and the antibacterial effect was similar to that of Van itself. Instead of adding Van directly, the cluster-based nanodrug delivery system can release the appropriate amount of Van according to the bacterial content, effectively increasing drug utilization and reducing side effects [157].

8.4 Catalysis

In general, the smaller the size of the catalyst, the larger the specific surface area, the higher the catalytic activity, and the higher the atomic utilization. Atomic precise metal NCs are unique materials that consist of a few to a few hundred atoms and are typically less than 2 nm in size. Due to their ultrasmall size, unique structure, molecular-like discrete energy levels, rich unsaturated active sites and other characteristics, metal NCs show completely different physical and chemical properties from the larger metal NPs. Therefore, they exhibit high catalytic activity and selectivity in many catalytic reactions. These metal NCs not only have broad application prospects in industrial catalysis due to their excellent catalytic performance, but also can be used as model catalysts to explore the catalytic reaction mechanism and establish the relationship between catalyst structure and performance. The reasons for this benefit are: (1)The size of nanocluster is very small, only changing one atom will greatly affect the nature of it, so the atomic precise cluster with accurately controlled the atom type, number, and position is very necessary for the study of catalytic structure-activity relationship. (2) Due to its clear and manageable structure, and absolutely single dispersion, that is beneficial to the research of catalytic reaction path from the perspective of whether experiment or theoretical simulation (it will effectively avoid the interference factors from nanoparticles polydispersity, e.g., different positions of nanoparticles, such as platform, edge, kink, corner, have different catalytic ability). (3) In nanoparticle, the interface between the metal core and cover agent environment is not clear. In contrast, the surrounding environment of atomic precise nanocluster is established, the amount and type of surface ligand also can be precisely adjustable, which is helpful to study the surface/interface effect. (4) Profiting from its molecular level purity and highly uniformity, making it supported catalyst, which can play the role of bridge to connect homogeneous and heterogeneous catalysis and realize high activity, selectivity, stability, and cyclic utilization at the same time. In this part, we will focus on the applications of atomic precise NCs in catalysis, such as electrocatalysis, photocatalysis, and catalysis of organic reactions.

8.4.1 Electrocatalysis

8.4.1.1 Anode electrocatalytic reactions in fuel cells

Zhou et al. has prepared different atom content but same ligand protected Au NC (thiolate-stabilized Au_{25}, Au_{38}, Au_{144}, Au_{333}, $Au_{\sim520}$ and $Au_{\sim940}$) by wet chemical synthesis method and investigated the transition from metallic to molecular state influence to the ethanol oxidation reaction (EOR) (Fig. 8.26A–C). They used femtosecond transient absorption spectroscopy to distinguish three different states of Au particles: metallic (size larger than Au_{333}, that is, larger than 2.3 nm), transition regime (between Au_{333} and Au_{144}, that is, 2.3–1.7 nm) and nonmetallic or excitonic state (smaller than Au_{144}, i.e., smaller than 1.7 nm). Au_{144} has exhibited highest current density (114.7 mA/mg) at ethanol oxidation peaks among Au_{25}, Au_{38}, Au_{144} and Au_{333}. It reflects the size-dependent activity trend is consistent with the transition from the molecular to plasmoic state. They found the optimum size for best EOR activity located in transition state between the metallic to molecular-state, rather than all quantum-sized particles showing high activity [230].

Lu et al. has utilized Pt_1Au_{24} NCs as catalysts and obtained high mass activity (3.7 A mg Pt^+Au^{-1}) in electrocatalytic formic acid oxidation (FAO), which was ~12 and 34 times as high as that of Pt NC and commercial Pt/C. Although the single Pt atom was located in the center of the Au NC, it has improved the catalytic activity and high-efficiency atom utilization. What's more, it protected the NC from CO poisoning in order to achieve long-term stability, which was evidenced by the normalized J–E curves for FAO, CO stripping test and in situ FT-IR analysis [231].

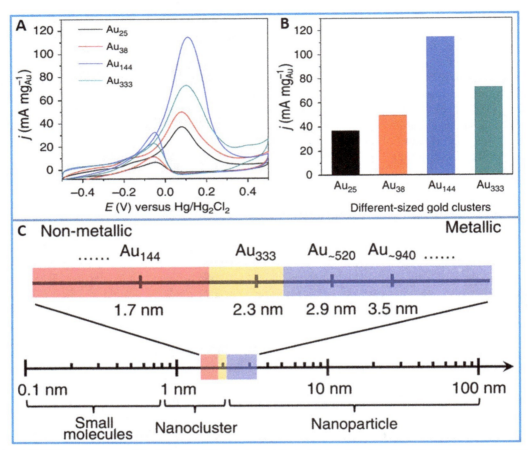

FIG. 8.26 (A and B) EOR performances of Au NCs with different atom numbers, (C) transition from the molecular to plasmonic state.

8.4.1.2 Cathode electrocatalytic reactions in fuel cells

Nesselberger et al. utilized ultrahigh vacuum with a laser ablation source to obtain a series of size-selected Pt clusters (Pt_{20}, Pt_{46}, $>Pt_{46}$) with precise control of size and coverage. The Pt_{20} and $Pt > 46$ have shown 2 and 3.5 times enhancements in surface-area-normalized specific activities and more than 2 and 6 times improvements in mass-normalized specific activities compared with the standard TKK catalysts, respectively, in electrocatalysis oxygen reduction reaction (ORR). Owing to the independent and precise regulation of the size and distribution of the clusters, they establish the structure-activity relationship. They exclude the effect factors such as cluster size, interparticle distance, specific anion adsorption, and the position of the metal d-band center. It is the electric field effect that the determinant factor for the electrode coverage with oxygenated species and ORR activity [232].

Weber et al. used the same approach to deposit mass-selected Pt_n (n from 1 to 14) on indium tin oxide (ITO) in ultrahigh vacuum. Among these Pt_n NCs, Pt_{10}/ITO exhibits the lowest mass activity, but its activity is still an order of magnitude higher than that of 5 nm Pt NPs/ITO. Different from the anticorrelation between EOR activity and Pt 4d binding energy in the previous work [233], there was no obvious relationship between Pt 4d binding energy and ORR activity. The authors disclosed another interesting rule about the size-dependent branching between H_2O and H_2O_2 production in ORR. The H_2O_2/H_2O ratio gradually decreased from 0.53 (for Pt_1/ITO) to 0.24 (for Pt_{14}/ITO) and to 0.04 (for Pt nano/ITO). The smaller cluster has not efficient appropriate sites to dissociate O_2, the ORR process two electrons pathway. The larger cluster has relatively efficient sites for O_2 dissociation, the ORR tends to process four electrons pathway. This suggests tuning the size of the active site is another useful strategy to control the catalytic selectivity [234].

Yamamoto et al. has synthesized a series of ultrafine Pt NCs with 12, 28, 60 atoms based on his spherical phenylazo-methine dendrimer templates. The three size Pt clusters were applied in ORR, the smallest clusters Pt_{12} shown highest catalytic activity with 13-folds enhancement factor compared with commercial Pt/C. Furthermore, they introduced a foreign metal (Sn) into Pt_n in synthesizing bimetallic clusters. Due to the synergistic effect induced d-band center energy

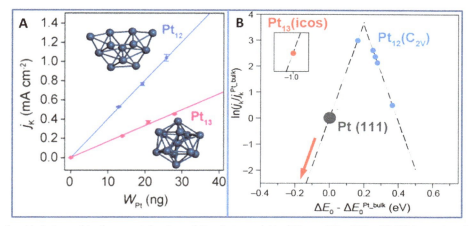

FIG. 8.27 (A) Relationship between kinetic current density and the cluster weight of Pt_{12} and Pt_{13} NCs; (B) "Volcano-shape" relationship between kinetic current density and oxygen adsorption energy of Pt_{12}, Pt_{13}, and Pt (111).

regulation, the $Pt_nSn_{(28-n)}$ exhibited higher onset potential and larger kinetic current compared with the mixture of Pt and Sn clusters [235].

Yamamoto et al. continued to apply the macromolecular template approach to synthesize two kinds of Pt clusters with different atom numbers (one is magic number 13, the other is misshapen 12). Although there was only one atom difference, Pt_{12} showed better activity than Pt_{13} with twice enhancement in mass activity (Fig. 8.27). Besides, Pt_{13} showed the lowest activity in comparison to the activities of Pt_{28} and Pt_{60} reported previously. Therefore, the high ratio of surface atoms theory was not applicable here any longer. Here, the ORR activity is influenced by the specific geometric structure of the cluster. Pt_{13} has a high symmetry and stability of icosahedral structure, while, Pt_{12} has different atomic coordination structures from Pt_{13}. Due to the structure difference, Pt_{12} has an ideal ΔE_0 (0.2–0.3 eV) in the known "volcano-shape" relationship, and Pt_{13} presents a strong oxygen binding energy ($\Delta E_0 = -1.0$ eV) which is not beneficial to ORR [236]. They also synthesized two different dendrimer stabilized Pt_n NCs to investigate the coordination environment effect. Due to the tight surrounding, the clusters surrounded by polyamidoamine dendrimers (PAMAM) generally showed lower activities than those protected by fourth-generation phenylazomethine dendrimer with a triphenylpyridylmethane core (DPAG$_4$-PyTPM) [237].

Chen et al. prepared a series of Au clusters ($Au_{11, 25, 55, 140}$) by solution-phase thiolate ligand protecting method and applied to ORR. Because a large number of surface Au atoms have low coordination number in smaller clusters, Au_{11} showed the best activity in ORR among the size series, and the ORR activity decreased with the increase in core size. Besides, theoretical calculations demonstrated that the smaller Au clusters, the narrower d bands, the closer to Fermi levels, the easier O_2 adsorption, which is beneficial to ORR [238].

Dass and Chakraborty et al. also investigated the size effect of NCs on the ORR activity. They used $Au_{28}(TBBT)_{20}$, $Au_{36}(TBBT)_{24}$, $Au_{133}(TBBT)_{52}$, and $Au_{279}(TBBT)_{84}$ NCs (TBBT, 4-tert-butylbenzenethiol) to study the kinetics of electrocatalytic ORR in alkaline condition. Unexpectedly, the smallest Au_{28} NCs exhibited the lowest activity with the highest overpotential to reach the same current and the worst selectivity. The reactivity trend is $Au_{36} > Au_{133} > Au_{279} > Au_{28}$ which is qualitatively correlated to the thermochemical stability trends of the NCs [239]. In addition, they expanded bulky t-butyl thiolated Au NCs, and obtained two new and larger NCs, $Au_{46}(S\text{-}tBu)_{24}$, and $Au_{65}(S\text{-}tBu)_{29}$. They investigated the core size effect of this series ($Au_n(S\text{-}tBu)_m$, $n, m = 23, 16; 30, 18; 46, 24; 65, 29$) on the activity for ORR. The ORR catalytic results indicated that the largest size has the highest activity. Au_{65} NCs exhibited the highest selectivity with 80% OH^- production and the smallest one, Au_{23} NCs produced 53% OH^-. Au_{65} only need 80 mV overpotential can reach -1 mA/cm^2 while Au_{23} need 680 mV overpotential [240].

Kauffman et al. investigated the effect of charge state of cluster on the electrocatalytic ORR activity. Three different charge state $Au_{25}{}^q$ NCs but with identical surface structure produced equivalent electron transfer numbers of 3.0 ± 0.3 e$^-$ in ORR and showed the ORR activity trend as $Au_{25}{}^- > Au_{25}{}^0 > Au_{25}{}^+$. There was an antidependence relationship between the calculated OH^- binding energy and the measured turnover frequency (TOF) value, i.e., $Au_{25}{}^+$ showed the largest OH^- binding energy and the lowest ORR TOF [241]. Lu et al. also reported three charge states Au_{25} NCs for electrocatalytic ORR. Au_{25} showed similar catalytic activity trend as reported by Kauffman et al. The Au_{25}^- displayed the most effective two-electron reduction pathway with maximum H_2O_2 production [242]. This charge-induced catalytic activity opens up a new field of view for the study of the mechanism of this catalytic reaction.

Introducing good conductive material to form composite electrocatalysts can improve activity and stability. Kwak et al. prepared Au_{25} NC and reduced graphene oxide (rGO) composites and controlled the number of the Au_{25} layers from 0 to 15 by adjusting the ratio of NCs to rGO. In ORR, the Au_{25}-rGO provided a much more efficient four electron-transfer process and showed a more positive onset potential and a higher limiting current density compared with Au_{25} and rGO. Besides, with increase in the Au_{25} film thickness, the catalytic current and ORR electrocatalytic rate constant increased and the charge transfer resistance (Rct) decreased [243].

Introducing foreign atom is an effective way to improve the catalytic activity. Zhu et al. synthesized bimetallic $Au_1Ag_{21}(dppf)_3(SAdm)_{12}/C$ (dppf, 1,1′-bis-(diphenylphosphino)ferrocene) showed better activity in ORR than corresponding $Ag_{22}(dppf)_3(SAdm)_{12}/C$ due to the synergy effect between the M_{13} kernel and dppf ligand in M_1Ag_{21}. The two clusters are both two-electron transfer process [244].

8.4.1.3 Electrocatalytic hydrogen evolution reaction

Zhao et al. loaded Au_{25} NCs on MoS_2 nanosheets to improve the HER activity of MoS_2. Furthermore, they investigated the interfacial effect by comparing the HER activities between $Au_{25}(SCH_2CH_2Ph)_{18}/MoS_2$ and $Au_{25}(SePh)_{18}/MoS_2$. Due to weak electron relaying capability, Au_{25} with -SePh ligand showed worse HER activity than that with -SCH_2CH_2Ph [245]. Gao et al. synthesized $Pd_6(SC_{12}H_{26})_{12}$ NCs and loaded it on active carbon (AC). They investigated the ligand effect on the HER activity by annealing the catalyst at 200°C. Pd_6/AC ligand-off showed a closer onset potential to that of commercial Pt/C (−0.043 V) and exhibited a 30 mV positive shift compared with that with ligand-on composite, which suggested ligand-off is beneficial to HER activity [246].

So far, Pt-based materials are still the most advanced catalyst for HER. However, the high price and scarce reserves of Pt prevent it from achieving large-scale commercial application. On the premise of ensuring the catalytic activity, the strategies to reduce the use of Pt are proposed. One is to reduce the particle size to expose more active sites, and the other is to introduce relatively cheap metals to form the alloy with enhanced activity. Kwak et al. reported that the $Pt_1Au_{24}(SC_6H_{13})_{18}$ showed highly efficient electrocatalytic HER activity in comparison with $Au_{25}(SC_6H_{13})_{18}$ [247].

Considering the synergistic effect in the compositions of bimetallic NCs, Zhu et al. synthesized Au_2Pd_6 NCs as HER catalysts and compared it with Pd_3 and Au_2 NCs. The structure of Au_2Pd_6 NC can be regarded as two Pd_3 triangles linked by an Au_2 unit. They deposited NCs on MoS_2, a popular HER electrocatalyst. Due to the electronic interaction between NCs and MoS_2, the NCs greatly improved the HER performance of MoS_2, which was confirmed by the XPS analysis. Au_2Pd_6 NCs caused a larger shift than Pd_3 NCs in the Mo 3d and S 2p binding energies of MoS_2. Besides, the DFT calculation also demonstrated that Au_2Pd_6/MoS_2 has the best ideal ΔG_H^* for the best HER activity among Pd_3/MoS_2, Au_2/MoS_2, MoS_2. Furthermore, the composite Au_2Pd_6/MoS_2 has more HER active sites than a single component of the composite (Fig. 8.28) [248].

8.4.1.4 Electrocatalytic oxygen evolution reaction

Gao et al. also reported that the Pd_6 NC was highly active for oxygen evolution reaction (OER). In contrast to the activity trend in HER, the ligand-on NCs show better OER activity than that without ligand. The electron cloud density of Pd_6/AC is higher than that without ligand, indicating that oxygen atom is easier to desorption, which is conducive to OER [246]. Zhao et al. deposited precise Au_{25}, Au_{144}, Au_{333} NC on functional $CoSe_2$ nanosheets, obtained high OER activity, and presented the correlation between cluster structure and OER activity (Fig. 8.29). They found the OER activity enhanced with NC size increase. XPS and Raman analysis indicate a significant synergistic electronic interaction between NC and support. DFT calculation demonstrated the acceleration in the *OOH (an important intermediate for OER) production [249].

8.4.1.5 Electrocatalytic CO₂ reduction

Kauffman et al. utilized negatively charged $Au_{25}(SC_2H_4Ph)_{18}^-$ to electrocatalysis CO_2 reduction, and CO and H_2 were the only products. Au_{25}^- showed a smaller onset overpotential of CO formation and a higher TOF than that observed with 2–5 nm larger Au NPs. The negative charge of Au_{25}^-, appropriate CO_2 adsorption on Au_{25}^-, and the reactive site on Au_{25}^- promoting C=O bond activation and Hads formation are the main reason for highly efficient CO_2 reduction [250]. They further investigate the effect of cluster charge state on the activity of electrocatalytic CO_2 reduction. For all three different charge states of Au_{25} NCs, the kind and amount of catalytic products, the onset potentials, and the Tafel slopes were almost the same. However, Au_{25}^- showed higher cathodic current density, TOF, and FE than those of Au_{25}^+ and Au_{25}^0. These results indicate that the three charge states of Au_{25} had the same catalytic pathway, but the catalytic rate of Au_{25}^- was the highest [241]. Moreover, they utilized cost-effective consumer-grade renewable energy (i.e., using solar cell and rechargeable battery to simulate day and night renewable energy source) and Au_{25} catalysts to drive CO_2 electroconversion on a large scale (Fig. 8.30) [251]. Zhao et al. utilized two different structures of Au_{25} NCs (nanosphere and nanorod) as a model to investigate geometric structure effect of cluster on the activity of electrocatalytic

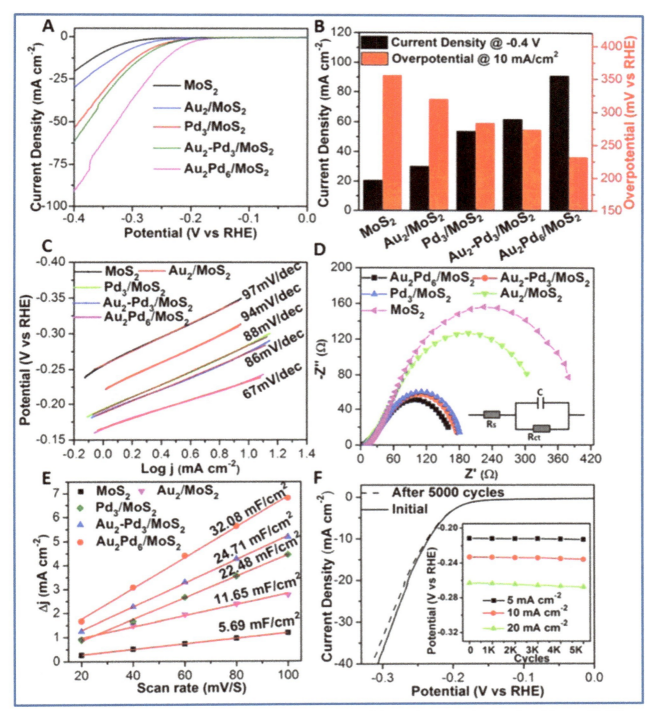

FIG. 8.28 (A) LSV curves of samples for HER; (B) current density and overpotential comparison among these samples; tafel curves (C), EIS curves (D) and electron double capacity calculation (E) of these samples; (F) stability tests for Au_2Pd_6/MoS_2.

CO_2 reduction. Au_{25} nanospheres have higher electrocatalytic CO_2 reduction activity than Au_{25} nanorods, reflecting that the total current density, FE, and CO formation rate of Au_{25} nanospheres are higher than that of Au_{25} nanorods at all potentials. The reasons why sphere-shaped Au_{25} had a better activity than nanorod-shaped Au_{25} are that sphere-shaped Au_{25} easily loses one ligand to expose an active site, stabilizes *COOH well, and is more electron rich than Au_{25} nanorod [252]. Wu et al. utilized two-phase antigalvanic reduction (AGR) method to synthesize $Au_{47}Cd_2(TBBT)_{31}$ NCs. $Au_{47}Cd_2$ NCs showed high FE up to 96% for electrocatalytic CO_2 reduction [253].

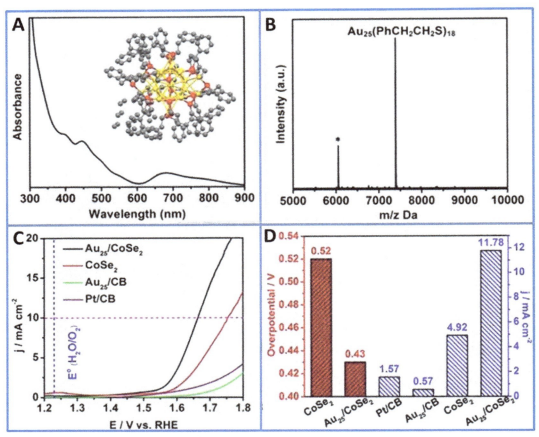

FIG. 8.29 (A) UV-vis. Absorption spectrum and (B) MALDI-TOF-MS spectrum of the NC Au_{25}; inset is its crystal structure. (C) LSVs for OER of these samples, and (D) comparison of the overpotential and current density of these samples.

8.4.2 Photocatalysis

Metal NCs with unique optical properties have unexpected interesting photoreactivity related to their structure. For example, Kauffman et al. has found that Au_{25}^- cannot be spontaneously oxidized by O_2, except under room light irradiation, electron transfer can occur between Au_{25}^- and O_2 [254]. Kawasaki et al. found that when Au_{25} is excited by visible/near-infrared (NIR) light, it can generate $^1O_2$ regardless of the solubility and charge status of Au_{25}. However, the triplet energy of Au_{38} is low, so it cannot produce $^1O_2$ under light irradiation [255]. Negishi et al. have reported that Au_{25} NC protected by azobenzene thiolate (S-Az)-can switch back and forth between *cis* and *trans* forms with 100% efficiency under photoirradiation, which provides a potential opportunity for photoinduced chiral catalysis [256]. Stamplecoskie et al. have investigated the size effect of the GSH-protected Au NC on its light harvesting ability. The larger the Au NC size, the less dominant the short lifetime relaxation component. As an efficient light collecting antenna, NC with long excited state is beneficial to the photocatalytic reaction [257]. These results demonstrate that NCs may play specific roles in photocatalytic reactions.

8.4.2.1 Photocatalytic decomposition of pollutants

Yu et al. deposited the $Au_{25}(SCH_2CH_2Ph)_{18}$ NCs on TiO_2 and applied it into photocatalytic methyl orange degradation, and they found it displayed light-dependent catalytic activity. Due to the introduction of Au_{25}, the light absorption range of the composite is extended from ultraviolet to near infrared. Under UV and visible light irradiation, the decomposition ratio (DR) in methyl orange degradation of the composite (with only 0.94% NC loading amount) was 69%, while the DR of bare TiO_2 with only 27%, indicating a higher catalytic activity of the composite [258]. Hu et al. reported similar results in their study on the photocatalytic degradation of rhodamine B (RhB) by using Au_{25}/porous ZnO as catalysts. In order to balance the separation efficiency of e^-—h^+ and the active surface area, they optimized the loading amount of Au_{25} in

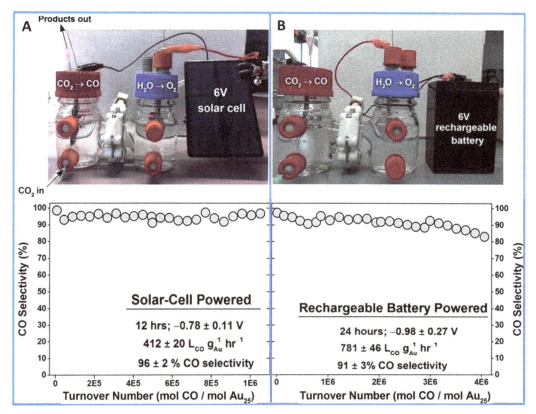

FIG. 8.30 The electrocatalytic CO_2 reduction instruments (on the top) by using (A) solar cell and (B) rechargeable battery, and corresponding CO selectivity and TON (on the bottom).

composite. When the loading amount is 1.16%, the catalytic activity of the composite is the highest, and the DR can reach 95% within 15 min [259].

Lee et al. investigated the size effect of NCs on the photocatalytic activity and electron transfer. They deposited three different sizes GSH-protected Au NCs (1.1 nm Au_{25}, 1.6 nm Au_{144}, 2.8 nm Au_{807}) on ZnO surface and used it to degrade thionine (TH). To reach 50% DR, ZnO-Au_{25} needs longest photolysis time-112 s while for ZnO-Au_{144} is 47 s, and for ZnO-Au_{807} is 34 s. The introduction of the Au NC does not expand light harvesting, but act as a charge-separation role. Larger Au NC size caused shorter lifetime and higher electron transfer efficiency, therefore induced higher photocatalytic activity [260].

Liu et al. prepared the 6-mercaptohexanoic acid (MHA)-protected Au_{25} NC as photosensitizer and loaded it on highly ordered TiO_2 nanotube arrays (TNAs) for photocatalytic decomposition of antibiotic tetracycline. The composite has expanded light absorption from UV to NIR range by introducing Au_{25}. Besides, Au_{25} NC can facilitate charge transfer and suppress charge recombination. Therefore, compared with bare TNAs, the activity of Au_{25}-TNAs was increased by 1.6 times, and the kinetic constants were increased by 1.67–1.75 times [261].

Weng et al. used positive charged poly-ethylenimine (BPEI) as the surface modifier to modify a series of metal oxide (i.e., SiO_2, TiO_2, ZnO, ZrO_2, etc.) and assemble negatively charged $Au_{25}(SG)_{18}$ NC. BPEI is not only used as a linker, but also as a reducing agent and stabilizer. The multifunction of BPEI protects the ligand of Au NCs from being oxidized, so that Au NCs remain unchanged under light irradiation. Therefore, the composite is stable and the degradation rate did not loss in 10 cycles [262].

8.4.2.2 Photocatalytic water splitting

Berr et al. utilized an ultrahigh-vacuum approach to decorate CdS nanorods with Pt NCs of different sizes. By keeping the cluster coverage of the system consistent (~23 NCs/CdS nanorod), they studied the effect of cluster size on the photocatalytic hydrogen evolution reaction (HER) activity. The trend of photocatalytic activity was $Pt_8 \approx Pt_{22} < Pt_{34} < Pt_{68} < Pt_{46}$. The LUMO position of the cluster must be lower than the CB position of the semiconductor and higher than the reduction potential of H^+/H_2 at the same time. Among these NCs, Pt_{46} has the most suitable LUMO position, thus showing the highest quantum efficiency (QE)

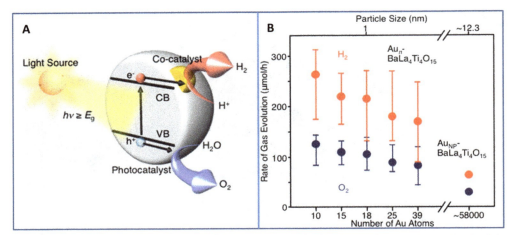

FIG. 8.31 (A) Scheme of photocatalytic water splitting using Au NCs as cocatalysts. (B) Relationship between the size of the Au NCs and the activity of H_2 and O_2 production.

[263,264]. In addition to the size of NCs, its valence state also plays a key role in photocatalytic hydrogen evolution. Li et al. have found that Pt NCs with higher oxidation state effectively suppress the side reaction-hydrogen oxidation reaction (HOR) which is because that O_2 dissociation on its surface was much more difficult than that on metallic state [265].

Negishi et al. prepared size uniform NC-based catalysts using the front-synthesis approach. They controlled the loading amount ($\sim 0.1\%$) and deposited a series of GSH-protected Au NCs ($Au_{10}(SG)_{10}$, $Au_{15}(SG)_{13}$, $Au_{18}(SG)_{14}$, $Au_{22}(SG)_{16}$, $Au_{25}(SG)_{18}$, $Au_{29}(SG)_{20}$, $Au_{33}(SG)_{22}$, and $Au_{39}(SG)_{24}$) on $BaLa_4Ti_4O_{15}$. After removing the ligand by calcination at 300°C in vacuo, for group 1 (Au_{10}, Au_{15}, Au_{18}, Au_{25}, and Au_{39}), the size of the Au NCs displayed no significant change with slight increase, while, for group 2 (Au_{22}, Au_{29}, and Au_{33}) these Au NCs showed clear size increase and a distinct plasmonic peak at ~ 520 nm. The Au NC-$BaLa_4Ti_4O_{15}$ system with more stable Au NCs (those in group 1) was used to reveal the size effect of Au NCs in photocatalytic water splitting activity. All Au NC-$Bala_4Ti_4O_{15}$ composites produced H_2 and O_2 in the ratio of 2:1, but the catalytic activity increased with the decrease in NCs size, which was due to the sharp increase of surface atoms with the size decrease (Fig. 8.31) [266].

Du et al. investigated the heteroatom effect on the photocatalytic activity by comparing the $Ag_{25}(SPhMe_2)_{18}PPh_4$ and $[Pt_1Ag_{24}(SPhMe_2)_{18}](PPh_4)_2$) NCs. Pt_1Ag_{24}/g-C_3N_4 showed the highest activity for photocatalytic H_2 generation with fourfold enhancements compared with Ag_{25}/g-C_3N_4. The origin of the better activity of the Pt_1Ag_{24}-g-C_3N_4 is (1) it showed a smaller barrier for photocarrier transfer; (2) it displayed improved charge separation ability; and (3) it showed a longer photocarrier lifetime [267].

8.4.2.3 Photocatalytic organic transformation reactions

Kawasaki et al. utilized Au_{25} NC as a photosensitizer to generate highly reactive $^1O_2$ by direct visible/NIR photoexcitation. The photoexcited produced $^1O_2$ was then used as an oxidant and further applied in selective photocatalytic oxidation of sulfides. Under visible/NIR light irradiation, Au_{25} catalyzed sulfide to convert to sulfoxide with nearly 100% selectivity [255]. Li et al. also utilized $^1O_2$ generated by photoexcited Au NC to selectively oxidize sulfides and amines. They synthesized S-Adm-protected Au_{38} with a larger HOMO-LUMO gap (~ 1.57 eV), which was appropriate to generate $^1O_2$. Au_{38} showed a higher activity than Au_{25} in selective photocatalytic sulfide oxidation because of its higher efficiency of $^1O_2$ production [268]. Zhu et al. utilized $Au_{25-x}Ag_x(SCH_2CH_2Ph)_{18}^-$ alloy NCs to achieve 100% conversion and selectivity in the catalytic transformation of benzylamine to N-benzylidenebenzylamine under visible light irradiation [269]. Chen et al. utilized Au_{25}/TiO_2 composites to realize highly efficient (1522 h^{-1} TOF) and selective amines oxidation under mild conditions. The yield of bare TiO_2 (both anatase and rutile phase) is only 12%–14%. After the deposition of Au_{25}, the conversion of rutile TiO_2 and anatase TiO_2 are significantly increased to 45% and 80%, respectively [270].

8.4.3 Catalytic selective oxidation

8.4.3.1 CO oxidation

As early as 1999, Heiz and Landman et al. have utilized size selected Au_n clusters ($n \leq 20$) to catalyze CO oxidation. They found Au_8 was the minimum sized specie that can be active in this reaction [271]. Soon afterwards, Anderson et al. used Au_n

of different sizes ($n=1, 2, 3, 4, 7$) to study the size effect. When $n \geq 3$, the clusters become active in CO oxidation [272]. Until 2008, Kiely and Hutchings used aberration-corrected STEM to identify the really active catalyst that was ~0.5 nm Au NCs [273]. Recently, they used a new counting protocol to examine the size distribution and activity contribution of various Au species. In addition to 0.5 nm Au NCs, 1–3 nm Au NCs were also active sources [274]. Besides, other factors were also carefully studied, including the composition and thickness of the support [275], charging effect [276], structural fluxionality [277], and the water promotion effect [278]. Besides Au NCs, size-selected Pd_n and Pt_n NCs were also used to catalyze CO oxidation. A lot of experiments and theoretical studies were systematically conducted to investigate the effects of the electronic structure [279], size [280–282], support [283–285], oxidation state [286,287], and geometric structure [288]. CO oxidation catalyzed by these size-selected clusters prepared in ultrahigh vacuum was already summarized by Corma et al. [289] Here, we mainly introduce the NCs with defined crystal structures prepared by liquid chemical methods.

Jin et al. reported the $Au_{25}(SR)_{18}/CeO_2$ has good activity in CO oxidation, while $Au_{25}(SR)_{18}/CeO_2$ is inactive for the same reaction. The strong influence of the support indicated that the NC-support interface should be the catalytic active site. The O_2-pretreated $Au_{25}(SR)_{18}/CeO_2$ catalyst is advantageous to the CO conversion, especially after joining the water steam into the feed gas. It shows that the system has high catalytic activity is not due to ligand desorption, but in the pretreatment process of O_2 including O_2 adsorption and activation on the catalyst and active oxygen species generation in the perimeter and low-coordinated corner of the Au sites [290].

Nie et al. further confirmed the O_2 pretreatment function by comparing the activity of $Au_{38}(SC_{12}H_{25})_{24}/CeO_2$ catalysts with and without ligand [291]. Though O_2 pretreated, the catalyst with ligand shows improved catalytic activity, while the one without ligand exhibit lower activity. In addition, the one with ligand are stable and show long durability for 20 h, but without ligand the NCs are unstable. DFT theoretical has demonstrated the self-promoting CO oxidation mechanism in $Au_{25}(SCH_2CH_2Ph)_{18}/CeO_2$ and $Au_{38}(SC_{12}H_{25})_{24}/CeO_2$ systems, the key point is the triangular Au_3 sites in the NCs [292]. The mild O_2 pretreatment process should involve O_2 adsorption and activation on the catalyst surface, generating active oxygen species, and O_2 from the feed gas can readily follow the "footprints" of the pretreatment-generated active oxygen species to replenish active oxygen, hence, offering high catalytic activity under lower reaction temperature. According to Pei's theoretical work, the dihapto peroxo complex is most probably formed by oxidative addition of the O—O bond at the superoxo stage [292].

Overbury et al. proposed a different view about the role of ligands through various technologies, such as reaction kinetic tests, in situ IR and X-ray absorption spectroscopy, DFT calculation: thiolate ligands act as a double-edged sword for the Au_{25} NCs for CO oxidation [293]. The intact catalyst $Au_{25}(SR)_{18}$ has no catalytic activity to CO oxidation due to the lack of the ability of all Au sites to adsorb CO, while the naked Au_{25} without protective ligand only being active at high temperature. Isotopic labeling experiments clearly indicate that CO oxidation on the $Au_{25}(SR)_{18}/CeO_2$ catalyst proceeds predominantly via the MvK (Mars-Van Krevelen) mechanism rather than L-H (Languir-Hinshewood) mechanism (Fig. 8.32) under lower temperature, that is CO adsorbed on the Au sites of the $Au_{25}(SR)_{18}/CeO_2$ catalyst reacts with the lattice oxygen of CeO_2 to form CO_2, while the O_2 replenishes the consumed lattice oxygen.

Jin et al. reported a mild activation strategy for CO oxidation. By adding reductive gas such as O_2, CO, O_2/H_2, or O_2/CO to enhance the O_2-pretreatment effect, they obtained a higher efficient $Au_{144}(CH_2CH_2Ph)_{60}$ catalyst. Furthermore, they have found that the activation of the catalyst is closely related to the production of active oxygen species on CeO_2 by Raman spectroscopy and pulse experiments [294].

Similarly, Spivey et al. reported that the reductively pretreated $Au_{38}(SC_{12}H_{25})_{24}/TiO_2$ composite is an efficient catalyst for CO oxidation, while the air-dried $Au_{38}(SC_{12}H_{25})_{24}/TiO_2$ catalyst showed almost no activity for CO oxidation under the same conditions. The H_2/He treated Au_{38}/TiO_2 material was no sulfur present in the catalyst, by the tests of FTIR, EXAFS and XPS, which indicates thiol ligand prevents the contact between Au sites and molecular CO. In addition, they reported the electropositive sulfide species strongly interacted with Au (+1), likely at the Au-TiO_2 support interface, suggesting that the interfacial sites play an important role in catalyzing CO oxidation [295].

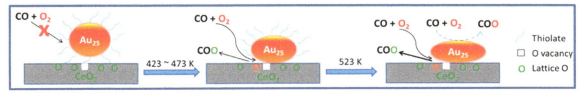

FIG. 8.32 Proposed oxidation mechanism of carbon monoxide by intact, partially and completely dethiolated $Au_{25}(SR)_{18}/CeO_2$ Catalysts.

Different with the removal of ligand to enhance activity, Overbury et al. proposed creating in situ cus Au atoms to improve the catalytic activity of intact NCs. They found a ligand-protected intact $Au_{22}(L_8)_6$ NC (L = 1,8-bis(diphenylphosphino) octane) with eight the cus Au atoms, exhibits high activity for low-temperature catalytic CO oxidation without the need of ligand removal. The eight cus Au atoms are close to neutral charge and barely affected by the charge transfer from the phosphine ligands to the Au NCs, which is quite different from that of the cus sites on slightly dethiolated Au_{25} clusters [296].

Wu et al. investigated the support shape effect on the catalytic activity by using rod and cube CeO_2 to support the $Au_{22}(L_8)_6$ (L = 1,8-bis(diphenylphosphino) octane) NC. In CO oxidation reaction, the rod CeO_2 supported $Au_{22}(L_8)_6$ NCs show higher activity than that of cube CeO_2-supported NCs, due to the difference in the nature and quantity of surface exposure of Au sites as well as the efficiency of organic ligand dissociation [297].

Li et al. found the electronic or steric effects of the protecting ligands has great influence to the catalytic performance of the NCs. CeO_2-supported different ligand protected $Au_{25}(SR)_{18}$ NC (SR = SPh, S-Nap and SC_2H_4Ph), they showed different CO oxidation catalytic activity. Besides, the heteroatom-doped can also modify the catalytic performance of the NCs. $M_nAu_{25-n}(SCH_3)_{18}$ (M = Cu, Ag) exhibited various catalytic performances in CO oxidation [298,299]. Furthermore, they proposed that the coordination of the lattice oxygen with the CO is the rate limiting step in CO oxidation by Au_{38}/CeO_2 catalyst, which is identified by regulating the Ce^{3+}/Ce^{4+} ratios [300].

Zhu et al. obtained $Au_8(DPPF)_4$ NCs with the help of the ferrocene ligand to direct the precise formation and solidify the structural pattern of the NCs. Compared with the Au_8 protected by PPh_3, the $Au_8(DPPF)_4$ cluster exhibits improved catalytic activity for the CO oxidation reaction, due to its resistance to aggregation into large NPs during the reactions. DFT studies suggested that the enhanced activity and the excellent stability of $Au_8(DPPF)_4$ is due to the homolytic phosphine dissociation nature and the post dissociation reconstruction effect induced by Fe [301].

8.4.3.2 Alcohol oxidation

Alcohol oxidation is important in industry because its products, aldehyde, ketone or acid, are important raw materials for the synthesis of fine chemicals. Tsukuda et al. reported that the ultrasmall Au: PVP (PVP = poly(N-vinyl-2-pyrrolidone)) NCs exhibits excellent activity in the aerobic oxidation of alcohol. They found an interesting phenomenon, which is the catalytic performance of Au NCs is facilitated with increasing electron density on the Au core, which is confirmed by XPS, FTIR and EXAFS tests. The key step for aerobic oxidation catalyzed by Au:PVP is the activation of O_2 by anionic Au cores, which can produce superoxo or peroxi-like species [302].

They further proposed two strategies for improving the catalytic performance of Au_{25} NCs, one is doping single Pd atom ($Au_{24}Pd_1$) [303] and the other is removing the protected ligands by high-temperature calcination (Fig. 8.33) [304]. The Au_{25} NCs fully protected by the thiolates were inert in the aerobic oxidation, while $Au_{24}Pd_1$ NCs and ligand-off Au_{25} NCs exhibit highly-efficient activity in catalytic oxidized benzyl alcohol. The heteroatom doping effect improving activity is relying on the electron transfer from Pd to Au, and the high-temperature calcination benefiting the activity is due to the creation of more empty sites available. Even so far, the two methods have been recognized as the effective method to facilitate the catalytic performance of NC catalysts. Besides, Zhu et al. investigated the effect of exact site-specific atom on the catalytic performance of benzyl alcohol oxidation. They found Au_8Pd_1 exhibited higher activity than Au_9 because the Pd atom provides active sites to adsorb and active O_2. In addition, the Au_8Pd_1 NCs can mediate the electrons and holes of the adsorbates, which is also beneficial to the high catalytic activity [305].

Zheng et al. synthesized three novel $Au_{13}Cu_x$ (x = 2, 4, 8) bimetallic NCs and investigated their catalytic activity for aerobic oxidation of alcohol. The bimetallic NCs show higher activity than Au_{25} NCs and pretreated $Au_{13}Cu_8$ NCs are more active than unpretreated one [306]. Meanwhile, Tuel et al. further demonstrated that the calcinated $Au_{25}(SPh-pNH_2)_{17}$@SBA-15 are active in aerobic benzyl alcohol oxidation. The catalytic activity increased upon calcination and reached the top at 400°C, when all the ligands were thermally removed [307]. Zhu et al. utilized metal-exchange method to synthesize cadmium-doped NC $Cd_1Au_{24}(SR)_{18}$, which show much higher activity than Au_{25} NC in benzyl alcohol oxidation [308]. Tsukuda's group prepared Pd_1Au_{33} and Pd_1Au_{43} NCs by coreduction of Au and Pd precursor ions. The doped Pd atom was located at the exposed surface of the Au:PVP NC, and the single Pd atom doping improved the activity of oxidation reactions and enhanced the selectivity for hydrogenation of the C=C bond over the C=O bond [309]. These experimental results fully confirmed heteroatom doping and high-temperature pretreatment can improve the catalytic performance.

Zhu et al. prepared two kinds of atomically precise composites, Au_{11}:PPh_3 NC@ZIF-8 and $Au_{13}Ag_{12}$:PPh_3@MIL-101, by a one-step using MOFs as the size-confining templates. They first combined structural precise MOF and metal NC to form well-defined composite. The confined effect of porous MOF solve the aggregation, sintering and structural change problem caused by high temperature calcination. The composites exhibited remarkable stability, favorable catalytic activity and excellent selectivity in the oxidation of benzyl alcohol [310].

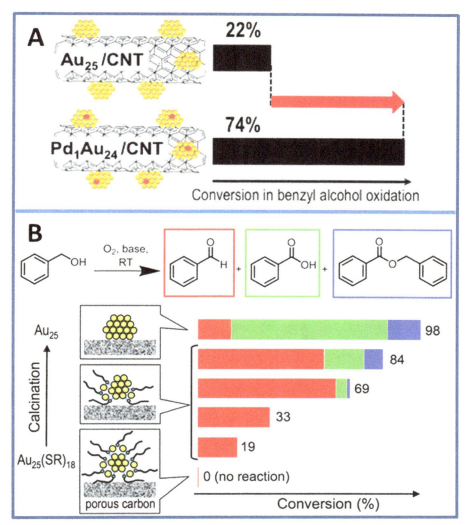

FIG. 8.33 Two strategies for improving the catalytic performance of the Au$_{25}$ NC, including (A) doping with a single Pd atom (Au$_{24}$Pd$_1$) and (B) removing the protected ligands by high-temperature calcination.

Zhang et al. used water-soluble Au$_{25}$ NCs as precursor to fabricate a serious of layered double hydroxides (LDH) supported Au NCs catalysts, including Au$_{25}$/M(Ni, Co)$_3$Al-LDH [311], Au$_{25}$/Ni$_x$Al-LDH [312] and γ-Fe$_2$O$_3$@M$_3$Al-LDH@Au$_{25-x}$ (x represents determined mass of gold in wt%) [313]. Due to the synergistic interaction between Au$_{25}$ NCs and LDH, and the abundant Ni-OH basic sites in LDH, the composite display unprecedentedly high activity. Among them, the γ-Fe$_2$O$_3$@Ni$_3$Al-LDH@Au$_{25-0.053}$ with calcination exhibits the highest activity (TOF: 112498 h^{-1}) for the aerobic oxidation under 1 atm O$_2$, without basic additives, solvent-free conditions.

Optimizing metal-support interactions (MSI) is an effective method to modify the electronic structure for the catalysts to obtain desired properties. Zhu et al. firstly synthesized an excellent catalyst Pd$_3$Cl/TNT (Pd$_3$Cl = [Pd$_3$(PPh$_2$)$_2$(PPh3)$_3$Cl]$^+$, TNT = titanate nanotubes). Due to the MSI effect boosting the key β-H elimination step from both kinetic and thermodynamic aspects, the catalyst exhibits good conversion (68.4%–99.3%) with 100% selectivity toward the aldehyde at 30°C with 1 atm O$_2$ in the aerobic oxidation [314].

8.4.3.3 Epoxidation of alkenes

Epoxides is an important commercially product, which is synthesized by the chlorohydrin routes. However, this route has its own limitation, such as the production of undesired byproducts or high cost of the H$_2$O$_2$ reactant [315]. Lee et al. reported the catalytic activity of subnanometer gold clusters (Au$_6$-Au$_{10}$) prepared by laser ablation method for direct propene epoxidation [316]. The model catalysts (Au$_6$-Au$_{10}$) are active for the partial oxidation of propene, and this is the first catalyst for the C$_3$H$_6$/O$_2$ system using Al$_2$O$_3$ as support instead of the commonly TiO$_2$ support.

In order to gain deep insight into the mechanism, Jin et al. exploits the catalytic capability of thiolate-capped $Au_{25}(SR)_{18}$, $Au_{38}(SR)_{24}$, and $Au_{144}(SR)_{60}$ nanoclusters for styrene oxidation with O_2 [317]. They found the activity and chemoselectivity of $Au_n(SR)_m$ catalysts has a strong connection with particle size (n), and the protected ligands do not influence the activity and selectivity. The smallest $Au_{25}(SR)_{18}$ NCs shows the highest conversion of styrene. The high activity is dependent on the partial positive charged of the NCs and the electron-rich kernel [318]. They further compared $Au_{25}(SR)_{18}/TiO_2$ and $Pt_1Au_{24}(SR)_{18}/TiO_2$ for the selective oxidation of styrene with $PhI(OAc)_2$ as the oxidant. Due to the heteroatom-doped effect, the $Pt_1Au_{24}(SR)_{18}/TiO_2$ exhibited greatly enhanced stability and catalytic activity [319].

Many studies have been performed on styrene oxidation by thiolate-protected Au NCs, and most catalytic systems focused on Au_{25} NCs as catalysts using O_2, tert-butyl hydroperoxide (TBHP) or $PhI(OAc)_2$ as the oxidants. Among them, the catalyst $0.5Au_{25}$-HAP (HAP=hydroxyapatite) prepared by Tsukuda et al. by calcinating the composite $0.5Au_{25}$: SG-HAP at 300°C showed 100% conversion and 92% selectivity in styrene epoxidation with anhydrous TBHP as the oxidant. The catalyst system presented the highest activity reported so far [320].

Zhu et al. succeeded in the preparation of the largest AgAu alloy NC, $[Ag_{46}Au_{24}(SR)_{32}](BPh_4)_2$, with a doping shell that consisted of a multilayer $Ag_2@Au_{18}@Ag_{20}$ core protected by a chiral $Ag_{24}Au_6(SR)_{32}$ shell. They compared the catalytic performance of a series of homometallic and alloy NCs to obtain the insight into the surface structure-catalytic property relationship. Compared with the homosilver NCs Ag_{44}/carbon nanotube (CNT), the surface doping $Au_{24}Ag_{46}$/CNT exhibits improved activity and selectivity for epoxides. In addition, the surface doped catalyst $Au_{24}Ag_{46}$/CNT shows better selectivity than the core-shell catalyst $Ag_{32}Au_{12}$/CNT (Fig. 8.34) [321]. Zhu et al. further investigated the water effect in styrene oxidation catalyzed by NCs. The XPS results indicated that water adsorption on the surface of Ag_{44} and $Au_{25-x}Ag_x$ influenced the valence state of Ag using mixed solvents (toluene and water), whereas the corresponding valence state of Au in Au_{25} and $Au_{25-x}Ag_x$ remained the same. The isotope labeling experiment confirmed electron transfer from H_2O to Ag and the reaction mechanism involve two processes for different products [322].

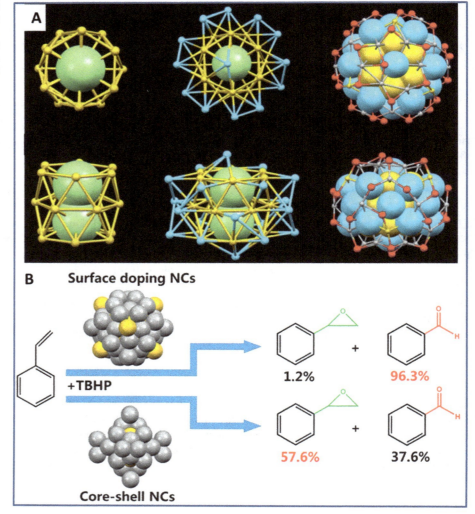

FIG. 8.34 (A) Core-shell structure of $[Ag_{46}Au_{24}(SR)_{32}]^{2+}$. Color label: *Light green/blue/gray* for silver; *yellow* for gold; *red* for sulfur. (B) Both surface-doped NC and core-shell structured NC as catalysts in the styrene oxidization with various selectivities.

8.4.3.4 Sulfide selective oxidation

Sulfoxide is a valuable and efficient reagent. At present, the most direct and effective method is sulfide oxidation by using H_2O_2, O_2 and odosylbenzene (PhIO) as oxidants [323]. Jin et al. investigated the catalytic capability of $Au_{25}(SCH_2CH_2Ph)_{18}$ for the selective sulfide oxidation by using PhIO as the oxidant for the first time [324]. The $Au_{25}(SR)_{18}/TiO_2$ catalysts gives an increase to excellent conversion (97%) with 92% selectivity for sulfoxide. The supports strongly influence the catalytic activity of $Au_{25}(SR)_{18}$ NCs, and the sequence of the performance of $Au_{25}(SR)_{18}$/support is as follows: $TiO_2 > Fe_2O_3 > CeO_2 > MgO$. A plausible mechanism of the reaction of sulfide oxidation by PhIO oxidant is like this: firstly, PhIO is proposed to be adsorbed onto the pocket site of the $Au_{25}(SR)_{18}$ cluster [325], and then the sulfide should coordinate with the Au atom of $Au_{25}(SR)_{18}$ cluster, finally, the activated oxidant transfers an oxygen atom to the sulfide, giving an increase to the sulfoxide product.

By using the best support, Li et al. further exploited the sulfide oxidation by different sizes Au nanoclusters. The $Au_{144}(SCH_2Ph)_{60}/TiO_2$ catalyst exhibits the highest catalytic performance (92% conversion of methyl phenyl sulfide with 99% selectivity for sulfoxide). Interestingly, the unusual size dependence of the gold nanocluster catalyst is observed: $Au_{144} > Au_{102} > Au_{99} > Au_{38} > Au_{25}$ [326,327].

8.4.3.5 Hydrocarbons oxidation

Hydrocarbon oxidation has been extensively studied as a prototype reaction because it is an important industrial process for the removal of volatile organic compounds. Tsukuda et al. prepared a series of Au_n/HAP-300 ($n=10, 18, 25, 39$) catalysts to reveal the size effect on their catalytic performance of cyclohexane oxidation [328]. They also synthesized a larger Au clusters (1.4 nm) by a conventional adsorption method, which is referred to as Au_{85} according to the average diameter of the Au clusters. The TOF increased with the size of Au nanoclusters, reaching the maximum values (18,500 h^{-1} Au atom^{-1}) at $n=39$. Since then, the values have started to drop with a further increase in size.

Barrabes et al. investigate the MSI effect of $Au_{38}(SC_2H_4Ph)_{24}$ nanocluster with two diverse supports CeO_2 and Al_2O_3. They found that Au_{38}/CeO_2 showed higher activity and selectivity than Au_{38}/Al_2O_3 after 300°C pretreatment, which may be caused by the state of Au (cationic and metallic) on the surface of Au_{38}/CeO_2 and Au_{38}/Al_2O_3, respectively. It is unexpected that cyclohexanethiol is formed especially in Au_{38}/Al_2O_3 system [329].

Wang and Zheng et al. reported that during the hydrolytic oxidation of triethylsilane, $Au_{34}Ag_{28}$ NC with acetylene ligand fully covered showed better activity than those with partially or completely removed surface ligands (Fig. 8.35). The untreated $Au_{34}Ag_{28}$/XC-72 catalyst shows high conversion (40% in 4 min and 100% in 12 min) for the hydrolytic oxidation of organosilanes to silanols, while 200°C-calcinated $Au_{34}Ag_{28}$/XC-72 catalyst gave conversion below 3% after 30 min. It is, for the first time, testify the promoting effect of surface ligands on catalysis [330].

Zhu et al. proposed a shutting an atom into and out of Au NCs strategy to achieve reversible switching of catalytic activity. Au_{24} is considered as Au_{25} NCs with one central Au atom loss. Compared with Au_{25}, Au_{24} showed better activity in the methane oxidation toward methanol. Besides, the activation and deactivation can be controlled and reversibly switched by one central atom removal and addition, which effectively avoids the irreversibility of catalytic capability and improves the durability [331].

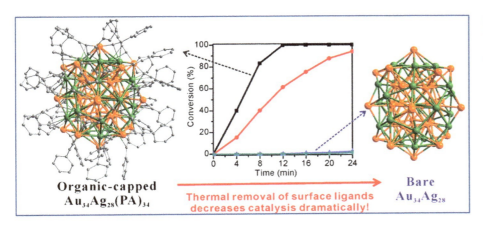

FIG. 8.35 Comparison of catalytic performances of $Au_{34}Ag_{28}$/XC-72 before and after thermal treatment for the hydrolytic oxidation of triethylsilane.

Yamamoto et al. synthesized a series of finely controlled NC catalysts (alloy NC around 1 nm in diameter) by using a macromolecular template with coordination sites on a gradient of alkalinity. The catalytic performance of $Cu_{32}Pt_{16}Au_{12}$@TPM-DPA G4/GMC (TPM-DPA G4 = fourth-generation polyphenylazomethine dendrimer with a tetraphenylmethane core, GMC = graphitized mesoporous carbon) is 24 times greater than that of a commercially available Pt catalyst for aerobic oxidation of hydrocarbons. The high activity of $Cu_{32}Pt_{16}Au_{12}$ was attributed to synergistic effects on the interface between Cu(0)/Cu(I) and other precious metals in the conversion from the peroxide intermediate to ketone [332]. In addition, Vajda and Anderson et al. used size-selected Pt clusters made in ultrahigh vacuum to investigate the size effect in propane and ethylene dehydrogenation reaction [333,334].

8.4.3.6 D-glucose oxidation

Li et al. studied the catalytic properties of a series of $Au_n(PET)_m$ with different size for D-glucose oxidation [335]. The catalytic activity of the gold NCs was significantly dependent on the sizes, following the order $Au_{144}(PET)_{60}$/AC > $Au_{38}(PET)_{24}$/AC > $Au_{25}(PET)_{18}$/AC. The unusual size dependence of the gold NC catalyst was also found in selective sulfide oxidation reactions [326]. Noteworthy, oxide supports (e.g., TiO_2 and ZrO_2) were also applied and the corresponding $Au_n(PET)_m$ nanocluster/oxides were almost inactivity (<5% conversion) under the identical conditions in the D-glucose oxidation reactions.

8.4.4 Catalytic selective reduction

Selective reduction reactions play an important role in industrial manufacturing, including carbon-carbon multiple-bond reduction, carboxide reduction, nitro derivative reduction and single-electron transfer reduction. Since 2010, Zhu and Jin have developed $Au_{25}(SR)_{18}$-catalyzed selective reductions of cyclic ketones and α,β-unsaturated ketones [325,336]. The high selectivity of $Au_{25}(SR)_{18}$ facilitated the investigation of other NCs as model catalysts for reduction reactions. Zhu's group has summarized the reduction reaction catalyzed by quasi-homogeneous Au NCs in the liquid phase [337]. Here, we will review some advances on supported and unsupported NCs catalysts in selective reduction.

8.4.4.1 Carbon-carbon multiple-bond reduction

Jin et al. investigated the catalytic performance of spherical Au_{25} and rod-shaped Au_{25} nanoclusters for the semihydrogenation of alkynes (Fig. 8.36) [338]. They found that above two ligand-on catalysts show excellent activities for semihydrogenation of terminal alkynes. Of note, the internal alkynes cannot be catalyzed by the ligand-on nanoclusters, whereas the ligand-off nanoclusters are found to be capable of catalyzing the semihydrogenation of internal alkynes. Therefore, a new deprotonation activation of terminal alkynes was proposed, which is verified by FT-IR spectroscopy. This pathway was in contrast with the adsorption of alkynes (Ru-vinylidene interfacial bond) on the conventional Ru colloid [339].

Based on the special activity characteristics of thiol-protected Au_{25} nanoclutens for alkynes semihydrogenation, Wang et al. synthesized two kinds of isostructural Au_{38} nanoclusters, $[Au_{38}(L)_{20}(Ph_3P)_4]^{2+}$ (L = alkynyl or thiolate) for the

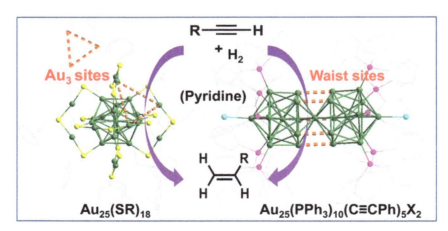

FIG. 8.36 Proposed schematic diagram for semihydrogenation of terminal alkynes catalyzed by $Au_{25}(SR)_{18}$ and $Au_{25}(PPh_3)_{10}(CCPh)_5X_2$ (X = Cl, Br) under H_2 (20 bar).

determination of the ligand effect on catalytic performance. Of note, the alkynyl-protected Au_{38} are very active (>97%) in the semihydrogenation of alkynes (including terminal and internal ones) to alkenes, whereas the thiol-protected Au_{38} showed a very low conversion (<2%). This fact suggests that the protecting ligands play an important role in semihydrogenation of alkynes [340].

In addition, Wu et al. synthesized the alloyed $Au_{20}Cd_4$-$(SH)(SR)_{19}$ nanocluster by way of antigalvanic reaction and explored its catalytic performance for the semihydrogenation of acetylene in an ethylene-rich stream [341]. The doping of Cd in Au_{23} can improve the selectivity for acetylene, although it decreases the activity. Tsukuda group successfully synthesized alloy Au_n: PVP NC with a single Rh or Pd atom doping [309,342]. Interestingly, the way of doping Au_{34} NC with Rh and Pd into was different. The Rh atom was added to Au_{34} yielding the $Au_{34}Rh_1$ NC, whereas the Pd atom replaced an Au atom of Au_{34} yielding the $Au_{33}Pd_1$ NC. Besides, $Au_{34}Rh_1$ exhibited much higher catalytic activity than $Au_{33}Pd_1$ for olefin hydrogenation.

Crampton et al. utilized a series of size-selected Pt_n ($n=7$–40) clusters prepared by physical method to investigate whether the ethylene hydrogenation reaction is structure sensitivity or not. By temperature programmed reaction (TPR) experiments, they found the reaction is structure sensitive when the cluster size is below 1 nm region (the most active one is Pt_{13}), while for those NPs with larger sizes, the reaction is structure insensitive. By combining with the IR absorption spectroscopy study, the carbon species formation is considered as the critical reason why the reaction is insensitive for the catalysts in the large size regime [343,344]. Besides, they studied the support effect by using three kinds of SiO_2 support (silicon-rich film, oxygen-rich film, and stoichiometric film) to deposit the Pt_{13} cluster to check the reactivity of Pt cluster [345].

Yamamoto et al. prepared Pt_{12}@TPM G4/GMC by dendrimer template method. The Pt_{12} showed higher activity (TOF = 1350 atom $(Pt)^{-1}$ h^{-1}) in olefin hydrogenation reaction than that of Pt NPs [346]. The Pt_{12} NC also exhibited higher activity and better poison tolerance to amines in the aldehyde amination reduction reaction in comparison with Pt NPs [347]. Thanks to dendrimer template method, other precise metal clusters such as Pd [348], Rh [349], Cu NCs [350] were successfully synthesized to be used as highly effective catalysts in olefin and nitroaren hydrogenations.

Corma et al. utilized the tunable penetrability of the well-defined pore structures to easily control the particular reactants in and out of the reactive center. They loaded Pt species on the MCM-22 zeolite (Pt@MCM-22 and Pt/MCM-22-imp are represented Pt NCs or Pt NPs in the internal space or on the external surface of MCM-22, respectively), and used the composites to catalyze olefin hydrogenation [351]. Because of the difference in the diffuse ability of olefins with different length carbon chain, propene can get into the internal space of MCM-22 and be catalyzed by Pt NCs, while isobutene only be catalyzed by Pt NPs at the outer surface of MCM-22. The comparison of catalytic propene and isobutene hydrogenation activity by two kinds of Pt species/MCM-22 composites has suggested Pt NCs showed higher intrinsic activity than Pt NPs, and Pt NCs in the MCM-22 can be used as size-selective catalysts. Besides combination with support materials, introducing foreign metal element is another effective way to improve the catalytic performance. Thomas et al. have prepared a series of bimetallic and trimetallic clusters to achieve highly efficiency in catalytic a lot of single-step hydrogenation reactions [352,353].

8.4.4.2 Carboxide reduction

Aldehyde or ketones reduction

Jin et al. first reported the stereoselective hydrogenation (almost completely 100%) of cyclic ketone to one specific isomer of cyclic alcohol by using $Au_{25}(SR)_{18}$ nanocluster [325]. The catalytic performance is related to atomic space structure and the core-shell properties of the $Au_{25}(SR)_{18}$. In $Au_{25}(SR)_{18}$ NCs, the electron enrichment Au_{13} core play the role of C=O bond activation, and the electron-deficient Au_{12} shell is responsible for providing active sites for H_2 adsorption and dissociation. Furthermore, they found that $Au_{25}(SR)_{18}/CeO_2$ catalyst shows good catalytic activity and selectivity in the aldehyde hydrogenation reactions under the condition of adding Lewis acid and base [354]. They proposed that the removal of an Au-SR fragment or the formation of an adduct exposes open metal atoms to the reactants. DFT calculations prove that the surface gold atoms on the Au_{13}-core or staple shell or the Lewis acids bound on the Au_{13}-core are the catalytic active sites for hydrogenation. The $Au_{24}(SR)_{17}$ species generated from removal of a "Au-SR" fragment in the presence of Lewis acids and ammonia is considered as the real catalytic activity intermediates for the aldehyde reduction based on the mass spectrometry results.

α, β-unsaturated ketones selective reduction

Jin et al. first reported that $Au_{25}(SR)_{18}$ NCs show 100% selectivity for the chemoselective hydrogenation of α,β-unsaturated ketones and aldehydes to unsaturated alcohols [336]. Based on the electronic structure of the $Au_{25}(SR)_{18}$ nanocluster and previous DFT calculations, they proposed a mechanism, that is the eight Au_3 faces of $Au_{25}(SR)_{18}$ adsorb and activate the C=O group, H_2 is dissociated on the Au atoms of the exterior shell and form two nearly symmetrical Au-H-Au bridges. The special structure of the $Au_{25}(SR)_{18}$ and its unique electronic characteristics are main causes for the observed extraordinary selectivity. Jin et al. also demonstrated that the NC $Pd_1Au_{24}(SR)_{18}$ showed similar selectivity for the hydrogenation of α,β-unsaturated ketone to unsaturated alcohols [355]. These results supported the fact that $Pd_1Au_{24}(SR)_{18}$ and $Au_{25}(SR)_{18}$ both have core-shell structures.

8.4.4.3 Reduction of nitroarene derivatives

Nitrophenol reduction

Scott et al. first probes the reactivity of the monolayer alkanethiolate-protected Au clusters for the reduction of 4-nitrophenol (4-NP) to 4-aminophenol (4-AP) in the presence of $NaBH_4$ [356]. Due to lack of available Au surface sites, all the monolayer alkanethiolate-protected Au NCs (i.e., $C_{18}SH$, $C_{12}SH$, C_6SH and so on) had lower catalytic activities than PVP stabilized Au NPs despite their smaller particle sizes. In contrast, the $Au_{25}(SR)_{18}$ (SR: several alkanethiols and phenylethanethiol) showed higher catalytic activity, that is because the active sites of the $Au_{25}(SR)_{18}$ is retaining open for 4-NP reduction reaction, which is attributed to the specific core-shell structure leading to leaves eight facets unblocked by any thiolates [357,358]. The hydrogenation of 4-NP by $NaBH_4$ to 4-AP was used as a model reaction to value the exposure of metal sites in the atomic precise nanoclusters, such as identification the accessibility of active sites in three $Au_{13}Cu_x$ ($x=2$, 4, 8) bimetallic nanoclusters. In spite of the presence of twelve 2-pyridine thioctic acid ligands in $Au_{13}Cu_8$, it is still leaved open space for small molecules to access its Au sites; therefore, $Au_{13}Cu_8$ can completely catalyze the 4-NP reduction in 10 min, while $Au_{13}Cu_2$ or $Au_{13}Cu_4$ clusters cannot catalyze the reaction [306].

From this point on, a great deal of work concentrates on the catalytic activity of atomic precise NCs for nitrophenol reduction catalysis [359–367]. Among them, Farrag et al. utilized microwave treatment to remove the ligands from the monolayer protected NCs without any agglomeration [365]. The ligand-off Au_{25}/Al_2O_3 (or TiO_2) catalysts showed higher catalytic activity than ligand-on catalysts in reduction of 4-NP. Wu et al. synthesized a new trimetallic NC, $Au_{16.8}Ag_{7.2}Hg_1(PET)_{18}$, by the "bi-antigalvanic reduction" method [362]. $Au_{24-x}Ag_xHg_1$ exhibited a remarkably higher catalytic activity than the Hg (or Ag)-doped Au_{25} NCs in the reduction of 4-NP. Jin et al. investigated the catalytic activities of three nearly identical sizes nanoclusters ($[Au_{23}(S\text{-}c\text{-}C_6H_{11})_{16}]^-$, $Au_{24}(SCH_2pHtBu)_{20}$ and $[Au_{25}(SCH_2CH_2pH)_{18}]^-$) for reduction of 4-nitrobenzene. The $[Au_{23}(S\text{-}c\text{-}C_6H_{11})_{16}]^-$ nanocluster shows the highest activity, suggesting the atomic packing mode and electronic structure play a key role in determining the catalytic performance [367].

Ultrasmall nanoclusters supported on oxides usually undergo sintering at high temperature, which makes them unsuitable catalysts for high-temperature reactions. Wang and Scott et al. prepared two novel silica-enclosed $Au_{25}[SC_3H_6Si(OCH_3)_3]_{18}$ and $Au_{25}(11\text{-}MUA)_{18}$, respectively ($Au_{25}@SiO_2$ as short for the two composites). The $Au_{25}@SiO_2$ composites are stable even at 400°C and they have excellent activity and recyclability for reduction of 4-NP [360,366]. Gao et al. fabricated a hyperstar polymer coating $Au_{25}(SR)_{18}$ nanocomposites by ligand exchange between thiolates of $Au_{25}(SR)_{18}$ NCs and disulfide groups of hyperstar polymers, which also efficient catalyze the reduction of 4-nitrophenol with great stability (no size change even after 3 months) [364].

Xie et al. successively reported two articles about the ligand effects in the NC catalysts $Au_{25}(SR)_{18}$ of 4-NP reduction [361,363]. The main findings were as follows: (i) the activity of the shorter-chain ligand-protected Au_{25} NCs was higher than those bearing a longer chain; (ii) the aromatic ligand was particularly effective in increasing the accessibility of Au_{25} NCs; (iii) the ligands with amine groups had detrimental impacts on the catalytic activity of Au_{25} NCs; and (iv) the ligand was also used to modulate the catalytically active sites of Au_{25} NCs, direct the catalytic reaction pathway, and tailor the selectivity induced by ligand-protected Au_{25} NCs.

Shi et al. developed two kinds of NCs/ZIF-8 composites to reduce NC aggregation with the help of the confinement effect of MOF [368]. One type is $Au_{25}(SG)_{18}$@ZIF-8, for which $Au_{25}(SG)_{18}$ was uniformly encapsulated into a ZIF-8 by coordination-assisted self-assembly. The other kind is $Au_{25}(SG)_{18}$/ZIF-8, in which $Au_{25}(SG)_{18}$ was impregnated onto the outer surface of ZIF-8. The catalytic activity was related to the distribution of $Au_{25}(SG)_{18}$ inside or outside the ZIF-8 matrix. $Au_{25}(SG)_{18}$@ZIF-8 took 12 h to complete the 4-NP reduction reaction, indicating it was a diffusion-controlled process for NCs encapsulated inside the MOF matrix. On the contrary, $Au_{25}(SG)_{18}$ anchored on the outer surface of ZIF-8 finished 4-NP reduction within 10 min.

Selective reduction of nitroarene derivative

Jin et al. explored the catalytic performance of $Au_{99}(SR)_{42}$ NC for chemoselective reduction for nitrobenzaldehyde derivatives [369]. Surprisingly, the thermally robust $Au_{99}(SR)_{42}$ NC was completely selective for the hydrogenation of aldehyde groups, which was different from the reduction of nitro groups by traditional gold nanocatcalysts. Due to the acido-basic properties of the oxide supports, the catalytic activities of Au$_{99}$(SR)$_{42}$/oxides were much higher than that of pure Au$_{99}$(SR)$_{42}$. They also investigated the catalytic properties of water-soluble Au$_{15}$(SG)$_{13}$, Au$_{18}$(SG)$_{14}$, Au$_{25}$(SG)$_{18}$, Au$_{38}$(SG)$_{24}$, and Au$_{25}$(Capt)$_{18}$ for chemoselective hydrogenation of 4-nitrobenzaldehyde [370]. The catalytic performances were enhanced with an increasing core size. In addition, less bulky ligand protected nanoclusters gives high activity: Au$_{25}$(Capt)$_{18}$ > Au$_{25}$(SG)$_{18}$.

Li et al. synthesized a basic PPh$_2$Py ligand-protected Au$_{11}$(PPh$_2$Py)$_7$Br$_3$ nanocluster, which efficiently catalyzed chemoselective hydrogenation of nitrobenzaldehyde without need of amine additive [371]. However, the corresponding Au$_{11}$(PPh$_3$)$_7$Cl$_3$ NC had no catalytic activity under the same conditions. Based on the experimental data and DFT simulations, they proposed the functional ligand (PPh$_2$Py) facilitation catalytic mechanism.

Zhang et al. achieved the chemoselective hydrogenation of 3-nitrostyrene to 3-vinylaniline by Au$_{25}$ nanoclusters supported on ZnAl-hydrotalcite [372]. The Au$_{25}$/ZnAl-HT-300 (300°C heat treatment) catalyst showed excellent selectivity (>99%) to 3-vinylaniline, and complete conversion of 3-nitrostyrene over broad reaction duration and temperature windows. The high chemoselectivity of hydrogenation of the nitro group over the catalyst Au$_{25}$/ZnAl-HT-300 was due to the fact that only the nitro group reduction was activated while the vinyl group was not. The inertia of vinyl over the Au catalyst was further proved by attenuated total reflection infrared spectra (ATR-IR) spectroscopy analysis.

8.4.4.4 Single-electron transfer reaction

The simplest chemical process, single electron transfer (SET) between an electron donor and an acceptor, profoundly affects catalytic reactions [373]. Murray's group conducted electrochemical and optical spectroscopic analysis of three stable Au$_{25}$ NCs under the three charge states −1, 0, and +1 (although this NC was considered as Au$_{38}$ at that time) [374–377]. They obtained the rate constant and activation energy barrier of the Au$_{25}^{0/-1}$ electron self-exchange process by NMR analysis, and found an important inner-sphere reorganization energy confirmed by Raman stretch shift of the Au—S bonds [378]. Then, they further investigated the temperature and size dependent electron transfer (ET) dynamics of the NC in solid-state cases [379].

Maran's group used Au$_{25}^{-1}$ as electron donor and polyfluoroperoxide as oxidant to synthesize Au$_{25}^{+1}$ by two dissociative ET steps. The charge dependent NMR spectra of Au$_{25}$ were fully monitored [380]. They further used Au$_{25}$ as an ET catalyst to take part in a series of peroxide reduction reaction, and the corresponding ET rate was measured by CV [381]. Besides the −1, 0, and +1, the three common valences of Au$_{25}$, they characterized three other charge states of Au$_{25}$, −2, +2, and +3 that were produced irreversibly. The ET rate constants of each of these electrode processes were determined by cyclic voltammetry [382]. The effect of the substituent length on the lifetime of Au$_{25}$ in the different charged states was also investigated by electrochemistry [383].

In addition to the electrochemistry method of interconverting the charge states of Au$_{25}$, the use of redox reagents also leads to stable Au$_{25}$ in various charge states. Tsukuda group obtained Au$_{25}$ with 0 and +1 charge using the strong oxidant Ce(SO$_4$)$_2$ upon reaction with Au$_{25}^{-1}$ and used ESI-MS to characterize the charge of the product [384]. Zhu et al. utilized trace oxygen to convert Au$_{25}^{-1}$ to Au$_{25}^{0}$ and obtained the crystal structure and optical spectrum of Au$_{25}^{0}$ [385]. They also used H$_2$O$_2$ and NaBH$_4$ to achieve the reversible conversion between Au$_{25}^{0}$ and Au$_{25}^{-1}$ and monitored the process by EPR [386]. They further investigated the reaction between Au$_{25}^{0}$ and two different salts (tetraoctylammonium halide (TOAX) and NaX, X being a halogen element), both halide salts converted Au$_{25}^{0}$ to Au$_{25}^{-1}$, but only TOA$^+$ relatively stabilized the cluster, while Na$^+$ did not provide protection [387]. Lee's group found that Au$_{25}$ gave its electron to fluorescence chromophores by interfacial ET process [388].

Zhu's group utilized two organic single electron oxidants, 2,2,6,6-tetramethylpiperidin-1-oxoammonium cation (TEMPO$^+$) and the donor-acceptor complex phenothiazine-tetrachloro-p-benzoquinone (PTZ-TCBQ) to investigate the single ET reaction between them and Au$_{25}^{-1}$ [389,390]. First, it was estimated whether the ET process occur by using CV to determine the redox potential of the single electron oxidant. For TEMPO, the half-wave potential $E_{1/2}$ (TEMPO$^{+1/0}$) was equal to 0.650 V, which is higher than that of $E_{1/2}$ (Au$_{25}^{+1/0}$) and $E_{1/2}$ (Au$_{25}^{0/-1}$). The potential difference between the redox peaks of the two Au$_{25}$ redox couples was 73 and 61 mV, respectively, which was similar to the value of the peak potential difference (59 mV) for a single ET process. Therefore, it was predicted that Au$_{25}^{-1}$ could transfer one or two electrons to TEMPO$^+$. For PTZ-TCBQ, the situation was similar [390]. The NMR and UV-vis. Absorption spectra were used to monitor in real time the effect of the added amount of oxidant on the Au$_{25}$ charge state and its chemical environment. EPR analysis was performed to confirm the single ET process. The radical concentration linearly increased with the addition of oxidant, until 2 equiv. oxidant, and then the EPR signal no longer increased [389]. This finding showed that Au$_{25}$ generated a single radical that was an active intermediate for many organic reactions [389,390].

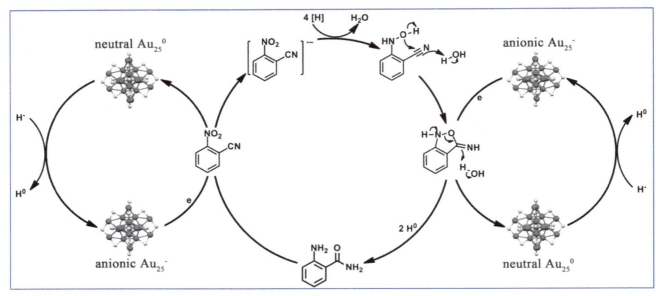

FIG. 8.37 Single-electron transfer mechanism of an intramolecular cascade reaction catalyzed by the Au_{25} NC.

Based on the characteristic of the NC Au_{25} mentioned above, Zhu's group first developed the ET catalysis of Au_{25} NCs in intramolecular cascade reactions of 2-nitrobenzonitrile (Fig. 8.37). This reduction reaction using excess $NaBH_4$ and catalyzed by Au_{25} led to 2-amniobenzamide with high conversion and selectivity. A detailed mechanism involving ET was proposed and confirmed by UV, cyclic voltammograms and EPR. First, 2-nitrobenzonitrile accepted an electron from Au_{25}^- with the formation of N-centered free radicals. Then, a five-member ring intermediate was formed by intramolecular cascade reaction and removal of an electron from Au_{25}^-. Finally, 2-amniobenzamide was obtained together with H_2O. Overall, Au_{25} served as the ET mediator to transfer one or two electrons from $NaBH_4$ to the organic substrate. This study provided ideas for the application of NCs as ET catalysts in a wide range of organic reactions [391].

8.4.5 Catalysis of coupling reactions

8.4.5.1 Ullmann coupling reaction

Jin et al. reported the $Au_{25}(SR)_{18}/CeO_2$ nanocluster catalyst show high activity in the Ullmann homocoupling of aryl iodides at 130°C for 2 days [392]. It is worth to note that these supports (CeO_2, TiO_2, Al_2O_3 and SiO_2) were all capable of promoting gold catalyst's activity in the coupling reaction compared with the pure $Au_{25}(SR)_{18}$, and the type of support has no significant difference on the catalytic activity.

Li and Jin et al. synthesized an aromatic-thiolate-protected $TOA^+[Au_{25}(S-Nap)_{18}]^-$ NC and investigated its catalytic properties in Ullmann heterocoupling reaction in order to understand the distinct effects of the aromatic ligands [393]. Compared with the $Au_{25}(SCH_2CH_2Ph)_{18}$, the obtained $Au_{25}(SNap)_{18}$ nanoclusters exhibit significantly improved catalytic activity (91% conversion) and selectivity (82% for the heterocoupling product) in the Ullmann heterocoupling reactions, indicating that the chemical nature of the protecting ligands exerts a major influence on the catalytic properties of the nanoclusters. The catalytic activity of $Au_{25}(SNap)_{18}/CeO_2$ was proved to be arisen from the nanoclusters by the blank experiment with only CeO_2 and control experiment with Au(I)-SNap polymer on CeO_2. But its recycle property is not good enough, the conversion and selectivity of nitroiodobenzene dropped to 79% and 47% (second cycle), 50% and 15% (third cycle), respectively.

8.4.5.2 Suzuki cross-coupling reaction

Jin et al. explored firstly the facilitating effect of ionic liquids (ILs) in $Au_{25}(SR)_{18}$ nanocluster catalyzed the Suzuki cross-coupling reaction at the atomic level [394]. The addition of imidazolium-based ILs could significantly improve product conversion. The cation of BMIM·X acted as an excellent catalytic promoter, whereas the anions of BMIM·X only have a minor effect. The process of imidazolium cation-induced removal of thiolate ligands (-SR) and "-SR-Au-" units from the protected Au_{25} nanocluster is studied by UV-vis spectroscopy and MALDI-TOF-MS. They proposed that the reactant p-iodoanisole is adsorbed and activated by the generated low-coordinated gold atoms [$Au_{25-n}(SR)_{18-n}$] for the subsequent Suzuki cross coupling. It is believed that the imidazolium cations of the ionic liquids engender strong interactions with the surface thiolate ligands on the cluster, which is the key to the promotional role of the imidazolium cations.

Corma et al. found that the three or four Pd atom clusters formed from various palladium source (including salts, complexes or NPs) were considered to be active catalysts in cross-coupling reactions (Heck, Sonogashira, Stille, and Suzuki) [395]. Zhu et al. designed and synthesized the trinuclear [$Pd_3Cl(PPh_2)_2(PPh_3)_3$]$^+$[SbF_6]$^-$ NC (Pd_3Cl for short) as an intermediate case between "classic" Pd(0)Ln complex and Pd NP for efficient catalysis and mechanistic investigations (Fig. 8.38) [396]. Pd_3Cl NCs show surprisingly activity in the Suzuki-Miyaura C-C cross coupling of a variety of aryl bromides and arylboronic acids with high yield under ambient aerobic conditions. The mechanism of Pd_3Cl in this reaction is different from traditional Pd catalyst. By monitoring the reaction process with ESI-MS and the extended EXAFS of Pd K-edge, they proposed the mechanism is that Pd_3Cl first reacts with phenylboronic acid to generate an intermediate [$Pd_3Ar(PPh_2)_2(PPh_3)_3$]$^+$ (Pd_3Ar for short) and the real catalyst was the Pd_3Ar intermediate, while Pd_3Cl only act as a precatalyst.

8.4.5.3 C-N cycloisomerization coupling reactions

Liu et al. synthesized [$Pd_3(C_7H_7)_2(MeCN)_3$]$^{2+}$ [BF_4^-]$_2$ ([$Pd_3(C_7H_7)_2$]$_2^+$ for short,) and used it as the active catalytic species for the cycloisomerization reaction of 2-phenylethynylaniline [397]. The catalyst exhibited high activity (>99% yield), whereas Pd(0) complexes, Pd(II) salts and the salt (C_7H_7)BF_4 were inactive under the same conditions. In addition, the catalyst is unusually stable and maintained its structural integrity during the catalytic process over many catalytic cycles, indicating that the [$Pd_3(C_7H_7)_2$]$_2^+$ is the catalytically active species. The Maestri and Malacria group reported the synthesis of other trinuclear Pd clusters, [((SAr')(PAr_3)$Pd)_3$]$^+$ (Pd_3 for short), involving activation of the C—S bond of isothioureas [398]. This group developed a new assembly method to synthesize Pd_3, Pt_3 and mixed Pd/Pt analogues [399]. The Pd_3 clusters showed high activity and selectivity for the semireduction of internal alkynes to (Z)-alkenes [400,401]. These results indicated that the Pd_3 catalyst operated through mechanisms that differed from those operating by traditional complexes, highlighting the great potential of "all-metal aromatics" in catalysis. In addition, Zheng et al. reported N-heterocyclic carbine stabilized Au_{25} NCs showed higher thermal and air stabilities than that with phosphines or thiols ligands for cycloisomerization of alkynyl amines to indoles [402].

8.4.5.4 Sonogashira cross-coupling reaction

Jin et al. loaded $Au_{25}(CH_2CH_2Ph)_{18}$ nanocluster on oxides to be served as effective catalysts for the Sonogashira cross-coupling reaction with excellent conversion (96.1%) and high selectivity for 1-methoxy-4-(2-phenylethynyl)benzene (88.1%) [403]. The two open facets of $Au_{25}(CH_2CH_2Ph)_{18}$ nanocluster are quite attractive to both reactants and likely to be the active site for the coupling reaction. The Ceria-supported Au_{25} catalyst presented a significant loss in selectivity (from 88.1% to 64.5% after 5 cycles). The reduction of selectivity is due to the gradual degradation of the nanoclusters (ligand-protected removing) in many cyclic experiments, as the larger Au particles have lower selectivity.

Li et al. used bimetallic $M_xAu_{25-x}(CH_2CH_2Ph)_{18}$ nanoclusters (M = Pt, Cu, Ag) to catalyze the carbon-carbon coupling reaction between p-iodoanisole and phenylacetylene to investigate the doping metal atoms effect. The conversion of p-iodoanisole was greatly influenced by the electronic effects in the bimetallic 13-atom cored NCs, and that the selectivity mainly depended on the type of atoms on the M_xAu_{12-x} shell.

8.4.5.5 A3-coupling reaction

Jin et al. first investigated the active sites of $Au_{38}(SC_2H_4Ph)_{24}$ for the catalysis of the A3-coupling reaction to connect the structure of the catalyst and its catalytic performance [404]. They compared the catalytic activities of ligand-off Au_{38} nanocluster, an Au(I)-SC_2H_4Ph complex and another atomically well-defined $Au_{25}(SC_2H_4Ph)_{18}$ (all supported on CeO_2) for the A3-coupling reaction. The results indicated that the importance of the intact structure of $Au_{38}(SC_2H_4Ph)_{24}$ for the catalytic performance, because both the $Au^{\sigma+}$ sites on the NC surface and the Au^0 atoms in the core were key to enhance the catalytic performance. Similarly to $Au_{25}(SR)_{18}/CeO_2$ for Ullmann coupling [393], the drawback of the catalyst was that the conversion decreased from 98% to 68% after 3 cycles. The drop in conversion was caused by the gradual degradation of the NCs in the multiple recycling reactions.

Wu et al. synthesized cadmium-doped gold NC, $Au_{26}Cd_5$, by peeling and doping from $Au_{25}(SCH_2CH_2Ph)_{18}$ [405]. Due to the cooperation between the cadmium atoms and the neighbor gold atoms on the surface of the Au_{13} icosahedron, the $Au_{26}Cd_5$ NC showed high catalytic activity, good recyclability and substrate tolerance for the A3-coupling reaction, whereas $Au_{25}(SCH_2CH_2Ph)_{18}$ was inactive under the same conditions.

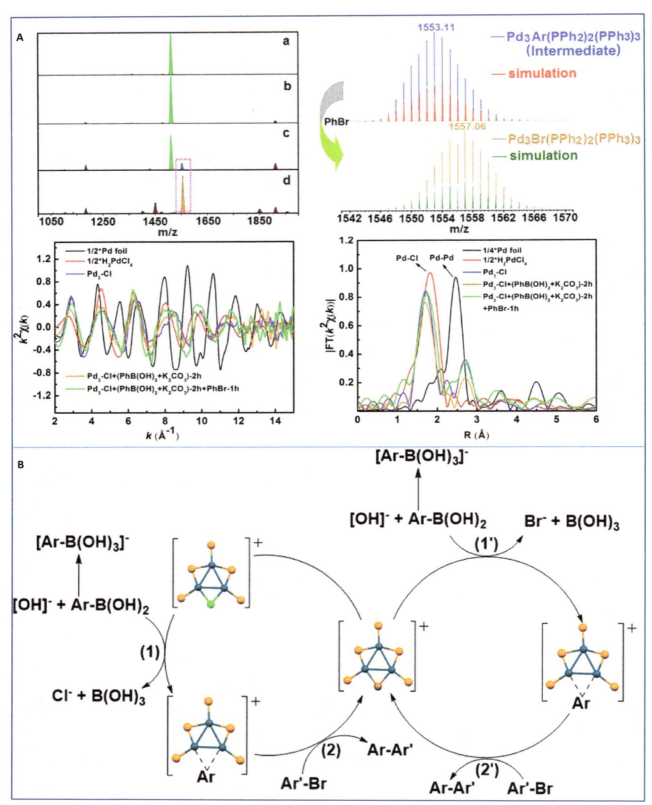

FIG. 8.38 (A) Experimental evidence for the odd mechanism: (a) ESI-MS tracking of the reaction. (b) Evolution of Pd$_3$Ar to Pd$_3$Br. (c) EXAFS $k^2\chi(k)$ oscillation functions, (d) their Fourier transform spectra for the reaction of Pd$_3$Cl with phenylboronic acid; (B) Odd mechanism of the Suzuki coupling reaction catalyzed by the Pd$_3$Cl cluster.

8.4.5.6 The carboxylation reaction of CO_2

Zhu's group investigated the influence of a foreign atom (Au, Pd and Pt) doping of Ag_{25} NCs on the catalytic performance for the conversion of phenylacetylene to phenylpropiolate [406]. The activity of $M_1@Ag_{24}$ is greatly influenced by the doping atom, following the order: $Au_1@Ag_{24} > Pd_1@Ag_{24} \approx Pt_1@Ag_{24} > Ag_1@Ag_{24}$. Furthermore, they achieved high efficient Chemical Fixation of CO_2 by precisely tuning surface Ag site in Au NCs. They compared the catalytic activity of three Au NCs ($Au_{19}Ag_4$(S-Adm)$_{15}$, $Au_{20}Ag_1$(S-Adm)$_{15}$ and Au_{21}(SAdm)$_{15}$) for the cycloaddition of CO_2 to epoxides. Due to the completely open surface Ag site facilitating the ring opening of epoxides and sequent CO_2 insertion, $Au_{19}Ag_4$ exhibited the best activity and stability. $Au_{20}Ag_1$ with partially exposed Ag site showed weak affinity for epoxides and poor efficiency of CO_2 capture, therefore, its catalytic performance is least efficient [407].

Zhu et al. have established an "electrostatic attraction" strategy for the synthesis of NCs@MOF composites (Fig. 8.39) [408]. They successfully synthesized eight NCs@MOF catalysts with reproducible high yields, which proved the universality of the method. The composite $Au_{12}Ag_{32}(SR)_{30}$@ZIF-8 presents excellent performance (TON = 18,164) in the carboxylation reaction of CO_2 with terminal alkyne under ambient CO_2 pressure. Due to the confined effect of MOF, the $Au_{12}Ag_{32}(SR)_{30}$@ZIF-8 can be reused five times without loss of catalytic activity. They further utilized coordination assisted self-assembly method to synthesize NCs-decorated MOF sandwich materials (i.e., ZIF-8@Au_{25}@ZIF-67 and ZIF-8@Au_{25}@ZIF-8) with well-defined structures and interface. The composites exhibited higher activity than Au_{25}/ZIF-8 and Au_{25}@ZIF-8 for both 4-nitrophenol reduction and terminal alkyne carboxylation with CO_2 [409].

8.4.6 Other catalytic reactions

Wu's group utilized $Au_{25}(SR)_{18}$ to catalyze the formation of α,β-unsaturated ketones or aldehydes from propargylic acetates with high yields [410]. Since the reaction conditions involved an alkaline atmosphere, the tandem hydrolysis-Meyer-Schuster rearrangement route was not reasonable. They speculated the reaction mechanism was that the $Au_{25}(SR)_{18}$ promotes the SN_2 reaction by attracting conjugate electrons of the C≡C bond and providing an appropriate location for the reaction.

Zhu et al. utilized Au_{25} and Au_{38} NCs to catalyze the dehydration of boronic acid to boroxine [411]. Au_{38} was found to catalyze the complete conversion of phenylboric acid into boroxine, whereas in comparison Au_{25} only accounted for 68% conversion. The 5-nm Au NPs did not achieve the dehydration. The high activities of Au_{38} and Au_{25} here were determined by the unique electronic properties of the NCs. The positive Au atoms in the shell were electrophilic, leading to the activation of the nucleophilic oxygen atoms in phenylborate. The electron rich Au core promoted the activation of H atoms.

Tsukuda et al. synthesized Ag_{44}/mesoporous carbon (MPC) as catalyst to catalyze dehydrogenation of NH_3BH_3. In this reaction, Ag_{44}/MPC exhibited high efficiency in the air or pure O_2 [412]. The dehydrogenation by Ag_{44} proceeded by a different mechanism from those previously reported with metal NPs.

Zhu et al. utilized bimetallic $Au_1Ag_{24}(SR)_{18}$, $Au_{25-x}Ag_x(SR)_{18}$ ($x = 2$–6), Ag_{25}(SPhMe$_2$)$_{18}$, and Au_{25}(SCH$_2$CH$_2$Ph)$_{18}$ NCs to investigate the synergistic effect in the alkynylation of CF_3-ketone (Fig. 8.40). The four NCs all have a similar sphere M25 structure with a 13-atom kernel ($M_1@M_{12}$) encapsulated in $M_{12}(SR)_{18}$ motifs. The Ag kernel shell provided the active site, and the Au center played the crucial role for stability. Therefore, the Au@Ag kernel was the essential structural feature for high activity and super stability. Based on this finding, they predicted that the bimetallic $Au_{12}Ag_{32}(SR)_{30}$

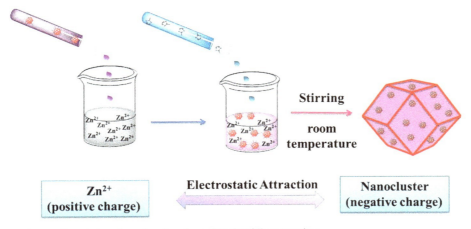

FIG. 8.39 An electrostatic attraction strategy for rational synthesis NCs@MOFs composites.

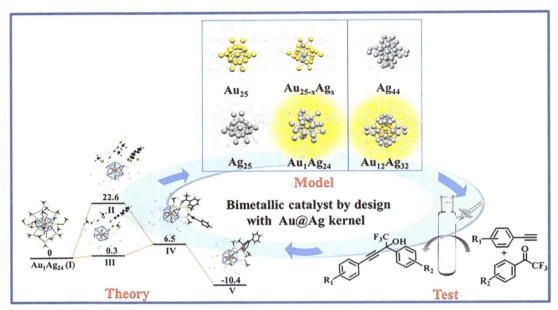

FIG. 8.40 Illustration for evaluating and predicting synergistic effect of bimetallic Au—Ag catalyst at atomic level.

NC may be the ideal catalyst for this reaction. $Au_{12}Ag_{32}(SR)_{30}$ NC exhibited the best performance (1790 h^{-1} TOF and at least 5 cycles without activity loss) [413].

Zhu et al. utilized Au_4Cu_4/CNT to achieve highly efficient CuAAC reaction of both terminal alkynes and internal alkynes. On the contrary, Cu_{11}/CNT can catalyze the AAA reaction of terminal alkyne on the basis of the deprotonation mechanism. Au_{11}/CNT and Cu_{11}/CNT catalysts exhibited no activity for the AAC reaction of internal alkynes. The catalytic mechanism of Au_4Cu_4 NCs is different from the classic alkyne deprotonation to σ,π-alkynyl intermediate mechanism (Fig. 8.41). The novel mechanism is that alkyne was activated by π-complexation with Au_4Cu_4, which is confirmed by the capture of three nanocluster—π-alkyne intermediates [Au_4Cu_4 (π-CH≡C-p-C_6H_4R)], R = H, Cl, and CH_3 [414]. They synthesized $Au_{24}Cu_6$(SPhtBu)$_{22}$ NCs via in situ two-phase ligand exchange method. Due to the exposure of the planar $Cu_3(SR)_3$ motif, $Au_{24}Cu_6$ NCs exhibited higher catalytic activity than homometallic Au_{25} NCs and $Au_{38-x}Cu_x$ alloy NCs for the epoxide ring opening reaction. The unique π conjugation in Cu(d)-S(p)π-bonding is the key feature to facilitating the reaction [415].

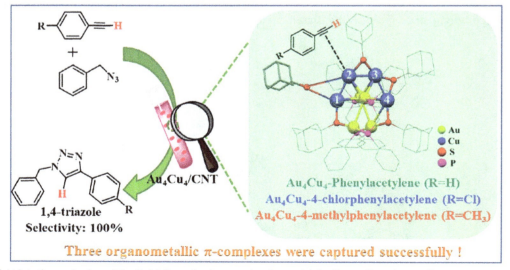

FIG. 8.41 Insight into the mechanism of the CuAAC reaction by capturing the crucial Au_4Cu_4-π-alkyne intermediate.

In order to investigate the influence of atomic structure on the distinct reaction channels of CO_2 hydrogenation, Zhu et al. utilized three nonmetallic Au NCs ($[Au_9(PPh_3)_8](NO_3)_3$, $[Au_{11}(PPh_3)_8Cl_2]Cl$, and $Au_{36}(TBBT)_{24}$) as catalysts for the hydrogenation reaction of CO_2. Using nonmetallic metal clusters can achieve controllable Conversion of CO_2. Different NCs selectively toward different target products: Au_9 for methane, Au_{11} for ethanol, Au_{36} for formic acid, this is totally different from Au NPs [416]. They further compared the catalytic performance of CO_2 hydrogenation of two NCs, Au_{24} and Au_{25} NCs. Au_{24} with internal vacancy displayed much more structural flexibility, can resist the aggregation and postpone the deactivation, therefore Au_{24} showed higher activity than Au_{25} without internal vacancy [417]. Besides, Au_{24} exhibits better activity than Au_{25} in the intramolecular hydroamination of alkynes [418]. They used two isomers of Au_{28} NCs ($Au_{28}(SPh-Bu)_{20}$ and $Au_{28}(SC_6H_{11})_{20}$) to catalyze the elective hydrogenation of CO_2. The two isomers have the same fcc Au_{20} kernel but different surface arrangements. In $Au_{28}(SPh-Bu)_{20}$, the remaining eight gold atoms and twelve ligands form two trimetric staples, while in $Au_{28}(SC_6H_{11})_{20}$, the eight gold and twelve ligands are arranged into two trimeric staples and two monomeric staples. Therefore, the two Au_{28} NCs showed different catalytic behaviors [419].

In addition, they found fully covered $Ni_6(SR)_{12}$ exhibited improved catalytic activity for nitriles hydrogenation toward primary amines with the help of NH_3. This method does not need to remove the ligand under thermal treatment. The ligand-shielding effect can easily suppress by upon the NH_3 [420]. They used FCC periodic series of the $Au_{8n+4}(SR)_{4n+8}$ ($n = 3, 4, 5$) NCs to investigate the evolvement in the catalytic activity accompanied by the periodic change in the inner sites. The series NCs have same surface motifs but their kernels are periodically changed. These NCs exhibited different catalytic activity in the hydration of phenylacetylene. The catalytic performance is as this trend: $Au_{28}(TBBT)_{20} > Au_{36}(TBBT)_{24} > Au_{44}(TBBT)_{28}$ [421].

8.5 Conclusions and outlooks

Thanks to various synthesis methods, lots of different kinds of metal NCs (i.e., monometallic NCs, alloy NCs, noble NCs, and nonnoble NCs) have been prepared with unique properties. Benefiting from the excellent optical properties, such as large stokes shift, long lifetime, and tunable emission colors, these metal NCs can be used to achieve sensitive sensing and imaging. But, the low QY and the weak affinity to a specific target is still challenging problem. Therefore, we need to synthesize metal NCs with high OY by introducing foreign metal atoms or changing the type of the ligand. In addition, although some protein or DNA-stabilized metal NCs are multifunctional with targeting, sensing, bioimaging and therapy functions integrated together. However, these NCs require high cost proteins or DNA chemicals. Besides, these NCs are unstable in extreme conditions; the performance of these NC is easily disturbed by external factors, such as solution pH, ionic strength, and temperature. Therefore, we have to optimize these factors when applying protein or DNA- stabilized NCs into sensing or imaging in real samples. Besides, most sensing strategies using metal NCs are based on fluorescent quenching. The quenching sensing mode has some drawbacks such as low sensitivity, poor selectivity, and strong background signals. Fluorescent enhancement sensing mode and FRET based sensing approach are more and more sensitive and selective. Therefore, we need to design metal NCs based sensing or imaging system to construct platform more suitable for clinical applications. We can combine other functional materials or modify the metal NCs to construct multimode sensing or imaging, such as MRI/PL coimaging as well. For therapy, we can also develop multimode therapy to improve the therapy effect by combining various features of metal NCs, such as its photodynamic, photothermal therapy ability, drug, gene carry ability, and radiation ability.

For catalysis application, metal NCs not only serve as efficient catalysts, but also act as model catalysts to figure out the catalytic mechanism and correlate the structure and catalytic performance based on the precise structure and composition. Although lots of meaningful work have been done, there are still many problems need to be solved. In the future, we need to investigate the composition effect by synthesizing multimetallic NCs. Ligands on the outside of the NCs, are the first to contact the reactants, so rational designing ligands is important to optimize catalytic activity and selectivity. Considering the structure of NCs, constructing inherent uncoordinated active sites in NCs is also an efficient strategy to prepare high active NCs catalyst. In addition, support can stabilize NCs and it is required in heterogeneous catalyst. Besides this, the synergetic effect between support and NCs can be rationally utilized to facilitate the catalytic reaction. Therefore, we should pay more attention to the functional supports. Furthermore, there are lots of chiral NCs have been synthesized recently, and chiral NCs catalysis is a region worthy to be explored. To build the relationship between the structure and activity of NCs, it is needed to monitor the whole catalysis process and capture the intermediates in the reaction by advanced techniques such as in situ PTIR, EXANES, EXAFS, MS, etc. Then, combined with more realistic theoretical simulations, the catalytic mechanism will be more solid. Moreover, for now, most NCs catalysts are on the basis of noble metal. To realize large-scale industrial applications in the future, non-noble metal NCs need to be developed.

References

[1] J.S. Lee, M.S. Han, C.A. Mirkin, Colorimetric detection of mercuric ion (Hg2+) in aqueous media using DNA-functionalized gold nanoparticles, Angew. Chem. Int. Ed. 46 (22) (2007) 4093–4096, https://doi.org/10.1002/anie.200700269.

[2] X. Xue, F. Wang, X. Liu, One-step, room temperature, colorimetric detection of mercury (Hg2+) using DNA/nanoparticle conjugates, J. Am. Chem. Soc. 130 (11) (2008) 3244–3245, https://doi.org/10.1021/ja076716c.

[3] G.K. Darbha, A.K. Singh, U.S. Rai, E. Yu, H. Yu, P.C. Ray, Selective detection of mercury (II) ion using nonlinear optical properties of gold nanoparticles, J. Am. Chem. Soc. 130 (25) (2008) 8038–8043, https://doi.org/10.1021/ja801412b.

[4] C.C. Huang, Z. Yang, K.H. Lee, H.T. Chang, Synthesis of highly fluorescent gold nanoparticles for sensing mercury(II), Angew. Chem. Int. Ed. 46 (36) (2007) 6824–6828, https://doi.org/10.1002/anie.200700803.

[5] J. Xie, Y. Zheng, J.Y. Ying, Protein-directed synthesis of highly fluorescent gold nanoclusters, J. Am. Chem. Soc. 131 (3) (2009) 888–889, https://doi.org/10.1021/ja806804u.

[6] J. Xie, Y. Zheng, J.Y. Ying, Highly selective and ultrasensitive detection ofHg2+ based on fluorescence quenching of au nanoclusters by Hg2+–Au+ interactions, Chem. Commun. 46 (6) (2010) 961–963, https://doi.org/10.1039/B920748A.

[7] D. Hu, Z. Sheng, P. Gong, P. Zhang, L. Cai, Highly selective fluorescent sensors for Hg2+ based on bovine serum albumin-capped gold nanoclusters, Analyst 135 (6) (2010) 1411–1416, https://doi.org/10.1039/c000589d.

[8] W. Jiang, B. Rao, Q. Li, et al., Fluorescence signal amplification of gold nanoclusters with silver ions, Anal. Methods 10 (43) (2018) 5181–5187, https://doi.org/10.1039/c8ay01955j.

[9] H. Kawasaki, K. Yoshimura, K. Hamaguchi, R. Arakawa, Trypsin-stabilized fluorescent gold nanocluster for sensitive and selective Hg2+ detection, Anal. Sci. 27 (6) (2011) 591, https://doi.org/10.2116/analsci.27.591.

[10] H. Wei, Z. Wang, L. Yang, S. Tian, C. Hou, Y. Lu, Lysozyme-stabilized gold fluorescent cluster: synthesis and application as Hg2+ sensor, Analyst 135 (6) (2010) 1406–1410, https://doi.org/10.1039/c0an00046a.

[11] L.Y. Chen, C.M. Ou, W.Y. Chen, C.C. Huang, H.T. Chang, Synthesis of photoluminescent Au ND-PNIPAM hybrid microgel for the detection of Hg2+, ACS Appl. Mater. Interfaces 5 (10) (2013) 4383–4388, https://doi.org/10.1021/am400628p.

[12] L. Shang, L. Yang, F. Stockmar, et al., Microwave-assisted rapid synthesis of luminescent gold nanoclusters for sensing Hg2+ in living cells using fluorescence imaging, Nanoscale 4 (14) (2012) 4155–4160, https://doi.org/10.1039/c2nr30219e.

[13] Y.H. Lin, W.L. Tseng, Ultrasensitive sensing of Hg2+ and CH3Hg+ based on the fluorescence quenching of lysozyme type VI-stabilized gold nanoclusters, Anal. Chem. 82 (22) (2010) 9194–9200, https://doi.org/10.1021/ac101427y.

[14] A. Senthamizhan, A. Celebioglu, T. Uyar, Flexible and highly stable electrospun nanofibrous membrane incorporating gold nanoclusters as an efficient probe for visual colorimetric detection of Hg(II), J. Mater. Chem. A 2 (32) (2014) 12717–12723, https://doi.org/10.1039/C4TA02295E.

[15] C. Guo, J. Irudayaraj, Fluorescent Ag clusters via a protein-directed approach as a Hg(II) ion sensor, Anal. Chem. 83 (8) (2011) 2883–2889, https://doi.org/10.1021/ac1032403.

[16] W. Guo, J. Yuan, E. Wang, Oligonucleotide-stabilized ag nanoclusters as novel fluorescence probes for the highly selective and sensitive detection of the Hg2+ ion, Chem. Commun. 23 (2009) 3395–3397, https://doi.org/10.1039/b821518a.

[17] B. Adhikari, A. Banerjee, Facile synthesis of water-soluble fluorescent silver nanoclusters and Hg II sensing, Chem. Mater. 22 (15) (2010) 4364–4371, https://doi.org/10.1021/cm1001253.

[18] I. Chakraborty, T. Udayabhaskararao, T. Pradeep, Luminescent sub-nanometer clusters for metal ion sensing: a new direction in nanosensors, J. Hazard. Mater. 211–212 (2012) 396–403, https://doi.org/10.1016/j.jhazmat.2011.12.032.

[19] G.V. Ramesh, T.P. Radhakrishnan, A universal sensor for mercury (Hg, Hg I, Hg II) based on silver nanoparticle-embedded polymer thin film, ACS Appl. Mater. Interfaces 3 (4) (2011) 988–994, https://doi.org/10.1021/am200023w.

[20] E. Sumesh, M.S. Bootharaju, P.T. Anshup, A practical silver nanoparticle-based adsorbent for the removal of Hg2+ from water, J. Hazard. Mater. 189 (1–2) (2011) 450–457, https://doi.org/10.1016/j.jhazmat.2011.02.061.

[21] B. Zheng, J. Zheng, T. Yu, et al., Fast microwave-assisted synthesis of AuAg bimetallic nanoclusters with strong yellow emission and their response to mercury(II) ions, Sens. Actuators B Chem. 221 (2015) 386–392, https://doi.org/10.1016/j.snb.2015.06.089.

[22] N. Xia, J. Yang, Z. Wu, Fast, high-yield synthesis of amphiphilic Ag nanoclusters and the sensing of Hg2+ in environmental samples, Nanoscale 7 (22) (2015) 10013–10020, https://doi.org/10.1039/c5nr00705d.

[23] K.Y. Pu, Z. Luo, K. Li, J. Xie, B. Liu, Energy transfer between conjugated-oligoelectrolyte-substituted POSS and gold nanocluster for multicolor intracellular detection of mercury ion, J. Phys. Chem. C 115 (26) (2011) 13069–13075, https://doi.org/10.1021/jp203133t.

[24] H. Liu, X. Zhang, X. Wu, L. Jiang, C. Burda, J.J. Zhu, Rapid sonochemical synthesis of highly luminescent non-toxic AuNCs and Au@AgNCs and Cu (II) sensing, Chem. Commun. 47 (14) (2011) 4237–4239, https://doi.org/10.1039/c1cc00103e.

[25] G. Zhang, Y. Li, J. Xu, et al., Glutathione-protected fluorescent gold nanoclusters for sensitive and selective detection of Cu2+, Sens. Actuators B Chem. 183 (2013) 583–588, https://doi.org/10.1016/j.snb.2013.04.023.

[26] Y. Chen, Y. Wang, C. Wang, et al., Papain-directed synthesis of luminescent gold nanoclusters and the sensitive detection of Cu2+, J. Colloid Interface Sci. 396 (2013) 63–68, https://doi.org/10.1016/j.jcis.2013.01.031.

[27] L. Shang, S. Dong, Silver nanocluster-based fluorescent sensors for sensitive detection of Cu(II), J. Mater. Chem. 18 (39) (2008) 4636–4640, https://doi.org/10.1039/b810409c.

[28] S. Wang, X. Meng, Y. Feng, H. Sheng, M. Zhu, An anti-galvanic reduction single-molecule fluorescent probe for detection of Cu(II), RSC Adv. 4 (19) (2014) 9680–9683, https://doi.org/10.1039/c3ra46877a.

[29] H. Ding, C. Liang, K. Sun, et al., Dithiothreitol-capped fluorescent gold nanoclusters: an efficient probe for detection of copper(II) ions in aqueous solution, Biosens. Bioelectron. 59 (2014) 216–220, https://doi.org/10.1016/j.bios.2014.03.045.

[30] A. George, E.S. Shibu, S.M. Maliyekkal, M.S. Bootharaju, T. Pradeep, Luminescent, freestanding composite films of Au 15 for specific metal ion sensing, ACS Appl. Mater. Interfaces 4 (2) (2012) 639–644, https://doi.org/10.1021/am201292a.

[31] P.L. Xavier, K. Chaudhari, P.K. Verma, S.K. Pal, T. Pradeep, Luminescent quantum clusters of gold in transferrin family protein, lactoferrin exhibiting FRET, Nanoscale 2 (12) (2010) 2769–2776, https://doi.org/10.1039/c0nr00377h.

[32] J. Zhang, Y. Yuan, G. Liang, et al., A microwave-facilitated rapid synthesis of gold nanoclusters with tunable optical properties for sensing ions and fluorescent ink, Chem. Commun. 51 (52) (2015) 10539–10542, https://doi.org/10.1039/c5cc03086b.

[33] C.V. Durgadas, C.P. Sharma, K. Sreenivasan, Fluorescent gold clusters as nanosensors for copper ions in live cells, Analyst 136 (5) (2011) 933–940, https://doi.org/10.1039/c0an00424c.

[34] Z. Lin, F. Luo, T. Dong, et al., Recyclable fluorescent gold nanocluster membrane for visual sensing of copper(II) ion in aqueous solution, Analyst 137 (10) (2012) 2394–2399, https://doi.org/10.1039/c2an35068h.

[35] X. Yang, L. Yang, Y. Dou, S. Zhu, Synthesis of highly fluorescent lysine-stabilized Au nanoclusters for sensitive and selective detection of Cu2+ ion, J. Mater. Chem. C 1 (41) (2013) 6748–6751, https://doi.org/10.1039/c3tc31398k.

[36] M. Zhang, H.N. Le, X.Q. Jiang, S.M. Guo, H.J. Yu, B.C. Ye, A ratiometric fluorescent probe for sensitive, selective and reversible detection of copper (II) based on riboflavin-stabilized gold nanoclusters, Talanta 117 (2013) 399–404, https://doi.org/10.1016/j.talanta.2013.09.034.

[37] Y.Q. Wang, T. Zhao, X.W. He, W.Y. Li, Y.K. Zhang, A novel core-satellite CdTe/Silica/Au NCs hybrid sphere as dual-emission ratiometric fluorescent probe for Cu2+, Biosens. Bioelectron. 51 (2014) 40–46, https://doi.org/10.1016/j.bios.2013.07.028.

[38] Y.T. Su, G.Y. Lan, W.Y. Chen, H.T. Chang, Detection of copper ions through recovery of the fluorescence of DNA-templated copper/silver nanoclusters in the presence of mercaptopropionic acid, Anal. Chem. 82 (20) (2010) 8566–8572, https://doi.org/10.1021/ac101659d.

[39] E.S. Shibu, T. Pradeep, Quantum clusters in cavities: trapped Au15 in cyclodextrins, Chem. Mater. 23 (4) (2011) 989–999, https://doi.org/10.1021/cm102743y.

[40] P.C. Chen, T.Y. Yeh, C.M. Ou, C.C. Shih, H.T. Chang, Synthesis of aluminum oxide supported fluorescent gold nanodots for the detection of silver ions, Nanoscale 5 (11) (2013) 4691–4695, https://doi.org/10.1039/c3nr00713h.

[41] Y. Yue, T.Y. Liu, H.W. Li, Z. Liu, Y. Wu, Microwave-assisted synthesis of BSA-protected small gold nanoclusters and their fluorescence-enhanced sensing of silver(I) ions, Nanoscale 4 (7) (2012) 2251–2254, https://doi.org/10.1039/c2nr12056a.

[42] H.W. Li, Y. Yue, T.Y. Liu, D. Li, Y. Wu, Fluorescence-enhanced sensing mechanism of BSA-protected small gold-nanoclusters to silver(I) ions in aqueous solutions, J. Phys. Chem. C 117 (31) (2013) 16159–16165, https://doi.org/10.1021/jp403466b.

[43] Z. Wu, M. Wang, J. Yang, et al., Well-defined nanoclusters as fluorescent nanosensors: a case study on Au25(SG)18, Small 8 (13) (2012) 2028–2035, https://doi.org/10.1002/smll.201102590.

[44] T.Y. Zhou, L.P. Lin, M.C. Rong, Y.Q. Jiang, X. Chen, Silver-gold alloy nanoclusters as a fluorescence-enhanced probe for aluminum ion sensing, Anal. Chem. 85 (20) (2013) 9839–9844, https://doi.org/10.1021/ac4023764.

[45] S. Roy, G. Palui, A. Banerjee, The as-prepared gold cluster-based fluorescent sensor for the selective detection of As III ions in aqueous solution, Nanoscale 4 (8) (2012) 2734–2740, https://doi.org/10.1039/c2nr11786j.

[46] J.A. Annie Ho, H.C. Chang, W.T. Su, DOPA-mediated reduction allows the facile synthesis of fluorescent gold nanoclusters for use as sensing probes for ferric ions, Anal. Chem. 84 (7) (2012) 3246–3253, https://doi.org/10.1021/ac203362g.

[47] H. Cao, Z. Chen, H. Zheng, Y. Huang, Copper nanoclusters as a highly sensitive and selective fluorescence sensor for ferric ions in serum and living cells by imaging, Biosens. Bioelectron. 62 (2014) 189–195, https://doi.org/10.1016/j.bios.2014.06.049.

[48] H. Zhang, Q. Liu, T. Wang, et al., Facile preparation of glutathione-stabilized gold nanoclusters for selective determination of chromium (III) and chromium (VI) in environmental water samples, Anal. Chim. Acta 770 (2013) 140–146, https://doi.org/10.1016/j.aca.2013.01.042.

[49] S. Liu, F. Lu, J.J. Zhu, Highly fluorescent Ag nanoclusters: microwave-assisted green synthesis and Cr3+ sensing, Chem. Commun. 47 (9) (2011) 2661–2663, https://doi.org/10.1039/c0cc04276e.

[50] N. Goswami, A. Giri, M.S. Bootharaju, P.L. Xavier, T. Pradeep, S.K. Pal, Copper quantum clusters in protein matrix: potential sensor of Pb 2+ ion, Anal. Chem. 83 (24) (2011) 9676–9680, https://doi.org/10.1021/ac202610e.

[51] H. Kawasaki, K. Hamaguchi, I. Osaka, R. Arakawa, Ph-dependent synthesis of pepsin-mediated gold nanoclusters with blue green and red fluorescent emission, Adv. Funct. Mater. 21 (18) (2011) 3508–3515, https://doi.org/10.1002/adfm.201100886.

[52] Z. Yuan, M. Peng, Y. He, E.S. Yeung, Functionalized fluorescent gold nanodots: synthesis and application for Pb2+ sensing, Chem. Commun. 47 (43) (2011) 11981–11983, https://doi.org/10.1039/c1cc14872a.

[53] Y. Liu, K. Ai, X. Cheng, L. Huo, L. Lu, Gold-nanocluster-based fluorescent sensors for highly sensitive and selective detection of cyanide in water, Adv. Funct. Mater. 20 (6) (2010) 951–956, https://doi.org/10.1002/adfm.200902062.

[54] D. Lu, L. Liu, F. Li, et al., Lysozyme-stabilized gold nanoclusters as a novel fluorescence probe for cyanide recognition, Spectrochim. Acta A Mol. Biomol. Spectrosc. 121 (2014) 77–80, https://doi.org/10.1016/j.saa.2013.10.009.

[55] C. Zong, L.R. Zheng, W. He, X. Ren, C. Jiang, L. Lu, In situ formation of phosphorescent molecular gold(I) cluster in a macroporous polymer film to achieve colorimetric cyanide sensing, Anal. Chem. 86 (3) (2014) 1687–1692, https://doi.org/10.1021/ac403480q.

[56] H. Liu, G. Yang, E.S. Abdel-Halim, J.J. Zhu, Highly selective and ultrasensitive detection of nitrite based on fluorescent gold nanoclusters, Talanta 104 (2013) 135–139, https://doi.org/10.1016/j.talanta.2012.11.020.

[57] B. Unnikrishnan, S.C. Wei, W.J. Chiu, J. Cang, P.H. Hsu, C.C. Huang, Nitrite ion-induced fluorescence quenching of luminescent BSA-Au25 nanoclusters: mechanism and application, Analyst 139 (9) (2014) 2221–2228, https://doi.org/10.1039/c3an02291a.

[58] W.Y. Chen, C.C. Huang, L.Y. Chen, H.T. Chang, Self-assembly of hybridized ligands on gold nanodots: tunable photoluminescence and sensing of nitrite, Nanoscale 6 (19) (2014) 11078–11083, https://doi.org/10.1039/c4nr02817a.

[59] J. Zhang, C. Chen, X. Xu, X. Wang, X. Yang, Use of fluorescent gold nanoclusters for the construction of a NAND logic gate for nitrite, Chem. Commun. 49 (26) (2013) 2691–2693, https://doi.org/10.1039/c3cc38298b.

[60] M.L. Cui, J.M. Liu, X.X. Wang, et al., A promising gold nanocluster fluorescent sensor for the highly sensitive and selective detection of S2, Sensors Actuators B Chem. 188 (2013) 53–58, https://doi.org/10.1016/j.snb.2013.05.098.

[61] W.Y. Chen, G.Y. Lan, H.T. Chang, Use of fluorescent DNA-templated gold/silver nanoclusters for the detection of sulfide ions, Anal. Chem. 83 (24) (2011) 9450–9455, https://doi.org/10.1021/ac202162u.

[62] Z. Yuan, M. Peng, L. Shi, et al., Disassembly mediated fluorescence recovery of gold nanodots for selective sulfide sensing, Nanoscale 5 (11) (2013) 4683–4686, https://doi.org/10.1039/c2nr33202g.

[63] M. Wang, Z. Wu, J. Yang, G. Wang, H. Wang, W. Cai, Au25(SG)18 as a fluorescent iodide sensor, Nanoscale 4 (14) (2012) 4087–4090, https://doi.org/10.1039/c2nr30169e.

[64] F. Qu, N.B. Li, H.Q. Luo, Polyethyleneimine-templated Ag nanoclusters: a new fluorescent and colorimetric platform for sensitive and selective sensing halide ions and high disturbance-tolerant recognitions of iodide and bromide in coexistence with chloride under condition of high ionic strength, Anal. Chem. 84 (23) (2012) 10373–10379, https://doi.org/10.1021/ac3024526.

[65] A. Mathew, P.R. Sajanlal, T. Pradeep, Selective visual detection of TNT at the sub-zeptomole level, Angew. Chem. Int. Ed. 51 (38) (2012) 9596–9600, https://doi.org/10.1002/anie.201203810.

[66] Z. Chen, S. Qian, X. Chen, W. Gao, Y. Lin, Protein-templated gold nanoclusters as fluorescence probes for the detection of methotrexate, Analyst 137 (18) (2012) 4356–4361, https://doi.org/10.1039/c2an35786k.

[67] K.S. Park, S.S. Oh, H.T. Soh, H.G. Park, Target-controlled formation of silver nanoclusters in abasic site-incorporated duplex DNA for label-free fluorescence detection of theophylline, Nanoscale 6 (17) (2014) 9977–9982, https://doi.org/10.1039/c4nr00625a.

[68] N. Zhang, Y. Si, Z. Sun, et al., Lab-on-a-drop: biocompatible fluorescent nanoprobes of gold nanoclusters for label-free evaluation of phosphorylation-induced inhibition of acetylcholinesterase activity towards the ultrasensitive detection of pesticide residues, Analyst 139 (18) (2014) 4620–4628, https://doi.org/10.1039/c4an00855c.

[69] G. Guan, Y. Cai, S. Liu, et al., High-level incorporation of silver in gold nanoclusters: fluorescence redshift upon interaction with hydrogen peroxide and fluorescence enhancement with herbicide, Chem. Eur. J. 22 (5) (2016) 1675–1681, https://doi.org/10.1002/chem.201504064.

[70] X. Wang, P. Wu, Y. Lv, X. Hou, Ultrasensitive fluorescence detection of glutaraldehyde in water samples with bovine serum albumin-Au nanoclusters, Microchem. J. 99 (2) (2011) 327–331, https://doi.org/10.1016/j.microc.2011.06.004.

[71] Y. Yong, C. Li, X. Li, T. Li, H. Cui, S. Lv, Ag 7 Au 6 cluster as a potential gas sensor for CO, HCN, and NO detection, J. Phys. Chem. C 119 (13) (2015) 7534–7540, https://doi.org/10.1021/acs.jpcc.5b02151.

[72] P.H. Li, J.Y. Lin, C.T. Chen, et al., Using gold nanoclusters as selective luminescent probes for phosphate-containing metabolites, Anal. Chem. 84 (13) (2012) 5484–5488, https://doi.org/10.1021/ac300332t.

[73] K. Selvaprakash, Y.C. Chen, Using protein-encapsulated gold nanoclusters as photoluminescent sensing probes for biomolecules, Biosens. Bioelectron. 61 (2014) 88–94, https://doi.org/10.1016/j.bios.2014.04.055.

[74] J. Sun, F. Yang, D. Zhao, X. Yang, Highly sensitive real-time assay of inorganic pyrophosphatase activity based on the fluorescent gold nanoclusters, Anal. Chem. 86 (15) (2014) 7883–7889, https://doi.org/10.1021/ac501814u.

[75] P. Zhang, Y. Wang, Y. Chang, Z.H. Xiong, C.Z. Huang, Highly selective detection of bacterial alarmone ppGpp with an off-on fluorescent probe of copper-mediated silver nanoclusters, Biosens. Bioelectron. 49 (2013) 433–437, https://doi.org/10.1016/j.bios.2013.05.056.

[76] F. Wen, Y. Dong, L. Feng, S. Wang, S. Zhang, X. Zhang, Horseradish peroxidase functionalized fluorescent gold nanoclusters for hydrogen peroxide sensing, Anal. Chem. 83 (4) (2011) 1193–1196, https://doi.org/10.1021/ac1031447.

[77] X. Jia, X. Yang, J. Li, D. Li, E. Wang, Stable Cu nanoclusters: from an aggregation-induced emission mechanism to biosensing and catalytic applications, Chem. Commun. 50 (2) (2014) 237–239, https://doi.org/10.1039/C3CC47771A.

[78] L. Shang, S. Dong, Sensitive detection of cysteine based on fluorescent silver clusters, Biosens. Bioelectron. 24 (6) (2009) 1569–1573, https://doi.org/10.1016/j.bios.2008.08.006.

[79] X. Yuan, Y. Tay, X. Dou, Z. Luo, D.T. Leong, J. Xie, Glutathione-protected silver nanoclusters as cysteine-selective fluorometric and colorimetric probe, Anal. Chem. 85 (3) (2013) 1913–1919, https://doi.org/10.1021/ac3033678.

[80] C.P. Liu, T.H. Wu, C.Y. Liu, S.Y. Lin, Live-cell imaging of biothiols via thiol/disulfide exchange to trigger the photoinduced electron transfer of gold-nanodot sensor, Anal. Chim. Acta 849 (2014) 57–63, https://doi.org/10.1016/j.aca.2014.08.022.

[81] M.L. Cui, J.M. Liu, X.X. Wang, et al., Selective determination of cysteine using BSA-stabilized gold nanoclusters with red emission, Analyst 137 (22) (2012) 5346–5351, https://doi.org/10.1039/c2an36284h.

[82] S. Ge, M. Yan, J. Lu, et al., Electrochemical biosensor based on graphene oxide-Au nanoclusters composites for l-cysteine analysis, Biosens. Bioelectron. 31 (1) (2012) 49–54, https://doi.org/10.1016/j.bios.2011.09.038.

[83] B. Han, E. Wang, Oligonucleotide-stabilized fluorescent silver nanoclusters for sensitive detection of biothiols in biological fluids, Biosens. Bioelectron. 26 (5) (2011) 2585–2589, https://doi.org/10.1016/j.bios.2010.11.011.

[84] Z. Huang, F. Pu, Y. Lin, J. Ren, X. Qu, Modulating DNA-templated silver nanoclusters for fluorescence turn-on detection of thiol compounds, Chem. Commun. 47 (12) (2011) 3487–3489, https://doi.org/10.1039/c0cc05651k.

[85] Y. He, X. Wang, J. Zhu, S. Zhong, G. Song, Ni2+−modified gold nanoclusters for fluorescence turn-on detection of histidine in biological fluids, Analyst 137 (17) (2012) 4005–4009, https://doi.org/10.1039/c2an35712g.

[86] M. Santhosh, S.R. Chinnadayyala, A. Kakoti, P. Goswami, Selective and sensitive detection of free bilirubin in blood serum using human serum albumin stabilized gold nanoclusters as fluorometric and colorimetric probe, Biosens. Bioelectron. 59 (2014) 370–376, https://doi.org/10.1016/j.bios.2014.04.003.

[87] B. Aswathy, G. Sony, Cu2+ modulated BSA-Au nanoclusters: a versatile fluorescence turn-on sensor for dopamine, Microchem. J. 116 (2014) 151–156, https://doi.org/10.1016/j.microc.2014.04.016.

[88] Y. Tao, Y. Lin, J. Ren, X. Qu, A dual fluorometric and colorimetric sensor for dopamine based on BSA-stabilized Au nanoclusters, Biosens. Bioelectron. 42 (1) (2013) 41–46, https://doi.org/10.1016/j.bios.2012.10.014.

[89] Y. Teng, X. Jia, J. Li, E. Wang, Ratiometric fluorescence detection of tyrosinase activity and dopamine using thiolate-protected gold nanoclusters, Anal. Chem. 87 (9) (2015) 4897–4902, https://doi.org/10.1021/acs.analchem.5b00468.

[90] S.S. Kumar, K. Kwak, D. Lee, Amperometric sensing based on glutathione protected Au25 nanoparticles and their pH dependent electrocatalytic activity, Electroanalysis 23 (9) (2011) 2116–2124, https://doi.org/10.1002/elan.201100240.

[91] L. Li, H. Liu, Y. Shen, J. Zhang, J.J. Zhu, Electrogenerated chemiluminescence of Au nanoclusters for the detection of dopamine, Anal. Chem. 83 (3) (2011) 661–665, https://doi.org/10.1021/ac102623r.

[92] J.M. Liu, J.T. Chen, X.P. Yan, Near infrared fluorescent trypsin stabilized gold nanoclusters as surface plasmon enhanced energy transfer biosensor and in vivo cancer imaging bioprobe, Anal. Chem. 85 (6) (2013) 3238–3245, https://doi.org/10.1021/ac303603f.

[93] M. Zhuang, C. Ding, A. Zhu, Y. Tian, Ratiometric fluorescence probe for monitoring hydroxyl radical in live cells based on gold nanoclusters, Anal. Chem. 86 (3) (2014) 1829–1836, https://doi.org/10.1021/ac403810g.

[94] T. Chen, Y. Hu, Y. Cen, X. Chu, Y. Lu, A dual-emission fluorescent nanocomplex of gold-cluster-decorated silica particles for live cell imaging of highly reactive oxygen species, J. Am. Chem. Soc. 135 (31) (2013) 11595–11602, https://doi.org/10.1021/ja4035939.

[95] E. Ju, Z. Liu, Y. Du, Y. Tao, J. Ren, X. Qu, Heterogeneous assembled nanocomplexes for ratiometric detection of highly reactive oxygen species in vitro and in vivo, ACS Nano 8 (6) (2014) 6014–6023, https://doi.org/10.1021/nn501135m.

[96] X. Xia, Y. Long, J. Wang, Glucose oxidase-functionalized fluorescent gold nanoclusters as probes for glucose, Anal. Chim. Acta 772 (2013) 81–86, https://doi.org/10.1016/j.aca.2013.02.025.

[97] A.M.P. Hussain, S.N. Sarangi, J.A. Kesarwani, S.N. Sahu, Au-nanocluster emission based glucose sensing, Biosens. Bioelectron. 29 (1) (2011) 60–65, https://doi.org/10.1016/j.bios.2011.07.066.

[98] L.L. Wang, J. Qiao, H.H. Liu, et al., Ratiometric fluorescent probe based on gold nanoclusters and alizarin red-boronic acid for monitoring glucose in brain microdialysate, Anal. Chem. 86 (19) (2014) 9758–9764, https://doi.org/10.1021/ac5023293.

[99] L.V. Nair, D.S. Philips, R.S. Jayasree, A. Ajayaghosh, A near-infrared fluorescent nanosensor (AuC@Urease) for the selective detection of blood urea, Small 9 (16) (2013) 2673–2677, https://doi.org/10.1002/smll.201300213.

[100] H.H. Deng, G.W. Wu, Z.Q. Zou, et al., pH-sensitive gold nanoclusters: preparation and analytical applications for urea, urease, and urease inhibitor detection, Chem. Commun. 51 (37) (2015) 7847–7850, https://doi.org/10.1039/c5cc00702j.

[101] B. Hemmateenejad, F. Shakerizadeh-Shirazi, F. Samari, BSA-modified gold nanoclusters for sensing of folic acid, Sens. Actuators B Chem. 199 (2014) 42–46, https://doi.org/10.1016/j.snb.2014.03.075.

[102] J. Zhang, M. Sajid, N. Na, L. Huang, D. He, J. Ouyang, The application of Au nanoclusters in the fluorescence imaging of human serum proteins after native PAGE: enhancing detection by low-temperature plasma treatment, Biosens. Bioelectron. 35 (1) (2012) 313–318, https://doi.org/10.1016/j.bios.2012.03.010.

[103] L.Y. Chen, C.C. Huang, W.Y. Chen, H.J. Lin, H.T. Chang, Using photoluminescent gold nanodots to detect hemoglobin in diluted blood samples, Biosens. Bioelectron. 43 (1) (2013) 38–44, https://doi.org/10.1016/j.bios.2012.11.034.

[104] C.C. Huang, C.K. Chiang, Z.H. Lin, K.H. Lee, H.T. Chang, Bioconjugated gold nanodots and nanoparticles for protein assays based on photoluminescence quenching, Anal. Chem. 80 (5) (2008) 1497–1504, https://doi.org/10.1021/ac701998f.

[105] C.C. Huang, C.T. Chen, Y.C. Shiang, Z.H. Lin, H.T. Chang, Synthesis of fluorescent carbohydrate-protected Au nanodots for detection of Concanavalin A and Escherichia coli, Anal. Chem. 81 (3) (2009) 875–882, https://doi.org/10.1021/ac8010654.

[106] C.C. Huang, Y.L. Hung, Y.C. Shiang, et al., Photoassisted synthesis of luminescent mannose-au nanodots for the detection of thyroglobulin in serum, Chem. Asian J. 5 (2) (2010) 334–341, https://doi.org/10.1002/asia.200900346.

[107] J. Yoshimoto, A. Sangsuwan, I. Osaka, et al., Optical properties of 2-methacryloyloxyethyl phosphorylcholine-protected Au 4 nanoclusters and their fluorescence sensing of C-reactive protein, J. Phys. Chem. C 119 (25) (2015) 14319–14325, https://doi.org/10.1021/acs.jpcc.5b03934.

[108] C.T. Chen, W.J. Chen, C.Z. Liu, L.Y. Chang, Y.C. Chen, Glutathione-bound gold nanoclusters for selective-binding and detection of glutathione S-transferase-fusion proteins from cell lysates, Chem. Commun. (48) (2009) 7515–7517, https://doi.org/10.1039/b916919a.

[109] G.Y. Lan, W.Y. Chen, H.T. Chang, Characterization and application to the detection of single-stranded DNA binding protein of fluorescent DNA-templated copper/silver nanoclusters, Analyst 136 (18) (2011) 3623–3628, https://doi.org/10.1039/c1an15258k.

[110] Z. Yuan, Y. Du, Y.T. Tseng, et al., Fluorescent gold nanodots based sensor array for proteins discrimination, Anal. Chem. 87 (8) (2015) 4253–4259, https://doi.org/10.1021/ac5045302.

[111] H. Kong, Y. Lu, H. Wang, F. Wen, S. Zhang, X. Zhang, Protein discrimination using fluorescent gold nanoparticles on plasmonic substrates, Anal. Chem. 84 (10) (2012) 4258–4261, https://doi.org/10.1021/ac300718p.

[112] L. Hu, S. Han, S. Parveen, Y. Yuan, L. Zhang, G. Xu, Highly sensitive fluorescent detection of trypsin based on BSA-stabilized gold nanoclusters, Biosens. Bioelectron. 32 (1) (2012) 297–299, https://doi.org/10.1016/j.bios.2011.12.007.

[113] J. Zhang, Z. Zhang, X. Nie, et al., A label-free gold nanocluster fluorescent probe for protease activity monitoring, J. Nanosci. Nanotechnol. 14 (6) (2014) 4029–4035, https://doi.org/10.1166/jnn.2014.8873.

[114] J. Sharma, H.C. Yeh, H. Yoo, J.H. Werner, J.S. Martinez, Silver nanocluster aptamers: in situ generation of intrinsically fluorescent recognition ligands for protein detection, Chem. Commun. 47 (8) (2011) 2294–2296, https://doi.org/10.1039/c0cc03711g.

[115] H.C. Yeh, J. Sharma, J.J. Han, J.S. Martinez, J.H. Werner, A DNA-silver nanocluster probe that fluoresces upon hybridization, Nano Lett. 10 (8) (2010) 3106–3110, https://doi.org/10.1021/nl101773c.

[116] L. Zhang, J. Zhu, Z. Zhou, et al., A new approach to light up DNA/Ag nanocluster-based beacons for bioanalysis, Chem. Sci. 4 (10) (2013) 4004–4010, https://doi.org/10.1039/c3sc51303c.

[117] J. Li, X. Zhong, H. Zhang, X.C. Le, J.J. Zhu, Binding-induced fluorescence turn-on assay using aptamer-functionalized silver nanocluster DNA probes, Anal. Chem. 84 (12) (2012) 5170–5174, https://doi.org/10.1021/ac3006268.

[118] C. Zheng, A.-X. Zheng, B. Liu, et al., One-pot synthesized DNA-templated Ag/Pt bimetallic nanoclusters as peroxidase mimics for colorimetric detection of thrombin, Chem. Commun. 50 (86) (2014) 13103–13106, https://doi.org/10.1039/C4CC05339G.

[119] Y. Qian, Y. Zhang, L. Lu, Y. Cai, A label-free DNA-templated silver nanocluster probe for fluorescence on-off detection of endonuclease activity and inhibition, Biosens. Bioelectron. 51 (2014) 408–412, https://doi.org/10.1016/j.bios.2013.07.060.

[120] M. Zhang, S.M. Guo, Y.R. Li, P. Zuo, B.C. Ye, A label-free fluorescent molecular beacon based on DNA-templated silver nanoclusters for detection of adenosine and adenosine deaminase, Chem. Commun. 48 (44) (2012) 5488–5490, https://doi.org/10.1039/c2cc31626a.

[121] K. Zhang, K. Wang, M. Xie, et al., DNA-templated silver nanoclusters based label-free fluorescent molecular beacon for the detection of adenosine deaminase, Biosens. Bioelectron. 52 (2014) 124–128, https://doi.org/10.1016/j.bios.2013.08.049.

[122] W. Song, Y. Wang, R.P. Liang, L. Zhang, J.D. Qiu, Label-free fluorescence assay for protein kinase based on peptide biomineralized gold nanoclusters as signal sensing probe, Biosens. Bioelectron. 64 (2015) 234–240, https://doi.org/10.1016/j.bios.2014.08.082.

[123] Q. Wen, Y. Gu, L.J. Tang, R.Q. Yu, J.H. Jiang, Peptide-templated gold nanocluster beacon as a sensitive, label-free sensor for protein post-translational modification enzymes, Anal. Chem. 85 (24) (2013) 11681–11685, https://doi.org/10.1021/ac403308b.

[124] Y. Wang, Y. Wang, F. Zhou, P. Kim, Y. Xia, Protein-protected Au clusters as a new class of nanoscale biosensor for label-free fluorescence detection of proteases, Small 8 (24) (2012) 3769–3773, https://doi.org/10.1002/smll.201201983.

[125] C. Shen, X. Xia, S. Hu, M. Yang, J. Wang, Silver nanoclusters-based fluorescence assay of protein kinase activity and inhibition, Anal. Chem. 87 (1) (2015) 693–698, https://doi.org/10.1021/ac503492k.

[126] Y. Gu, Q. Wen, Y. Kuang, L. Tang, J. Jiang, Peptide-templated gold nanoclusters as a novel label-free biosensor for the detection of protease activity, RSC Adv. 4 (27) (2014) 13753–13756, https://doi.org/10.1039/C4RA00096J.

[127] Y. Chen, H. Zhou, Y. Wang, et al., Substrate hydrolysis triggered formation of fluorescent gold nanoclusters—a new platform for the sensing of enzyme activity, Chem. Commun. 49 (84) (2013) 9821–9823, https://doi.org/10.1039/c3cc45494k.

[128] W.Y. Chen, L.Y. Chen, C.M. Ou, C.C. Huang, S.C. Wei, H.T. Chang, Synthesis of fluorescent gold nanodot-liposome hybrids for detection of phospholipase C and its inhibitor, Anal. Chem. 85 (18) (2013) 8834–8840, https://doi.org/10.1021/ac402043t.

[129] Y. Zhang, Y. Cai, Z. Qi, L. Lu, Y. Qian, DNA-templated silver nanoclusters for fluorescence turn-on assay of acetylcholinesterase activity, Anal. Chem. 85 (17) (2013) 8455–8461, https://doi.org/10.1021/ac401966d.

[130] X. Liu, F. Wang, A. Niazov-Elkan, W. Guo, I. Willner, Probing biocatalytic transformations with luminescent DNA/silver nanoclusters, Nano Lett. 13 (1) (2013) 309–314, https://doi.org/10.1021/nl304283c.

[131] Y. Tao, Y. Lin, Z. Huang, J. Ren, X. Qu, DNA-templated silver nanoclusters-graphene oxide nanohybrid materials: a platform for label-free and sensitive fluorescence turn-on detection of multiple nucleic acid targets, Analyst 137 (11) (2012) 2588–2592, https://doi.org/10.1039/c2an35373c.

[132] X. Liu, F. Wang, R. Aizen, O. Yehezkeli, I. Willner, Graphene oxide/nucleic-acid-stabilized silver nanoclusters: functional hybrid materials for optical aptamer sensing and multiplexed analysis of pathogenic DNAs, J. Am. Chem. Soc. 135 (32) (2013) 11832–11839, https://doi.org/10.1021/ja403485r.

[133] L. Zhang, J. Zhu, S. Guo, T. Li, J. Li, E. Wang, Photoinduced electron transfer of DNA/Ag nanoclusters modulated by G-quadruplex/hemin complex for the construction of versatile biosensors, J. Am. Chem. Soc. 135 (7) (2013) 2403–2406, https://doi.org/10.1021/ja3089857.

[134] A.L. West, M.H. Griep, D.P. Cole, S.P. Karna, DNase 1 retains endodeoxyribonuclease activity following gold nanocluster synthesis, Anal. Chem. 86 (15) (2014) 7377–7382, https://doi.org/10.1021/ac5005794.

[135] J.T. Petty, B. Sengupta, S.P. Story, N.N. Degtyareva, DNA sensing by amplifying the number of near-infrared emitting, oligonucleotide-encapsulated silver clusters, Anal. Chem. 83 (15) (2011) 5957–5964, https://doi.org/10.1021/ac201321m.

[136] Y. Xiao, Z. Wu, K.Y. Wong, Z. Liu, Hairpin DNA probes based on target-induced in situ generation of luminescent silver nanoclusters, Chem. Commun. 50 (37) (2014) 4849–4852, https://doi.org/10.1039/c4cc01154f.

[137] Y. Zhang, C. Zhu, L. Zhang, et al., DNA-templated silver nanoclusters for multiplexed fluorescent DNA detection, Small 11 (12) (2015) 1385–1389, https://doi.org/10.1002/smll.201402044.

[138] L. Liu, Q. Yang, J. Lei, N. Xu, H. Ju, DNA-regulated silver nanoclusters for label-free ratiometric fluorescence detection of DNA, Chem. Commun. 50 (89) (2014) 13698–13701, https://doi.org/10.1039/C4CC04615C.

[139] W. Guo, J. Yuan, Q. Dong, E. Wang, Highly sequence-dependent formation of fluorescent silver nanoclusters in hybridized DNA duplexes for single nucleotide mutation identification, J. Am. Chem. Soc. 132 (3) (2010) 932–934, https://doi.org/10.1021/ja907075s.

[140] X. Jia, J. Li, L. Han, J. Ren, X. Yang, E. Wang, DNA-hosted copper nanoclusters for fluorescent identification of single nucleotide polymorphisms, ACS Nano 6 (4) (2012) 3311–3317, https://doi.org/10.1021/nn3002455.

[141] H.C. Yeh, J. Sharma, I.M. Shih, D.M. Vu, J.S. Martinez, J.H. Werner, A fluorescence light-up Ag nanocluster probe that discriminates single-nucleotide variants by emission color, J. Am. Chem. Soc. 134 (28) (2012) 11550–11558, https://doi.org/10.1021/ja3024737.

[142] X. Qiu, P. Wang, Z. Cao, Hybridization chain reaction modulated DNA-hosted silver nanoclusters for fluorescent identification of single nucleotide polymorphisms in the let-7 miRNA family, Biosens. Bioelectron. 60 (2014) 351–357, https://doi.org/10.1016/j.bios.2014.04.040.

[143] S.W. Yang, T. Vosch, Rapid detection of microRNA by a silver nanocluster DNA probe, Anal. Chem. 83 (18) (2011) 6935–6939, https://doi.org/10.1021/ac201903n.

[144] Y.Q. Liu, M. Zhang, B.C. Yin, B.C. Ye, Attomolar ultrasensitive microRNA detection by DNA-scaffolded silver-nanocluster probe based on isothermal amplification, Anal. Chem. 84 (12) (2012) 5165–5169, https://doi.org/10.1021/ac300483f.

[145] F. Xu, H. Shi, X. He, et al., Concatemeric dsDNA-templated copper nanoparticles strategy with improved sensitivity and stability based on rolling circle replication and its application in microRNA detection, Anal. Chem. 86 (14) (2014) 6976–6982, https://doi.org/10.1021/ac500955r.

[146] Y. Zhang, Z. Chen, Y. Tao, Z. Wang, J. Ren, X. Qu, Hybridization chain reaction engineered dsDNA for Cu metallization: an enzyme-free platform for amplified detection of cancer cells and microRNAs, Chem. Commun. 51 (57) (2015) 11496–11499, https://doi.org/10.1039/c5cc03144c.

[147] H. Dong, S. Jin, H. Ju, et al., Trace and label-free microRNA detection using oligonucleotide encapsulated silver nanoclusters as probes, Anal. Chem. 84 (20) (2012) 8670–8674, https://doi.org/10.1021/ac301860v.

[148] Y. Si, Z. Sun, N. Zhang, et al., Ultrasensitive electroanalysis of low-level free micrornas in blood by maximum signal amplification of catalytic silver deposition using alkaline phosphatase-incorporated gold nanoclusters, Anal. Chem. 86 (20) (2014) 10406–10414, https://doi.org/10.1021/ac5028885.

[149] Y.C. Shiang, C.A. Lin, C.C. Huang, H.T. Chang, Protein A-conjugated luminescent gold nanodots as a label-free assay for immunoglobulin G in plasma, Analyst 136 (6) (2011) 1177–1182, https://doi.org/10.1039/c0an00889c.

[150] H. Liu, X. Wu, X. Zhang, C. Burda, J.J. Zhu, Gold nanoclusters as signal amplification labels for optical immunosensors, J. Phys. Chem. C 116 (3) (2012) 2548–2554, https://doi.org/10.1021/jp206256j.

[151] M.C. Alonso, L. Trapiella-Alfonso, J.M.C. Fernández, R. Pereiro, A. Sanz-Medel, Functionalized gold nanoclusters as fluorescent labels for immunoassays: application to human serum immunoglobulin E determination, Biosens. Bioelectron. 77 (2016) 1055–1061, https://doi.org/10.1016/j.bios.2015.08.011.

[152] G.H. Yang, J.J. Shi, S. Wang, et al., Fabrication of a boron nitride-gold nanocluster composite and its versatile application for immunoassays, Chem. Commun. 49 (91) (2013) 10757–10759, https://doi.org/10.1039/c3cc45759a.

[153] J. Peng, L.N. Feng, K. Zhang, X.H. Li, L.P. Jiang, J.J. Zhu, Calcium carbonate-gold nanocluster hybrid spheres: synthesis and versatile application in immunoassays, Chem. Eur. J. 18 (17) (2012) 5261–5268, https://doi.org/10.1002/chem.201102876.

[154] Y.N. Chen, P.C. Chen, C.W. Wang, et al., One-pot synthesis of fluorescent BSA–Ce/Au nanoclusters as ratiometric pH probes, Chem. Commun. 50 (62) (2014) 8571–8574, https://doi.org/10.1039/c4cc03949a.

[155] L. Shang, F. Stockmar, N. Azadfar, G.U. Nienhaus, Intracellular thermometry by using fluorescent gold nanoclusters, Angew. Chem. Int. Ed. 52 (42) (2013) 11154–11157, https://doi.org/10.1002/anie.201306366.

[156] X. Chen, J.B. Essner, G.A. Baker, Exploring luminescence-based temperature sensing using protein-passivated gold nanoclusters, Nanoscale 6 (16) (2014) 9594–9598, https://doi.org/10.1039/c4nr02069c.

[157] Q. Li, Y. Pan, T. Chen, et al., Design and mechanistic study of a novel gold nanocluster-based drug delivery system, Nanoscale 10 (21) (2018) 10166–10172, https://doi.org/10.1039/c8nr02189a.

[158] C.L. Liu, M.L. Ho, Y.C. Chen, et al., Thiol-functionalized gold nanodots: two-photon absorption property and imaging in vitro, J. Phys. Chem. C 113 (50) (2009) 21082–21089, https://doi.org/10.1021/jp9080492.

[159] L. Polavarapu, M. Manna, Q.H. Xu, Biocompatible glutathione capped gold clusters as one- and two-photon excitation fluorescence contrast agents for live cells imaging, Nanoscale 3 (2) (2011) 429–434, https://doi.org/10.1039/c0nr00458h.

[160] L. Shang, R.M. Dörlich, S. Brandholt, et al., Facile preparation of water-soluble fluorescent gold nanoclusters for cellular imaging applications, Nanoscale 2009 (2011), https://doi.org/10.1039/c0nr00947d.

[161] X. Yang, L. Gan, L. Han, D. Li, J. Wang, E. Wang, Facile preparation of chiral penicillamine protected gold nanoclusters and their applications in cell imaging, Chem. Commun. 49 (23) (2013) 2302–2304, https://doi.org/10.1039/c3cc00200d.

[162] X. Le Guével, B. Hötzer, G. Jung, M. Schneider, NIR-emitting fluorescent gold nanoclusters doped in silica nanoparticles, J. Mater. Chem. 21 (9) (2011) 2974–2981, https://doi.org/10.1039/c0jm02660c.

[163] C. Wang, C. Wang, L. Xu, H. Cheng, Q. Lin, C. Zhang, Protein-directed synthesis of pH-responsive red fluorescent copper nanoclusters and their applications in cellular imaging and catalysis, Nanoscale 6 (3) (2014) 1775–1781, https://doi.org/10.1039/c3nr04835g.

[164] X. Le Guével, C. Spies, N. Daum, G. Jung, M. Schneider, Highly fluorescent silver nanoclusters stabilized by glutathione: a promising fluorescent label for bioimaging, Nano Res. 5 (6) (2012) 379–387, https://doi.org/10.1007/s12274-012-0218-1.

[165] S. Wang, X. Meng, A. Das, et al., A 200-fold quantum yield boost in the photoluminescence of silver-doped Ag x Au 25 − x nanoclusters: The 13th silver atom matters, Angew. Chem. Int. Ed. 53 (9) (2014) 2376–2380, https://doi.org/10.1002/anie.201307480.

[166] X. Wu, X. He, K. Wang, C. Xie, B. Zhou, Z. Qing, Ultrasmall near-infrared gold nanoclusters for tumor fluorescence imaging in vivo, Nanoscale 2 (10) (2010) 2244–2249, https://doi.org/10.1039/c0nr00359j.

[167] S.L. Raut, R. Fudala, R. Rich, et al., Long lived BSA Au clusters as a time gated intensity imaging probe, Nanoscale 6 (5) (2014) 2594–2597, https://doi.org/10.1039/c3nr05692a.

[168] J. Zhang, Y. Fu, C.V. Conroy, et al., Fluorescence intensity and lifetime cell imaging with luminescent gold nanoclusters, J. Phys. Chem. C 116 (50) (2012) 26561–26569, https://doi.org/10.1021/jp306036y.

[169] L. Shang, N. Azadfar, F. Stockmar, et al., One-pot synthesis of near-infrared fluorescent gold clusters for cellular fluorescence lifetime imaging, Small 7 (18) (2011) 2614–2620, https://doi.org/10.1002/smll.201100746.

[170] Q. Zhuang, H. Jia, L. Du, et al., Targeted surface-functionalized gold nanoclusters for mitochondrial imaging, Biosens. Bioelectron. 55 (2014) 76–82, https://doi.org/10.1016/j.bios.2013.12.003.
[171] Z. Sun, Y. Wang, Y. Wei, et al., Ag cluster-aptamer hybrid: specifically marking the nucleus of live cells, Chem. Commun. 47 (43) (2011) 11960–11962, https://doi.org/10.1039/c1cc14652a.
[172] Y. Wang, Y. Cui, Y. Zhao, et al., Bifunctional peptides that precisely biomineralize Au clusters and specifically stain cell nuclei, Chem. Commun. 48 (6) (2012) 871–873, https://doi.org/10.1039/c1cc15926g.
[173] Y. Wang, Y. Cui, R. Liu, et al., Blue two-photon fluorescence metal cluster probe precisely marking cell nuclei of two cell lines, Chem. Commun. 49 (91) (2013) 10724–10726, https://doi.org/10.1039/c3cc46690f.
[174] V. Venkatesh, A. Shukla, S. Sivakumar, S. Verma, Purine-stabilized green fluorescent gold nanoclusters for cell nuclei imaging applications, ACS Appl. Mater. Interfaces 6 (3) (2014) 2185–2191, https://doi.org/10.1021/am405345h.
[175] Y. Yang, S. Wang, S. Chen, Y. Shen, M. Zhu, Switching the subcellular organelle targeting of atomically precise gold nanoclusters by modifying the capping ligand, Chem. Commun. 54 (66) (2018) 9222–9225, https://doi.org/10.1039/c8cc04474k.
[176] X.L. Guével, N. Daum, M. Schneider, Synthesis and characterization of human transferrin-stabilized gold nanoclusters, Nanotechnology 22 (27) (2011), https://doi.org/10.1088/0957-4484/22/27/275103.
[177] Y. Wang, J.T. Chen, X.P. Yan, Fabrication of transferrin functionalized gold nanoclusters/graphene oxide nanocomposite for turn-on near-infrared fluorescent bioimaging of cancer cells and small animals, Anal. Chem. 85 (4) (2013) 2529–2535, https://doi.org/10.1021/ac303747t.
[178] Y. Pan, Q. Li, Q. Zhou, et al., Cancer cell specific fluorescent methionine protected gold nanoclusters for in-vitro cell imaging studies, Talanta 188 (2018) 259–265, https://doi.org/10.1016/j.talanta.2018.05.079.
[179] Y. Kong, J. Chen, F. Gao, et al., Near-infrared fluorescent ribonuclease-A-encapsulated gold nanoclusters: preparation, characterization, cancer targeting and imaging, Nanoscale 5 (3) (2013) 1009–1017, https://doi.org/10.1039/c2nr32760k.
[180] A. Retnakumari, S. Setua, D. Menon, et al., Molecular-receptor-specific, non-toxic, near-infrared-emitting Au cluster-protein nanoconjugates for targeted cancer imaging, Nanotechnology 21 (5) (2010), 055103, https://doi.org/10.1088/0957-4484/21/5/055103.
[181] M.A.H. Muhammed, P.K. Verma, S.K. Pal, et al., Luminescent quantum clusters of gold in bulk by albumin-induced core etching of nanoparticles: metal ion sensing, metal-enhanced luminescence, and biolabeling, Chem. Eur. J. 16 (33) (2010) 10103–10112, https://doi.org/10.1002/chem.201000841.
[182] J. Qiao, X. Mu, L. Qi, J. Deng, L. Mao, Folic acid-functionalized fluorescent gold nanoclusters with polymers as linkers for cancer cell imaging, Chem. Commun. 49 (73) (2013) 8030–8032, https://doi.org/10.1039/c3cc44256j.
[183] Y. Wang, C. Dai, X.-P. Yan, Fabrication of folate bioconjugated near-infrared fluorescent silver nanoclusters for targeted in vitro and in vivo bioimaging, Chem. Commun. 50 (92) (2014) 14341–14344, https://doi.org/10.1039/C4CC06329E.
[184] P. Zhang, X.X. Yang, Y. Wang, N.W. Zhao, Z.H. Xiong, C.Z. Huang, Rapid synthesis of highly luminescent and stable Au20 nanoclusters for active tumor-targeted imaging in vitro and in vivo, Nanoscale 6 (4) (2014) 2261–2269, https://doi.org/10.1039/c3nr05269a.
[185] M.A.H. Muhammed, P.K. Verma, S.K. Pal, et al., Bright, NIR-emitting Au23 from Au25: characterization and applications including biolabeling, Chem. Eur. J. 15 (39) (2009) 10110–10120, https://doi.org/10.1002/chem.200901425.
[186] S.Y. Lin, N.T. Chen, S.P. Sun, et al., The protease-mediated nucleus shuttles of subnanometer gold quantum dots for real-time monitoring of apoptotic cell death, J. Am. Chem. Soc. 132 (24) (2010) 8309–8315, https://doi.org/10.1021/ja100561k.
[187] C. Sun, H. Yang, Y. Yuan, et al., Controlling assembly of paired gold clusters within apoferritin nanoreactor for in vivo kidney targeting and biomedical imaging, J. Am. Chem. Soc. 133 (22) (2011) 8617–8624, https://doi.org/10.1021/ja200746p.
[188] E. Oh, F.K. Fatemi, M. Currie, et al., PEGylated luminescent gold nanoclusters: synthesis, characterization, bioconjugation, and application to one- and two-photon cellular imaging, Part. Part. Syst. Charact. 30 (5) (2013) 453–466, https://doi.org/10.1002/ppsc.201200140.
[189] J. Li, X. Zhong, F. Cheng, J.R. Zhang, L.P. Jiang, J.J. Zhu, One-pot synthesis of aptamer-functionalized silver nanoclusters for cell-type-specific imaging, Anal. Chem. 84 (9) (2012) 4140–4146, https://doi.org/10.1021/ac3003402.
[190] J. Yin, X. He, K. Wang, et al., One-step engineering of silver nanoclusters-aptamer assemblies as luminescent labels to target tumor cells, Nanoscale 4 (1) (2012) 110–112, https://doi.org/10.1039/c1nr11265a.
[191] J. Li, J. You, Y. Zhuang, et al., A "light-up" and "spectrum-shift" response of aptamer-functionalized silver nanoclusters for intracellular mRNA imaging, Chem. Commun. 50 (54) (2014) 7107–7110, https://doi.org/10.1039/c4cc00160e.
[192] J. Yu, S.A. Patel, R.M. Dickson, In vitro and intracellular production of peptide-encapsulated fluorescent silver nanoclusters, Angew. Chem. Int. Ed. 46 (12) (2007) 2028–2030, https://doi.org/10.1002/anie.200604253.
[193] J. Wang, G. Zhang, Q. Li, et al., In vivo self-bio-imaging of tumors through in situ biosynthesized fluorescent gold nanoclusters, Sci. Rep. 3 (2013) 1157, https://doi.org/10.1038/srep01157.
[194] S. Gao, D. Chen, Q. Li, et al., Near-infrared fluorescence imaging of cancer cells and tumors through specific biosynthesis of silver nanoclusters, Sci. Rep. 4 (2014), https://doi.org/10.1038/srep04384.
[195] H. Hu, P. Huang, O.J. Weiss, et al., PET and NIR optical imaging using self-illuminating 64Cu-doped chelator-free gold nanoclusters, Biomaterials 35 (37) (2014) 9868–9876, https://doi.org/10.1016/j.biomaterials.2014.08.038.
[196] G. Sun, L. Zhou, Y. Liu, Z. Zhao, Biocompatible GdIII-functionalized fluorescent gold nanoclusters for optical and magnetic resonance imaging, New J. Chem. 37 (4) (2013) 1028–1035, https://doi.org/10.1039/c3nj00052d.
[197] S.K. Sun, L.X. Dong, Y. Cao, H.R. Sun, X.P. Yan, Fabrication of multifunctional Gd2O3/Au hybrid nanoprobe via a one-step approach for near-infrared fluorescence and magnetic resonance multimodal imaging in vivo, Anal. Chem. 85 (17) (2013) 8436–8441, https://doi.org/10.1021/ac401879y.

[198] D.H. Hu, Z.H. Sheng, P.F. Zhang, et al., Hybrid gold-gadolinium nanoclusters for tumor-targeted NIRF/CT/MRI triple-modal imaging in vivo, Nanoscale 5 (4) (2013) 1624–1628, https://doi.org/10.1039/c2nr33543c.

[199] X. Wu, C. Li, S. Liao, et al., Silica-encapsulated Gd3+−aggregated gold nanoclusters for in vitro and in vivo multimodal cancer imaging, Chem. Eur. J. 20 (29) (2014) 8876–8882, https://doi.org/10.1002/chem.201403202.

[200] E.S. Shibu, S. Sugino, K. Ono, et al., Singlet-oxygen-sensitizing near-infrared-fluorescent multimodal nanoparticles, Angew. Chem. Int. Ed. 52 (40) (2013) 10559–10563, https://doi.org/10.1002/anie.201304264.

[201] A. Retnakumari, J. Jayasimhan, P. Chandran, et al., CD33 monoclonal antibody conjugated Au cluster nano-bioprobe for targeted flow-cytometric detection of acute myeloid leukaemia, Nanotechnology 22 (28) (2011), 285102, https://doi.org/10.1088/0957-4484/22/28/285102.

[202] J. Yin, X. He, K. Wang, et al., Label-free and turn-on aptamer strategy for cancer cells detection based on a DNA-silver nanocluster fluorescence upon recognition-induced hybridization, Anal. Chem. 85 (24) (2013) 12011–12019, https://doi.org/10.1021/ac402989u.

[203] Y. Tao, Y. Lin, Z. Huang, J. Ren, X. Qu, Incorporating graphene oxide and gold nanoclusters: a synergistic catalyst with surprisingly high peroxidase-like activity over a broad pH range and its application for cancer cell detection, Adv. Mater. 25 (18) (2013) 2594–2599, https://doi.org/10.1002/adma.201204419.

[204] D. Hu, Z. Sheng, S. Fang, et al., Folate receptor-targeting gold nanoclusters as fluorescence enzyme mimetic nanoprobes for tumor molecular colocalization diagnosis, Theranostics 4 (2) (2014) 142–153, https://doi.org/10.7150/thno.7266.

[205] X. Zhang, F.G. Wu, P. Liu, N. Gu, Z. Chen, Enhanced fluorescence of gold nanoclusters composed of HAuCl4 and histidine by glutathione: glutathione detection and selective cancer cell imaging, Small 10 (24) (2014) 5170–5177, https://doi.org/10.1002/smll.201401658.

[206] H. Chen, B. Li, X. Ren, et al., Multifunctional near-infrared-emitting nano-conjugates based on gold clusters for tumor imaging and therapy, Biomaterials 33 (33) (2012) 8461–8476, https://doi.org/10.1016/j.biomaterials.2012.08.034.

[207] X. Zan, Q. Li, Y. Pan, et al., Versatile ligand-exchange method for the synthesis of water-soluble monodisperse AuAg nanoclusters for cancer therapy, ACS Appl. Nano Mater. 1 (12) (2018) 6773–6781, https://doi.org/10.1021/acsanm.8b01559.

[208] C. Wang, J. Li, C. Amatore, Y. Chen, H. Jiang, X.M. Wang, Gold nanoclusters and graphene nanocomposites for drug delivery and imaging of cancer cells, Angew. Chem. Int. Ed. 50 (49) (2011) 11644–11648, https://doi.org/10.1002/anie.201105573.

[209] A.K. Sahoo, S. Banerjee, S.S. Ghosh, A. Chattopadhyay, Simultaneous RGB emitting Au nanoclusters in chitosan nanoparticles for anticancer gene theranostics, ACS Appl. Mater. Interfaces 6 (1) (2014) 712–724, https://doi.org/10.1021/am4051266.

[210] Y. Wang, J. Chen, J. Irudayaraj, Nuclear targeting dynamics of gold nanoclusters for enhanced therapy of HER2 + breast cancer, ACS Nano 5 (12) (2011) 9718–9725, https://doi.org/10.1021/nn2032177.

[211] S. Su, H. Wang, X. Liu, Y. Wu, G. Nie, IRGD-coupled responsive fluorescent nanogel for targeted drug delivery, Biomaterials 34 (13) (2013) 3523–3533, https://doi.org/10.1016/j.biomaterials.2013.01.083.

[212] R. Gui, A. Wan, X. Liu, H. Jin, Retracted article: intracellular fluorescent thermometry and photothermal-triggered drug release developed from gold nanoclusters and doxorubicin dual-loaded liposomes, Chem. Commun. 50 (13) (2014) 1546–1548, https://doi.org/10.1039/c3cc47981a.

[213] T. Chen, S. Xu, T. Zhao, et al., Gold nanocluster-conjugated amphiphilic block copolymer for tumor-targeted drug delivery, ACS Appl. Mater. Interfaces 4 (11) (2012) 5766–5774, https://doi.org/10.1021/am301223n.

[214] D. Chen, Z. Luo, N. Li, J.Y. Lee, J. Xie, J. Lu, Amphiphilic polymeric nanocarriers with luminescent gold nanoclusters for concurrent bioimaging and controlled drug release, Adv. Funct. Mater. 23 (35) (2013) 4324–4331, https://doi.org/10.1002/adfm.201300411.

[215] P. Huang, J. Lin, S. Wang, et al., Photosensitizer-conjugated silica-coated gold nanoclusters for fluorescence imaging-guided photodynamic therapy, Biomaterials 34 (19) (2013) 4643–4654, https://doi.org/10.1016/j.biomaterials.2013.02.063.

[216] L.V. Nair, S.S. Nazeer, R.S. Jayasree, A. Ajayaghosh, Fluorescence imaging assisted photodynamic therapy using photosensitizer-linked gold quantum clusters, ACS Nano 9 (6) (2015) 5825–5832, https://doi.org/10.1021/acsnano.5b00406.

[217] Y. Yang, S. Wang, C. Xu, A. Xie, Y. Shen, M. Zhu, Improved fluorescence imaging and synergistic anticancer phototherapy of hydrosoluble gold nanoclusters assisted by a novel two-level mesoporous canal structured silica nanocarrier, Chem. Commun. 54 (22) (2018) 2731–2734, https://doi.org/10.1039/c8cc00685g.

[218] Y. Wang, Q. Yuan, D. An, J. Li, X. Gao, The Au clusters induce tumor cell apoptosis via specifically targeting thioredoxin reductase 1 (TrxR1) and suppressing its activity, Chem. Commun. 50 (74) (2014) 10687–10690, https://doi.org/10.1039/c4cc03320e.

[219] X.D. Zhang, J. Chen, Z. Luo, et al., Enhanced tumor accumulation of Sub-2 nm gold nanoclusters for cancer radiation therapy, Adv. Healthcare Mater. 3 (1) (2014) 133–141, https://doi.org/10.1002/adhm.201300189.

[220] X.D. Zhang, Z. Luo, J. Chen, et al., Ultrasmall Au10-12(SG)10-12 nanomolecules for high tumor specificity and cancer radiotherapy, Adv. Mater. 26 (26) (2014) 4565–4568, https://doi.org/10.1002/adma.201400866.

[221] Y. Tao, Z. Li, E. Ju, J. Ren, X. Qu, Polycations-functionalized water-soluble gold nanoclusters: a potential platform for simultaneous enhanced gene delivery and cell imaging, Nanoscale 5 (13) (2013) 6154–6160, https://doi.org/10.1039/c3nr01326j.

[222] J. Li, W. Wang, D. Sun, et al., Aptamer-functionalized silver nanoclusters-mediated cell type-specific siRNA delivery and tracking, Chem. Sci. 4 (9) (2013) 3514–3521, https://doi.org/10.1039/c3sc51538a.

[223] Y. Tao, E. Ju, Z. Li, J. Ren, X. Qu, Engineered CpG-antigen conjugates protected gold nanoclusters as smart self-vaccines for enhanced immune response and cell imaging, Adv. Funct. Mater. 24 (7) (2014) 1004–1010, https://doi.org/10.1002/adfm.201302347.

[224] Y. Tao, Y. Zhang, E. Ju, H. Ren, J. Ren, Gold nanocluster-based vaccines for dual-delivery of antigens and immunostimulatory oligonucleotides, Nanoscale 7 (29) (2015) 12419–12426, https://doi.org/10.1039/c5nr02240a.

[225] R. Lu, W. Zou, H. Du, J. Wang, S. Zhang, Antimicrobial activity of Ag nanoclusters encapsulated in porous silica nanospheres, Ceram. Int. 40 (2) (2014) 3693–3698, https://doi.org/10.1016/j.ceramint.2013.09.055.

[226] I. Chakraborty, T. Udayabhaskararao, G.K. Deepesh, T. Pradeep, Sunlight mediated synthesis and antibacterial properties of monolayer protected silver clusters, J. Mater. Chem. B 1 (33) (2013) 4059–4064, https://doi.org/10.1039/c3tb20603c.

[227] X. Yuan, M.I. Setyawati, A.S. Tan, C.N. Ong, D.T. Leong, J. Xie, Highly luminescent silver nanoclusters with tunable emissions: cyclic reduction-decomposition synthesis and antimicrobial properties, NPG Asia Mater. 5 (2) (2013), https://doi.org/10.1038/am.2013.3.

[228] X. Yuan, M.I. Setyawati, D.T. Leong, J. Xie, Ultrasmall Ag+−rich nanoclusters as highly efficient nanoreservoirs for bacterial killing, Nano Res. 7 (3) (2014) 301–307, https://doi.org/10.1007/s12274-013-0395-6.

[229] W.Y. Chen, J.Y. Lin, W.J. Chen, L. Luo, E. Wei-Guang Diau, Y.C. Chen, Functional gold nanoclusters as antimicrobial agents for antibiotic-resistant bacteria, Nanomedicine 5 (5) (2010) 755–764, https://doi.org/10.2217/nnm.10.43.

[230] M. Zhou, C. Zeng, Y. Chen, et al., Evolution from the plasmon to exciton state in ligand-protected atomically precise gold nanoparticles, Nat. Commun. 7 (2016), https://doi.org/10.1038/ncomms13240.

[231] Y. Lu, C. Zhang, X. Li, et al., Significantly enhanced electrocatalytic activity of Au25 clusters by single platinum atom doping, Nano Energy 50 (2018) 316–322, https://doi.org/10.1016/j.nanoen.2018.05.052.

[232] M. Nesselberger, M. Roefzaad, R. Fayçal Hamou, et al., The effect of particle proximity on the oxygen reduction rate of size-selected platinum clusters, Nat. Mater. 12 (10) (2013) 919–924, https://doi.org/10.1038/nmat3712.

[233] A. Von Weber, E.T. Baxter, S. Proch, et al., Size-dependent electronic structure controls activity for ethanol electro-oxidation at Ptn/indium tin oxide (n = 1 to 14), Phys. Chem. Chem. Phys. 17 (27) (2015) 17601–17610, https://doi.org/10.1039/c5cp01824b.

[234] A. Von Weber, E.T. Baxter, H.S. White, S.L. Anderson, Cluster size controls branching between water and hydrogen peroxide production in electrochemical oxygen reduction at Ptn/ITO, J. Phys. Chem. C 119 (20) (2015) 11160–11170, https://doi.org/10.1021/jp5119234.

[235] K. Yamamoto, T. Imaoka, W.J. Chun, et al., Size-specific catalytic activity of platinum clusters enhances oxygen reduction reactions, Nat. Chem. 1 (5) (2009) 397–402, https://doi.org/10.1038/nchem.288.

[236] T. Imaoka, H. Kitazawa, W.J. Chun, S. Omura, K. Albrecht, K. Yamamoto, Magic number Pt13 and misshapen Pt12 clusters: which one is the better catalyst? J. Am. Chem. Soc. 135 (35) (2013) 13089–13095, https://doi.org/10.1021/ja405922m.

[237] T. Imaoka, H. Kitazawa, W.J. Chun, K. Yamamoto, Finding the most catalytically active platinum clusters with low atomicity, Angew. Chem. Int. Ed. 54 (34) (2015) 9810–9815, https://doi.org/10.1002/anie.201504473.

[238] W. Chen, S. Chen, Oxygen electroreduction catalyzed by gold nanoclusters: strong core size effects, Angew. Chem. Int. Ed. 48 (24) (2009) 4386–4389, https://doi.org/10.1002/anie.200901185.

[239] L. Sumner, N.A. Sakthivel, H. Schrock, K. Artyushkova, A. Dass, S. Chakraborty, Electrocatalytic oxygen reduction activities of thiol-protected nanomolecules ranging in size from Au28(SR)20 to Au279(SR)84, J. Phys. Chem. C 122 (43) (2018) 24809–24817, https://doi.org/10.1021/acs.jpcc.8b07962.

[240] T.C. Jones, L. Sumner, G. Ramakrishna, et al., Bulky t-butyl thiolated gold nanomolecular series: synthesis, characterization, optical properties, and electrocatalysis, J. Phys. Chem. C 122 (31) (2018) 17726–17737, https://doi.org/10.1021/acs.jpcc.8b01106.

[241] D.R. Kauffman, D. Alfonso, C. Matranga, et al., Probing active site chemistry with differently charged Au25q nanoclusters (q = −1, 0, +1), Chem. Sci. 5 (8) (2014) 3151–3157, https://doi.org/10.1039/c4sc00997e.

[242] Y. Lu, Y. Jiang, X. Gao, W. Chen, Charge state-dependent catalytic activity of [Au 25 (SC 12 H 25) 18] nanoclusters for the two-electron reduction of dioxygen to hydrogen peroxide, Chem. Commun. 50 (62) (2014) 8464–8467, https://doi.org/10.1039/C4CC01841A.

[243] K. Kwak, U.P. Azad, W. Choi, K. Pyo, M. Jang, D. Lee, Efficient oxygen reduction electrocatalysts based on gold nanocluster–graphene composites, ChemElectroChem 3 (8) (2016) 1253–1260, https://doi.org/10.1002/celc.201600154.

[244] X. Zou, S. He, X. Kang, et al., New atomically precise M1Ag21 (M = Au/Ag) nanoclusters as excellent oxygen reduction reaction catalysts, Chem. Sci. 12 (10) (2021) 3660–3667, https://doi.org/10.1039/d0sc05923d.

[245] S. Zhao, R. Jin, Y. Song, et al., Atomically precise gold nanoclusters accelerate hydrogen evolution over MoS2 nanosheets: the dual interfacial effect, Small 13 (43) (2017), https://doi.org/10.1002/smll.201701519.

[246] X. Gao, W. Chen, Highly stable and efficient Pd6(SR)12 cluster catalysts for the hydrogen and oxygen evolution reactions, Chem. Commun. 53 (70) (2017) 9733–9736, https://doi.org/10.1039/c7cc04787h.

[247] K. Kwak, W. Choi, Q. Tang, et al., A molecule-like PtAu24(SC6H13)18 nanocluster as an electrocatalyst for hydrogen production, Nat. Commun. 8 (2017).

[248] Y. Du, J. Xiang, K. Ni, et al., Design of atomically precise Au 2 Pd 6 nanoclusters for boosting electrocatalytic hydrogen evolution on MoS2, Inorg. Chem. Front. 5 (11) (2018) 2948–2954, https://doi.org/10.1039/c8qi00697k.

[249] S. Zhao, R. Jin, H. Abroshan, et al., Gold nanoclusters promote electrocatalytic water oxidation at the nanocluster/cose2 interface, J. Am. Chem. Soc. 139 (3) (2017) 1077–1080, https://doi.org/10.1021/jacs.6b12529.

[250] D.R. Kauffman, D. Alfonso, C. Matranga, H. Qian, R. Jin, Experimental and computational investigation of Au 25 clusters and CO2: a unique interaction and enhanced electrocatalytic activity, J. Am. Chem. Soc. 134 (24) (2012) 10237–10243, https://doi.org/10.1021/ja303259q.

[251] D.R. Kauffman, J. Thakkar, R. Siva, et al., Efficient electrochemical CO2 conversion powered by renewable energy, ACS Appl. Mater. Interfaces 7 (28) (2015) 15626–15632, https://doi.org/10.1021/acsami.5b04393.

[252] S. Zhao, N. Austin, M. Li, et al., Influence of atomic-level morphology on catalysis: the case of sphere and rod-like gold nanoclusters for CO2 electroreduction, ACS Catal. 8 (6) (2018) 4996–5001, https://doi.org/10.1021/acscatal.8b00365.

[253] S. Zhuang, D. Chen, L. Liao, et al., Hard-sphere random close-packed Au47Cd2(TBBT)31 nanoclusters with a faradaic efficiency of up to 96% for electrocatalytic CO2 reduction to CO, Angew. Chem. Int. Ed. 59 (8) (2020) 3073–3077, https://doi.org/10.1002/anie.201912845.

[254] D.R. Kauffman, D. Alfonso, C. Matranga, G. Li, R. Jin, Photomediated oxidation of atomically precise Au25(SC 2H4Ph)18- nanoclusters, J. Phys. Chem. Lett. 4 (1) (2013) 195–202, https://doi.org/10.1021/jz302056q.

[255] H. Kawasaki, S. Kumar, G. Li, et al., Generation of singlet oxygen by photoexcited Au25(SR) 18 clusters, Chem. Mater. 26 (9) (2014) 2777–2788,- https://doi.org/10.1021/cm500260z.

[256] Y. Negishi, U. Kamimura, M. Ide, M. Hirayama, A photoresponsive Au25 nanocluster protected by azobenzene derivative thiolates, Nanoscale 4 (14) (2012) 4263–4268, https://doi.org/10.1039/c2nr30830d.

[257] K.G. Stamplecoskie, P.V. Kamat, Size-dependent excited state behavior of glutathione-capped gold clusters and their light-harvesting capacity, J. Am. Chem. Soc. 136 (31) (2014) 11093–11099, https://doi.org/10.1021/ja505361n.

[258] C. Yu, G. Li, S. Kumar, H. Kawasaki, R. Jin, Stable Au 25 (SR) 18 /TiO 2 composite nanostructure with enhanced visible light photocatalytic activity, J. Phys. Chem. Lett. 4 (17) (2013) 2847–2852, https://doi.org/10.1021/jz401447w.

[259] J. Hu, H. Yuan, P. Li, et al., Synthesis and photocatalytic activity of ZnO-Au25 nanocomposites, SCIENCE CHINA Chem. 59 (3) (2016) 277–281,- https://doi.org/10.1007/s11426-015-5487-6.

[260] J. Lee, H.S. Shim, M. Lee, J.K. Song, D. Lee, Size-controlled electron transfer and photocatalytic activity of ZnO-Au nanoparticle composites, J. Phys. Chem. Lett. 2 (22) (2011) 2840–2845, https://doi.org/10.1021/jz2013352.

[261] Y. Liu, Q. Yao, X. Wu, et al., Gold nanocluster sensitized TiO2 nanotube arrays for visible-light driven photoelectrocatalytic removal of antibiotic tetracycline, Nanoscale 8 (19) (2016) 10145–10151, https://doi.org/10.1039/c6nr01702a.

[262] B. Weng, K.Q. Lu, Z. Tang, H.M. Chen, Y.J. Xu, Stabilizing ultrasmall Au clusters for enhanced photoredox catalysis, Nat. Commun. 9 (1) (2018), https://doi.org/10.1038/s41467-018-04020-2.

[263] M.J. Berr, F.F. Schweinberger, M. Döblinger, et al., Size-selected subnanometer cluster catalysts on semiconductor nanocrystal films for atomic scale insight into photocatalysis, Nano Lett. 12 (11) (2012) 5903–5906, https://doi.org/10.1021/nl3033069.

[264] F.F. Schweinberger, M.J. Berr, M. Döblinger, et al., Cluster size effects in the photocatalytic hydrogen evolution reaction, J. Am. Chem. Soc. 135 (36) (2013) 13262–13265, https://doi.org/10.1021/ja406070q.

[265] Y. Hang Li, J. Xing, Z. Jia Chen, et al., Unidirectional suppression of hydrogen oxidation on oxidized platinum clusters, Nat. Commun. (2013), https://doi.org/10.1038/ncomms3500.

[266] Y. Negishi, Y. Matsuura, R. Tomizawa, et al., Controlled loading of small Au n clusters (n = 10–39) onto BaLa 4 Ti 4 O 15 photocatalysts: toward an understanding of size effect of cocatalyst on water-splitting photocatalytic activity, J. Phys. Chem. C 119 (20) (2015) 11224–11232, https://doi.org/10.1021/jp5122432.

[267] X.L. Du, X.L. Wang, Y.H. Li, et al., Isolation of single Pt atoms in a silver cluster: forming highly efficient silver-based cocatalysts for photocatalytic hydrogen evolution, Chem. Commun. 53 (68) (2017) 9402–9405, https://doi.org/10.1039/c7cc04061j.

[268] Z. Li, C. Liu, H. Abroshan, D.R. Kauffman, G. Li, Au38S2(SAdm)20 photocatalyst for one-step selective aerobic oxidations, ACS Catal. 7 (5) (2017) 3368–3374, https://doi.org/10.1021/acscatal.7b00239.

[269] C. Hanbao, G. Guiqi, C. Jinsong, et al., Photoinduced oxidation catalysis by Au25-xAgx(SR)18 nanoclusters, ChemNanoMat 4 (5) (2018) 482–486,- https://doi.org/10.1002/cnma.201700336.

[270] H. Chen, C. Liu, M. Wang, et al., Visible light gold nanocluster photocatalyst: selective aerobic oxidation of amines to imines, ACS Catal. 7 (5) (2017) 3632–3638, https://doi.org/10.1021/acscatal.6b03509.

[271] A. Sanchez, S. Abbet, U. Heiz, et al., When gold is not Noble: nanoscale gold catalysts, J. Phys. Chem. A 103 (48) (1999) 9573–9578, https://doi.org/10.1021/jp9935992.

[272] S. Lee, C. Fan, T. Wu, S.L. Anderson, CO oxidation on Aun/TiO2 catalysts produced by size-selected cluster deposition, J. Am. Chem. Soc. 126 (18) (2004) 5682–5683, https://doi.org/10.1021/ja049436v.

[273] A.A. Herzing, C.J. Kiely, A.F. Carley, P. Landon, G.J. Hutchings, Identification of active gold nanoclusters on iron oxide supports for CO oxidation, Science 321 (5894) (2008) 1331–1335, https://doi.org/10.1126/science.1159639.

[274] Q. He, S.J. Freakley, J.K. Edwards, et al., Population and hierarchy of active species in gold iron oxide catalysts for carbon monoxide oxidation, Nat. Commun. 7 (2016), https://doi.org/10.1038/ncomms12905.

[275] C. Harding, V. Habibpour, S. Kunz, et al., Control and manipulation of gold nanocatalysis: effects of metal oxide support thickness and composition, J. Am. Chem. Soc. 131 (2) (2009) 538–548, https://doi.org/10.1021/ja804893b.

[276] B. Yoon, H. Häkkinen, U. Landman, et al., Charging effects on bonding and catalyzed oxidation of CO on Au8 clusters on MgO, Science 307 (5708) (2005) 403–407, https://doi.org/10.1126/science.1104168.

[277] M. Arenz, U. Landman, U. Heiz, CO combustion on supported gold clusters, ChemPhysChem 7 (9) (2006) 1871–1879, https://doi.org/10.1002/cphc.200600029.

[278] U. Landman, B. Yoon, C. Zhang, U. Heiz, M. Arenz, Factors in gold nanocatalysis: oxidation of CO in the non-scalable size regime, Top. Catal. 44 (1–2) (2007) 145–158, https://doi.org/10.1007/s11244-007-0288-6.

[279] W.E. Kaden, T. Wu, W.A. Kunkel, S.L. Anderson, Electronic structure controls reactivity of size-selected pd clusters adsorbed on tio2 surfaces, Science 326 (5954) (2009) 826–829, https://doi.org/10.1126/science.1180297.

[280] S. Kunz, F.F. Schweinberger, V. Habibpour, et al., Temperature dependent CO oxidation mechanisms on size-selected clusters, J. Phys. Chem. C 114 (3) (2010) 1651–1654, https://doi.org/10.1021/jp911269z.

[281] U. Heiz, A. Sanchez, S. Abbet, W.D. Schneider, Catalytic oxidation of carbon monoxide on monodispersed platinum clusters: each atom counts, J. Am. Chem. Soc. 121 (13) (1999) 3214–3217, https://doi.org/10.1021/ja983616l.

[282] S. Bonanni, K. Aït-Mansour, W. Harbich, H. Brune, Reaction-induced cluster ripening and initial size-dependent reaction rates for CO oxidation on Pt n/TiO 2 (110)-(1 × 1), J. Am. Chem. Soc. 136 (24) (2014) 8702–8707, https://doi.org/10.1021/ja502867r.

[283] M.D. Kane, F.S. Roberts, S.L. Anderson, Effects of alumina thickness on CO oxidation activity over Pd20/Alumina/Re(0001): correlated effects of alumina electronic properties and Pd20 geometry on activity, J. Phys. Chem. C 119 (3) (2015) 1359–1375, https://doi.org/10.1021/jp5093543.

[284] S. Bonanni, K. Aït-Mansour, W. Harbich, H. Brune, Effect of the TiO 2 reduction state on the catalytic CO oxidation on deposited size-selected Pt clusters, J. Am. Chem. Soc. 134 (7) (2012) 3445–3450, https://doi.org/10.1021/ja2098854.

[285] C. Yin, F.R. Negreiros, G. Barcaro, et al., Alumina-supported sub-nanometer Pt10 clusters: amorphization and role of the support material in a highly active CO oxidation catalyst, J. Mater. Chem. A 5 (10) (2017) 4923–4931, https://doi.org/10.1039/c6ta10989f.

[286] M. Moseler, M. Walter, B. Yoon, et al., Oxidation state and symmetry of magnesia-supported Pd 13 O x nanocatalysts influence activation barriers of CO oxidation, J. Am. Chem. Soc. 134 (18) (2012) 7690–7699, https://doi.org/10.1021/ja211121m.

[287] H. Jeong, J. Bae, J.W. Han, H. Lee, Promoting effects of hydrothermal treatment on the activity and durability of Pd/CeO2 catalysts for CO oxidation, ACS Catal. 7 (10) (2017) 7097–7105, https://doi.org/10.1021/acscatal.7b01810.

[288] Y. Watanabe, X. Wu, H. Hirata, N. Isomura, Size-dependent catalytic activity and geometries of size-selected Pt clusters on TiO2(110) surfaces, Catal. Sci. Technol. 1 (8) (2011) 1490–1495, https://doi.org/10.1039/c1cy00204j.

[289] L. Liu, A. Corma, Metal catalysts for heterogeneous catalysis: from single atoms to nanoclusters and nanoparticles, Chem. Rev. 118 (10) (2018) 4981–5079, https://doi.org/10.1021/acs.chemrev.7b00776.

[290] X. Nie, H. Qian, Q. Ge, H. Xu, R. Jin, CO oxidation catalyzed by oxide-supported Au 25(SR) 18 nanoclusters and identification of perimeter sites as active centers, ACS Nano 6 (7) (2012) 6014–6022, https://doi.org/10.1021/nn301019f.

[291] X. Nie, C. Zeng, X. Ma, et al., CeO2-supported Au38(SR)24 nanocluster catalysts for CO oxidation: a comparison of ligand-on and -off catalysts, Nanoscale 5 (13) (2013) 5912–5918, https://doi.org/10.1039/c3nr00970j.

[292] C. Liu, Y. Tan, S. Lin, et al., CO self-promoting oxidation on nanosized gold clusters: triangular Au 3 active site and CO induced O-O scission, J. Am. Chem. Soc. 135 (7) (2013) 2583–2595, https://doi.org/10.1021/ja309460v.

[293] Z. Wu, D.E. Jiang, A.K.P. Mann, et al., Thiolate ligands as a double-edged sword for CO oxidation on CeO 2 supported Au25(SCH2CH2Ph) 18 nanoclusters, J. Am. Chem. Soc. 136 (16) (2014) 6111–6122, https://doi.org/10.1021/ja5018706.

[294] W. Li, Q. Ge, X. Ma, et al., Mild activation of CeO2-supported gold nanoclusters and insight into the catalytic behavior in CO oxidation, Nanoscale 8 (4) (2016) 2378–2385, https://doi.org/10.1039/c5nr07498c.

[295] S. Gaur, J.T. Miller, D. Stellwagen, A. Sanampudi, C.S.S.R. Kumar, J.J. Spivey, Synthesis, characterization, and testing of supported Au catalysts prepared from atomically-tailored Au 38(SC 12H 25) 24 clusters, Phys. Chem. Chem. Phys. 14 (5) (2012) 1627–1634, https://doi.org/10.1039/c1cp22438g.

[296] Z. Wu, G. Hu, D.E. Jiang, et al., Diphosphine-protected Au22 nanoclusters on oxide supports are active for gas-phase catalysis without ligand removal, Nano Lett. 16 (10) (2016) 6560–6567, https://doi.org/10.1021/acs.nanolett.6b03221.

[297] Z. Wu, D.R. Mullins, L.F. Allard, Q. Zhang, L. Wang, CO oxidation over ceria supported Au22 nanoclusters: shape effect of the support, Chin. Chem. Lett. 29 (6) (2018) 795–799, https://doi.org/10.1016/j.cclet.2018.01.038.

[298] J. Lin, W. Li, C. Liu, et al., One-phase controlled synthesis of Au25 nanospheres and nanorods from 1.3 nm Au : PPh3 nanoparticles: the ligand effects, Nanoscale 7 (32) (2015) 13663–13670, https://doi.org/10.1039/c5nr02638e.

[299] W. Li, C. Liu, H. Abroshan, et al., Catalytic CO oxidation using bimetallic M x Au 25− x clusters: a combined experimental and computational study on doping effects, J. Phys. Chem. C 120 (19) (2016) 10261–10267, https://doi.org/10.1021/acs.jpcc.6b00793.

[300] J. Good, P.N. Duchesne, P. Zhang, W. Koshut, M. Zhou, R. Jin, On the functional role of the cerium oxide support in the Au38(SR)24/CeO2 catalyst for CO oxidation, Catal. Today 280 (2017) 239–245, https://doi.org/10.1016/j.cattod.2016.04.016.

[301] S.H. Li, X. Liu, W. Hu, M. Chen, Y. Zhu, An Au8Cluster fortified by four ferrocenes, J. Phys. Chem. A 124 (29) (2020) 6061–6067, https://doi.org/10.1021/acs.jpca.0c03366.

[302] H. Tsunoyama, N. Ichikuni, H. Sakurai, T. Tsukuda, Effect of electronic structures of Au clusters stabilized by poly(N -vinyl-2-pyrrolidone) on aerobic oxidation catalysis, J. Am. Chem. Soc. 131 (20) (2009) 7086–7093, https://doi.org/10.1021/ja810045y.

[303] S. Xie, H. Tsunoyama, W. Kurashige, Y. Negishi, T. Tsukuda, Enhancement in aerobic alcohol oxidation catalysis of Au 25 clusters by single Pd atom doping, ACS Catal. 2 (7) (2012) 1519–1523, https://doi.org/10.1021/cs300252g.

[304] T. Yoskamtorn, S. Yamazoe, R. Takahata, et al., Thiolate-mediated selectivity control in aerobic alcohol oxidation by porous carbon-supported Au25 clusters, ACS Catal. 4 (10) (2014) 3696–3700, https://doi.org/10.1021/cs501010x.

[305] J. Xu, S. Xu, M. Chen, Y. Zhu, Unlocking the catalytic activity of an eight-atom gold cluster with a Pd atom, Nanoscale 12 (10) (2020) 6020–6028, https://doi.org/10.1039/c9nr10198e.

[306] H. Yang, Y. Wang, J. Lei, et al., Ligand-stabilized Au13Cux (x = 2, 4, 8) bimetallic nanoclusters: ligand engineering to control the exposure of metal sites, J. Am. Chem. Soc. 135 (26) (2013) 9568–9571, https://doi.org/10.1021/ja402249s.

[307] C. Lavenn, A. Demessence, A. Tuel, Au25(SPh-pNH2)17 nanoclusters deposited on SBA-15 as catalysts for aerobic benzyl alcohol oxidation, J. Catal. 322 (2015) 130–138, https://doi.org/10.1016/j.jcat.2014.12.002.

[308] H. Deng, S. Wang, S. Jin, et al., Active metal (cadmium) doping enhanced the stability of inert metal (gold) nanocluster under O2 atmosphere and the catalysis activity of benzyl alcohol oxidation, Gold Bull. 48 (3–4) (2015) 161–167, https://doi.org/10.1007/s13404-015-0174-0.

[309] S. Hayashi, R. Ishida, S. Hasegawa, S. Yamazoe, T. Tsukuda, Doping a single palladium atom into gold superatoms stabilized by PVP: emergence of hydrogenation catalysis, Top. Catal. 61 (1–2) (2018) 136–141, https://doi.org/10.1007/s11244-017-0876-z.

[310] L. Liu, Y. Song, H. Chong, et al., Size-confined growth of atom-precise nanoclusters in metal-organic frameworks and their catalytic applications, Nanoscale 8 (3) (2016) 1407–1412, https://doi.org/10.1039/c5nr06930k.
[311] L. Li, L. Dou, H. Zhang, Layered double hydroxide supported gold nanoclusters by glutathione-capped Au nanoclusters precursor method for highly efficient aerobic oxidation of alcohols, Nanoscale 6 (7) (2014) 3753–3763, https://doi.org/10.1039/c3nr05604j.
[312] S. Wang, S. Yin, G. Chen, L. Li, H. Zhang, Nearly atomic precise gold nanoclusters on nickel-based layered double hydroxides for extraordinarily efficient aerobic oxidation of alcohols, Catal. Sci. Technol. 6 (12) (2016) 4090–4104, https://doi.org/10.1039/c6cy00186f.
[313] S. Yin, J. Li, H. Zhang, Hierarchical hollow nanostructured core@shell recyclable catalysts γ-Fe2O3@LDH@Au25-: X for highly efficient alcohol oxidation, Green Chem. 18 (21) (2016) 5900–5914, https://doi.org/10.1039/c6gc01290f.
[314] Y. Yun, H. Sheng, J. Yu, et al., Boosting the activity of ligand-on atomically precise Pd3Cl cluster catalyst by metal-support interaction from kinetic and thermodynamic aspects, Adv. Synth. Catal. 360 (24) (2018) 4731–4743, https://doi.org/10.1002/adsc.201800603.
[315] T.A. Nijhuis, M. Makkee, J.A. Moulijn, B.M. Weckhuysen, The production of propene oxide: catalytic processes and recent developments, Ind. Eng. Chem. Res. 45 (10) (2006) 3447–3459, https://doi.org/10.1021/ie0513090.
[316] S. Lee, L.M. Molina, M.J. López, et al., Selective propene epoxidation on immobilized Au6-10 clusters: the effect of hydrogen and water on activity and selectivity, Angew. Chem. Int. Ed. 48 (8) (2009) 1467–1471, https://doi.org/10.1002/anie.200804154.
[317] Y. Zhu, H. Qian, M. Zhu, R. Jin, Thiolate-protected Aun nanoclusters as catalysts for selective oxidation and hydrogenation processes, Adv. Mater. 22 (17) (2010) 1915–1920, https://doi.org/10.1002/adma.200903934.
[318] Y. Zhu, H. Qian, R. Jin, An atomic-level strategy for unraveling gold nanocatalysis from the perspective of Aun(SR)m nanoclusters, Chem. Eur. J. 16 (37) (2010) 11455–11462, https://doi.org/10.1002/chem.201001086.
[319] H. Qian, D.E. Jiang, G. Li, et al., Monoplatinum doping of gold nanoclusters and catalytic application, J. Am. Chem. Soc. 134 (39) (2012) 16159–16162, https://doi.org/10.1021/ja307657a.
[320] Y. Liu, H. Tsunoyama, T. Akita, T. Tsukuda, Efficient and selective epoxidation of styrene with TBHP catalyzed by Au25clusters on hydroxyapatite, Chem. Commun. 46 (4) (2010) 550–552, https://doi.org/10.1039/b921082b.
[321] S. Wang, S. Jin, S. Yang, et al., (BPh4)2 nanocluster and its structure-related catalytic property, Sci. Adv. 1 (2015).
[322] J. Chai, H. Chong, S. Wang, S. Yang, M. Wu, M. Zhu, Controlling the selectivity of catalytic oxidation of styrene over nanocluster catalysts, RSC Adv. 6 (112) (2016) 111399–111405, https://doi.org/10.1039/c6ra23014h.
[323] E. Wojaczyńska, J.E. Wojaczyński, Synthesis, sulfoxides, Chem. Rev. 110 (2000) 4303–4356.
[324] G. Li, H. Qian, R. Jin, Gold nanocluster-catalyzed selective oxidation of sulfide to sulfoxide, Nanoscale 4 (21) (2012) 6714–6717, https://doi.org/10.1039/c2nr32171h.
[325] Y. Zhu, Z. Wu, C. Gayathri, H. Qian, R.R. Gil, R. Jin, Exploring stereoselectivity of Au25 nanoparticle catalyst for hydrogenation of cyclic ketone, J. Catal. 271 (2) (2010) 155–160, https://doi.org/10.1016/j.jcat.2010.02.027.
[326] C. Liu, C. Yan, J. Lin, C. Yu, J. Huang, G. Li, One-pot synthesis of Au 144 (SCH 2 Ph) 60 nanoclusters and their catalytic application, J. Mater. Chem. A 3 (40) (2015) 20167–20173, https://doi.org/10.1039/C5TA05747G.
[327] Y. Chen, J. Wang, C. Liu, Z. Li, G. Li, Kinetically controlled synthesis of Au102(SPh)44 nanoclusters and catalytic application, Nanoscale 8 (19) (2016) 10059–10065, https://doi.org/10.1039/c5nr08338a.
[328] Y. Liu, H. Tsunoyama, T. Akita, S. Xie, T. Tsukuda, Aerobic oxidation of cyclohexane catalyzed by size-controlled au clusters on hydroxyapatite: size effect in the sub-2 nm regime, ACS Catal. 1 (1) (2011) 2–6, https://doi.org/10.1021/cs100043j.
[329] B. Zhang, S. Kaziz, H. Li, et al., Modulation of active sites in supported Au38(SC2H4Ph)24 cluster catalysts: effect of atmosphere and support material, J. Phys. Chem. C 119 (20) (2015) 11193–11199, https://doi.org/10.1021/jp512022v.
[330] Y. Wang, X.K. Wan, L. Ren, et al., Atomically precise alkynyl-protected metal nanoclusters as a model catalyst: observation of promoting effect of surface ligands on catalysis by metal nanoparticles, J. Am. Chem. Soc. 138 (10) (2016) 3278–3281, https://doi.org/10.1021/jacs.5b12730.
[331] X. Cai, G. Saranya, K. Shen, et al., Reversible switching of catalytic activity by shuttling an atom into and out of gold nanoclusters, Angew. Chem. Int. Ed. 58 (29) (2019) 9964–9968, https://doi.org/10.1002/anie.201903853.
[332] M. Takahashi, H. Koizumi, W.J. Chun, M. Kori, T. Imaoka, K. Yamamoto, Finely controlled multimetallic nanocluster catalysts for solvent-free aerobic oxidation of hydrocarbons, Sci. Adv. 3 (7) (2017), https://doi.org/10.1126/sciadv.1700101.
[333] S. Vajda, M.J. Pellin, J.P. Greeley, et al., Subnanometre platinum clusters as highly active and selective catalysts for the oxidative dehydrogenation of propane, Nat. Mater. 8 (3) (2009) 213–216, https://doi.org/10.1038/nmat2384.
[334] E.T. Baxter, M.A. Ha, A.C. Cass, A.N. Alexandrova, S.L. Anderson, Ethylene dehydrogenation on Pt4,7,8 clusters on Al2O3: strong cluster size dependence linked to preferred catalyst morphologies, ACS Catal. 7 (5) (2017) 3322–3335, https://doi.org/10.1021/acscatal.7b00409.
[335] J. Zhang, Z. Li, J. Huang, et al., Size dependence of gold clusters with precise numbers of atoms in aerobic oxidation of d-glucose, Nanoscale 9 (43) (2017) 16879–16886, https://doi.org/10.1039/c7nr06566c.
[336] Y. Zhu, H. Qian, B.A. Drake, R. Jin, Atomically precise Au25(SR)18 nanoparticles as catalysts for the selective hydrogenation of α,β-unsaturated ketones and aldehydes, Angew. Chem. Int. Ed. 49 (7) (2010) 1295–1298, https://doi.org/10.1002/anie.200906249.
[337] H. Chong, M. Zhu, Catalytic reduction by quasi-homogeneous gold nanoclusters in the liquid phase, ChemCatChem 7 (15) (2015) 2296–2304,- https://doi.org/10.1002/cctc.201500247.
[338] G. Li, R. Jin, Gold nanocluster-catalyzed semihydrogenation: a unique activation pathway for terminal alkynes, J. Am. Chem. Soc. 136 (32) (2014) 11347–11354, https://doi.org/10.1021/ja503724j.
[339] X. Kang, N.B. Zuckerman, J.P. Konopelski, S. Chen, Alkyne-functionalized ruthenium nanoparticles: ruthenium-vinylidene bonds at the metal-ligand interface, J. Am. Chem. Soc. 134 (3) (2012) 1412–1415, https://doi.org/10.1021/ja209568v.

[340] X.K. Wan, J.Q. Wang, Z.A. Nan, Q.M. Wang, Ligand effects in catalysis by atomically precise gold nanoclusters, Sci. Adv. 3 (10) (2017), https://doi.org/10.1126/sciadv.1701823.

[341] M. Zhu, P. Wang, N. Yan, et al., The fourth alloying mode by way of anti-galvanic reaction, Angew. Chem. Int. Ed. 57 (17) (2018) 4500–4504,- https://doi.org/10.1002/anie.201800877.

[342] S. Hasegawa, S. Takano, S. Yamazoe, T. Tsukuda, Prominent hydrogenation catalysis of a PVP-stabilized Au34 superatom provided by doping a single Rh atom, Chem. Commun. 54 (46) (2018) 5915–5918, https://doi.org/10.1039/c8cc03123a.

[343] A.S. Crampton, M.D. Rötzer, C.J. Ridge, et al., Structure sensitivity in the nonscalable regime explored via catalysed ethylene hydrogenation on supported platinum nanoclusters, Nat. Commun. 7 (2016), https://doi.org/10.1038/ncomms10389.

[344] A.S. Crampton, M.D. Rötzer, C.J. Ridge, et al., Assessing the concept of structure sensitivity or insensitivity for sub-nanometer catalyst materials, Surf. Sci. 652 (2016) 7–19, https://doi.org/10.1016/j.susc.2016.02.006.

[345] A.S. Crampton, M.D. Rötzer, F.F. Schweinberger, B. Yoon, U. Landman, U. Heiz, Controlling ethylene hydrogenation reactivity on Pt13Clusters by varying the stoichiometry of the amorphous silica support, Angew. Chem. Int. Ed. 55 (31) (2016) 8953–8957, https://doi.org/10.1002/anie.201603332.

[346] M. Takahashi, T. Imaoka, Y. Hongo, K. Yamamoto, Formation of a Pt 12 cluster by single-atom control that leads to enhanced reactivity: hydrogenation of unreactive olefins, Angew. Chem. Int. Ed. 52 (29) (2013) 7419–7421, https://doi.org/10.1002/anie.201302860.

[347] M. Takahashi, T. Imaoka, Y. Hongo, K. Yamamoto, A highly-active and poison-tolerant Pt12 sub-nanocluster catalyst for the reductive amination of aldehydes with amines, Dalton Trans. 42 (45) (2013) 15919–15921, https://doi.org/10.1039/c3dt52099d.

[348] Z. Maeno, T. Kibata, T. Mitsudome, T. Mizugaki, K. Jitsukawa, K. Kaneda, Subnanoscale size effect of dendrimer-encapsulated Pd clusters on catalytic hydrogenation of olefin, Chem. Lett. 40 (2) (2011) 180–181, https://doi.org/10.1246/cl.2011.180.

[349] I. Nakamula, Y. Yamanoi, T. Yonezawa, T. Imaoka, K. Yamamoto, H. Nishihara, Nanocage catalysts—rhodium nanoclusters encapsulated with dendrimers as accessible and stable catalysts for olefin and nitroarene hydrogenations, Chem. Commun. 44 (2008) 5716–5718, https://doi.org/10.1039/b813649a.

[350] P. Maity, S. Yamazoe, T. Tsukuda, Dendrimer-encapsulated copper cluster as a chemoselective and regenerable hydrogenation catalyst, ACS Catal. 3 (2) (2013) 182–185, https://doi.org/10.1021/cs3007318.

[351] L. Liu, U. Díaz, R. Arenal, G. Agostini, P. Concepción, A. Corma, Generation of subnanometric platinum with high stability during transformation of a 2D zeolite into 3D, Nat. Mater. 16 (1) (2017) 132–138, https://doi.org/10.1038/nmat4757.

[352] J.M. Thomas, B.F.G. Johnson, R. Raja, G. Sankar, P.A. Midgley, High-performance nanocatalysts for single-step hydrogenations, Acc. Chem. Res. 36 (1) (2003) 20–30, https://doi.org/10.1021/ar990017q.

[353] A.B. Hungria, R. Raja, R.D. Adams, et al., Single-step conversion of dimethyl terephthalate into cyclohexanedimethanol with Ru5PtSn, a trimetallic nanoparticle catalyst, Angew. Chem. Int. Ed. 45 (29) (2006) 4782–4785, https://doi.org/10.1002/anie.200600359.

[354] G. Li, H. Abroshan, Y. Chen, R. Jin, H.J. Kim, Experimental and mechanistic understanding of aldehyde hydrogenation using Au25 nanoclusters with Lewis acids: unique sites for catalytic reactions, J. Am. Chem. Soc. 137 (45) (2015) 14295–14304, https://doi.org/10.1021/jacs.5b07716.

[355] Q. Huifeng, B. Ellen, Z. Yan, J. Rongchao, Doping 25-atom and 38-atom gold nanoclusters with palladium, Acta Phys. -Chim. Sin. (2011) 513–519,- https://doi.org/10.3866/pku.whxb20110304.

[356] M. Dasog, W. Hou, R.W.J. Scott, Controlled growth and catalytic activity of gold monolayer protected clusters in presence of borohydride salts, Chem. Commun. 47 (30) (2011) 8569–8571, https://doi.org/10.1039/c1cc11813g.

[357] H. Yamamoto, H. Yano, H. Kouchi, Y. Obora, R. Arakawa, H. Kawasaki, N,N-Dimethylformamide-stabilized gold nanoclusters as a catalyst for the reduction of 4-nitrophenol, Nanoscale 4 (14) (2012) 4148–4154, https://doi.org/10.1039/c2nr30222e.

[358] A. Shivhare, S.J. Ambrose, H. Zhang, R.W. Purves, R.W.J. Scott, Stable and recyclable Au25 clusters for the reduction of 4-nitrophenol, Chem. Commun. 49 (3) (2013) 276–278, https://doi.org/10.1039/c2cc37205c.

[359] J. Fang, J. Li, B. Zhang, et al., The support effect on the size and catalytic activity of thiolated Au25 nanoclusters as precatalysts, Nanoscale 7 (14) (2015) 6325–6333, https://doi.org/10.1039/c5nr00549c.

[360] V. Sudheeshkumar, A. Shivhare, R.W.J. Scott, Synthesis of sinter-resistant Au@silica catalysts derived from Au25 clusters, Catal. Sci. Technol. 7 (1) (2017) 272–280, https://doi.org/10.1039/c6cy01822j.

[361] J. Li, R.R. Nasaruddin, Y. Feng, J. Yang, N. Yan, J. Xie, Tuning the accessibility and activity of Au25(SR)18Nanocluster catalysts through ligand engineering, Chem. Eur. J. 22 (42) (2016) 14816–14820, https://doi.org/10.1002/chem.201603247.

[362] N. Yan, L. Liao, J. Yuan, et al., Bimetal doping in nanoclusters: synergistic or counteractive? Chem. Mater. 28 (22) (2016) 8240–8247, https://doi.org/10.1021/acs.chemmater.6b03132.

[363] R.R. Nasaruddin, T. Chen, J. Li, et al., Ligands modulate reaction pathway in the hydrogenation of 4-Nitrophenol catalyzed by gold nanoclusters, ChemCatChem 10 (2) (2018) 395–402, https://doi.org/10.1002/cctc.201701472.

[364] D. Hu, S. Jin, Y. Shi, et al., Preparation of hyperstar polymers with encapsulated Au25(SR)18 clusters as recyclable catalysts for nitrophenol reduction, Nanoscale 9 (10) (2017) 3629–3636, https://doi.org/10.1039/c6nr09727h.

[365] M. Farrag, Microwave-assisted synthesis of ultra small bare gold clusters supported over Al2O3 and TiO2 as catalysts in reduction of 4-nitrophenol to 4-aminophenol, Microporous Mesoporous Mater. 232 (2016) 248–255, https://doi.org/10.1016/j.micromeso.2016.06.032.

[366] H. Chen, C. Liu, M. Wang, C. Zhang, G. Li, F. Wang, Thermally robust silica-enclosed Au25 nanocluster and its catalysis, Cuihua Xuebao 37 (10) (2016) 1787–1793, https://doi.org/10.1016/S1872-2067(16)62478-6.

[367] S. Zhao, A. Das, H. Zhang, R. Jin, Y. Song, R. Jin, Mechanistic insights from atomically precise gold nanocluster-catalyzed reduction of 4-nitrophenol, Prog. Nat. Sci. Mater. Int. 26 (5) (2016) 483–486, https://doi.org/10.1016/j.pnsc.2016.08.009.

[368] Y. Luo, S. Fan, W. Yu, et al., Fabrication of Au25(SG)18–ZIF-8 nanocomposites: a facile strategy to position Au25(SG)18 nanoclusters inside and outside ZIF-8, Adv. Mater. 30 (2018), https://doi.org/10.1002/adma.201704576.
[369] G. Li, C. Zeng, R. Jin, Thermally robust Au99(SPh)42 nanoclusters for chemoselective hydrogenation of nitrobenzaldehyde derivatives in water, J. Am. Chem. Soc. 136 (9) (2014) 3673–3679, https://doi.org/10.1021/ja500121v.
[370] G. Li, D.E. Jiang, S. Kumar, Y. Chen, R. Jin, Size dependence of atomically precise gold nanoclusters in chemoselective hydrogenation and active site structure, ACS Catal. 4 (8) (2014) 2463–2469, https://doi.org/10.1021/cs500533h.
[371] C. Liu, H. Abroshan, C. Yan, G. Li, M. Haruta, One-pot synthesis of Au11(PPh2Py)7Br3 for the highly chemoselective hydrogenation of Nitrobenzaldehyde, ACS Catal. 6 (1) (2016) 92–99, https://doi.org/10.1021/acscatal.5b02116.
[372] Y. Tan, X.Y. Liu, L. Zhang, et al., ZnAl-hydrotalcite-supported Au25 nanoclusters as precatalysts for chemoselective hydrogenation of 3-nitrostyrene, Angew. Chem. Int. Ed. 56 (10) (2017) 2709–2713, https://doi.org/10.1002/anie.201610736.
[373] D. Astruc, Electron Transfer and Radical Processes in Transition Metal Chemistry, 1995.
[374] D. Lee, R.L. Donkers, G. Wang, A.S. Harper, R.W. Murray, Electrochemistry and optical absorbance and luminescence of molecule-like Au38 nanoparticles, J. Am. Chem. Soc. 126 (19) (2004) 6193–6199, https://doi.org/10.1021/ja049605b.
[375] J. Kim, K. Lema, M. Ukaigwe, D. Lee, Facile preparative route to alkanethiolate-coated Au38 nanoparticles: postsynthesis core size evolution, Langmuir 23 (14) (2007) 7853–7858, https://doi.org/10.1021/la700753u.
[376] S. Antonello, A.H. Holm, E. Instuli, F. Maran, Molecular electron-transfer properties of Au38 clusters, J. Am. Chem. Soc. 129 (32) (2007) 9836–9837, https://doi.org/10.1021/ja071191+.
[377] B.M. Quinn, P. Liljeroth, V. Ruiz, T. Laaksonen, K. Kontturi, Electrochemical resolution of 15 oxidation states for monolayer protected gold nanoparticles, J. Am. Chem. Soc. 125 (22) (2003) 6644–6645, https://doi.org/10.1021/ja0349305.
[378] J.F. Parker, J.P. Choi, W. Wang, R.W. Murray, Electron self-exchange dynamics of the nanoparticle couple [Au25(SC2Ph)18]0/1- by nuclear magnetic resonance line-broadening, J. Phys. Chem. C 112 (36) (2008) 13976–13981, https://doi.org/10.1021/jp805638x.
[379] T.M. Carducci, R.W. Murray, Kinetics and low temperature studies of electron transfers in films of small (<2 nm) Au monolayer protected clusters, J. Am. Chem. Soc. 135 (30) (2013) 11351–11356, https://doi.org/10.1021/ja405342r.
[380] A. Venzo, S. Antonello, J.A. Gascón, et al., Effect of the charge state (z = −1, 0, +1) on the nuclear magnetic resonance of monodisperse Au 25 [S(CH 2) 2Ph] 18z clusters, Anal. Chem. 83 (16) (2011) 6355–6362, https://doi.org/10.1021/ac2012653.
[381] S. Antonello, M. Hesari, F. Polo, F. Maran, Electron transfer catalysis with monolayer protected Au25 clusters, Nanoscale 4 (17) (2012) 5333–5342, https://doi.org/10.1039/c2nr31066j.
[382] S. Antonello, N.V. Perera, M. Ruzzi, J.A. Gascón, F. Maran, Interplay of charge state, lability, and magnetism in the molecule-like Au25(SR)18 cluster, J. Am. Chem. Soc. 135 (41) (2013) 15585–15594, https://doi.org/10.1021/ja407887d.
[383] S. Antonello, T. Dainese, F. Maran, Exploring collective substituent effects: dependence of the lifetime of charged states of Au25(SCnH2n+1)18 nanoclusters on the length of the thiolate ligands, Electroanalysis 28 (11) (2016) 2771–2776, https://doi.org/10.1002/elan.201600323.
[384] Y. Negishi, N.K. Chaki, Y. Shichibu, R.L. Whetten, T. Tsukuda, Origin of magic stability of thiolated gold clusters: a case study on Au25(SC6H13) 18, J. Am. Chem. Soc. 129 (37) (2007) 11322–11323, https://doi.org/10.1021/ja073580+.
[385] M. Zhu, W.T. Eckenhoff, T. Pintauer, R. Jin, Conversion of anionic [Au25(SCH2CH 2Ph)18]- cluster to charge neutral cluster via air oxidation, J. Phys. Chem. C 112 (37) (2008) 14221–14224, https://doi.org/10.1021/jp805786p.
[386] M. Zhu, C.M. Aikens, M.P. Hendrich, et al., Reversible switching of magnetism in thiolate-protected Au25 superatoms, J. Am. Chem. Soc. 131 (7) (2009) 2490–2492, https://doi.org/10.1021/ja809157f.
[387] M. Zhu, G. Chan, H. Qian, R. Jin, Unexpected reactivity of Au25(SCH2CH 2Ph)18 nanoclusters with salts, Nanoscale 3 (4) (2011) 1703–1707, https://doi.org/10.1039/c0nr00878h.
[388] M.S. Devadas, K. Kwak, J.W. Park, et al., Directional electron transfer in chromophore-labeled quantum-sized Au 25 clusters: Au25 as an electron donor, J. Phys. Chem. Lett. 1 (9) (2010) 1497–1503, https://doi.org/10.1021/jz100395p.
[389] Z. Liu, M. Zhu, X. Meng, G. Xu, R. Jin, Electron transfer between [Au25(SC2H4Ph)18]—TOA + and oxoammonium cations, J. Phys. Chem. Lett. 2 (17) (2011) 2104–2109, https://doi.org/10.1021/jz200925h.
[390] Z. Liu, Q. Xu, S. Jin, S. Wang, G. Xu, M. Zhu, Electron transfer reaction between Au25 nanocluster and phenothiazine-tetrachloro-p-benzoquinone complex, Int. J. Hydrogen Energy 38 (36) (2013) 16722–16726, https://doi.org/10.1016/j.ijhydene.2013.06.030.
[391] H. Chong, P. Li, S. Wang, et al., Au25 clusters as electron-transfer catalysts induced the intramolecular cascade reaction of 2-nitrobenzonitrile, Sci. Rep. 3 (2013), https://doi.org/10.1038/srep03214.
[392] G. Li, C. Liu, Y. Lei, R. Jin, Au25 nanocluster-catalyzed ullmann-type homocoupling reaction of aryl iodides, Chem. Commun. 48 (98) (2012) 12005–12007, https://doi.org/10.1039/c2cc34765b.
[393] G. Li, H. Abroshan, C. Liu, et al., Tailoring the electronic and catalytic properties of Au25 nanoclusters via ligand engineering, ACS Nano 10 (8) (2016) 7998–8005, https://doi.org/10.1021/acsnano.6b03964.
[394] H. Abroshan, G. Li, J. Lin, H.J. Kim, R. Jin, Molecular mechanism for the activation of Au25(SCH2CH2Ph)18 nanoclusters by imidazolium-based ionic liquids for catalysis, J. Catal. 337 (2016) 72–79, https://doi.org/10.1016/j.jcat.2016.01.011.
[395] A. Leyva-Pérez, J. Oliver-Meseguer, P. Rubio-Marqués, A. Corma, Water-stabilized three- and four-atom palladium clusters as highly active catalytic species in ligand-free C-C cross-coupling reactions, Angew. Chem. Int. Ed. 52 (44) (2013) 11554–11559, https://doi.org/10.1002/anie.201303188.
[396] F. Fu, J. Xiang, H. Cheng, et al., A robust and efficient Pd3 cluster catalyst for the Suzuki reaction and its odd mechanism, ACS Catal. 7 (3) (2017) 1860–1867, https://doi.org/10.1021/acscatal.6b02527.

[397] C. Lv, H. Cheng, W. He, et al., Pd3 cluster catalysis: compelling evidence from in operando spectroscopic, kinetic, and density functional theory studies, Nano Res. 9 (9) (2016) 2544–2550, https://doi.org/10.1007/s12274-016-1140-8.

[398] S. Blanchard, L. Fensterbank, G. Gontard, E. Lacôte, G. Maestri, M. Malacria, Synthesis of triangular tripalladium cations as noble-metal analogues of the cyclopropenyl cation, Angew. Chem. Int. Ed. 53 (1987).

[399] Y. Wang, P.A. Deyris, T. Caneque, et al., A simple synthesis of triangular all-metal aromatics allowing access to isolobal all-metal heteroaromatics, Chem. Eur. J. 21 (35) (2015) 12271–12274, https://doi.org/10.1002/chem.201501239.

[400] P.A. Deyris, T. Cañeque, Y. Wang, et al., Catalytic semireduction of internal alkynes with all-metal aromatic complexes, ChemCatChem 7 (20) (2015) 3266–3269, https://doi.org/10.1002/cctc.201500729.

[401] A. Monfredini, V. Santacroce, P.A. Deyris, et al., Boosting catalyst activity in cis -selective semi-reduction of internal alkynes by tailoring the assembly of all-metal aromatic tri-palladium complexes, Dalton Trans. 45 (40) (2016) 15786–15790, https://doi.org/10.1039/c6dt01840h.

[402] H. Shen, G. Deng, S. Kaappa, et al., Highly robust but surface-active: An N-heterocyclic carbene-stabilized Au25 nanocluster, Angew. Chem. Int. Ed. 58 (49) (2019) 17731–17735, https://doi.org/10.1002/anie.201908983.

[403] G. Li, D.E. Jiang, C. Liu, C. Yu, R. Jin, Oxide-supported atomically precise gold nanocluster for catalyzing Sonogashira cross-coupling, J. Catal. 306 (2013) 177–183, https://doi.org/10.1016/j.jcat.2013.06.017.

[404] Q. Li, A. Das, S. Wang, Y. Chen, R. Jin, Highly efficient three-component coupling reaction catalysed by atomically precise ligand-protected Au38(SC2H4Ph)24 nanoclusters, Chem. Commun. 52 (99) (2016) 14298–14301, https://doi.org/10.1039/c6cc07825g.

[405] M.B. Li, S.K. Tian, Z. Wu, Improving the catalytic activity of Au25 nanocluster by peeling and doping, Chin. J. Chem. 35 (5) (2017) 567–571,- https://doi.org/10.1002/cjoc.201600526.

[406] Y. Liu, X. Chai, X. Cai, et al., Central doping of a foreign atom into the silver cluster for catalytic conversion of CO 2 toward C − C bond formation, Angew. Chem. Int. Ed. 57 (31) (2018) 9775–9779, https://doi.org/10.1002/anie.201805319.

[407] G. Li, X. Sui, X. Cai, et al., Precisely constructed silver active sites in gold nanoclusters for chemical fixation of CO 2, Angew. Chem. 133 (19) (2021) 10667–10670, https://doi.org/10.1002/ange.202100071.

[408] L. Sun, Y. Yun, H. Sheng, et al., Rational encapsulation of atomically precise nanoclusters into metal–organic frameworks by electrostatic attraction for CO2 conversion, J. Mater. Chem. A 6 (31) (2018) 15371–15376, https://doi.org/10.1039/c8ta04667k.

[409] Y. Yun, H. Sheng, K. Bao, et al., Design and remarkable efficiency of the robust Sandwich cluster composite nanocatalysts ZIF-8@Au25@ZIF-67, J. Am. Chem. Soc. 142 (9) (2020) 4126–4130, https://doi.org/10.1021/jacs.0c00378.

[410] M.B. Li, S.K. Tian, Z. Wu, Catalyzed formation of α,β-unsaturated ketones or aldehydes from propargylic acetates by a recoverable and recyclable nanocluster catalyst, Nanoscale 6 (11) (2014) 5714–5717, https://doi.org/10.1039/c4nr00658e.

[411] P. Huang, Z. Jiang, G. Chen, Y. Zhu, Y. Sun, Dehydration of phenylboronic acid to boroxine catalyzed by Aun nanoclusters with atom packing core-shell structure, J. Nanosci. Nanotechnol. 13 (7) (2013) 5088–5092, https://doi.org/10.1166/jnn.2013.7577.

[412] M. Urushizaki, H. Kitazawa, S. Takano, R. Takahata, S. Yamazoe, T. Tsukuda, Synthesis and catalytic application of Ag44 clusters supported on mesoporous carbon, J. Phys. Chem. C 119 (49) (2015) 27483–27488, https://doi.org/10.1021/acs.jpcc.5b08903.

[413] L. Sun, K. Shen, H. Sheng, et al., Au-Ag synergistic effect in CF3-ketone alkynylation catalyzed by precise nanoclusters, J. Catal. 378 (2019) 220–225, https://doi.org/10.1016/j.jcat.2019.08.043.

[414] Y. Fang, K. Bao, P. Zhang, et al., Insight into the mechanism of the CuAAC reaction by capturing the crucial Au4Cu4-π-alkyne intermediate, J. Am. Chem. Soc. 143 (4) (2021) 1768–1772, https://doi.org/10.1021/jacs.0c12498.

[415] J. Chai, S. Yang, Y. Lv, H. Chong, H. Yu, M. Zhu, Exposing the delocalized Cu − S π bonds on the Au24Cu6(SPhtBu)22 nanocluster and its application in ring-opening reactions, Angew. Chem. Int. Ed. 58 (44) (2019) 15671–15674, https://doi.org/10.1002/anie.201907609.

[416] D. Yang, W. Pei, S. Zhou, J. Zhao, W. Ding, Y. Zhu, Controllable conversion of CO2 on non-metallic gold clusters, Angew. Chem. Int. Ed. 59 (5) (2020) 1919–1924, https://doi.org/10.1002/anie.201913635.

[417] X. Cai, W. Hu, S. Xu, et al., Structural relaxation enabled by internal vacancy available in a 24-atom gold cluster reinforces catalytic reactivity, J. Am. Chem. Soc. 142 (9) (2020) 4141–4153, https://doi.org/10.1021/jacs.9b07761.

[418] L. Wang, K. Shen, M. Chen, Y. Zhu, One-core-atom loss in a gold nanocluster promotes hydroamination reaction of alkynes, Nanoscale 11 (29) (2019) 13767–13772, https://doi.org/10.1039/c9nr04219a.

[419] J. Xu, M. Chen, Y. Zhu, Selective hydrogenation of CO2 dictated by isomers in Au28(SR)20 nanoclusters: which one is better? Chem. A Eur. J. 25 (39) (2019) 9185–9190, https://doi.org/10.1002/chem.201901698.

[420] X. Chai, T. Li, M. Chen, R. Jin, W. Ding, Y. Zhu, Suppressing the active site-blocking impact of ligands of Ni6(SR)12 clusters with the assistance of NH3 on catalytic hydrogenation of nitriles, Nanoscale 10 (41) (2018) 19375–19382, https://doi.org/10.1039/c8nr03700k.

[421] Y. Sun, E. Wang, Y. Ren, et al., The evolution in catalytic activity driven by periodic transformation in the inner sites of gold clusters, Adv. Funct. Mater. 29 (38) (2019) 1904242, https://doi.org/10.1002/adfm.201904242.

Chapter 9

Summary and perspectives

Manzhou Zhu and Xi Kang

9.1 Summary of this book

Nanocluster materials have unique characterizations such as atomic precision, molecular purity, and hence, they resemble molecular compounds. In this book, we first introduce the controllable preparation and structural determination of metal nanoclusters. Then, the mechanisms of structural evolutions and the origins of chemical-physical properties of metal nanoclusters are disclosed. Furthermore, the theoretical approaches to analyze nanoclusters are reviewed. Moreover, the assembly of nanocluster compounds and the effect on properties are summarized. Finally, the application status of nanoclusters and cluster-based nanomaterials are presented. Indeed, tremendous researching efforts have been made on understanding, manipulating, and applying metal nanoclusters in these years, especially in the past two decades. Through these advances, it is anticipated that metal nanoclusters have opened up a new era of colloidal nanoparticle research.

9.2 Personal perspectives to the nanocluster science

Although tremendous advances have been made to the nanocluster science, much work remains to be done. In regard to future perspectives, we briefly comment on the following issues.

9.2.1 New synthetic methodologies

The preparation and purification are the basis of the subsequent analyses and applications of metal nanoclusters. So far, although a wide series of gold, silver, and their homologous alloy nanoclusters with different nuclearities have been controllably synthesized and structural determined, the reports on other metal-based nanoclusters (such as Pd, Pt, Cu, Fe, and Ni nanoclusters) are limited. Considering the different electronic structures between Au/Ag and other metals, new synthetic methodologies may be necessary for the preparation of Pd, Pt, Cu, Fe, and Ni nanoclusters. Significant improvement in synthetic methodology is needed in order to increase the diversity of metal nanoclusters, which is favorable to the in-depth understanding of properties and the wide applications of cluster-based nanomaterials.

9.2.2 Preparing more novel alloy nanoclusters

Alloying has proven to be a versatile strategy for improving the chemical and physical performance of metal materials. For example, alloy nanoclusters exhibit improved stability, enhanced fluorescence, and intriguing chirality relative to the corresponding monometallic nanoclusters, which are expected to benefit the applications of such nanomaterials [1]. Besides, alloying in nanoclusters is of major significance as this strategy can be utilized to increase the diversity of nanoclusters, improve their functionality and broaden their applications [1].

Besides, the alloying of metal nanoclusters has some implications for the wider scientific community: (i) the alloying of different heterometals into a nanocluster template follows distinct modes in terms of doping locations, numbers, and pathways, which leads to the structural and compositional diversities of cluster-based nanomaterials; (ii) atomically precise alloy nanoclusters constitute a unique platform for investigating intermetallic synergism at the unprecedented atomic level, which is not possible to accomplish with conventional (larger) nanoparticles. It is anticipated that atomic level understanding of such synergisms will guide the preparation of other types of metal nanomaterials with controllable compositions and desirable properties; (iii) alloy nanoclusters often exhibit enhanced optical, electrochemical, and catalytic properties compared with monometallic nanoclusters; and alloy nanocluster-based materials are being widely used in bio-labeling, imaging, sensing, catalysis, to name a few. All these benefits drive us to continually prepare and research alloy nanoclusters.

9.2.3 Ligand effect on nanoclusters

The ligand effect was largely overlooked in earlier research since the long-chain HS-C_n H_{2n+1}, HS-C_2H_4Ph, and GSH ligands behave quite similarly in the preparation of $Au_{25}(SR)_{18}$ nanoclusters [2]. Along with the preparation of more nanoclusters with different nuclearities, the ligand effect in controlling the structures and compositions of metal nanoclusters is disclosed. The most evident of the ligand effect is manifested in the ligand exchange-induced structure transformation of metal nanoclusters, such as from $Au_{25}(S-C_2H_4Ph)_{18}$ to $Au_{28}(S-PhtBu)_{20}$ [3], and from $Au_{144}(S-C_2H_4Ph)_{60}$ to $Au_{133}(S-PhtBu)_{52}$ [4].

The ligand effect has also been exploited in dictating the chemical-physical properties of metal nanoclusters. For example, the LMCT (ligand-to-metal charge transfer) mechanism has been proposed to explicate the origin of nanocluster photoluminescence [5]. From this point of view, the emission of several nanoclusters (e.g., Au_{25} or Pt_1Ag_{28}) has been regulating by controlling the electron-donating and -absorbing ability of ligands [5,6]. In this context, the requirement of highly emissive nanoclusters calls for the development of more ligands with strong electron-donating ability.

Besides, joint efforts between experiment and theory are still needed to map out more details of the ligand effect at the atomic level, which will lead to the discovery of more nanoclusters with novel sizes, structures, and properties.

9.2.4 Cocrystallization of metal nanoclusters

Of the great class of material crystallization, the "cocrystallization" has gained a significant presence. The term "cocrystallization" refers to the crystalline system consisting of two or more different components that form a unique crystalline structure. The assembled structures endow cocrystallized materials with novel physical-chemical properties that differ from individual components [7]. In this context, the cocrystal engineering is closely relevant to the production of functionalized compounds.

The cocrystallization of heterogeneous nanoclusters has emerged as a new "growth point" in nanocluster science since it allows for the full understanding of both intracluster structures and intercluster packing modes [8]. However, although thousands of crystal structures of metal nanoclusters have been reported, the research of cocrystallization in metal nanoclusters is still in its infancy and approximately 10 cases of nanocluster cocrystallization systems have been reported up to the present [8]. The absence of the nanocluster cocrystallization hinders both the fundamental research and the practical application of cluster-based nanomaterials. In this context, the further development of the nanocluster cocrystallization calls for more efforts.

9.2.5 Periodicity of nanoclusters

Nanocluster researchers always have a dream to arrange nanocluster entities into a periodic table by referring to their regular structures and compositions, which is like the periodic table of elements [9]. However, the reported nanocluster sizes appear to be discrete, and their structures appear to have no mutual relation. The Jin group reported identified the first magic series of Au nanoclusters protected by the same HS-PhtBu ligands, with a general formula of $Au_{8n+4}(S-PhtBu)_{4n+8}$ ($n=3-6$) [3,10–13]. These gold nanoclusters exhibited a uniform evolution in atomic structures, optical spectra, and HOMO-LUMO gaps (HOMO = the highest occupied molecular orbital and LUMO = lowest unoccupied molecular orbital) after every growth of the $Au_8(S-PhtBu)_4$ unit [10,12,13].

The research of the periodicity of nanoclusters allows for an in-depth understanding of growth modes of metal nanoclusters and nanoparticles as well as the mechanism of their material properties. Future work could identify more magic series and reveal deeper relationships among the sizes of nanoclusters. The periodicity of nanoclusters is an exciting direction for future research.

9.2.6 Seeing the transformation of nanoclusters

From crystal structures of nanocluster precursors and their transforming counterparts, one can only gain information about the final conversion sites. However, the atomically precise transformation routes remain mysteries. For example, as to the mono- and central doping of Au in $Ag_{25}(SPhMe_2)_{18}$, Bakr et al. proposed that the introduced Au^+ was first reduced to Au(0), replacing one Ag on the icosahedral kernel shell, and then the Au(0) diffused to the center due to thermodynamic favorability [14]. Xie and coworkers further demonstrated this Au-doping process based on alloying $Ag_{25}(MHA)_{18}$ into $Au_1Ag_{24}(MHA)_{18}$ (MHA = 6-mercaptohexanoic acid) by analyzing the mass spectrometry of the transformation [15].

Besides, ligand exchange serves as an efficient approach to transform the nanocluster structures and to dictate their properties [16,17]. Previous works can only propose the possible transformations by examining the time-dependent mass spectrometry or performing the theoretical means; however, to develop an intuitive technique to directly "see" these transformations remains challenging. Future work is still necessary to provide more intuitive perspectives in many other nanocluster transformations. In situ EXAFS and the progress of microscope technology may provide precise information on these.

9.2.7 Fixing the boundary between quantum-sized nanoclusters and metallic-state nanoparticles

Colloid nanoparticles with large sizes are more like bulk metal with continuously electronic bands, whereas the small-sized metal nanoclusters exhibit discrete electronic energy levels [18]. The fixing of the boundary between quantum-sized nanoclusters and metallic-state nanoparticles remains a key topic in the nanocluster science to explore the mechanism of material properties of both nanoclusters and nanoparticles.

To date, the largest Au nanocluster with structure elucidation reported is $Au_{279}(SR)_{84}$, and a sharp transition from a plasmonic to molecular state occurs between $Au_{246}(SR)_{80}$ and $Au_{279}(SR)_{84}$ nanoclusters [19,20]. On the other hand, the "transition region" of Ag nanoclusters moves to a much smaller size [21]. Thus, alloying is expected to alter the boundary between metallic and nonmetallic states. Alloy nanoclusters with large sizes, $Au_{130-x}Ag_x(SR)_{50}$ and $Au_{52}Cu_{72}(SR)_{55}$, still show nonmetallic properties, although Ag or Cu atoms outnumber Au in these cases [22,23]. Whether $Au_{144}(SR)_{60}$ is metallic or nonmetallic is still controversial based on femtosecond time-resolved transient absorption spectroscopy [24]. Upon doping, $Au_{144-x}Ag_x(SR)_{60}$ shows nonmetallic optical absorptions, while that of $Au_{144-x}Cu_x(SR)_{60}$ is plasmon-like [25,26]. Therefore, doping/alloying on nanoclusters of large sizes is worth studying in the future.

9.2.8 Mechanisms of properties of nanoclusters

Owing to the discrete electron energy levels and strong quantum size effect, metal nanoclusters prone to display structure-dependent chemical-physical properties, and even a single-atom variation on nanocluster structures will remarkably influence their properties. In this context, nanocluster materials exhibit abundant material properties, such as fluorescence, chirality, magnetism, and electrochemical and electroluminescence properties, and so on, which provide a material library for applications of cluster-based nanomaterials.

However, the meticulous mechanism of these properties remains to be elucidated. Taking the fluorescence as an example, geometric structure factors (e.g., the restriction of intramolecular motion) and electronic structure factors (e.g., the ligand-to-metal charge transfer, the relativistic effect, or the superatomic $P \leftarrow D$ transition) are often invoked to explain the emission of a specific nanocluster [5,27,28]. In this context, the emission generation (or the emission enhancement) of nanoclusters involves multiple factors, rather than any single factor, which calls for more efforts to dig out. Future work is expected to reveal more insight into the emission issue (and also for the mechanism of other properties).

9.2.9 Future theoretical works

Although density functional theory (DFT) is very powerful and is currently the primary method, the simulated electronic structures and optical absorption spectra still not sufficiently match the experimental ones. Besides, the computational demands rise sharply with the increasingly reported $M_n(SR)_m$ nanoclusters, especially for large-size nanoclusters (i.e., with kernel metals of more than 100 atoms). In this context, it is highly desirable to develop more efficient DFT methods.

Also, DFT calculations are essential to understand the cluster evolution and transformation, such as the early stages of the nanocluster nucleation and growth and the ligand exchange-induced structural transformation, while such information is impossible to capture with the current level of instrumentation. However, the modeling is essential for the corresponding DFT calculation, which is subjective in a certain sense. The approaches to perfectly combine the experimental results (e.g., mass spectrometry) and the DFT calculations call for more effort.

9.2.10 Nanocluster-based assemblies

The rich chemistry of peripheral ligands enables nanoclusters to self-assemble into cluster-based nanomaterials, or serve as hosts/guests to produce cluster-based hybrid nanomaterials [29]. Specifically, the intercluster assembly of nanoclusters belongs to the realm of supramolecular chemistry, such as cluster-based crystal lattices, metal-organic frameworks, and aggregation-induced emission materials. Cluster-based host-guest nanosystems also belong to the supramolecular chemistry of nanoclusters. Nanoclusters can not only serve as hosts to absorb functional molecular guests, but also act as guests to be anchored onto/into the 1D nanowires, 2D nanosheets, and 3D nanostructures. Such assemblies render nanoclusters promising nanomaterials to be applied in catalysis, chemical sensor, gas adsorption, and so on [29].

Although the assembly of cluster building blocks into cluster-based metal-organic frameworks (MOFs) is an efficient method to tailor the chemical-physical properties of nanoclusters, only Ag-N based cluster MOFs have been reported [30,31]. M-O and M-S are also robust and common in the nanocluster research, but cluster-based MOFs based on these interactions have not been reported. Moreover, cluster-based aggregation-induced emission (AIE) materials have shown improved optical properties relative to nonemissive and weakly emissive individual clusters [32,33]. However, research into these cluster-based AIE materials is lacking.

Compared with the intercluster assembly, research on cluster-based host-guest nanosystems are still in its infancy. Although the interactions between hosts and guests can tune both the geometric/electronic structures and the chemical/physical properties of nanoclusters, only a few cases have been reported [29]. Further work should consider more types of interactions and arrange nanocluster guests into/onto different nanostructure basements.

9.2.11 Nanocluster crystals as materials

The nanocluster molecules are assembled into the crystalline lattice due to the intercluster interactions, i.e., noncovalent interactions among adjacent cluster molecules, such as hydrogen bonding, $\pi \cdots \pi$, $C-H \cdots \pi$, metal-ligand coordination, and so on. Crystal packing of nanoclusters is also directed by the crystal lattice energy, which includes a lot of parameters (e.g., size, charge, and geometric/electronic configuration of nanoclusters) [29]. In this context, the control of these intercluster interactions and environment parameters enables the regulation of the material properties of these crystals.

However, although several nanocluster crystals with different packing modes and material properties have been reported, nanocluster crystals as materials have not been examined. While crystal structures have been studied, research have not gone beyond to determine the material properties of these crystals, such as the optical property, the catalytic ability, the gas adsorption capacity, the electrical conductivity, etc. A rich area of nanocluster crystals as materials is yet to be explored.

9.2.12 Grant applications of cluster-based nanomaterials

The rich chemical-physical properties of metal nanoclusters render them promising nanomaterials for applications [9,34]. As ongoing improvement is being made in terms of control over the structures and compositions of nanoclusters which are closely related to their properties, one future direction is to move on to precise tailoring of properties of nanoclusters for applications in various fields.

At present metal nanoclusters have been widely exploited in catalysis, chemical and biological sensors, bio-imaging, therapeutic applications, antimicrobial application, energy storage application, and so on [9,34]. Besides, applications of cluster-based materials in terms of gas separation, gas storage, solar cells, logic gates, and drug and gene delivery have also appeared. However, potentially useful technologies which could impact industry have not come yet. The grant applications of cluster-based nanomaterials still call for more efforts, and their use in large-scale heterogeneous catalysts or test kits for detections may have promising prospect.

In closing, we believe that research into metal nanoclusters can lead to novel, smart nanomaterials that will open up new horizons in nanoscience and nanotechnology and offer solutions to problems related to the environment, biology, and human health.

References

[1] X. Kang, Y. Li, M. Zhu, R. Jin, Atomically precise alloy nanoclusters: syntheses, structures, and properties, Chem. Soc. Rev. 49 (17) (2020) 6443–6514, https://doi.org/10.1039/c9cs00633h.

[2] X. Kang, H. Chong, M. Zhu, $Au_{25}(SR)_{18}$: the captain of the great nanocluster ship, Nanoscale 10 (23) (2018) 10758–10834, https://doi.org/10.1039/c8nr02973c.

[3] C. Zeng, T. Li, A. Das, N.L. Rosi, R. Jin, Chiral structure of thiolate-protected 28-gold-atom nanocluster determined by X-ray crystallography, J. Am. Chem. Soc. 135 (27) (2013) 10011–10013, https://doi.org/10.1021/ja404058q.

[4] C. Zeng, Y. Chen, K. Kirschbaum, K. Appavoo, M.Y. Sfeir, R. Jin, Structural patterns at all scales in a nonmetallic chiral $Au_{133}(SR)_{52}$ nanoparticle, Sci. Adv. 1 (2) (2015), https://doi.org/10.1126/sciadv.1500045.

[5] Z. Wu, R. Jin, On the ligand's role in the fluorescence of gold nanoclusters, Nano Lett. 10 (7) (2010) 2568–2573, https://doi.org/10.1021/nl101225f.

[6] X. Kang, X. Wei, S. Wang, M. Zhu, Controlling the phosphine ligands of $Pt_1Ag_{28}(S-Adm)_{18}(PR_3)_4$ nanoclusters, Inorg. Chem. 59 (13) (2020) 8736–8743, https://doi.org/10.1021/acs.inorgchem.0c00350.

[7] M.N. O'Brien, M.R. Jones, B. Lee, C.A. Mirkin, Anisotropic nanoparticle complementarity in DNA-mediated co-crystallization, Nat. Mater. 14 (8) (2015) 833–839, https://doi.org/10.1038/nmat4293.

[8] X. Kang, M. Zhu, Cocrystallization of atomically precise nanoclusters, ACS Mater. Lett. 2 (10) (2020) 1303–1314, https://doi.org/10.1021/acsmaterialslett.0c00262.

[9] R. Jin, C. Zeng, M. Zhou, Y. Chen, Atomically precise colloidal metal nanoclusters and nanoparticles: fundamentals and opportunities, Chem. Rev. 116 (18) (2016) 10346–10413, https://doi.org/10.1021/acs.chemrev.5b00703.

[10] M. Zhou, C. Zeng, M.Y. Sfeir, et al., Evolution of excited-state dynamics in periodic Au_{28}, Au_{36}, Au_{44}, and Au_{52} nanoclusters, J. Phys. Chem. Lett. 8 (17) (2017) 4023–4030, https://doi.org/10.1021/acs.jpclett.7b01597.

[11] C. Zeng, H. Qian, T. Li, et al., Total structure and electronic properties of the gold nanocrystal $Au_{36}(SR)_{24}$, Angew. Chem. Int. Ed. 51 (52) (2012) 13114–13118, https://doi.org/10.1002/anie.201207098.

[12] C. Zeng, Y. Chen, C. Liu, K. Nobusada, N.L. Rosi, R. Jin, Gold tetrahedra coil up: kekule-like and double helical superstructures, Sci. Adv. 1 (9) (2015), https://doi.org/10.1126/sciadv.1500425.

[13] C. Zeng, Y. Chen, K. Iida, et al., Gold quantum boxes: on the periodicities and the quantum confinement in the Au-28, Au-36, Au-44, and Au-52 magic series, J. Am. Chem. Soc. 138 (12) (2016) 3950–3953, https://doi.org/10.1021/jacs.5b12747.

[14] M.S. Bootharaju, C.P. Joshi, M.R. Parida, O.F. Mohammed, O.M. Bakr, Templated atom-precise galvanic synthesis and structure elucidation of a $Ag_{24}Au(SR)_{18}$ nanocluster, Angew. Chem. Int. Ed. 55 (3) (2016) 922–926, https://doi.org/10.1002/anie.201509381.

[15] K. Zheng, V. Fung, X. Yuan, D. Jiang, J. Xie, Real time monitoring of the dynamic intracluster diffusion of single gold atoms into silver nanoclusters, J. Am. Chem. Soc. 141 (48) (2019) 18977–18983, https://doi.org/10.1021/jacs.9b05776.

[16] C. Zeng, Y. Chen, A. Das, R. Jin, Transformation chemistry of gold nanoclusters: from one stable size to another, J. Phys. Chem. Lett. 6 (15) (2015) 2976–2986, https://doi.org/10.1021/acs.jpclett.5b01150.

[17] X. Kang, M. Zhu, Transformation of atomically precise nanoclusters by ligand-exchange, Chem. Mater. 31 (24) (2019) 9939–9969, https://doi.org/10.1021/acs.chemmater.9b03674.

[18] R. Jin, T. Higaki, Open questions on the transition between nanoscale and bulk properties of metals, Commun. Chem. 4 (1) (2021), https://doi.org/10.1038/s42004-021-00466-6.

[19] N.A. Sakthivel, S. Theivendran, V. Ganeshraj, A.G. Oliver, A. Dass, Crystal structure of faradaurate-279: $Au_{279}(SPh-tBu)_{84}$ plasmonic nanocrystal molecules, J. Am. Chem. Soc. 139 (43) (2017) 15450–15459, https://doi.org/10.1021/jacs.7b08651.

[20] T. Higaki, M. Zhou, K.J. Lambright, K. Kirschbaum, M.Y. Sfeir, R. Jin, Sharp transition from nonmetallic Au-246 to metallic Au-279 with nascent surface plasmon resonance, J. Am. Chem. Soc. 140 (17) (2018) 5691–5695, https://doi.org/10.1021/jacs.8b02487.

[21] Y. Song, K. Lambright, M. Zhou, et al., Large-scale synthesis, crystal structure, and optical properties of the $Ag_{146}Br_2(SR)_{80}$ nanocluster, ACS Nano 12 (9) (2018) 9318–9325, https://doi.org/10.1021/acsnano.8b04233.

[22] Y. Song, Y. Li, H. Li, et al., Atomically resolved $Au_{52}Cu_{72}(SR)_{55}$ nanoalloy reveals marks decahedron truncation and penrose tiling surface, Nat. Commun. 11 (1) (2020), https://doi.org/10.1038/s41467-020-14400-2.

[23] S. Sharma, W. Kurashige, K. Nobusada, Y. Negishi, Effect of trimetallization in thiolate-protected $Au_{24-n}CuPd$ clusters, Nanoscale 7 (24) (2015) 10606–10612, https://doi.org/10.1039/c5nr01491c.

[24] M. Zhou, T. Higaki, Y. Li, et al., Three-stage evolution from nonscalable to scalable optical properties of thiolate-protected gold nanoclusters, J. Am. Chem. Soc. 141 (50) (2019) 19754–19764, https://doi.org/10.1021/jacs.9b09066.

[25] C. Kumara, A. Dass, $(AuAg)_{144}(SR)_{60}$ alloy nanomolecules, Nanoscale 3 (8) (2011) 3064–3067, https://doi.org/10.1039/c1nr10429b.

[26] A.C. Dharmaratne, A. Dass, $Au_{144-x}Cu_x(SC_6H_{13})_{60}$ nanomolecules: effect of Cu incorporation on composition and plasmon-like peak emergence in optical spectra, Chem. Commun. 50 (14) (2014) 1722–1724, https://doi.org/10.1039/c3cc47060a.

[27] X. Kang, S. Wang, M. Zhu, Observation of a new type of aggregation-induced emission in nanoclusters, Chem. Sci. 9 (11) (2018) 3062–3068, https://doi.org/10.1039/c7sc05317g.

[28] K.L.D.M. Weerawardene, P. Pandeya, M. Zhou, Y. Chen, R. Jin, C.M. Aikens, Luminescence and electron dynamics in atomically precise nanoclusters with eight superatomic electrons, J. Am. Chem. Soc. 141 (47) (2019) 18715–18726, https://doi.org/10.1021/jacs.9b07626.

[29] X. Kang, M. Zhu, Intra-cluster growth meets inter-cluster assembly: the molecular and supramolecular chemistry of atomically precise nanoclusters, Coord. Chem. Rev. 394 (2019) 1–38, https://doi.org/10.1016/j.ccr.2019.05.015.

[30] R.-W. Huang, Y.-S. Wei, X.-Y. Dong, et al., Hypersensitive dual-function luminescence switching of a silver-chalcogenolate cluster-based metal-organic framework, Nat. Chem. 9 (7) (2017) 689–697, https://doi.org/10.1038/nchem.2718.

[31] M. Cao, R. Pang, Q.-Y. Wan, et al., Porphyrinic silver cluster assembled material for simultaneous capture and photocatalysis of mustard-gas simulant, J. Am. Chem. Soc. 141 (37) (2019) 14505–14509, https://doi.org/10.1021/jacs.9b05952.

[32] N. Goswami, Q. Yao, Z. Luo, J. Li, T. Chen, J. Xie, Luminescent metal nanoclusters with aggregation-induced emission, J. Phys. Chem. Lett. 7 (6) (2016) 962–975, https://doi.org/10.1021/acs.jpclett.5b02765.

[33] X. Kang, M. Zhu, Tailoring the photoluminescence of atomically precise nanoclusters, Chem. Soc. Rev. 48 (8) (2019) 2422–2457, https://doi.org/10.1039/c8cs00800k.

[34] I. Chakraborty, T. Pradeep, Atomically precise clusters of noble metals: emerging link between atoms and nanoparticles, Chem. Rev. 117 (12) (2017) 8208–8271, https://doi.org/10.1021/acs.chemrev.6b00769.

Index

Note: Page numbers followed by *f* indicate figures, and *b* indicate boxes.

A

Absolute quantum yield, 47
Acetycholinesterase (AChE), 304, 319
A3-coupling reaction, 353–354
Adenosine aptamer, 316–318
Adenosine deaminase (ADA), 316–318
Adenosine-5-triphosphate (ATP), 306–307
Adiabatic integration method, 204
Ag-Ag scrambling, 118–119, 119*f*
Ag_{12} clusters, solvent induced isomerization, 129–130, 130–131*f*
Aggregation-induced emission (AIE), 165–168, 166*f*, 168*f*, 275, 322
  in Au-thiolate nanoclusters, 166
  Cu nanoclusters, 167
  for luminescent Ag nanoclusters, 166–167
Aggregation-induced emission luminogens (AIEgens), 275–277
Aggregative growth model, 80–81, 81*f*
Ag^+ induced size-growth, 128, 128*f*
$AgNO_3$ induced size-growth, 126, 127*f*
Ag^+ sensing, 296, 297–298*f*
Alcohol oxidation, 344–345, 345*f*
Aldehyde reduction, 349
Alkaline phosphatase (ALP), 318
Alkenes, epoxidation of, 345–346, 346*f*
Alkynyl ligands, 62
Alloy nanoclusters, 34–38
  chirality of, 177
  co-reduction method, 34–35
  intercluster reaction method, 37–38, 37*f*
  ligand exchange method, 35–36, 35*f*
  method exchange method, 36–37
  photoluminescence of, 162–165
α-cyclodextrin (α-CD), 179, 179*f*, 235
α-thrombin, 316
Amine ligands, 64
Anion sensing, 298–303, 301–302*f*
Anode electrocatalytic reactions, in fuel cells, 335, 336*f*
Antibonding orbital, 45
Antigalvanic reduction method, 36
Antigalvanic theory, 296
Antimicrobial agents, 333–335
Applications, of metal nanoclusters, 289
  biological application, 323–335
    antimicrobial agents, 333–335
    biolabeling/imaging, 323–329
    disease diagnostics and therapy, 329–333
    self-vaccine, 333
  catalysis, 335–357
    catalytic selective oxidation, 342–348
    catalytic selective reduction, 348–352
    coupling reactions, 352–355
    electrocatalysis, 335–339
    photocatalysis, 340–342
    reactions, 355–357, 356*f*
  overview, 289
  sensors, 289–323
    biological sensor, 306–322
    chemical sensor, 289–306
    pH meters, 322, 323*f*
    thermometers, 322–323
Assembly of metal nanoclusters
  cluster-based host-guest nanosystem, 235–245
  cluster-based metal-organic framework via linker, 263–279, 264*f*, 266–271*f*, 273–281*f*
  crystal lattice assembly for assemble nanomaterial via supramolecular interactions, 245–263, 246*f*, 248–250*f*, 252–263*f*
  overview, 280–281
Atomic absorption spectrophotometric (AAS) method, 295
Atomically precise metal clusters
  ligand etching induced size-reduction of, 110–115, 111–115*f*
  reduction/oxidation induced size-conversion of, 83–97
  size-maintained ligand exchange of, 105–110, 107–110*f*
Atomically precise structures
  kernels in nanoclusters, 65–73
    body-centered cubic (BCC) stacking, 68–69
    decahedron, 71–73
    FCC kernels, 65–68, 66–67*f*
    HCP kernels, 68
    icosahedron, 69–71
  ligands and metal-ligand interfaces, 59–65
Au-Ag alloy nanocluster, 20
Au-Ag bimetallic nanoclusters, 34
Au-Ag exchange, 120, 120*f*
Au(I) complex induced size-growth, 122–123, 123*f*
AuCu alloy nanocluster
  magnetic properties of, 174, 174*f*
  photoluminescence of, 167, 168*f*
Au-LA-DEDA, ECL intensity of, 190
Au_{25} nanoclusters, 27
  $AgNO_3$ induced size-growth of, 126, 127*f*
  magnetic properties of, 170–172, 170–171*f*
  phenylselenol ligand-protected, 29, 30*f*
  thiolated, 28–29, 28*f*
Au_{144} nanoclusters, 27, 28*f*
  phenylselenol ligand-protected, 29, 30*f*
Au_{21}, photoluminescence of, 162, 162*f*
AuSAdm induced size-conversion, 121–122, 122*f*
Au-tetraalkylammonium complex, 23–24
$Au_{23}X_{15}$ clusters, 128–129, 129*f*
Average binding energy (ABE), 145*b*, 146

B

Basis sets, 204–205
Battery, 184
β-cyclodextrin, 235
Bi-antigalvanic reduction method, 350
Bilirubin sensing, 308–309
Biological application, 323–335
  antimicrobial agents, 333–335
  biolabeling/imaging, 323–329
  disease diagnostics and therapy, 329–333
  self-vaccine, 333
Biological sensor, 306–322
Biomolecule sensing, 306–322
  bilirubin, 308–309
  biothiol, 308
  DNA, 319–320
  dopamine (DA), 309, 310*f*
  enzyme, 315–319, 317*f*
  folic acid (FA), 312
  glucose, 312, 313*f*
  heparin, 309, 311*f*
  H_2O_2, 307, 307*f*
  microRNAs (miRNAs), 320–322, 321*f*
  phosphate-containing metabolites, 306–307, 306*f*
  protein, 312–315, 314*f*
  reactive oxygen species (ROS), 310–312
  urea, 312
Biothiol sensing, 308
Body-centered cubic (BCC) stacking, 68–69
Bombardment, fast atom, 13–14, 14*f*

Bonding interaction analysis, 224–226, 225–226f
4-Bromobenzenethiol (BBT), 158
Brust-Schiffrin method (BSM), 21, 23, 24f
  synthetic route for, 24–25, 25f
  two-phase, 5
Brust synthesis, 20–25
  mechanism of, 81–82, 82f
  thiol capped gold nanoparticles, 21f

C

Calixarenes, 64
Cancer
  diagnosis and detection, 329–330, 330f
  therapy, 331–332
Carbon-carbon multiple-bond reduction, 348–349, 348f
Carboxide reduction, 349–350
Carboxylic acid, 181–182
Catalysis, 335–357
Catalytic reactions, 355–357, 356f
Catalytic selective oxidation, 342–348
  alcohol oxidation, 344–345, 345f
  CO oxidation, 342–344, 343f
  D-glucose oxidation, 348
  epoxidation of alkenes, 345–346, 346f
  hydrocarbons oxidation, 347–348, 347f
  sulfide selective oxidation, 347
Catalytic selective reduction, 348–352
  carbon-carbon multiple-bond reduction, 348–349, 348f
  carboxide reduction, 349–350
  reduction of nitroarene derivatives, 350–351
  single-electron transfer reaction, 351–352, 352f
Cathode electrocatalytic reactions, in fuel cells, 336–338, 337f
Cerenkov resonance energy transfer (CRET), 327–329
Characterization techniques
  circular dichroic spectroscopy, 48–49
  dynamic light scattering, 53–54
  electron paramagnetic resonance, 55–56, 56f
  Fourier transform infrared (FTIR) spectrometer, 47–48
  mass spectrometry, 50–52
  nuclear magnetic resonance spectroscopy, 54–55, 55f
  photoluminescence spectroscopy, 45–47
  single crystal X-ray diffraction, 58–59
  thermogravimetric analysis, 54
  transient absorption (TA) spectroscopy, 49
  transmission electron microscope, 52–53
  UV-vis absorption spectroscopy, 45
  X-ray absorption fine structure, 57–58
  X-ray photoelectron spectroscopy, 56–57
Chemical sensing, 303–306, 305f
Chemical sensor, 289–306
Chemical shift, 56–57
Chemical synthesis, in solution
  Brust synthesis, 20–25, 21f
  controlled conversion method, 30–33
  inorganic anions as synthetic templates, 33–34
  kinetically controlled synthesis, 25–26, 25f
  metal complex reduction method, 19–20
  one-pot synthesis, 26–30, 28f, 30f
Chemotherapy, phototriggered, 332
Chirality, 175–184
  definition of, 175
  identification through crystal structure, 175–178, 176–177f
Chiral ligands/compounds, 179–182, 181f
Chiral separation, 178–179, 178–179f
Circular dichroic spectroscopy, 48–49
Circularly polarized luminescence (CPL), 182–183
  copper nanoclusters, 182, 183f
  superatom complex inorganic framework material, 182, 183f
Cluster-based host-guest nanosystem, for assemble nanomaterial, 235–245
  cluster compound, 240–245
  supramolecular interactions of NCs as hosts
    with cyclodextrins, 235–237, 235–236f
    with fullerene, 237–239, 238f
    with other molecules, 239–240, 240f
Cluster-based metal-organic framework, via linker, 263–279, 264f, 266–271f, 273–281f
Cluster-based nanomaterials, applications of, 376
Cluster-CD hybrid materials, 236–237, 236f
Cluster-cluster reactions, 105–106, 107f
Cluster compound, 240–245
Cluster-ligand reactions, 105–106, 107f
Cluster of clusters, 20, 69–70
Cluster self-assembly material, 29–30
C-N cycloisomerization coupling reactions, 353
CO_2, carboxylation reaction of, 355, 355f
Cocrystallization, 374
Cold reflective discharge ion source (CORDIS), 13–14
Colloidal gold, 18–19
Complete cubic evolution model, 68
Connectivity, 145b
Constrained structural optimization, 211
Controlled conversion method, 30
  ligand exchange method, 32
  two-phase ligand exchange method, 30–31, 31–32f
CO oxidation, 342–344, 343f
Copper (Cu) nanoclusters
  aggregation-induced emission of, 167
  chirality of, 176, 176f
  circularly polarized luminescence of, 182, 183f
  icosahedral kernel in, 69
  magnetic properties of, 174
Co-reduction method, 34–35
Core etching method, 296–297
Coupling reactions, catalysis of, 352–355
  A3-coupling reaction, 353–354
  carboxylation reaction of CO_2, 355, 355f
  C-N cycloisomerization coupling reactions, 353
  Sonogashira cross-coupling reaction, 353
  Suzuki cross-coupling reaction, 352–353, 354f
  Ullmann coupling reaction, 352
CPL. See Circularly polarized luminescence (CPL)
C-reactive protein (CRP), 315
Critical-sized aggregates, 80–81
Crystal, 167
Crystal-induced emission (CIE), 167
Crystal lattice, assembly in, 245–263, 246f, 248–250f, 252–263f
Crystallization induced emission enhancement (CIEE), 260–263
Cu^{2+} sensing, 294–296, 294–295f
Cyclodextrins (CDs), supramolecular interactions of NCs as hosts with, 235–237, 235–236f
Cyclopentanethiol (CPT), 158
Cysteine (Cys), 308

D

Decahedral M_7-centered shell-by-shell kernels, 72
Decahedron kernels, 71–73
Denatured bovine serum albumin (dBSA), 292
Density functional theory (DFT), 375
  application of, 205–226
  calculations, 305–306, 375
  functionals, 201–205
  time-dependent, 205
D-glucose oxidation, 348
Diamagnetic gold nanoclusters, 173, 173f
Dirac equation, 205
Dissolution/concentration induced size conversion, 130–131, 131f
DNA sensing, 319–320
Dopamine (DA) sensing, 309, 310f
Doping, metal, 117–121, 119–121f
  photoluminescence and, 162–165
Double nanocluster ion compound (DNIC), 253
Dynamic light scattering (DLS), 53–54, 298

E

ECL. See Electrochemiluminescence (ECL)
Effective core potentials (ECPs), 205
Electrocatalysis, 335–339
  electrocatalytic CO_2 reduction, 338–339, 341f
  electrocatalytic hydrogen evolution reaction, 338, 339f
  electrocatalytic oxygen evolution reaction, 338, 340f
  fuel cells
    anode electrocatalytic reactions in, 335, 336f
    cathode electrocatalytic reactions in, 336–338, 337f
Electrocatalytic CO_2 reduction, 338–339, 341f
Electrocatalytic hydrogen evolution reaction, 338, 339f
Electrocatalytic oxygen evolution reaction, 338, 340f
Electrochemical analysis, 185–187, 185f, 188f
Electrochemical/chemo-redox strategy, 86, 86f
Electrochemical energy gap, 184
Electrochemical property, 184–191
Electrochemiluminescence (ECL), 187–191, 189–190f, 309

Electrochemistry, 184
Electrode, 184
Electrolyte, 184
Electron circular dichroism (ECD), 49
Electronic state elucidation, 213–216, 214f, 216–217f
Electron nuclear double resonance (ENDOR), 171–172
Electron paramagnetic resonance (EPR), 55–56, 56f
   of AuCu-I and AuCu-II, 174, 174f
   of $[Au_{25}(SBu)_{18}0]_n$, 171–172, 171f
   of $Au_{25}(PET)_{18}$ nanocluster, 170–171, 170f
Electron spin resonance (ESR). See Electron paramagnetic resonance (EPR)
Electron-transfer transformations, 84–87, 88f
Electrospray ionization mass spectrometry (ESI-MS), 26, 82–83, 89–90, 90–91b, 91f
Electrospray ionization (ESI)-TOF-MS, 51–52, 52f
Electrostatic attraction strategy, 355
Energy fragment decomposition analysis, 214–215
Environmental protection agency (EPA), 289
Enzyme sensing, 315–319, 317f
Epoxidation, of alkenes, 345–346, 346f
EPR. See Electron paramagnetic resonance (EPR)
Esterase, 318
Ethanol oxidation reaction (EOR), 335
Evaporation, 15–16
Excited state absorption (ESA), 49
Extended X-ray absorption fine structure (EXAFS), 57–58, 58f, 300

F

Face-centered cubic (FCC) kernels, 65–68, 66–67f
Fast atom bombardment, 13–14, 14f
FCC kernels. See Face-centered cubic (FCC) kernels
FCC-packed gold clusters, size-evolution of, 140–141, 141f, 146, 146f
FCC-packed silver clusters, size-evolution of, 142–144, 142–144f
Femtosecond time-resolved transient absorption spectroscopy, 154
Ferromagnetic gold nanoclusters, 173, 173f
Field effect transistor (FET), 257–259
Fluorescence quenching mechanism, 295
Fluorescence resonance energy transfer (FRET), 293
Fluorescence spectrophotometer, 46
Fluorescent alloy nanoclusters, 165
4-Fluorothiophenol (FBT), 158
Folic acid (FA) sensing, 312
Four-circle diffractometer, 59
Fourier transform extended X-ray absorption fine-structure (FT-EXAFS), 255
Fourier transform infrared (FTIR) spectrometer, 47–48, 48f
Frequency calculation, 212–213
Fuel cells
   anode electrocatalytic reactions in, 335, 336f
   cathode electrocatalytic reactions in, 336–338, 337f
Fullerene, supramolecular interactions of NCs as hosts with, 237–239, 238f

G

Galvanic synthesis, alloy nanoclusters, 36
γ-cyclodextrin, 235
Gas condensation, 15–16
Gas phase nanoclusters, 15
   evaporation and gas condensation, 15–16
   laser vaporization, 17–19, 17–18f
Gauge-including atomic orbitals (GIAOs) formalism, 224
Generalized gradient approximations (GGA), 203
Gene therapy, 333, 334f
Geometry optimization, 206–212, 208–209f
Glucose sensing, 312, 313f
Glutathione (GSH), 308
Gold atom insertion-thiolate group elimination mechanism, 219, 219f
Gold(0)-gold(I) polynuclear complexes, 19
Gold (Au) nanoclusters, 19
   chiral, 175–176
   circularly polarized luminescence of, 182
   electrochemical properties of, 185, 185f, 188f
   electrochemiluminescence of, 188–189, 189–190f
   electron paramagnetic resonance, 56, 56f
   FCC kernel in, 65
   FCC-packed, size-evolution of, 140–141, 141f, 146, 146f
   fluorescence spectra of, 156, 157f
   grand unified model for, 138–140, 139f
   GS-protected, 45, 46f
   kinetically controlled synthesis, 25–26
   light-induced transformation of, 33
   magnetic properties of, 172–173, 173f
   monodisperse, 82–83, 83f
   one-pot synthesis, 29, 29f
   phosphine protected
      ligand exchange induced size-growth, 105, 106f
      redox-induced size-conversion, 92–95, 92–95f
   two-phase ligand exchange method, 30, 31f
Gold nanoparticles
   Brust-Schiffrin synthesis, 23, 24f
   polydisperse, size-conversion of, 82–83, 83f
   thiol capped, 21–22f
Gold(I)-thiolate complex, 21–22, 22f
Grand unified model, 138–140, 139f
Ground state bleaching (GSB), 49

H

Halide ions, 303
Halogen ions, 62–63
Hartree Fock equation, 202
Heating induced size-conversion, 133–135, 134–135f
Heavy metal ion sensing, 289–298
Hemoglobin (Hb), 313–314
Hemolytic anemia, 316–318
Heparin sensing, 309, 311f
Heteroatom doping, 186
Hexagonal close-packed (HCP) kernels, 68
Hg^{2+} sensing, 290–293, 291–292f
Highest occupied molecular orbital (HOMO), 184
High-performance liquid chromatography (HPLC), 304
High resolution transmission electron microscopy (HRTEM), 293
Homocysteine (Hcy), 308
HOMO-LUMO gap, 184, 187
Hopping mechanism, reduction induced $2e^-$, 90–92
H_2O_2 sensing, 307, 307f
H-SAdm etching induced size-growth, 98, 98f
Hydride ions, 63–64
Hydrocarbons oxidation, 347–348, 347f

I

Icosahedral metal nanoclusters, size-evolution of, 140, 140–141f
Icosahedron, 140
Icosahedron-based assembly kernels, 69–70
Icosahedron-centered shell-by-shell kernels, 70–71
Icosahedron kernels, 69–71
Immunoassays, 322
Indium tin oxide (ITO), 336
Ino decahedral M_{13}-centered shell-by-shell kernels, 72–73
Ino decahedron, 71–72
Inorganic anions, as synthetic templates, 33–34
Inorganic ions, as capping ligands, 64–65
In situ two-phase ligand exchange method, 31, 32f
Instantaneous/burst nucleation, 80
Intercluster compounds, 239
Intercluster reaction method, 37–38, 37f
Interconversion
   induced by extraligands, 115–117, 115–117f
   ligand exchange induced, 107–108, 108–109f, 112f
Ion mobility mass spectrometry (IM-MS), 236–237
Isolated icosahedral kernels, 69
Isomerization
   ligand-exchange induced, 107–108
   solvent-triggered, 128–132, 129–131f
Isotopic labeling, 118–119

J

Jellium model, 137–138, 138f

K

Kernels, in nanoclusters, 65
   body-centered cubic (BCC) stacking, 68–69
   decahedron, 71–73
   FCC kernels, 65–68, 66–67f

Index

Kernels, in nanoclusters *(Continued)*
  HCP kernels, 68
  icosahedron, 69–71
Ketones reduction, 349
Kinetically controlled synthesis, 25–26, 25f

L

Lambert-Beer Law, 87–88b
Lamer model, 79–80, 80f
Lamer theory, 80
Langmuir-Hinshelwood (L-H) mechanism, 343
Laser vaporization, 17–19, 17–18f
Ligand etching induced size-growth, 98–105, 98–103f, 106f
Ligand etching induced size-reduction, 110–115, 111–115f
Ligand exchange, 32
  alloy nanoclusters, 35–36, 35f
    antigalvanic reduction method, 36
    galvanic synthesis, 36
  size-maintained, 104f, 105–110, 107–110f
  two-phase, 30–31, 31f
  in situ, 31, 32f
  via cluster-complex reaction, 121, 121f
Ligands, chiral, 179–182, 181f
Ligand-to-metal charge transfer (LMCT), 158, 374
Limit of the detection (LOD), 290
Liquid metal ion source (LMIS), 14–15
Local density approximation (LDA), 203
Lowest unoccupied molecular orbital (LUMO), 184
Luminescence.
  See also Electrochemiluminescence (ECL); Photoluminescence (PL)
  aggregation-induced, 165–166
  circularly polarized, 182–183, 183f

M

Magnetic circular dichroism (MCD) spectroscopy, 174
Magnetic resonance imaging (MRI), 327–329
Magnetism, 170, 172–175, 173–174f
  Au_{25}-based metal nanoclusters, 170–172, 170–171f
Marks decahedron, 71–72
Mars-Van Krevelen (MvK) mechanism, 343
Mass spectrometry (MS), 50–52
  ESI-TOF-MS, 51–52, 52f
  MALDI-TOF-MS, 50–51, 51f
Matrix assisted laser desorption ionization (MALDI) TOF-MS, 50–51, 51f
Mercaptan, 20–21
Metal complex induced size-growth, 121–128
  ligand exchange/addition via cluster-complex reaction, 121, 121f
  via extra-metal component incorporation, 121–128, 122–127f
Metal complex reduction method, 19–20
Metal exchange method, 35–36
Metal ion sensing, 296–298, 299–300f
Metallic-state nanoparticles, 375
Metal-ligand interfaces, 59–65
Metal nanoclusters
  chirality of, 175–184
  definition of, 1
  electrochemical property of, 184–191
  electronic structure evolution of, 137–140
  history of, 1–4
  magnetism of, 170–175
  metal complex catalyzed size-growth of, 128, 128f
  optical absorption of, 153–154, 154f
  photoluminescence of, 155–170
  property of, 3–4, 3f
  research status of, 4, 4f
    controllable preparation, 5
    potential application, 6–7
    property regulation, 6
    structure elucidation, 5
  size-growth of, 79–82
  size of, 2, 2f
  structures of, 2–3, 3f
Metal-organic frameworks (MOFs), 263–264
  Ag NC-based, 277–279, 279f
  cluster-based, 263–279
  with intense luminescent properties, 275–277, 278f
Metal-support interactions (MSI), 345
Metal to ligand charge transfer (MLCT), 158
Method exchange method, 36–37
Methotrexate, 304
(2-(3-Methylpyrazinyl)-diphenylphosphine), 160
Michelson interferometer, 47–48
MicroRNAs (miRNAs) sensing, 320–322, 321f
Minimal bactericidal concentration (MBC), 333
Minimum inhibitory concentration (MIC), 333
Mn-doped Au_{25} nanoclusters, 171
Modified Kirby-Bauer method, 333
MOFs@nanocluster, 243–245, 243–245f
Molecular dynamic (MD) calculations, 227
Molecular dynamic (MD) simulations, 237
Monodisperse gold nanoclusters, 82–83, 83f
Monotonic size-growth, 90–91, 91f
Motif, 59–60
Multimodal imaging, 327–329, 328f

N

Nanocluster-based assemblies, 375–376
Nanocluster crystals, 376
Nanocluster science, 373–376
  applications of cluster-based nanomaterials, 376
  boundary between quantum-sized nanoclusters and metallic-state nanoparticles, 375
  cocrystallization of metal nanoclusters, 374
  future theoretical works, 375
  ligand effect, 374
  mechanisms of, 375
  nanocluster-based assemblies, 375–376
  nanocluster crystals as materials, 376
  new synthetic methodologies, 373
  novel alloy nanoclusters, 373
  periodicity of, 374
  transformation of, 374–375
Nanoelectrochemistry, 184–185
Nanomaterials
  assembly of, 233–235
  cluster-based, applications of, 376
Nanoparticle@nanocluster, 242, 242f
Nanoparticles
  metallic-state, 375
  optical gap of, 184
Nanoscale synthesis methods, 233–234
Nanowire@nanocluster, 241, 241f
Natural bond orbital (NBO) charge analysis, 216–217
Near-infrared electrochemiluminescence (NIR-ECL), 189–190
Near-infrared photoluminescence (NIR-PL), 188–189
Neurotransmitter, 309
N-heterocyclic carbenes (NHCs), 61–62
Nickel nanoclusters, magnetic properties of, 173–174
Nitrite, 301–303
Nitroarene derivatives
  reduction of, 350–351
  selective reduction of, 350–351
Nitrophenol reduction, 350
NMR analysis, in size-conversion, 132–133b, 133f
NMR spectra, calculations on, 223–224
Noble metals, 11
Noncovalent interactions, 226
Novel alloy nanoclusters, 373
Nuclear magnetic resonance (NMR) spectroscopy, 54–55, 55f

O

One-pot synthesis, 26–30, 28f, 30f
Optical absorption, 153–154, 154–155f
Organelle-targeted imaging, 324–325
Organophosphorus compounds (OPs), 304
Oscillation, 33
Oxidation induced size-conversion, 92–93, 93f
Oxidation induced size-growth, $[Au_{23}(SR)_{16}]^-$ clusters, 96–97, 97f

P

Paramagnetic gold nanoclusters, 173, 173f
Partial optimization, 211–212, 212f
Perfluoroglutarate ligands, 64
Phenylacetylene induced size-reduction, 117, 118f
2-Phenylpropane-1-thiol, 180
pH-induced size-conversion, 135–137, 136–137f
pH meters, 322, 323f
Phosphate-containing metabolites sensing, 306–307, 306f
Phosphine ligands, 61–62
Phospholipase C (PLC), 318–319
Photocatalysis, 340–342
  photocatalytic decomposition of pollutants, 340–341

photocatalytic organic transformation reactions, 342
photocatalytic water splitting, 341–342, 342f
Photocatalytic decomposition, of pollutants, 340–341
Photocatalytic organic transformation reactions, 342
Photocatalytic water splitting, 341–342, 342f
Photodynamic therapy (PDT), 332
Photoelectric conversion, 191
Photoinduced electron transfer (PET) process, 308
Photoluminescence (PL), 45–46, 155, 168–170, 169f.
    See also Electrochemiluminescence (ECL)
    aggregation-induced emission, 165–168, 166f, 168f
    inherent geometry and electronic structure, 156–158, 157f
    ligand effect on, 158–162
        structural changes inducted by ligand exchange, 160–162, 160–162f
        structure maintenance by ligand substitution, 158–159, 159f
    in metal doping and alloying, 162–165, 163f, 165f
Photoluminescence spectroscopy, 45–47, 47f
Photon correlation spectroscopy (PCS). See Dynamic light scattering (DLS)
Photo-oxidation induced size-growth, 97, 97f
Photothermal therapy (PTT), 332
Phototriggered chemotherapy, 332
Physical-chemical properties
    chirality, 175–184
    magnetism, 170–175
    optical absorption, 153–154, 154–155f
    photoluminescence, 155–170
Platelet-derived growth factor (PDGF), 314–315
Polydisperse Au nanoparticles, size-conversion of, 82–83, 83f
Polyethyleneimine (PEI), 333
Polyhedral oligomeric silsesquioxane (POSS), 239
Polyoxometalates (POMs), 33–34
Potable gold, 1
Proteases, 315
Protein posttranslational modifications (PTMs), 318
Protein sensing, 312–315, 314f
$Pt_1Ag_{31}(SR)_{16}(DPPM)_3Cl_3$ nanocluster, 160–161, 161f
Pump-probe technique, 49

Q
"Quadruple-gear" meshing mechanism, 247
Quantum confinement, 45
Quantum mechanics-molecular mechanics (QM-MM) calculations, 226–228, 227f
Quantum-sized nanoclusters, 375
Quasielastic scattering. See Dynamic light scattering (DLS)

R
Reaction mechanism
    analysis, 216–221
    of metal nanocluster catalysis and conversion, 219–221, 220f, 222f
Reactive oxygen species (ROS) sensing, 310–312
Redox induced ligand addition/removal reactions, 88–90, 88–89f
Redox induced single electron oxidation/reduction, 85, 86f
Redox-induced size-conversion, phosphine protected gold nanoclusters, 92–95, 92–95f
Reduction induced $2e^-$ hopping mechanism, 90–92
Reduction/oxidation induced size-conversion, 83–97
    Au_{25} and Au_{23} clusters, 95–96, 95–96f
    $[Au_{23}(SR)_{16}]^-$ clusters, 96–97, 97f
    electron-transfer transformations, 84–87
Relativistic effects, 205
Root-mean-square (RMS), 207
Runge–Gross theorem, 205

S
Sacrificial ligand induced size-conversion, 117, 118f
Scanning electron microscopy (SEM), 293
Schrödinger equation
    time-dependent, 201–202
    time-independent, 201–202
Seed NPs, 98
Selenophenol ligand induced size-reduction, 115, 115f
Self-assembly, 233
Self-vaccine, 333
Semiconductor clusters, 18
Sensors, 289–323
Serum proteins, 312–313
Severe combined immunodeficiency disease (SCID), 316–318
Shell-by-shell kernels
    decahedral M_7-centered, 72
    icosahedron-centered, 70–71
    Ino decahedral M_{13}-centered, 72–73
Silver cluster-assembled material, 265, 267f
    1D, 265–267, 268f
    2D, 272–273, 274f
    3D, 274, 275–276f
Silver (Ag) nanoclusters, 33
    chirality of, 176
    circularly polarized luminescence of, 182
    FCC-packed, size-evolution of, 142–144, 142–144f
    FCC patterns in, 67
    photoluminescence of, 156–157, 157f
    synthesis method for, 38, 38f
Single atom effect, 219–220
Single crystal X-ray diffraction (SC-XRD), 5, 58–59
Single doping, 120, 120f
Single electron transfer (SET) reactions, 84, 84f, 351–352, 352f
Single nanocluster ion compound (SNIC), 253
Single nucleotide polymorphisms (SNPs), 320
Single-occupied molecular orbital (SOMO), 171
Single point energy calculations, 205–206
6H left-handed helix (6HLH) arrangement, 251
Size-conversion, 79
    aggregative growth model, 80–81, 81f
    heating induced, 133–135, 134–135f
    interconversion induced by extraligands, 115–117, 115–117f
    Lamer model, 79–80, 80f
    ligand etching induced size-reduction, 110–115, 111–115f
    mechanism of, 79
        Brust synthesis, 81–82, 82f
        ligand exchange, 98–117
        metal complex induced size-growth, 121–128
        metal doping, 117–121
        molecular-level, 82–137
    pH-induced, 135–137, 136–137f
    polydisperse Au nanoparticles, 82–83, 83f
    reduction/oxidation induced, 83–97
    sacrificial ligand induced, 117, 118f
    size-growth induced by ligand etching, 98–105, 98–102f, 106f
    solvent-triggered isomerization of metal clusters, 128–132, 129–131f
Size-evolution, 79
    electronic structure evolution, 137–140
    grand unified model, 138–140, 139f
    Jellium model, 137–138, 138f
    FCC-packed gold clusters, 140–141, 141f
    FCC-packed silver clusters, 142–144, 142–144f
    of icosahedral metal nanoclusters, 140, 140–141f
    principals, 144–146, 145–146f
Size-maintained ligand exchange, 105–110, 107–110f
Slater type orbitals (STO), 204
Solid-state method, silver nanoclusters, 38, 38f
Solvent evaporation technique, 252
Solvent microphase separation, 234–235
Solvent-triggered isomerization, 128–132, 129–131f
Sonogashira cross-coupling reaction, 353
Sputtering, 12–13, 12f
"Staple-motif-growth" pathway, 144–146
Stimulated emission (SE), 49
Structure-property correlation, 224–226
Sulfide ions, 303
Sulfide selective oxidation, 347
Sulfur ions, 62–63
Superatom complex inorganic framework material, CPL spectrum of, 182, 183f
Superatom complex model (SACM), 69
Superatom electronic theory, 213, 228
Superatom-network model, 228
Super valence bond model, 228

Supramolecular assembly strategy, α-CD based, 179, 179f
Surface-motif-exchange (SME) mechanism, 124
Surface plasmon enhanced energy transfer (SPEET) sensor, 309
Surface plasmon resonance (SPR), 290
Suzuki cross-coupling reaction, 352–353, 354f

T

Target-assisted isothermal exponential amplification (TAIEA), 320
Targeted imaging, in vitro and in vivo, 326–327
Theophylline, 304
Theoretical simulation methods
  density functional theory
    application of, 205–226
    basis sets, 204–205
    electronic state elucidation, 213–216, 214f, 216–217f
    frequency calculation, 212–213
    functionals, 201–205
    geometry optimization, 206–212, 208–209f
    NMR spectra, calculations, 223–224, 224f
    optical properties, 221–223, 223f
    reaction mechanism analysis, 216–221
    relativistic effects, 205
    single point energy calculations, 205–206
    structure-property correlation, 224–226
    time-dependent, 205
  foundation of, 201–205
  quantum mechanics-molecular mechanics (QM-MM) calculations, 226–228, 227f
Thermodynamic analysis, 216–218, 218f
Thermogravimetric analysis (TGA), 54
Thermometers, 322–323
Thiolate ligand etching induced size-reduction, 113–114
Thiolate ligands, 60–61, 61f
Thiol capped gold nanoparticles, 21–22f
Thyroglobulin (Tg), 315
Time-correlated single photon counting (TCSPC), 47
Time-dependent density functional theory (TDDFT), 201, 205
Time-dependent Schrödinger equation, 201–202
Time-independent Schrödinger equation, 201–202
Time-of-flight mass analyzer (TOF), 50
Transient absorption (TA) spectroscopy, 49, 50f
Transition state, location of, 209–211, 210–211f
Transmission electron microscope (TEM), 52–53
Trinitrotoluene (TNT), 303–304
Twist-and-lock mechanism, 255–257
Two-phase ligand exchange induced size-conversion, 104f
Two-phase ligand exchange method, 30–31, 31f
  in situ, 31, 32f

U

Ullmann coupling reaction, 352
Ultrafast absorption spectra, 154, 155f
α, β-Unsaturated ketones selective reduction, 349–350
Untargeted imaging, 324, 325f
Urea sensing, 312
UV-vis absorption spectroscopy, 45
  Au_{38Q} and Au_{38T}, 45, 46f
  GS-protected Au NCs, 45, 46f
  of neutral $[Au_{25}(PET)_{18}]^0$ or anionic $[Au_{25}(PET)_{18}]^{1-}$, 85, 85f
  on size-conversion, 87f, 87–88b
  pH-regulated, 135–136, 136f

V

Vacuum synthesis, 11–15
  device in, 12f
  fast atom bombardment, 13–14, 14f
  liquid metal ion source, 14–15
  sputtering, 12–13
van der Waals forces, 18–19
Vaporization, laser, 17–19, 17–18f
Vibrational circular dichroism (VCD), 49, 184
Volcano-shape size-growth, 90–91, 91f
Voltammogram, 186

W

Water splitting, photocatalytic, 341–342, 342f

X

X-ray absorption fine structure (XAFS), 5, 57–58, 58f
X-ray absorption near edge structure (XANES), 57–58, 300
X-ray absorption spectroscopy (XAS), 5
X-ray diffraction (XRD), 293
X-ray magnetic circular dichroism (XMCD), 172
X-ray photoelectron spectroscopy (XPS), 56–57, 293
  applications, 57
  characteristics, 57

Z

Zero order regular approximation (ZORA), 205

Printed in the USA
CPSIA information can be obtained
at www.ICGtesting.com
LVHW011540011123
762789LV00016B/369